Springer Proceedings in Complexity

Springer Proceedings in Complexity publishes proceedings from scholarly meetings on all topics relating to the interdisciplinary studies of complex systems science. Springer welcomes book ideas from authors. The series is indexed in Scopus.

Proposals must include the following:

- name, place and date of the scientific meeting
- a link to the committees (local organization, international advisors etc.)
- scientific description of the meeting
- list of invited/plenary speakers
- an estimate of the planned proceedings book parameters (number of pages/articles, requested number of bulk copies, submission deadline)

Submit your proposals to: Hisako.Niko@springer.com

Anna Visvizi · Orlando Troisi · Mara Grimaldi
Editors

Research and Innovation Forum 2022

Rupture, Resilience and Recovery in the Post-Covid World

Editors
Anna Visvizi
SGH Warsaw School of Economics
Warsaw, Poland

Orlando Troisi
University of Salerno
Fisciano, Italy

Mara Grimaldi
University of Salerno
Fisciano, Italy

ISSN 2213-8684 ISSN 2213-8692 (electronic)
Springer Proceedings in Complexity
ISBN 978-3-031-19562-4 ISBN 978-3-031-19560-0 (eBook)
https://doi.org/10.1007/978-3-031-19560-0

This Springer imprint is published by the registered company Springer Nature Switzerland AG
The registered company address is: Gewerbestrasse 11, 6330 Cham, Switzerland

Organization

The Rii Forum Chairs

Anna Visvizi, Ph.D., Associate Professor, SGH Warsaw School of Economics, Warsaw, Poland
Orlando Troisi, Ph.D., Associate Professor, University of Salerno, Fisciano, Italy

The Rii Forum Steering Committee

Marek Bodziany, Ph.D. (dr hab.), Associate Professor, Colonel of the Polish Armed Forces, Deputy Dean of the Faculty of Security Sciences at The General Kościuszko Military University of Land Forces (AWL), Wrocław, Poland
Higinio Mora, Ph.D., Associate Professor, Computer Technology Department, University of Alicante, Alicante, Spain
Raheel Nawaz, Ph.D., Professor, Pro Vice Chancellor (Digital Transformation), Staffordshire University, United Kingdom; Professor of AI, Professor of Digital Education, PFHEA, NTF, Staffordshire University, United Kingdom
Francesco Polese, Prof. Eng., Director of Simas—Inter Department Centre for Innovation Systems and Health Management, Director of the Ph.D. Program 'Big Data Management' of DISA MIS (Department of Business Sciences—Management & Innovation Systems), Professor of Business Management—DISA MIS (Department of Business Sciences—Management & Innovation Systems), Professor of Health Management—Department of Medicine and Surgery, University of Salerno, Fisciano, Italy
Orlando Troisi, Ph.D., Associate Professor, University of Salerno, Fisciano, Italy

Anna Visvizi, Ph.D. (dr hab.), Associate Professor, SGH Warsaw School of Economics, Warsaw, Poland & Deree College—The American College of Greece, Athens, Greece
Yenchun Jim Wu, Ph.D., Distinguished Professor, Graduate Institute of Global Business and Strategy, National Taiwan Normal University, Taiwan

The Rii Forum Academic Committee

Antonio Botti, Ph.D., Professor, Management & Innovation Systems (DISA-MIS), University of Salerno, Fisciano, Italy
Antonella Carbonaro, Ph.D., Professor, Department of Computer Science and Engineering University of Bologna, Italy
Michał Choraś, Ph.D., D.Sc., Full Professor, Bydgoszcz University of Science and Technology, Poland
Vincenzo Corvello, Ph.D., Assistant Professor, University of Calabria, Cosenza, Calabria, Italy; Editor-in-Chief, European Journal of Innovation Management (EJIM)
Howard Duncan, Ph.D., Chair Emeritus, International Metropolis Project, Former Editor of *International Migration* (Wiley), Adjunct Research Professor, Carleton University, Ottawa, Canada
Mara Grimaldi, Ph.D., Assistant Professor, Management & Innovation Systems (DISA-MIS), University of Salerno, Fisciano, Italy
Jari Jussila, Ph.D., Principal Research Scientist at HAMK Smart Research Unit, Häme University of Applied Sciences, Finland
Krzysztof Kozłowski, Ph.D., Associate Professor, Vice-Rector for Students and Teaching, SGH Warsaw School of Economics, Warsaw, Poland
Amina Muazzam, Ph.D., Director of Research (DOR), Lahore College for Women University, Lahore, Pakistan
Raquel Perez-del-Hoyo, Ph.D., Associate Professor, Building Sciences and Urbanism Department, Urban Design and Regional Planning Unit, University of Alicante, Spain
Wim Naudé, Ph.D., Professor, RWTH Aachen University, Aachen, Germany
Placido Pinheiro, Ph.D., Professor, Graduate Program in Applied Informatics, University of Fortaleza, Brazil
Shahira Assem Abdel Razek, Ph.D., Assistant Professor of Architecture, Delta University for Science and Technology, Egypt
Lucie Rohlikova, Ph.D., Associate Professor, University of Western Bohemia, Plzen, Czechia
Kawther Saeedi, Ph.D., Assistant Professor of Information Systems, Faculty of Computing and Information Technology, King Abdulaziz University, Saudi Arabia
Yves Wautelet, Ph.D., Research Centre for Information Systems Engineering (LIRIS), Brussels Campus, KU Leuven, Belgium

The Rii Forum Support Team

Vilma Çekani, Ph.D. Candidate, Management & Innovation Systems (DISA-MIS), University of Salerno, Fisciano, Italy
Gianluca Maria Guazzo, Ph.D. Candidate, Management & Innovation Systems (DISA-MIS), University of Salerno, Fisciano, Italy
Atheena Gouviotis, Research & Innovation Institute (Rii)

Contact at: info@rii-forum.org
Website: https://rii-forum.org

Research & Innovation Forum 2022

Where science & innovation meet

Athens, April 20-22, 2022 |on-site & on-line | rii-forum.org

Rupture, Resilience & Recovery in the Post-Covid World

Preface

The **Research and Innovation Forum** (Rii Forum) is an annual conference that brings together researchers, academics, and practitioners from all over the world to engage in conceptually sound inter- and multi-disciplinary, empirically driven debate on key issues influencing the dynamics of social interaction today. Technology, innovation, and education define the rationale behind the Rii Forum and are at the heart of all discussions held during the conference. As the COVID-19 had defined the tone of our lives in the past two years, the key theme underlying the Rii Forum 2022 was related to it. Specifically, the Rii Forum 2022 was devoted to the broadly defined question of ways of overcoming **rupture** caused by the pandemic, of **resilience** (of societies, the business sector, higher education, academia and other), and of fostering **recovery,** hence the so called "three Rs".

COVID-19 and its corollaries brought rupture into nearly every aspect of our lives and into every dimension of our social, political and economic systems, locally, internationally and globally. The need to address the implications revived the discussion on resilience at the level of families, communities, organizations (public and private), including international organizations, and governments. In these complex and uncertain contexts, the debate on means and strategies aimed at fostering recovery has become an imperative. The Rii Forum 2022 focus on advances in sophisticated information and communication technology (ICT) was meant to highlight that considerate (or smart) and efficient use of ICT offers ways of by-passing and mitigating some of the challenges pertaining to the "three Rs". Indeed, artificial intelligence (AI), big data, cognitive computing, cloud computing, blockchain, the internet of things (IoT), social networking sites, 5G, and others equip us with a new set of tools, techniques and approaches by which to view, interpret, explain, and respond to the plethora of challenges our societies face today. However, ICT remains but one factor that shapes contemporary affairs. A variety of other factors weigh in the equation of contemporary society and its ability to adapt and respond to change amidst volatility, uncertainty, complexity, and ambiguity (VUCA).

In detail, the Rii Forum 2022 encouraged debate on ways of utilizing ICT as a means of boosting resilience and the prospect of recovery across diverse, frequently overlapping levels, issues, and domains. This includes resilience and antifragility at the community and organization (public, private, voluntary) levels. It also includes approaches and strategies implemented in such domains as education, public policy, economic policy, security and defense, and, for instance, healthcare. An equally important aspect of the discussion on ICT in connection with rupture, resilience and recovery, comprises the question of ICT and its local, regional, international and global implications. While digital transformation is a powerful term to account for the impact of ICT in several of these domains, there is much more to it. The Rii Forum 2022 sought to delineate and explore this yet unclaimed area of research.

The Rii Forum 2022 served as a virtual arena for debate on how sophisticated information and communication technology (ICT), including artificial intelligence (AI), blockchain, big data, cloud and edge computing, 5G, Internet of Things (IoT), and social networking, could help stakeholders to manage continuity, innovation, and change in the post-Covid world. As an inter- and multi-disciplinary conference, the Rii Forum 2022 welcomed insights from a range of research domains and academic disciplines. To this end, the conceptual foundations of the Rii Forum 2022 were drawn from social sciences, management science, computer science, as well as education science and humanities.

The objective of Rii Forum 2022 was to discuss questions of rupture, resilience and recovery in the post-Covid worlds by focusing on such technology-mediated domains as education, smart cities, the business sector and management strategies, safety and security, but also questions closer culture and politics. Accordingly, the discussions during Rii Forum 2022 were organized along the following themes (tracks):

- Track 1: Technology-enhanced teaching and learning: from distance teaching to remote learning and beyond
- Track 2: Critical insights into smart cities and smart villages: beyond the ICT-hype, toward new avenues of research
- Track 3: Managing rupture, resilience and recovery in organizations
- Track 4: ICT, safety, and security in the digital age: bringing the human factor back into the analysis
- Track 5: Lessons from the past: technology and cultural heritage: preserving, restoring, and emulating cultural heritage in today's societies
- Track 6: Government and governance and the VUCA tale.

Even if the veil of uncertainty related to the pandemic was still present in the run-up to Rii Forum 2022, it was possible at last to organize the conference, at least partially, on-site. Clearly, as the dynamics of the pandemic was not linear and Rii Forum is a truly international conference, due to travel restrictions still in place, several participants could participate only online. The conference venue was Innovathens, part of Technopolis, i.e., former Athens' gasworks company, which powered the city from the mid-19th century throughout the 1980s. Following a thorough refurbishment and adaptation, today, Technopolis serves as a hub of innovation, debate, and entertainment. The Rii Forum 2022 conference attracted delegates from nearly all continents, including North and South America, Asia, the Arab Peninsula, Europe and Sub-Saharan Africa. The conference was held in Athens on April 20–22, 2022, and consisted of twenty panel sessions. In addition, two keynote speeches were delivered, a workshop was organized, and a roundtable discussion was held.

Regarding the **Rii Forum Keynote**, on April 20, 2022, the Rii Forum 2022 delegates had the opportunity to listen to Professor Krzysztof Kozłowski, Ph.D. (dr hab.) Vice-Rector for Teaching and Student Affairs, Director of the Institute of International Affairs, SGH Warsaw School of Economics, Warsaw, Poland. The title of the keynote speech was "The Past That's Still Here, just unevenly distributed: the war in Ukraine & the challenges for Higher Education Institutions". On April 21, 2022, the Rii Forum 2021 delegates were invited to join the keynote speech, titled "Accelerating the recovery by collaborating with startups: the new paradigm of innovation in the post-Covid era", by Professor Vincenzo Corvello, Editor in Chief of European Journal of Innovation Management, Department of Engineering, University of Messina, Italy.

The **Rii Forum Workshop**, titled "Immerse yourself in a virtual world", was organized by Professors Lucie Rohlıkova and Petr Hořejší, from the University of West Bohemia, the Czech Republic. It was held on April 21, 2022. Virtual-reality (VR)-based solutions and applications are used across issues and domains, including healthcare, education, tourism, engineering, and many others. The objective of the multidisciplinary and practice-oriented workshop was to give the Rii Forum 2022 participants, and especially Ph.D. students attending the conference, the opportunity to engage in a hands-on (including VR head-set testing) conversation on new research topics in the domain of VR research.

The **Rii Forum Round Table** was held on Friday, April 22, 2022. The topic of the roundtable discussion was "Higher Education at a Turn". The rationale behind the debate was the recognition that the COVID-19 pandemic revealed that HEIs around the world had been challenged by an unprecedented number of factors. The thrust of the challenge pertained not only to the capacity of HEIs to embark on remote learning but also to the administrative and managerial capacity to effectively anticipate and respond to developments taking place in their external environment. The objective of the Rii Forum Round Table was to dwell on that multi-scalar challenge. The speakers included Professor Christelle Scharff, Associate Dean, Computer Scientist & Entrepreneur, Pace University, New York; Professor Lloyd George Waller, The University of the West Indies (Mona Campus), Kingston, Jamaica, and Prof. Enric Serradell-López, Associate Dean, Director of the MBA Program, Open University of Catalonia. The roundtable discussion was moderated by Professor Anna Visvizi, SGH Warsaw School of Economics, Warsaw. The Rii Forum Roundtable debate served as an opportunity to present two recently published book on this topic, including the monography by Lloyd G. Waller (2022) Digital Transformation for Higher Education Institutions: A Framework for the Future, Ian Randle Publishers [1], and the edited book by Linda Daniela and Anna Visvizi. A. (eds.) (2021) Remote Learning in Times of Pandemic: Issues, Implications and Best Practice, Routledge [2].

124 extended paper proposals were submitted in response to the initial call for papers. Off these proposals, in course of a double-blind review process, 74 papers

were accepted to be included in the proceedings. This suggests an acceptance-rejection ratio of 59:41. All papers were subjected to several rounds of a double-blind peer review process (at least two reviews in each round of the review process, followed by a metareview). Reflecting the inter- and multi-disciplinary focus and scope of the Rii Forum 2022, the papers included in this volume are grouped into five broad thematic sections. These include

- Innovation & Technology Transfer: developments & Issues
- Smart cities: rupture & resilience building
- Education & Technology enhanced learning: the post-Covid "new normal"
- Business sector response to Covid-19: strategic recovery
- Covid-19 & the society: resilience & recovery.

The papers included in these Rii Forum 2022 proceedings serve as evidence that in times as unpredictable and filled with inconsistencies as today, it is mandatory to identify the drivers for the recovery of businesses, community and society that can support decision-making and policy-making processes in the attainment of resilience in VUCA times. This book offers a unique view not only on the scope of the rupture the COVID-19 pandemic inflicted on our societies and political and economic systems, but also on how amidst uncertainty resilience can be built and recovery can be attained.

We remain grateful to the Rii Forum Steering Committee and the Rii Forum Program Committee for their commitment, sound judgment and hard work in the process of organizing the Rii Forum 2022 and then successfully organizing the conference in a hybrid mode, i.e., onsite and online. We would like to say 'thank you' to all contributing authors. We remain indebted to the reviewers who worked around the clock to evaluate papers and revisions submitted to this book. Finally, we would like to express our gratitude to the entire Springer team and the Editors of Complexity for their continued support and guidance.

As our societies succeed in building resilience and the massive project of delivering the COVID-19 vaccine to our societies begins yielding results, we are hopeful that also the forthcoming Rii Forum 2023 will be held onsite, and that the online mode of the conference delivery will be just an option. Rii Forum 2023 will be held in Cracow, Poland, on April 12–14, 2023. Please, check the Rii Forum website (https://rii-forum.org) for updates. We hope to meet you in Kraków, a city beloved by so many, including Polish poets, including Gałczynski [3].

Warsaw, Poland
Fisciano, Italy

Anna Visvizi
Orlando Troisi
Rii Forum 2022 Chairs

References

1. Waller, L.G. (2022) Digital Transformation for Higher Education Institutions: A Framework for the Future, Ian Randle Publishers.
2. Daniela, L., Visvizi. A. (eds) (2021) Remote Learning in Times of Pandemic: Issues, Implications and Best Practice, Routledge.
3. Gałczynski, Konstanty Ildefons, And Furdyna A. (1975) The Enchanted Doroshka (Zaczarowana Dorożka), *The Polish Review*, vol. 20, no. 4, pp. 109–12. *JSTOR*, http://www.jstor.org/stable/25777341. Accessed 13 Jul. 2022.

Acknowledgments

As the Rii Forum 2022 Chairs, we would like to extend our "thank you" to everyone who contributed to the conference's success. The Rii Forum 2022 was an instance of exemplary teamwork, problem solving, co-creation, knowledge sharing, learning, and adding to the broader debate on key issues that shape our world today. While technology and its diverse facets remained the key interpretive lens guiding the conversations at the conference, it was the social impact thereof that was of primary concern to the Rii Forum 2022 participants. We would like to thank everyone who participated in this process.

We are grateful to the Rii Forum 2022 keynote speakers, i.e., Prof. Vincenzo Corvello, Editor in Chief of European Journal of Innovation Management (EJIM), Department of Engineering, University of Messina, and Professor Krzysztof Kozłowski, Ph.D. (dr hab.), Vice-Rector for Teaching and Student Affairs, Director of the Institute of International Affairs, SGH Warsaw School of Economics, Warsaw, Poland. The points they raised in their talks reverberated throughout the conference. Thank you!

We would also like to thank our dear colleagues and friends, who responded to the initial call for panels and papers, thus making the Rii Forum 2022 a truly inter- and multi-disciplinary conference. Even more so, we would like to thank all Rii Forum 2022 delegates for creating a friendly and welcoming atmosphere at the conference. Indeed, the Rii Forum, each subsequent edition, prides itself on the quality of presentations, on ensuring an atmosphere filled with respect, self-discipline, and respect for others (mind that presentation time is respected, and a discussion always follows the presentations!), and finally, on critical and constructive feedback. We are grateful to all participants for that.

We would also like to extend our gratitude to the Rii Forum 2022 reviewers. To review all papers submitted to the Rii Forum 2022 was a challenge. Thank you all, who joined the process, for your hard work.

Finally, we would like to thank the Publisher, Springer Nature, and its entire team for granting us the opportunity to collaborate on this volume. We praise your professionalism. We appreciate the privilege of collaborating with you.

Anna Visvizi
Orlando Troisi
Rii Forum 2022
Chairs

Contents

Contributors

Opeoluwa Aiyenitaju Department of Operations, Technology, Events and Hospitality Management, Manchester Metropolitan University, Manchester, UK

Sanaa AL-Kesmi Master of Science in Urban Design, College of Architecture and Design, Effat University, Jeddah, Saudi Arabia

Suaad Alarifi Information Systems Department, King Abdulaziz University, Jeddah, Saudi Arabia

Muzoon Alasbali Department of Operations and Supply Chain Management, Kingdom of Saudi Arabia, Effat University, Jeddah, Saudi Arabia

Omnia Sameer Alghazi Electrical and Computer Engineering Department, Effat University, Jeddah, Saudi Arabia

Salem AlJanah College of Computer and Information Sciences, Imam Mohammad Ibn Saud Islamic University (IMSIU), Riyadh, Saudi Arabia

Naif Radi Aljohani King Abdulaziz University, Jeddah, Saudi Arabia

Haya Almagwashi Information Systems Department, King Abdulaziz University, Jeddah, Saudi Arabia

Abdulaziz T. Almaktoom Department of Operations and Supply Chain Management, Kingdom of Saudi Arabia, Effat University, Jeddah, Saudi Arabia

Muhammad ALSurf Regional Manager-Market Dev.-Green Business Certification Inc, Jeddah, Saudi Arabia

Mohammed Alzibak Department of Computer Science Technology and Computation, University of Alicante, Alicante, Spain

Nada T. Bakri Effat University, Jeddah City, KSA, Saudi Arabia

Guillermo Balastegui-García Department of Computer Science Technology and Computation, University of Alicante, Alicante, Spain

Rico Baldegger School of Management Fribourg, Fribourg, Switzerland

Giovanni Baldi University of Salerno, Fisciano, SA, Italy

Vincenzo Basile Department of Economics, Management and Institutions, University of Naples 'Federico II', Naples, Italy

Krishna M. Bathula School of Computer Science and Information Systems, Pace University, New York, NY, USA

Miroslav Bednář Faculty of Engineering, University of West Bohemia, Pilsen, Czech Republic

Marek Bodziany The General Kosciuszko Military University of Land Forces, Wroclaw, Poland

Ander Dorado Bole Department of Computer Science Technology and Computation, University of Alicante, Alicante, Spain

Antonio Botti Università degli Studi di Salerno, Salerno, Italy; Ipag Business School, Parigi, France

Marek Bureš Faculty of Engineering, University of West Bohemia, Pilsen, Czech Republic

Nancy Capobianco Department of Economics, Management and Institutions, University of Naples 'Federico II', Naples, Italy

Luca Carrubbo University of Salerno, Fisciano, SA, Italy

Jackie Carter Manchester Metropolitan University, Manchester, UK

Gian Luca Casali Queensland University of Technology, Brisbane, QLD, Australia

Ylenia Cavacece University of Cassino and Southern Lazio, Cassino, FR, Italy

Vilma Çekani University of Salerno, Fisciano, Salerno, Italy

Ricky Celenta Università degli Studi di Salerno, Salerno, Italy

Karel Chadt The Institute of Hospitality Management and Economics, Prague, Czech Republic

Štěpán Chalupa The Institute of Hospitality Management and Economics, Prague, Czech Republic

Michał Choraś Bydgoszcz University of Science and Technology, Bydgoszcz, Poland

Maria Vincenza Ciasullo Department of Management and Innovation Systems, University of Salerno, Fisciano, Italy; Faculty of Business, Design and Arts, Swinburne University of Technology, Kuching, Malaysia; Department of Management, University of Isfahan, Isfahan, Iran

Rafael Comin-Nunes University Center, Fortaleza, Brazil

Vincenzo Corvello Department of Mechanical, Energy and Management Engineering, University of Calabria, Rende, Italy

Valentina Cucino Scuola Superiore Sant'Anna, Pisa, Italy

Ciro Clemente De Falco University of Naples Federico II, Naples, Italy; University of Salerno, Fisciano, Salerno, Italy

María de los Ángeles Tárraga Sánchez CEIP Bec de L'Águila, Alicante, Spain

María del Mar Ballesteros García CEIP Bec de L'Águila, Alicante, Spain

Maddalena della Volpe University of Salerno, Salerno, Italy

Claudio Del Regno University of Salerno, Fisciano, SA, Italy

Babacar Diop mJangale, Coding School, Thiès, Senegal

Václav Duffek Centre of Biology, Geoscience and Environmental Education, University of West Bohemia, Pilsen, Czech Republic

Jacek Dworzecki AMBIS University, Prague, Czech Republic

Małgorzata Dymyt General Tadeusz Kosciuszko Military University of Land Forces, Wroclaw, Poland

Małgorzata Dziembała The Department of International Economic Relations, The University of Economics in Katowice, Katowice, Poland

Aya Elouali Department of Computer Science Technology and Computation, University of Alicante, Alicante, Spain

Benedetta Esposito Department of Business Studies Management and Innovation Systems, University of Salerno, Salerno, Italy

Aldrin Espín-León Faculty of Sociology and Social Work, Central University of Ecuador, Quito, Ecuador

Laura Ferrando-Martínez Department of Building Sciences and Urbanism, University of Alicante, Alicante, Spain

Miriana Ferrara Department of Management and Innovation Systems, University of Salerno, Fisciano, Italy

Raquel Ferreras-Garcia Universitat Oberta de Catalunya, Barcelona, Spain

Jan Fiala Department of Computer Science and Educational Technology, University of West Bohemia, Pilsen, Czech Republic

Kata Fredheim Stockholm School of Economics in Riga, Riga, Latvia; Baltic International Centre for Economic Policy Studies, Riga, Latvia

Pablo Garrido Universidad Miguel Hernandez de Elche, Elche, Spain

Janusz Gierszewski Pomeranian University in Slupsk, Slupsk, Poland

Mara Grimaldi Department of Business Science, Management & Innovation Systems (DISA-MIS), University of Salerno, Fisciano, SA, Italy

Gianluca Maria Guazzo Department of Business Science – Management and Innovation System, University of Salerno, Fisciano, SA, Italy

Diane Hart Faculty of Business and Law, Department of Operations, Technology, Events and Hospitality Management, Manchester Metropolitan University, Manchester, UK

Mariyam Hassan Information Systems Department, King Abdulaziz University, Jeddah, Saudi Arabia

Saeed-Ul Hassan Department of Computing and Mathematics, Management Manchester Metropolitan University, Manchester, UK

Kristýna Havlíková University of West Bohemia, Plzen, Czech Republic

Andrés Henao-Rosero Catholic University of Pereira, Pereira, Colombia

Samedi Heng HEC Liège, Université de Liège, Liège, Belgium

Ana Beatriz Hernández-Lara Department of Business Management, Universitat Rovira i Virgili, Avinguda Universitat, Reus, Spain

Lucía Martínez Hernández Department of Computer Science Technology and Computation, University of Alicante, Alicante, Spain

Raimir Holanda Filho University of Fortaleza, Fortaleza, Brazil

Marianne Honkasaari Häme University of Applied Sciences, Hämeenlinna, Finland

Petr Hořejší Department of Industrial Engineering and Management, University of West Bohemia, Pilsen, Czech Republic

Klaudia Gabriella Horváth National University of Public Service, Budapest, Hungary

Wojciech Horyn The General Kosciuszko Military University of Land Forces, Wroclaw, Poland

Jan Husák Center for Digitalization and Educational Technologies (CEDET), Czech Institute of Informatics, Robotics, and Cybernetics (CIIRC), Czech Technical University in Prague (CTU), Dejvice, Czech Republic

Xhimi Hysa POLIS University, Tirana, Albania

Jan Hán The Institute of Hospitality Management and Economics, Prague, Czech Republic

Asmaa Ibrahim Department of Architecture, College of Architecture and Design, Effat University, Jeddah, Saudi Arabia; Effat University, Jeddah City, KSA, Saudi Arabia

Alexandra Jaramillo-Gutiérrez University of Salerno, Salerno, Italy

Jiřina Jenčková Perfect Hotel Concept, Prague, Czech Republic

Antonio Jimeno-Morenilla Department of Computer Technology, University of Alicante, Alicante, Spain

Jari Jussila Häme University of Applied Sciences, Hämeenlinna, Finland

Anne-Mari Järvenpää Häme University of Applied Sciences, Hämeenlinna, Finland

Kaleemunnisa School of Computer Science and Information Systems, Pace University, New York, NY, USA

Hadi Karami Faculty of Business and Law, Department of Operations, Technology, Events and Hospitality Management, Manchester Metropolitan University, Manchester, UK

Ilona Kačerová Faculty of Engineering, University of West Bohemia, Pilsen, Czech Republic

Mehr-un-Nisa Khalid Information Technology University, Lahore, Pakistan

Aroob Khashoggi Effat University, Jeddah, Saudi Arabia

Jana Kleinová Faculty of Engineering, University of West Bohemia, Pilsen, Czech Republic

Robert Kocur General Tadeusz Kosciuszko Military University of Land Forces, Wroclaw, Poland

Pavel Kopeček University of West Bohemia, Plzen, Czech Republic

Olli Koskela Häme University of Applied Sciences, Hämeenlinna, Finland

Martin Kotek Center for Digitalization and Educational Technologies (CEDET), Czech Institute of Informatics, Robotics, and Cybernetics (CIIRC), Czech Technical University in Prague (CTU), Dejvice, Czech Republic

Michal Kotek Center for Digitalization and Educational Technologies (CEDET), Czech Institute of Informatics, Robotics, and Cybernetics (CIIRC), Czech Technical University in Prague (CTU), Dejvice, Czech Republic

Agnieszka Maria Koziel National Central University, Taoyuan, Taiwan

Rafał Kozik Bydgoszcz University of Science and Technology, Bydgoszcz, Poland

Krzysztof Kozłowski SGH Warsaw School of Economics, Warsaw, Poland

David Krákora Faculty of Engineering, University of West Bohemia, Pilsen, Czech Republic

Jan Kubr Faculty of Engineering, University of West Bohemia, Pilsen, Czech Republic

Saikat Kundu Department of Engineering, Manchester Metropolitan University, Manchester, UK

Iivari Kunttu Häme University of Applied Sciences, Hämeenlinna, Finland

Pablo Lara-Navarra Universitat Oberta de Catalunya. Tibidabo, Barcelona, Spain; Open University of Catalonia, Barcelona, Spain

Muhammad Latif Department of Engineering, Manchester Metropolitan University, Manchester, UK

Angela Lehr Creighton University, Omaha, NE, USA

Giulia Leoni University of Bologna, Bologna, Italy

Francesca Loia Department of Economics, Management and Institutions, University of Naples 'Federico II', Naples, Italy

Otoniel López Granado Department of Computer Engineering, Miguel Hernández University, ElcheAlicante, Spain

Gennaro Maione University of Salerno, Fisciano, Italy

Ornella Malandrino Department of Business Studies Management and Innovation Systems, University of Salerno, Salerno, Italy

Radosław Malik SGH Warsaw School of Economics, Warsaw, Poland

Alessandro Manetti European Institute of Design, Barcelona, Spain; Open University of Catalonia, Barcelona, Spain

Dorien Martinet KU Leuven, Leuven, Belgium

Miguel Martinez-Rach Universidad Miguel Hernandez de Elche, Elche, Spain

Justyna Matkowska Department of History, University at Albany, State University of New York, Albany, NY, USA; Centre for Migration Studies, Adam Mickiewicz University of Poznan, Poznan, Poland

Luca Mazzara University of Bologna, Bologna, Italy

Antonietta Megaro University of Salerno, Fisciano, SA, Italy

Julio C. Mendoza-Tello Faculty of Engineering and Applied Sciences, Central University of Ecuador, Quito, Ecuador

Tatiana Mendoza-Tello Insurance Consultant, Quito, Ecuador

Pavel Mentlík Centre of Biology, Geoscience and Environmental Education, University of West Bohemia, Pilsen, Czech Republic

Héctor Migallón Department of Computer Engineering, Miguel Hernández University, ElcheAlicante, Spain

Ivan A. Mikhaylov RUDN University, Moscow, Russian Federation

Abeer Samy Yousef Mohamed Effat University, Jeddah, Saudi Arabia; Tanta University, Tanta, Egypt

Mady Mohamed Architecture Department, College of Architecture and design, Effat University, Jeddah, Saudi Arabia; Architecture Department, College of Engineering, Zagazig University, Zagazig, Egypt

Mohammed F. M. Mohammed Effat University, Jeddah, Saudi Arabia

Serafina Montefresco Department of Mechanical, Energy and Management Engineering, University of Calabria, Rende, Italy

Raffaella Montera Department of Management and Innovation Systems, University of Salerno, Fisciano, Italy; Department of Economics and Business, University of Florence, Florence, Italy

Higinio Mora Department of Computer Science Technology and Computation, University of Alicante, Alicante, Spain

Ali Mohamad Mouazen Management and International Management Department, Lebanese International University, Beirut International University, Beirut, Lebanon

Rania Nasreldin Department of Architecture, Faculty of Engineering, Cairo University, Giza, Egypt

Aitana Jiménez Navarro Department of Computer Science Technology and Computation, University of Alicante, Alicante, Spain

Raheel Nawaz Staffordshire University, United Kingdom

Ndeye Massata Ndiaye Pôle Science, Technologie et Numérique, Université Virtuelle du Sénégal, Dakar, Senegal

Anastasija Nikiforova Faculty of Science and Technology, Institute of Computer Science, Chair of Software Engineering, University of Tartu, Tartu, Estonia

Ifra Nisar Information Technology University, Lahore, Pakistan

Izabela Nowicka Military University of the Land Forces, Wroclaw, Poland

Marina-Paola Ojan Istituto Europeo di Design. Biada, Barcelona, Spain

Aitana Pastor Osuna Department of Computer Science Technology and Computation, University of Alicante, Alicante, Spain

Mandy Parkinson Manchester Metropolitan University, Manchester, UK

Christian Pauletto International University in Geneva, Geneva, Switzerland

Aleksandra Pawlicka University of Warsaw, Warsaw, Poland

Marek Pawlicki Bydgoszcz University of Science and Technology, Bydgoszcz, Poland

Antonio Soriano Payá Department of Computer Science Technology and Computation, University of Alicante, Alicante, Spain

Marco Pellicano University of Salerno, Fisciano, SA, Italy

Mirko Perano University of Salerno, Fisciano, SA, Italy

Maria Lúcia D. Pereira Ceara State University, (L.I.C.C.P.), Fortaleza, Brazil

Manuel Perez-Malumbres Universidad Miguel Hernandez de Elche, Elche, Spain

María Luisa Pertegal-Felices Developmental and Educational Psychology Department, University of Alicante, Alicante, Spain

Martina Perutková Center for Digitalization and Educational Technologies (CEDET), Czech Institute of Informatics, Robotics, and Cybernetics (CIIRC), Czech Technical University in Prague (CTU), Dejvice, Czech Republic

Andrzej Pieczywok Kazimierz Wielki University, Bydgoszcz, Poland

Wojciech Pietrzyński Pomeranian University in Slupsk, Slupsk, Poland

Luana Ibiapina C. C. Pinheiro Ceara State University, (L.I.C.C.P.), Fortaleza, Brazil

Pedro Gabriel Calíope Dantas Pinheiro University of Fortaleza, Fortaleza, Brazil; Estacio University Center, Fortaleza, Brazil

Plácido Rogerio Pinheiro University of Fortaleza, Fortaleza, Brazil; Ceara State University, (L.I.C.C.P.), Fortaleza, Brazil

Pablo Piñol Universidad Miguel Hernandez de Elche, Elche, Spain

Francesco Polese Department of Business Science – Management and Innovation System, University of Salerno, Fisciano, SA, Italy

Ferenc Pongrácz Tungsram Operations Kft, Budapest, Hungary

Tomáš Průcha Department of Computer Science and Educational Technology, University of West Bohemia, Pilsen, Czech Republic

Raquel Pérez-delHoyo Department of Building Sciences and Urbanism, University of Alicante, Alicante, Spain

Saeed Mian Qaisar Electrical and Computer Engineering Department, Effat University, Jeddah, Saudi Arabia

Laura-Diana Radu Faculty of Economics and Business Administration, Department of Business Information Systems, Alexandru Ioan Cuza University of Iasi, Iasi, Romania

Tarek Ragab College of Architecture and Design, Effat University, Jeddah, Saudi Arabia

Sema Refae Architecture Department, Dar AlHekma University, Jeddah, Saudi Arabia

Marina S. Reshetnikova RUDN University, Moscow, Russian Federation

Lucie Rohlíková Department of Computer Science and Educational Technology, University of West Bohemia, Pilsen, Czech Republic; Department of Computer Science and Educational Technology, University of West, Plzen, Czech Republic

Emilia Romeo University of Naples Federico II, Naples, Italy; University of Salerno, Fisciano, Salerno, Italy

Marlena Rybczyńska The General Tadeusz Kościuszko Military University of Land Forces, Wroclaw, Poland

Filip Rybnikár Faculty of Engineering, University of West Bohemia, Pilsen, Czech Republic

Elvi Mihai Sabau Sabau Department of Computer Science Technology and Computation, University of Alicante, Alicante, Spain

Kawther Saeedi Information Systems Department, King Abdulaziz University, Jeddah, Saudi Arabia

Daisy Valdivieso Salazar Faculty of Sociology and Social Work, Central University of Ecuador, Quito, Ecuador

Jordi Sales-Zaguirre Institut Químic de Sarrià, Universitat Ramon Llull, Barcelona, Spain

Haitham Samir College of Architecture and Design, Effat University, Jeddah, Saudi Arabia; School of Engineering, New Giza University, Cairo, Egypt

Pedro José Leal Santiago University of Fortaleza, Fortaleza, Brazil

Christelle Scharff School of Computer Science and Information Systems, Pace University, New York, NY, USA; Seidenberg School of Computer Science and Information Systems, Pace University, New York, USA

Adja Codou Seck Department of Computer Science, Université Alioune Diop de Bambey, Bambey, Senegal

Enric Serradell-López Open University of Catalonia, Barcelona, Spain

Maria Rosaria Sessa Department of Business Studies Management and Innovation Systems, University of Salerno, Salerno, Italy

Ali Shahid Information Technology University, Lahore, Pakistan

Chien-wen Shen National Central University, Taoyuan, Taiwan

Daniela Sica Department of Business Studies Management and Innovation Systems, University of Salerno, Salerno, Italy

María Teresa Signes-Pont Department of Computer Science Technology and Computation, University of Alicante, Alicante, Spain

Michal Šimon Faculty of Engineering, University of West Bohemia, Pilsen, Czech Republic

Carlo Alessandro Sirianni Department of Business Science – Management and Innovation System, University of Salerno, Fisciano, SA, Italy

Asamaporn Sitthi Department of Geography, Faculty of Social Science, Srinakharinwirot University, Bangkok, Thailand

Jakub Stejskal Center for Digitalization and Educational Technologies (CEDET), Czech Institute of Informatics, Robotics, and Cybernetics (CIIRC), Czech Technical University in Prague (CTU), Dejvice, Czech Republic

Fernando Tafalla-Porto Department of Building Sciences and Urbanism, University of Alicante, Alicante, Spain

Andrea Moretta Tartaglione University of Cassino and Southern Lazio, Cassino, FR, Italy

Jin Yee Tay Department of Computing and Mathematics, Manchester Metropolitan University, Manchester, UK

Sofiane Tebboune Faculty of Business and Law, Department of Operations, Technology, Events and Hospitality Management, Manchester Metropolitan University, Manchester, UK

Pin Shen Teh Department of Operations, Technology, Events and Hospitality Management, Manchester Metropolitan University, Manchester, UK

Magdalena Tomala Jan Kochanowski University in Kielce, Kielce, Poland

Orlando Troisi Department of Business Science, Management & Innovation Systems (DISA-MIS), University of Salerno, Fisciano, SA, Italy

Konstantinos Tsilionis KU Leuven, Leuven, Belgium

Zane Varpina Stockholm School of Economics in Riga, Riga, Latvia; Baltic International Centre for Economic Policy Studies, Riga, Latvia

Galina A. Vasilieva RUDN University, Moscow, Russian Federation

Susie Vaughan Susie Vaughan & Company, McKinney, TX, USA

Saverino Verteramo Department of Mechanical, Energy and Management Engineering, University of Calabria, Rende, Italy

Anna Visvizi SGH Warsaw School of Economics, Warsaw, Poland; Effat University, Jeddah, Saudi Arabia

Roberto Vona Department of Economics, Management and Institutions, University of Naples 'Federico II', Naples, Italy

Pavel Vránek Faculty of Engineering, University of West Bohemia, Pilsen, Czech Republic

Hajra Waheed Information Technology University, Lahore, Pakistan

Yves Wautelet KU Leuven, Leuven, Belgium

Marta Wincewicz-Bosy General Tadeusz Kosciuszko Military University of Land Forces, Wroclaw, Poland

Tso-hsuan Yeh National Central University, Taoyuan, Taiwan

Miroslav Zíka Department of Computer Science and Educational Technology, University of West Bohemia, Pilsen, Czech Republic

Innovation and Technology Transfer: Developments and Issues

Data Security as a Top Priority in the Digital World: Preserve Data Value by Being Proactive and Thinking Security First

Anastasija Nikiforova

Abstract Today, large amounts of data are being continuously produced, collected, and exchanged between systems. As the number of devices, systems and data produced grows up, the risk of security breaches increases. This is all the more relevant in times of Covid-19, which has affected not only the health and lives of human beings' but also the lifestyle of society, i.e. the digital environment has replaced the physical. This has led to an increase in cyber security threats of various nature. While security breaches and different security protection mechanisms have been widely covered in the literature, the concept of a "primitive" artifact such as data management system seems to have been more neglected by researchers and practitioners. But are data management systems always protected by default? Previous research and regular updates on data leakages suggest that the number and nature of these vulnerabilities are high. It also refers to little or no DBMS protection, especially in case of NoSQL, which are thus vulnerable to attacks. The aim of this paper is to examine whether "traditional" vulnerability registries provide a sufficiently comprehensive view of DBMS security, or they should be intensively and dynamically inspected by DBMS owners by referring to Internet of Things Search Engines moving towards a sustainable and resilient digitized environment. The paper brings attention to this problem and makes the reader think about data security before looking for and introducing more advanced security and protection mechanisms, which, in the absence of the above, may bring no value.

Keywords Data · Database · Internet of things · NoSQL · Security · Vulnerability

---

A. Nikiforova (✉)
Faculty of Science and Technology, Institute of Computer Science, Chair of Software Engineering, University of Tartu, 51009 Tartu, Estonia
e-mail: Nikiforova.Anastasija@gmail.com; anastasija.nikiforova@ut.ee

A. Visvizi et al. (eds.), *Research and Innovation Forum 2022*,
Springer Proceedings in Complexity, https://doi.org/10.1007/978-3-031-19560-0_1

1 Introduction

Today, in the age of information and Industry 4.0, billions of data sources, including but not limited to interconnected devices (sensors, monitoring devices) forming Cyber-Physical Systems (CPS) and the Internet of Things (IoT) ecosystem, continuously generate, collect, process, and exchange data [1]. With the rapid increase in the number of devices (smart objects or "things", e.g., smartphones, smartwatches, intelligent vehicles etc.) and information systems in use, the amount of data is increasing. Moreover, due to the digitization and variety of data being continuously produced and processed with a reference to Big Data, their value, is also growing and as a result, the risk of security breaches and data leaks, including but not limited to users' privacy [2]. The value of data, however, is dependent on several factors, where data quality and data security that can affect the data quality if the data are accessed and corrupted, are the most vital. Data serve as the basis for decision-making, input for AI- (Artificial Intelligence) and ML- (Machine Learning) driven models, forecasts, simulations etc., which can be of high strategical and commercial/business value.

This has become even more relevant in terms of Covid-19 pandemic, when in addition to affecting the health, lives, and lifestyle of billions of citizens globally, making it even more digitized, it has had a significant impact on business [3]. This is especially the case because of challenges companies have faced in maintaining business continuity in this so-called "new normal". However, in addition to those cybersecurity threats that are caused by changes directly related to the pandemic and its consequences, many previously known threats have become even more desirable targets for intruders, hackers. Every year millions of personal records become available online [4–6]. Lallie et al. [3] have compiled statistics on the current state of cyber-security horizon during the pandemic, which clearly indicate a significant increase of such. As an example, Shi [7] reported a 600% increase in phishing attacks in March 2020, just a few months after the start of the pandemic, when some countries were not even affected. Miles [8], however, reported that in 2021, there was a record-breaking number of data compromises, where "the number of data compromises was up more than 68% when compared to 2020", when LinkedIn was the most exploited brand in phishing attacks, followed by DHL, Google, Microsoft, FedEx, WhatsApp, Amazon, Maersk, AliExpress and Apple. And while [5] suggests that vulnerability landscape is returning to normal, there is another trigger closely related to cyber-security that is now affecting the world - geopolitical upheaval.

Recent research demonstrated that weak data and database protection in particular is one of the key security threats [4, 6, 9–11]. This poses a serious security risk, especially in the light of the popularity of search engines for Internet connected devices, also known as Internet of Things Search Engines (IoTSE), Internet of Everything (IoE) or Open Source Intelligence (OSINT) Search Engines such as Shodan, Censys, ZoomEye, BinaryEdge, Hunter, Greynoise, Shodan, Censys, IoTCrawler. While these tools may represent a security risk, they provide many positive and security-enhancing opportunities. They provide an overview on network security, i.e., devices connected to the Internet within the company, are useful for market research and adapting business strategies, allow to track the growing number of smart devices representing the IoT world, tracking ransomware—the number and nature of devices affected by it, and therefore allow to determine the appropriate actions to protect yourself in the light of current trends. However, almost every of these white hat-oriented objectives can be exploited by black-hatters. The popularity of IoTSE decreased a level of complexity of searching for connected devices on the internet and easy access even for novices due to the widespread popularity of step-by-step guides on how to use IoT search engine to find and gain access if insufficiently protected to webcams, routers, databases and in particular non-relational (NoSQL) databases, and other more «exotic» artifacts such as power plants, wind turbines or refrigerators. They provide service- and country- wised exposure dashboards, TOP vulnerabilities according to CVE, statistics about the authentication status, Heartbleed, BlueKeep–a vulnerability revealed in Microsoft's Remote Desktop Protocol that has become even more widely used during pandemics, port usage and the number of already compromised databases. Some of these data play a significant role for experienced and skilled attackers, making these activities even less resource-consuming by providing an overview of the ports to be used to increase the likelihood of faster access to the artifact etc.

In the past, vulnerability databases such as CVE Details were considered useful resources for monitoring the security level of a product being used. However, they are static and refer to very common vulnerabilities in the product being registered when a vulnerability is detected. Advances in ICT, including the power of the IoTSE, require the use of more advanced techniques for this purpose.

The aim of this paper is to examine both current data security research and to analyse whether "traditional" vulnerability registries provide a sufficient insight on DBMS security, or they should be rather inspected by using IoTSE-based and respective passive testing, or dynamically inspected by DBMS holders conducting an active testing. As regards the IoTSE tool, this study refers to Shodan- and Binary Edge-based vulnerable open data sources detection tool—ShoBeVODSDT—proposed by Daskevics and Nikiforova [9].

The paper is structured as follows: Sect. 2 provides the reader with a background, including a brief overview of data(base) security research, Sect. 3 gives an overview of database security threats according to the CVE Details, Sect. 4 provides a comparative analysis of the results extracted from the CVE database with the results obtained as a result of the application of the IoTSE-based tool. Section 5 summarizes the study, making call for "security first" principle.

2 Rationale of the Study

Data security, and therefore database security, should be a priority for IT management as an extremely valuable asset for any organization [10, 11]. Failure to comply with the requirements for security and protection of data and sensitive data in particular can lead to significant damage and losses of commercial, reputation, operational etc. nature. Recent research, however, often point out the problems associated with meeting even the most trivial requirements. A Data Breach Investigations Report [6] revealed that one of the most prominent and growing problems is the misconfiguration of DBMS. This is even more the case for NoSQL. Given that in case of NoSQL there is less focus on the security mechanism (i.e., it was not their priority), some research such as Fahd et al. [11] do not recommended to directly expose them to an open environment, where untrusted clients can directly access them. This refers to a frequently observed highly vulnerable combination of data (such as relational or document databases or cloud file storage) placed on the Internet without controls, combined with security researchers looking for them [6]. These rather undesirable combinations have been on the rise for the past few years constituting the concept of "open database". The term does not refer to data storage facilities that have been assigned the Open Database License (ODL), whereby they are knowingly made available to users to be freely used, shared, modified while maintaining the same freedom for others, thus contributing to the openness paradigm. Instead, it refers to insecure and unprotected database that can be accessed by any stakeholder despite the agreement of the data holder, which poses a serious risk. This was studied in [9], revealing that while there are databases that can be considered as open databases being accessible via Internet, the difference between NoSQL and relational database management systems (RDBMS) is not as obvious with some weak results demonstrated by SQL databases.

All in all, research aimed at identifying security threats and vulnerabilities in databases is relatively limited, with especially little research on NoSQL security, despite the vulnerabilities in NoSQL database systems is a well-known problem [9, 11]. Moreover, when examining recent database research, other aspects not related to security - their performance and efficiency in certain scenarios, scalability (sharding), availability (replication), dynamism (no rigid schema) etc. appear to be more popular with much less attention paid to security [11]. An analysis of a set of key security features offered by four NoSQL systems - Redis, Cassandra, MongoDB and Neo4j [11], however, concluded that NoSQL is characterized by mostly low level of both built-in security, encryption, authentication and authorization, and auditing, while most NoSQL lack them.

The next Section is intended to provide an insight on database security provided by CVE Details – probably the most widely known registry of vulnerabilities.

3 CVE Security Vulnerability Database

3.1 CVE Scope and Classification of the Vulnerabilities

The CVE security vulnerability database is a free source of information providing details on disclosed cybersecurity vulnerabilities and exploits constituting a catalogue of over 172 thousand entries. These records are added to the database through a six-step process, where each interested party can contribute to the database and, if the identified vulnerability is approved, the relevant information will become part of the registry with the priority and risk assigned as a result of its discovery, where the verification is performed by the CVE participant, thereby making this list authoritative.

CVE registry divide vulnerabilities into 13 types:

1. bypass something, e.g., restrictionss
2. cross-site scripting known as XSS
3. denial of service (DoS),
4. directory traversal,
5. code execution (arbitrary code on vulnerable system),
6. gain privileges
7. HTTP response splitting
8. memory corruption
9. gain / obtain information
10. overflow
11. cross site request forgery (CSRF)
12. file inclusion
13. SQL injection.

3.2 CVE Statistics of Most Popular Databases

For the purposes of this study, the most popular databases were selected for their further analysis. The list was formed based on the results of the DB-Engines Ranking,[1] presenting data on March 2022. In addition to the TOP-10 most popular databases CouchDB, Memcached and Cassandra were selected based on their popularity in other lists. Table 1 lists them along with their type and basic statistics on their vulnerabilities. The latter is retrieved from the CVE registry, where the date of the 1st and last reported vulnerability is recorded to determine if the registry provides continuous and up-to-date data, the total number of vulnerabilities reported between these dates, the most frequently reported vulnerability, and a list of 3 most popular vulnerabilities in recent 5 years. This is intended to provide some general statistics and point to current trends and whether they have changed, i.e., whether the most popular vulnerability over the years is still a key threat or the developers managed to resolve it.

Despite the undeniable popularity of NoSQL databases, relational databases remain popular, and TOP-5 consists of 4 RDBMS and MongoDB. However, at the same time, it should be noted that all the most popular relational DBMS, taking the highest places are multi-model, i.e., adapted to current trends. For example, Oracle has proven to be the most popular, using a relational DBMS as its primary database model, while secondary models include document store, graph DBMS, RDF store,

[1] https://db-engines.com/en/ranking.

Table 1 General DB-wised statistics of their vulnerability [author, based on CVE Details]

Database	Type of database	1st vulnerabilityregistered	last vulnerability registered	Total # of vulnerabilities	Most popular vulnerability	TOP-3 vulnerabilities in 2018–2022
Oracle	Relational, multi-model	2008	2021	44	DoS	DoS, code execution, gain information
MySQL	Relational, multi-model	2001	2015	152	DoS	–
Microsoft SQL Server	Relational, multi-model	1999	2021	87	Code execution	Code execution
PostgreSQL	Relational, multi-model	1999	2022	134	DoS	Code execution, overflow, Sql injection
MongoDB	Document, multi-model	2013	2022	38	DoS	DoS, Code execution, overflow, bypass something
Redis	Key-value, multi-model	2015	2021	23	Overflow	Overflow, code execution, memory corruption, bypass something
IBM Db2	Relational, multi-model	2004	2021	106	DoS	Code execution, overflow, gain information
Elasticsearch	Search engine, multi-model	2018	2022	22	Gain information	Gain information, DoS, Gain privilege, Code execution
Microsoft Access	Relational	1999	2020	17	Code execution	Code execution, Overflow

(continued)

Table 1 (continued)

Database	Type of database	1st vulnerabilityregistered	last vulnerability registered	Total # of vulnerabilities	Most popular vulnerability	TOP-3 vulnerabilities in 2018–2022
SQLite	Relational	2009	2022	48	DoS	Code execution, DoS, Overflow
Cassandra	Wide column store	2015	2022	6	Code execution	Code execution, DoS, Bypass Something
Memcached	Key-value store	2013	2020	14	DoS	DoS, overflow
CouchDB	Document, multi-model	2010	2021	15	Code execution	Code execution, bypass something, gain privileges

and Spatial DBMS. Similarly, MySQL secondary database models are represented by document store and spatial database, as are PostgreSQL and Microsoft SQL, although the latter uses graph DBMS in addition to the above.

The highest number of discovered vulnerabilities are in MySQL, although this is the only database for which data are no longer provided, i.e., the last vulnerability was registered in 2015. MySQL is followed by PostgreSQL and IBM Db2, with Cassandra, Memcached, CouchDB, Microsoft Access, Elasticsearch and Redis reporting the fewest vulnerabilities. Although the number of revealed vulnerabilities does not necessarily mean that the level of the relevant databases is definitely higher or lower, which may depend on the popularity of these databases, users and community involvement, this suggests such an assumption. In some cases, these statistics play a decisive role in choosing a database giving the impression of a higher "security-by-design" level. At the same time, the aforementioned databases with fewer reported vulnerabilities have come under the spotlight in some of recent data leakages, with Elasticsearch dominating [12, 13], from which data on unique 1.2 billion people was leaked in 2019, making this one of the largest data leaks from a single source organization in history. This also applies to perhaps the most provocative database—MongoDB, whose low security level has been widely discussed and because of which it is very often the object of IoT search engines "trainings", for which step-by-step guides are provided. Ferrari et al. [14], however, inspected compromised databases, where Redis dominated with about 30% of databases were compromised, followed by Elasticsearch (13%) and MongoDB (8%). In most cases, this was caused by misconfiguration of these databases.

For the most common and major vulnerabilities encountered over time, most of them are DoS, although code execution is also a widespread vulnerability. A database-wised analysis of the most frequently reported vulnerabilities over the past 5 years demonstrate that Code Execution is the most common and is in the TOP-3 for 11 databases, followed by overflow (7), DoS (6), bypassing something (4), gaining information (3).

While these data are very general, Fig. 1 shows data on vulnerabilities reported in 2021 and their scores, i.e., risk level (red bars indicate the highest scores or the highest number of high-risk level vulnerabilities). While for some of them data for 2022 is also provided, the purpose of this study requires to focus on 2021, when IoTSE-driven analysis by Daskevics and Nikiforova [9] took place, thereby allowing for more consistent comparison of results. At the same time, there data on Elasticsearch, Microsoft Access and Memcached for 2021 are not available. In terms of their vulnerability in 2020, however, Elasticsearch suffered most from XSS with information obtaining and DoS, Microsoft Access – code execution, where one was combined with an overflow, while for Memcached DoS was registered.

For the most popular vulnerability, the same trend is observed, i.e., 8 of 43 vulnerabilities refer to DoS, another 8 – to code execution, followed by 7 cases of overflow and 5 information leaks. Some vulnerabilities may overlap, which explains the inconsistencies in the data obtained from the registry, i.e., 31 of 43 have been supplied with corresponding detail. This, however, together with [5], according to which VulnDB has identified many thousands of vulnerabilities that were not registered in

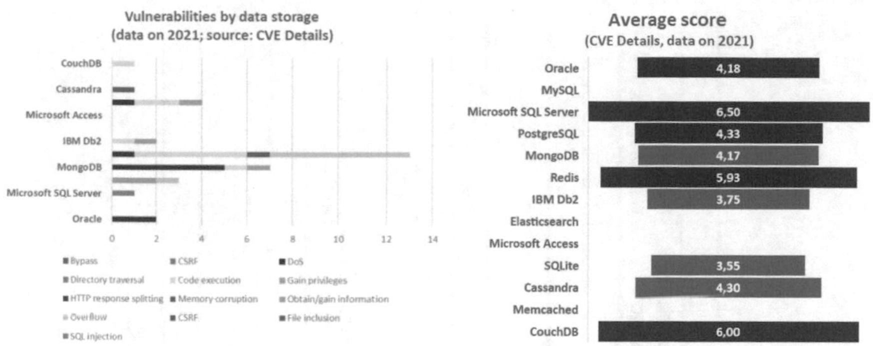

Fig. 1 Vulnerability of data storages in 2021 and their score (based on CVE Details) [author]

the CVE Details database, puts into question the completeness and accuracy of the CVE registry in regard to the actual state of the art.

MongoDB and Oracle have the most reported vulnerabilities, followed by Redis and PostgreSQL. Again, despite the widespread discussion about the highest risk of vulnerabilities for NoSQL compared to SQL, this is not so obvious since RDBMS are also at risk. The level of risk of registered vulnerabilities for Microsoft SQL is the highest.

Otherwise, very obvious and strong conclusions cannot be drawn from the data provided. However, it can be speculated that MongoDB is weak against DoS, but Redis against code execution and overflow. To get more supported results, this paper addresses the call made in [9] and maps the results obtained in that study to the data obtained from CVE Details.

4 CVE Registry Versus IoTSE-Based Testing Results

In this section, a brief overview of the results will be given, and more specifically their comparison with the results obtained by IoTSE-based tool ShoBeVODSDST that conducts a penetration testing [9]. In addition to providing both results and ranking databases by overall results, the focus will be on "Gain Information" category, as it corresponds well to the aspect inspected by ShoBeVODSDST - "managed to connect and gather sensitive data" (4th of 5 risk levels). In [9], it was expected that a correlation will be determined, allowing assumption to be made about less secure data sources.

Table 2 shows the statistics of the services under consideration, as well as their total number of vulnerabilities and the percentage of the "Gain Information" vulnerability. It should be noted that ShoBeVODSDST inspects a limited list of predefined data sources (8) with the possibility of enriching it, as its source code is publicly available.

According to Table 2, MySQL, the data on which is not updated by CVE, accounts more than half of all databases found on the Internet. However, the number of instances that it was able to connect to is not very high for MySQL representing

Table 2 CVE Details- and IoTSE- statistics on database vulnerability

CVE				IoTSE tool			
Database	Total # of vulnerabilities	Total registered	Ratio (Info gained/total)	Total DBMS found	# DBMS connected	Gathered data or compromised	Ratio (Info gained/connected)
Oracle	**11**	**2**	0%	–	–	–	–
MySQL	–	0	0%	**13,452**	0,13%	0%	0%
Microsoft SQL server	1	1	0%	–		–	–
PostgreSQL	**5**	**3**	**67%**	**1187**	0,17%	0%	0%
MongoDB	**13**	**7**	14%	**177**	8%	**79%**	7%
Redis	**8**	**13**	0%	122	10%	**83%**	**83%**
IBM Db2	2	2	**50%**	–		–	–
Elasticsearch	–	0	0%	86	**90%**	**27%**	**9%**
Microsoft Access	–	0	0%	–		–	–
SQLite	2	1	**50%**	–		–	–
Cassandra	1	1	0%	7	**14%**	0%	0%
Memcached	–	0	0%	116	**80%**	26%	**24%**
CouchDB	1	1	0%	14	0	0	0

18 databases, which is similar to PostgreSQL where the number of found databases is 1187 with only 2 databases could be connected. However, there were 5 vulnerabilities in PostgreSQL registered by CVE Details, with 2 of them related to information gaining that was not found by ShoBeVODSDT. At the same time, the absolute leader in this negative trend is Memcached, where it was possible to connect to 93 of 116 databases with more than 20% of the databases, from which data have been either gathered or they were found to be already compromised. Similar results were obtained for Elasticsearch, where it was possible to connect to 90% of all databases found, and 27% of the databases are already compromised or data could be gathered from them. Similarly, CVE Details does not provide details of its vulnerabilities in 2021. MongoDB and Redis showed the worst results for both data sources, where MongoDB was inferior to data gatherings and has a large number of compromised databases according to ShoBeVODSDT and is subject to both DoS, code execution and data gatherings according to CVE Details. Redis, however, with being relatively difficult to connect to, where every 10th database is inferior to this, is characterized by a high ratio of information gatherings. According to CVE Details, both DoS, code execution, overflow, and memory corruption have been detected for it. Additionally, Oracle was one of the most frequently reported databases in CVE Details, with 10 vulnerabilities in total, while only two of them have a comprehensive description—both related to DoS.

All in all, the results in most cases are rather complimentary, and one source cannot completely replace the second. This is not only due to scope limitations of both sources—CVE Details cover some databases not covered by ShobeVODSDT, while not providing the most up-to-date information with a very limited insight on MySQL.

At the same time, there are cases when both sources refer to a security-related issues and their frequency, which can be seen as a trend and treated by users respectively taking action to secure the database that definitely do not comply with the "secure by design" principle. This refers to MongoDB, PostgreSQL and Redis. CouchDB, however, can be considered relatively secure by design, as is less affected, as evidenced by both data sources, where only one vulnerability was reported in CVE Details in 2021, while it was the only data source, to which ShoBeVODSDT was not able to connect. The latter, however, could be because CouchDB proved to be less popular, with only 14 of nearly 15 000 instances found.

5 Conclusions

Obviously, data security should be the top priority of any information security strategy. Failure to comply with the requirements for security and protection of data can lead to significant damage and losses of commercial, reputation, operational etc. nature [11]. However, despite the undeniable importance of data security, the current level of data security is relatively low—data leaks occur regularly, data become corrupted in many cases remaining unnoticed for IS holders. According to

Risk Based Security Monthly Newsletter, 73 million records were exposed in March 2022, and 358 vulnerabilities were identified as having a public exploit that had not yet been provided with CVE IDs.

This study provided a brief insight of the current state of data security provided by CVE Details—the most widely known vulnerability registry, considering 13 databases. Although the idea of CVE Details is appealing, i.e., it supports stakeholder engagement, where each person or organization can submit a report about a detected vulnerability in the product, it is obviously not sufficiently comprehensive. It can be used to monitor the current state of vulnerabilities, but this static approach, which sometimes provides incomplete or inconsistent information even about revealed vulnerabilities, must be complemented by other more dynamic solutions. This includes not only the use of IoTSE-based tools, which, while providing valuable insight into unprotected databases seen or even accessible from outside the organization, are also insufficient.

The paper shows an obvious reality, which, however, is not always visible to the company. In other words, while this may seem surprisingly in light of current advances, the first step that still needs to be taken thinking about date security is to make sure that the database uses the basic security features: authentication, access control, authorization, auditing, data encryption and network security [11, 15, 16]. Ignorance or non-awareness can have serious consequences leading to data leakages if these vulnerabilities are exploited. Data security and appropriate database configuration is not only about NoSQL, which is typically considered to be much less secured, but also about RDBMS. This study has shown that RDBMS are also relatively inferior to various types of vulnerabilities. Moreover, there is no "secure by design" database, which is not surprising since absolute security is known to be impossible. However, this does not mean that actions should not be taken to improve it. More precisely, it should be a continuous process consisting of a set of interrelated steps, sometimes referred to as "reveal-prioritize-remediate". It should be noted that 85% of breaches in 2021 were due to a human factor, with social engineering recognized as the most popular pattern [6]. The reason for this is that even in the case of highly developed and mature data and system protection mechanism (e.g., Intrusion Detection System (IDS)), the human factor remains very difficult to control. Therefore, education and training of system users regarding digital literacy and raising Information Security Awareness (ISA), as well as the definition, implementation and maintaining security policies and risk management strategy, must complement technical advances.

Acknowledgements This research has been funded by European Social Fund via IT Academy programme (University of Tartu)

References

1. Pevnev, V., Kapchynskyi, S.: Database security: threats and preventive measures (2018)
2. Himeur, Y., Sohail, S. S., Bensaali, F., Amira, A., Alazab, M.: Latest trends of security and privacy in recommender systems: a comprehensive review and future perspectives. Comput. Secur. 102746 (2022)
3. Lallie, H.S., Shepherd, L.A., Nurse, J.R., Erola, A., Epiphaniou, G., Maple, C., Bellekens, X.: Cyber security in the age of COVID-19: A timeline and analysis of cyber-crime and cyber-attacks during the pandemic. Comput. Secur. **105**, 102248 (2021)
4. Risk Based Security, Talentbuddy.co / Talentguide.co database exposed, company reacts swiftly (2016). https://www.riskbasedsecurity.com/2016/05/06/talentbuddy-co-talentguide-co-database-exposed-company-reacts-swiftly/. Last Accessed 31 Mar 2022
5. Risk Based Security and Flashpoint 2021. Year end report vulnerability quickview (2021)
6. Verizon. Data Breach Investigations Report (DBIR), p. 119 (2021). https://www.verizon.com/business/resources/reports/2021/2021-data-breach-investigations-report.pdf. Last Accessed 31 Mar 2022
7. Shi, F.: Threat spotlight: coronavirus-related phishing. Barracuda Networks (2020). https://blog.barracuda.com/2020/03/26/threat-spotlight-coronavirus-related-phishing. Last Accessed 31 Mar 2022
8. Miles B.: How to minimize security risks: Follow these best practices for success (2022). https://www.techrepublic.com/article/minimizing-security-risks-best-practices/?utm_source=email&utm_medium=referral&utm_campaign=techrepublic-news-special-offers
9. Daskevics, A., Nikiforova, A.: ShoBeVODSDT: Shodan and binary edge based vulnerable open data sources detection tool or what Internet of Things Search Engines know about you. In: 2021 Second International Conference on Intelligent Data Science Technologies and Applications (IDSTA), pp. 38–45. IEEE (2021)
10. Li, L., Qian, K., Chen, Q., Hasan, R., Shao, G.:. Developing hands-on labware for emerging database security. In: Proceedings of the 17th Annual Conference on Information Technology Education, pp. 60–64 (2016)
11. Fahd, K., Venkatraman, S., Hammeed, F.K.: A comparative study of NoSQL system vulnerabilities with big data. Int. J. Manage. Inf. Tech **11**(4), 1–19 (2019)
12. Tunggal, A.: The 61 biggest data breaches (2021). https://www.upguard.com/blog/biggest-data-breaches. Last Accessed 31 Mar 2022
13. Panda Security.: Over 1 billion people's data leaked in an unsecured server (2019). https://www.pandasecurity.com/en/mediacenter/news/billion-consumers-data-breach-elasticsearch/
14. Ferrari, D., Carminati, M., Polino, M., Zanero, S.: NoSQL breakdown: a large-scale analysis of misconfigured NoSQL services. In: Annual Computer Security Applications Conference, pp. 567–581 (2020)
15. Teimoor, R.A.: A review of database security concepts, risks, and problems. UHD J. Sci. Technol. **5**(2), 38–46 (2021)
16. Malik, M., Patel, T.: Database security-attacks and control methods. Int.J. Inf. **6**(1/2), 175–183 (2016)

Design Thinking Innovation and Trends Foresighting

Alessandro Manetti, Pablo Lara-Navarra, and Enric Serradell-López

Abstract During the 20th Century, designers mainly focused on products' form and function to improve their value on markets. In recent times, design has becoming a discipline more and more used to tackle more complex problems and to produce and promote strategic innovation not only inside products but on a wider scale. The systematization of the design process in the concept of Design Thinking widen the application of this methodology in a big scale of sectors and departments. In the last two decades, society has been transforming itself dramatically due to the pace of technology, demography and globalization with an accelerating speed never experienced in the recent past. It is for this reason that analyzing and foreseeing macrotrends has become a strategic key factor for their development and adaptation to a complete mutating environment. Design Thinking methodologies have common phases, patterns and key points that can be considered as a user centered creative processes that lead to an innovation of the status quo. This process emphasizes observation, collaboration, fast learning, visualization of ideas, rapid concept prototyping, and concurrent business analysis, which ultimately influences innovation and business strategy. The idea of the research paper is matching three territories of research (Design Thinking Process, Trends Foresighting and Analysis, and Visual Mapping) to justify that the foresighting and analysis of a macrotrend—together with a visual representation of a system of complex phenomena—plays a strategic role in orientating and facilitating design process methodologies aimed to produce better design driven innovation inside organizations.

Keywords Design thinking · Forecasting · Visualization

A. Manetti
European Institute of Design, Carrer Biada. 11, 0812 Barcelona, Spain

A. Manetti (✉) · P. Lara-Navarra · E. Serradell-López
Open University of Catalonia, Av. Tibidabo. 39, 08018 Barcelona, Spain
e-mail: amanetti@uoc.edu

P. Lara-Navarra
e-mail: plara@uoc.edu

E. Serradell-López
e-mail: eserradell@uoc.edu

A. Visvizi et al. (eds.), *Research and Innovation Forum 2022*,
Springer Proceedings in Complexity, https://doi.org/10.1007/978-3-031-19560-0_2

1 Introduction

In recent times, design has become a discipline increasingly used to address complex problems [10] and to produce and promote strategic innovation not only in products, but in decision-making within organizations [2]. This has been possible thanks to the systematization of the design process, as it allows the application of this methodology on a large scale in different sectors and organizations [5]. Design Thinking has been definitely gaining attention over the past decade with more and more companies seeking to use it as a human-centered problem-solving approach that can create novel solutions and foster (radical) innovation [8].

In addition, in contemporary interconnected world, the availability of large amount of data across organizations' activities, operations, strategies and decision-making redefined dramatically the dynamics of economic, political, social and companies' systems [14].

The need of radical innovation has been accentuated in the last decade, society has been drastically transformed due to changes in technology, demographics and globalization, an accelerated speed never experienced in the recent past. It is for this main reason that applying design thinking to analyze and predict trends and their impact on the structure and evolution of organizations has become a strategic factor for a changing and unstable environment.

The idea of the research paper is to work with the areas of knowledge of Design thinking Process and Visual Mapping to justify decision making based on Foresighting and trend analysis. Therefore, the analysis of trends integrated in a visual representation together with the design thinking process offers a new research perspective and a new tool for designers and innovation managers to respond to new strategic challenges driven by design.

2 Methodology

The framework is to understand whether visualization and design are methodologies that help to make trend projections for strategic decision-making, specifically, in the field of future studies. In this sense, the work establishes as an objective the creation of visual tools of futurization to innovate in the creation of products and services. To answer the research, the following questions were asked:

(A) Is Design Thinking a valid methodology for innovating in futurization?
(B) How does visualization help you understand the complexity of trends?
(C) Are Automated tools based on design and supported by visualization techniques useful for understanding the future?

In the following sections we present the main characteristics of the research design, in addition to raising the foundations and describing the context of the debate developed in the generation of the trend visualization tool that we call the Deflexor map.

2.1 *Participants*

The study involved 18 professionals with a transdisciplinary and transgenerational profile, prioritizing gender equality. Finally, nine women and nine men were selected, with an age range between 23 and 65 years, all of them linked to the different disciplines of design.

2.2 *Knowledge Context*

The option chosen for the dynamics of study and exchange of knowledge is the face-to-face workshop with systematic, orderly, transparent, transferable and iterative procedures typical of design and which was based on the competences of research, critical thinking, creative capacity, and collaborative work.

2.3 *Data Sources and Instruments*

Regarding the use of data collection tools in the workshops, the Worth Global Style Network (WGSN) trend materials from the Istituto Europeo di Design (IED) library were consulted. In addition, the library contains a specific bibliographic collection on trends, stored in a database and in physical copies, and has the archive of all the projects and works of students and teachers.

Regarding data collection instruments, different techniques were used. On the one hand, the Deflexor map conceived as a physical trend map based on the design process that is used as a creative methodology to help, promote and apply innovation in companies and organizations was used. This tool gives us the possibility to generate new knowledge based on design (Fig. 1).

In addition, for the development of the research, the mind map tool with the use of post-it virtual labels [3] supported by the Miro platform was used. This digital software is a flexible canvas set that allows teams to build accessible and dynamic visual maps that can be developed collaboratively, distributed, anywhere, anytime.

3 Conceptual Framework, Design and Foresight

Design thinking has been defined and studied as the cognitive process of designers [4, 7]. Born during the 90 s, the research movement in Design Thinking methods is still in continuous development, trying to identify the fundamental reasoning patterns and concretizing the core of the practice of design thinking [6].

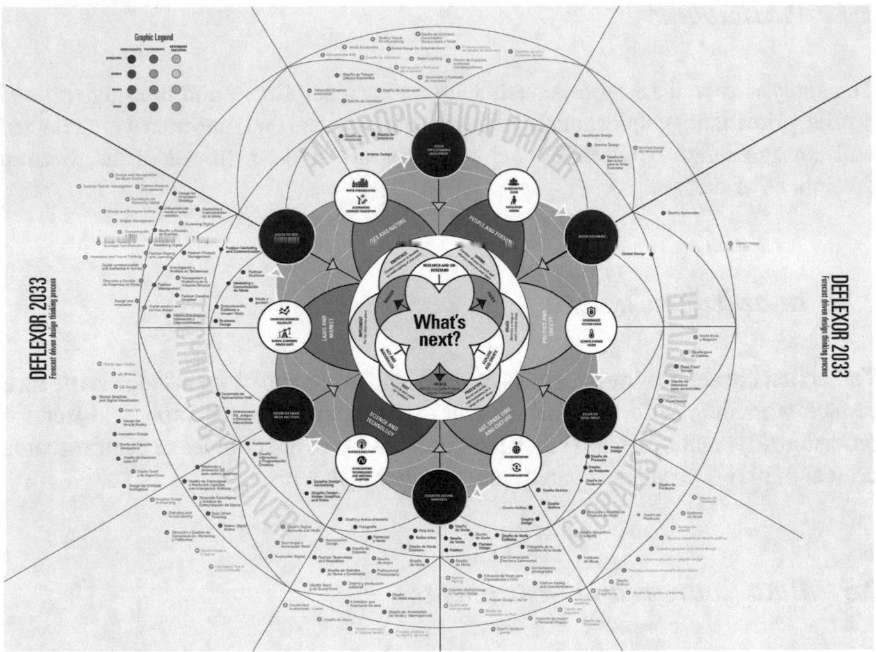

Fig. 1 Deflexor map with the different macrofields and megatrends components. *Source* Alessandro Manetti IED Barcelona

Other researchers attempt to build a bridge between design thinking as a cognitive process and design thinking as an innovation process where designers meet professionals from other disciplinary fields [11]. In this last meaning, the relevance of user-centered research [1], prototyping, testing, interdisciplinarity, the iteration process, visual results and teamwork [2] around complex issues are mentioned as important aspects of the Design Thinking innovation methodology.

3.1 Conceptual Notes Between Prediction and Forecasting

Before going into the presentation of the results, we believe it is necessary to make some annotations on the concepts of prediction and forecasting. The generic concept of prediction refers to the analysis of phenomena or evolutions that have implications for the future. However, it should be noted that after the 1990s, the term has been distorted and used in the most disparate contexts, even in a totally inadequate way, erroneously approaching both prospective practice and other methods of financial planning, for example, the budgetary. For this reason, we must clearly determine the concept of trend prediction, identifying its context and distancing itself from the term forecast, above all, from the forecasting concept of an organization.

To avoid any contextualization error, it is necessary to delimit the use of the terms forecast and prediction (foresighting). In the business environment, forecasting means a set of forecasts supported by quantitative and qualitative data, to establish strategic planning. Forecasting was defined by Martin and Irvine [13] as the exercise of establishing with high probability and accuracy that a future event will happen. Uncertainty is introduced in the definition as a constituent element of the prediction study, in fact, it is the intrinsic nature of each of these studies, to prepare indicators that reliably quantify the risk and then guide it so that the strategic planning process is effective.

The forecast of organizations differs significantly from the prediction of trends. The definition of the term foresight is more complex, foresighting would be described as a cognitive capacity of the human being that allows us to think about the future, consider, model, create and respond to future eventualities, and again, where foresighting is not the ability to predict the future, it is an attribute that allows us to weigh pros and cons, evaluate different options to use as decision aids.

In the representation (Fig. 2), it can be seen how prediction (foresight) needs a greater understanding of highly volatile, complex and uncertain future environments and potentially generating strong impacts on the organization (attraction of the future); while forecasting is born from the push of elements of the present typically linked to the production, economic and financial of the organization (thrust of the present). On the other hand, there is another way to futurize, in organizations, from reading and understanding phenomena, through a historical analysis, looking for comparison with similar events in the past [12]. This modality can generate the replication of effective patterns in the past to realize plausible futures strategies with minimal changes in your strategy. The three forces are usually used in different proportions in the construction of a global strategy for the future.

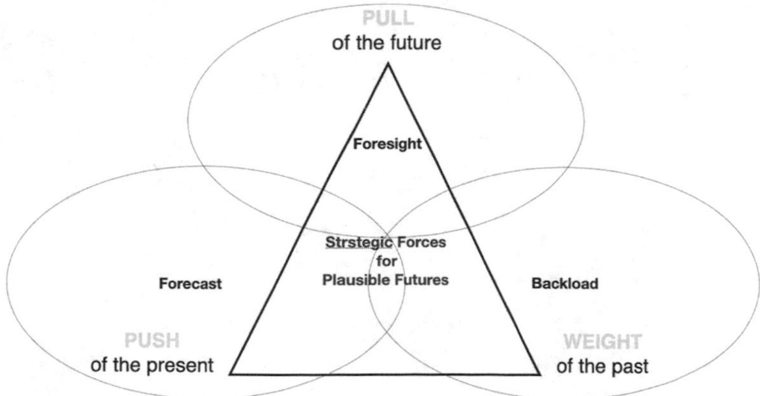

Fig. 2 Strategic forces in building plausible futures. *Source* of own development inspired by Future Platform Ebook 2021

4 Results

This part relates the application of the proposed methodology as part of a research project carried out at the IED design school in Madrid for a backpack production company whose objective is to validate strategic decision-making based on defining new users considering the future evolution of different design scenarios.

The company is aimed with its product at a child audience, specifically primary and secondary school students who use the product as a container for books and other materials related to their student life and classroom activity. The briefing proposes to explore other possible situations of use and asks what other audiences can be users of your product. All this based on strategic decisions that are motivated by a description of future scenarios.

The project was developed with six working groups using the technological tool Miro, loaded with the Deflexor map. Each of the groups was assigned a design scenario alongside a pair of predetermined megatrends on the bending map. Each group appointed a project leader and facilitator and through the post-it technique, a research session was started among the different trends present on the map (Fig. 3).

The first phase closed with a vote using the post-it tool on the Miro platform prioritizing five macrotrends for each of the 6 scenarios.

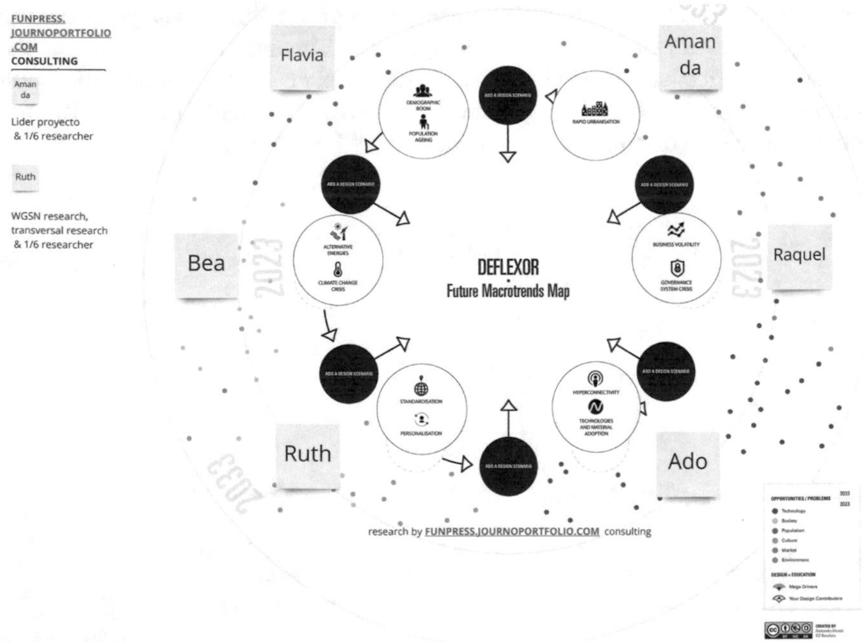

Fig. 3 Miro canvas with deflexor map with six selected megatrends and scenarios. *Source* Alessandro Manetti IED Barcelona

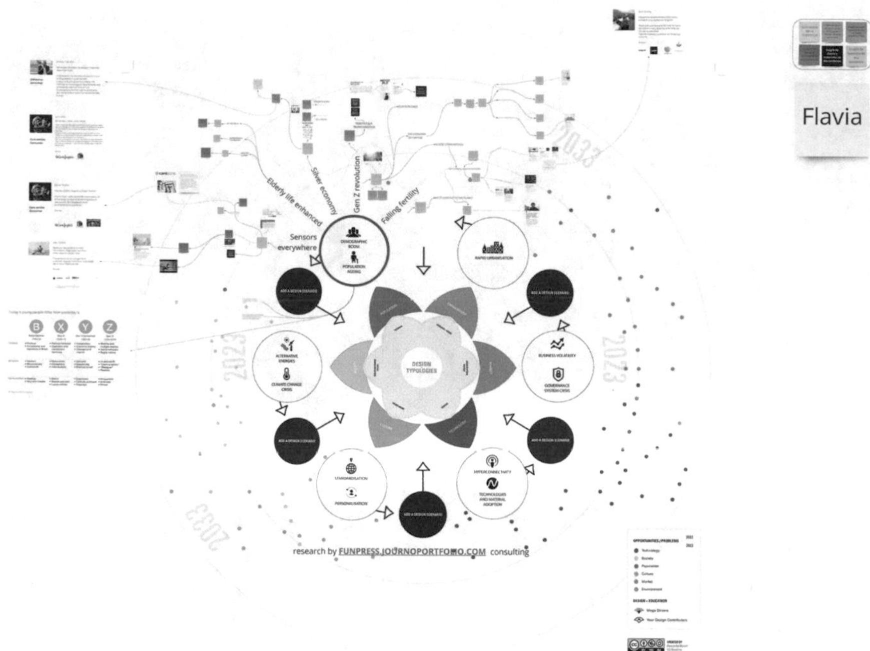

Fig. 4 Miro canvas with deflexor map with design for humans scenario. *Source* Alessandro Manetti IED Barcelona

The first group worked on the Design for Humans scenario related to Population Ageing and Demographic Boom. Five trends were selected: Sensors Everywhere; Elderly Life Enhanced; Silver Economy; Gen Z Revolution; Falling Fertility. For each of the trends, it was built a visual map connected to research content on digital portals and through the WGSN portal. The first group presented 5 new user persona's profiles based on reflections inspired by the selected trends (Fig. 4).

The second group worked on the scenario of the Design for New Business Models associated with the megatrends of Business Volatility and Governance System Crisis selecting four particularly relevant trends: Segmented Tourism, Global Middle Class Growth, Pandemic Risk Response, Individualism.

From these selected trends, seven new profiles of user personas were built (Fig. 5).

The third group worked on Design for Cultural Complexity scenario with Standardization and Personalization megatrends. Six trends were selected: Economy for The Common Good; Spiritual and Holistic Lifestyles; Premiumization; Slow Food Culture; New Sobriety and Simplicity Needs; 3D/4D Do It Yourself. Processing these trends, the group identified seven new user persona profiles (Fig. 6).

The fourth group worked on the scenario of Design for Sustainable Development associated with the Rapid Urbanization megatrend selecting seven relevant trends: New Urban Mindset, Green Conscience Diffusion, Smart City Systems, Mobility Revolution, Smart Retail & Logistic Revolution, Smart Working Places, Smart Home

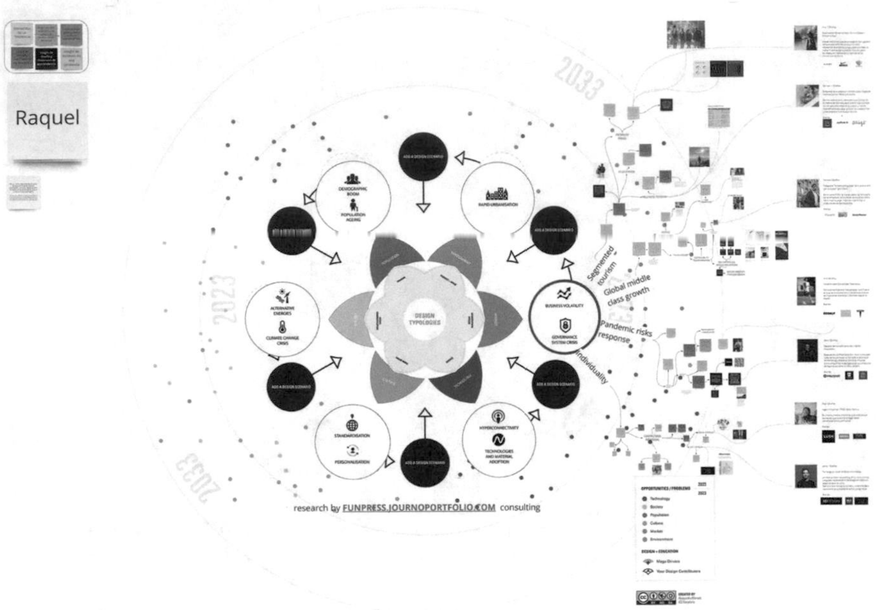

Fig. 5 Miro canvas with deflexor map with design for new business models scenario. *Source* Alessandro Manetti IED Barcelona

Solutions. From these trends, five new user persona's profiles of potential users for the company were generated (Fig. 7).

The fifth group worked on the scenario of The Design for Social Impact associated with the megatrends of Climate Change Crisis and Alternative Energies Transition selecting seven trends: Sport Attitude; Green Conscious Diffusion; Cradle to Cradle; Smart Energies; Waste management. From these trends, four new user persona's profiles of potential users were generated (Figs. 8 and 9).

The sixth group worked on the stage of the Design for Smart Technology and Media associated with the megatrends of Hyperconnectivity and Accelerating Technologies and Materials Adoption, selecting four trends: 3d/4D Printing Tech; Virtual Reality & Augmented Reality Applications; Remix Culture; Waste Reduction. From these trends, 3 new profiles of potential users were generated.

5 Conclusions

The first conclusion: the association between a visualization tool and design thinking process can be considered a valid methodology to innovate in futurization. From the results obtained, it can be confirmed that the Deflexor methodology based on

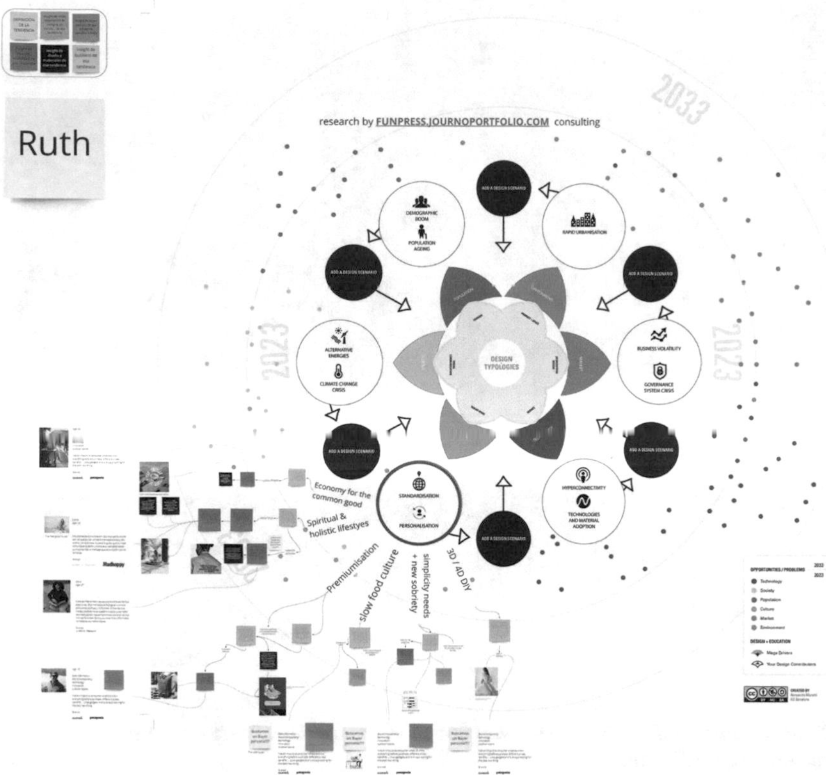

Fig. 6 Miro canvas with deflexor map with design for cultural complexity scenario. *Source* Alessandro Manetti IED Barcelona

design thinking supported by trends allows to generate detailed profiles of users, whose behavior pattern is not only a brief description of the individual but also draws the interaction with social, economic, technological, environmental, cultural and interpersonal environments. On the other hand, visualization helps to understand the complexity of trends and globally understand their interrelationships in order to facilitate and establish future scenarios generation.

The association between a visualization tool and design thinking process helps furthermore to bypass the limitation of the technology-driven approach, so designers and managers should employ an integrated perspective that combines digital, knowledge-based, human centered approach and social orientation in the implementation of innovation strategies [16].

In this sense, working with automated tools based on design and supported by visualization techniques are helpful for understanding the future. Even if, technology per se does not allow the automatic attainment of innovation and it is only through the right application of flexible skills and knowledge that innovation can be realized successfully [15], these platforms allow to set faster strategic decision processes to

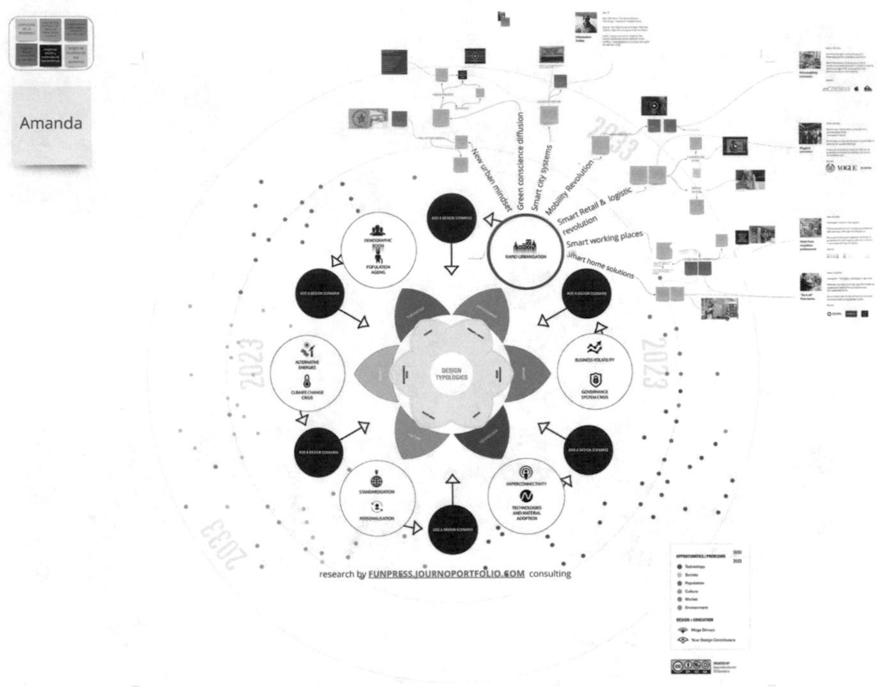

Fig. 7 Miro canvas with deflexor map with design for sustainable development scenario. *Source* Alessandro Manetti IED Barcelona

be implemented in new products. In addition, these tools allow to contextualize the different users them inside the universe of trends, macro trends and megatrends activating trends driven strategic reflections in order to promote and implement innovation.

The integration of design thinking inside the visual trends foresighting mapping is particularly useful at different levels:

- It produces new knowledge about the impact and the characteristics of the trends most relevant to the company's activity cycle on short, medium and long term
- It generates insights about the product placement and distribution spaces and about their different elements: communication systems, brand display elements, materials, etc.
- It offers insights about user persona's behavioral patterns, such as the emerging needs during the shopping experience in the physical and digital point of sale, together with the interactions between spaces, product systems and other different typologies of users and potential consumers; not limiting the exploration to the single moment of purchase and consumption but extending to the characteristics of people in their set of sensations, emotions and lifestyles

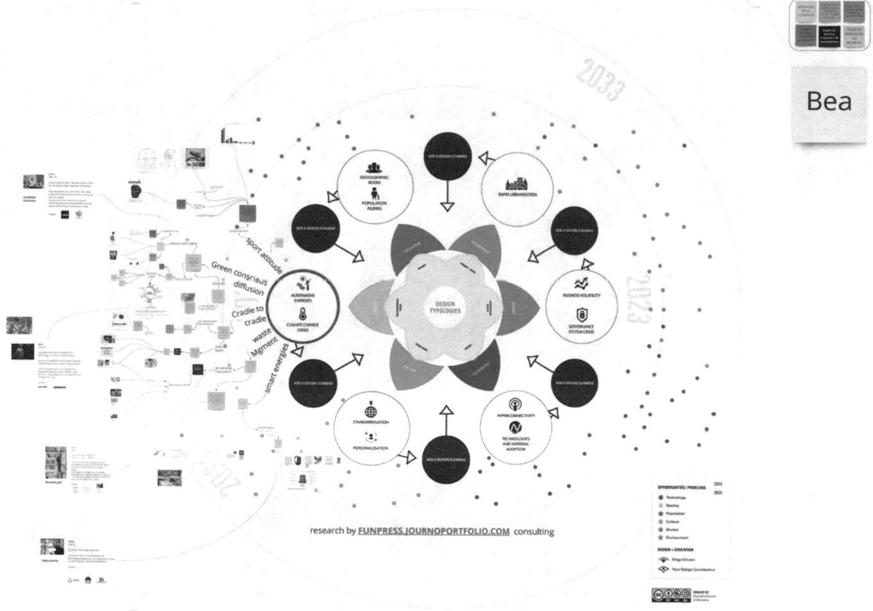

Fig. 8 Miro canvas with deflexor map with design for social impact scenario. *Source* Alessandro Manetti IED Barcelona

- It generates personal insights of the user persona at different levels at the time of purchase: ergonomic, relational, psychological, emotional, etc.
- insights of lifestyle clusters linked to different users
- insights of design style features (biomorphic, angled, retro, futuristic, etc.) and material ranges
- provides a powerful tool to foster change in organizations,
- insights about business model alternatives related to the selected trends and to potential buyers interested to follow these trends, in other words, this methodology provides visions to gain resilience for future developments and disruptions [9]

The use of tools such as the Miro platform during the workshop, has improved the active participation of designers who, due to their characteristics, correspond by themselves to potential new users of the category of the products investigated, amplifying the dimension of collaboration and co-creation throughout the whole process.

Finally, the use of the Deflexor Map methodology opens a new path of research that goes beyond the activity of defining new potential users for a company, since it can be used in other design driven innovation dynamics such as:

- generating and developing focused and mixed trends and design driven new ideas for products and services;

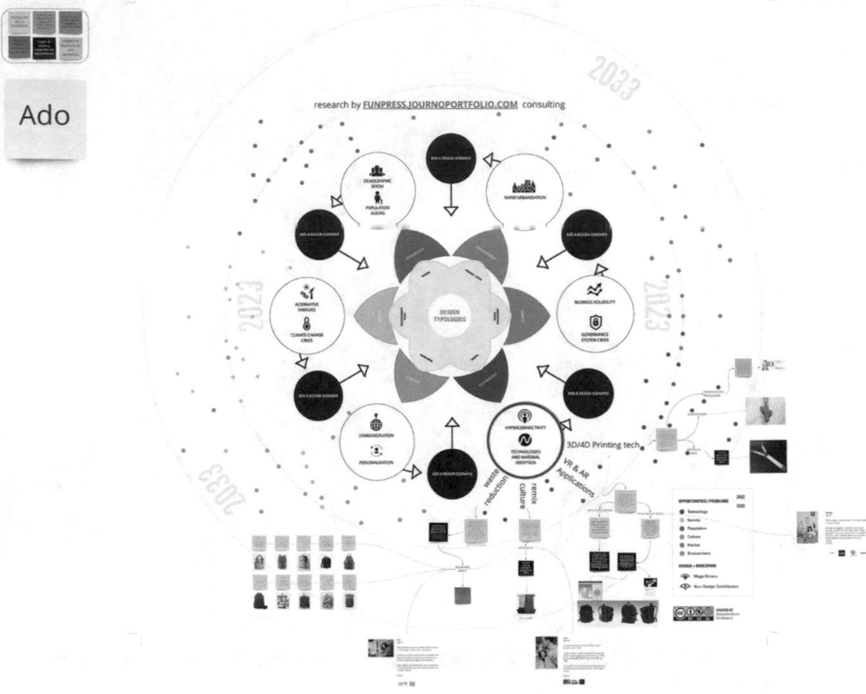

Fig. 9 Miro canvas with deflexor map with design for smart technology and media scenario. *Source* Alessandro Manetti IED Barcelona

- rethinking organization's models, gaining resilience, in order to manage more efficiently the big challenges produced by the impact of macrotrends in the different fields of the contemporary world;
- Reducing the risk of taking decisions in the products and services innovation process in a future environment full of uncertainty;
- orientating and guiding new offering of products and services.

References

1. Brown, T.: Change by Design: How Design Thinking Transforms Organizations and Inspires Innovation. New York (2009)
2. Bucolo, S., Wrigley, C.: Design-Led Innovation: Overcoming Challenges to Designing Competitiveness to Succeed in High Cost Environments. IGI Global (2014)
3. Buzan, T.: How to Mind Map: the Ultimate Thinking Tool that will Change Your Life. Thorsons Publishers, London (2022)
4. Cross, N., Roozenburg, N.: Modelling the design process in engineering and in architecture. J. Eng. Des. **3**(4), 325–337 (1992)

5. Daalhuizen, J.: Method Usage in Design: How Methods Function as Mental Tools for Designers. Delft University of Technology (2014)
6. Dorst, K.: Notes on Design: How Creative Practice Works. BIS Publishers, Amsterdam (2017)
7. Eastman, C., Newstetter, W., McCracken, M.: Design Knowing and Learning: Cognition in Design Education. Elsevier (2001)
8. Eisenbart, B., Bouwman, S., Voorendt, J., McKillagan, S., Kuys B., Ranscombe C.: Implementing design thinking to drive innovation in technical design. Int. J. Des. Creat. Innov. (2022)
9. Habicher, D., Erschbamer, G., Pechlaner, H., Ghirardello, L., Walder, M.: Transformation and design thinking: perspectives on sustainable change, company resilience and democratic leadership in SMEs. Leadership Educ. Pers. Interdiscipli-Nary J. **3**(2), 145–156 (2021)
10. Julier, G.: The Culture of Design (2008)
11. Kimbell, L.: Rethinking Design Thinking: Part II Design and Culture (2012)
12. Laloux, F.: Reinventing Organizations. Nelson Parker, Brussels (2014)
13. Martin, B.R., Irvine, J.: Research Foresight: Priority-Setting in Science. Pinter, London (1989)
14. Troisi, O., Grimaldi, M.: Guest editorial: Data-driven orientation and open innovation: the role of resilience in the (co-)development of social changes. Transform. Gov. Ment: People Process. Pol. **16**(2), 165–171 (2022)
15. Troisi O., Visvizi A., Grimaldi M.: The different shades of innovation emergence in smart service systems: the case of Italian cluster for aerospace technology. J. Bus. Ind. Mark (2021)
16. Visvizi, A., Troisi, O., Grimaldi, M., Loia, F.: Think human, act digital: activating data-driven orientation in innovative start-ups. Eur. J. Innov. Manag. **25**(6), 452–478 (2022)

Fake News and Threats to IoT—The Crucial Aspects of Cyberspace in the Times of Cyberwar

Aleksandra Pawlicka, Marek Pawlicki, Rafał Kozik, and Michał Choraś

Abstract There has been an ongoing debate whether the possibility of cyberspace becoming both a weapon and a battlefield should be treated as an act of war. Although NATO recognized in 2014 that an armed response can be invoked following a cyber-attack, the anonymous nature of the Internet makes it extremely difficult to attribute the attack to a specific nation-state, which makes the chances of retaliation slim. The events of the war in Ukraine have proved that yet again. On top of that, there emerged new ways of weaponizing cyberspace. In this context, the current concept of cybersecurity must be revised and enhanced. This situation emphasizes the necessity to put a spotlight on the threats posed to the Internet of Things devices, due to their ubiquity in the modern world and the possibility of weaponizing them. Online disinformation campaigns being employed as part of warfare in Ukraine have also inspired considerations on broadening the concept of cybersecurity, by acknowledging the significance of fake news. This paper analyses both aspects from the perspective of threat actors and their motivations.

Keywords Cybersecurity · Cyberwar · Fake news · Internet-of-Things

---

A. Pawlicka (✉) · M. Pawlicki · R. Kozik · M. Choraś
ITTI Sp. z o.o., Poznań, Poland
e-mail: apawlicka@itti.com.pl

M. Pawlicki
e-mail: mpawlicki@itti.com.pl

R. Kozik
e-mail: rkozik@itti.com.pl

M. Choraś
e-mail: mchoras@itti.com.pl

A. Pawlicka
University of Warsaw, Warsaw, Poland

M. Pawlicki · R. Kozik · M. Choraś
Bydgoszcz University of Science and Technology, Bydgoszcz, Poland

A. Visvizi et al. (eds.), *Research and Innovation Forum 2022*,
Springer Proceedings in Complexity, https://doi.org/10.1007/978-3-031-19560-0_3

1 Introduction

Although Arquilla and Ronfeldt wrote that "cyberwar is coming" almost three decades ago [1], there has been an intense debate on whether the possibility of cyberspace becoming both a weapon and a battlefield means it should be treated as an act of war. In 2014, NATO acknowledged that an armed collective defence response can be invoked following a cyberattack. Two years later, the Allies recognized cyberspace to be a domain of military operations; this resulted in giving even greater priority to the cyber defence [2]. Yet, due to the anonymous and dispersed nature of the Internet, attributing a cyberattack to a specific nation-state tends to prove impossible, and thus, the chances of retaliation are slim. The recent events of the war in Ukraine have proved yet again that although it is fair to suppose who the perpetrator is, answering to a cyberthreat is a very complicated matter. Even more so, as the actions taken in cyberspace do not only affect the sides of the conflict; conversely, they often influence otherwise unrelated countries and entities, proving that "cyberwar has no borders [3]". As a consequence, the current state of cybersecurity needs to be given careful consideration.

This paper puts the spotlight on the two new challenges to ensuring cybersecurity in today's world—attacks on the IoT devices, and weaponizing disinformation. In Sect. 2, the current image of cybersecurity is concisely discussed. Against this background, Sects. 3 and 4 introduce both threats and highlight the actors who pose them.

2 The Current Vision of Cybersecurity

In a general sense, cybersecurity deals with providing protection for each element constituting cyberspace—computers, networks, systems and software. Typically, the threat is posed by a wide range of illegal actions and attacks, the purpose of which being extorting money, stealing or tampering with users' data, interfering with the functioning of businesses and organisations, etc. At an individual level, cyberattacks can lead to loss of data, money and other assets; attacks at a greater scale, for example, targeted at critical infrastructure, may disrupt the functioning of a society [4]. Thus, the main objective of cybersecurity is to make cyberspace more secure, by means of uncovering and ridding of vulnerabilities of systems and networks, as well as tackling cyberattacks in a proactive way. Thus, by the name "cybersecurity" one usually means a wide array of technologies, tools and methods, as well as policies, concepts and best practices; the term also encompasses all the relevant assets—infrastructure, data, programs, or even personnel [5, 6]. The level of cybersecurity of a given network is proportional to the way the security goals are met. Traditionally, the goals were presented as protecting the Confidentiality, Integrity and Availability of a system. In other words, in line with these goals, a system is secure when information is disclosed exclusively to the appropriate entities and processes, it is not modified without notice and that it is available when expected [7, 8]. These goals constitute the so-called CIA triad [9]. In addition, the triad has often been supplemented with additional three A-goals, namely Accountability, Authenticity and Access control. This means that a system may be perceived to be secure when its actions are not deniable, its entities are credible and verifiable, and the access is possible only to the authorized entities [7, 8].

The malicious actors who pose a threat to cyberspace are no longer stereotypical hooded hackers; just the opposite—the range of threat actors is vast and diverse. Besides the "script kiddies", some actors who play major roles in cybersecurity dynamics are:

- nation states—sophisticated, with generous budgets, substantial capabilities and expertise; funded, directed and coordinated by governments [10, 11]
- cyberterrorists—either state-sponsored or not, driven by ideology [12], using cyberattacks to inspire terror and fear, sow panic and disinformation, to jeopardise the well-being and safety of citizens [13]
- hacktivists—hackers-activists, such as Anonymous, who engage hacking skills in activism against governments, institutions and individuals, to express their discontent and bring about change [14, 15]
- trolls—bored pranksters, provokers and cyberbullies, as well as sponsored political trolls, all of them trying to sow deception or controversy, or simply hurt other people [16]
- organized crime—including scammers and the criminals of all sorts, interested in making financial gain by means of exploiting vulnerabilities of cyberspace [17]
- cyber militia—groups of organized volunteers who use hacking skills for achieving political goals [14].

Yet, the recent events, as well as the state-of-the-art literature, have highlighted two more aspects of cybersecurity that were lacking from the "classical" image of cybersecurity, but need to be properly addressed as soon as possible, namely weaponizing the Internet-of-Things (IoT) devices and online disinformation, a.k.a. fake news.

3 The New Threat—Compromising IoT Devices

The Internet of Things is the network of devices which are capable of connecting with systems, as well as other devices, by means of the Internet. The objects which comprise the IoT range from small household items to large industrial machines. The network has been constantly growing, with the number of devices expected to reach 22 billion in 2025 [18]. It has also attracted all sorts of cyberspace wrongdoers.

According to ENISA [19], the catalogue of threats to IoT is very vast, comprising:

- Nefarious activity/abuse.
- Outages.

- Physical attacks.
- Disasters.
- Eavesdropping/interception/hijacking.
- Damage/loss (IT assets).
- Failures/malfunctions [19].

The threat actors who present threat to IoT have been scrutinized by Bugeja, Jacobsson and Davidsson [20]. According to them, the major forces who are capable of harming citizens by means of compromising IoT are nation-states, cyberterrorists, competitors and organized criminals, hacktivists, individual thieves, and other malicious hackers. A close analysis of the motivations behind exploiting the IoT has uncovered that they vary and are often closely connected with the nature of a particular threat actors. The attacks on IoT may be motivated by:

- national/political interest—where the IoT devices may be employed to spy on people, or to perform attacks on a large scale
- terrorism—destruction, blackmail or revenge, e.g., by means of inducing epileptic seizures by means of hacked smart lightning patterns [21]; no significant attack on IoT has been performed yet which could be pinpointed to terrorist motivations, though [11]
- personal gain—understood as money, knowledge, data, or even just peer recognition
- curiosity, intellectual challenge or the wish to create confusion [20, 22].

Experts argued that a large-scale attack on IoT as a means of a military conflict was not a question of "if" but of "when" ([23–25], etc.). The war in Ukraine showed that this time has come. It was found that a new botnet, potentially compromising the availability of Ukrainian IoT devices, was used to attack government and financial websites before the invasion [26]. Then, a long-planned attack of Russian military hackers on the power grid, targeted at millions of Ukrainians was foiled at the end of March [27].

Thus, from the cybersecurity defenders' standpoint, the highest priority should be given to ridding of vulnerabilities, not leaving any device unsecured, and preventing cyberattacks in a proactive manner; only then will the adversaries not be given the whole networks of devices on a silver platter.

4 Weaponizing Online Disinformation

In the recent years, the ways information is spread have changed drastically. For decades, or even centuries, they were the professionals, such as journalists and writers, who were responsible for creating reliable reports on recent events. The content was under close quality control, by experts in the field and editors, who were able to cease the publication of a piece of information. If the news was not trustworthy or did not meet proper standards, the reputation of a publisher would suffer [28]. With the introduction of the Internet and social media, this dynamics have changed. Today, anyone can create, publish and disseminate content, with no quality control or censorship whatsoever. What is more, they may remain fully anonymous if they please. Lastly, once a piece of news goes online, there is hardly any possibility to stop its dissemination; detecting it may prove to be challenging, too [29, 30]. This shift in the way information is created gave rise to the phenomenon of fake news, the term referring to deliberate disinformation, i.e., verifiably false information, which appears to be news and is spread by means of the Internet [31, 32]. There is a lot of controversy surrounding this term, partly because it has become an umbrella name, applied to various forms of fabricated content of varying degree of maliciousness and intention [33], and partly because the term was misused in the public discourse and may have lost its original meaning [34, 35].

Skilfully crafted and planted disinformation is a powerful tool, or even weapon. Over the years, malicious, false information has undermined individuals' credibility and ruined their reputation and careers. However, the damaging potential of fake news reaches significantly further—from influencing the results of democratic elections [36], skewing people's attitudes towards vaccination [37], giving bogus, deadly dangerous health advice [38] and inspiring riots, to shaking the foundations of democracy [39, 40].

The main threat actor who resort to spreading fake news as their weapon of choice are nation states [41]. In addition, hacktivists and cyberterrorists have been known to spread disinformation as well, either to create confusion, or create the atmosphere of panic, anxiety and distrust [42].

The question arises of why threat actors employ online disinformation. In their study, Kim, Xiong, Lee and Han [43] have identified three motives for creating fake news. Firstly, disinformation is spread for ideological and political purposes, in order to enhance polarization and biases, and as a result influence public opinions. This kind of fake news campaigns, especially if coordinated and well-funded, have the power to affect elections and voting, like in the case of 2016 US Presidential Election [36] or Brexit [43].

Another factor which tends to strongly motivate the creators of fake news is monetary gain. Some websites spreading mass-produced fake news often rely heavily on advertising for profit; thus, the titles of news items are "clickbaity", provocative and attention-grabbing.

Lastly, fake news is employed to prey on people's fears and anxiety. For example, after the onset of the coronavirus pandemic, the Internet exploded with false, misleading health advice which, if followed, could prove fatal. On the other hand, some news items are crafted in such a way that they give false hope to people or play down the severity of a situation [43].

Massive Russian online disinformation and propaganda campaigns were conducted even before the physical outbreak of war on Ukraine. They aimed at creating chaos, making fun of the Ukrainian President, weakening morale, or even implying hostile actions against Russia [44]. After the outbreak, the rhetoric has changed—whilst for Ukraine and other related countries (e.g., Poland) the news was intended to confuse and create panic [45], in Russia it carefully painted the picture of success of their troops in the "special military operation" [46]. Although the media have tried to raise the awareness of the disinformation campaigns and even debunked some of the news items, people have kept believing the fake news, for example by rushing to petrol stations in order to panic buy fuel [45].

5 Conclusions

Over the last few months, the world has witnessed cyberspace indeed becoming a battlefield of a hybrid war, to an unprecedented scale. This requires rethinking what we so far considered to be the pillars and dynamics of ensuring cybersecurity. This paper has emphasized the significance of not letting the Internet of Things devices become a weapon. Similarly, as fake news has taken a prominent role of arms in the cyberwar, efforts must be made in order to both detect it and raise people's awareness of their being the targets of malicious disinformation.

As both aforementioned aspects of the war fought in cyberspace bring severe consequences in the real world, from now on, they must be included as an inseparable element of any cybersecurity policy, method and best practice.

Acknowledgements This work is partially funded under the ELEGANT project, which has received funding from the European Union's Horizon 2020 research and innovation programme under grant agreement No. 957286.

This publication is cofinanced by the National Center for Research and Development within INFOSTRATEG program, number of application for funding: INFOSTRATEG-I/0019/2021-00.

References

1. Arquilla, J., Ronfeldt, D.: Cyberwar is coming! Comp. Strateg. **12** (1993). https://doi.org/10.1080/01495939308402915
2. Brent, L.: NATO's role in cyberspace. Nato Rev. (2019)
3. Ballmer, D.: Ukraine Proves Cyber War Has No Borders. BlackBerry (2022)
4. Cisco: What Is Cybersecurity?. https://www.cisco.com/c/en/us/products/security/what-is-cybersecurity.html
5. Frank, I., Odunayo, E.: Approach to cyber security issues in nigeria: challenges and solution. Int. J. Cogn. Res. Sci. Eng. Educ. **1** (2013)
6. Pawlicka, A., Choraś, M., Pawlicki, M., Kozik, R.: A $10 million question and other cybersecurity-related ethical dilemmas amid the COVID-19 pandemic. Bus. Horiz. **64**, 729–734 (2021). https://doi.org/10.1016/j.bushor.2021.07.010
7. Wustrich, L., Pahl, M.-O., Liebald, S.: Towards an Extensible IoT Security Taxonomy. In: 2020 IEEE Symposium on Computers and Communications (ISCC). pp. 1–6. IEEE (2020). https://doi.org/10.1109/ISCC50000.2020.9219584
8. International Telecommunications Union (ITU).: ITU-TX.1205: series X: data networks, open system communications and security: telecommunication security: overview of cybersecurity 2008 (2008)

9. Qadir, S., Quadri, S.M.K.: Information Availability: An Insight into the Most Important Attribute of Information Security. J. Inf. Secur. **07**, 185–194 (2016). https://doi.org/10.4236/jis.2016.73014
10. Gargano, F.: Three Common Threat Actors and the One You Might Not Know About.
11. Pawlicka, A., Choraś, M., Pawlicki, M.: Cyberspace threats: Cyberspace threats: not only hackers and criminals. Raising the awareness of selected unusual cyberspace actors - cybersecurity researchers' perspective. In: Proceedings of the 15th International Conference on Availability, Reliability and Security. pp. 1–11. ACM, New York, NY, USA (2020). https://doi.org/10.1145/3407023.3409181
12. Ahmad, R., Yunos, Z.: A Dynamic Cyber Terrorism Framework. Int. J. Comput. Sci. Inf. Secur. **10**(2) (2012)
13. Dziundziuk, D.V.: Stopping Cyber terror Countries must work together to thwart efforts of internet Criminals. Per Concordiam, J. Eur. Secur. Def. 2
14. Pawlicka, A., Choraś, M., Pawlicki, M.: The stray sheep of cyberspace a.k.a. the actors who claim they break the law for the greater good. Pers. Ubiquitous Comput. **25**, 843–852 (2021). https://doi.org/10.1007/s00779-021-01568-7
15. Ohlin, D., Govern, K., Oxford, C.F.: Nicolò Bussolati " The Rise of Non-State Actors in Cyberwarfare (2015)
16. Gorwa, R.: Computational Propaganda in Poland: False Amplifiers and the Digital Public Sphere (2017)
17. Sigholm, J.: Non-State Actors in Cyberspace Operations. J. Mil. Stud. **4**, 1–37 (2013). https://doi.org/10.1515/jms-2016-0184
18. Oracle: What is IoT? Oracle India (2019)
19. ENISA: Baseline Security Recommendations for IoT (2017)
20. Bugeja, J., Jacobsson, A., Davidsson, P.: An analysis of malicious threat agents for the smart connected home. In: 2017 IEEE International Conference on Pervasive Computing and Communications Workshops (PerCom Workshops). pp. 557–562. IEEE (2017). https://doi.org/10.1109/PERCOMW.2017.7917623
21. Ronen, E., Shamir, A., Weingarten, A.-O., O'Flynn, C.: IoT Goes Nuclear: Creating a Zigbee Chain Reaction. IEEE Secur. Priv. **16**, 54–62 (2018). https://doi.org/10.1109/MSP.2018.1331033
22. CSA.: Future-proofing the Connected World: 13 Steps to Developing Secure IoT Products (2016)
23. Stavridis, J., Weinstein, D.: The Internet of Things Is a Cyberwar Nightmare. Foreign Policy (2016)
24. Bobbio, A., Campanile, L., Gribaudo, M., Iacono, M., Marulli, F., Mastroianni, M.: A cyber warfare perspective on risks related to health IoT devices and contact tracing. Neural Comput. Appl. (2022). https://doi.org/10.1007/s00521-021-06720-1
25. CISOMAG.: Cybercriminals Will leverage IoT and 5G for Large-Scale Attacks. Cisomag (2022)
26. Slaney, R.: SecurityScorecard Discovers new botnet, 'Zhadnost,' responsible for Ukraine DDoS attacks. SecurityScorecard (2022)
27. Bajak, F.: Ukraine says potent Russian hack against power grid thwarted. AP News (2022)
28. Turk, Z.: Technology as enabler of fake news and a potential tool to combat it. Eur. Parliam. (2018)
29. Kozik, R., Kula, S., Choraś, M., Woźniak, M.: Technical solution to counter potential crime: text analysis to detect fake news and disinformation. J. Comput. Sci. **60**, 101576 (2022). https://doi.org/10.1016/j.jocs.2022.101576
30. Choraś, M., Demestichas, K., Giełczyk, A., Herrero, Á., Ksieniewicz, P., Remoundou, K., Urda, D., Woźniak, M.: Advanced Machine Learning techniques for fake news (online disinformation) detection: a systematic mapping study. Appl. Soft Comput. **101**, 107050 (2021). https://doi.org/10.1016/j.asoc.2020.107050
31. Allcott, H., Gentzkow, M.: Social media and fake news in the 2016 election. J. Econ. Perspect. **31**, 211–236 (2017). https://doi.org/10.1257/jep.31.2.211

32. Cambridge Dictionary: Fake News
33. Wardle, C.: Fake news. It's complicated. First Draft (2017)
34. Lind, D.: President Donald Trump finally admits that "fake news" just means news he doesn't like. https://www.vox.com/policy-and-politics/2018/5/9/17335306/trump-tweet-twitter-latest-fake-news-credentials (2018)
35. Giuliani-Hoffman, F.: "F*** News" should be replaced by these words, Claire Wardle says. CNN Bus. (2017)
36. Allcott, H., Gentzkow, M.: Social Media and Fake News in the 2016 Election. J. Econ. Perspect. (2017)
37. Belluz, J.: Research fraud catalyzed the anti-vaccination movement. Let's not repeat history (2019)
38. Islam, M.S., Sarkar, T., Khan, S.H., Mostofa Kamal, A.-H., Hasan, S.M.M., Kabir, A., Yeasmin, D., Islam, M.A., Amin Chowdhury, K.I., Anwar, K.S., Chughtai, A.A., Seale, H.: COVID-19–related infodemic and its impact on public health: a global social media analysis. Am. J. Trop. Med. Hyg. **103**, 1621–1629 (2020). https://doi.org/10.4269/ajtmh.20-0812
39. Curtin, L.: The 'fake' news effect. Am. Nurse. (2020)
40. Choraś, M., Pawlicka, A., Kozik, R., Woźniak, M.: How machine learning may prevent the breakdown of democracy by contributing to fake news detection. IT Prof. (2022)
41. Alba, D., Satariano, A.: At Least 70 Countries have had disinformation campaigns, study finds. New York TImes (2019)
42. Sługocki, W., Sowa, B.: Disinformation as a threat to national security on the example of the COVID-19 pandemic. Secur. Def. Q. (2021) https://doi.org/10.35467/sdq/138876
43. Kim, B., Xiong, A., Lee, D., Han, K.: A systematic review on fake news research through the lens of news creation and consumption: research efforts, challenges, and future directions. PLoS ONE **16**, e0260080 (2021). https://doi.org/10.1371/journal.pone.0260080
44. Thomsen, I.: Russian disinformation is part of its war effort in Ukraine. How can the West respond? News@Northeastern (2022)
45. Martewicz, M.: War in Ukraine sparks panic buying at polish fuel stations. bloomberg (2022)
46. Jazeera, A.: Do not call Ukraine invasion a 'war', Russia tells media, schools. Al Jazeera. https://www.aljazeera.com/news/2022/3/2/do-not-call-ukraine-invasion-a-war-russia-tells-media-schools (2022)

Information Security Risk Awareness Survey of Non-governmental Organization in Saudi Arabia

Mariyam Hassan, Kawther Saeedi, Haya Almagwashi, and Suaad Alarifi

Abstract Nowadays, the adoption of digital technology for NGOs (Non-Governmental Organizations) has become essential and unavoidable no matter how small the NGO are. Using technology come with its risk and opportunities; to control the risk an appropriate information security risk assessment methodology should be adopted. It helps to assess the risks and identify security requirements to protect information as well as maintain its confidentiality, integrity and availability. Furthermore, discovering vulnerabilities on the systems and defending threats help reduce risks and control the level of uncertainty NGOs facing. However, complying with information system risk assessments standards requires knowledge and experience on information security management which is a challenge for most of NGOs in Saudi Arabia as often lack resources. This paper contributes to demonstrates an analysis approach providing insight into the current awareness of information security risks in NGOs in Saudi Arabia. A survey was conducted on a sample of 168 NGOs accredited by the Ministry of Human Resources and Social Development (MHRSD) in Saudi Arabia were selected using a multi-stage stratified sampling approach. The results show a lack of security awareness in terms of protecting information security and the need for a straightforward tool to help assisting the information security risk with limited expertise and recourses.

Keywords Risk assessment · Information security · Non-governmental Organizations · Risk management

---

M. Hassan (✉) · K. Saeedi · H. Almagwashi · S. Alarifi
Information Systems Department, King Abdulaziz University, Jeddah, Saudi Arabia
e-mail: mhassan0056@stu.kau.edu.sa

K. Saeedi
e-mail: ksaeedi@kau.edu.sa

A. Visvizi et al. (eds.), *Research and Innovation Forum 2022*,
Springer Proceedings in Complexity, https://doi.org/10.1007/978-3-031-19560-0_4

1 Introduction

Non-Governmental Organizations (NGOs) are defined as the third sector (voluntary or civil society) which are supported by their members to solve some social problems and issues in various fields such as health, education, relief, and other [1]. NGOs are not-for-profit organizations, and any surplus is invested in supporting the activities of the organization. Ideally, NGOs are not government entities even if they receive financial support from the government [2]. The most important characteristic of NGOs is their reliance on service innovation, passion and commitment of stakeholders combined with limited resources [3]. Even with limited resources, NGOs still have to comply with regulations and governance which is the main driver of change in NGOs [3]. In Saudi Arabia, NGOs are accredited as an independent agency and authorized by the Ministry of Human Resource and Social Development (HRSD). NGOs are classified by HRSD according to their activities including education, health, social services and others [4].

As technology and information systems have evolved to become increasingly reliable as an enabler for organizations, the adoption of digital technology for NGOs has become essential [3]. Governments are also moving towards e-government to improve the quality and effectiveness of services, forcing some Non-Governmental Organizations to link their systems with government systems to facilitate data exchange and service delivery [5]. Also, some governments are demanding Non-Governmental Organizations to comply with the standards set for information security management in all sectors of the state, for example, in Saudi Arabia, the National Cybersecurity Authority [6] has established a set of controls and guidelines that all organizations must comply with in order to implement e-governance; where e-governance is considered one of the basics that support the development of smart cities. Smart cities are multidimensional systems that rely on technology to improve the lives of citizens and improve the quality and effectiveness of public services [7].

The development of technology has contributed to the increase of risks to information security, forcing organizations to adopt information security risk assessment methodologies to control the risks and maintain business continuity. Complying with information security policies can provide a level of security and assurance to organizations, however, risk management processes are the bases of security and assurance according to best practices [8, 9]. Usually, organizations choose the appropriate information security risk assessment methodology that helps them identify their information security requirements and assess the assets to be protected against possible threats. There are many information security management standards and approaches available to organizations such as the National Institute of Standards and Technology (NIST), International Organization for Standardization [10], Organization for Economic Co-operation and Development (OECD) and others. These standards focus on protecting the enterprise's assets from threats that might exploit vulnerabilities exist on the enterprise assets [10–12]. NIST guidelines are considered one of the more detailed guidelines in terms of the steps for risk management compared to other standards [13].

NGOs are facing some obstacles in adopting the existing information security management standards. Lack of resources, lack of experienced staff, and limited budget are considered factors obstructing the development of security plan and other information security risk assessment procedures. Another contributing factor is that some NGOs are not considering information security as a priority [14, 15]. Moreover, with the adoption of e-government, it can be difficult for NGOs with their limited resources to assess risks and predict threats that may impact their information security. Third sector is part of the economic system that is connected to the e-government and private sector; it is of no exception in complying with regulations [2, 5].

To sum up, NGOs have to follow regulations and perform risk assessment, however, the available risk assessment approaches are considerably advance and complicated for the unarmed NGOs to follow [1, 14]. Security requirements of NGOs are different from regular (for-profit) organizations due to the difference in assets, goals, strategic plans, and the extended effect of any reputation damage on the support provided by donors and volunteers. All these factors make NGOs different than regular organizations in regard of security requirements [14].

The goal of this study is to examine the necessity for NGOs to have a suitable information security risk assessment methodology that provides an effective way for assessing information security risk while considering NGOs' unique security requirements. To do this, we created a survey that assesses the internal and security situation in non-governmental organizations. In choosing the most relevant areas to be covered in the survey, we relied on related research that discussed cybersecurity risks of NGOs. We used the multi-stage stratified sampling approach to sample the questionnaire because it takes into account the sample conditions accessible from non-governmental organizations that have been approved by the Ministry of Human Resources and Social Development. After reviewing the findings, we concluded that NGOs require a cybersecurity risk assessment approach that takes into consideration their security requirements and needs. Thus, the research contributed to providing insight into the current awareness of information security risks in NGOs in Saudi Arabia.

The rest of the paper is organized as following: next section presents the security challenges of NGOs in addition to the existing practices and approaches of information security risk assessment for NGOs. The third section discusses the research methodology. The section after, discusses the results of data analysis. Conclusion are presented in the last section.

2 Theoretical Framework

This section presents two aspects: first, the basic concepts of information security risk assessment and define the terms that we will mention. Second, studies that discussed cybersecurity issues for NGOs.

2.1 Basic Concepts

Since the main objective of this research is to evaluate the current situation of NGOs about awareness of information security risks and to verify the need of NGOs for an appropriate information security risk assessment approach; therefore, it is important to define these terms.

According to NIST [16]:

- Risk management can be defined as the overall process of identifying, controlling and mitigating risks related to an information system. Risk management consist of many procedures including risk assessment.
- Risk assessment is the first step of risk management procedures, which identifies the extent of the potential threat and risks associated with information systems, as well as their impact on the organization.
- Risk is the possibility that a threat source may exploit a potential vulnerability, resulting in an adverse impact on the organization.
- Threat refers to the possibility that a particular source may exploit a particular vulnerability.

Risk assessment encompasses nine stages include: (1) define system characterization. (2) Threat Identification. (3) Vulnerability Identification. (4) Control Analysis. (5) Likelihood Determination. (6) Impact Analysis. (7) Risk Determination. (8) Control Recommendations. (9) Results Documentation [16]. The members' awareness of the importance of comply with the implemented controls that maintain the security of their organization is one of the criteria that determines the success of an organization risk management program [16].

2.2 Related Work

This section presents studies related to security issues of NGOs in recent years. The issues discussed in most studies are centered on two aspects: one is the security challenges of NGOs and its effect on risk level. Second, methods and practices that is used to enhance NGOs security.

NGOs Security Challenges

NGOs available resources are usually invested in their core competencies which are provided to the community in terms of education, care, health, environmental protection, and other different services [14]. Since these organizations focus on helping the community instead of making profits through selling goods and services, they usually suffer from a lack of resources and depend on government funding as the main source of income [3]. The largest share of income is allocated to support the services provided by these organizations to the community and ensure their continuity, while Minimal budgets are allocated to IT management and support [14]. It is not unusual that non-IT employee manages the IT department in these types of organizations [14], Also, their assets, goals and strategic plans differ from regular (for profit) organizations and do not focus on information security as much as it focuses on providing services to society despite the scarcity of resources. These factors make them vulnerable to attacks that may exploit their vulnerabilities in computers, networks, as well as human factors [14]. With the rapid development of technology and communications, NGOs are one of the targets that criminal hackers exploit to cause damage, whether financial, physical or other [17].

Al Achkar [18] discusses the state of digital insecurity that NGOs may face with their dependence on technology in their daily operations. Digital insecurity means the absence of strong data protection policies and practices, weak encryption protocols and infrastructure defects, in addition to sharing sensitive data inadvertently. The researcher discussed the impact of the state of digital insecurity on NGOs, which makes them vulnerable to the risk of data dissemination and exploitation, spreading rumors and misleading statements about the organization, which negatively affects its reputation. Furthermore, Mois [19] mentioned that one of the most prominent challenges facing NGOs from the use of mass media and social media is cyber-attacks such as phishing attacks.

Carey-Smith et al. [14] has used a case study of two NGOs to discover the challenges facing these organizations and determine their security needs. Both organizations did not have risk assessment or information security management, but one of them had information security practices outdated and not fully implemented. Furthermore, the main concepts of security management were not well understood in addition to poor planning, poor communication within the team, and lack of information security management resources. All these factors may lead to the loss of significant assets to these NGOs. This study enables us to visualize the general security situation in NGOs and the security challenges they face; in addition to understand the points that must be taken into consideration when discussing security issues related to NGOs.

Rice [20] also has used a case study of a non-profit contracting company called Sequoia, to determine the challenges of information technology infrastructure that affect information security. The risks facing Sequoia is that their data might be accessed or changed by attackers via the Internet. Also, the spread of viruses between devices is a major threat to the organization as well as infrastructure defects that cause network outages or congestion. Furthermore, Kolb and Abdullah [21] listed some risks that NGOs may face in the event of security breaches, including financial loss, reputation loss, damage staff morale, and donor loss. The threats and risks that NGOs may face have been summarized in Rice et al. study, and this gives us a greater understanding of the extent of the losses these organizations may experience if they are exposed to these risks.

Mierzwa and Scott [12] conducted a survey directed to technical groups focusing on NGOs to discover security controls, assessments, and programs that are employed to manage information security in their organizations. According to the survey results, half of the NGOs do not have information security units nor security experts personnel. Moreover, 86% of these organizations do not have plans to create information security units or hire experts in cybersecurity. Regarding the applied information security frameworks and methodologies, it was found that the vast majority use an internally developed framework, while about 32% reported that they follow NIST guidelines. Moreover, 50% of these NGOs have been subjected to ransomware attacks, most of whom have recovered by restoring backups. Although this survey lacks some detailed questions regarding the frameworks used to manage information security, it provided a general sense of what NGOs need.

Another survey conducted by Imboden [22] for a wide range of individuals who hold management and technical positions in NGOs. According to the survey, nearly half of these NGOs do not have information security policies, while two-thirds of these NGOs have been exposed to security incidents.

The most prominent finding from previous studies is the difference between the statistics of regular organizations and non-governmental organizations, the level of awareness of information security and developing security plans to address risks in NGOs is considered low compared to regular organizations. For example, the results of a survey conducted on Small to Medium-sized Enterprises (SMEs) showed that about 81% of the organizations participating in the survey consider information security as a priority to be taken into consideration, 77% of these organizations issued a set of their own security policies. Where 50% of them have human errors as a main source of vulnerabilities [23]. As the results show, most SMEs have their own security policies while the majority of NGOs do not have any security plans or policies (Table 1).

We conclude from the previous studies that NGOs are security unconscious to a large extent although they are at risk. Risk that might affect confidentiality, privacy, and integrity of data without having an effective recovery plan is a major threat facing NGOs. Viruses and the damage they cause to the organization's devices is also one of the concerns that must be considered by NGOs. Moreover, network or server disruption due to infrastructure defects may pose a threat to the NGO. All these threats if not properly managed will inevitably lead to damage to the NGO reputation, loss of financial resources, and possibly loss of donor confidence. These reasons call for considering an appropriate approach for NGOs to information security risks assessment designed with the consideration of NGOs nature and special needs.

Existing practices and approaches to manage information security in NGOs

Most of the recommended practices and approaches in this section are domain-agnostic and not limited to NGOs, but we chose to present them because they are suggested and recommended for the NGO field by researchers. Most studies in this section focused on the importance of awareness programs directed to employees to make them aware of the importance of information security risk assessment and how to comply with best practices that contribute to security management in NGOs. Kolb and Abdullah [21] mentioned that most NGOs focus on fundraising and operations management rather than information security management and risk assessment. They relied on NIST guidelines to provide recommendations for an information security awareness program targeting employees. Recommendations include approval and

Table 1 Summarize the security challenges of NGOs

Author	Aim of study	Finding
Al Achkar [18]	To discuss the impact of digital threats on NGOs and suggest approaches that address digital security concerns	NGOs vulnerable to the risk of data dissemination and exploitation, spreading rumors and misleading statements about the organization, which negatively affects its reputation
Mois [19]	To discuss how NGOs use different means of communication for social change	NGOs are vulnerable to cyber attacks such as phishing attacks
Carey-Smith et al. [1]	To discover the challenges facing NGOs and determine their security needs	Both organizations did not have risk assessment or information security management, there was no full awareness about the management of information security, poor communication within the team, limited time and lack of resources
Rice [20]	To determine the challenges of information technology infrastructure that affect information security	Data can be accessed or changed by the public via the Internet, the spread of viruses between devices considered a potent threat, infrastructure defects cause network outages or congestion
Kolb and Abdullah [21]	To identifies the potential information security risks faced by NGOs	Financial losses, loss of reputation, damage staff morale, and donor loss
Mierzwa and Scott [12]	To learn about the controls, assessments, and programs that are employed to manage information security in NGOs	Half of the NGOs do not have information security units or experts 86% of NGOs do not have plans to create information security units or hire experts in cybersecurity The majority use an internally developed framework, while about 32% follow NIST guidelines 50% of these NGOs have been subjected to ransomware attacks
Imboden [22]	To explore the adoption as well as attitudes regarding information security policies at NGOs	Nearly half of NGOs do not have information security policies Two-thirds of NGOs have been exposed to security incidents

support from top management, which will ensure adequate resources. Also, assemble a team with different expertise, including legal and human resources, in addition to information technology, then assess the environment and review policies and amendments to what needs to be changed. Survey employees to measure their level of knowledge of policies and started training sessions to enhance security awareness.

Furthermore, Al Achkar [18] discusses some strategies that would help in addressing the digital risks that may face NGOs. One of the proposed strategies is to build internal capabilities and train employees to avoid practices that lead to digital risks in addition to train them to spot weaknesses within the capabilities available to the organization. The researcher also suggests that NGOs share good practices and lessons learned with local communities. Moreover, having a network of experts to work on developing minimum acceptable standards that can be expanded to include the largest number of NGOs can improve their digital risk environment.

Carey-Smith et al. [1] set out an approach for improving information security management in NGOs. They choose action research as a methodology and the technology acceptance model as part of the theoretical framework underpinning the research project. Their framework aims to help NGOs employees to continue improving their skills and capabilities without needing the help of an external "expert." This framework corresponds to the attributes of NGOs that are consensus-based decision making and feeling of informality in relationships between members. These features distinguish them from small enterprises. However, the researchers could not get results because of some difficulties that occurred in conducting the research with NGO.

Ghani et al. [24] discussed the impact of employees' understanding of risk assessment process on risk management in NGOs. They used surveys to prove their hypotheses that employee's knowledge and understanding of risk identification, assessment, and management will contribute to risk control. The researchers stated that the results might not be realistic due to small sample, only one NGO has included. As a result of this study, the researchers believe that there is an urgent need to provide awareness programs directed to employees to improve their knowledge of risk management.

Carey-Smith et al. [14] described the applicability of ISO27001 in NGOs and stated that this standard is not compatible with NGOs because it is relatively expensive and difficult to understand for those who lack knowledge and experience in information security management. He suggested some recommendations to help NGOs in the management of information security, included the effective management of finance and human resources, hiring people with the necessary expertise and skills to manage information security, and creating awareness programs that emphasize the importance of information security risk assessment. Moreover, encouraging students and graduates to train and work in NGOs to prepare them to solve the real problems facing information security.

Along with staff awareness programs on information security risk assessment, some research has discussed the need to develop an information security risk management plan. Rice [20] stressed the importance of developing a security plan containing an accurate risk assessment of the NGOs. Although it is difficult for NGOs to develop a such plan, it is essential to protect critical assets and data and all departments and staff must prioritize information security risk assessment. Also, Mierzwa and Scott [12] believes that because of some limitations for NGOs in the face of lack of resources, it is possible to use existing information security standards such as ISO 27001 and NIST guidelines and develop a security plan that includes steps to be taken when there are security concerns in addition to enhancing awareness for IT staff to achieve the information security risk assessment activities in the NGOs.

Imboden [22] stated that budget is the first factor determining whether NGOs are prepared to adopt information security management approach or not. Even if the organization holds sensitive data, the budget remains the factor that determines whether to consider security or not. They also mentioned that there are ready-made and reliable security templates and standards that could be enough with routine training and discussion of information security policies and practices among staff.

On the other hand, Lin [5] suggested some practices for governments that adopt e-government in dealing with NGOs, which include improving online services by adding open platforms that help NGOs, improving data management and sharing with NGOs, protecting information security and actively working on enhancing the advantages of e-government in various departments.

Previous studies have discussed the aspect of improving the information security awareness among employees by providing training programs which we believe will require resources NGOs might not have. Also, some research discussed the importance of employing people with expertise in cybersecurity management and developing a security plan to deal with risks that may face the NGOs or rely on existing standards of information security risks assessment such as ISO 27001 and NIST guidelines; while we believe NGOs lack the ability to recruit security experts. Although some researchers suggest using the existing standards of information security risks assessment, but we did not come across research that tests the suitability or the methodology of complying with these standards for NGOs (Table 2).

Previous studies helped to understand the nature of NGOs, the challenges or security concerns they face in general, and what recommendations are proposed to reduce them. Our research will focus on studying the information risk assessment for NGOs in Saudi Arabia aiming to assist their awareness and capture the challenges faced to comply with information security standards. Then outline an appropriate solution to assist them in assessing information security risks.

Based on previous research into the extent to which NGOs are aware of cybersecurity risks and their lack of interest in this field, we've determined how these organizations should be addressed and what the most critical aspects of our survey should be covered. Because of the lack of awareness of cyber security among these organizations, we utilized simple vocabulary that is more easily understood by the public. Furthermore, the best practices from the research aided us in crafting questions that allowed us to better analyze NGOs' internal and security conditions.

Table 2 Summarize the information security best practices and approaches in NGOs

Author	Aim of study	Finding
Kolb and Abdullah [21]	To provides recommendations to implement an information security awareness program	NIST Guidelines are appropriate for making recommendations for an information security awareness program targeting employees
Al Achkar [18]	To discuss the impact of digital threats on NGOs and suggest approaches that address digital security concerns	Build internal capabilities and train employees to avoid practices that lead to digital risks, share good practices and lessons learned by NGOs with local communities. Moreover, having a network of experts to work on developing minimum acceptable standards that can be expanded to include the largest number of NGOs can improve their digital risk environment
Carey-Smith et al. [14]	To develop a framework aims to help NGOs employees to continue improving their skills and capabilities without needing the help of an external "expert."	There are some frustrations that occurred in conducting research with NGO and the researchers could not get results
Ghani et al. [24]	To discuss the impact of employees' understanding of the risk assessment process on risk management in NGOs	There is an urgent need to provide awareness programs directed to employees to improve their knowledge about risk assessment
Carey-Smith et al. [1]	To discover the challenges facing NGOs and determine their security needs	ISO27001standard is not compatible with NGOs There is a need for the effective management of finance and human resources, hiring people with the necessary expertise and skills, in addition, creating awareness programs for employees and encouraging students and graduates to train and work in NGOs
Rice [20]	To determine the challenges of information technology infrastructure that affect information security	There is a need for developing a security plan containing an accurate risk assessment of the NGOs It takes some time to implement the security in Sequoia

(continued)

Table 2 (continued)

Author	Aim of study	Finding
Mierzwa and Scott [12]	To learn about the controls, assessments, and programs that are employed to manage information security in NGOs	It is possible to use standards such as ISO 27001 and NIST guidelines to develop a security plan There is a need for enhancing awareness for IT staff in NGOs
Imboden [22]	To explore the adoption as well as attitudes regarding information security policies at NGOs	The budget is the first factor that determines whether NGOs arc prepared to adopt information security management Routine training and discussion of information security policies and practices among staff could be useful
Lin [5]	To propose a more complete NGOs management model based on government	There are some practices for governments that adopt e-governance in dealing with NGOs, which include improving online services and data management and protecting information security and actively working on enhancing the advantages of e-government in various departments

Compliance approaches with existing standards of information security management

In the previous section, we discussed the most critical practices and approaches that may help NGOs to manage the information security, essential points they focused on is staff awareness programs and use of available information security management guidelines and standards such as ISO27001 and NIST. Mark Carey-Smith stated that there are many efforts to produce standards for information security management; however, there is no literature examining the suitability of these standards for NGOs [14].

Small and medium-sized enterprises (SMEs) have some characteristics similar to NGOs, as they have limited resources in terms of assets and workforce in addition to financing. The strategic plans of regular organizations including SMEs differ from NGOs, and therefore they differ in some characteristics, such as being dependent on reaping their profits mainly from selling products or services; since profit is an essential factor in the business continuity of these organizations [25]. Otherwise, NGOs depend for their survival on government funding, donations, and volunteer individuals as a workforce. Reputation is one of the fundamental intangible assets of NGOs, as it affects the set of variables that determine business continuity which includes the confidence of donors and their continued support to the organization's activities, as well as the volunteers continuing to work alongside the organization [26].

Limited resources in both SMEs and NGOs affect the level of maturity of information technology and attention to information security. Lack of awareness and the absence of effective policies to manage information security increases the vulnerability of these organizations to attacks and intrusions. Nevertheless, awareness about the importance of information security risk assessment may help SMEs to think seriously about allocating a reasonable portion of their budget to training their employees on the most important practices and policies that help in reaching the lowest level of information security risk assessment. These decisions approved by senior management and all employees can be obligated to it. Also, existing standards for information security management such as ISO 27001 and NIST may be effective with SMEs because it required the support of senior management.

As for NGOs, they face a greater risk if they are attacked or infiltrated, as this will negatively affect their reputation and consequently the continuity of their business. Nevertheless, awareness of the importance of information security is not enough, because they are unable to use strict recruitment practices or enforce training opportunities for their volunteers.

It may be useful to consider how SMEs comply with existing information security management standards as they are similar to NGOs in case of limited resources [1] There are some studies discussed the SMEs compliance with information security standards. Susanto and Almunawar [23] proposed an integrated solutions system (ISF) to measure the level of compliance with the ISO27001 standard. The system contains two sections: electronic assessment and electronic monitoring, and has been tested in terms of reliability, performance, and usability on large, medium, and small organizations. However, the evaluation results and details of the proposed system are not included in the declared part of the study.

Montenegro and Moncayo [27] have discussed the obstacles that faced SMEs in applying risk assessment approaches to information security and proposed the development of a hybrid model that combines different risk assessment approaches including ISO 27005, OCTAVE-S and MAGERIT. Their model is applicable and involves managers and operators with limited IT experience and can contribute to reducing information security risks to these enterprises. Also, Valdevit and Mayer [28] believes that SMEs face problems when implementing an information security risk assessment system. They proposed a gap analysis tool to analyze the compliance level between the current situation of SMEs and the requirements of the ISO27001 standard. Their research methodology involves modeling the requirements of ISO27001 by simplifying its structure, then designing the assessment tool, which is in the form of a hierarchical questionnaire, then the testing phase. The results showed that this tool improves the efficiency of the gap analysis task.

Moreover, given that risk assessment is a necessary and arduous process as it contains many elements, Valdevit and et al. [29] have developed an implementation guide for SMEs to deploying an ISMS, which can help them to get ISO 27001 certificate through simplifying the information security management procedures while considering the lack of resources and ignore the unnecessary processes. In the same context, Richard Henson et al. [30] have developed the IASME standard for SMEs, which uses the risk profile that contains the classification of risk in order to prioritize controls according to the risk profile; therefore, the number of controls to be taken into account will be lower than that of ISO 27001 implementation procedures.

Furthermore, Ponsard and Grandclaudon [31] compared a set of standards and approaches to information security management in terms of its suitability for SMEs. They considered that the ISO27001 standard was not suitable for SMEs because of its complexity to contain many directives, while the NIST standard was appropriate.

However, to the author's knowledge, some researchers suggest using the existing standards of information security management, but there are no researches study the suitability of compliance or the methodology of compliance with existing standards of information security management for NGOs. Nevertheless, given the fact the NIST guidelines are suitable for SMEs, they may also be suitable for NGOs due to their similarities in terms of limited resources and limited experience.

3 Research Methodology

Since previous research assumed that the security requirements of NGOs differ than regular organizations, the research used the quantitative and qualitative approach to evaluate the current situation of NGOs in Saudi Arabia and measure the suitability of the of information security risk assessment standards for them. We used the survey tool because it is one of the approved methods for data collection in both quantitative and qualitative approaches through diversifying between the use of closed and open questions [32]. There are also many similar studies that have used the survey for their research methodology [33, 34]. The survey questions were developed and structured based on the findings drawn from related work (Fig. 1).

3.1 Sampling

The survey was conducted on a sample of accredited NGOs by the Ministry of Human Resources and Social Development (MHRSD) in Saudi Arabia. The survey sample frame is MHRSD's directory of accredited NGOs which is available on MHRSD's website [4]. The number of NGOs is 648, most of them are classified as social, health and educational organizations according to the directory until the date of 2014, which is the latest version available on the website, until the date of access in 2021.

The sample design used in this research is Stratified multi-stages Sampling. This methodology is best suited for this research due to the heterogeneity of the study population. The concept of this methodology is dividing the sample into several homogeneous stages and each stage is divided into a group of homogeneous sub-stages [35, 36].

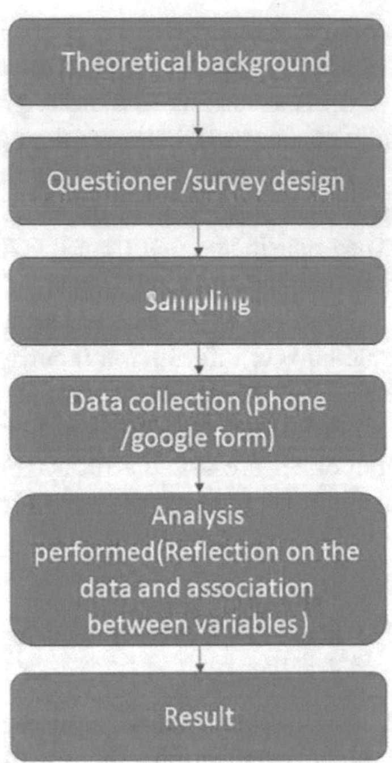

Fig. 1 Methodology steps

In this research, the sample was divided into homogeneous groups according to the administrative regions in Saudi Arabia, each group known as a stratum. Each stratum is divided into subgroups according to the size of the NGO (Micro, small, Medium, Large). Size description is taken from the official definition of the organization's sizes issued by the MHRSD and The General Authority for Small and Medium Enterprises "Monshaat" [4, 37]. The advantage of using the stratified sample is to obtain results at the levels of regions and provinces. Also, it is important for the research sample to be representative of the Saudi NGOs community [38].

To determine sample size of research, Cochran formula and Finite Population Correction for Proportions formula have been adopted [39, 40]. To determine the sample size in each stratum the proportional distribution method was applied [41, 42].

For the research sample, the acceptable margin of error is 5%, while the confidence level is 95%. According to a group of studies, these percentages are acceptable [33, 34, 43].

Due to the lack of updates on the NGO Directory issued in 2014 by MHRSD, many information are inaccurate or missed. Therefore, some measures have been taken to enhance the quality of the used sample frame and increase the response rate. The NGOs not registered with a phone number or registered with wrong phone numbers and cannot be contacted through other methods have been excluded from the sample frame. Also, other resources such as the organizations' accounts on social media sites have been adapted to obtain phone numbers and fill some gaps in the sample frame. These steps have been recommended in many previous studies [33, 44, 45].

Table 3 Sample frame

Region	Micro (1–5 full-time employees)	Small (6–49 full-time employees)	Medium (50–249 full-time employees)	Large (250 and more full-time employees)	Total	Percentage (%)
Riyadh	28	32	2	2	64	22
Makkah	28	32	6	0	66	23
Madina	7	9	2	0	18	6
Qassim	9	12	2	2	25	9
Eastern Region	4	14	4	0	22	8
Asir	18	9	2	0	29	10
Tabuk	2	10	0	0	12	4
Hail	9	10	0	0	19	7
Northern Borders Region	2	8	0	0	10	3
Jazan	6	2	0	0	8	3
Najran	2	1	0	0	3	1
Al-Baha	6	3	0	0	9	3
Al-Jouf	1	4	0	0	5	2
Total	**122**	**146**	**18**	**4**	**290**	**100**

After updating the sample frame, approximately 53% of the original sample frame. in other words, 290 out of 550 NGOs were included in the sample frame (Table 3), the rest considered unusable and have been excluded. Thus, the number of responses required for the survey is 165 responses (Table 4). The response rate with the updated sample frame was 75%, and the researcher was able to reach 168 responses.

3.2 Data Collection

We used an online survey through Google forms to conduct the survey with the executives or staff responsible for cybersecurity and information technology initiatives in NGOs. The survey was conducted through two ways: either by calling the individual concerned from the NGO and asking him the questions by phone, or by sending the survey to the person concerned via e-mail and following up with him/her through reminder messages. The first technique was used to decrease the non-response bias as well as to facilitate for respondents who have some difficulties in using technology and the Internet.

To minimize the outliers and improve sample viability, the iterative questioning methodology was used to validate some of the answers that may be unreal or not

Table 4 The sample required for the survey

Region	Micro (1–5 full-time employees)	Small (6–49 full-time employees)	Medium (50–249 full-time employees)	Large (250 and more full-time employees)	Total	Percentage (%)
Riyadh	16	18	1	1	37	22
Makkah	16	18	3	0	38	23
Madina	4	5	1	0	10	6
Qassim	5	7	1	1	14	9
Eastern Region	2	8	2	0	13	8
Asir	10	5	1	0	17	10
Tabuk	1	6	0	0	7	4
Hail	5	6	0	0	11	7
Northern Borders Region	1	5	0	0	6	3
Jazan	3	1	0	0	5	3
Najran	1	1	0	0	2	1
Al-Baha	3	2	0	0	5	3
Al-Jouf	1	2	0	0	3	2
Total	**70**	**83**	**10**	**2**	**165**	**100**

representative of reality. The proportion of answers that indicated outliers were very small and were excluded [46].

The contacting procedure was as follows: the person involved is called via his telephone number to receive his approval for the survey conducting through telephone or through an online survey. If a telephone survey is selected by the respondent, the respondent will be asked to decide the appropriate date. Then, he will be called later on the specified date. In case he does not reply on the specified date, the call will be repeated at subsequent times, or he will be asked to fill out the online survey. If the respondent prefers the online survey, the link will be sent to him and he will be reminded in subsequent periods to fill it out. Most of the survey calls lasted between 15 to 20 min per person. As for the reminder and follow-up, it was done every two days for a week, and then the organization would be excluded if its members did not respond.

The data collected from NGOs included information about the size of the organization and its branches, in addition to the most important practices used to manage and assess cybersecurity risks, as well as a set of questions that measured the level of awareness of cybersecurity issues, these survey elements were applied in another study [33]. The respondents' opinions on the most important characteristics that should be present in any proposed framework for assessing cybersecurity risks were included in the survey. Survey questions are attached in the appendix section.

3.3 *Ethical Consideration*

All the organizations that took part in the survey consented to participate on the condition of maintaining confidentiality, not disclosing any personally identifying information, and not disclosing the identity of the people who took part in the study. As a result, the information gathered will only be used for research and analysis and will not be made public.

4 Results and Discussion

A quantitative approach was used to analyze the data collected from the survey. SPSS software was used to perform the descriptive analysis and conduct the correlation tests to interpret the associations between the variables.

4.1 *Descriptive Analysis*

As mentioned previously, according to the sampling methodology the number of responses to the survey amounted to 168, the largest percentage of the sample was for micro and small-sized NGOs since they constitute the largest percentage of NGOs in Saudi Arabia (Table 5).

Cybersecurity functions in NGOs

According to the survey statistics, the vast majority of the respondents NGOs (74.5%) specifically micro and small-sized NGOs, do not have dedicated departments or personnel responsible for IT functions as well as for cybersecurity functions, whereas 23.6% of NGOs have a department or staff to handle the IT functions including cybersecurity functions. A very small percentage equal to 1.2 percent of respondents, opted to handle IT and cybersecurity activities by contracting with a third party (Table 6).

Table 5 Distribution of NGOs according to size

NGO size	Frequency	Percent
Large (250 and more full-time employees)	2	1.2
Medium (50–249 full-time employees)	9	5.4
Micro (1–5 full-time employees)	71	42.3
Small (6–49 full-time employees)	86	51.2

Table 6 The responsible for the cybersecurity functions

The responsible for the cybersecurity functions	Frequency	Percent
Contracting with a third party	2	1.2
Cybersecurity department	1	0.6
IT staff, one or more	39	23.6
Not set	123	74.5

Compliance with information security risk management measures

To assess the current situation in NGOs and predict the security risks that may face them, they were asked about the percentage of their commitment to implement the most important measures to manage information security risks. Regarding the existence of applied information security risk management policies, (32.2%) of respondents have applied policies for information security risk management, while (69.8%) do not have any policies. Most of those with policies in place were in the category of large and medium organizations.

Respondents were asked about the extent of their commitment to policies and to implement a set of basic and routine procedures in managing cybersecurity. The percentage of commitment to implementing the following measures was more than 75% which is high compared to others: maintaining customer privacy, as the percentage of NGOs committed to applying this measure reached about 79.8%. Also, applying some access control procedures such as securing NGOs devices with passwords (59.5%). In addition to taking regular backups of data (54.8%). On the other hand, the following procedures have witnessed varying proportions and are considered less common than the procedures mentioned previously, such as obligating employees not to share personal information with unauthorized individuals and obligating them to log out of their accounts upon completion of work. Moreover, activating anti-virus programs in addition to activating automatic updates for systems and software. There are some procedures that were considered uncommon and were not applied in large percentages, such as setting up firewalls and adopting a policy to change passwords periodically, as the non-application rates for these two measures reached 45.2% (Table 7).

To find out the reasons behind the NGOs' failure to apply all basic information security risk management procedures, the respondents were asked about the reasons behind their point of view, the largest percentage (37.8%) of them answered that the lack of experience in information technology is the main reason. Lack of financial resources (22.3%) and a lack of human cadres (20.4%) are also considered possible causes according to the respondents. However, some replied that information security is not a priority for the organization or there are no threats that affect information security for NGOs (Table 8).

Furthermore, respondents were asked whether their NGOs compatible with standard for information security risks assessment; the vast majority (92.4%) do not have any standard or approach for information security risks assessment, whereas 7.6% of the respondents adopt a standard and the most of them have adopted the guidelines of The National Cybersecurity Authority (ECC-2018).

Table 7 Compliance with information security risk management measures

	Applied, and the commitment ratio is 0%	Applied, and the commitment ratio is greater than 75%	Applied, and the commitment ratio is less than 25%	Applied, and the commitment ratio is less than 50%	Applied, and the commitment ratio is less than 75%	Not applied (%)
Set up firewalls	4.2	29.8	6.0	3.0	11.9	45.2
Applying access control measures to the organization's equipment, for example securing computers with a password	3.0	59.5	4.2	10.7	19.6	3.0
Activating antiviruses programs	4.2	42.9	7.7	17.3	19.6	8.3
Backup data periodically	2.4	54.8	5.4	10.7	16.7	10.1
Update programs automatically for all organization's devices	3.6	17.9	12.5	24.4	23.2	18.5
Set up a policy to change passwords periodically	3.0	19.6	9.5	6.5	16.1	45.2
Obliging employees not to share personal information or login data and passwords with unauthorized employees	3.6	45.2	2.4	7.7	22.0	19.0

(continued)

Table 7 (continued)

	Applied, and the commitment ratio is 0%	Applied, and the commitment ratio is greater than 75%	Applied, and the commitment ratio is less than 25%	Applied, and the commitment ratio is less than 50%	Applied, and the commitment ratio is less than 75%	Not applied (%)
Requiring employees to log out of their accounts upon completion of work	3.0	39.9	8.3	10.1	17.3	21.4
Employees are aware of the risk of downloading files attached to email messages that may lead to an email phishing attack	9.5	24.4	13.1	16.1	32.1	4.8
Applying procedures to maintain customer privacy	1.8	79.8	1.8	3.6	10.1	3.0

Table 8 The reasons for not adhering to policies for information security management

Respondents' opinions	Frequency	Percent
Information security management is not a priority for the organization	34	10.4
There are no threats affecting the information security	30	9.1
Lack of financial resources	73	22.3
Lack of human resources	67	20.4
Lack of experience in information security	124	37.8

To investigate the extent of compliance or knowledge of the respondents with the guidelines of the National Cybersecurity Authority (NCA), they were asked about the status of their compliance with the guidelines. More than half of the respondents (56.7%) reported that they had not heard of these controls before. While about 34 percent plan to comply with it in the future. For the remaining percentages, they are distributed among the compatible NGOs, whether they are highly compatible or partially compatible with the NCA guidelines (Table 9).

Table 9 Compliance with the National Cybersecurity Authority (NCA) controls

Status of compatibility	Frequency	Percent
Complaint >90%	8	6.3
Currently planning the compliance with it	34	26.8
Never heard of the controls and guidelines	72	56.7
Relatively complaint >75%	13	10.2

As for the obstacles to compliance with NCA guidelines, half of the respondents (50%) stated that the lack of experienced human cadres is one of the most important obstacles that face NGOs, and the lack of financial resources comes second (31.3%). 15.5% divided between the lack of an adequate information security risk assessment methodology and the difficulty of executing the existing approaches. Whereas individuals who consider information security not a priority for the NGOs considered that if they chose to comply with the controls, there were no obstacles.

Information security awareness

Awareness is one of cybersecurity's core elements, providing information security awareness programs for employees is an efficient way of gaining expertise that helps individuals concentrate on information security concerns. In addition, providing training programs for employees contributes to raising awareness and influencing their behavior to be consistent with the imposed security policies. Given the importance of the information security awareness aspect, the respondents were asked whether they have information security awareness programs for employees. According to the results, 15.1% have awareness or training programs directed to employees while the remaining percentage of respondents do not have any awareness or training programs to increase employees' awareness of information security (Table 10).

The proposed approach

At the end of the survey, respondents were asked about the specifications that they believe are important to be available in any proposed framework for information

Table 10 Providing information security awareness programs for employees * NGO size Crosstabulation

Providing information security awareness programs for employees	Large (250 and more full-time employees)	Medium (50–249 full-time employees)	Small (6–49 full-time employees)	Micro (1–5 full-time employees)
No	0.0	4.3	56.0	39.7
Yes	8.0	12.0	28.0	52.0

Table 11 The relationship between compliance with NCA controls and other variables

Compliance with NCA controls			
Variables	Fisher exact test	P	Phi coefficient
Providing information security awareness programs for employees	9.558	0.014	0.302
Applying policies for information security management	9.526	0.018	0.313
Adopting information security risk assessment standard	10.888	0.006	0.332

security risks assessment for NGOs. The feature that received the largest percentage of votes was defining the gaps affecting the compliance with the NCA guidelines, followed by the rest of the specifications, which includes considering financial resources and assets and proposing mitigation measures for information security risks.

4.2 Associations Between Variables

In this part, the relationship between the variables will be investigated to find out the factors affecting the NGO's compatibility with NCA controls. To study the relationship between variables chi-square and the Fisher Exact test were used to verify whether variables are related or not. Also, the Phi coefficient test was used to measure the effect size of the chi-square test.

After performing the test, we found that there are three variables associated with the NCA compliance, which are information security awareness for employees, applying policies for information security management and adopting information security risk assessment standards. The availability of all these factors or variables increases the possibility of compliance with NCA controls. For example, the organization's adoption of any information security risk assessment standard increases the likelihood of compliance with the NCA controls. The relationship between the compliance with NCA controls and other variables can be seen in (Fig. 2), where the p-value defines whether there is an association between variables. A p-value less than 0.05 indicates a correlation. The effect size and correlation details can be seen in (Table 11).

5 Conclusion

The main conclusion we reached from the previous survey is that the majority of NGOs in Saudi Arabia are considered small organizations. The most significant challenge confronting NGOs in general, and tiny and micro organizations in particular, is a lack of experience and resources, both financial and human, in addition

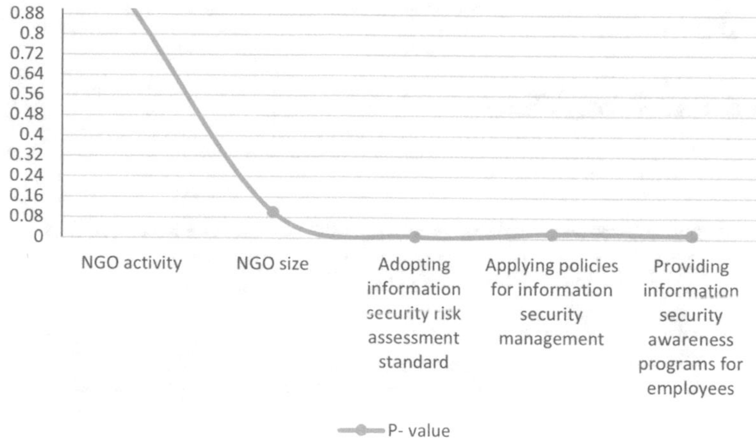

Fig. 2 The relationship between compliance with NCA controls and other variables

to lack of security awareness. Therefore, we find that many organizations are not committed to implementing basic information security management procedures on a regular basis. We can conclude that the current standards and approaches of information security risks assessment are considered insufficient for NGOs; since NGOs face many difficulties applying basic information security management procedures, due to the voluntary nature of NGOs and that all of their resources are used to improve the quality of provided to community services. Therefore, information security risks assessment was not a priority for the NGOs due to the belief of some NGOs that applying some basic measures such as securing devices with a password or installing anti-virus programs, even if they did not update it, is sufficient to protect the information security of the organization. While some other NGOs are aware of the information security concerns that may threaten the organization, but the lack of experienced human cadres in the first place and the lack of financial resources impede the organization in adopting steps that would provide more protection for information security in the organization. Also, the awareness aspect plays an important role in improving the current situation. As shown by the survey results, most NGOs do not have information security awareness programs for employees. While compatibility with NCA controls may become mandatory, particularly with the implementation of e-government, many NGOs do not have any background on what cybersecurity is and how to comply with NCA controls.

The purpose of this research is to assess the current situation of information security in NGOs located in Saudi Arabia and to study the security requirements and challenges facing them, which provides an added value to the local case and some benefits to the general community. For future, we will work to develop an information security risk assessment approach designed specifically for NGOs that is a comprehensive and meets their special security requirements.

Appendix

Appendix A: Survey Questions

Survey of the information security situation at NGOs

This survey is part of a research for a master's degree and is directed at Non-Governmental Organizations (NGOs) for the purpose of determining the state of cybersecurity for NGOs, in addition to measuring awareness and adherence to privacy and cybersecurity policies. Also, define the most important specifications to be met in any proposed framework to define cybersecurity strategies and risks.

mhassan0056@stu.kau.edu.sa (لا تتم مشاركته) تبديل الحساب

*مطلوب

* ?The name of NGO

إجابتك

What is your role in your organization and in which department are you .1
* ?working

Manager

IT employee

Consultant

أخرى:

* ?What kind of activity does your organization do .2

Culture and recreation

Education and research

Health

Social services

Environment

housing

Law, advocacy and politics

Voluntarism promotion

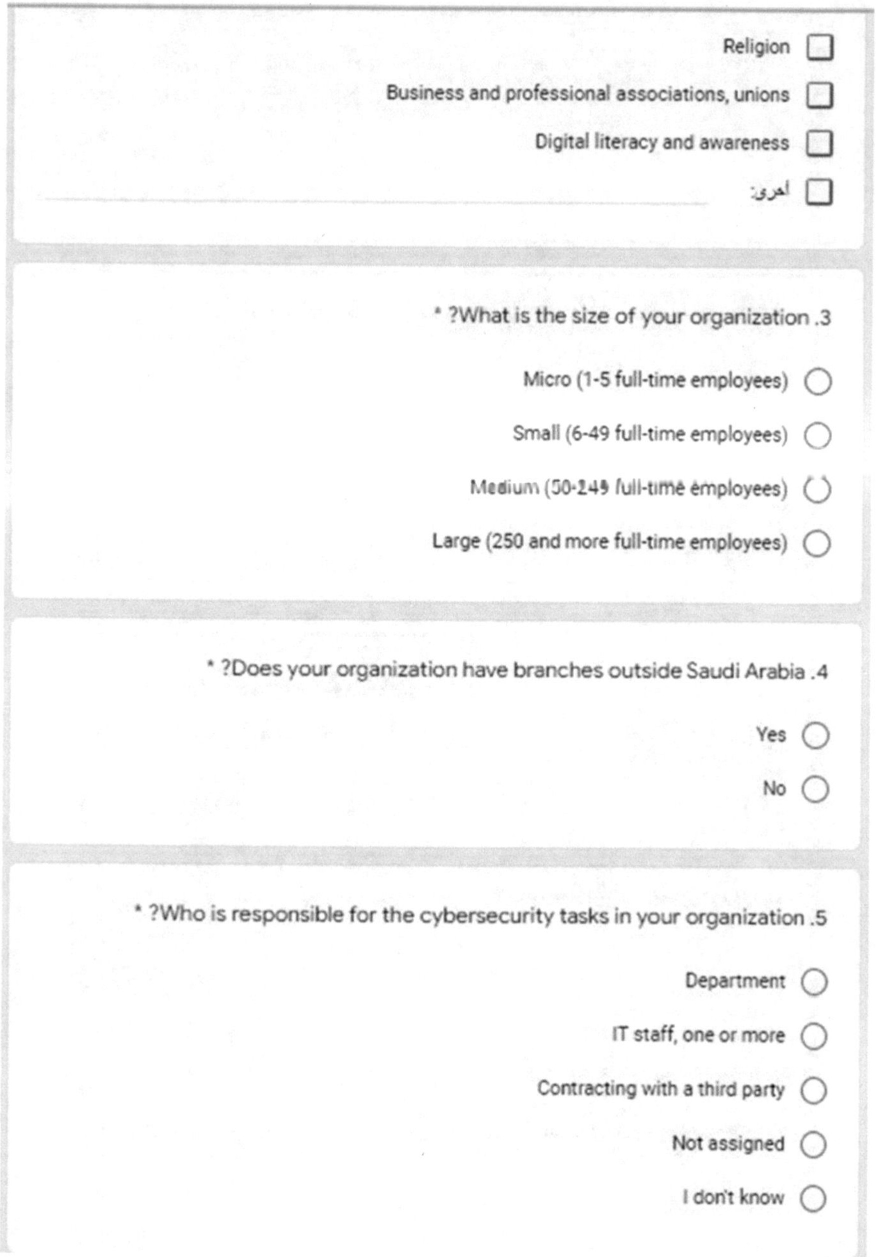

Religion ☐

Business and professional associations, unions ☐

Digital literacy and awareness ☐

أخرى: ☐

* ?What is the size of your organization .3

Micro (1-5 full-time employees) ○

Small (6-49 full-time employees) ○

Medium (50-249 full-time employees) ○

Large (250 and more full-time employees) ○

* ?Does your organization have branches outside Saudi Arabia .4

Yes ○

No ○

* ?Who is responsible for the cybersecurity tasks in your organization .5

Department ○

IT staff, one or more ○

Contracting with a third party ○

Not assigned ○

I don't know ○

If you have employees who are responsible for managing cybersecurity, are .6 ?they volunteers or are they paid employees

Volunteers ◯

Paid employees ◯

Is there an applied standard or methodology for cybersecurity risk assessment .7 * ?in your organization

Yes ◯

No ◯

I don't know ◯

What is the applied standard or methodology for cybersecurity risk .8 ?assessment in your organization (in case you have one)

Essential Cybersecurity Controls (ECC-2018) ☐

SP800-30 Risk Assessment Guide from National Institute of Standards and Technology (NIST) ☐

The International Organization for Standardization Risk assessment guide as part o- ISO27001 ☐

Organization for Economic Co-operation and Development (OECD) Recommendation for Digital Security Risk Management ☐

OCTAVE-S ☐

I don't know ☐

أخرى: ☐

* ?Are there any policies in place to manage the cybersecurity in your NGO .9

Yes ◯

No ◯

I don't know ◯

Which of the following tasks are routinely performed as basic procedures for .10 managing cybersecurity in your organization? Clarify the percentage of * .employees' commitment to implement it

Not ɔplied	Applied, and the commitment ratio is greater than 75%	Applied, and the commitment ratio is less than 75%	Applied, and the commitment ratio is less than 50%	Applied, and the commitment ratio is less % than 25	Applied, and the commitment ratio is 0%	
○	○	○	○	○	○	Set up firewalls
○	○	○	○	○	○	Applying measures of access control to the organization's hardware and resources, for example securing computers with a password
○	○	○	○	○	○	Activating antiviruses programs
○	○	○	○	○	○	Backup data periodically
○	○	○	○	○	○	Update programs automatically for all organization's devices
○	○	○	○	○	○	Set up a policy to change passwords periodically
○	○	○	○	○	○	Obliging employees not to share personal information or login data and passwords with

○	○	○	○	○	○	Requiring employees to log out of their accounts upon completion of work
○	○	○	○	○	○	Employees are aware of the risk of downloading files attached to email messages that may lead to an email phishing attack
○	○	○	○	○	○	Applying procedures to maintain customer privacy

In your opinion, what are the reasons for not adopting or adhering to .11
* cybersecurity policies in general? (You can select more than one)

- ☐ Cybersecurity management is not a priority for the organization ☐
- ☐ There are no threats affecting the cybersecurity ☐
- Lack of financial resources ☐
- Lack of human resources ☐
- Lack of experience in information technology ☐
- I don't know ☐
- أخرى: ☐

Does your organization have cybersecurity awareness programs for their .12
* ?employees

Yes

No

I don't know

To what extent is your organization compliant with the National Cybersecurity .13
Authority (NCA) relevant regulations and guidelines (such as Essential
* ?Cybersecurity Controls (ECC-2018)

Complaint >90%

Relatively complaint >75%

Currently planning the compliance with it

Never heard of the regulations and guidelines.

I don't know about the status of my NGO compliance

In your opinion, what are the obstacles that could the organization face when .14
adopting a cybersecurity risk assessment standard or methodology? (you can
* choose more than one)

Difficult implementation and frequent procedures

Lack of financial resources

Lack of experienced human resources

I don't think there are any obstacles

أخرى:

If there is a proposed solution within which you can assess and manage risks .15 in cybersecurity, what are the specifications that should be considered essential * ?for this framework to be executable in your NGO

	I do not know	Not important at all	Slightly important	Important	Very important
The tool can be operated by inexperienced members in the cybersecurity	○	○	○	○	○
Identify the cybersecurity risks that threaten the organization easily	○	○	○	○	○
Propose measures that are specifically designed for NGOs to mitigate the cybersecurity risks	○	○	○	○	○
Identify gaps that impede the organization in complying with the regulations of the National Cybersecurity Authority	○	○	○	○	○
Consider existing human resources	○	○	○	○	○
Consider existing budget	○	○	○	○	○

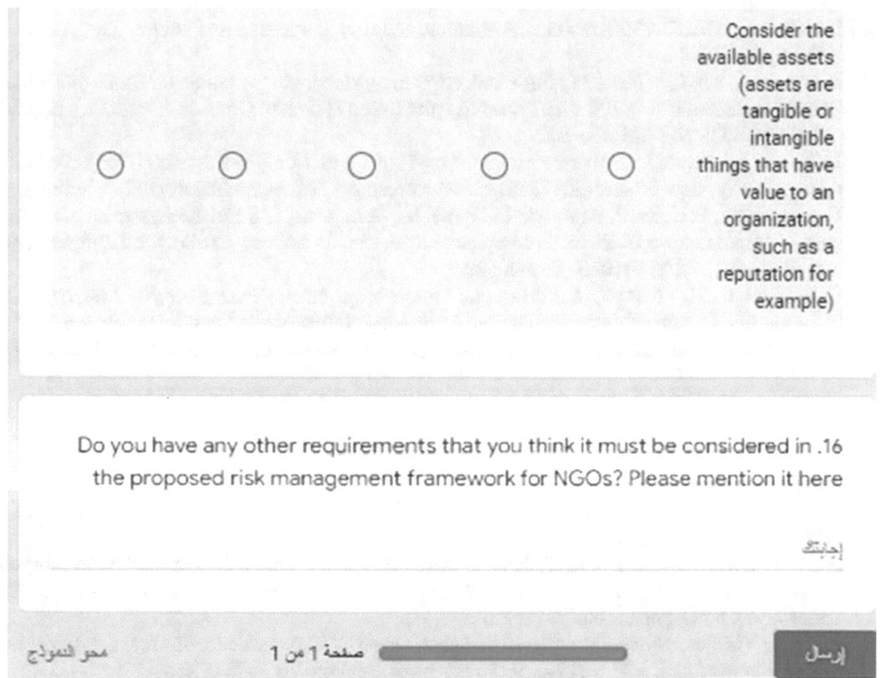

References

1. Carey-Smith, M., Nelson, K., May, L.: Improving information security management in nonprofit organisations with action research. In: Proceedings of the 5th Australian Information Security Management Conference, pp. 38–46 (2007b).
2. Nações Unidas.: Handbook on Non-profit institutions in the System of National Accounts (2003). http://unstats.un.org/unsd/publication/seriesf/seriesf_91e.pdf
3. Akingbola, K., Rogers, S.E., Baluch, A.: Change management in nonprofit organizations. In: Change Management in Nonprofit Organizations (2019). https://doi.org/10.1007/978-3-030-14774-7
4. Development, M. of H. R. and S.: Ministry of Human Resources and Social Development (2019). https://hrsd.gov.sa/
5. Lin, Y.: Government Management Model of Non-profit Organizations Based on E-government, pp. 164–168 (2019). https://doi.org/10.1145/3348445.3348464
6. Authority, N.C. (n.d.).: National Cybersecurity Authority. Retrieved November 14, 2019. https://nca.gov.sa/en/index.html
7. Anthopoulos, L.G.: Smart government: a new adjective to government transformation or a trick? In: Public Administration and Information Technology, vol. 22, pp. 263–293. Springer. https://doi.org/10.1007/978-3-319-57015-0_6
8. Bernardo, D.V.: Security risk assessment: Toward a comprehensive practical risk management. Int. J. Inf. Comput. Secur. **5**(2), 77–104 (2012). https://doi.org/10.1504/IJICS.2012.051775
9. Wangen, G., Hallstensen, C., Snekkenes, E.: A framework for estimating information security risk assessment method completeness: core unified risk framework CURF. Int. J. Infor. Secur. **17**(6), 681–699 (2018). https://doi.org/10.1007/s10207-017-0382-0

10. ISO/IEC.: ISO/IEC 27001:2005, Information security management systems-requirements. Infor. Syst. (2005)
11. Bowen, P., Hash, J., Wilson, M.: Information Security Handbook: A Guide for Managers NIST Special Publication 800–100. NIST Special Publication 800–100, October, 137 (2006). https://doi.org/10.6028/NIST.SP.800-100
12. Mierzwa, S., Scott, J.: Cybersecurity in Non-Profit and Non-Governmental Organizations Cybersecurity View project (2017). https://www.researchgate.net/publication/314096686
13. Ngamboé, M., Berthier, P., Ammari, N., Dyrda, K., Fernandez, J.M.: Risk assessment of cyber-attacks on telemetry-enabled cardiac implantable electronic devices (CIED). Int. J. Infor. Secur. https://doi.org/10.1007/s10207-020-00522-7
14. Carey-Smith, M., Nelson, K., May, L.: Improving Information Security Management in Nonprofit Organisations with Action Improving Information Security Management in Nonprofit Organisations with Action Research (2007a). https://doi.org/10.4225/75/57b52bb243e30
15. ENISA. (n.d.).: ENISA. Retrieved February 11, 2020. https://www.enisa.europa.eu/
16. Stoneburner, G., Goguen, A., Feringa, A.: Risk Management Guide for Information Technology Systems Recommendations of the National Institute of Standards and Technology (2002)
17. Tufan, E., Tezcan, C., Acartürk, C.: Anomaly-based intrusion detection by machine learning: a case study on probing attacks to an institutional network. IEEE Access **9**, 50078–50092 (2021). https://doi.org/10.1109/ACCESS.2021.3068961
18. Al Achkar, Z.: Achieving Safe Operations through Acceptance: challenges and opportunities for security risk management Digital Risk: How New Technologies Impact Acceptance and Raise New Challenges for NGOs (2021)
19. Moist, R.: Giuliana Sorce (Ed.). Global perspectives on NGO communication for social change. Studies Commu. Sci. **22**(1), 277–279 (2022). https://doi.org/10.24434/j.scoms.2022.01.042
20. Rice, L.E.: Non-profit organizations' need to address security for effective government. **4**(4), 53–71 (2012)
21. Kolb, N., Abdullah, F.: Developing an information security awareness program for a non-profit organization. Int. Manag. Rev. **5**(2), 103 (2009)
22. Imboden, T.R.: How are nonprofit organizations influenced to create and adopt information security policies? Issues Infor. Syst. **14**(2), 166–173 (2013)
23. Yeniman Yildirim, E., Akalp, G., Aytac, S., Bayram, N.: Factors influencing information security management in small- and medium-sized enterprises: a case study from Turkey. Int. J. Inf. Manage. **31**(4), 360–365 (2011). https://doi.org/10.1016/j.ijinfomgt.2010.10.006
24. Ghani, E.K., Hassin, N.H.N., Muhammad, K.: Effect of employees' understanding on risk management process on risk management: a case study in a non-profit organisation. Int. J. Finan. Res. **10**(3), 144–152 (2019). https://doi.org/10.5430/ijfr.v10n3p144
25. Of, I., By, A.: V Oluntary D Isclosure of S Ales By S Mall and M Edium S Ized E Nterprises : I (2002)
26. Sarstedt, M., Schloderer, M.P.: Developing a measurement approach for reputation of non-profit organizations. Inter. J. Nonprofit Voluntary Sector Marketing), 276–299 (2010). 15(January 2009. https://doi.org/10.1002/nvsm
27. Montenegro, C., Moncayo, D., Provemovil, S.A.: Information Security Risk in SMEs : a Hybrid Model compatible with IFRS Evaluation in two Ecuadorian SMEs of Automotive Sector Information Security Risk in SMEs : a Hybrid Model compatible with IFRS Evaluation in two Ecuadorian SMEs of Automotive Sector. October 2016 (2017). https://doi.org/10.1109/INFOCOMAN.2016.7784226
28. Valdevit, T., Mayer, N.: A gap analysis tool for SMES targeting ISO/IEC 27001 compliance. In: ICEIS 2010 - Proceedings of the 12th International Conference on Enterprise Information Systems, 3 ISAS, pp. 413–416. https://doi.org/10.5220/0002865504130416
29. Valdevit, T., Mayer, N., Barafort, B.: Tailoring ISO/IEC 27001 for SMEs: a guide to implement an information security management system in small settings. Commun. Comp. Infor. Sci. **42**, 201–212 (2009). https://doi.org/10.1007/978-3-642-04133-4_17

30. Richard Henson, W.B.S., Daniel Dresner, H.I.A.N., David Booth, I.S.C. (n.d.).: IASME: Information Security Management evolution for SMEs
31. Ponsard, C., Grandclaudon, J.: Survey and Guidelines for the Design and Deployment of a Cyber Security Label for SMEs. Springer International Publishing (2019). https://doi.org/10.1007/978-3-030-25109-3_13
32. Kumar, R.: Research methodology a step-by-step guide for beginners. In: Acta Universitatis Agriculturae et Silviculturae Mendelianae Brunensis, vol. 53, Issue 9 (2015). http://publications.lib.chalmers.se/records/fulltext/245180/245180.pdf
33. Department for Digital, Culture, M. & S. (DCMS).: Cyber security skills in the UK labour market 2020 (2020). https://www.gov.uk/government/publications/cyber-security-skills-in-the-uk-labour-market-2020/cyber-security-skills-in-the-uk-labour-market-2020
34. Fatokun Faith, B., Hamid, S., Norman, A., Fatokun Johnson, O., Eke, C.I.: Relating factors of tertiary institution students' cybersecurity behavior. In: 2020 International Conference in Mathematics, Computer Engineering and Computer Science, ICMCECS 2020, pp. 0–5. https://doi.org/10.1109/ICMCECS47690.2020.246990
35. Jarques, C.: Chapter 4 stratified sampling. Stratified Sampling, pp. 1–27 (2014)
36. Ronald N. Forthofer, Eun Sul Lee, M.H.: Biostatistics (2544)
37. Monshaat. (n.d.).: Monshaat. Retrieved December 2, 2020. https://www.monshaat.gov.sa/
38. Singh, A.S., Masuku, M.B.: Fundamentals of applied research and sampling techniques. Int. J. Medical Appl. Sci. **2**(4), 124–132 (2013)
39. Bartlett II, J.E., Kotrlik, J.W., Higgins, C.C.: Determing appropriate sample size in survey research. Infor. Technol. Learning Perform J. **19**(1), 43–50 (2001). https://www.opalco.com/wp-content/uploads/2014/10/Reading-Sample-Size1.pdf
40. Size, D.S. (n.d.).: Using Published Tables Using Formulas To Calculate A Sample Size Using A Census For Small Populations
41. Kanpur, I.: Chapter 10 two stage sampling (subsampling). Sampling Theory, Two Stage Sampling, pp. 1–21 (2013)
42. Pandey, R., Verma, M.R.: Samples allocation in different strata for impact. Rev. Bras. Biom. **26**(4), 103–112 (2008). http://jaguar.fcav.unesp.br/RME/fasciculos/v26/v26_n4/A7_Artigo_Verma.pdf
43. Norris, D.F., Mateczun, L., Joshi, A., Finin, T.: Cyberattacks at the grass roots: american local governments and the need for high levels of cybersecurity. Public Adm. Rev. **79**(6), 895–904 (2019). https://doi.org/10.1111/puar.13028
44. Six, M.: Quality in Multisource Statistics Quality Guidelines for. 07112, 1–93
45. Sudman, S., Lessler, J.T., Kalsbeek, W.D.: Nonsampling error in surveys. J. Mark. Res. **30**(3), 392 (1993). https://doi.org/10.2307/3172891
46. Shenton, A.K.: Strategies for ensuring trustworthiness in qualitative research projects.: University of Liverpool Library. Educ. Infor. **22**, 63–75. https://pdfs.semanticscholar.org/cbe6/70d35e449ceed731466c316cd273032b28ca.pdf

Digital Transformation in Tourism Ecosystems: What Impact on Sustainability and Innovation?

Orlando Troisi, Mara Grimaldi, and Anna Visvizi

Abstract The study seeks to address the following research objectives: (1) to reveal how smart tourism ecosystems (STE) can be redefined through a data-driven approach for digital transformation; (2) to assess the impact of data-driven approach on the development of sustainability and innovation. **Methodology**: The empirical research is based on a case study methodology performed through the technique of qualitative content analysis. The key data-driven strategies and practices implemented through the Project "Smart Tourism" are analyzed. **Findings**: The findings reveal that data-driven smart tourism ecosystems can create innovation and sustainability based on the activation of a data culture, of different kinds of resources and digital skills and on users' participation. **Originality**: The identification of the drivers for the digital redefinition of smart tourism ecosystems can be useful for researchers and managers that should face the acceleration of digitalization processes caused by Covid-19.

Keywords Digital transformation · Smart tourism ecosystem (STE) · Data data-driven orientation · Sustainable Development Goals (SDGs) · Innovation · Sustainability-oriented innovation (SOI)

O. Troisi · M. Grimaldi (✉)
University of Salerno, Via Giovanni Paolo II, 132, 84014 Fisciano, SA, Italy
e-mail: margrimaldi@unisa.it

O. Troisi
e-mail: otroisi@unisa.it

A. Visvizi
Institute for Hellenic Growth & Prosperity (IHGP), The American College of Greece, 6 Gravias Street, 153-42, Aghia Paraskevi, Athens, Greece

Effat University, 8482 Qasr Khouzam, Al-Nazlah Al-Yamaniyah, Jeddah 22332, Saudi Arabia

A. Visvizi et al. (eds.), *Research and Innovation Forum 2022*,
Springer Proceedings in Complexity, https://doi.org/10.1007/978-3-031-19560-0_5

1 Introduction

Over the last decades, digitalization redefined the rules of competition in markets by forcing companies to reread their strategies, processes and business models [1] through the application of technology [2]. The adoption of smart technologies has brought disruption in many sectors and also in services which are grounded on the key role of human interactions, such as tourism and hospitality [3, 4].

The pervasive role of technologies in tourism service has led to the redefinition of tourism businesses as smart tourism ecosystems (STE), interconnected networks that create, manage and deliver intelligent touristic experiences through technology-mediated information sharing and value co-creation [5].

By exploring how contemporary tourism ecosystems are complying with the demands of digital transformation can permit to analyze the new ways of creating value in tourism service and the ways in which this value can give birth to innovation [6, 7].

The redesign of business orientation through new technologies requires that a strategic data management is implemented to improve decision making processes and prompt innovative insights. In this regard, data-driven decision making (DDDM) suggests the need to reframe and rethink business strategies and practices [8, 9]. Big data analysis allows companies at increasing their value: then, the digital transformation of contemporary companies implies the consideration of data as a strategic resource for the improvement of value creation. By influencing value creation, big data can also have an impact on sustainability [10, 11] and on the development of innovation [12, 13]. Hence, the exploration of how businesses can be reinterpreted through data-driven approach can help identify the mechanisms that can foster sustainability and innovation.

The adoption of data-driven principles in business orientation is expected to play a significant role in improved future development of tourism ecosystems. However, the nexus between data-driven orientation and tourism sector has been until now neglected in research. Few studies investigate the adoption of data-driven approach in tourism and the enabling factors for the development of innovation [14] and for the promotion of inclusive and sustainable growth [15]. Previous research scarcely analyze the use of big data, analytics and blockchain technologies in tourism and hospitality as a technological means rather than as a source of value and innovation [16].

For this reason, this study aims at: (i) detecting how data-driven orientation can be useful when managing diverse aspects of tourism; (ii) analyzing the impact of data-driven orientation on sustainability and innovation.

2 Theoretical Background. Data-Driven Orientation, Innovation and Sustainability: What Link?

Big Data and the adoption of data-driven orientation encompass all industries but can have a particular impact on knowledge-intensive business services, such as tourism. In fact, tourism sector is essentially grounded on the provision of a totalizing services for consumers and on the capabilities of companies to establish multiple touchpoints with travellers. Hence, big data is the key lever for the digital redefinition of tourism [17] and can enable knowledge exchanges as key levers for innovation and for the competitiveness of destinations. In service literature, contemporary tourism companies are reinterpreted as smart tourism ecosystems (STE), that are a set of services that offer tourists an integrated experience [18]. The adoption of a data-driven mindset can increase value generation in tourism in terms of transparency and immediate feedback, customers segmentation, decision-making and innovation.

The adoption of big data analytics cannot only increase competitiveness and improve the development of innovation but can also have an impact on the social side of businesses. A smart data management can turn big data into Social Big Data [19], a new source of relevant knowledge that can enhance the effectiveness of decision and the growth of companies as well as in terms of sustainability.

Sustainability is a crucial issue in contemporary agenda, from political and social science to management and economics. According to the Sustainable Development Goals (SDG) developed by United Nations in 2016, sustainable development is the result of the harmonization of three dimensions: economic growth, social inclusion and environmental protection. Hence, in management studies, sustainability is considered as a key driver for competitive advantage [20]. The increasing diffusion of studies on sustainability leads scholars to analyze the relevance of this concept in tourism sector. In contemporary context, a tourism ecosystem cannot attain competitiveness if it cannot pursue sustainability [21].

According to Buhalis [3] tourism ecosystems can develop competitiveness not only through the offering of unique experiences, but also through the satisfaction of multiple benefits for stakeholders in the local community and the territory, by enhancing economic and social growth. Moreover, the relevance of sustainability in tourism sector is emphasized also by World Tourism Organization that defines tourism development as the balance between economic, social and environmental dimensions [22]. Smart tourism ecosystems can benefit from the application of big data analytics (to collect, integrate and analyze data) for the development of social connections and social needs, by potentially improving sustainability [23] and sustainability- oriented innovation [SOI, 11, 24].

Thus, the recognizes relevance of big data analytics and of data-driven orientation in the development of innovation and sustainability should be further explored. For this reason, this paper aims to address the following research questions:

RQ1: which are the key enablers of data-driven orientation in smart tourism ecosystems (STE)?

RQ2: which is the role of data-driven orientation in the development of sustainability and innovation?

3 Methodology

To address the two research questions introduced above, the empirical research analyzes the data-driven practices implemented by the project "Smart Tourism", funded by the Program "Interreg Italia Francia Marittimo[1]", that provides tourism companies with digital services for competitiveness of transnational tourism business and valuechain. The key services offered are: (1) access to technology for digital transformation; (2) enhancement of networks and integrated supply systems; (3) development of a market of qualified suppliers able to support the innovation processes.

This case has been selected since the project involves a network of tourism companies, therefore a tourism ecosystem, and aims at transforming them digitally (RQ1) by promoting innovation and sustainable development (RQ2).

The work adopts an exploratory qualitative approach based on a content analysis as inquiry [25, 26], performed through the collection of textual data on the activities implemented by the Project through the gathering of information from the official website of "Smart Tourism" and of the specific events and programs launched.

Therefore, based on the key dimensions of data-driven orientation and of the key enablers of innovation and sustainability identified in literature, the content analysis sketch presented in Table 1 has been designed.

4 Findings

The content analysis permits to inspect textual data based on some key categories identified in literature. Then, after the interpretation of the findings, some categories are revised, deleted or enrich to identify new conceptualizations.

4.1 RQ1: Data-Driven Orientation in the STE

As for the RQ1, the key enablers of data-driven orientation in smart tourism ecosystems are identified. To support the competitiveness (dimension 1) of small and

[1] The 2014–2020 Interreg Italy-France Maritime Program is a cross-border program co-financed by the European Regional Development Fund (ERDF) within the European Territorial Cooperation (CTE) to promote the competitiveness of businesses in cross-border supply chains and to improve the social connections and sustainability of territories.

Table 1 Variables and content analysis sketch

Variables	Content analysis sketch
RQ1: Data-driven orientation dimensions	– Which are the key technological tools implemented to analyze data? – Is there a strategic planning for data collection in line with company's objectives? – Is there an integrated strategy of actions for the training of employees and the enhancement of digital skills? – Is there a coherent data culture that includes data management in business' strategies? Are the digital culture of community and the access to technology improved after the development of the smart projects? – Is the data collected stored for continuous improvement?
RQ2: Enablers for innovation	– Did the smart projects and the use of data analytics and technology promote the development of innovation opportunities? – Did the development and updating of companies' digital culture and skills contribute to the development of innovation?
RQ2: Enablers for sustainability	– Did the smart projects and the use of data analytics and technology promote economic well-being by changing the way of doing business in the nations engaged? – Did the smart projects and the use of data analytics and technology promote social inclusion and users' participation? – Did the smart projects and the use of data analytics and technology promote the environmental growth of the territory?

medium enterprises operating in the tourism supply chain, the Project offers companies the assistance of freelancers and consultants with specialized skills for the provision of a catalog of qualified 4.0 services. In particular, the experts make accurate assessment of the digitization level of the companies and define a personalized digital roadmap to help companies embark on the digital transformation journey. The digital consulting companies offer also free access to smart technologies for the digitalization of internal processes, organizational structures, and for the improvement of technology adoption. The experts not only support the growth of the supply chain through big data analysis and greater openness to innovation, integration and synergy between the actors, but also promote the change in the attitude toward smart technology.

The support to the organizational and digital change is realized also through the enhancement of company's capabilities (dimension 2). Different education and training activities (such as workshop, traineeships, open day with universities and research centers) are organized to raise the digital competencies of SME's managers and employees and to increase the awareness of the opportunities and challenges of the fourth industrial revolution.

In relation to the third dimension, i.e. data-driven culture and strategy, the SMEs are provided with the values for a digital redefinition of attitude. To ensure a successful digital transformation, the experts need to ensure that the selected technologies are tailor-cut to a given business' values, mission, culture, and processes. Notably, successful implementation of selected technologies, frequently remains a function of the support obtained from the digital transformation partner.

In context of tourism and travel, a unified ICT-enhanced, smart system, composed of smart visas, borders, security processes, and infrastructure, may substantially improve the customer's perception of the service provided, therefore also the customer's satisfaction. This in turn, can create incentives for and enhance the digital transformation of companies. More specifically, an integrated smart tourism and travel system might improve the travel experience by detecting travel distortions and pre-empting the customer being exposed to them. Another, yet, related example, includes the case of online ticket purchase platforms, online check-in and connected mobile devices, e.g. smart phones, which again facilitate the process of detecting and pre-empting travel distortions. These methods greatly the overall quality and perceived utility of the travel experience, as well as safety and security, both for the individual customer/traveller and for the customers/travellers at large.

4.2 RQ2: The Development of Innovation and Sustainability

To address RQ2, some of the data-driven enabling dimensions of innovation and sustainability are detected.

The key technologies implemented to pursue innovation are Internet of Things, location-based services, artificial intelligence, augmented and virtual reality, and blockchain technology that help companies implement an efficient, inclusive, and economically, socially, and environmentally sustainable tourism offering.

The experts offer SMEs support for the innovation of the offer and training for the development of corporate innovation strategies through the enrichment of skills of stakeholders in the use of tools such as Big Data, Smart Data, applications, devices, etc. To enhance digital skills and the creation of new knowledge, participation in Workshops with other professionals and specialized consultants and Living Labs can help co-develop innovative ideas and stimulate opportunities for innovation.

Smart tourism build, with the active involvement of experienced companies and suppliers, participatory process of co-planning of qualified services in which the beneficiary companies will participate and involve users in the co-development of innovative solutions.

The set of digital tools provided has a positive impact on enhance innovation, sustainability, accessibility, and inclusion strategy across the tourism cycle and customer journey, i.e. before, during, and after the visit. Educated tourism professionals on environmental sustainability and accessibility. The project promotes the use of big data to facilitate the process of selected tourism regions to embark on future-oriented development. The challenge is to enable a digital transformation of these regions in a manner that will enable leveraging digitalization and data-driven technologies in a sustainable manner.

As for economic sustainability, the greatest impact of digital change pursued is the enhancement of industry's workforce. The automation of processes impacts jobs currently done manually. This may include either modifying the process underpinning these professions or making them obsolete. The assumption is, however, that digitally-enabled growth will create many new job opportunities, creating space for the renovation of traditional professions.

As for social inclusion, users are engaged in the co-development of tourism offers through an integrated set of tools based on AI, fuzzy logic, machine learning. Visitors are engaged in the improvement of the service with location-based communications and personalized recommendations based on their buying behaviors. Moreover, a network of social connections is created by leveraging on Artificial Intelligence, Big Data analytics, Mobile apps, Web, Wi-Fi Networks, Geo-location, IoT sensors, and other contextual information to deliver personalized communications and experiences to people based on their profiles, time and context in the physical world.

Regarding environmental sustainability, AI and agent-based modelling/simulation are used to enhance resilience and sustainability through flow control. The active control and management of visitor flows based on the intelligent use of data can help to sustainably alter the behaviour of tourists and lead to a better balance of capacities in the long run.

Moreover, through analytics and artificial intelligence smart solutions created by software development company can predict the amount of food needed to feed all guests, highlight the most and least popular dishes to make purchasing strategy more rational, analyze the kinds of wasted food and the frequency of throwing it away (for example, determine some specific occasions and patterns) in order to make appropriate changes in the strategy as well.

5 Conclusions

The analysis performed confirms that the application of data-driven orientation consists in the integration of technology with human skills and with a managerial dimension, based on the adoption of a proactive mind-set to strategic data management. The activities implemented in the project "Smart tourism" shows that the application of an integrated set of smart technologies and analytics in tourism ecosystems can enhance users' knowledge, develop innovation and increase economic, social and environmental sustainability.

Detecting the disruptive changes introduced by digital transformation and the adoption of data-driven orientation in smart tourism ecosystems allows at identifying the enablers for sustainability and innovation. In particular, the drivers for innovation and sustainability identified through content analysis are: (i) the creation of a smart system that support customer's journey through IoT, AI, virtual reality, etc.; (ii) the improvement of digital entrepreneurial orientation and of digital literacy; (iii) the engagement of users in the co-development of innovative solutions.

The findings of the study can help managers and scholars understand: (1) how smart projects, digital technologies and big data analytics can help tourism companies transform their business digitally; (2) how ecosystem's adaptation and change can lead to the introduction of new practices that can create innovation and transform durably the culture of community, by enhancing well-being. The findings of the research can help pinpoint how data-driven orientation can enable the attainment of social, economic and environmental goals through the achievement of continuous transformations within ecosystems and through the renewal of innovation processes over time [7]. Moreover, the exploration of how data-driven orientation can boost humans' relationships and social connections can address a gap in literature related to the absence of studies that investigate the role of technologies in reframing social connections within a community (Lytras and Visvizi 2018).

Further studies can start from the enablers of sustainability and innovation identified in this study to create a conceptual framework that can be validated through qualitative research (observation, semi-structured interviews) or quantitative research (regression or structural equation modeling) which can allow researchers at exploring how the adoption of a new data culture and the enrichment of knowledge can develop data-driven innovation and, in turn, encourage sustainability-oriented innovation by shedding light on the different innovation outcomes generated.

References

1. Zott, C., Amit, R.: Business model innovation: how to create value in a digital world. NIM Marketing Intell. Rev. **9**(1), 18 (2017)
2. Corvello, V., De Carolis, M., Verteramo, S., Steiber, A.: The digital transformation of entrepreneurial work. Int. J. Entrep. Behav. Res. **28**(5), 1167–1183 (2021)
3. Buhalis, D., Law, R.: Progress in tourism management: twenty years on and 10 years after the internet: the state of eTourism research. Tour. Manag. **29**, 609–663 (2008)
4. Gretzel, U., Sigala, M., Xiang, Z., Koo, C.: Smart tourism: foundations and developments. Electron. Mark. **25**(3), 179–188 (2015)
5. Gretzel, U., Werthner, H., Koo, C., Lamsfus, C.: Conceptual foundations for understanding smart tourism ecosystems. Comput. Hum. Behav. **50**, 558–563 (2015)
6. Barile, S., Ciasullo, M.V., Troisi, O., Sarno, D.: The role of technology and institutions in tourism service ecosystems: Findings from a case study. TQM J. **29**(6), 811–833 (2017)
7. Polese, F., Botti, A., Grimaldi, M., Monda, A., Vesci, M.: Social innovation in smart tourism ecosystems: How technology and institutions shape sustainable value co-creation. Sustainability **10**(1), 140 (2018)
8. LaValle, S., Lesser, E., Shockley, R., Hopkins, M.S., Kruschwitz, N.: Big data, analytics and the path from insights to value. MIT Sloan Manag. Rev. **52**(2), 21–32 (2011)
9. McAfee, A., Brynjolfsson, E., Davenport, T.H., Patil, D.J., Barton, D.: Big data: the management revolution. Harv. Bus. Rev. **90**(10), 60–68 (2012)

10. Dubey, R., Gunasekaran, A., Childe, S.J., Papadopoulos, T., Luo, Z., Wamba, S.F., Roubaud, D.: Can big data and predictive analytics improve social and environmental sustainability? Technol. Forecast. Soc. Chang. **144**, 534–545 (2019)
11. Etzion, D., Aragon-Correa, J.A.: Big data, management, and sustainability: strategic opportunities ahead. Organ. Environ. **29**(2), 147–155 (2016)
12. Trabucchi, D., Buganza, T.: Data-driven innovation: switching the perspective on Big Data. Eur. J. Innov. Manag. **22** (2018)
13. Troisi, O., Grimaldi, M., Loia, F.: Redesigning business models for data-driven innovation: a three-layered framework. In: The International Research & Innovation Forum, (pp. 421–435). Springer, Cham (2020)
14. Visvizi, A.: Artificial intelligence (AI) and sustainable development goals (SDGs): exploring the impact of AI on politics and society. Sustainability **14**(3), 1730 (2022)
15. Boluk, K. A., Cavaliere, C. T., Higgins-Desbiolles, F.: A critical framework for interrogating the United Nations Sustainable Development Goals 2030 Agenda in tourism. J. Sustain. Tourism. **27**(2), 847–864 (2019).
16. Camilleri, M.A.: The use of data-driven technologies in tourism marketing. In: Entrepreneurship, Innovation and Inequality, 182–194. Routledge (2019).
17. Buhalis, D., Amaranggana, A.: Smart tourism destinations. In: Information and communication technologies in tourism, 553–564. Springer, Cham (2014).
18. Buhalis, D., Harwood, T., Bogicevic, V., Viglia, G., Beldona, S., Hofacker, C.: Technological disruptions in services: lessons from tourism and hospitality. J. Serv. Manage. **30**(4), 484–506 (2019).
19. Bello-Orgaz, G., Jung, J.J., Camacho, D.: Social big data: recent achievements and new challenges. Inform. Fusion. **1**(28), 45–59, (2016).
20. Hall, J., Wagner, M.: Integrating sustainability into firms' processes: performance effects and the moderating role of business models and innovation. Bus. Strategy. Environ. **21**(3), pp.183–196 (2012).
21. Iunius, R.F., Cismaru, L., Foris, D.: Raising competitiveness for tourist destinations through information technologies within the newest tourism action framework proposed by the European commission. Sustainability **7**, 12891–12909 (2015)
22. World Tourism Organization (UNWTO); United Nations Environmental Programme (UNEP). Making Tourism More Sustainable: A Guide for Policy-Makers; World Tourism Organization Publications: Madrid, Spain (2005)
23. Del Vecchio, P., Mele, G., Ndou, V., Secundo, G.: Open innovation and social big data for sustainability: evidence from the tourism industry. Sustainability **10**(9), 3215 (2018)
24. Akhtar, P., Tse, Y.K., Khan, Z., & Rao-Nicholson, R.: Data-driven and adaptive leadership contributing to sustainability: global agri-food supply chains connected with emerging markets. Int. J. Prod. Econ. **181**, 392–401, (2016).
25. Losito, G.: L'analisi del contenuto nella ricerca sociale. FrancoAngeli, Milano, (1996).
26. Krippendorff, K.: Reliability in content analysis: some common misconceptions and recommendations. Hum commun. Res. **30**(3), pp. 411–433, (2004).
27. Lytras, M.D., Visvizi, A.: Who uses smart city services and what to make of it: toward interdisciplinary smart cities research. Sustainability, **10**(6), pp.1998, (2018).

Science, Technology and Innovation Policy in the USA During the Covid-19 Pandemic

Małgorzata Dziembała

Abstract The Covid-19 pandemic caused some long-term consequences in various fields, including science, technology and innovation (STI). The main objective of the article is to present the position of the United States during the pandemic in the light of the Global Innovation Index and to present the STI policy measures implemented in the USA to deal with the consequences of the Covid-19 pandemic. The analysis shows that the United States is doing relatively well in terms of STI, which is reflected in its position in the GII. Still, in some areas, STI outperforms its main competitor, the EU. However, it cannot be ignored that the effects of a pandemic may be seen over a longer period of time due to the nature of innovation processes. In the USA, during the Covid-19 pandemic, more than 40 STI policy initiatives were implemented with a focus on drug development and others to maintain the stability of the STI system during turbulent times with multiple pandemic shocks.

Keywords STI Policy · Covid-19 pandemic · Innovation system · Research and development

1 Introduction

The Covid-19 pandemic has not only had a short-term impact on socio-economic activities, which varies from country to country and from region to region, but its impact is also manifested in certain long-term changes in various fields, including science, technology and innovation (STI). Possible implications of the pandemic for the STI ecosystem include, on the one hand, new ways of doing things, including more efficient and effective use of digital tools, but on the other hand, there is a risk of reduced funding for innovation by companies, government and other actors in the innovation system. Therefore, STI policy should actively support the promotion of innovation through the use of appropriate tools and budget. Otherwise the innovation achievement gap may widen and, in addition, some actors or employees may be

M. Dziembała (✉)
The Department of International Economic Relations, The University of Economics in Katowice, 1 Maja 50, 40-287 Katowice, Poland
e-mail: malgorzata.dziembala@ue.katowice.pl

A. Visvizi et al. (eds.), *Research and Innovation Forum 2022*,
Springer Proceedings in Complexity, https://doi.org/10.1007/978-3-031-19560-0_6

excluded. However, there are some challenges for future STI spending by businesses, as business R&D spending is procyclical and new innovative solutions to cope with the pandemic need to be developed, which also requires a significant demand for R&D funding. During a pandemic, public funding may even increase due to the effective contribution of STIs to counter the effects of Covid-19. There are many other implications that could potentially affect the STI ecosystem, and policies may undergo some changes to effectively respond to the challenges of Covid-19 [20].

The long term impact of Covid 19 can be discussed from different perspectives: STI funding, digital and artificial intelligence adoption. The openness of STI ecosystems seen with open science and open data initiatives, as well as the inclusiveness of STI and the agility of this ecosystem are another channel through which the potential impact can be seen. Finally, due to the increase in international collaboration that can be enhanced, an efficient global STI system can emerge [15]. However, nowadays the short-term effects of the Covid-19 pandemic on the STI and the STI policy of particular countries are observed.

In major economies such as the USA, some STI-related programs focused on vaccine discovery have been initiated, but also aimed at stabilizing the innovation system and adapting it to upcoming challenges. However, there are not many studies analyzing the implications of the Covid-19 pandemic on a country's STI policy and STI system as reflected in specific indicators.

The following research questions were formulated: what is the USA position in terms of innovation during the Covid-19 pandemic compared to other countries? How is the USA STI policy responding to the Covid-19 pandemic?

The main objective of the article is to present the position of the USA during the pandemic in the light of the Global Innovation Index and to present the STI policy measures implemented in the USA to deal with the implications of Covid-19. A literature review was used as research methods to substantiate the arguments.

The article consists of two parts. The first discusses the position of the USA during the Covid-19 pandemic and the second focuses on presenting some of the measures and actions taken in this country in the wake of the Covid-19 pandemic to stimulate innovation.

2 Methodology

The Covid-19 pandemic has short- and long-term effects on the economy [1]. However, they vary from country to country and different policy measures have been implemented, including STI policies. Therefore, it is worth examining the impact of the Covid-19 pandemic on innovation leaders, and the USA serves as a good example.

From the above discussion, it is clear that in the short term, the impact of the pandemic on innovation in the USA economy varies. Therefore, only by having a long series of data could the complex impact of the pandemic be assessed. However, it is now important to mitigate the short-term effects, and the role of economic policy, including STI policy, is important in this regard.

Therefore, it is worth investigating what is the position of the USA in terms of innovation during the Covid-19 pandemic compared to other countries, also compared to the situation in the pre-pandemic period? Another research question was posed: How is STI policy in the USA responding to the Covid-19 virus pandemic? Thus, it was found that public funding for research plays a fundamental role in STI policy during a pandemic. This was illustrated using the example of the USA.

A review of STI policy and its instruments under pandemic conditions in the USA was used to answer the above questions as well as GII 2021 was conducted.

3 The Position of the USA in Terms of Innovation in the Light of the Global Innovation Index 2021

The Covid-19 pandemic has diverse repercussions in different spheres of the economy and its actors. In order to overcome the negative effects of the pandemic, some actions and policies have also been taken under the STI policy to enhance its innovation. Strengthening innovation should help to cope with the impact of the pandemic and to build resilience in the economy.

There are many individual indices that reflect the innovativeness of a country's economy and can be categorized as input and/or output indicators. However, the complex indices capture the diverse dimensions of a country's innovation and provide an overall picture.

The current situation of leading countries in terms of innovation capacity is represented by the Global Innovation Index (GII), which ranks 132 economies in terms of their innovation capacity, but also allows the examination of the factors that influence their position. The construction of the index is based on 81 factors [4].

The analysis of the ranking shows that the first place in the list is occupied by Switzerland, followed by Sweden, and the USA occupies the third position. After that, the following countries: the United Kingdom, the Republic of Korea, the Netherlands, Finland, Singapore, Denmark, and Germany round out the top ten innovation leaders according to the GII 2021 ranking.

It is worth mentioning that the top-ranked countries, namely Switzerland, Sweden, the USA and the UK, have been ranked among the top five countries in terms of innovation over the past three years. The results of the aforementioned GII 2021 study show that the Covid-19 pandemic has had a varied impact on innovation across sectors and regions. Investment in innovation has proven resilient, as contrary to earlier expectations of a reduction in this type of investment in innovation (R&D spending), the indicator for 2020 is promising–continuing its growth. Global business R&D spending is estimated to have increased by about 10%. Some other indicators related to innovation have seen growth in 2020, such as showcasing scientific output–scientific publications, intellectual property filings, venture capital deals [6].

The GII 2021 includes the following pillars that frame this index. Namely, related to innovation inputs–institutions, human capital and research, infrastructure, market sophistication, business sophistication, knowledge and technology performance, and creative performance [6]. The position of the USA in the various pillars of this index varies, with the highest in terms of market sophistication and business sophistication (related to business conditions, including knowledge workers, innovation linkages, knowledge absorption)–it ranked second, while in terms of knowledge and technology performance–it ranked third. However, in terms of the pillar Infrastructure, the country ranks only 23rd. The Infrastructure pillar includes: ICT, General Infrastructure and Environmental Sustainability. The pillar Institutions includes the following dimensions–Policy Environment, Regulatory Environment, and Business Environment, for which the USA ranks 12th [6]. The particular dimensions of GII 2021 for the USA and Germany are shown in Fig. 1.

A comparison of the GII 2021 and the GII 2019 for the United States shows that their innovation performance with respect to some dimensions of the presented index varies. It means that some dimensions have improved while others have deteriorated. During the period of instability due to the pandemic and the countermeasures taken, the investment dimension, for example, remains stable and the Education pillar has improved, as have its composite indices. Analyzing the situation of the USA and

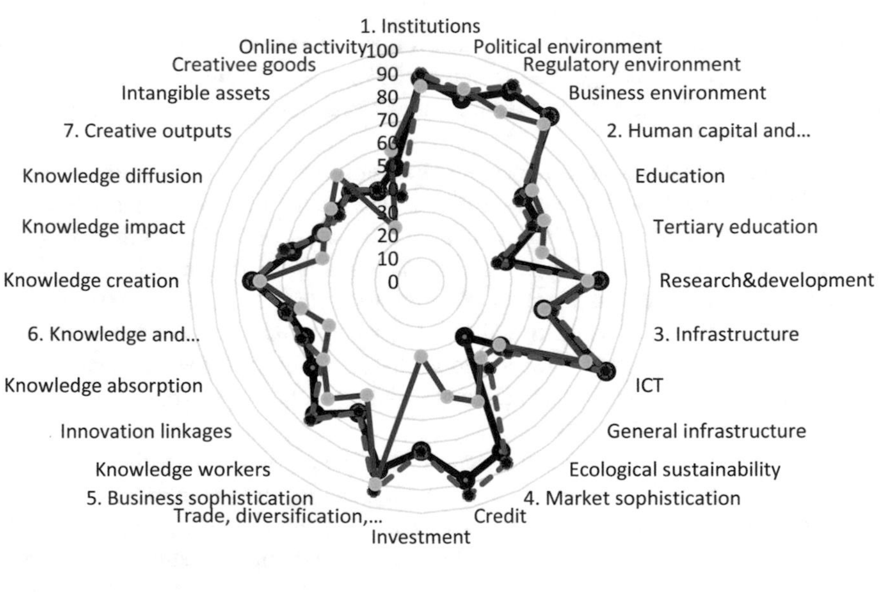

Fig. 1 The pillars and subpillars of GII 2021 for the USA and Germany and GII 2019 for the USA. *Source* own elaboration based on [6, 7]

Germany, however, it can be concluded that the USA is doing quite well in most dimensions of the GII 2021.

The USA is a leader in innovation, ahead of its major competitors, including the EU. The USA innovation performance compares relatively well with that of the EU in the following areas: the business sector in terms of R&D expenditure and product/process innovators (Table 1). According to the European Innovation Scoreboard 2021, the USA innovation performance improved from 2014 to 2021, but the rate was higher for the USA than for the EU, 16% and 12.5%, respectively [3].

Recent analyses show relatively strong the USA innovation performance in recent years, spanning the period before and just at the beginning of the Covid-19 pandemic. However, given the nature of innovation processes, the consequences of a pandemic can be seen in the long-term process, including the negative ones. The negative impact of the Covid-19 pandemic could have been mitigated by certain actions taken in the area of innovation.

Table 1 Performance of the USA in 2014 and 2021 relative to the EU in 2014

Indicators	2014	2021	2014–2021
Doctorate graduates	67.2	82.7	15.5
Tertiary education	125.5	127.9	2.4
International co-publications	129.1	115.8	−13.3
Most cited publications	149.7	135.1	−14.6
R&D expenditure public sector	98.8	93.7	−5.0
Government funding business R&D	185.9	119.0	−66.9
R&D expenditure business sector	151.8	155.7	3.9
Employment in ICT	121.7	119.2	−2.4
Product/process innovators	67.5	150.3	82.8
Marketing/organisational innovators	n/a	n/a	n.a
Innovation co-operation	n/a	n/a	n.a
Public–private co-publications	148.4	126.1	−22.3
PCT patent applications	100.3	114.7	14.4
Trademark applications	62.8	66.3	3.6
Design applications	25.6	34.6	9.0
Medium & high-tech product exports	94.4	94.1	−0.4
Knowledge-intensive services exports	109.3	105.3	−4.0
Air pollution by fine particulate matter	175.3	175.6	0.2
Environment-related technologies	91.4	87.2	−4.2

Source [4]

4 The STI Policy Measures Implemented in the USA During the Covid-19 Pandemic

An innovation policy is one that influences innovation, so it consists not only of different policies, but also of policy instruments with different motivations (using different labels). Three types of innovation policy are recognized: mission-oriented, which refers to offering solutions to challenges that are placed on the policy agenda, invention-oriented with a narrow focus, and system-oriented policy, which refers to policies at the system level and the national innovation system [3, pp. 3–5]. The rationale for this policy is the "market failure" approach to innovation policy with three types of instruments: investing in public knowledge production, subsidizing R&D in private firms, and strengthening incomplete property rights. An innovation policy approach based on an innovation system that is described by a set of factors, and when this system fails to provide the factors–a "system failure" can be identified [3, pp. 6–11].

More recently, attention has been paid to the innovation policy mix approach, which consists of four dimensions: the domain areas, the rationale for the policy, the strategic tasks and the instruments used. However, a different approach to the policy mix can be applied, emphasizing the interactions and balances within each dimension [14, pp. 251–279].

Innovation policy is necessary to create an environment for business innovation through appropriate regulations. These are regulations that can also affect the innovation activities of companies. There are also government support programmes that include transfer (direct or indirect to companies). The main types of policy instruments to promote innovation include: grants, equity and debt financing, guarantees for debt financing, payments for goods and services, tax incentives, infrastructure and services that are dedicated to innovation activities for companies. There is also public infrastructure and a conducive macroeconomic environment. The social context for innovation is also important [5, 9, pp. 156–159; p. 3]. Given these considerations, it is useful to examine STI policy in the US from an innovation policy perspective.

STIs responded quickly to the situation triggered by Covid-19, which was related to vaccine development. However, the functioning of innovation systems was disrupted due to the aftermath of the pandemic, including lockdowns, which was reflected in the effects on innovation capacity, including the sectors that were most affected. It is important to consider that business investment in R&D is pro-cyclical, so different responses to the pandemic can be observed in leading and lagging sectors, in large and small firms, and in different geographic areas. The result is a widening productivity gap [15].

Therefore, there is a need to rethink STI policy, which should be clearly articulated, taking into account the role that science and innovation should play in the transition towards "more sustainable, equitable and resilient societies". Accordingly, innovation efforts should be directed to those areas where they are most needed. Policies that support business R&D should be effective, so well-designed measures, direct and indirect, should be created to stimulate business to innovate [15]. The OECD study shows that the response of research and innovation systems has been observed to be very impressive, both from the public and private sectors. These actions were reflected in the development of vaccines, more intensive use of digital technologies, and the stimulation of observed trends, namely the promotion of open access, the use of digital tools, the intensification of STI collaboration internationally, and the promotion of public–private partnerships. However, some elements of the innovation system have suffered, namely small and medium-sized early stage start-ups, young researchers, and women. The role of governments in the short term is to mitigate the negative effects of the pandemic, and to develop solutions [2, 15]. However, the impact on the innovation ecosystem varies from country to country and depends on the innovation capacity and measures that have been taken through economic policies to counter the negative impacts, not only with the support of the government but also private businesses.

Therefore, it is worth mentioning the USA response to the pandemic in relation to the country's innovation system. The USA innovation system is diverse and large, consisting of a variety of actors. These include federal and state governments and their branches, public agencies, universities, and the private sector. It also includes public agencies, various organizations: non-profit and intermediary. The country is one of the leading countries in terms of innovation in the international arena, which also reflects the efforts and potential of the country's innovation system based on high R&D investments and market orientation. The federal government primarily funds basic research, also providing infrastructure development, some framework activities, and support for some research conducted by individual executive agencies. As highlighted, there are multiple actors in this system–the federal level is represented by the White House and the Office of Science and Technology Policy (OSTP), whose purpose is also to coordinate the activities of the executive agencies. There are expert committees that provide advice on innovation. There are a number of federal agencies that deal with innovation policy, programs, and there are White House departments that also deal with innovation that are responsible for certain agencies, such as the Census Bureau. As mentioned earlier, there are some federal agencies including the National Science Foundation, the National Institutes of Health with a significant R&D budget, the Department of Defense Small Business Administration which supports programs related to innovation, namely the Small Business Innovation Research (SBIR) program and the Small Business Technology Transfer (STTR) program. The USA Congress is responsible for legislation related to innovation. The authority of the federal government is shared with state and local governments. There are also other entities, including private industry, that play a special role in the innovation system [12].

As stated earlier, the USA innovation system is complex and composed of many actors. As a result, there are many initiatives to respond to the impact of the pandemic at different levels and targeting different groups. There are some courses of action to address the challenges of Covid-19, including not only spending on health or certain economic measures, but also spending on R&D and strengthening innovation capacity. However, these actions vary from country to country and depend on their specific characteristics, including financial capacity.

The USA has implemented certain packages of health and economic measures, including spending related to R&D, covering:

- Coronavirus Preparedness and Response Supplemental Appropriations Act, March 6, 2020 (USD 836 million for Covid-19 virus research for one of the National Institutes of Health (NIH); USD 3.1 billion for the Emergency Fund for Health and Social Services (Department of Health), including spending for vaccine development and directed among various institutions to the Biomedical Advanced Research and Development Authority (BARDA)),
- Families First Coronavirus Response Act, 18 March, 2020 covering sanitary and health care measures,
- Cares Act, 27 March 2020, predicting 1 trillion USD for new resources in the economy, USD 500 billion for direct public spending, USD 500 billion for credit companies [11].

As of April 4, 2022, there were 42 policy initiatives to combat Covid-19 with STIs in the USA. It includes policy initiatives that were reported in the OECD database for 2020–2021. The largest number of initiatives were for institutional funding for public research (7), followed by grants for public research projects (5). However, it is worth mentioning that 4 public research grants have a value of more than EUR 500 million (Table 2).

There are many institutions responsible for certain initiatives in the field of STI to counteract the implications of the pandemic (Table 3).

The analysis shows that institutional funding for public research dominates. It is also worth noting the scale of policy support in financial terms. In terms of subject matter, funding focuses on diagnosing and stimulating the STI system. However, other areas should not be neglected. The pandemic underlines the role of ICT in stimulating innovation. The smart cities which extensively use digital tools help to counteract the negative consequences of pandemic [18, 19], where social challenges will have to be addressed [10]. However there are other tools to counteract the effect of pandemic like artificial intelligence [8, 16, 17].

Table 2 Policy instruments applied within the STI policy in the USA as a response to the Covid-19 pandemic (number of initiatives, million EUR)

Instruments	1–5 M	20–50 M	50–100 M	100–500 M	More than 500 M	Unknown	Not applicable
Creation or reform of governance structure or public body						3	
Dedicated support to research infrastructures					1		

(continued)

Table 2 (continued)

Instruments	1–5 M	20–50 M	50–100 M	100–500 M	More than 500 M	Unknown	Not applicable
Fellowships and postgraduate loans and scholarships					1		
Formal consultation of stakeholders or experts						1	
Grants for business R&D and innovation					1		
Horizontal STI coordination bodies					1		
Information services and access to datasets					1	1	
Institutional funding for public research	1	1	2	1	2		
Intellectual property regulation and incentives						1	
Procurement programmes for R&D and innovation				1	1		
Project grants for public research			1		4		
Public awareness campaigns and other outreach activities						1	1
Science and innovation challenges, prizes, awards					1		

(continued)

Table 2 (continued)

Instruments	1–5 M	20–50 M	50–100 M	100–500 M	More than 500 M	Unknown	Not applicable
Standards and certification for technology development			1				
Strategies, agendas and plans					1		
Technology extension and business advisory services			1				

Source own elaboration based on [13]

Table 3 Institutions responsible for initiatives with regard to innovation in the USA as a response to the Covid-19 pandemic

Organisation	Number of initiatives
National Institutes of Health (NIH)	5
Center for Disease Control and Prevention (CDP)	5
Biomedical Advanced Research and Development Authority (BARDA)	5
Food and Drug Administration (FDA)	5
Department of Health and Human Services (HHS)	3
National Science Foundation (NSF)	3
Office of Management and Budget (OMB)	3
Department of Defense (DOD)	3
US Patent and Trademark Office (USPTO)	2
Midwest Big Data Innovation Hub	1
Federal Emergency Management Agency (FEMA)	1
Northeast Big Data Innovation Hub	1
Department of Energy (DOE)	1
Department of Agriculture (USDA)	1
Environmental Protection Agency (EPA)	1
National Center for Advancing Transnational Sciences (NCATS)	1
Department of Veteran Affairs (VA)	1
The White House	1
Council on Governmental Relations (COGR)	1

(continued)

Table 3 (continued)

Organisation	Number of initiatives
National Library of Medicine	1
Critical Path Institute	1
West Big Data Innovation Hub	1
South Big Data Innovation Hub	1
Columbia University	1

Source own elaboration based on [13]

5 Conclusions

The COVID-19 pandemic was an unprecedented phenomenon, and its repercussions and impact have been noted in various spheres, including STIs. The consequences of the pandemic can be analyzed from a short-term and long-term perspective. Short-term consequences can be observed at the moment, but they are not evenly distributed when it comes to STIs. The United States is doing relatively well in terms of STIs, which is reflected in its GII ranking. However, it still outperforms its main competitor, the EU, in some areas of STI.

However, it cannot be ignored that the effects of a pandemic can be seen over a longer period of time due to the nature of innovation processes. Therefore the appropriate policy mix could be applied which should be adjusted and take into considerations the changes in the economic environment due the pandemic which is reflected in the recovery packages. The role of STI policy is even greater nowadays when it is more and more involved in the health protection and also the public institutions and organisations are increasing its contribution to NSI and as a result to innovativeness of this country.

Consequently, an appropriate policy mix can be applied, which should be adapted and take into account the changes in the economic environment caused by the pandemic, as reflected in the recovery packages. The role of STI policy is even greater now that it is increasingly involved in health care, and public institutions and organizations are increasing their contribution to NSI and, as a result, to the country's innovation.

The USA NSI is composed of many different actors who can contribute to the system, i.e., the creation and diffusion of innovation. Innovation appears to be even more important during a pandemic. Therefore, funding initiatives focus on developing drugs and other products, as well as maintaining the stability of the STI system during the turbulent times of COVID-19. In the United States, the scale of funding seems to be a challenge. However, the broad scope of the impact of Covid-19 on STIs in the USA is still fully unknown, also limited by data availability.

Future research should focus on the effects of the pandemic on the national innovation system and the evaluation of the measures used and implemented. The Covid-19 pandemic stimulates the STI transition process, so the effectiveness of the USA STI system during the Covid-19 pandemic requires further research and analysis.

References

1. Baldwin, R., Weder di Mauro, B. (eds.): Economics in the Time of COVID-19, CEPR Press, London (2020). https://voxeu.org/system/files/epublication/COVID-19.pdf. Accessed 3 July 2022
2. Ciasullo, M.V., Troisi, O., Grimaldi, M., Leone, D.: Multi-level governance for sustainable innovation in smart communities: an ecosystems approach. Int. Entrep. Manag. J. Springer **16**(4), 1167–1195 (2020). https://doi.org/10.1007/s11365-020-00641-6.. (Springer, Dec 2020)
3. Edler, J., Fagerberg J.: Innovation policy: What, why, and how, TIK Working Paper on Innovation Studies No. 20161111, pp. 1–27. https://www.sv.uio.no/tik/InnoWP/tik_working_paper_20161111.pdf. Accessed 15 June 2022
4. European Innovation Scoreboard 2021, Office for Official Publications of the European union, Luxembourg (2021). https://ec.europa.eu/docsroom/documents/46013. Accessed 05 Apr 2022
5. Frascati Manual 2015: Guidelines for Collecting and Reporting Data on Research and Experimental Development, The Measurement of Scientific, Technological and Innovation Activities, OECD Publishing, Paris (2015). https://doi.org/10.1787/9789264239012-en
6. Global Innovation Index 2021. Tracking innovation through the COVID-19 crisis, WIPO Geneva (2021). https://www.wipo.int/edocs/pubdocs/en/wipo_pub_gii_2021.pdf. Accessed 05 Apr 2022
7. Global Innovation Index 2019. Creating Healthy Lives—The Future of Medical Innovation, WIPO Geneva (2019). https://www.wipo.int/edocs/pubdocs/en/wipo_pub_gii_2019.pdf. Accessed 20 Jun 2022
8. Maione, G., Loia, F.: New shades on the smart city paradigm during covid-19: a multiple case study analysis of Italian local governments. In: Visvizi, A., Troisi, O., Saeedi, K. (eds.). Research and Innovation Forum 2021. RII FORUM 2021. Springer Proceedings in Complexity. Springer, Cham (2021). https://doi.org/10.1007/978-3-030-84311-3_22
9. Oslo Manual 2018: Guidelines for Collecting, Reporting and Using Data on Innovation, 4th Edition, The Measurement of Scientific, Technological and Innovation Activities, OECD, Publishing, Paris/Eurostat, Luxembourg. OECD/Eurostat (2018) https://doi.org/10.1787/9789264304604-en
10. Polese, F., Troisi, O., Grimaldi, M., Loia, F.: Reinterpreting governance in smart cities: an ecosystem-based view. In: Smart Cities and the UN SDGs, pp. 71–89. Elsevier (2021)
11. Public policies for research and innovation in the face of the Covid-19 crisis, https://www.ipea.gov.br/cts/en/all-contents/articles/articles/203-public-policies-for-research-and-innovation-in-the-face-of-the-covid-19-crisis-3. Accessed 05 Apr 2022
12. Sapira, P., Youtie, J.: The Innovation System and Innovation Policy in the United States, University of Manchester (2010). https://www.euussciencetechnology.eu/assets/content/documents/InnovationSystemInnovationPolicyUS.pdf. Accessed 05 Apr 2022
13. STIP COVID-19 Tracker, 6 April 6, 2022. https://stip.oecd.org/covid/countries/UnitedStates. Accessed 20 Apr 2022
14. The Innovation Policy Mix: in OECD Science, technology and industry outlook 2010, OECD (2010). https://doi.org/10.1787/sti_outlook-2010-en
15. Times of crisis and opportunity. OECD Science, technology and innovation outlook 2021, OECD (2021)
16. Troisi, O., Visvizi, A., Grimaldi, M.: The different shades of innovation emergence in smart service systems: the case of Italian cluster for aerospace technology. J. Bus. Ind. Market. Vol. ahead-of-print No. ahead-of-print (2021). https://doi.org/10.1108/JBIM-02-2020-0091

17. Visvizi, A., Bodziany, M. (eds.) Artificial Intelligence and Its Contexts, Advanced Sciences and Technologies for Security Applications (2021). https://doi.org/10.1007/978-3-030-88972-2_2
18. Visvizi, A., Lytras, M. (eds.) Smart cities: Issues and challenges: Mapping political, social and economic risks and threats. Elsevier (2019).
19. Visvizi, A., Lytras, M.D., Aljohani, N.R. (eds.) Research and Innovation Forum 2020: Disruptive Technologies in Times of Change. Springer Nature (2021).
20. What future for science, technology and innovation after COVID-19? OECD Science, Technology and Industry Policy Papers, OECD, April, no. 107 (2021).

An Application of Machine Learning in the Early Diagnosis of Meningitis

Pedro Gabriel Calíope Dantas Pinheiro, Luana Ibiapina C. C. Pinheiro, Raimir Holanda Filho, Maria Lúcia D. Pereira, Plácido Rogerio Pinheiro, Pedro José Leal Santiago, and Rafael Comin-Nunes

Abstract Meningitis is an infectious disease that can lead to neurocognitive impairments due to an inflammatory process in the meninges caused by various agents, mainly viruses and bacteria. Early diagnosis, especially when dealing with bacterial meningitis, reduces the risk of complications and mortality. Able to identify the most relevant features in the early diagnosis of bacterial meningitis. The model is designed to explore the prediction of specific data through the Logistic Regression, K Nearest Neighbour, and Random Forest algorithms. Early identification of the patient's clinical evolution, cure, or death is essential to offer more effective and agile therapy. Random Forest Algorithm is the best performing algorithm with 90.6% accuracy, Logistic Regression with 90.3% performance and KNN with 90.1%. The most relevant characteristics to predict deaths are low education level and red blood cells in the CSF, suggesting intracranial haemorrhage. The best-performing algorithm will predict the evolution of the clinical condition that the patient will present at the end of hospitalisation and help health professionals identify the most relevant characteristics capable of predicting an improvement or worsening of your general clinical condition early on.

Keywords Meningitis · Logistic regression · Random forest · K nearest neighbour

P. G. C. D. Pinheiro · R. Holanda Filho · P. R. Pinheiro (✉) · P. J. L. Santiago
University of Fortaleza, Fortaleza, Brazil
e-mail: placido@unifor.br

P. G. C. D. Pinheiro
e-mail: pedrogcdpinheiro@edu.unifor.br

R. Holanda Filho
e-mail: raimir@unifor.br

L. I. C. C. Pinheiro · M. L. D. Pereira · P. R. Pinheiro
Ceara State University, (L.I.C.C.P.), Fortaleza, Brazil
e-mail: luana.ibiapina@aluno.uece.br

P. G. C. D. Pinheiro
Estacio University Center, Fortaleza, Brazil

R. Comin-Nunes
University Center, September 7, Fortaleza, Brazil

A. Visvizi et al. (eds.), *Research and Innovation Forum 2022*,
Springer Proceedings in Complexity, https://doi.org/10.1007/978-3-031-19560-0_7

1 Introduction

The use of machine learning algorithms is a technological strategy capable of predicting early diagnoses of diseases and assisting health professionals in decision making in the most targeted, agile, and effective clinical treatment. An early approach is essential to prevent death and neurological sequelae caused by diseases directly affecting the Central Nervous System (CNS), such as bacterial meningitis [1].

Bacterial meningitis is considered a public health problem due to its high mortality and morbidity rates. The mortality rate can vary from 17 to 40% depending on the pathogen and the country's socio-economic condition in cognitive damage. It can affect up to 50% of patients [2]. Regarding the epidemiological context of bacterial meningitis, it is known that its incidence is 0.8 and 2.6 cases per 100,000 adults in developed countries and may be up to 10 times higher in developing countries [3].

The Infectious Diseases Society guides the initiation of therapy if the case is considered probable. Such guidelines are essential since the late initiation of antibiotic treatment may be associated with an unfavourable outcome [4, 5].

Data say that this type of meningeal disease can lead to death within 24 h. After infection [6–8]. In a global context, meningitis is still a health problem, causing 300,000 deaths per ye and the sequelae left by the disease in patients who survived [8].

In 2021, the World Health Organization launched the project "Defeating Meningitis by 2030: The Global Roadmap", which sets out an action plan to eradicate meningitis by 2030, explaining the importance of mass vaccination as the primary form of prevention. In Brazil, during the year 2021, 831 confirmed cases of meningitis were reported by the Information System of Notifiable Diseases, of which 92 died as an outcome, and about 24% of patients who underwent death were infected with meningitis bacteria. The mortality rate from meningitis is currently among patients younger than one year and in adults aged 40–49 years [9, 10]. The early diagnosis performed by the health professional favours the rapid initiation of treatment and a possible reduction in the risk of neurological impairment due to greater control over the evolution of the disease [11, 12]. Seeking to support the decision making of health professionals, Artificial Intelligence (AI) can act as a link between the large amount of data allied to medical devices to obtain accurate and agile diagnoses [13–16].

This article aims to predict the clinical outcome for bacterial meningitis, identifying the main clinical and laboratory characteristics through a better-performing algorithm in machine learning.

1.1 Machine Learning Algorithms

The application of machine learning and Artificial Intelligence algorithms to predict the predominant characteristics in the clinical outcome of patients diagnosed with bacterial meningitis. The database used in this research is in the public domain and was obtained through the Brazilian Information System on Notifiable Diseases.

The data analysis approach is made quantitatively of patients diagnosed with Bacterial Meningitis from January 2019 to December 2021 in Brazil. Initially, health professionals classified the data with clinical and academic experience in infectious diseases. Then, the Machine Learning algorithms (Logistic Regression, Random Forest, and KNN) were tested and evaluated for their performance.

2 Orange Canvas

The research results were processed through the Orange Canvas and Google Collaboratory tools (https://colab.research.google.com/), using the Python language Machine Learning animals: Logistic Regression, Random Forest, and KNN were implemented and tested, approaching the evolution of the case (high 1.0; death 2.0). As can be seen in Fig. 1.

2.1 Random Forest

The Random Forest algorithm is a learning method used for classification and regression. It overcomes the overfitting problem and has high robustness to noise and outliers o scalability and parallelism for classifying high dimensional data [17].

The central concept of the algorithm is the combination of several decision tree classifiers, being able to combine Bagging and random subspace for decision making and obtain the result [18]. The construction process is shown in Fig. 2.

The process is divided into two steps: the voting process and the tree's growth, divided into three points: the random selection of the training set, the construction of random forests, and the split node. When splitting the nodes, the feature with the

```
BaseDados ='./BaseDados/'  #Caminho para o arquivo CSV (dataset)
Meningite = pd.read_csv(BaseDados+"Meningite.csv" , sep = ';')#Dataset
.
.
.
# Logistic Regression
modelo = LogisticRegression(max_iter=10000)
ModeloGenerico(modelo, 'Logistic Regression')

# Random Forest
modelo = RandomForestClassifier()
ModeloGenerico(modelo, 'Random Forest Classifier')

# KNN
modelo = KNeighborsClassifier()
ModeloGenerico(modelo, 'KNearestNeighbor')
```

Fig. 1 Excerpt algorithm used

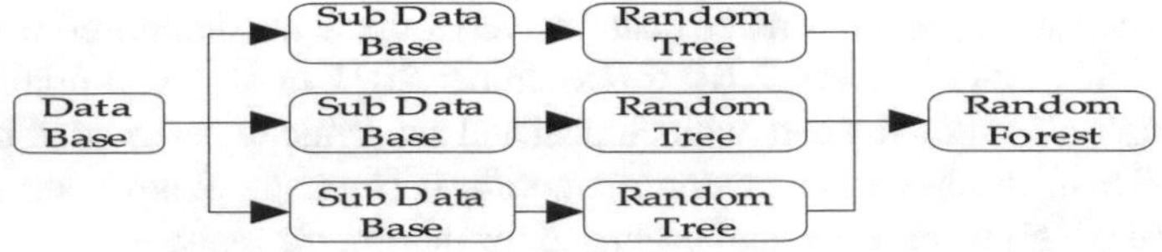

Fig. 2 Structural map of random forest

lowest Gini coefficient is selected as the splitting feature [18]. The Gini coefficient calculation steps as shown in Eq. 1, where the probability of sample set j in sample set S:

$$Gini(s) = 1 - \sum_{i=1}^{m} P_i^2 \tag{1}$$

2.2 *K-Nearest-Neighbors*

It is a machine learning algorithm used for classification and regression, where the learning is structured in the similarity of one data to another, based on instances of input k to learn a specific function to produce the desired output, having as the most straightforward case k = 1 [19]. KNN assumes similarity between existing data to generate the classification of classes. The objective is to determine the sample belonging to a specific class by analysing neighbouring samples (k-neighbor). The appropriate choice of k has a significant impact on the performance of the algorithm, which can improve or worsen the accuracy [20] given by Eq. 2, where x is the point to be labelled, and Y is the closest point to x:

$$C_n^{1nn}(x) = Y \tag{2}$$

Since x and y are close enough, the probability that x and y belong to the same class is high. However, the higher the value of k, the more significant the reduction in the risk of overfitting due to noise in the training data. Mainly we can consider the most popular distance measure. The Euclidean distance is given by Eq. 3:

$$D_{euclidiana} = \sqrt{(p_1 - q_1)^2 + \cdots + (p_n - q_n)^2} = \sqrt{\sum_{i=1}^{n} (p_n - q_n)^2} \tag{3}$$

2.3 Logistic Regression

Linear Regression uses regression analysis to determine a quantitative relationship between variables. Considering X as the independent variable and Y as the dependent variable, a model of the relationship between the independent and dependent variable can be established through the least-squares function of the equation when the variables present a linear relationship [21]. The objective is to find more suitable parameters and use a line to adjust the points, as shown in Fig. 3 [22].

The Logistic Regression Algorithm is a type of linear regression of simple complexity, being one of the most used algorithms for binary classification (0 or 1), based on Machine learning for predictive analysis and classification structured in probability theory [21], can be applied to areas of knowledge such as Economics [23], Finance [24] and Medicine [25–27].

The algorithm performs the prediction, given in Eq. 4, of the probability of a specific event occurring, where the probability is considered a dependent variable.

$$ln\left(\frac{P}{1-p}\right) = ln\left[\frac{\sum_{i=1}^{j} P_i}{1-\sum_{i=1}^{j} P_1}\right] = \beta_0 + \sum_{i=1}^{n} \beta_i x_i \tag{4}$$

where:

$J = 1, 2, ..., K - 1$
β_0: is the intercept
β_i: Partial regression coefficient
P: Probability of event occurrence
$\frac{P}{1-p}$: Probability of result
X: preditor.

The curve or logistic function is a sigmoid curve S is applied to produce the continuous probability distribution over bounded output classes between 0 and 1 generated by equation five and can be seen in Fig. 4:

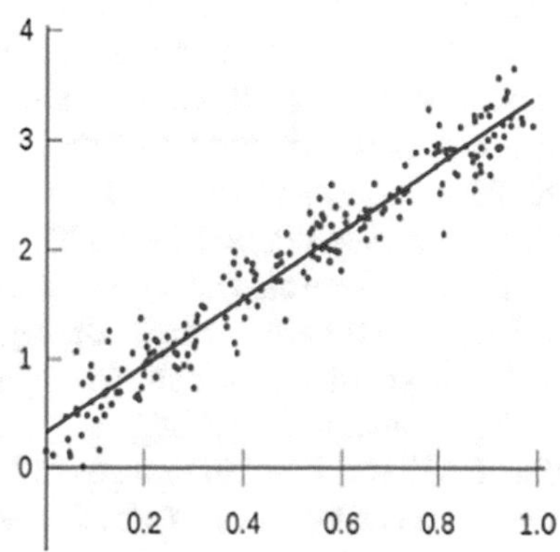

Fig. 3 Linear regression diagram

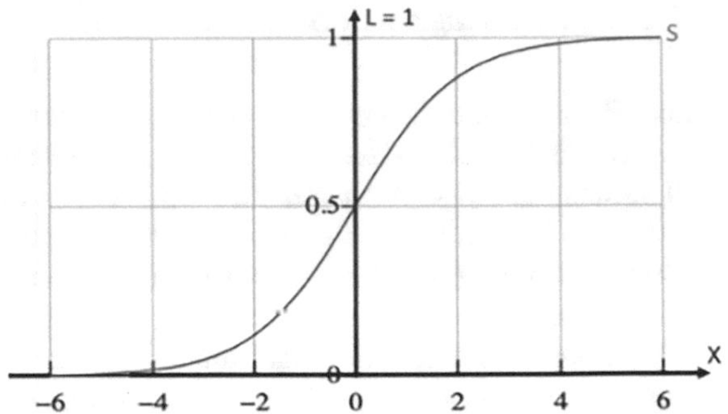

Fig. 4 Sigmoid S curve as defined in Eq. 2

$$f(x) = \frac{L}{1 + e^{-k(x - x_0)}} \tag{5}$$

where:

the x_0= value of the midpoint of the sigmoid,
$\boldsymbol{L}$ = maximum value of the curve,
$\boldsymbol{k}$ = the logistic growth rate or slope of the curve.

The curve S shown in Fig. 3 is obtained through the graph f as it approaches L as x approaches $+\infty$ and approaching 0 (zero) when x comes $-\infty$ [21].

3 Results

This section presents the results obtained in this research using a machine with a Windows operating system, CPU Intel(R) Core (TM) i7-4500U and 16 GB of RAM. Three artificial intelligence algorithms were applied to 36 features of meningitis. After testing the three machine learning algorithms, the Random Forest algorithm showed a more extended training and testing time of 4.914 and 0.54 s. However, it presented performance to predict discharge or death from meningitis, with an accuracy of 90.6% and F1 of 95.1%. Logistic Regression showed 90.3% accuracy, 90.8% accuracy and 99.4% F1. The KNN, on the other hand, showed an accuracy of 90.1%, precision of 90.5% and F1 of 94.8%. In Table 1.

The data were also analysed using the area under the Receiver Operating Characteristic (ROC) curve, which corresponds to the relationship between valid positive and false-positive values in the range from 0 to 1, as shown in Fig. 5.

With Random Forest in green, obtaining a value of 0.701, Logistic Regression in lilac 0.632 and KNN in orange 0.502. The model generated by the best-performing algorithm can be viewed from the tree represented in Fig. 6.

Table 1 Results avaliation

Results evaluation							
Model	Train time [s]	Test time [s]	AUC	CA	F1	Precision	Recall
Random forest	4,914	0,54	0,701	0,906	0,951	0,906	1
Logistic regression	1,092	0,154	0,632	0,903	0,949	0,908	0,994
KNN	1,004	0,546	0,502	0,901	0,948	0,905	0,994

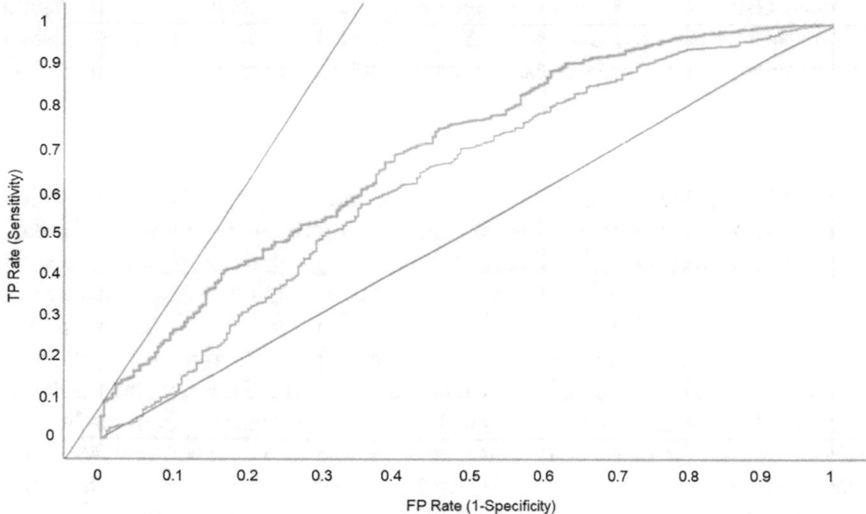

Fig. 5 Area under the ROC curve

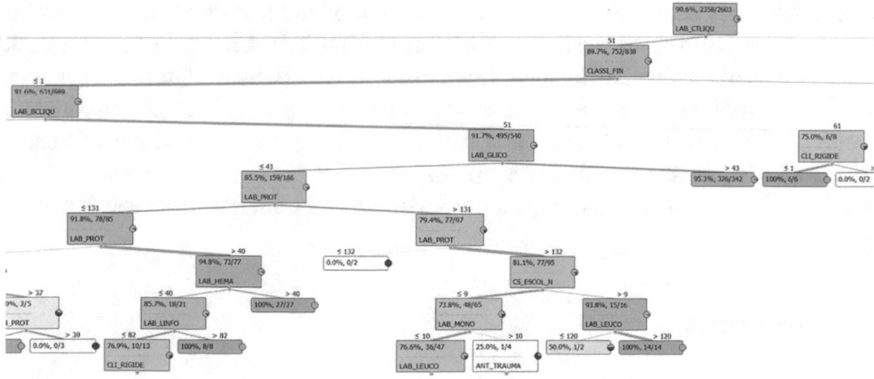

Fig. 6 Decision tree

It can be identified that the main characteristics for predicting cure in a patient with bacterial meningitis are related to the cytochemical parameters of the CSF.

4 Discussion

Data were analysed through cross-validation, of which 70% were used for testing and 30% for training. As for the characteristics identified, it is known that among the main tests for the evaluation of meningitis is the collection of cerebrospinal fluid, capable of revealing the main etiological agents of bacterial meningitis, which are Haemophilus Influenzae, Neisseria meningitides, and Streptococcus pneumonia came to be considered the principal causative agent of the disease. Then, from identifying the disease, the most effective medical conduct can be defined by choosing specific drug therapy to avoid indiscriminate use of drug requests for more targeted exams with the patient's clinical condition. Thus, favouring a good health condition. To diagnose whether the patient has bacterial meningitis, the proliferation of a positive bacterial culture present in the CSF must be evaluated, whose parameter considered normal is cellularity of fewer than ten cells per cubimillimetrer of blood (mm^3); an increase considered significant is if its quantity is greater than 200 cells/mm^3. The values obtained about the outcome of death, the cellularity of 51 cells/mm^3 was obtained, which is considered a moderate increase. However, only 9.4% of patients initially evaluated with this same value were discharged from the hospital (cure). However, the model identified that all people who died had erythrocytes at a count greater than 2130 erythrocytes/mm^3, whereas those who were cured had a reduced value, being less than 132 erythrocytes/mm^3. To confirm the cause of red blood cells, tests should be performed: to evaluate clot formation and xanthochromia in the three tubes. Thus, immunisation prevents tuberculous meningitis. Regarding vaccination, it is known that despite the high fatality rate from meningitis, Pentavalent vaccines can be prevented, including preventing infections such as conjugated meningococcal meningitis Haemophilus influenze type B, as well as the prevention of children by Neisseria Meningitidis serogroup C. It is essential to recognise that the health professional needs to early assess the various signs and conditions of meningitis and differentiate them since therapeutic measures are appropriate, especially in its initial phase, to reduce the impact of morbidity and mortality from the disease.

5 Conclusion

Identifying the characteristics that predict the clinical evolution of the patient diagnosed with Bacterial Meningitis, considering the clinical and socio-economic situation of the patient, is a complex process. Among the machine learning algorithms, the one that presented the best performance was Random Forest with 90.6% accuracy, followed by Logistic Regression with 90.3% and KNN with 90.1%. Thus, the most relevant characteristics resulting from evaluating parameters to predict the outcome of death from bacterial meningitis were low education level and red blood cells in the CSF, suggesting intracranial haemorrhage. The underreporting of information by health professionals on the Notification Form was a limitation of the research, being necessary to treat the database, eliminating invalid data through a filter, the need to expand the database and the lack of information limited the number of functionalities to be served. The data were also analysed using the area under the Receiver Operating Characteristic (ROC) curve, which corresponds to the relationship between valid positive and false-positive values from 0 to 1. Random Forest obtained a value of 0.701, Logistic Regression 0.632 and KNN 0.502. Future work is applying hybrid models to expand the universe of classifications and risk. Thus, it intends to incorporate the Multicriteria Decision Analysis, with the Multiattribute Utility of the decision to the bases presented.

References

1. Teixeira, D.C., Diniz, L.M., Guimarães, N.S., Moreira, H.M., Teixeira, C.C., Romanelli, R.M.: Risk factors associated with theoutcomes of pediatric bacterial meningitis: a systematic review. J Pediatr. **96**, 159–167 (2020)
2. Christo, P.P.: "Time is brain" also for bacterial meningitis. Arq Neuropsiquiatr. **77**(4), 221–223 (2019)
3. Lucas, M.J., Brouwer, M.C., Van de Beek, D.: Sequelas neurológicas da meningite bacteriana. J Infectar. **73**(1), 1827 (2016)
4. Costerus, J.M., Brouwer, M.C., Bijlsma, M.W., Beek, D.: Meningite bacteriana adquirida na comunidade. Curr Opin Infect Dis. **30**(1), 135–141 (2017)
5. Veronesi, R., Focaccia R.: Tratado de Infectologia. 5ª ed. São Paulo: Ed. Atheneu (2015)
6. BRASIL. Meningite. Disponível em: https://www.gov.br/saude/pt-br/assuntos/saude-de-a-a-z/m/meningite (2020)
7. Teixeira, A.B., Cavalcante, J.C.V., Moreno, I.C., Soares, I.A., Holanda, F.O.A.: Meningite bacteriana: uma atualização. RBAC **50** (4), 327-329 (2018)
8. WHO guidelines on self-care interventions for health and well-being. Geneva: World Health Organization (2021)
9. Datasus. Available online: http://tabnet.datasus.gov.br/cgi/deftohtm.exe?sinannet/cnv/meninbr.def. Accessed 28 Dec 2021
10. CEARÁ, Governo do Estado do. Secretaria de Saúde do Estado do Ceará. Célula de Imunização (CEMUN). Boletim Epidemiológico: Meningite. Ceará (2020)
11. Teixeira, D.C., et al.: Risk factors associated with the outcomes of pediatric bacterial meningitis: a systematic review. J Pediatr. **96**, 159–167 (2020)
12. Castro, A.K.A., Pinheiro, P.R., Pinheiro, M.C.D., Tamanini, I.: Towards the applied hybrid model in decision making: a neuropsychological diagnosis of alzheimer's disease study case. Int. J. Comput. Intell. Syst. 89–99 (2011)
13. Andrade, E.C., Pinheiro, P.R., Holanda Filho, R., Nunes, L.C., Pinheiro, M.C.D., Abreu, W.C., Simão Filho, M., Pinheiro, L.I.C.C., Pereira, M.L.D., Pinheiro, P.G.C.D.: Comin-Nunes, R.: Application of machine learning to infer symptoms and risk factors of covid-19. Springer Proc Compl 13–24 (2021)
14. Ara, A.: Case study: Integrating IOT, streaming analytics and machine learning to improve intelligent diabetes management system. In: International Conference on Energy, Communication, Data Analytics and Soft Computing. IEEE, pp 3179–3182 (2017)

15. Andrade, E., Portela, S., Pinheiro, P.R., Nunes, L.C., Simão Filho, M., Costa, W.S., Pinheiro, M.C.D.: A protocol for the diagnosis of autism spectrum disorder structured in machine learning and verbal decision analysis. Comput. Mathat. Meth. Med. (2021)
16. Santos, H.G.: Machine Learning para Análises Preditivas em Saúde: Exemplo de Aplicação para Predizer Óbitos em Idosos de São Paulo, Cad. Saúde Pública **35**(7) (2019)
17. Guo, Y., Zhou, Y., Hu, X., Cheng, W.: Research on recommendation of insurance products based on random forest. In: International Conference on Machine Learning, Big Data and Business Intelligence (2019)
18. Lan, H., Pan, Y.A.: crowdsourcing quality prediction model based on random forests. In: 18th International Conference on Computer and Information Science (ICIS) (2019)
19. Fukunaga, K., Narendra, P.M.A.: branch and bound algorithm for computing k-nearest neighbours. IEEE Trans. Comput. **100**(7) (1975)
20. Altman, N.S.: An introduction to kernel and nearest-neighbour nonparametric regression. Am. Statist. **46**, 175–185 (1992)
21. Pinheiro, L.I.C.C., Pereira, M.L.D., Andrade, E.C.D., Nunes, L.C., Abreu, W.C.D., Pinheiro, P.G.C.D., Holanda Filho, R., Pinheiro, P.R.: An intelligent multicriteria model for diagnosing dementia in people infected with human immunodeficiency virus. Appl. Sci. (2021)
22. Zou, X., Hu, Y., Tian, Z., Shen, K.: Logistic regression model optimization and case analysis. In: IEEE 7th International Conference on Computer Science and Network Technology (2019)
23. Kurdyś-Kujawskaa, A., Zawadzkaa, D.: applying logistic regression models to assess domestic financial decisions relating to debt. Proc. Comput. Sci. **176**, 3418–3427 (2020)
24. Han, D., Ma, L., Yu, C.: Financial prediction: application of logistic regression with factor analysis. In: 4th International Conference on Wireless Communications (2008)
25. Pinheiro, L.I.C.C., Pereira, M.L.D.P., Fernandez, M.P., Vieira Filho, F.M., Abreu, W.J.C.P, Pinheiro, P.G.C.D.: Application of data mining algorithms for dementia in people with HIV/AIDS. Comput. Mathemat. Meth. Med. (2021)
26. Araújo de Castro A.K., Pinheiro P.R., Dantas Pinheiro M.C.: Applying a decision-making model in the early diagnosis of alzheimer's disease, vol. 4481. LNCS, Springer (2007)
27. Araújo de Castro, A.K., Pinheiro, P.R., Pinheiro, M.C.D.: A hybrid model for aiding in decision making for the neuropsychological diagnosis of Alzheimer's disease, vol. 5009 (pp 495- 504). LNCS, Springer (2008)

Research Development on Assistive Technology: A Network and Concept-Linking Analysis

Chien-wen Shen, Agnieszka Maria Koziel, and Tso-hsuan Yeh

Abstract This study aims to provide an in-depth understanding of assistive technology (AT) research development by investigating relationships between related keywords and concepts and uncovering how trends change over time. This study employed network analysis to identify the relationships between frequently cited papers and keywords in the AT literature and concept-linking analysis to uncover the key concepts and classifies them into clusters determining the changing trends in AT research. The network analysis results on author keywords co-occurrence and citation indicate that the development trend of AT is primarily observed in medicine and is related to medical devices used in rehabilitation or available for disabled and people for well-being improvement. Meanwhile, the concept linking analysis identifies seven groups of key AT concepts, including technical issues, education, health and disability, policy, user/people, and ways/medium. We also depict the yearly changes of the key concepts in AT research development. This research extracts the strongest key concepts, the most productive authors and countries, and the connecting relationship between authors and literature within the big data. It provides a comprehensive view of temporal patterns of AT developments and their development in terms of human–computer interactions.

Keywords Assistive technology · Network analysis · Concept-linking analysis · Disability

C. Shen · A. M. Koziel (✉) · T. Yeh
National Central University, Taoyuan, Taiwan
e-mail: aga.koziel@g.ncu.edu.tw

C. Shen
e-mail: cwshen@ncu.edu.tw

T. Yeh
e-mail: tsohsuan.yeh@g.ncu.edu.tw

A. Visvizi et al. (eds.), *Research and Innovation Forum 2022*,
Springer Proceedings in Complexity, https://doi.org/10.1007/978-3-031-19560-0_8

1 Introduction

Assistive technologies (ATs) support people in managing everyday tasks, particularly helping older adults, disabled people, and people with health issues improve their quality of life [1]. People who need AT are not limited to those with physical disabilities but also those with visual, hearing, or cognitive impairment, deafblindness, or learning disability [2–4]. In addition, AT provides useful devices based on human–computer relations for aging people [5, 6]. Adequate assistive technology should address people with disabilities and elders with specific needs and other health conditions. Hence it is crucial to provide the technologies which better fit the needs of the disabled and older people [7]. People with visual and hearing impairment face particular challenges in terms of AT development [8]. Research shows that providing an appropriate AT for different diseases and disabilities can strongly affect people's well-being. Moreover, establishing a good environment with AT for this group of people can help them improve their life satisfaction [9]. Innovative technology can provide support for older adults and reduce the burden on their future lives. New and innovative assistive technologies are constantly developing across the globe to enable older adults, disabled people, and people with health issues to enjoy an unprecedented quality of life. What's unique about innovations in AT development is that devices target specific challenges experienced by people in need. Innovative AT devices also endeavor to promote greater inclusivity within society by improving each user's health and well-being [10].

The subject of AT has been explored in specific research domains concerning disabled persons with visual disorders [3, 11, 12], dementia [13–15], brain disorders [4, 16, 17], learning disabilities [18–20] or people with mobility impairments [21–23]. However, these studies have predominantly discussed AT only in terms of people with specific kinds of impairments or with special needs. The recently published research provides a theoretical overview of the AT device design and features, but only with a focus on older people [7]. Therefore, the trajectory of AT is not clear [11, 21]. Hence research on AT development trends as an innovative interactive system can clarify where it aims to go. This research broadens the view on AT usage before the Covid-19 pandemic started and allows us to understand better the direction of AT development to fulfill certain groups of people's needs and improve their well-being. Besides, to further understand the development patterns, relationships, and critical concepts in the different research domains, this study conducts network and concept-linking analyses [24]. These methods are used to comprehensively analyzes the literature on AT development and thus, identify the research trends, relationship networks, and critical concepts. Analyzing the collective data can help reveal the AT development trends from different aspects. Thus, this study's findings are expected to provide readers with an in-depth understanding of AT-related research and clearly show the potential directions for further AT development.

2 Materials and Methods

2.1 *Data Collection and Descriptive Analysis*

The scope of this study includes research articles retrieved from the Scopus database published between 2000 and 2018. The database searching process was specified according to three available categories chosen in search engine filters - titles, abstracts, and author key concepts, with "assistive technology" included as the main keyword. Scopus was selected as the source of data collection because it is widely accepted as the largest abstract and citation database in the peer-reviewed literature [24]. Compared to other databases, Scopus was best suited for this study. Scopus is the largest abstract and citation database of peer-reviewed literature and the most comprehensive database with indexed content from more than 20,000 journals [25]. The database was searched for literature related to the main keyword, and the key concept's development trend in academic papers was investigated. Searches were conducted in the database, using the main keyword "assistive technology" and the terms: "assistive products," "assistive devices," "assistive equipment," "assistive instruments," and "assistive software" as keywords in studies related to AT Only articles published in English language articles were selected. After applying the criteria listed above, 12,582 articles from 135 countries with 18,452 author keywords were retrieved.

2.2 *Network Analysis*

A network map was carried out through VOS viewer, a software that provides network visualization of cluster analysis. The network map is created based on the bibliometric data extracted in the data collection phase [26]. These data were used to construct co-occurrence and citation network maps that provide a quick and accurate reading of the information and help identify the emerging trends in advance [27]. The co-occurrence network helps determine the link between the two terms that occur together in the same publication [28]. The citation network map was constructed according to the number of citations and the relations between all items. In the created network visualization, items are shown along the time axis, with colors indicating the clusters they belong to. Visualization provides an overview of the most frequently used keywords in the publications, cited publications in a citation network, the relation of citations in these publications, and the clusters to which the publications belong [29]. The network construction process concerns three phases normalization, mapping, and clustering. In the normalization process, the strength for differences between the nodes is calculated. In the mapping phase, the algorithm is used to position the node in the network in a two-dimensional space. Finally, in the third phase, the clustering algorithm assigns nodes to clusters. The network analysis itself can reveal historical patterns in citation data, and to detect recent trends in the AT development literature, a rudimental concept-linking analysis in data mining was performed.

2.3 Concept-Linking Analysis

In this study, the concept-linking analysis was adopted to perform literature mining to identify the emerging key concepts in AT. Moreover, the occurrence of key concepts was used to understand the key concepts' trends across the time frame. The concept-linking analysis focuses on the title and abstract fields of the extracted data, where the year is used as a dimension [30]. The SAS Text Miner was used to analyze the whole data set. The data mining process flow includes text parsing and filtering; and concept linking, clustering, and strength between key terms and highly associated terms. The text parsing stage extracts, clean, and creates a dictionary of words from the documents. This step involves parsing the extracted words to identify entities, removing stop words, and spell-checking. The text filtering is next used to reduce the total number of parsed terms, recognize irrelevant terms that are not directly related to the research, and eliminate them [31]. Further concept linking analysis included linking the concepts to identify other terms that are highly associated with a key concept. The key concept is shown at the center of a link diagram. The terms with the key concept most often circle the central term. Key concept links are represented by the strength of the relationship with the central term, which is assistive technology. The strength of the connection between terms is calculated using the binomial distribution [32]. The concept-linking analysis is reviewed according to the first tier of key concepts. It provides eight links of the terms most related to the central term, assistive technology. A key concept-linking table was developed to depict the changes in the key concept strength and key concept clusters through the years. The strongest related terms to key concepts were classified into clusters by grouping a set of objects with similar content into the same cluster.

3 Results

3.1 Descriptive Analysis

Research shows, that the highest annual publications distribution growth rate was recorded in 2002 and 2006 at 47.6% and 43.4%, respectively. Moreover, the number of publications until 2006 remained below 500 per year. Later, between 2007 and 2015, the number grew by 100 every alternate year. In the following years, approximately 100 more papers were published each year. At the end of 2018, 1331 publications containing author keywords were recorded. The author's keyword analysis showed the frequency of the appearance of each keyword extracted each year from

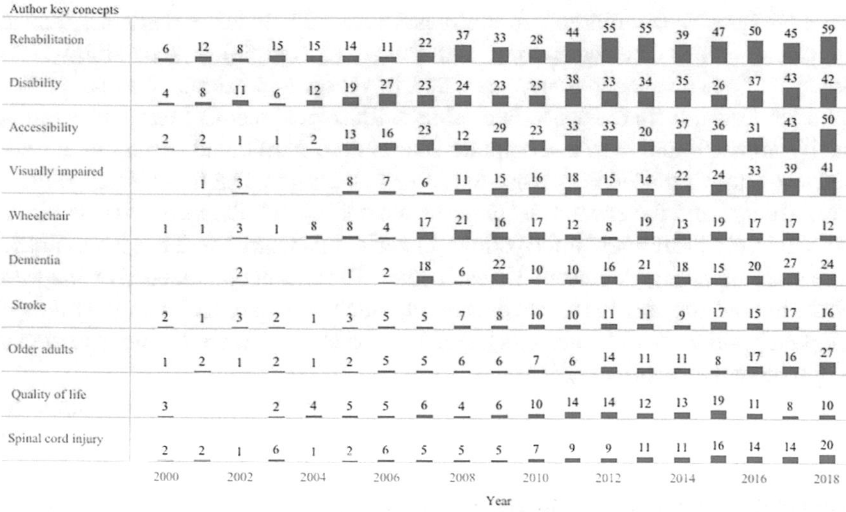

Author key concepts	2000		2002		2004		2006		2008		2010		2012		2014		2016		2018
Rehabilitation	6	12	8	15	15	14	11	22	37	33	28	44	55	55	39	47	50	45	59
Disability	4	8	11	6	12	19	27	23	24	23	25	38	33	34	35	26	37	43	42
Accessibility	2	2	1	1	2	13	16	23	12	29	23	33	33	20	37	36	31	43	50
Visually impaired		1	3			8	7	6	11	15	16	18	15	14	22	24	33	39	41
Wheelchair	1	1	3	1	8	8	4	17	21	16	17	12	8	19	13	19	17	17	12
Dementia			2			1	2	18	6	22	10	10	16	21	18	15	20	27	24
Stroke	2	1	3	2	1	3	5	5	7	8	10	10	11	11	9	17	15	17	16
Older adults	1	2	1	2	1	2	5	5	6	6	7	6	14	11	11	8	17	16	27
Quality of life	3			2	4	5	5	6	4	6	10	14	14	12	13	19	11	8	10
Spinal cord injury	2	2	1	6	1	2	6	5	5	5	7	9	9	11	11	16	14	14	20

Year

Fig. 1 Author keywords with yearly frequency

18,452 keywords between 2000 and 2018. Figure 1 illustrates the keywords with a high yearly frequency. The most repeated term was "rehabilitation," followed by "disability" and "accessibility," which directly reveals the characteristics of the target group associated with AT. Overall, most keywords' usage typically increased each year, except for "accessibility," "visually impaired," "dementia," and "quality of life" (Fig. 1). From 2000 to 2004, only one or two related articles defined "accessibility" as an author keyword, although the number of annual publications increased in 2005 and kept growing afterward. Between 2000 and 2006, the key concept of "dementia" appeared in approximately six publications, and the publications' number drastically increased in 2007. The keyword "quality of life" was not discussed in 2001 and 2002. After two years, however, the number of publications with this key concept slowly started to rise.

3.2 *Network Analysis Results in Co-occurrence and Direct Citation*

A co-occurrence network was built with 1318 author keywords. In order to make the visualization map (see Fig. 2) easier to perceive and interpret, the top 100 words with the highest co-occurrence value were extracted. The author's keyword with the highest number of co-occurrences was "assistive technology," which also had the highest link strength, followed by "rehabilitation," "disability," and "accessibility." In this analysis, the top 100 keywords were grouped into five clusters. Five large clusters appeared in different colors, as follows: cluster #1 (red) = rehabilitation, cluster #2 (green) = children's disabilities, cluster #3 (blue) = dementia, cluster #4 (yellow) = assistive technology, and cluster #5 (purple) = accessibility. The main term, "assistive technology," appeared in yellow and exerted absolute control over other terms. It shared space in clusters with several medical terms concerning visually impaired and human–computer interaction. Purple and green colors indicated two clusters placed very close to the yellow cluster. The first cluster refers to "accessibility" and the second to "children's disabilities." Between "accessibility" and "children's disabilities" is the yellow cluster's main word, "assistive technology." These three are strongly connected to each other. The cluster commanded by the word "rehabilitation" appears in the red cluster, in which "assistive technology" holds an important position. A fifth cluster is formed by the term "dementia" and is related to aging society and technology.

Another network-analysis map was constructed based on the number of citations and the relations among different items (see Fig. 2). The node is labeled by the author of the cited document and year of publication. The most cited articles mostly concern technology, rehabilitation, disabilities, and older people. In terms of the research topic, the authors were grouped into clusters to indicate who published papers using a similar key concept. The clusters were named based on the majority of references belonging to them, and are represented by six colors: light blue = disability and rehabilitation, yellow = education and technology, red = aged population, green = mobility and acceptability, dark blue = children and purple = impact of AT. Although the red cluster is the central one, the biggest nodes appear in yellow and light blue clusters, in which the most frequently cited articles are included.

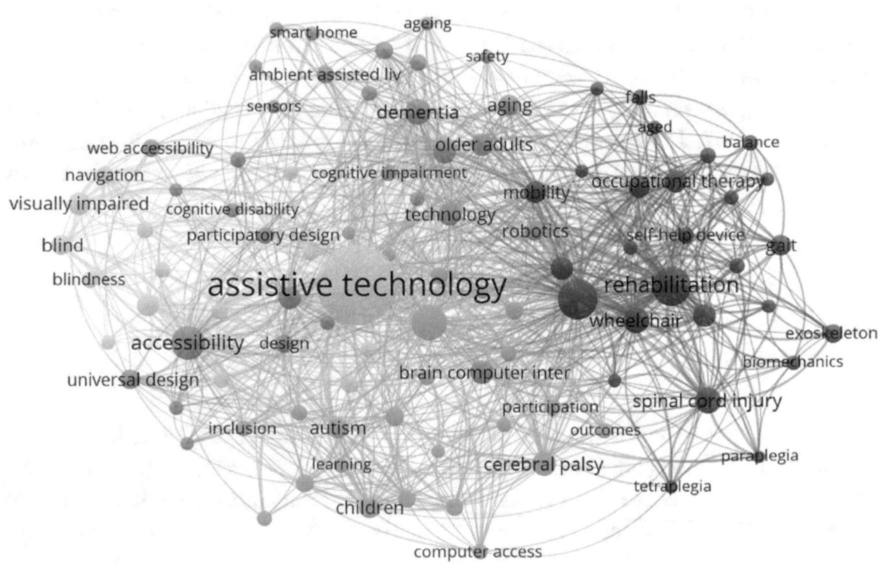

Fig. 2 Author keywords co-occurrence network map

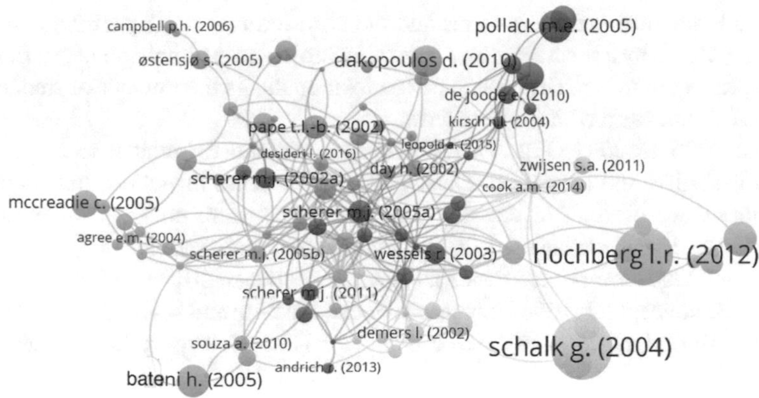

Fig. 3 Citation network map

The publication retrieved as top1 is "BCI2000: a general-purpose brain-computer interface system," which was published in 2004 by Schalk regarding the education and technology concept [33]. This paper is highly related to innovative interactive systems based on the brain-computer interface. The following two papers with full citations belong to the light blue cluster and are related to the disability and rehabilitation concept. One of the papers was published by Hochberg in 2012, where the author investigated a neural interface system that could restore mobility and independence in people with paralysis [34]. The second paper classified in the light blue cluster published in 2002, examined a system to provide computer access for people with severe disabilities [35]. A citation network map discovered among six different clusters indicated that they had issues in common (Fig. 3).

3.3 Results About Concept-Linking Analysis by Year

The concept-linking diagram reveals eight links between the most related terms and the given term "assistive technology." The analysis aimed to identify the eight terms with the maximum connection with assistive technology each year and the type of group to which the words belong. In the concept-linking analysis for 2000–2018, seven groups of key concepts were determined. The recognized terms include technical issues (yellow), education (green), health and disability (blue), policy (orange), user/people (pink), ways/medium (grey), and miscellaneous (white). The most frequently used keywords belong to the technical issues groups, followed by health and disability, ways/medium, and education. As AT is commonly used in medical issues, another group of common key concepts that appeared in the documents is group blue, where the term "disability" has constantly appeared since 2009. Figures 4, 5, 6 and 7 summarize the uncovered terms in the concept link diagram of each year from 2000 to 2018. The strength value under each term shows the strength of the link between particular terms and the core term (AT). The attribute has two values separated by a slash sign. The latter sign indicates the total number of frequencies appearing in the specific year, and the former shows the number of frequencies arising in connection with the core term.

From 2000 to 2004 (Fig. 4), the concepts are mostly related to technology, health/disability, and education. During those five years in publications appeared ideas that were difficult to classify. Therefore they were collected together into one group called miscellanios. During the next four years (2005–2008), more technology and education-related concepts (Fig. 5). Between 2010 and 2009, and 2013 (Fig. 6), the concepts classified into the groups: education and health/disability significantly increased, however in the next five years (Fig. 7), terms related to education

Year	2000	2001	2002	2003	2004
Term 1 Strength	technology (29/45)	technology (11/48)	technology (60/95)	technology (9/103)	technology (76/112)
Term 2 Strength	provision (4/5)	disability (16/40)	Life (17/30)	disability (31/61)	provision (8/9)
Term 3 Strength	user (14/18)	student (7/10)	provision (10/14)	family (9/14)	Life (16/27)
Term 4 Strength	development (7/13)	community (6/11)	family (5/5)	delivery (6/8)	people (34/63)
Term 5 Strength	approach (7/11)	alternative (6/10)	person (20/37)	policy (8/10)	project (14/22)
Term 6 Strength	application (6/9)	Integration (4/6)	field (10/13)	strategy (9/16)	form (7/9)
Term 7 Strength	issue (5/8)	employment (4/5)	direction (5/5)	consumer (7/8)	teacher (6/7)
Term 8 Strength	software (5/6)	nature (4/5)	developer (5/5)	loan (5/5)	Innovation (3/7)

LEGEND:	TECHNOLOGY	HEALTH/ disability	USER/PEOPLE	Ways/medium
	EDUCATION	POLICY	Misc	

Fig. 4 Concept linking analysis by year groped into 7 clusters (1)

Year	2005	2006	2007	2008
Term 1 Strength	technology (85/143)	technology (111/181)	technology (157/283)	technology (156/269)
Term 2 Strength	disability (40/87)	student (23/34)	disability (62/142)	education (20/35)
Term 3 Strength	environment (25/48)	engineering (15/19)	engineering (11/19)	service (33/66)
Term 4 Strength	delivery (8/10)	curriculum (9/9)	Individual (35/79)	adaptation (10/11)
Term 5 Strength	accommodation (8/11)	structure (13/18)	iOS (33/74)	learning (13/19)
Term 6 Strength	contribution (7/8)	value (12/15)	assessment (24/51)	web (14/21)
Term 7 Strength	technical (6/7)	frequency (9/10)	experience (23/51)	communication technology (9/9)
Term 8 Strength	outcome (5/23)	microcontroller (8/8)	quality (22/47)	Isolation (6/6)

LEGEND:	TECHNOLOGY	HEALTH/ disability	USER/PEOPLE	Ways/medium
	EDUCATION	POLICY	Misc	

Fig. 5 Concept linking analysis by year groped into 7 clusters (2)

Year	2009	2010	2011	2012	2013
Term 1 Strength	technology (192/336)	technology (204/348)	technology (234/438)	technology (223/420)	technology (295/497)
Term 2 Strength	disability (71/180)	disability (72/172)	disability (90/222)	disability (70/187)	disability (109/229)
Term 3 Strength	education (21/37)	education (28/46)	education (31/60)	education (34/55)	education (38/66)
Term 4 Strength	service (40/82)	environment (47/103)	adaptation (16/27)	student (33/60)	student (41/73)
Term 5 Strength	Interaction (24/47)	science (19/31)	policy (22/39)	learning (27/42)	autonomy (14/18)
Term 6 Strength	issue (24/47)	professional (18/31)	accessibility (27/54)	curriculum (11/11)	adoption (13/19)
Term 7 Strength	opportunity (22/40)	theory (11/16)	law (11/15)	application (54/125)	educator (9/10)
Term 8 Strength	collaboration (11/14)	cognition (9/11)	clinician (11/16)	mobile (11/17)	special education (8/9)

LEGEND: TECHNOLOGY, HEALTH/ disability, USER/PEOPLE, Ways/medium, EDUCATION, POLICY, Misc

Fig. 6 Concept linking analysis by year groped into 7 clusters (3)

Year	2014	2015	2016	2017	2018
Term 1 Strength	technology (334/541)	technology (316/569)	technology (379/619)	technology (364/608)	technology (409/685)
Term 2 Strength	disability (95/215)	disability (91/228)	dementia (25/37)	disability (104/258)	disability (120/282)
Term 3 Strength	Life (61/130)	dementia (13/21)	Life (79/178)	education (47/93)	education (51/101)
Term 4 Strength	service (52/101)	participation (21/41)	student (45/88)	accessibility (70/135)	communication (68/148)
Term 5 Strength	participation (31/54)	accessibility (26/53)	science (23/38)	community (36/77)	innovation (21/28)
Term 6 Strength	impact (32/58)	web accessibility (23/45)	observation (17/27)	concern (19/32)	collaboration (20/32)
Term 7 Strength	methodology (27/47)	screen reader (8/10)	employment (12/16)	reader (18/25)	government (12/14)
Term 8 Strength	emotion (11/14)	media (6/7)	communication technology (10/13)	important role (13/16)	older user (7/7)

LEGEND: TECHNOLOGY, HEALTH/ disability, USER/PEOPLE, Ways/medium, EDUCATION, POLICY, Misc

Fig. 7 Concept linking analysis by year grouped into 7 clusters (4)

decreased, while the health-related concepts increased. The concept "technology" appears every year and the terms classified in the same group are spread throughout the entire period. However, literature published in the first two years 2000 and 2001, and also in the last years 2017 and 2018 indicate strong relations between "technology", "disability", "community", and "education".

4 Conclusion

Aiming to better understand the way and direction the AT is developing in order to fulfill certain groups of people's needs and improve their well-being, this paper is a literature analysis of the papers related to AT in different domains. Our study provided an overall review and extracted the strongest key concepts, the most productive authors and countries, and the connecting relationship between authors and literature within the considerable data. Although, the increased number of publications with the main keyword "assistive technology" is considered an obvious conclusion, that also indicates the importance of AT. Moreover, it is necessary to research AT itself to better understand its development as innovative technology. Keywords such as "rehabilitation," "disability," and "accessibility" were used most frequently in the last decades, and other keywords related to specific diseases, disabilities, devices, or groups of people appeared progressively. This proves an increasing in-depth analysis of AT and the growing interest in those subjects. Moreover, the most cited articles concerned topics such as technology, rehabilitation, disabilities, education, and older people. Thus, the development trend of AT is primarily observed in medicine and is related to different, innovative medical devices used in rehabilitation or available for disabled and older people, to improve their well-being. It is also very important to pay attention to the words related to education, which started to appear more often since 2010 and increase connection to the main concept.

Our study results would help researchers in further exploring the topic of AT in terms of networks and key concepts. It may also help policymakers in better understanding AT development and the relations between policy terms in the last few years. Besides this research is a good attempt to provide a deep overview of key concepts related to assistive technology before the Covid-19 pandemic. It might be a good start for future research to discover changes in AT research between pre and the post-pandemic world and better understand possible world resilience in connection to disabled and aged people using AT in daily life.

References

1. Tao, G., et al.: Evaluation tools for assistive technologies: a scoping review. Arch. Phys. Med. Rehab. **101**(6), 1025–1040 (2020)
2. O'Neill, S.J., et al.: Assistive technology: understanding the needs and experiences of individuals with autism spectrum disorder and/or intellectual disability in Ireland and the UK. Assist. Technol. **32**, 1–9 (2019)
3. Perfect, E., Jaiswal, A., Davies, T.C.: Systematic review: Investigating the effectiveness of assistive technology to enable internet access for individuals with deafblindness. Assist. Technol. **31**(5), 276–285 (2019)
4. Tofani, M., et al.: The psychosocial impact of assistive device scale: Italian validation in a cohort of nonambulant people with neuromotor disorders. Assist. Technol. **32**(1), 54–59 (2020)
5. ATIA.: What is assistive technology. Available from: https://www.atia.org/at-resources/what-is-at/

6. Dudgeon, B.J., et al.: Managing activity difficulties at home: a survey of medicare beneficiaries. Arch. Phys. Med. Rehabil. **89**(7), 1256–1261 (2008)
7. Iancu, I., Iancu, B.: Designing mobile technology for elderly. A theoretical overview. Technol. Forecast. Social Change **155**, 119977 (2020)
8. Hermawati, S., Pieri, K.: Assistive technologies for severe and profound hearing loss: beyond hearing aids and implants. Assist. Technol. **32**(4), 182–193 (2020)
9. Özsungur, F.: A research on the effects of successful aging on the acceptance and use of technology of the elderly. Assist. Technol. **34**, 1–14 (2019)
10. Oomens, I.M.F., Scholten, C.: Inclusion in social innovation through the primary and secondary use of technology: a conceptual framework. Int. Rev. Appl. Econ. **34**(5), 672–686 (2020)
11. Bhowmick, A., Hazarika, S.M.: An insight into assistive technology for the visually impaired and blind people: state-of-the-art and future trends. J. Mult. User Inter. **11**(2), 149–172 (2017)
12. Elmannai, W., Elleithy, K.: Sensor-based assistive devices for visually-impaired people: current status, challenges, and future directions. Sensors (Basel) **17**(3), 565 (2017)
13. Asghar, I., Cang, S., Yu, H.: Assistive technology for people with dementia: an overview and bibliometric study. Health Info. Libr. J. **34**(1), 5–19 (2017)
14. Brims, L., Oliver, K.: Effectiveness of assistive technology in improving the safety of people with dementia: a systematic review and meta-analysis. Aging Ment. Health **23**(8), 942–951 (2019)
15. Gibson, G., et al.: Personalisation, customisation and bricolage: how people with dementia and their families make assistive technology work for them. Ageing Soc. **39**(11), 2502–2519 (2019)
16. de Oliveira, J.M., et al.: REHAB FUN: an assistive technology in neurological motor disorders rehabilitation of children with cerebral palsy. Neural Comput. Appl. **32**(15), 10957–10970 (2020)
17. Geman, O., et al.: An intelligent assistive tool using exergaming and response surface methodology for patients with brain disorders. IEEE Access **7**, 21502–21513 (2019)
18. Abbott, C., et al.: Emerging issues and current trends in assistive technology use 2007–2010: practising, assisting and enabling learning for all. Disabil. Rehabil. Assist. Technol. **9**(6), 453–462 (2014)
19. Alnahdi, G.: Assistive technology in special education and the universal design for learning. Turkish Online J. Educ. Technol. **13** (2014)
20. Atanga, C., et al.: Teachers of students with learning disabilities: assistive technology knowledge, perceptions, interests, and barriers. J. Special Educ. Technol. **35**(4), 0162643419864858 (2019).
21. Cowan, R., et al.: Recent trends in assistive technology for mobility. J. Neuroeng. Rehabil. **9**, 20 (2012)
22. Ehlers, S.G., Field, W.E.: Accessing and operating agricultural machinery: advancements in assistive technology for users with impaired mobility. Assist. Technol. **31**(5), 251–258 (2019)
23. Remillard, E.T., Fausset, C.B., Fain, W.B.: aging with long-term mobility impairment: maintaining activities of daily living via selection, optimization, and compensation. Gerontologist **59**(3), 559–569 (2017)
24. Gavel, Y., Iselid, L.: Web of Science and Scopus: a journal title overlap study. Online Inf. Rev. **32**(1), 8–21 (2008)
25. Agarwal, A., et al.: Bibliometrics: tracking research impact by selecting the appropriate metrics. Asian J. Androl. **18**(2), 296–309 (2016)
26. van Eck, N.J., Waltman, L.: Software survey: VOSviewer, a computer program for bibliometric mapping. Scientometrics **84**(2), 523–538 (2010)
27. Persson, O.: Identifying research themes with weighted direct citation links. J. Informet. **4**(3), 415–422 (2010)
28. Kong, X., et al.: Academic social networks: Modeling, analysis, mining and applications. J. Netw. Comput. Appl. **132**, 86–103 (2019)
29. Leung, X.Y., Sun, J., Bai, B.: Bibliometrics of social media research: a co-citation and co-word analysis. Int. J. Hosp. Manag. **66**, 35–45 (2017)

30. Nishanth, K.J., et al.: Soft computing based imputation and hybrid data and text mining: the case of predicting the severity of phishing alerts. Expert Syst. Appl. **39**(12), 10583–10589 (2012)
31. SAS, I.I.: Getting Started with SAS® Text Miner 12.1. 2012: Cary, NC: SAS Institute Inc.
32. SAS, SAS® Text Miner 15.1: Strength of Association for Concept Linking (2020)
33. Schalk, G., et al.: BCI2000: a general-purpose brain-computer interface (BCI) system. IEEE Trans. Biomed. Eng. **51**(6), 1034–1043 (2004)
34. Hochberg, L.R., et al.: Reach and grasp by people with tetraplegia using a neurally controlled robotic arm. Nature **485**(7398), 372 (2012)
35. Betke, M., Gips, J., Fleming, P.: The camera mouse: visual tracking of body features to provide computer access for people with severe disabilities. IEEE Trans. Neural Syst. Rehabil. Eng. **10**(1), 1–10 (2002)

Intelligent Screening from X-Ray Digital Images Based on Deep Learning

Aitana Jiménez Navarro, Lucía Martínez Hernández, Aya Elouali, Higinio Mora, and María Teresa Signes-Pont

Abstract Coronavirus disease (Covid-19) is an infectious respiratory disease caused by SARS-CoV-2. Among the symptoms, the respiratory system of the sufferer is affected. This respiratory condition suggests that the chest imaging plays a key role in the diagnosis of the disease. Several pre-trained deep learning models have been developed to detect Covid-19 through chest radiographs. These models provide high precision for binary detection, however, when combined with other diseases such as pneumonia that also affect the respiratory system and lungs, they offer poorer quality performance and reduce screening performance. In this study, we analyze some neural networks models for binary and multiclass classification of X-ray images in order to find out the best accuracy of classification. The models are based on deep learning methodology to learn and classify images. They are extracted from well-known repositories such as Kaggle. The conducted experiments compare their performance in several scenarios: a multiclass classification model versus the combination of several binary classification models. Two methods for combining the output of the binary models are proposed to achieve the best performance. The results show that the best results are obtained with a well-trained multiclass model. However, a preliminary screening can be obtained from the binary models without creating and training a more complex model.

Keywords Image screening · Covid-19 · Deep learning · Neural network

A. J. Navarro · L. M. Hernández · A. Elouali · H. Mora (✉) · M. T. Signes-Pont
Department of Computer Science Technology and Computation, University of Alicante, Alicante, Spain
e-mail: hmora@ua.es

A. J. Navarro
e-mail: ajn9@alu.ua.es

L. M. Hernández
e-mail: lmh29@alu.ua.es

A. Elouali
e-mail: ae56@alu.ua.es

M. T. Signes-Pont
e-mail: teresa@ua.es

A. Visvizi et al. (eds.), *Research and Innovation Forum 2022*,
Springer Proceedings in Complexity, https://doi.org/10.1007/978-3-031-19560-0_9

1 Introduction

Coronavirus disease 2019, commonly known as Covid-19, is an infectious disease caused by the SARS-CoV-2 virus [1]. This disease originated on 2019, in Wuhan City, China was detected when the World Health Organization (WHO) began receiving reports of an epidemic outbreak of pneumonia, of unknown cause and origin in this city [2]. In January of the same year, the authorities of this country identified the cause as a new strain. From Wuhan, the disease has spread to other continents such as Europe, America and Asia, and is now recognized as a pandemic by the WHO.

Transmission of the virus occurs when there is direct person-to-person contact through small droplets or microdroplets that are emitted by the infected person (who may not have symptoms of the disease) when talking, sneezing or coughing, and passed directly to the other person through inhalation or touch contact.

Most people infected by the virus may experience different symptoms ranging from mild to moderate respiratory illnesses, such as a common cold (fever, cough, dyspnea, myalgia and fatigue); to more serious illnesses, such as pneumonia, severe acute respiratory syndrome (SARS), Middle East respiratory syndrome (MERS), sepsis, septic shock, and other respiratory illnesses [1]. These symptoms usually appear between two and fourteen days (incubation period) after exposure to the virus. This disease can be fatal and for the treatment of this disease there is no specific method, being the therapeutic measures to alleviate the symptoms and maintain vital functions [3].

Covid-19 mainly affects respiration, i.e. the lungs. Being such a contagious and serious disease in some cases it is crucial to detect it as early as possible. In general, Covid-19 detection methods include the following methods [4]:

- PCR (Polymerase Chain Reaction). It is a technique that aims to obtain a large number of viral RNA sequences to detect Covid.
- Antigen test: Results are obtained in about 15 min. They detect virus proteins.
- Antibody test: in this case it determines whether the individual has antibodies generated as a result of the infection.

These tests mentioned above usually take some time to produce results: 6–48 h in the case of PCRs, up to 15 min in case of antigen/antibody test. In addition to the previous methods, Covid detection by image analysis also provides accuracy and fast response [5, 6].

Radiological image processing and analysis can be a crucial element in the accurate and early detection of Covid-19. Image analysis consists of extracting features from images for subsequent study. In this case, the characteristics of interest to be extracted from radiological images are the possible stains that Covid-19 can generate in the lungs.

At the moment, neither radiographs nor computed tomography are recommended diagnostic tools for the detection of Covid, however it can be faster and less expensive method of detection than the usual diagnostics. Well trained models with deep learning neural networks to detect Covid-19 through chest X-rays have high accuracy.

Among the proposed models to carry out this task there are two sets: (a) binary models which classify the images between a respiratory disease and healthy images; (b) multiclass models which classify the images between several types of respiratory diseases such as pneumonia, bronchial asthma, Covid and others.

Results probe binary models have higher accuracy than multiclass ones. In the neural networks available in the literature 95% accuracy is achieved for multiclass neural networks and near to 99% accuracy for binary neural networks [7].

However, a multiclass model needs a large dataset to learn from images, thus, it requires too much data, takes more time to be developed, and in addition, training process could take a long time. These are the issues and the main problem addressed in this work.

The number of the images needed for training a Deep Learning model is related to the number of different classes the model has to classify [8]. A binary classification model is able to classify the images in two sets [9], while a multiclass model classifies the images in several sets.

The work hypothesis is to find a suitable method for multiclass classifications based on the algorithmic combination of simpler or binary classification models. In this hypothesis, the "suitable" term refers to the acceptable lack of precision of this combined model with regard to precision of the multiclass one.

The contribution of this work is twofold: in first place, to probe the feasibility of combination of deep learning binary models to obtain a multiclass model in the case of X-Ray Digital Images screening, and in second place, to propose a method to perform a multiclass classification of images for several diseases (Covid-19, pneumonia) by means the combination of binary models. As a result, this work aims to create a multiclass expert system composed of several binary neural networks models and thus be able to run them obtaining a good accuracy instead of training a multiclass model.

The methodology of this work according to the scientific method starts with the proposal of a hypothesis. After that, several models have been reviewed in the literature, combination methods have been proposed, and finally, the proposals have been tested with experiments. In this way, binary and multiclass neural networks models for classification of X-ray images are analyzed and compared in order to find out the best accuracy of classification.

This work is organized as follows: First, we review some advanced techniques for intelligent screening of images and present both binary and multiclass neural networks models for radiological images to compare their results and performance. We focus on recent well-known models with high rate of accuracy. Secondly, the proposed combination methods of the binary models are proposed, and then the results of the experimentations are described. Finally, some conclusions and future lines are drawn.

2 Techniques for Intelligent Screening

In recent times, advances techniques for data analysis have been developed in order to get knowledge and find out information on Covid-19 to provide solutions in monitoring, detection, diagnosis and treatment of diseases associated with this virus [10, 11]. Some of the most important techniques and related research projects are explained below.

2.1 Big Data

Big data represents a set of computational methods used to analyze large volumes of data, which are very difficult to process with traditional methods; with the objective of obtaining information containing relevant patterns or relationships [12].

The big data trend for the coming years will be the integration with biometric systems, through the use of thermal cameras; as well as the early identification of a possible infection through biosensors, to help the healthcare world [13].

Big Data is characterized by five adjectives for research development (the 5Vs): Volume, Variety, Velocity, Veracity, and Value. Using the 5Vs of big data must guarantee reliable and truthful information in order to implement AI systems with deep learning or machine learning.

Using Big Data for the study of Covid-19 allows to focus the study through indicators and trends over time, which includes future predictions. These techniques are commonly included in data science scope. Data science employs statistical and mathematical techniques characterized in the study variables that allow for the extension of techniques and models represented as clusters through data patterns or correlations, which in turn can be integrated with AI to improve results and anticipate a solution. Therefore, for this type of development, Big Data is supported by AI through various algorithms applied in deep learning and machine learning.

Some examples of these techniques are included in Table 1. Such classifications use both Big Data and AI to thus achieve a better diagnosis of Covid-19 disease using X-Ray images.

2.2 Artificial Intelligence

There are Artificial Intelligence models that can process a large amount of data in a short time for decision making. In this field, AI plays an important role in addressing different types of image classification problems.

Table 1 Neural networks models for screening of respiratory diseases based on digital images

Model	Input data	Main result
Comparing different classification models [14]	305 chest radiographs	The highest accuracy is the binary classification of Normal/Covid-19 with a value of 97.4%
Research on classification of covid-19 based on DL [15]	4,099 chest x-rays	The accuracy of the Normal/Covid-19 binary classification model is 96%
Detecting coronavirus from chest x-rays using transfer learning [16]	15,000 chest x-rays	Binary classification accuracy: 99.62% Multi-class classification accuracy: 95.48%
Digital image analysis based on DL [17]	1800 images	Multi-class classification accuracy: 97.80%

An example is the Covid-net system [18], which aims to detect Covid-19 infection in a patient from X-rays and chest CT (Computed Tomography) scans. These studies are able to detect Covid-19 infection and also other diseases that cause respiratory problems. In addition, it can be considered whether these diseases derive from Covid-19 infection or not. The common IA techniques used to analyze these data are deep learning and machine learning [19].

Machine learning (ML). Machine learning (ML) is wide term that refers to a broad range of algorithms that perform intelligent predictions based on data [20].

In the case of Covid-19, machine learning is used for multiple purposes. For example, to identify and diagnose the population at highest risk of infection through predictive analysis of protein-drug interactions.

In machine learning, predictive algorithms are needed, including decision trees, regression for statistical and predictive analysis and neural networks, among others [21].

The information retrieved through ML analysis of clinical images enables a wide range of predictions including the prediction of infection risk based on a person's characteristics such as age, hygiene habits, geographic location, weather condition and others. Using these data, a comprehensive predictive model can be created about the probability of contracting this disease and how it can affect the individual who contracts it.

Deep Learning (DL). Deep learning is essentially a neural network with several internal layers. Usually, deep learning outperforms other machine learning techniques for number of tasks in radiology [22]. This AI technique specialty helps to minimize the diagnostic problems, and it is able to obtain results in a very short time compared to biological test.

DL networks learn by detecting complex structures in the images. By creating computational models composed of several layers of processing, networks can create several levels of abstraction that represent the data. These computational models, known as Convolutional Neural Networks, give very good results over other techniques [23, 24]. This type of neural network typically learns from the pixels contained in the images provided.

These techniques perform morphological feature extraction processes to obtain the main characteristics and patterns to identify the diseases. In Deep Learning analysis, the input data are digital images. Thus, deep learning analysis is performed through graphical information. However, the effectiveness of this method is conditioned by the availability of examples in order to provide the system with a large number of images with the cases to be classified, so that the system can learn the relevant features autonomously.

There are various types of detection algorithms using X-ray images, tomography and ultrasound, including the identification of the pulmonary anomalies that can be caused, based on the virulence and transfer rate detected.

Thanks to DL methods, it has been possible to evaluate how many people have been infected by Covid and other diseases, since there are models able to use the DNA mutation rate to interpolate how many people were able to pass the virus, by means of a branching model [25].

A combined system of deep learning models provides a facility for detection of various pulmonary anomalies that can additionally obtain information about the risk of the disease in the patient by means of probabilistic models.

Next table shows some neural networks models for screening of respiratory diseases based on digital images. Models used in this research are conditioned by their availability in public repositories and data compatibility. Of course, some preprocessing and adaptation work could be made in order to increase the interoperability of the models. However, this work focusses on compatible models in order to probe the hypothesis.

3 Combination Method Proposal

Having several (two or more) models $M_1, \ldots, M_n$ each one classifies the same input I into a different set of classes (Eq. 1):

$$\begin{aligned} &M1: I \rightarrow \{c_{11}, \ldots, c_{1j}\} \\ &M2: I \rightarrow \{c_{21}, \ldots, c_{2j}\} \\ &\ldots \\ &Mn: I \rightarrow \{c_{n1}, \ldots, c_{nj}\} \end{aligned} \tag{1}$$

We aim to create a model M that classifies the input to the union of these classes. This model is obtained without a training phase, just by combining the available previously trained models (Eq. 2):

$$M: M1 \cup M2 \cup \ldots \cup Mn: I \rightarrow \{c_{11}, \ldots, c_{1j}, \ldots, c_{n1}, \ldots, c_{nj}\} \tag{2}$$

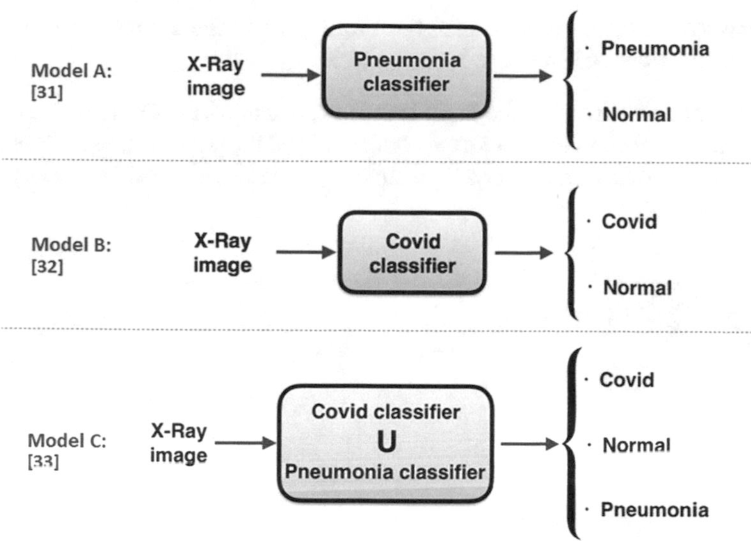

Fig. 1 X-Ray images ML classifiers

The idea is to compare this model created by combination of several models to a model trained properly on classifying the union of classes.

In this section, we test this idea on two models that classify chest X-ray images: one of them classifies into Covid or Normal and the other into Pneumonia or Normal. Therefore, the objective is to combine those pretrained models to get a model that classifies X-ray images into Covid, Pneumonia or Normal. Figure 1 schematically shows this idea.

We used well-know and pretrained CNN model for pneumonia image classification (Model A) [26], a VGG16 model for Covid image classification (Model B) [27], and a multiclass model trained on classifying an image to Pneumonia, Covid and Normal (Model C) [28].

In this work, we proposed to perform the combination of the results of those two binary models (Model A and Model B) and compare the classification made with the results produced by a multiclass model (Model C).

3.1 Combination Method #1

This method is by passing the input to both of binary classification models and then classify the image according to both outputs. This method considerers only the cases with no contradictory disease classification. Each classification result has a score of the accuracy of the result. In each model, the result is assigned to Covid or Pneumonia when the score exceeds a predefined threshold (50%). There are three possibilities as described below and shown in Figs. 2 and 3:

(a) Both models agree that the input is normal, doesn't have Covid or pneumonia.
(b) Covid model classifies the image as normal, and the pneumonia model classifies it as pneumonia then the class is pneumonia since Covid classifier does not detect pneumonia.

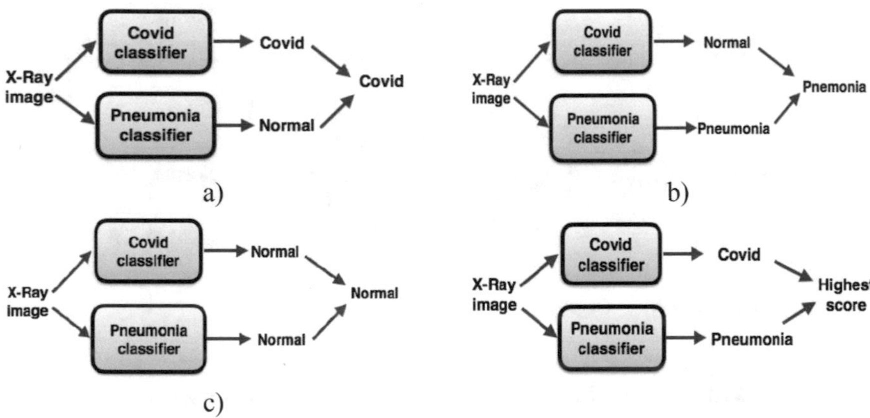

Fig. 2 Possibilities of combination for parallel classification

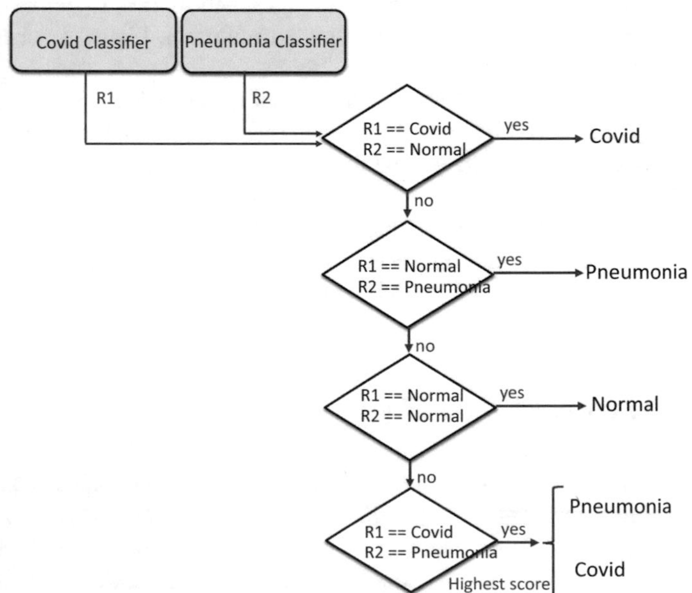

Fig. 3 Flowchart of parallel classification

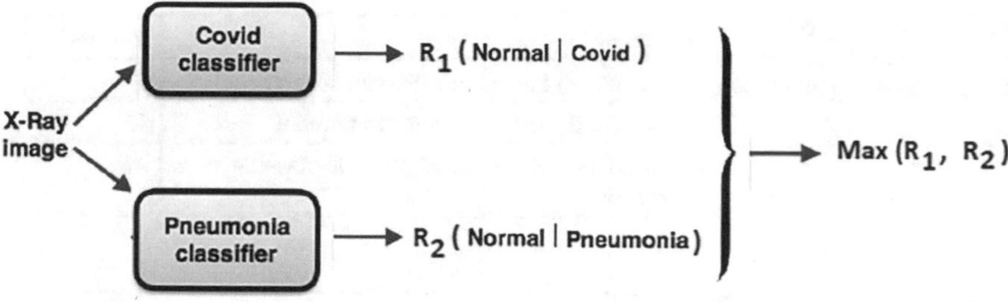

Fig. 4 Best score classification

(c) For the same reason if the pneumonia model classifies it as normal and Covid model as Covid the class will be Covid.
(d) Otherwise, if Covid classifier detects Covid and pneumonia classifier detects pneumonia then the class is the one with the highest score (where the model is surer).

Next figure shows the flowchart of this combination method.

3.2 Combination Method #2

The input image is also passed to both models. Each classification has a score that marks the reliability of it according to each model, then the class considered is the one with the highest score for each image, that is, the model that is surer about its results. Figure 4 schematically represents this case.

This method is similar to the fourth case of the previous combination method. in this case, the scores is applied in all cases.

3.3 Experiments

In our experiments, the Covid-19 Radiography Database is used [29–31]. It is one of the largest independent datasets publicly available. This dataset consists of a large set of chest X-ray images for Covid-19 positive cases along with Normal and Viral Pneumonia images and it is the winner of the COVID-19 Dataset Award by Kaggle Community (Kaggle is an online community platform for data scientists and machine learning researches). This dataset contains a total of 33,920 chest X-ray gathered from different sources including 11,956 COVID-19, 11,263 Non-COVID infections (Viral or Bacterial Pneumonia), and 10,701 Normal.

Before combining the models, we tested the three models (Covid, pneumonia, Covid and pneumonia) on the dataset containing all the classes. The models are executed as they are in the repository. This work maintains defined parameters and the

Table 2 Accuracy of pretrained models on the dataset containing all classes

Model	Accuracy (%)
(Model A) Pneumonia classification model	63
(Model B) covid classification model	55
(Model C) covid and pneumonia classification model	98

Table 3 Combination results

Model	Accuracy (%)
First combination method	69
Second combination method	72

calibration of the original deep learning networks in order to make a fair compassion. The results are shown in Table 2.

Binary Pneumonia and Covid models had such low accuracy since the dataset contains images of a class they are not trained on.

Next, the results of combining the models are described in Table 3.

The combined model using the first combination method was able to classify images correctly with a percentage of 69% while the second method classified correctly 72% of the images. Both methos outperforms the results of the single binary models by themselves. However, no one of those methods gave comparable results to the model trained on the three classes (98%).

Current research in this topic is mainly focused on developing new DL models for improving detection and screening capabilities, and/or fine-tuning pre-trained CNNs for obtaining better classification results and training parameters for transfer learning [15, 19, 23, 24, 32, 33]. The hypothesis raised in this work is genuine since there is no other works exploring the possibilities of combination of models for enhancing Covid detection in digital images. Thus, there is not possible to carry out exact comparisons of the results reported herein and many other works.

4 Conclusions and Future Work

In this work some methods for Intelligent Screening from X-Ray Digital Images have been reviewed. Among these, DL methods are able to screen these images between different classes according to some respiratory disease.

There are specific binary classification models which classify images with a high rate of accuracy between normal and some disease, and there are also other more complex models able to classify the images between several classes. In general, binary models obtain higher accuracy since the CNN only focus on the features of a pair of patterns. For several classes, the network has to learn an increasing number of features that, in many cases, are separated by a fine line. For these reasons, multiclass models need huge datasets to train the neural networks and spend much more time to achieve results. As result, they are complex to design.

In this work, two methods for combining the high specific binary models have been proposed aiming to achieve the performance of multiclass models. The results obtained show that for a proper classification, training a complex model for all types of diseases is the best option. Thus, comparisons using the same datasets strongly suggest that the fine-tuned multiclass network achieved better performance than several other combinations of binary models in terms of classification accuracy.

However, in contexts where there is not enough time or resources for training, the proposal for combining existing models can be a solution for a quick first classification at low-cost.

These are still further research in this area. For future work, we aim to study the circumstances in which the combination of models can provide advantages, e.g. in case of heterogeneous images to analyse the complexity of binary and multi-class models in terms of cost, training time and data requirements, and to extend the study to other types of images of a different nature.

References

1. World Health Organization: Coronavirus disease (COVID-19). Available online at https://www.who.int/health-topics/coronavirus. Accessed 20 April 2022
2. Shi, Y., Wang, G., Cai, Xp. et al.: An overview of COVID-19. J. Zhejiang Univ. Sci. B. **21**, 343–360 (2020). https://doi.org/10.1631/jzus.B2000083
3. Park, M., Cook, A.R., Lim, J.T., Sun, Y., Dickens, B.L.: A systematic review of COVID-19 epidemiology based on current evidence. J. Clin. Med. **9**(4), 967 (2020). https://doi.org/10.3390/jcm9040967
4. Sun, J., He, W.-T., Wang, L., Lai, A., Ji, X., Zhai, X., Li, G., Suchard, M.A., Tian, J., Zhou, J., Veit, M., Su, S.: COVID-19: Epidemiology, evolution, and cross-disciplinary perspectives. Trends Mol. Med. **26**(5), 483–495 (2020). https://doi.org/10.1016/j.molmed.2020.02.008
5. Poly, T.N., et al.: (2021) Application of artificial intelligence for screening COVID-19 patients using digital images: meta-analysis. JMIR Med. Inform. **9**(4), e21394 (2021). https://doi.org/10.2196/21394
6. Kumar, A., Manikandan, R., Magesh, S., Patan, R., Ramesh, S., Gupta, D.: Image analysis and data processing for COVID-19. In: Data science for COVID-19. Computational Perspectives. https://doi.org/10.1016/B978-0-12-824536-1.00035-6
7. Badawi, A., Elgazzar, K.L.: Detecting coronavirus from chest X-rays using transfer learning. COVID 2021. **1**(1), 403–415 (2021) https://doi.org/10.3390/Covid1010034
8. Zaidia, S.S.A., Ansari, M.S., Aslam, A., Kanwal, N., Asghar, M., Lee, B.: A survey of modern deep learning based object detection models. Digit. Signal Process. **126**, 103514 (2022). https://doi.org/10.1016/j.dsp.2022.103514
9. Zerouaoui, H., Idri, A.: Deep hybrid architectures for binary classification of medical breast cancer images. Biomed. Signal Process. Control. **71** (Part B), 103226 (2022). https://doi.org/10.1016/j.bspc.2021.103226
10. Bragazzi, N.L., Dai, H., Damiani, G., Behzadifar, M., Martini, M., Wu, J.: How big data and artificial intelligence can help better manage the covid-19 pandemic. Int. J. Environ. Res. Public Health **17**(9), 3176 (2020). https://doi.org/10.3390/ijerph17093176
11. Lin, L., Hou, Z.: Combat COVID-19 with artificial intelligence and big data. J. Travel. Med. **27**(5), taaa080 (2020), https://doi.org/10.1093/jtm/taaa080

12. Leung, C.K, Chen, Y., Shang, S., Deng, D.: Big data science on COVID-19 data. In: IEEE International Conference on Big Data Science and Engineering (BigDataSE), https://doi.org/10.1109/BigDataSE50710.2020.00010
13. Mujawar, M.A., Gohel, H., Bhardwaj, S.K., Srinivasan, S., Hickman, N., Kaushik, A.: Nano-enabled biosensing systems for intelligent healthcare: towards COVID-19 management. Mater. Today Chem. **17** (2020). https://doi.org/10.1016/j.mtchem.2020.100306.
14. Mahmud, T., Rahman, A., Fattah, S.A.: CovXNet: A multi-dilation convolutional neural network for automatic COVID-19 and other pneumonia detection from chest X-ray images with transferable multi-receptive feature optimization. Comput. Biol. Med. **122**, (2020). https://doi.org/10.1016/j.compbiomed.2020.103869
15. Ji, D., Zhang, Z., Zhao, Y., Zhao, Q.: Research on classification of covid-19 chest x-ray image modal feature fusion based on deep learning. J. Healthc. Eng., (2021). Article ID 6799202. https://doi.org/10.1155/2021/6799202
16. Badawi, A., Elgazzar, K.: Detecting coronavirus from chest x-rays using transfer learning. COVID. **1**(1), 403–415 (2021). https://doi.org/10.3390/Covid1010034
17. Gámez Guerrero, M. A., Rocha Nava, S. L., Hernández Oropeza, J. I.: Sistema auxiliar para el diagnóstico de covid-19 mediante análisis de imágenes de cr torácica basado en deep learning. memorias del concurso lasallista de investigación, desarrollo e innovación. (in spanish). Available online at https://repositorio.lasalle.mx/handle/lasalle/2216. Accessed 21 April 2022
18. Centers for Disease Control and Prevention.: Coronavirus disease 2019 (COVID-19)-Associated hospitalization surveillance network (Covid-Net). Available online at https://www.cdc.gov/coronavirus/2019-ncov/Covid-data/Covid-net/purpose-methods.html. Accessed 21 April 2022
19. Rehman, A., Iqbal, M.A., Xing, H., Ahmed, I.: COVID-19 detection empowered with machine learning and deep learning techniques: a systematic review. Appl. Sci. **11**(8), 3414 (2021). https://doi.org/10.3390/app11083414
20. Nichols, J.A., Chan, H.W.H., Baker, M.A.B.: Machine learning: applications of artificial intelligence to imaging and diagnosis. Biophys. Rev. **11**(1), 111–118 (2019). https://doi.org/10.1007/s12551-018-0449-9
21. Syeda, H.B., Syed, M., Sexton, K.W., Syed, S., Begum, S., Syed, F., Prior, F., Yu, F.: Role of machine learning techniques to tackle the covid-19 crisis: systematic review. JMIR Med. Inform. **9**(1) (2021). https://doi.org/10.2196/23811
22. Chassagnon, G., Vakalopolou, M., Paragios, N., et al.: Deep learning: definition and perspectives for thoracic imaging. Eur Radiol **30**, 2021–2030 (2020). https://doi.org/10.1007/s00330-019-06564-3
23. Islam, M.R., Nahiduzzaman, M.: Complex features extraction with deep learning model for the detection of COVID19 from CT scan images using ensemble based machine learning approach. Expert. Syst. Appl. **195**, 116554 (2022). https://doi.org/10.1016/j.eswa.2022.116554
24. Khan, E., Rehman, M.Z.U., Ahmed, F., Alfouzan, F.A., Alzahrani, N.M., Ahmad, J.: Chest x-ray classification for the detection of covid-19 using deep learning techniques. Sensors **22**, 1211 (2022). https://doi.org/10.3390/s22031211
25. Pathan, R.K., Biswas, M., Khandaker, M.U.: Time series prediction of COVID-19 by mutation rate analysis using recurrent neural network-based LSTM model. Chaos, Solitons Fractals. **138** (2020). https://doi.org/10.1016/j.chaos.2020.110018
26. Pneumonia-detection-using-CNN.: Available online https://github.com/Yashwanth-23/Pneumonia-detection-using-CNN. Accessed 5 June 2022
27. COVID-19_Classification.: Available online https://github.com/hakantekgul/COVID-19_Classification. Accessed 5 June 2022
28. Skripsi-multiclass classification.: Available online: https://www.kaggle.com/code/badslam/skripsi/notebook. Accessed 5 June 2022.
29. COVID-19 Radiography Database-Kaggle.: Available online https://www.kaggle.com/tawsifurrahman/covid19-radiography-database. Accessed 5 June 2022

30. Chowdhury, M.E.H., Rahman, T., Khandakar, A., Mazhar, R., Kadir, M.A., Mahbub, Z.B., Islam, K.R., Khan, M.S., Iqbal, A., Al-Emadi, N., Reaz, M.B.I., Islam, M.T.: Can AI help in screening Viral and COVID-19 pneumonia? IEEE Access **8**, 132665–132676 (2020). https://doi.org/10.1109/ACCESS.2020.3010287
31. Rahman, T., Khandakar, A., Qiblawey, Y., Tahir, A., Kiranyaz, S., Kashem, S.B.A., Islam, M.T., Maadeed, S.A., Zughaier, S.M., Khan, M.S., Chowdhury, M.E.: Exploring the effect of image enhancement techniques on covid-19 detection using chest x-ray images. Comput. Biol. Med. **132**, 104319 (2021). https://doi.org/10.1016/j.compbiomed.2021.104319
32. Pham, T.D.: Classification of COVID-19 chest X-rays with deep learning: new models or fine tuning? Health Inf. Sci. Systems. **9**(1), 2 (2021). https://doi.org/10.1007/s13755-020-00135-3
33. Subramanian, N., Elharrouss, O., Al-Maadeed, S., Chowdhury, M.: A review of deep learning-based detection methods for COVID-19. Comput. Biol. Med. **143**, 105233 (2022). https://doi.org/10.1016/j.compbiomed.2022.105233

Addressing the Challenges of Biological Passport Through Blockchain Technology

Aitana Pastor Osuna, Mohammed Alzibak, Ander Dorado Bole, Antonio Soriano Payá, and Higinio Mora

Abstract A biological passport collects the physiological parameters of athletes, through various blood and urine tests over a period of time. This passport includes two modules: the hematological and the endocrine (which also includes the steroidal data). A mathematical algorithm is applied to these data to calculate the profile of the person. This profile establishes the range of values within the biological controls and tests should move and if there were any anomaly in order to detect doping since the normal limits are exceeded. This work describes a distributed registration method based on blockchain technology to provide a solution to the previous challenges with clear benefits for all participants. The purpose of the proposed blockchain network is to consolidate a tamper-proof and distributed registry database, where anti-doping processes can be implemented through the development of distributed applications and smart contracts. Blockchain technology is a promising disruptive technology to address government policies and access control rules by a consortium of sports organizations in order to conduct the continuous doping controls and health revisions of athletes.

Keywords Blockchain · Smart contracts · Biological passport

A. P. Osuna · M. Alzibak · A. D. Bole · A. S. Payá · H. Mora (✉)
Department of Computer Science Technology and Computation, University of Alicante, Alicante, Spain
e-mail: hmora@ua.es

A. P. Osuna
e-mail: apo24@alu.ua.es

A. D. Bole
e-mail: adb35@alu.ua.es

A. S. Payá
e-mail: soriano@ua.es

A. Visvizi et al. (eds.), *Research and Innovation Forum 2022*,
Springer Proceedings in Complexity, https://doi.org/10.1007/978-3-031-19560-0_10

1 Introduction

Doping is one of the main problems in sport. Unfortunately, science is one step behind those who resort to cheating and when a new substance or method is detected, there is already another on the market. To address this issue, World Anti-Doping Agency (WADA) has focused on the oversight and enforcement of anti-doping measures in sport. In an attempt to improve anti-doping controls, in 1999 WADA launched a new strategy for detecting the consumption of products that may be of natural origin, the so-called Biological Passport Program. In 2009 WADA published the Athlete Biological Passport Operating Guidelines [1].

The Athlete Biological Passport (ABP) is a powerful tool for doping control and monitoring in high-level athletes. It consists of an analysis of the blood and urine samples taken from the athlete, with the aim of identifying prohibited substances and/or their metabolites [2].

A biological passport is the compilation of the biological parameters of an athlete through various blood and urine tests over a period of time. These blood or urine samples are analysed in authorized specific laboratories.

The transportation of the samples to these laboratories can be subject to errors, and there is always the possibility that the athlete biological passport samples could be tampered with or changed. This process may have errors, security or trust issues, so an improved and more secure system is necessary. The process should guarantee the provenance of the samples to confirm their authenticity.

New disruptive technologies can be introduced to address these issues for improving that anti-doping control process. Blockchain is becoming as a key technology for decentralized information sharing. This technology goes beyond being used to create cryptocurrencies [3]. By means blockchain, there is not needed that a single entity controls and verify the data, since blockchain provides enough security and trust, and also simplicity, visibility, and automation to the global system. Thus, it is more difficult to modify or alter the data stored in a blockchain [4].

The implementation of the ABP based on a blockchain system would allow athletes' samples to be tracked from their collection in the laboratory of origin, to the laboratory where they are analysed, going through each of the stages in their transport. Likewise, it would allow complete visibility and transparency of the samples and results. The incorporation of blockchain technology to the ABP process would increase security and trust in the global system.

In this paper, we will first briefly describe ABP and blockchain technology. Secondly, we will propose a blockchain-based ABP implementation solution. Next, we will describe the implementation of the proposed system. Finally, we will discuss the potential of this system.

2 Background

2.1 Athlete Biological Passport

The concept of athlete biological passport was first born in the early 2000s when the preservation and shadowing of a longitudinal record of hematologic variables was accepted to be used as a means of defining an individual's hematologic profile [5]. Three different modules can be distinguished in the ABP: hematology module, steroid module, and endocrinology module. Through these profiles, parameter changes for each athlete can be monitored, helping to detect abnormal fluctuations that may be due to doping use, as well as other possible causes. In other words, the purpose of the Biological Passport is to control some biomarkers or biological parameters of the athlete, over an extended period of time, in such a way that, in the event of appreciating variation with respect to the profile created previously, it implies an indication of doping in the behavior of the athlete in question.

The way to implement the ABP is homogenized in guidelines published by the WADA at the beginning of December 2009 and updated in successive revisions. These guidelines establish action protocols in everything related to the collection, transport and analysis of samples, which are mandatory for those Anti-Doping Organizations that express their intention to use the biological passport (Fig. 1).

The ABP program is carried out through WADA's Anti-Doping Administration and Management System (ADAMS). This is a centralized online database management tool used by athletes and WADA-accredited laboratories to coordinate and simplify anti-doping activities. in their anti-doping operations [1].

The centralized database can be subject to attacks by hackers. It can also be blocked by system administrators, which would prevent users from accessing the data. The centralized application can also be subject to attacks, which could cause a service interruption. Some disadvantages of Centralized Database Management System are:

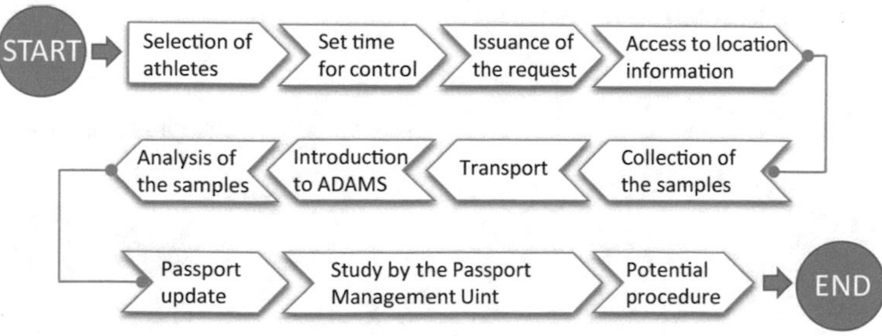

Fig. 1 Implementation of biological passport

- Because all the data is centralized at one location, it takes more time to search and access it. This may reduce the efficiency of the system. In addition, if the network is slow, this process takes even more time.
- There is a lot of data access traffic for the centralized database. This may create a bottleneck situation.

2.2 Overview of Blockchain

The blockchain has been defined as a decentralized ledger that manages transaction records on various computers at the same time thought a P2P (Peer-to-Peer) network. Blockchain technology is based on a type of decentralized database, which stores transactions permanently and immutably, that is, it cannot be manipulated. A blockchain is a growing list of records, called blocks, that are linked using advanced cryptography methods. Each block contains a cryptographic hash of the previous block, a timestamp, and transaction data [6, 7] as shown in Fig. 2.

Blockchain technology is mainly composed of a digital ledger where all the transactions are encoded in the form of blocks and are stored in a decentralized and open way. The information is distributed in several independent nodes that validate it without the need of a centralized authority. Different elements are identified in a blockchain [4, 8, 9]:

- *The ledger*. Accounting book or main book is the place where the transactions are located and each of the nodes are documented.
- *Node*. The block or node is the main part of the blockchain, it is where the previous and current information of the chain is registered.
- *Hash algorithm*. it is a mathematical algorithm whose objective is to convert a character into a code of a fixed length in a cryptographic way, any block of data will always be transformed into a mathematical code of a specific length.
- *Smart Contracts*. They are a program (script) that has the ability to execute autonomously and automatically, based on a series of programmed parameters and predefined conditions.

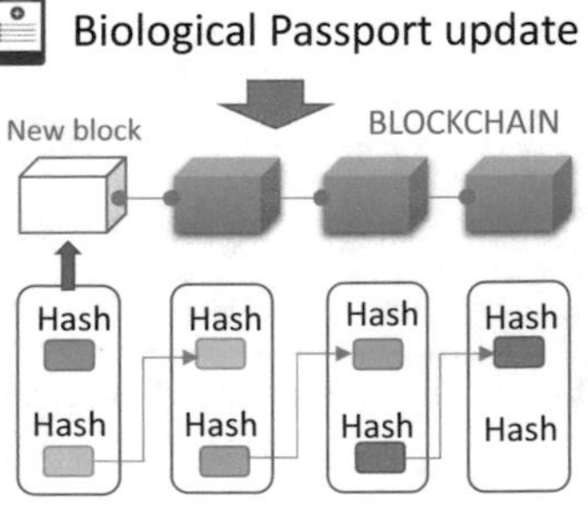

Fig. 2 Composition of Blockchain

- *Consensus.* Since blockchain is a distributed system, this technology needs some method to decide the correct state of the records after performing a transaction. This method is known as consensus algorithm. There are several accepted consensus algorithms in the blockchain system implementation: Proof-of-Work (PoW), Proof-of-Stake (PoS) or Byzantine are the most used.

The main characteristics of the blockchain technology are the following [10, 11]:

- *Distributed database.* In a blockchain, there is no authority or entity that has the entire database. All data is distributed along many nodes. Copies are stored and they are updated on the servers where the network is deployed.
- *Peer-to-Peer (P2P).* Communication is carried out between peers without the need for an intermediary. Each node stores and shares information with other nodes.
- *Unalterable Records.* This is an append-only structure. Once a transaction is in the database and has been validated; the information cannot be changed. Any change must be approved or validated by several nodes and stick to the end of the chain as a new block.
- *Transparency.* All transactions are visible to all users on the network. All information stored can be traced. Thus, it allows to verify the previous records of athletes at any time. The origin of the information will always be known by the signature of the node.

According to the access level of the user and to the visibility of the information there are several types of blockchain systems: public, private, consortium, and hybrid blockchain. Each type has its their own specific characteristics, level of transparency, efficiency, and immutability features [12, 13].

- *Public blockchain.* This blockchain network is accessible to any user. Thus, in public blockchain anyone on the internet have the right to read, make new transactions, and validate the nodes through the consensus algorithm. The best known public blockchain is the Bitcoin blockchain, which was the first to use this technology. Example: Bitcoin, Ethereum, IOTA, and Litecoin, among others.
- *Private blockchain.* This blockchain are not accessible to any user. There is a single entity in charge of maintaining the blockchain and granting permissions to the members to operate with the blockchain. This type of blockchain is usually used within an organization or reduced group of enterprises. Example: Multichain and Quorum, among others.
- *Consortium blockchain.* This blockchain are not accessible to any user, but in contrast to private blockchain, this type is governed and controlled by multiple organizations or enterprises. Examples: Hyperledger and Corda, among others.
- *Hybrid blockchain.* A hybrid blockchain combines features of a public and a private blockchain. Hybrid blockchains allow users to control who has access to information stored on the blockchain, making them ideal for privacy-sensitive applications such as financial transaction processing. Example: Ripple and XRP, among others.

In a blockchain, the block is the record, and each block contains a hash of the previous block, a timestamp, and a transaction data. Transaction data is, in general, any information that concerns the transaction, such as the amount of currency exchanged. This results in the creation of a chain of blocks from the genesis block to the current block.

Blockchain technology is used to address many challenges in which security, transparency, and traceability of data is needed. In this way, blockchain provides trust to the shared information system. Many applications are being developed in many scopes, such as smart cities [14], public administration [15–17], and others [18].

3 Proposed Solution

In this section, the main idea of this work is discussed. This is a conceptual proposal where the main advantages of using blockchain to address the anti-doping control process are described. Thus, this work just connects the features of the blockchain technology with the requirements of the current ABP to show the blockchain implementation is the best option to improve this process. In this way, blockchain can play a disruptive role in sharing information among the interested actors without the need of a central authority: athletes, laboratories, agencies, sport federations, etc.

In the next section, a prototype of the ABP process is implementing in a well-known current blockchain (Hyperledger Fabric) to show in a simple way the main components and participants involved. Of course, the World Anti-Doping Agency should update its procedures and technology support. We think this paper should serve as inspiration for them.

There are many contributions that blockchain technology can offer to the security of the biological passport. The traceability and immutability of the information provided by the blockchain together with smart IoT devices allow to collect product data.

Data of each sample are stored so that it can be verified that the temperature or relevant parameters at each stage of transport and storage were maintained under the appropriate conditions.

But in addition, the application of blockchain technology and IoT devices combined with the printing of security labels, not only ensures an adequate transport and conservation process, but also allows health professionals and authorities to know that the samples have not been manipulated.

By means of a QR code we will always have access to information on its proper transport, storage, and the different steps that it has followed from the moment of taking samples until they are already analyzed (Fig. 3).

The process to apply the athlete biological passport to a blockchain would be as follows. The blocks will contain all the critical information, laboratory results and the biological passport.

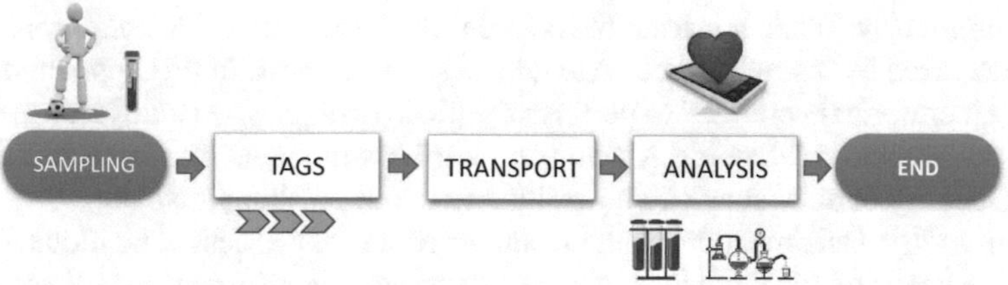

Fig. 3 Application of blockchain on biological passport system

- Blood and urine samples would be taken from the athlete.
- Samples are analyzed by mass spectrometry to identify prohibited substances and are stored in a secure database.
- The database is published on a blockchain.

Apart from this, personal data such as ID or passport number and name must be included so that there is no data manipulation.

Transactions will be carried out through the exchange of data between the participants on a blockchain. The transaction data is recorded in a block of data, which is then stored in the blockchain. No incentives are needed but they can be implemented as a reward when using the blockchain-based ADAMS application.

The database could be published on a blockchain through a special protocol, which establishes the rules for data access and modification. In this way, transparency and confidence in the results of the biological passport can be ensured.

4 Implementation

For this implementation, we use Hyperledger Fabric. Hyperledger Fabric is a modular open-source framework that allows deploying distributed applications on a permissioned blockchain system without systematic dependencies of a cryptocurrency [19].

4.1 Implementation Requirements

In this section we will analyze the necessary requirements to be able to implement a real solution to the case proposed in this study using a blockchain platform. The system has to provide features such as confidentiality, identifiability and availability of data [20]:

- *Confidentiality*: Working with athletes' private medical data requires that the data be accessible only by authorized entities. In addition, each piece of data must be tracked.

- *Identifiability*: When a private blockchain platform is used, this requirement can be acquired by restricting access to network participants. In this application case, the laboratories accredited to perform the biological passport analyses are chosen by the World Anti-Doping Agency and are identifiable at all times. In this way, the identity of an agency can be verified to avoid phishing attacks.
- *Availability*: Data must be available and have an easy access. The blockchain by nature keeps multiple copies of the data to ensure the information will not always available.

Other important feature inherited from the Hyperledger Fabric framework is scalability, since its modular design can add the necessary elements to handle the information processing needs [19].

4.2 Participants

In this proposal of ABP implementation there are four main actors. Three of them participate in the network with infrastructure and the fourth through their identity:

- *Global Anti-Doping Agency*: This is the agency in charge of setting up and administering the network. This organisation will be the administrator and will be responsible for generating credentials for all other participants. This organisation may add or revoke the credentials of other participants as defined in the network configuration. Despite being the administrator, the blockchain remains unchangeable, so data integrity is maintained.
- *Official laboratories*: The Global Anti-Doping Agency will add the laboratories to the network. Each laboratory will be responsible for storing the test results in an internal database.
- *National Anti-Doping Organisations*: These organisations will be added by the Global Anti-Doping Agency and are responsible for generating credentials for each athlete as well as storing the athlete's private data. The information can be verified at any time while maintaining confidentiality.
- *Athlete*: Relevant actor who for security and test integrity reasons does not participate in the network infrastructure. The athlete can request data from his/her biological passport (athlete does not have access to the most up-to-date information) and verify the veracity of this data [21].

Figure 4 shows the general architecture. Through this architecture we maintain only the public information of each athlete in the network and an immutability of the data that can be verifiable at all times. Private data is hidden for the rest of the organisations not involved in the sampling and analysis process in order to enhance privacy and confidentiality.

We achieve a high traceability of those who have requested private data through the use of a fully verifiable identity system. It allows the creation of multiple channels

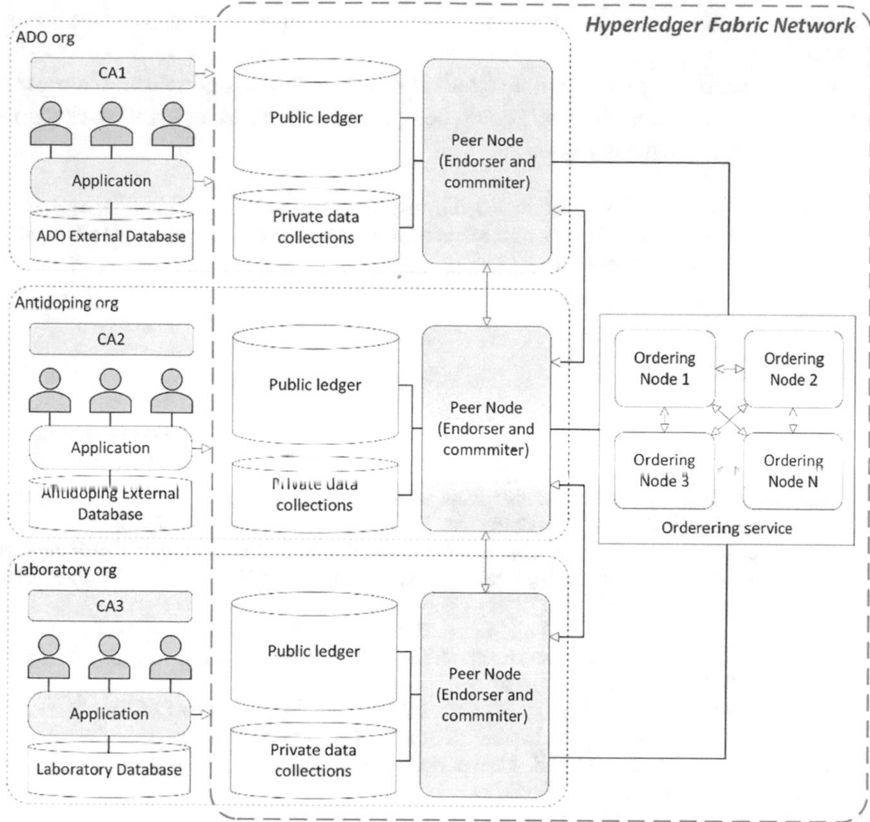

Fig. 4 Hyperledger fabric APB architecture

to the participating organisations in the network to maintain private transactions in case the network increases in number of participants.

5 Conclusions

In this work, blockchain technology has been proposed to support the supply chain of the ABP in order to improve security, interoperability, accessibility, automation and trust in the anti-doping control process. In this way, a decentralized technology is introduced, which means that there is no single entity that controls the data. Thus, it is more difficult to modify or alter the data stored in a blockchain.

An implementation of the ABP has been proposed that allows maintaining high information security by using cryptography with a system of public and private keys. In addition, once a piece of data has been saved or a transaction has been carried out, it cannot be removed, and it is very easy to realize that someone wants to modify the information.

As future work, we aim to explore regulation frameworks at national and European level, in order to propose how this could be updated to support an open blockchain-based anti-doping control process.

Acknowledgements This work was supported by the Conselleria of Participation, Transparency, Cooperation and Democratic Quality, of the Community of Valencia, Spain, under SDG oriented projects of the University of Alicante.

References

1. Athlete Biological Passport Operating Guidelines & Compilation of Required Elements. World Anti-Doping Agency. Available online at: https://www.wada-ama.org/sites/default/files/resources/files/WADA-ABP-Operating-Guidelines_v4.0-EN.pdf. Accessed 14 June 2022
2. Saugy, M., Lundby, C., Robinson, N.: Monitoring of biological markers indica-tive of doping: the athlete biological passport. Br. J. Sports Med. **48**(10), 827–832 (2014). https://doi.org/10.1136/bjsports-2014-093512
3. Mendoza-Tello, J.C., Mora, H., Pujol-López, F.A., et al.: Disruptive innovation of cryptocurrencies in consumer acceptance and trust. Inf. Syst. E-Bus Manag. **17**, 195–222 (2019). https://doi.org/10.1007/s10257-019-00415-w
4. Yaga, D., Mell, P., Roby, N., Scarfone, K.: Blockchain technology overview (2018). arXiv:1906.11078. https://nvlpubs.nist.gov/nistpubs/ir/2018/NIST.IR.8202.pdf. Accessed 1 May 2022
5. Sottas, P.-E., et al.: The athlete biological passport. Clin. Chem. **57**(7), 969–976 (2011). https://doi.org/10.1373/clinchem.2011.162271
6. Bhushan, B., Sinha, P., Martin Sagayam, K., Andrew, J.: Untangling blockchain technology: a survey on state of the art, security threats, privacy services, applications and future research directions, https://doi.org/10.1016/j.compeleceng.2020.106897
7. Sicilia, M., Visvizi, A.: Blockchain and OECD data repositories: opportunities and policy-making implications. Library Hi Tech **37**(1), 30–42 (2019). https://doi.org/10.1108/LHT-12-2017-0276
8. Casino, F., Dasaklis, T.K., Patsakis, C.: A systematic literature review of blockchain-based applications: current status, classification and open issues Telemat Inform **36**, 55–81 (2019)
9. Chukwu, E., Garg, L.: A systematic review of blockchain in healthcare: frameworks, prototypes, and implementations. IEEE Access **8**, 21196–21214 (2020)
10. Seebacher, S., Schüritz, R.: Blockchain technology as an enabler of service systems: a structured literature review. In: The 8th international conference on exploring service science, pp 12–23 (2017)
11. Yeoh, P.: Regulatory issues in blockchain technology. J Financ Regul Compl **25**(2), 196–208 (2017)
12. Monrat, A.A., Schelén, O.: Andersson K A survey of blockchain from the perspectives of applications, challenges, and opportunities. IEEE Access **7**, 117134–117151 (2019)
13. Ali, O., Jaradat, A., Kulakli, A., Abuhalimeh, A.: A comparative study: blockchain technology utilization benefits, challenges and functionalities. IEEE Access **9**, 12730–12749 (2021)
14. Mora, H., Mendoza-Tello, J.C., Varela-Guzmán, E.G., Szymanski, J.: Blockchain technologies to address smart city and society challenges. Comput. Hum. Behav. **122**, 106854 (2021). https://doi.org/10.1016/j.chb.2021.106854
15. Mora, H., Pérez-delHoyo, R., Sirvent, R.M., Gilart-Iglesias, V.: Management city model based on Blockchain and smart contracts technology. In: Visvizi, A., Lytras, M. (eds.) Research &

Innovation Forum 2019. RIIFORUM 2019. Springer Proceedings in Complexity. Springer, Cham (2019). https://doi.org/10.1007/978-3-030-30809-4_28
16. Barański, S., Szymański, J., Sobecki, A., Gil, D., Mora, H.: Practical I-Voting on Stellar Blockchain. Appl. Sci. **10**(21), 7606 (2020). https://doi.org/10.3390/app10217606
17. Mora, H., Pujol-López, F.A., Morales, M.R., Mollá-Sirvent, R.: Disruptive technologies for enabling smart government. In: Visvizi, A., Lytras, M.D., Aljohani, N.R. (eds.) Research and Innovation Forum 2020. RIIFORUM 2020. Springer Proceedings in Complexity. Springer, Cham (2021). https://doi.org/10.1007/978-3-030-62066-0_6
18. Mora, H., Morales-Morales, M.R., Pujol-López, F.A., Mollá-Sirvent, R.: Social cryptocurrencies as model for enhancing sustainable development. Kybernetes **50**(10), 2883–2916 (2021). https://doi.org/10.1108/K-05-2020-0259
19. Androulaki, E. et al.: Hyperledger fabric: a distributed operating system for permissioned blockchains. In: Proceedings of the thirteenth EuroSys conference, pp. 1–15 (2018)
20. Brandenburger, M., Cachin, C., Kapitza, R., Sorniotti, A.: Blockchain and trusted computing: Problems, pitfalls, and a solution for hyperledger fabric (2018). arXiv preprint arXiv:1805.08541. https://doi.org/10.48550/arXiv.1805.08541
21. Devriendt, T., Chokoshvili, D., Favaretto, M., Borry, P.: Do athletes have a right to access data in their Athlete Biological Passport? Drug Test. Anal. **10**(5), 802–806 (2018)

Impact of 5S Method in Apparel Industry

Muzoon Alasbali and Abdulaziz T. Almaktoom

Abstract This paper explores the 5S management concept, its advantages and the framework required to implement the system in the Apparel Industry. The 5S method is a Japanese system that involves workplace management improvements. It further analyses the effect of the 5S method in the apparel & textile industry as a case study and how its implementation directly impacts saving time and costs. 5S process helps to eliminate waste, reduce defects, and help in improving the productivity of the business. Further, reducing clutter and junk at the workplace is hygienic and ergonomic. This research paper identifies how the 5S method is implemented in the Apparel industry and how it has helped the industry improve its waste reduction targets, profitability, and productivity. The apparel industry uses the 5S to dispose of the wastage and use the factory production layout to smooth working, wastages are cleared, and regularly maintained space. This would help minimize handling time, moving part used in apparel from one process to another quickly, and final apparel can be stitched and packed in a faster time process. This will improve the business's profitability and enhance its competitiveness in the local and international market sectors.

Keywords Lean management · Quality improvement · 5S · Apparel industry

1 Introduction

The apparel industry in Saudi Arabia is one of the important contributors to the GDP. It exports about USD 5 billion and imports bout $52 billion, which indicates that the country is heavily dependent on imports from other countries. Textile Info Media [10]. The 5S method is a workplace organization tool used significantly by Japanese management to attain the highest level of safety, cleaning, sorting, arrangement, and movement of raw materials, work in progress and finished goods. This tool is used to achieve continuous improvement and quality management as part of the Kaizen

M. Alasbali · A. T. Almaktoom (✉)
Department of Operations and Supply Chain Management, Kingdom of Saudi Arabia, Effat University, Jeddah 21478, Saudi Arabia
e-mail: abalmaktoom@effatuniversity.edu.sa

A. Visvizi et al. (eds.), *Research and Innovation Forum 2022*,
Springer Proceedings in Complexity, https://doi.org/10.1007/978-3-031-19560-0_11

process. 5S method acts as a tool to accomplish the management in the workplace from a visual point of view so looks neat, clean, sorted and organized to make sure the usage of space is achieved.

This research is essential to improve the profitability and efficiency of the apparel industry in Saudi Arabia and make this industry contributing at an increased rate year on year to the country's GDP and its growth and reduce dependencies on imports from other countries.

1.1 Problem Statement

The problems which occur before the implementation of 5S in an organization are:

- Wastage of production time from searching materials, accessories as materials are located at temporary locations
- Wastage of production time from searching the tools, machinery and spares
- Storage area for materials, work-in-progress and finished goods not utilized properly
- A storage area used for unwanted materials and wastage
- Space for unwanted and rejected materials is not well defined.

1.2 Objectives

The main objective of this research paper is to carry out a theoretical review of the literature available in the field of the 5S method, its benefits with a focus on the apparel industry.

This study will look forward to answering the following:

(a) To understand the 5S workplace organization technique and its benefits.
(b) To carry out a critical review of the literature available about the 5S technique and its benefits to the Apparel industry.
(c) To understand the results from available literature and apply the techniques and develop suggestions to the Saudi Arabia Apparel industry.

2 Literature Review

2.1 Meaning and Definition

Five Japanese words make up 5S, and each word starts with S. These are seiri meaning sort, seiton meaning set in order, seiso meaning shine, seiketsu meaning standardize, and shitsuke meaning sustain. This technique has been practiced in Japan for quite

Table 1 5S Japanese word and their English meaning

Japanese word	English meaning
Siri	Structures, Sort, Shift, Cleaning up, Clearing out
Seiten	Straighten, Simplify, Set (in order), Configure
Seiso	Sanitize, Scrub, Shine, Sweep, Clean and check
Seiketsu	Standardize, Systemize, Conform
Shitsuke	Self-Discipline, Sustain, Custom and practice

Source Warwood and Knowles [11]

some time. Still, the original concept of 5S was developed by Takashi Osada in the 1980s and then later on Japanese research Hiroyuki Hirano who focused on sustainable, high-quality housekeeping at the workplace and productivity management [12]. On top of these 5S, today we have two more S, i.e. Safety and Security. The business must maintain the safety and security of the workplace, i.e. both factory and office place in the apparel industry. This technique is the starter of a productive life for all, including typical households, and is used for productivity improvement [1]. The successful implementation requires full support and commitment from all levels of the management especially Top-level management [5].

The meaning of the 5 Japanese words is shown in the Table 1.

The in-depth meaning of these five words is summarized below:

Seiri. It means meaning sorting or keeping things tidy and putting things in order. From an organizational perspective, it means to sort and organize the essential, vital, expensive, and frequently used items that are considered waste, useless, and of low value, which is not needed at a given point in time [7]. Unwanted items can be scrapped and sold. Critical things need to be kept very near to where the item is being used, and items that will not be utilized shortly should be taken off and stored correctly. The value of the article is based on the utility function and not cost function, which helps to reduce the time to search the product [7].

Seiton. It setting or straightening, i.e. maintaining proper arrangement and orderliness of the items at the workplace. It applies the principle of a place for everything and everything in its place, which indicates that things need to be kept back to their original position after usage is completed [4, 13]. This is mainly used for spares maintenance. Also, to manage orderliness, the business can use identification methods such as naming conventions, coloured tags, etc. [13].

Seiso. It means shining or ensuring Cleanliness at the workplace, free of wastage, oil spillage, and scrape. Free burrs by keeping the place neat and clean by inspecting areas regularly, and if unnecessary items are found, they need to be discarded as explained in 1 above or put in the right place as described in 2 above. In this inspection process, wear and tear of the machine and electrical components are made regularly. Management uses this stage to lubricate, tighten nuts and bolts, remove the dust on floors, windows, desks, tables, machines, etc. It is done at the macro level first and then at the individual micro-level [13].

Seiketsu. It means Standardization of the processes. This suggests that processes, if standardized and set in order, there will be clarity of thoughts in the employees' minds regarding what comes next, so there is no surprise and wastage is reduced [13]. The standards at work need to be communicated at all levels and make sure they are easy to understand, and participation from employees is required [7, 13].

Shitsuke. It means sustain, self-discipline and top management commitment to follow all the steps wholeheartedly. According to Ho and Cicmil [6], it means establishing and instilling the ability to do things the way they should be done by creating good habits, culture at work, and methodology to be shared in teaching training. Hence, everyone is aware of how things should be done.

2.2 Benefits of 5S Workplace Organization Technique

From the previous studies of various authors, namely Ho and Cicmil [6], Ho [5], Warwood and Knowles [11], Ablanedo-Rosas et al. [1], Yadav et al. [12], Gurel [4] and Young [13], Zagzoog et al. [14], Michalska and Szewieczek [8] the main benefits of using 5S workplace organization technique are summarized below:

- helps to provide a basis to implement a total quality environment
- increases the safety, Cleanliness and organization of material, things, spares, etc., in the workplaces, enhances the speed of products and services delivery, decreases customer waiting for the time for products and services to be delivered to them.
- reduces and minimizes the possibility of losses and failures of machines and operations
- improves the quality of the products and services as they are produced in clean, error-free and suitable machines devoid of any fault, which enhances the overall performance of organizations
- helps to improve the prevailing culture to the desired culture of quality management.
- enhances quality and efficiency at the workplace and ensures high-value spaces are used most optimally.
- helps reduce the overall costs of products, services, storage, maintenance, delivery, and operations.
- helps to build goodwill and build up the business's image as it helps to enhance customer satisfaction and improve teamwork and social culture at work.
- helps improve safety at work as machines are clean, working without problems, and accidents are reduced by regular maintenance.
- improves the worker's skills and makes them more innovative and creative.

2.3 *Apparel Factory and Production Processes*

The readymade garment and apparel industry is one of the most developing industries involving continuous changes, development, and quality improvement. A normal apparel production firm utilizes fabric storage, accessories storage, sampling unit, cutting department, stitching line, checking, and finishing department, Ironing team, packaging department and office & administration [2]. As the production processes change, customer requirements change, requiring overhaul in the production systems and techniques. To keep the costs of production low, the industry's operations need to be very effective, efficient and productive. It requires a massive round of efforts, future planning, and coordination of different production lines at one go and attains the highest level of productivity [3].

The apparel industry in Saudi Arabia has 101 factories with total finance employed 11,671.6 million riyals in 2016 with the employment of 14,297 [3]. Based on research by debes [3], few problems related to production which were noted is not able to provide materials to machines on time, lack of use of modern devices and use of outdated machinery requiring repairs and maintenance regularly, break down of machines at irregular intervals, lack of a culture of training and gaining new skills impacting the quality of product, lack of forwarding production planning and availability of ideal energies, spaces and high level of wastage in the industry.

3 Methodology

This research paper uses the qualitative research method to help understand the subject matter in-depth as part of the discussion with the operations managers of the Apparel industry in Saudi Arabia. As part of the study, the researcher visited one of the renowned apparel industries in Saudi Arabia and analyzed the production processes of the apparel industry. They have interviewed the operations manager of the company. This method was chosen to give an in-depth analysis of how the apparel industry works and how the 5S method could be helpful in the apparel industry. The research paper will use a deductive approach to draw inferences from available theories, concepts, and reports from the industry. The research philosophy used in this research is interpretivism using qualitative data and interpreting the data collected to inform discussion and analysis. The interview method was chosen as it provides in-depth analysis of the data of one particular business and then uses it as a case study to generalize the impact of the 5S method on the overall Apparel industry as the processes in the industry are standard and does not differ significantly.

4 Results, Discussion and Proposed Action Plan

4.1 Results

The interview with the apparel industry indicated that the company has a large amount of inventory which is not segregated between one which will be used currently or in the future, high value or low value, waste and unwanted or strategic, which makes the company and its employees to not able to meet the customer orders on time. Also, with a large amount of dust involved when cutting and stitching of the garments due to dust from the fabric, it is settled all over the factory, which makes the area dusty, untidy and unhygienic which makes the employees often go sick and take leaves which reduces the efficiency and productivity. At times, after cutting the parts of the fabric, some features go missing as they are not organized and stacked in an orderly manner which means that there is excess fabric usage to compensate the missing parts for stitching on time, and then this part is found, which increases the overall wastage and costs of the product and reduces the profitability. Further, as the machines are not cleaned and repaired on time, they get dusty, making the machine break down at irregular times, hampers the production process. Apparel industry processes are interlinked with one another, and if one step faces an issue with production due to breaking down, it has a trickling impact on the other methods. For example, e.g. If the stitching machine breaks down, workers cannot move the stitched pieces to the finishing stage, making the workers in other departments stay idle and reducing the overall productivity of the production system. The factory workers do not feel energetic working in dusty and unorganized environments, making them lazy and often making excuses to take leaves.

4.2 Discussion

From reviewing the literature and understanding the data available from the interview, the 5S method would be an appropriate workplace organization tool that the companies in the Apparel industry can use in the following manner.

S1 Seiri—Sort. The unnecessary items such as spare parts, needles, thread cones, and equipment such as fusing machines that are not regularly used can be sorted away. The company can define the items needed daily (thread cones, needles), weekly (label fusing machine), and others that are rarely used, which can be kept separate. The storage area can be adequately organized where fabric, accessories are sorted based on size, color and other specific features, which makes the identification and pick up more accessible and faster when the requirement of the fabric or other material comes up. This will help eliminate the wastage of time in searching current and new items that were previously kept haphazardly.

S2 Seiton—Set in order. Ensuring the order of the machines, materials are well labeled and kept organized, e.g. In the apparel industry. The devices are of different types that can be arranged in an orderly manner and named so that staff members in the factory know which machine they have to operate and rather than searching for the same. Also, the spares, accessories such as thread cones, needles, labels all can be organized in a manner using identified by a title and not mix two categories in one place. Creating the dividing lines, aisles mark and ensuring stairway and the common moving area is kept clean, appropriately lighted as per the required standard, which can quickly reduce injuries and movement of man and material. All accesses need to be accessible at all times, including emergency exits during working hours. In a warehouse or storage area, if the materials are kept orderly, workers can plan their visit predetermined with all their needs for the day and fill their bin, collect and take to their workstation and start working smoothly.

S3 Seiso—Shining. This step would involve keeping work areas, common areas, such as floors, stairs, walls, ceilings, racks, shelves, cabinets, furniture, machinery, equipment, tools, lighting all cleaned Machines, equipment, tools, stored items, materials, products and lighting are kept clean and arranged so that access is easy and less and less movement is required to pick up essential items. The cleaning department is appointed to take every kind of wastage away from the production and operation area, and checklists are filled regularly and made visible. Pest control should be scheduled to ensure apparel is not bitten by pests and making the place unclean.

S4 Seiketsu—Standardize. The apparel industry is one business in which most of the tasks are standardized, continuous and frequent. Hence, the apparel industry can use the display method, information placards, other visual means and markings available throughout the factory about processes, procedures, and use checklists to ensure work processes are carried out in a uniform and standardized manner which would avoid quality issues if processes are followed in line with standards. Also, employees at work need to be given formal training, and training needs to be refreshed at regular intervals for any change in processes, machinery or material.

S5 Shitsuke—Sustain. The employees need to be motivated as team and success stores to be displayed and shared with everyone and reward and recognition should be made an essential element of the work organization for, e.g. Target achievement, order completion before the due date, reduction in wastage or shortages below the target standards needs to be rewarded through financial and non-financial rewards. This will help the employees continue being motivated and believe that their work is being supported and recognized by top management.

4.3 *Proposed Action Plan for Implementation of 5S in Apparel Industry*

- Create Committee for implementation for 5S and include members from all levels
- Conduct 5S training for employees and encourage them to use checklists for each work procedure
- Take before and after photographs and post them on news boards
- Evaluate scores at regular intervals, maybe monthly or quarterly, so adjustments to procedures can be made quicker
- Create a maintenance chart for housekeeping and schedule the work for the cleaner and housekeeper, and regular audit to be carried out to ensure the job done is as per standards.
- Inter-departmental healthy competition to increase the interest and attention towards 5S.
- Periodic inspection and audit and evaluation by an external agency to ensure independent evaluation is carried out and formal reports without any bias are submitted to senior management.
- Statistics of data collected regarding fabric, quality control, records of inspection for printing, fusing, embroidery and other processes must be accompanied to the management before final dispatch of goods [9].

5 Conclusion

Defective and substandard quality products lead to significant scale rejection, rework, and loss of work for the apparel industry, increasing work time. Customer dissatisfaction rises for unnecessary reasons that could be easily avoided. To solve these issues, the 5S Workplace organization act as quality improvement technique which will help business to solve problems and minimize waste as the lean management tools are applied in the apparel industry.

Quality improvement is a continuous process and becomes very important from customer satisfaction as it improves the product's value. Quality product at reasonable prices becomes a value for the customer and enhances the company's brand image. Bringing the 5S method and implementing and making continuous improvements in the apparel industry is expected to reap solid benefits and help increase the business's sales and profits. It will also help to reduce the defects and minimize the losses.

The research objectives of understanding the 5S techniques and their critical benefits focusing on the apparel industry are successfully achieved as part of this research and based on case study analysis. A review of results obtained from an interview of one of Saudi Arabia Apparel industry will help the management of the company to strive towards implementation of the 5S method and motivate other players in the industry also to match the same standards and help Saudi Arabia Apparel industry to reach new heights in coming years and contribute significantly to the GDP.

References

1. Ablanedo-Rosas, J., Bahram, A., Juan, C.M., Javier, U.: Quality improvement supported by the 5S, an empirical case study of Mexican organizations. Int. J. Prod. Res. **48**(23), 7063–7087 (2010)
2. Al-Subiani, N.: Factors affecting production in the readymade garment industry in Jeddah. J. Specific Educ. Res. Mansoura University, Egypt (2011)
3. Debes, R.M.K.A.: Readymade garments sector in Saudi Arabia in light of vision 2030. Faculty of Human Sciences and Design, King Abdulaziz University, Jeddah, Saudi Arabia, http://vat.ft.tul.cz/2020/2/VaT_2020_2_6.pdf (2020)
4. Gurel, D.: A conceptual evaluation of 5S model in hotels. Afr. J. Bus. Manage. **7**(30), 3035–3042 (2013)
5. HO, S.: The 5S auditing. Managerial Aud. J. **14**(6), 294–301 (1999)
6. HO, S., Cicmil, S.: Japanese 5S practice. TQM Mag. **8**(1), 45–53 (1996)
7. Korkut, D.S., Cakicier, N., Erdinier, E.S., Ulay, E., Dogan, A.M.: 5S activities and its application at a simple company. African J. Biotechnol. **8**(8), 1720–1728 (2009)
8. Michalska, J., Szewieczek, D.: The 5S methodology as a tool for improving the organization. J. Achieve. Mater. Manuf. Eng. **24**(2), 211–214 (2007)
9. Tahiduzzaman, Md., Rahman, M., Dey, S.K., Kapuria, T.K.: Minimization of sewing defects of an apparel industry in Bangladesh with 5S & PDCA. Amer. J. Indus. Eng. **5**(1), 17–24 (2018). https://doi.org/10.12691/ajie-5-1-3
10. Textile Info Media Textile Industry in Saudi Arabia, https://www.textileinfomedia.com/textile-industry-in-saudi-arabia (n.d.)
11. Warwood, S., Knowles, G.: An investigation into Japanese 5S practice in UK industry. TQM Mag. **16**(5), 347–353 (2004)
12. Yadav, Yadav, Y., Chauhan, S.: Implementation of 5S in Banks. Int. J. Res. Commerce Econ. Manag **1**(2), 135–149 (2011)
13. Young, F.: The use of 5S in health care services: a literature review. Int. J. Bus. Soc. Sci. **5**(10), 240–248 (2014)
14. Zagzoog, G. W., Samkari, M. M., Almaktoom, A. T.: A case of eliminating wastes using 5S for a household electrical appliance warehouse. IEOM Soc. Int. (2019)

An Interactive Augmented Reality Tool to Enhance the Quality Inspection Process in Manufacturing—Pilot Study

Kristýna Havlíková, Petr Hořejší, and Pavel Kopeček

Abstract Quality inspection processes are an essential part of most industrial systems. These are repetitive and precise operations that are often very complex and require multiple steps to be performed correctly by different inspectors or operators. Augmented Reality (AR), one of the most promising and enabling technologies for assisting employees and engineers in the manufacturing workplace, has the potential to help operators to better focus on tasks while having virtual data at their disposal. Therefore, it is important to verify whether, with the support of this technology, it is possible to help workers perform these activities faster, more efficiently, and with less mental effort than with traditional paper-based documents. This paper describes a pilot interactive AR tool designed to support quality controllers in their inspection of welded products in the manufacturing environment. This tool is designed to guide the employee through the inspection process, provide them with all relevant information and help them find any discrepancies or deviations. The presented AR tool is created with the game engine Unity 3D and SDK Vuforia to assure compatibility with commonly used devices, such as tablets or smartphones. It does not use markers, but uses the object tracking method instead. A pilot study was conducted with a group of five probands to test the usability and functionality of the solution using SUS (System Usability Scale) standard questionnaires. The average SUS value rated by the probands was 78, confirming both a high level of usability and user satisfaction.

Keywords Augmented reality · Manufacturing · Quality control

K. Havlíková · P. Hořejší (✉) · P. Kopeček
University of West Bohemia, Univerzitni 8, 306 14 Plzen, Czech Republic
e-mail: tucnak@kpv.zcu.cz

K. Havlíková
e-mail: khavliko@kpv.zcu.cz

A. Visvizi et al. (eds.), *Research and Innovation Forum 2022*,
Springer Proceedings in Complexity, https://doi.org/10.1007/978-3-031-19560-0_12

1 Introduction

As manufacturing companies are forced to achieve ever-increasing efficiency and quality at the lowest cost, there is a growing interest in innovative approaches for rationalisation in both scientific and practical environments. In Industry 4.0, the relationship between digital and physical environments is transforming. Published research describes new systems to support human operators and improve their cognitive capabilities [1, 2]. The use of Augmented Reality (AR) techniques can serve as a supporting tool in many industries, including mechanical engineering. The rapid advances of AR in terms of resolution, computing power, and power consumptions have made AR accessible for research purposes and practical applications.

Quality control involves complex tasks requiring high levels of skill, experience, and concentration. They place a cognitive load on employees [3]. Therefore, it is important to verify whether AR can help workers perform these activities more efficiently compared to traditional paper-based documentation, without increasing their perceived mental workload.

The increasing amount of holistic data requires interoperability and changes in quality control methodology [4, 5]. Information and communication technologies play vital roles for modern manufacturing processes, monitoring and inspection systems [6]. To increase the flexibility, it is necessary to increase the availability of information and to link relevant quality characteristics [7, 8]. A growing number of publications devoted to AR in industry indicate that this technology is gaining interest in industrial processes [9].

AR as a tool to support workers in engineering processes has been addressed by a number of authors, e.g., [10, 11]. The authors discuss the possibilities of supporting workers by sharing expertise, increasing efficiency, and reducing the risk of errors stemming from lack of experience, distraction, or inattention. Inspection procedures in welding are addressed by the authors in [12] or [13].

In [14], a system displaying instructions after reading a QR code is presented. Similarly, [15] proposes an AR-based inspection system that displays 3D models on real objects. In [16], an AR tool for inspecting the installation of piping systems is introduced. A multiple marker system is used for object detection. Authors observed productivity parameters and were the first to investigate the system in terms of mental load perceived by the users. The above quoted references refer to AR technology used in industrial applications.

A literature review identified a research gap, which is the focus of this study. We investigate AR as a tool to systematically and interactively support quality controllers in their inspection of welded assemblies without the use of markers. With this in mind, **this paper presents AR software, its design, development, and a pilot study conducted with a group of probands to obtain feedback on functionality and usability.** This is the basis of large-scale research for monitoring the impact of AR-enabled quality control as a tool for increasing the efficiency of inspection while reducing the cognitive burden perceived by the worker.

This study focuses on a small-batch highly variable production line, where a worker encounters a large number of often relatively very similar products, with the probability of frequent technical changes, for which it is not possible or beneficial to consider full automation of the production or control. These objects must be correctly identified and assessed with a consistent level of concentration. For simplicity, this study considers welded industrial products of smaller dimensions.

In the first part of this article, the design and development of an interactive AR tool, the programmed functions, and the conditions of dataset creation are described. Then, the methodology of the research project is summarized and the progress of the pilot study is presented. Section 4 discusses the results and findings from the pilot study. Finally, the contributions of this study and directions for further research are summarized.

2 AR Software Design and Development

In relation to the findings obtained from the literature research, an interactive AR software is designed and described. Its purpose is to increase the productivity and the probability of detecting errors in comparison with using traditional paper documents. At the same time, this rationalization seeks to minimise the cognitive load perceived by workers in performing these activities, and to explore potential means for reducing this load. Also, the aim is to achieve a positive evaluation of usability of the software.

2.1 Designing the AR Software

The aim was to verify the functionality of the AR environment and to summarize comments and suggestions for further development of the software and the methodology of subsequent research.

The role of the inspector in this process is to verify that no welded component is missing. They must also assess the positioning of the welded components and the weld placement. As the complexity of the product increases, the likelihood of error rises. With the delay in catching an error, the cost of that error grows. Therefore, an AR solution is proposed that will interactively support the inspector in interpreting product geometry while being flexible for use directly at the inspection site. For this reason, a hand-held variant will be prioritised.

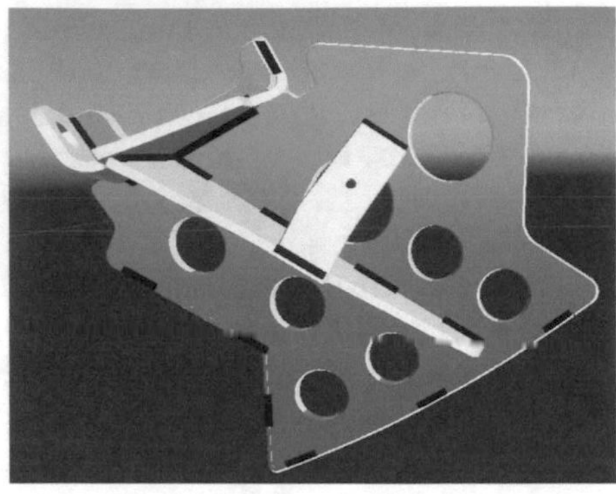

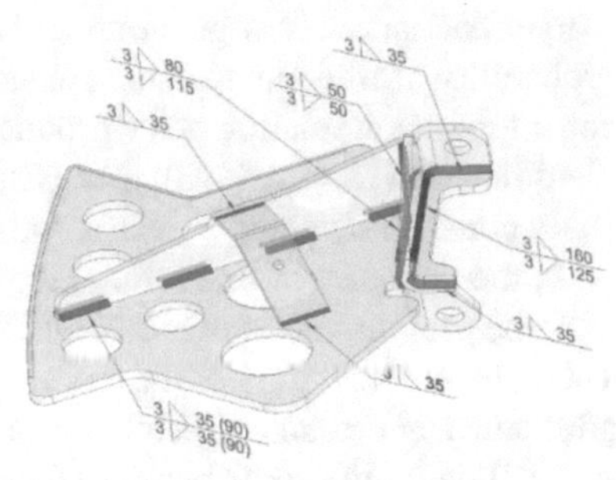

Fig. 1 Model input for pilot AR SW

2.2 Preparation of the Dataset

The AR tool was developed with Unity 3D and Vuforia SDK (Software Development Kit). Unity is a fully integrated multi-platform game engine. Vuforia was chosen due to its high compatibility with both Unity and portable devices.

Unlike similar studies, no 2D artificial markers will be used, instead the parts will be recognized using the marlerless object recognition method. This is a technique using a deep learning method. In the Vuforia environment, an AI database stored in the cloud is trained based on the given CAD model and defined parameters, which is then linked to the AR tool through Unity. This allows the inspected part itself to act as a marker. By pointing the camera of the device at the real part, the database is searched, a virtual model and programmed environment is displayed on the screen over the scanned part. This helps the user with interpretation of the standards. The variant of object tracking could contribute to reducing the negative influence of the lighting conditions and minimise the risk of misalignment of models.

The Vuforia Model Target Generator allows a 3D CAD model to be used instead of manually scanning the model, which can be laborious, not reliable for more complex parts and prone to inaccuracies depending on the lighting conditions. The model used for this pilot study is shown in Fig. 1. Input parameters are manually defined and by using deep learning algorithms, a trained database is generated and a model target is processed for intelligent object recognition. An example of a model configuration is shown in Fig. 2. Interactive content can then be programmed.

2.3 AR Software Development

The software serves as a support tool for checking the welded part. The design of the virtual environment and the interactive functions were created and modified based on interviews with workers with experience in the quality control of welded structures.

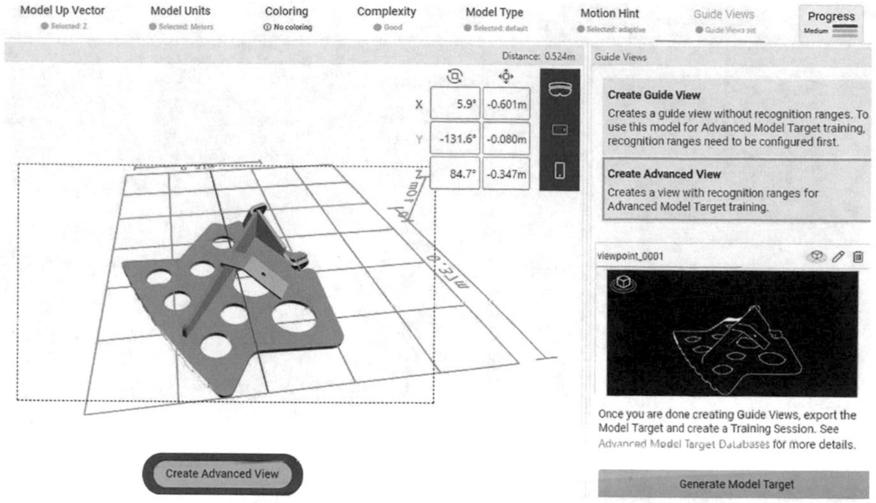

Fig. 2 Preparation of dataset for pilot AR software

The application can be built in Unity 3D for iOS, Android, and Universal Windows platforms. Smart handheld devices with the Android operating system were used in the pilot testing.

The basic feature is the recognition of the inspected part. Virtual content was created to intuitively guide the user through the entire inspection process. The main part of the application consists of a graphic comparison with the CAD model and informative panels interactively linked to the real object. The second part is the user interface. The data displayed can be selected using dynamic buttons. The application is complemented by an interactive checklist. The user can display or hide the help page with additional instructions.

Figure 3 shows the interactive AR interface. The inspector's task is to verify that all components are placed in the correct position, and that the length, number, and placement of the welded joints follow the designer's requirements.

The pilot tool presented was developed as a basis for future extended research, which will be described in more detail at the end of the paper.

3 Pilot Study: Methodology

The pilot study was conducted using the AR SW with a group of 5 probands to evaluate its functionality and usability. This section describes the results of the initial pilot study, the benefits and limitations of the AR solution, and plans for further research. A standardized System Usability Scale (SUS) questionnaire and interviews with probands were used to evaluate the interaction with the SW and to identify directions for further research and design of advanced AR SW.

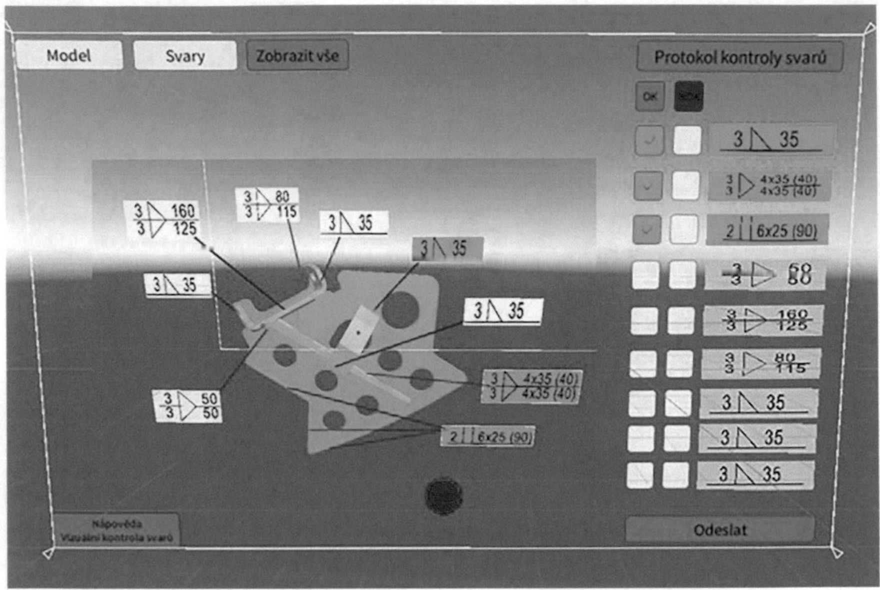

Fig. 3 Interactive interface of the software (interface in Czech language)

The pilot study was reduced to the inspection of one part. The aim was to verify the functionality and, if necessary, to summarize suggestions for further modifications. Volunteers working as quality inspectors in the welding industry with at least 1 year of working experience were selected as probands. The pilot group consisted of 2 women and 3 men. The women were aged 50 and 32, and the men were aged 27, 33, and 42. Testing was conducted at a quality control workplace in a welding hall.

The pilot study was conducted in several steps, including a brief introduction to the tool and its functionality, which took about 5 min for each proband. Each participant was then given a few minutes to familiarize themselves with the application before starting the actual test. None of the probands had any major problems in controlling the application, as they were using smart handheld devices.

After completing this phase, the probands were interviewed to obtain any ideas for improvements. Participants also completed the SUS usability evaluation questionnaire. The SUS includes 10 mixed tone items; odd numbers have a positive tone, even numbers have a negative tone. Responses are expressed on a 5-point Likert scale, where 5 is "Strongly agree" and 1 is "Strongly disagree". The responses are summed and converted using multiplicative coefficients. The resulting scores range from 0 to 100. This scale is divided into different levels of usability, which are defined from F for "completely unsatisfactory" to A+ for "completely satisfactory" [17, 18].

4 Results and Discussion

This section discusses the results of the pilot study. The tool was evaluated in terms of functionality and usability using the SUS questionnaire and interviews with probands.

Participants rate the usability positively, which is confirmed by the results of the SUS questionnaires and personal interviews. The resulting mean of the SUS scores was 78 with an average deviation of 3.7. The acceptability threshold of the SUS score is 70. The mean value, therefore, reached the level rated as "C", i.e., "Excellent" [19].

Users appreciate the good usability and practical functionality in the effective support of their inspection activities. Specifically, they appreciate "the ability to use a simple comparison method for the initial part assessment", "the visual display of welds that traditionally have to be searched for on several drawing sheets and in several views and sections" and last but not least "the continuous visual marking of already inspected areas, thus reducing the risk of missing some prescribed welds". Users commented positively on the optimal size of the selected part for handling and performing this form of inspection. Suggestions for improvement included, for example, the need to increase the contrast and adjust the transparency of some graphic elements. Users positively evaluated the smartphone option, which, while maintaining good legibility and clarity, allowed one hand to be free to manipulate the measuring tools or the controlled product.

Similar to [15, 16], this paper presents the design and development of an innovative AR tool for industrial applications. In [15], a framework for AR annotations is described, allowing useful information to be generated and made accessible and easy to interact with. They also use a markerless method, but unlike this paper, they use scanning. Relevant information about the parts is input by the users themselves, reducing the risk of redundant data. The study confirmed that AR content created in this way can be preserved and reused more efficiently than when using traditional information sources. However, the ability to scan a part is associated with inaccuracies, placing high demands on stable visual details during scanning and uniform lighting conditions during scanning and subsequent detection. In [16] a multi-marker system is used for AR assembly inspection. Also, in this case, the authors address the disadvantages of unfavourable lighting conditions. A study with end-users was also conducted. There were 2 groups of participants; production workers and designers. Among other things, usability was evaluated using a SUS, in which both groups rated the tool with an average value exceeding 85. All 3 studies designed an AR tool for hand-held devices, for different industries and applications. They have positive user evaluation results in terms of usability, and show potential for further development.

From the pilot study, it can be assumed that the designed AR SW could be successfully accepted for supporting quality control inspection and detecting manufacturing non-conformities. The user evaluation of the tool shows positive results for ease of use and intuitive usability. One factor contributing to this positive outcome was the involvement of the end-users in the design phase.

5 Conclusion and Future Work

In this paper, the development of AR software for supporting quality control in industrial processes is presented. Validation was performed in a pilot study with a group of probands evaluating usability with a SUS questionnaire. The results confirm a high level of usability. At the same time, suggestions for further modifications in the development of the advanced software and follow-up research methodology were collected. The AR SW was developed in the Unity 3D game engine environment in conjunction with the Vuforia SDK. Markerless object tracking is used to recognize the inspected object. Once the part is recognized, an interactive environment is automatically displayed in the user's field of view to guide the worker through the inspection process. The tool is designed for portable hand-held devices such as smartphones or tablets. The pilot study shows good readability and sufficient resolution even when using a smartphone. The aim was to verify the applicability of the AR software for further development of the tool in advanced research. The AR solution is designed for field research at a real inspection workplace, but in controlled conditions corresponding to the level of complexity of inspection activities usual for the process and the field.

Using the system will eliminate the need for the inspection department to maintain paper records, making everything easier to control and manage through the simple logging of history. The optimization of the process will thus lead to a reduction in the level of management of these departments.

A follow-up study will be conducted using a set of parts. With the high variability of very similar products, even experienced inspectors find it difficult to navigate inspection, part identification, and interpretation of requirements. These parts will randomly contain defects that include incorrect weld placement, incorrect or incomplete placement of welded components, or incorrect selection of these components.

In future research, welded products that intentionally contain defects related to component or weld placement will be tested in a large-scale study. The quality controllers will carry out the inspection process and objective metrics such as the speed or success rate of finding a defect will be observed. Users will evaluate the system in terms of usability with the SUS questionnaire and in terms of perceived stress using the NASA TLX questionnaire. The user study will include 2 groups of respondents; experienced welding quality control inspectors and operators with minimal experience.

The "Covid-19" pandemic has highlighted the need to digitise and optimise corporate knowledge. The system enables general digital support for commonly used control processes within the limiting factors mentioned above and fits precisely into the long-term digital transformation strategy of the European Union countries and other countries.

In general, the results of the pilot study are encouraging for further research into the possibilities of integrating the AR tool into industrial inspection processes.

Funding This work was supported by the Internal Science Foundation of the University of West Bohemia under Grant SGS-2021–028 'Developmental and Training Tools for the Interaction of Man and the Cyber–Physical Production System'.

References

1. Romero, D., Stahre, J., Taisch, M.: The Operator 4.0: Towards socially sustainable factories of the future. Comput. Indus. Eng., 139 (2020). https://doi.org/10.1016/J.CIE.2019.106128
2. Moeuf, A., Pellerin, R., Lamouri, S., Tamayo-Giraldo, S., Barbaray, R.: The industrial management of SMEs in the era of Industry 4.0. Int. J. Prod. Res. **56**, 1118–1136 (2018). https://doi.org/10.1080/00207543.2017.1372647
3. Wickens, C.D., Helton, W.S., Hollands, J.G., Banbury, S.: Engineering psychology and human performance. Eng. Psychol. Human Perform. (2021).https://doi.org/10.4324/9781003177616/ENGINEERING-PSYCHOLOGY-HUMAN-PERFORMANCE-CHRISTOPHER-WICKENS-WILLIAM-HELTON-JUSTIN-HOLLANDS-SIMON-BANBURY
4. Azamfirei, V., Granlund, A., Lagrosen, Y.: Multi-Layer Quality Inspection System Framework for Industry 4.0. https://doi.org/10.20965/ijat.2021.p0641
5. Imkamp, D., Berthold, J., Heizmann, M., Kniel, K., Manske, E., Peterek, M., Schmitt, R., Seidler, J., Sommer, K.D.: Challenges and trends in manufacturing measurement technology—the "industrie 4.0" concept. J. Sens. Sensor Syst. **5**, 325–335 (2016). https://doi.org/10.5194/jsss-5-325-2016
6. Phuyal, S., Bista, D., Bista, R.: Challenges, opportunities and future directions of smart manufacturing: a state of art review. Sustain. Futures **2**, 100023 (2020). https://doi.org/10.1016/J.SFTR.2020.100023
7. Germani, M., Mandorli, F., Mengoni, M., Raffaeli, R.: CAD-based environment to bridge the gap between product design and tolerance control. Precis. Eng. **34**, 7–15 (2010). https://doi.org/10.1016/J.PRECISIONENG.2008.10.002
8. Molleda, J., Carús, J.L., Usamentiaga, R., García, D.F., Granda, J.C., Rendueles, J.L.: A fast and robust decision support system for in-line quality assessment of resistance seam welds in the steelmaking industry. Comput. Ind. **63**, 222–230 (2012). https://doi.org/10.1016/J.COMPIND.2012.01.003
9. Bottani, E., Vignali, G.: Augmented reality technology in the manufacturing industry: A review of the last decade. IISE Trans. **51**, 284–310 (2019). https://doi.org/10.1080/24725854.2018.1493244
10. Palmarini, R., Erkoyuncu, J.A., Roy, R., Torabmostaedi, H.: A systematic review of augmented reality applications in maintenance. Robot. Comput. Integr. Manuf. **49**, 215–228 (2018). https://doi.org/10.1016/J.RCIM.2017.06.002
11. Bottani, E., Longo, F., Nicoletti, L., Padovano, A., Tancredi, G.P.C., Tebaldi, L., Vetrano, M., Vignali, G.: Wearable and interactive mixed reality solutions for fault diagnosis and assistance in manufacturing systems: Implementation and testing in an aseptic bottling line. Comput. Indus. 128 (2021)
12. Antonelli, D., Astanin, S.: Enhancing the quality of manual spot welding through augmented reality assisted guidance. Procedia CIRP. **33**, 556–561 (2015). https://doi.org/10.1016/j.procir.2015.06.076
13. Zhou, J., Lee, I., Thomas, B., Menassa, R., Farrant, A., Sansome, A.: Applying spatial augmented Reality to facilitate in-situ support for automotive spot welding inspection. Proceedings of VRCAI 2011: ACM SIGGRAPH Conference on Virtual-Reality Continuum and its Applications to Industry, pp 195–200 (2011). https://doi.org/10.1145/2087756.2087784
14. Ramakrishna, P., Hassan, E., Hebbalaguppe, R., Sharma, M., Gupta, G., Vig, L., Sharma, G., Shroff, G.: An AR inspection framework: feasibility study with multiple AR devices. Adjunct

Proceedings of the 2016 IEEE International Symposium on Mixed and Augmented Reality, ISMAR-Adjunct 2016, pp 221–226 (2017). https://doi.org/10.1109/ISMAR-ADJUNCT.2016.0080
15. He, F., Ong, S.K., Nee, A.Y.C.: A mobile solution for augmenting a manufacturing environment with user-generated annotations. Information **10**, Page 60. 10, 60 (2019). https://doi.org/10.3390/INFO10020060
16. Marino, E., Barbieri, L., Colacino, B., Fleri, A.K., Bruno, F.: An Augmented Reality inspection tool to support workers in Industry 4.0 environments. Comput. Indus. **127**, 103412 (2021). https://doi.org/10.1016/J.COMPIND.2021.103412
17. Lewis, J.R.: Usability testing. Handbook Human Factors Ergon., 1275–1316 (2006). https://doi.org/10.1002/0470048204.CH49
18. Lewis, J.R., Utesch, B.S., Maher, D.E.: Measuring perceived usability: the SUS, UMUX-LITE, and AltUsability. Int. J. Human-Comput. Interact. **31**, 496–505 (2015). https://doi.org/10.1080/10447318.2015.1064654
19. Bangor, A., Kortum, P., Miller, J.: Determining what individual SUS scores mean: adding an adjective rating scale. J. Usability Stud. **4**, 114–123 (2009)

Power Quality Disturbances Classification Based on the Machine Learning Algorithms

Omnia Sameer Alghazi and Saeed Mian Qaisar

Abstract This paper presents an approach for classification of the power quality disturbances (PQDs). The classification of real-time power quality disturbances (PQDs) is proposed in this work. The PQD signals are modelled based on the IEEE 1159–2019 standard. The outcome of the used PQD model is employed for analyzing the performance of suggested classification method. Firstly, the PQD signals are segmented and then each segment is further processed by machine learning based classifiers for identification of PQDs. The study is conducted for six major classes of the PQDs. The highest identification precision is secured by the Support Vector Machine classifier. It respectively attains the Accuracy = 94.32%, Precision = 84.55%, Recall = 84.33%, Specificity = 96.52%, F-measure = 84.19%, Kappa = 92.59%, and Area Under the Curve (AUC) = 97.83%.

Keywords Power quality disturbances · Machine learning · Classification · Evaluation measures

1 Introduction

The recent integration of renewable sources into the electricity grid via the distributed generation (DG) concept causes power quality disturbances (PQDs), system reliability reduction, and other concerns. To increase power quality, disturbances should first be identified using enormous power quality data. Power quality disturbance (PQD) auto categorization is a key aspect of power quality analysis and control [1–3].

Power quality disruptions (PQDs) has been investigated to ensure high-quality power generation in smart grids. Detecting PQ disturbances and effectively preventing them are crucial. This framework's essential components include a precise

O. S. Alghazi (✉) · S. M. Qaisar
Electrical and Computer Engineering Department, Effat University, Jeddah 22332, Saudi Arabia
e-mail: oalghazi@effat.edu.sa; omnia.sameer.oms@gmail.com

S. M. Qaisar
e-mail: sqaisar@effatuniversity.edu.sa

A. Visvizi et al. (eds.), *Research and Innovation Forum 2022*,
Springer Proceedings in Complexity, https://doi.org/10.1007/978-3-031-19560-0_13

knowledge as a first step, which may be followed by real-time therapy of PQ difficulties [4].

The IEEE suggested a practical approach for monitoring the electric PQ. The IEEE 1159–2019 standard includes PQD-related suggestions. This suggested standard practice offers suggestions for monitoring the electrical characteristics of single-phase and multiphase power systems. The recommended practice outlines the starting circumstances and abnormalities that resulted from the supply or load equipment, as well as interactions between the source and the load. This standard also covers power quality monitoring equipment, application techniques, and how to assess monitoring results. It also provides a set of terminology for detecting PQDs. The employment of technology that creates and is vulnerable to a variety of electromagnetic disturbances has increased interest in power quality [5, pp. 1159–2019].

Electrical energy is transported from transmission substations to diverse loads via distribution networks. Power is often transported from the transmission substation to the end consumers via distribution networks, which are normally operated radially. In fact, numerous power quality problems, such as harmonic distortion, might develop at any point in the power system. Disturbances, on the other hand, only become an issue at the interface between a network and its consumers, or at the equipment terminals. The equipment of energy efficient is a major source of power quality problems. Both variable speed motors and energy-saving bulbs are significant sources of waveform distortion and are susceptible to certain forms of power quality disturbances. While power quality become a barrier concerns to the broad adoption of sustainable sources and loads [6]. Due to recent advancements in power electronics, traction drives and other non-linear loads have grown in popularity. Electronics power based on converter power contributed to PQDs, resulting in higher power losses and severe economic losses. Every power issue represented in current, voltage, or frequency changes that results in destruction, or failure of consumer devices is considered as a PQDs problem [7].

The penetration of renewable energy (RE) sources to fulfill energy demands and achieve de-carbonization objectives is growing global concern over power quality. Because of the changing outputs and interface converters, power quality (PQ) disruptions are increasingly prevalent with RE penetration. To provide clean electricity to the consumer, it is necessary to identify and minimize PQDs [8]. Renewable energy is derived from natural sources that regenerate rapidly, such as photovoltaic (PV), a technology that turns solar energy into electrical energy. Wind power is another fast growing energy source. Wind energy was one of the most economically feasible solutions among the various clean energy sources [9]. Renewable energy is becoming an important contributor to our contemporary civilization but integrating it into the power system offers substantial technological problems. Power quality is a key part of integrating renewable energy [10]. Concerns about power quality are growing as renewable energy (RE) sources are increasingly used to meet energy demands and decarbonization targets. PQDs are becoming more common with RE penetration due to shifting outputs and interface converters. It is critical to identify and relieve PQDs in order to provide clean power to consumers [11].

This paper presents an approach for classification of the power quality disturbances (PQDs). The classification of instances power quality disturbances (PQDs) is proposed in this work. The PQD signals are modelled based on the IEEE 1159–2019 standard.

2 Literature Review

In recent decades, researchers have focused on the automated classification of PQD. To fulfill the power system's real-time demands, the process of identifying PQDs must be efficient while still achieving accuracy criteria. To fulfill the real-world standards, the effects of noisy signals must also be considered. The standard identification procedure includes two steps: feature extraction and pattern recognition.

A rigorous evaluation of methodologies for identifying and characterizing PQDs in utility grids with renewable energy penetration is provided. The researchers proposed several strategies for extracting characteristics in noisy situations in order to detect and characterize PQDs [8]. PQD feature extraction approaches include the "discrete wavelet transform" (DWT), the "S transform", the "wavelet packet decomposition" (WPD), the "Curvelet transform", and the "Hilbert transform" (HT). The Euler's rotation hypothesis enables two-dimensional curves to be translated into three dimensions instantly.

In [12], describes a novel approach for detecting various sorts of power quality disruptions (PQDs). In this article, adaptive chirp mode pursuit (ACMP) is used to extract important features, and the greedy algorithm is employed in the ACMP, which applies similar concepts to the matching pursuit approach. Furthermore, the ACMP makes use of sparse matrices, which reduces the computing cost. For the removal of incorrect features, the infinite feature selection as a filter-based method is used. In addition, the grasshopper optimization algorithm (GOA) is utilized to optimize the SVM as the classifier's parameters. The simulation and actual disturbance results demonstrate that the suggested method has excellent accuracy and speed, allowing it to be used in power quality measurement and analysis equipment. Furthermore, the suggested approach has noise rejection capabilities, which validates its usage in measuring and evaluating PQDs.

For the categorization of PQDs, three basic techniques are used: model-based, statistical, and data-driven. The data-driven method, which is based on machine and deep learning algorithms, is gaining popularity.

This study focuses on the classification of PQDs, as per definition of the IEEE 1159–2019 standard, in a hybrid grid by intelligently hybridizing the signal segmentation and machine learning algorithms.

The proposed methodology is described in Sect. 3. The empirical results are presented and discussed in Sect. 4. The conclusion is drawn in Sect. 5.

3 Methodology

There are several methods that may be taken to address power quality issues that create grid and power system difficulties. This study emphasizes the detection of PQDs. The first stage in this research is to model and simulate PQD signals in compliance with IEEE-1159 standards. By using real time data, each PQD signal classifiers use the acquired feature set to automate the classification of PQDs. The suggested PQDs identification technique is depicted in Fig. 1

3.1 *The Power Quality Disturbance (PQD) Model*

In the modeling and simulation of six types of PQD signals, the IEEE 1159–2019 standard is used. The following criteria are taken into account: pure signal, interruption, swell, sag, harmonics, and flicker [13, 14].

(1) **The pure signal:** It is a signal that is devoid of errors and signifies the execution of a faultless and steady supply. The signal can be quantitatively expressed using Eq. (1) [15, 16]. Whare, A represents the power signal amplitude. $\omega = 2.\ \pi.f$ and f is the frequency 50Hz and φ is the phase.

$$v(t) = Asin(wt - \varphi) \tag{1}$$

(2) **The interruption:** It is caused by a brief decrease of supply voltage for less than a minute. This reduces the root main square (RMS) voltage to less than 0.1 per unit (pu). The interruption impact on signal is represented by Eq. (2) [15].

$$v(t) = A(1 - \rho(x) \sin(wt - \varphi) \tag{2}$$

The parameters in Eq. (2) are defined below

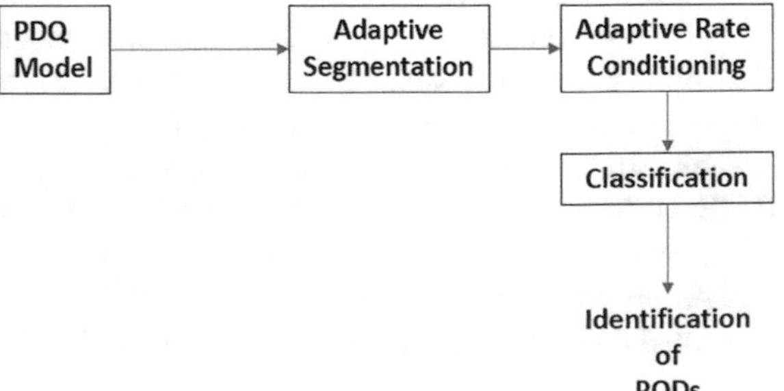

Fig. 1 The proposed PQDs classification system block diagram

$$x = u(t - t_1) - u(t - t_2))$$

$$-\pi \leq \varphi \leq \pi$$

$$u(t) = \begin{cases} 0 & t < 0 \\ 1 & t \geq 0 \end{cases}$$

$$T \leq t_2 - t_1 \leq (N - 1)T$$

$$T = \frac{1}{f}$$

$$0.9 \leq \rho \leq 1.0$$

(3) **The sag:** It is caused by brief short circuits and overloads. The RMS voltage will be reduced from 0.1 to 0.9 V. The singing impact on signal is represented by Eq. (3) [15].

$$\mathrm{v(t) = A(1 - \ y)sin(wt - \psi)} \tag{3}$$

The parameters in Eq. (3) are defined below

$$\mathrm{y} = \alpha(\mathrm{u(t} - t_1) - \mathrm{u(t} - t_2))$$

$$0.1 \leq \alpha \leq 0.9$$

$$-\pi \leq \varphi \leq \pi$$

$$u(t) = \begin{cases} 0 & t < 0 \\ 1 & t \geq 0 \end{cases}$$

$$T \leq t_2 - t_1 \leq (N - 1)T$$

$$T = \frac{1}{f}$$

(4) **The swell:** It occurs as a result of brief increases in power signal, which raises the RMS voltage between 1.1 and 1.8 pu. The swell signal is represented by Eq. (4) [15].

$$v(t) = A(1 + z)sin(wt - \varphi) \tag{4}$$

The parameters in Eq. (4) are defined below.

$$z = \beta(u(t - t_1) - u(t - t_2))$$

$$0.1 \le \beta \le 0.8$$

$$-\pi \le \varphi \le \pi$$

$$u(t) = \begin{cases} 0 & t < 0 \\ 1 & t \ge 0 \end{cases}$$

$$T \le t_2 - t_1 \le (N - 1)T$$

$$T = \frac{1}{f}$$

(5) **The harmonics:** It generally happens as a result of non-linear loads interfering with the supply device. The mathematical model of harmonics is described in Eq. (5) [15]. The parameters were defined as follows:

$$v(t) = A[sin(wt - \varphi) + H] \tag{5}$$

The parameters in Eq. (5) are defined below

$$H = \sum_{n=3}^{7} \alpha_n sin(nwt - \vartheta_n)$$

$$-\pi \le \varphi \le \pi$$

$$u(t) = \begin{cases} 0 & t < 0 \\ 1 & t \ge 0 \end{cases}$$

$$n = \{3, 5, 7\};\ 0.05 \le \alpha_n \le 0.15$$

$$-\pi \le \vartheta_n \le \pi$$

(6) **The flicker:** The waveform amplitude is modified at frequencies less than 25 Hz in this scenario. The flicker signal is defined by Eq. (6) [15]. The parameters were defined as follows:

$$v(t) = A[1 + \lambda \sin(w_f f)] \sin(wt - \varphi) \tag{6}$$

The parameters in Eq. (6) are defined below

$$0.05 \leq \lambda \leq 0.1$$

$$8 \leq f_f \leq 25Hz$$

$$w_f = 2\pi f_f$$

$$-\pi \leq \varphi \leq \pi$$

3.2 The Segmentation

The PQDs data is based on IEEE 1159–1195 standard. The data has been generated at a sampling rate of 4 kHz and by using a 16-Bit uniform based on quantization ADC. The data size for each class is 200 × 4000. Each instance is segmented for a 1-s length by using the rectangular window function.

3.3 The Classification

In this study the performance of known and robust supervised machine learning algorithms is compared for the identification of PQDs while using the mined features set. The considered algorithms are the "K-nearest neighbour" (KNN), "Support Vector Machine" (SVM), and Decision Tree (DT). This selection is made based on their frequent use in literature for the PQDs classification.

(1) **The KNN:** It is a simple and straightforward algorithm that used to solve both classification and regression problems. For dealing with categorization difficulties, it can be used in various configurations. In this study the fine KNN is used. The considered number of neighbours is one and the distance metric is Euclidean.
(2) **The decision tree (DT):** This algorithm is used for both the classification and regression problems. We explored deferent parameters of the DT and found that the best accuracy is secured by using the maximum number of splits as 100 and the spilt criterion Gini's diversity index.
(3) **The SVM:** The main purpose of SVM is to find the maximum range of samples data to reach for better results. The used kernel is quadratic SVM. The box constraint level is one. The used multiclass method is one-vs-one.

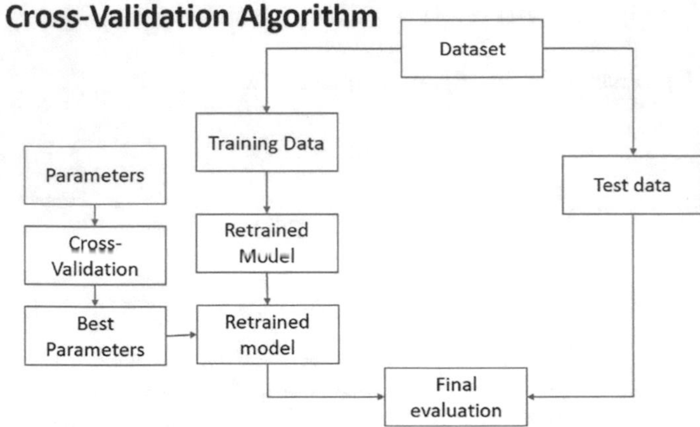

Fig. 2 The cross-validation algorithm

3.4 The Cross Validation (CV)

The CV is used in this study. It is a resampling method that tests and trains a model using different versions of the data on successive rounds. It is used to test machine learning models using a sample of data. To avoid overfitting, the testing data should be kept separate from the training data [17]. The Fig. 2 represents the cross-validation Algorithm.

3.5 The Confusion Matrix

In the realm of machine learning, a confusion matrix is a special table structure that allows visualization of the performance of an algorithm, usually a supervised learning algorithm, in the issue of statistical classification. Each row of the matrix represents the instances in an actual class, but each column, or vice versa, represents the instances in a forecast class. It is a one-of-a-kind contingency table with two dimensions: actual and expected, with the same set of classes in both [18]. The process is shown that the confusion matrix is made up of four main parameters that establish the classifier's measuring metrics:

1. A true positive (TP) is a test result that correctly detects the presence of a characteristic.
2. A test result that precisely indicates the lack of an attribute is known as a true negative (TN).
3. A test result that wrongly shows the presence of a characteristic is known as a false positive (FP).
4. A test result that mistakenly suggests the lack of a given trait is called a false negative (FN).

After that we calculated the measures, which are used to evaluate the accuracy of considered classifiers.

3.6 The Evaluation Measures

The following premeasured are considered to evaluate the accuracy of classification results [18]:

(1) **Accuracy (ACC):** It is determined as the proportion of each class's right prediction in the classifier results. The calculation of accuracy will be influenced by the confusion matrix idea. Equation (7) uses true TP, FP, FN, and TN values to find the accuracy in percentage

$$ACC = \frac{TP + TN}{TP + TN + FP + FN} \times 100\% \tag{7}$$

(2) **Precision (Pre):** It is the proportion of correctly predicted positive instances in a positive class to the total expected cases. The process may be expressed quantitatively in using mathematics Eq. (8). Where TN and TP are correct classifications. While FP and FN are wrong classification results.

$$\text{Precision} = \frac{TP}{(TP + FP)} \tag{8}$$

(3) **Specificity (Sp):** It indicates the fraction of correctly diagnosed negative situations, as illustrated in Eq. (9). Where true negatives (TN) and false positive (FP).

$$\text{Specificity} = \frac{TN}{(TN + FP)} \tag{9}$$

(4) **Recall:** It is also known as Sensitivity and is defined as the percentage of relevant documents retrieved successfully. Where do TP and FN. The recall is used to measure the percentage of accurately diagnosed positive instances as shown in Eq. (10).

$$\text{Recall} = \frac{TP}{(TP + FN)} \tag{10}$$

(5) **F-measure:** It is a strategy of being concerned about both measurements. This measure between recall and accuracy values represents the harmonic mean, as stated in Eq. (11).

$$F1 = \frac{2*precision*recall}{precision + recall} \tag{11}$$

(6) **The "area under the ROC Curve" (AUC-ROC):** To find AUC-ROC both TPR and FPR will be calculated as given by Eqs. (12) and (13).

$$\text{True Positive Rate (TPR)} = \text{Recall} = \frac{TP}{TP + FN} \tag{12}$$

$$\text{False Positive Rate (FPR)} = 1 - Percision = \frac{FP}{FP + TN} \tag{13}$$

(7) **The Kappa:** It is another way of describing the accuracy of the classifier. Its coefficient differences are employed in comparison analysis among unavoidable or accidental elements to discover the genuine characteristic by checking the number of correctives in the test results. The Kappa score measures inter-range reliability. Equations (14)–(16) is used to calculate Kappa:

$$kappa = 1 - \frac{1 - p_0}{1 - p_e} \tag{14}$$

$$p_o = \frac{(TP + TN)}{(TP + TN + FP + FN)} \tag{15}$$

$$p_e = \frac{(TP + TN)(TP + FN) + (FP + TN)(FP + FN)}{(TP + TN + FP + FN)^2} \tag{16}$$

4 Empirical Results and Discussion

In this study, the six PQDs classes have been considered namely, the pure signal (C1), flicker (C2), harmonics (C3), interruption (C4), sag (C5), and swell (C6). The MATLAB software is used to simulate and process the different PQDs signals. We investigated several classification algorithms. The evaluation of identifications is made and compared. The SVM, neural network and bagged trees secures the highest performance, and the findings are reported in the following Tables. The average evaluation measures for these classifiers are also displayed as following.

The confusion matrix is used to evaluate the evaluation measures of each classifier. The outcomes for Quadratic SVM are outlined in Table 1. We calculated the ACC, Pre, Sp, recall, F1, Kappa, and the AUC-ROC for each class. Then, the average of each evaluation measurement has been taken. The results show the average values of ACC = 92.08%, Pre = 81.51%, Recall = 73%, Sp. = 96.33%, F1 = 77.04%, kappa = 89.69%, and AUC-ROC = 97% respectively.

The confusion matrix is used to evaluate the evaluation measures of each classifier. The outcomes for wide neural network are outlined in Table 2. We calculated the

Table 1 Quadratic SVM results

Classes	Accuracy (%)	Precision (%)	Recall (%)	Specificity (%)	F1 (%)	Kappa (%)	AUC (%)
C1	95.56	80.97	100.00	94.53	89.49	94.34	100.00
C2	88.62	68.82	64.00	93.84	66.32	84.84	93.00
C3	98.54	100.00	92.50	100.00	96.10	98.17	100.00
C4	92.08	81.56	73.00	96.33	77.04	89.69	97.00
C5	99.31	99.49	97.00	99.88	98.23	99.14	100.00
C6	91.83	76.44	79.50	94.57	77.94	89.34	97.00
Average	94.32	84.55	84.33	96.52	84.19	92.59	97.83

Table 2 Wide neural network results

Classes	Accuracy (%)	Precision (%)	Recall (%)	Specificity (%)	F1 (%)	Kappa (%)	AUC (%)
C1	89.15	64.86	96.00	87.50	77.42	85.33	95.00
C2	86.71	64.25	66.50	91.41	65.36	81.76	81.00
C3	96.34	91.04	91.50	97.62	91.27	95.24	95.00
C4	89.84	76.67	69.00	94.90	72.63	86.33	79.00
C5	93.88	85.35	84.50	96.28	84.92	91.94	95.00
C6	89.49	88.98	52.50	98.43	66.04	85.83	87.00
Average	90.90	78.53	76.67	94.36	76.27	87.74	88.67

ACC, Pre, Sp, recall, F1, Kappa, and the AUC-ROC for each class. Then, the average of each evaluation measurement has been taken. The results show the average values of ACC = 90.90%, Pre = 78.58%, Recall = 76.67%, Sp. = 94.36%, F1 = 76.27%, kappa = 87.74%, and AUC-ROC = 88.67%.

The confusion matrix is used to evaluate the evaluation measures of each classifier. The outcomes for bagged tree are outlined in Table 3. We calculated the ACC, Pre, Sp, recall, F1, Kappa, and the AUC-ROC for each class. Then, the average of each evaluation measurement has been taken. The results show the average values of ACC = 88.61%, Pre = 75.24%, Recall = 70.5%, Sp. = 93.23%, F1 = 71.82%, kappa = 83.57%, and AUC-ROC = 91%.

The results indicate that the Quadratic SVM secures the highest performance in all evaluation measures while securing the highest accuracy as shown in Fig. 3. On average the accuracy secured by the Quadratic SVM for all 6 classes is 92.8% which is higher than wide neural network accuracy by 1.9%.

Table 3 Bagged trees results

Classes	Accuracy (%)	Precision (%)	Recall (%)	Specificity (%)	F1 (%)	Kappa (%)	AUC (%)
C1	75.27	38.53	65.50	77.38	48.52	62.68	82.00
C2	90.48	82.08	71.00	95.78	76.14	86.99	94.00
C3	93.90	96.18	75.50	99.14	84.59	91.80	94.00
C4	94.42	95.18	79.00	98.85	86.34	92.52	97.00
C5	98.49	96.52	97.00	98.94	96.76	98.01	100.00
C6	79.14	42.94	35.00	89.30	38.57	69.43	79.00
Average	88.61	75.24	70.50	93.23	71.82	83.57	91.00

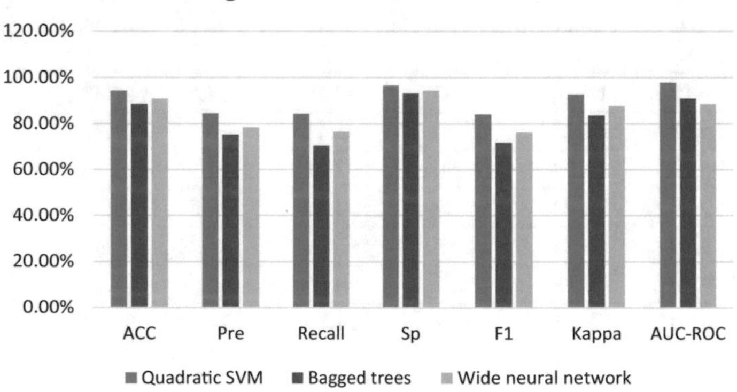

Fig. 3 The average accuracy percentage

5 Conclusion

Electrical gadgets have changed our lives to enhance our convenience. In accordance with the advancements in electrical equipment, there comes the need to implement efficient methods to detect any misgivings that may hinder the performance of such gadgets. Therefore, the only means of enabling the working of these electronics needs to be accurate and protected from any possible damage. In this paper a performance comparison is made among various machine learning algorithms for identification of the power quality disturbances while directly processing the PQD time-series and without any handcrafted features extraction. The results are encouraging and in future the performance of proposed method will be investigated while considering larger categories of the PQDs. These results will be used for future study by using extraction method. Moreover, investigation on the incorporation of ensemble learning techniques is another future research axis.

References

1. He, S., Li, K., Zhang, M.: A real-time power quality disturbances classification using hybrid method based on S-transform and dynamics. IEEE Trans. Instrum. Meas. **62**(9), 2465–2475 (2013). https://doi.org/10.1109/TIM.2013.2258761
2. Mian Qaisar, S., Alyamani, N., Waqar, A., Krichen, M.: Machine learning with adaptive rate processing for power quality disturbances identification. SN Comput. Sci. **3** (1), 1–6 (2022)
3. Mian Qaisar, S.: Signal-piloted processing and machine learning based efficient power quality disturbances recognition. PloS One. **16**(5), e0252104 (2021)
4. Zhong, T., Zhang, S., Cai, G., Li, Y., Yang, B., Chen, Y.: Power quality disturbance recognition based on multiresolution S-Transform and decision tree. IEEE Access **7**, 88380–88392 (2019). https://doi.org/10.1109/ACCESS.2019.2924918
5. Kipness, M.: IEEE SA—IEEE 1159–2019. SA Main Site. https://standards.ieee.org/ieee/1159/6124/. Accessed 05 Apr 2022
6. Bollen, M. H. J., Gu, I. Y. H.: Signal processing of power quality disturbances. John Wiley & Sons (2006)
7. A Comprehensive Survey on Different Control - ProQuest.: https://www.proquest.com/docview/2558798684. Accessed 05 Apr 2022
8. Chawda, G.S., et al.: Comprehensive review on detection and classification of power quality disturbances in utility grid with renewable energy penetration. IEEE Access **8**, 146807–146830 (2020). https://doi.org/10.1109/ACCESS.2020.3014732
9. Ray, P.K., Mohanty, S.R., Kishor, N.: Classification of power quality disturbances due to environmental characteristics in distributed generation system. IEEE Trans. Sustain. Energy **4**(2), 302–313 (2013). https://doi.org/10.1109/TSTE.2012.2224678
10. Liang, X.: Emerging Power Quality Challenges Due to Integration of Renewable Energy Sources. IEEE Trans. Ind. Appl. **53**(2), 855–866 (2017). https://doi.org/10.1109/TIA.2016.2626253
11. Khadem, S. K., Basu, M., Conlon, M.: Power quality in grid connected renewable energy systems: Role of custom power devices. Conf. Pap. (2010). [Online]. Available https://arrow.tudublin.ie/engscheleart/144
12. Motlagh, S. Z. T., Akbari Foroud, A.: Power quality disturbances recognition using adaptive chirp mode pursuit and grasshopper optimized support vector machines. Measurement. **168**, 108461 (2021). https://doi.org/10.1016/j.measurement.2020.108461
13. Qaisar, S.M., Aljefri, R.: Event-driven time-domain elucidation of the power quality disturbances. Procedia Comput. Sci. **168**, 217–223 (2020)
14. Qaisar, S. M., Alyamani, N.: Adaptive rate sampling and machine learning based power quality disturbances interpretation, pp. 1–6. (2021)
15. Igual, R., Medrano, C., Arcega,F. J., Mantescu, G.: Integral mathematical model of power quality disturbances. In: 2018 18th International conference on harmonics and quality of power (ICHQP), pp. 1–6 (2018). https://doi.org/10.1109/ICHQP.2018.8378902
16. Qaisar, S. M., Aljefri, R.: Time-domain identification of the power quality disturbances based on the event-driven processing, pp. 1–5. (2019)
17. Scikit-Learn Developers.: 3.1. Cross-validation: evaluating estimator performance. scikit-learn. https://scikit-learn/stable/modules/cross_validation.html. Accessed 20 Dec 2021
18. Kulkarni, A., Batarseh, F. A.: Confusion Matrix—an overview | ScienceDirect Topics. https://www.sciencedirect.com/topics/engineering/confusion-matrix. Accessed 20 Dec 2021

Smart Cities: Rupture and Resilience Building

Getting Things Right: Ontology and Epistemology in Smart Cities Research

Anna Visvizi, Orlando Troisi, Mara Grimaldi, and Krzysztof Kozłowski

Abstract The increasing pace of urbanization is posing multi-scalar challenges to cities and urban ecosystems worldwide. Sophisticated information and communication technology (ICT) has a pervasive influence on society, and ICT-enhanced solutions impact every aspect of our lives. In the urban context, this results in an, usually incremental, evolution of cities into smart cities. This, one of its kind, digital transformation, influences all facets of a city's functioning. Interestingly, in the otherwise rich body of research on smart cities, enquiries into business activity, innovation, competitiveness, seen especially as a function of economic growth, have been largely overlooked. Clearly, however, in a smart city there is a close positive relationship between ICT-based solutions, innovation, competitiveness, economic growth, and a thriving business sector. The ability to define and delineate the related to these concepts research field and the ability to recognize the mechanisms that underpin the relationships that unfold among the objects populating that field, bears the promise of identifying the gaps that exists in the literature, as well as the tools and strategies decision-makers may embark on to improve the functioning of a smart city. This paper delves into these issues. The value added of this paper is three-fold. First, it justifies the need to engage with the question of economic performance in the smart city. Second, it represents a rare attempt to add some metatheoretical order to the, at times unruly, smart city debate. Third, it highlights the new research avenues that need to be explored.

Keywords Smart city · Economic activity · Competitiveness · Business sector · Ontology · Epistemology

---

A. Visvizi (✉) · K. Kozłowski · K. Kozłowski
SGH Warsaw School of Economics, Warsaw, Poland
e-mail: avisvi@sgh.waw.pl

A. Visvizi
Effat University, Jeddah, Saudi Arabia

O. Troisi · O. Troisi · M. Grimaldi · M. Grimaldi
Department of Business Science, Management & Innovation Systems (DISA-MIS), University of Salerno, Fisciano, SA, Italy

A. Visvizi et al. (eds.), *Research and Innovation Forum 2022*,
Springer Proceedings in Complexity, https://doi.org/10.1007/978-3-031-19560-0_14

1 Introduction

Irrespective of the depth and breadth of smart-cities research, several questions remain unexplored. For instance, relatively little has been written about the economic aspects of the smart city reality. Notably, enquiries into business activity, innovation, competitiveness, seen especially as a function of economic performance, innovative capacity, business model innovation (BMI) etc., have been largely overlooked [1, 2]. Clearly, however, in a smart city, there is a close positive relationship between ICT-based solutions, innovation, competitiveness, economic growth, and a thriving business sector. In turn, these dimensions impact the well-being of the smart city inhabitants. Certainly, there is more to that. That is, advances in ICT influence the economy, including the forms and modes of economic exchange, models of economic collaboration (circular economy, sharing economy, platform), the velocity of economic exchange (including the payment systems), as well as modes and forms of communication between and among market agents. Moreover, given that the smart city represents not only a geographically delimited ecosystem, but also a nodal point where ICT and the society intersect, changes pertaining to the economy at large, have direct impact on the forms and modes of economic activity in the smart city. In other words, considering the increasing pace of urbanization, followed by the increasing velocity of economic exchange, and the increasing administrative autonomy of cities worldwide, it is imperative that developments in the economy at large are disaggregated to the smart city level; both in terms of concepts and data collection.

From a different angle, as cities and urban areas worldwide face a multitude of challenges, a targeted, well-thought-out use of technology for the city space, may mitigate several of these. As the economy and economic activity are the most efficient drivers of society's development, it is necessary to explore in more detail the connection between ICT-based solutions and economic activity/performance in the smart city, a topic largely overlooked in the literature. Accordingly, by means of bridging the gulf in the smart city debate, in this paper the relationship between ICT-enhanced services, applications and infrastructure, on the one hand, and business activity, innovation, competitiveness and economic growth on the other, are approximated through the lens of metatheory. To this end, the argument in this paper is structured as follows. In the next section, a brief review of the literature is offered to make a case that more attention needs to be paid to the economic aspects of smart cities functioning. To approach the question of the relationship between the ICT-based solutions and the economy in the smart city context, the concepts specific to the smart city debate are organized through the lens of ontology and epistemology. Discussion and conclusions follow.

2 Literature Review

Due to its origins in the diametrically opposed arms of, on the one hand, engineering and computer science, and, on the other, geography/spatial planning, until recently the smart-city domain exhibited the following weaknesses [1, 3]. Debate advanced by computer science and engineering communities overemphasized ICT-enhanced solutions. While technologically possible, these were not necessarily either feasible or implementable. As a result, studies addressed ever-more specialized problems, providing ever-more sophisticated responses to frequently peripheral issues. Mean while, the gap between society's needs, scientists' vision and research was widening. By contrast, the debate promoted by geographers viewed the (smart) city from a totally different perspective, focusing on the territorial, spatial and social aspects of the ICT driven transformation of the city and urban space. In this analysis, the role of ICT and the ICT-enhanced services and applications was diminished, thus widening the chasm running through the smart cities debate. The inroads of insights from social sciences exacerbated the chasm. The call for inter-disciplinary turn in smart-cities research was suggested as a way of bridging the chasm [4]. Simultaneously, it was cautioned that at least some consideration should be given to ontology and epistemology, if the resultant multi- and inter-disciplinary debate is to represent science proper, and the smart cities research can consolidate around a well-defined research agenda [1].

This plea notwithstanding, the debate on smart cities continues expanding regardless, and is thus populated by a variety of topics and methodological approaches. In this context, it is perhaps surprising that so little in the smart cities debate has been written on the impact of ICT enhanced applications and services on business activity, economic performance, innovation, competitiveness and eventually, economic growth [5–9]. The existing contributions tend to explore specific, important, and yet, minutiae of the smart city reality. These include: the circular economy [10, 11], services and the sharing economy [12], entrepreneurial opportunities [13–15], smart urban farming [16], entrepreneurship and start-ups [17, 18], the role of smart-city stakeholders in driving demand for specific services and technologies [19], smart services [20] and so on. Some research has been undertaken on the role of (smart) cities in the knowledge economy [21], also from the perspective of (global) talent management [22–25]. Smart-city performance, including innovativeness, has also been addressed [8, 26]; however, again, only as a part of the rather generic debate on smart-city policies and policy approaches.

The value of the contributions lies in their analytical categories, which are helpful to an understanding of the complexity of the issues and domains comprising a smart city. Nonetheless, the approaches fall short of offering a comprehensive insight into the mechanisms, or processes, that bring the smart city stakeholders together in advancing growth, development, and well-being. A definition of the domain's variables and their role in building, consolidating, and advancing a smart city's economic performance, including competitiveness is also absent in the existing research.

To be fair, a welcome trend in research on smart cities attempts to quantify and measure diverse aspects of smart cities reality [27–33]. A useful review of frameworks to assess smart cities has been conducted by Sharifi [27, 28], who developed a typology of the variables employed in the key approaches: economy; people; governance; environment; mobility; living; and data. And yet, more research is needed to offer conceptualizations of the processes and mechanisms underpinning respective areas of smart city performance [34], as well as to pinpoint which variables to employ, how to bypass the challenge of lack of data, comparability, and quality of data, and so on.

3 The Smart City Debate and Its Metatheoretical Dimensions

Typically, any discussion on ontology will start with question of the structure and the agency, and their relationship, whereas the conversation on epistemology will start with the question of how we can know about it. To understand how these issues play out in the context of smart cities research, this section dwells on the definitions of the smart city and thus extracts the cues pertaining to metatheory in the smart cities debate.

Given the diversity of approaches employed in the smart-city debate, several coexisting conceptualizations of the smart city exist. These are complementary; each offers an insight into an aspect of the smart-city reality. In this context, some argue that the smart city represents an urban development model geared to the utilization of human, collective and technological capital of urban agglomerations [35, 36]. In this view, the smart city is an "an ultra-modern urban area that addresses the needs of businesses, institutions, and especially citizens" [37]. From another angle, smart city is seen as a territorially limited and administratively delineated space whose ICT-enhanced services and applications aim to increase the city's, and its inhabitants' resilience to risks and threats and to promote sustainability [1, 38, 39].

The smart city has also been defined as "a place where the traditional networks and services are made more efficient through the use of digital and telecommunication technologies, for the benefit of its inhabitants and businesses" [40]. The definition constituent of the ISO family of standards related to smart cities places emphasis not only on the uses of "data information and modern technologies to deliver better services and quality of life to those in the city (residents, businesses, visitors)" [41], but also on the leadership and management practices in place. What brings these two definitions of smart cities together is the engagement with the end-users, i.e., citizens, businesses, and other stakeholders, an issue postulated in the literature for some time now [42–44]. The temporal dimension is very important in this context. That is, while there is a tendency in the literature to talk about smart cities' sustainability and resilience, references to a timeframe are absent. In contrast, the ISO standard is very explicit in this regard in that it defines the timeframe as "now and for the foreseeable future" [41].

Several rather fragmentary definitions of the smart city exist, and new terms are being coined, e.g. sustainable smart city [45, 46], smart sustainable city [47, 48] livable city [49], resilient city. Given this variety of terms and definitions, it is necessary to build a more comprehensive frame that rather than addressing solely the question of "the what?", would also highlight "the how?" and the "for what?". To this end, the three-tiered perspective on the smart city [43] that sees it, i.e. as a concept; as an artifact; and as a policymaking objective, is particularly useful. From this ontologically savvy conceptual, empirical and policymaking perspective, the smart city emerges as a key object of research and venue of research-driven policymaking. This way of viewing the smart city renders it possible to put (a) the subjects populating the smart city, i.e. its stakeholders; (b) the objects defining the smart city's structure, i.e. its tangible ICT-enhanced infrastructure; and (c) the outcomes of their relationship, beneath a spotlight to examine the mechanisms driving this relationship. This three-tiered approach focuses on the multiple stakeholders involved in a smart city, on its ICT-enhanced services and applications and on the modes of governing their relationships. Each of these categories of concepts retain their distinct nature and power and no conflation takes place.

Indeed, deriving from this approach, the smart city has been further conceptualized as a smart service system [44]. Here the weight of the argument falls on the synergies that emerge among people, technology, organizations, and information as they intertwine in the smart city. Viewing the latter from the perspective of smart services system [50–53], allows identifying the ways through which smart city stakeholders interact, and therefore co-create value and innovative solutions.

Who are the smart city stakeholders? There has been a tendency in the literature to include in the group of stakeholders the citizens, the authorities, and the business sector, including the providers of the ICT-services. The ISO definition of the smart city makes a clear reference to "visitors". The literature frequently refers to a broader category of 'end-users' [42]. The questions are (i) whether all these stakeholders can be considered agents endowed with purposive agency, and (ii) what kind of properties can be attributed to them. If so, then the next question that needs to be addressed is that of the structure in which they are embedded.

What is the structure in the smart city context? The simplicity of this question is tricky in that rarely in the debate on smart cities the question of the smart city ontology, or its constitutive parts, is explicitly posed. Indeed, in which ways a smart city is different from a city that is not 'smart' and what is it that it consists of? As argued elsewhere [34], it is the ICT component, and more specifically, the ICT-enhanced infrastructure, including wi-fi hotspots, smart traffic lights, e-services available to citizens, transport and traffic management related services, ICT-enhanced energy and water use systems, etc., that allows to distinguish between a non-smart city and a smart city. Hence, a case can be made that the smart city structure consists of tangible components, predominantly the built environment, the ICT-enhanced infrastructure, and their emergent properties [54], consistent of factors conditioning (enabling and constraining) that which is feasible in the smart city, e.g. that the agents, the stakeholders can act upon. the ICT-enhanced infrastructure and the ICT-conditioned components of the structure that most decisively distinguish the smart city from a non-smart city, in that they trigger the emergence of systemic properties that the agents can act upon [54].

The next question is what are the mechanisms that connect the stakeholders populating the smart city, or conversely, what is the relationship that unfolds between the agents and the structure, and consequently, also between/among the agents themselves. This is the classic question of the agent-structure relationship. In the approach proposed here, agents and the structure retain their ontologically distinct status and no conflation whatsoever is admitted. Nevertheless, the structure plays a powerful mediating role in enabling the emergence and in fostering the very specific, ICT-enabled, relationships between and among agents populating the smart city. A very good example in this context includes co-creation, co-production, but also and platform economy, circular economy, sharing economy. Notably, also more traditional forms of engagement and collaboration such as governance, leadership and management practices, (e-)service provision, collaboration, etc. acquire new facets when mediated by ICT.

The question of how epistemology informs research on smart cities at large proves to be an equally elusive a question. Consider this: there is no shortage of topics and issues in the smart-city debate. And yet, very little has been written about the smart city in a manner that would make the epistemological foundations of the argument abundantly clear. All too frequently, the debate on smart cities remains overly vague as regards the distinction between "what is", "how we know about it", "what it ought to be", and "how to make it happen", with relating categories of research questions remaining unconcise, overlapping, and leading essentially nowhere. Had it not been for the methodology requirement in most of the academic outlets, the situation would be worse. Consider the case of attempts at quantifying and measuring the 'smartness' of smart cities [27, 28]. Implicitly, these approaches adopt a positivist take on the smart city in that—as it transpires through the lines of the argument is—what exists, is only that which can be measured. Interpretive approaches abound in smart cities research too. Frequently, they merge with a very strong normative component, e.g. the use and application of ICT-based solutions is associated with positive outcomes. As a result, an entire discussion, a paper, will be then driven by that unspoken assumption, effectively excluding the possibility of discovering the contrary. The emergence of such terms as smart sustainable city, livable city, seems to attest to that. Obviously, knowledge is not neutral. Individual researchers' views and believes influence the specific research questions posed. The same applies to smart cities research. Consider the strand of research that views the smart city as a new name for a neoliberal agenda in urban space [cf. 55–57]. Examples abound. As questions pertaining to economic performance in the smart city are imminent, it is mandatory to bear these considerations in mind and be guided by them. The following section suggests how.

4 How to Explore the Smart City Performance? How to Get Things Right?

Against the backdrop of the points raised in the previous section, the objective of these paragraphs is to place respective concepts populating smart cities research in appropriate slots of the metatheoretical equation. By so doing, the ontological and epistemological foundations underpinning research on smart city performance will become more articulate. Correspondingly, the caveats besetting the issue and ways of navigating it will be identified.

In terms of *ontology*, research on smart city performance, needs to assume a holistic view on the smart city stakeholders, thus including citizens/end users, authorities (at all levels), and the business sector. It is important to stress the conceptual value of the term 'end-user' as compared to citizen, smart city inhabitant, or visitor. The term 'end user' is broader and, thus also includes those agents, who are not citizens, inhabitants etc. The term 'authorities' refers to the local (at all levels) and national authorities, who in their specific domains exert influence on other agents and condition the development/evolution of the structure. Similarly, the 'business sector' is employed as an aggregate term to account for individual business organizations inhabiting the smart city.

The smart city structure is defined as the set of factors related to the built environment, and the ICT-enhanced infrastructure, along with the structure of enabling/constraining properties emerging synergistically from the application of ICT to the domain of agential interaction. In practical terms, the ICT-enhanced infrastructure specific to the smart city exerts substantial impact on decision-making and management [57]. It influences how companies and organizations can address the needs of multiple actors, e.g. users, managers, policymakers, fosters value co-creation and supports the co-development of innovative solutions [58]. IoT, machine learning, cognitive computing, cloud computing systems applied in the smart city may increase end-users' participation by engaging them in the co-design of services and can boost end-users' experience by improving knowledge sharing [59]. Moreover, companies and organizations can benefit from the possibility to gather and analyze data on end-user's behavior, by gaining feedback to improve services and enhancing well-being and addressing more effectively the multiple needs of community [44].

In terms of the *agent-structure relationship*: Even if the structures are given, i.e. created by the previous generations of agents, the temporal dimension implicit in the debate on smart cities, consistent with "now and in foreseeable future" [41], suggests that the agents are not entirely deprived from the power to influence the structure. In fact, the agents are endowed with a degree of interpretive freedom to constantly seek to remold the structure [54]. As for the structure, its influence on the agents is decisive, in that it organizes the relationships among the agents, and defines that which is possible, including the structure of opportunities. Only through the proper combination of a given harmonized and strategically conceived structure (tangible and intangible dimensions) and of the uniqueness of agents (human component composed of attitude, know-how, beliefs, skills, willingness to engage), the ICT-mediated relationship can be fully exploited. The synergistic result of effective structure-agent relationship can help smart city develop competitiveness, and eventually resilience.

Regards the relationships between/among agents: Within the confines of the structure, the agents are endowed with agency, i.e. the capacity of purposeful action. The existence of the ICT component and the mediating role of the structure in this context, leads to the emergence of enablements and constraints that agents will consider when seeking to act. In the smart city context, action will include such modes as co-creation, co-production, as well as collaboration and interaction enabled by platform economy, circular economy, sharing economy. These mechanisms of collaboration add to the traditional modes of collaboration in the city context, and hence may be considered as distinct to the smart city. As such they may also be considered as distinct sophisticated sources of competitive advantage (smart) cities aim at.

In terms of epistemology, research on smart cities is vulnerable to two sets of challenges. First, it is frequent that the questions of "how can we know about it" and "what it ought to be" are conflated in research. Should it happen, and it frequently does, then entire sets of hypotheses leading to the construction of sophisticated surveys are driven by wishful thinking. As a result, the survey results are biased and so on. Second, naturally, all researchers bring their cognitive bias, i.e. the set of their specific views about the world, in the act of research. This is how questions of ideology and corresponding conceptualizations of smart cities barge in in the debate. Whereas ideologies are important, it is imperative that the research agenda is kept unbiased, and questions of "how we can know about" and "how we think it happened and what it represents" are not confused.

Mixed method approach [60], which combines qualitative approach (for the exploration of subdimensions of immaterial and complex constructs, such as competitiveness, innovation, ecosystems) and quantitative approach (to assess the statistical validity of the dimensions identified through qualitative approach) offers a way of navigating the challenge. Consider for instance, the Gioia methodology [61] that help to maneuver between generalizations and idiographic exploration of a context. Through inductive and deductive methods, the Gioia methodology suggests a continuous reproduction of the interpretative cycle of data-theory ("data collection"/ "reconnection of results to theory"), in which the data collected is compared with "old" knowledge and with existing theory, thus eventually helping in discovering complementary conceptualizations of a given reality.

That being said, the question is how does it all apply to the study of smart city performance? Throughout the paper it was argued that the economic aspects of the smart city remain underexplored in the literature. It was also argued that the smart cities research at large is filled with nonintentional metatheoretical inconsistencies. Hence, to dare to pose the question of the smart city performance, for instance of smart city competitiveness, it is necessary to understand the "who", the "what", and the "how". This paper suggested how to conceive of the structure in the smart city context, who are the agents populating that structure and what kind of relationship governs the agent-structure relationship in the smart city. These are important issues if we are to understand which factors and how drive economic performance, including competitiveness, in/of the smart city. This paper has also postulated a dose of self-reflection and an effort to adopt cognitive-bias free approaches to the study of the smart city. In brief, a case was made that only by identifying and naming the subjects of the analysis and by recognizing the causality inherent in their relationship it is possible to examine questions as complex as that of economic performance in the smart city. There is a beauty to setting things right in research, regardless of what we would like things to be.

5 Conclusions

Irrespective of the depth and breadth of smart-cities research, several questions remain unexplored, while the debate that unfolds bears a touch of unintended ontological and epistemological frivolity. Clearly, questions of metatheory rarely raise enthusiasm of the masses, and several emerging and consolidating research fields/approaches could be criticized on account of their vague metatheoretical foundations. Smart cities, as a field of research is no different. As the input economists to the smart-city debate is yet to leave its mark, it is imperative that a bit more attention is paid to ontology and epistemology in smart cities research. The challenge is twofold.

Since the vast range of topics that have been brought up by the emergence of the smart-city concept, require inter- and multidisciplinary perspectives [62–67], it is mandatory that at least a degree of awareness of matters pertaining to ontology and epistemology in smart cities research is in place. Only in this way, it is possible to recognize, name, identify that which is being explored (ontology), and distinguish between claims of what exists and how we can know about it (epistemology), and what ought to be. This paper did not seek to navigate all issues and questions pertaining to metatheory in smart cities research. Yet, it sought to showcase that metatheoretical considerations are being disregarded in the mainstream research on smart cities and that it is an issue.

Acknowledgements Research presented in this paper constitutes a part of the implementation of the following grant: 'Smart cities: Modelling, Indexing and Querying Smart City Competitiveness', National Science Centre (NCN), Poland, grant OPUS 20, Nr DEC-2020/39/B/HS4/00579.

References

1. Visvizi, A., Lytras, M.: Rescaling and refocusing smart cities research: from mega cities to smart villages. J. Sci. Technol. Policy Manag. (JSTPM) (2018). https://doi.org/10.1108/JSTPM-02-2018-0020
2. Appio, F.P., Lima, M., Paroutis, S.: Understanding smart cities: Innovation ecosystems, technological advancements, and societal challenges. Technol. Forecast. Soc. Chang. **142**, 1–14 (2019)

3. Visvizi, A., Lytras, M.D.: Smart cities research and debate: what is in there?. In: Visvizi, A., Lytras, M.D. (eds.) Smart Cities: Issues and Challenges: Mapping Political, Social and Economic Risks and Threats. Elsevier. ISBN: 9780128166390. (2019). https://www.elsevier.com/books/smart-cities-issues-and-challenges/lytras/978-0-12-816639-0
4. Visvizi, A., Mazzucelli, C.G., Lytras, M.: Irregular migratory flows: towards an ICTs' enabled integrated framework for resilient urban systems. J. Sci. Technol. Policy Manag **8**(2), 227–242 (2017). https://doi.org/10.1108/JSTPM-05-2017-0020
5. Tranos, E., Ioannides, Y.M.: ICT and cities revisited. Telematics Inf. **55**, 101439. ISSN: 0736-5853 (2020). https://doi.org/10.1016/j.tele.2020.101439
6. Vinod Kumar, T.M., Dahiya, B.: Smart economy in smart cities. In: Vinod Kumar, T. (eds.) Smart Economy in Smart Cities. Advances in 21st Century Human settlements. Springer, Singapore (2017). https://doi.org/10.1007/978-981-10-1610-3_1
7. Kim, K., Jung, J.-K., Choi, J.Y.: (2016) Impact of the smart city industry on the korean national economy: input-output analysis. Sustainability. **8**(7), 649 (2016)
8. Ferrara, R.: The smart city and the green economy in Europe: a critical approach. Energies **8**(6), 4724–4734 (2015). https://doi.org/10.3390/en8064724
9. Anttiroiko, A.V., Valkama, P., Bailey, S.J.: Smart cities in the new service economy: building platforms for smart services. AI Soc. **29**, 323–334 (2014). https://doi.org/10.1007/s00146-013-0464-0
10. Prendeville, S., Cherim, E., Bocken, N.: Circular cities: mapping six cities in transition, environmental innovation and societal transitions **26**, 171–194 (2018). ISSN: 2210-4224. https://doi.org/10.1016/j.eist.2017.03.002
11. Chauhan, A., Jakhar, S.K., Chauhan, C.H.: The interplay of circular economy with industry 4.0 enabled smart city drivers of healthcare waste disposal. J. Clean. Prod. **279**, 123854 (2021). ISSN: 0959-6526. https://doi.org/10.1016/j.jclepro.2020.123854
12. Rahman, M.A., Rashid, M.N., Hossain, M.S., Hassanain, E., Alhamid, M.F., Guizani, M.: Blockchain and IoT-based cognitive edge framework for sharing economy services in a smart city. IEEE Access **7**, 18611–18621 (2019). https://doi.org/10.1109/ACCESS.2019.2896065
13. Eichelberger, S., Peters, M., Pikkemaat, B., Chan, C.-S.: Entrepreneurial ecosystems in smart cities for tourism development: From stakeholder perceptions to regional tourism policy implications. J. Hosp. Tour. Manag. **45**, 319–329 (2020). ISSN: 1447-6770. https://doi.org/10.1016/j.jhtm.2020.06.011
14. Barba-Sánchez, V., Arias-Antúnez, E., Orozco-Barbosa, L.: Smart cities as a source for entrepreneurial opportunities: evidence for Spain, technological forecasting and social change **148**, 119713 (2019). ISSN: 0040-1625. https://doi.org/10.1016/j.techfore.2019.119713
15. Kummitha, R.K.R.: Smart cities and entrepreneurship: an agenda for future research. Technol. Forecast. Soc. Chang. **149**, 119763 (2019). ISSN: 0040-1625. https://doi.org/10.1016/j.techfore.2019.119763
16. Kullu, P., Majeedullah, S., Pranay, P.V.S., Yakub, B.: Smart urban farming (Entrepreneurship through EPICS). Procedia Comput. Sci. **172**, 452–459 (2020). ISSN: 1877-0509. https://doi.org/10.1016/j.procs.2020.05.098
17. Perng, S.Y., Kitchin, R., Mac Donncha, D.: Hackathons, entrepreneurial life and the making of smart cities. Geoforum. **97**, 189–197 (2018). ISSN: 0016-7185. https://doi.org/10.1016/j.geoforum.2018.08.024
18. Blanck, M., Duarte Ribeiro, J.L., Anzanello, M.J.: A relational exploratory study of business incubation and smart cities—findings from Europe, Cities, vol. 88, pp. 48–58 (2019). ISSN: 0264-2751. https://doi.org/10.1016/j.cities.2018.12.032
19. Kummitha, R.K.R., Crutzen, N.: Smart cities and the citizen-driven internet of things: a qualitative inquiry into an emerging smart city. Technol. Forecast. Soc. Change **140**, 44–53 (2019). ISSN: 0040-1625.https://doi.org/10.1016/j.techfore.2018.12.001
20. Malik, R., Visvizi, A., Troisi, O., Grimaldi, M.: Smart services in smart cities: insights from science mapping analysis. Sustainability **14**, 6506 (2022). https://doi.org/10.3390/su14116506
21. Penco, L., Ivaldi, E., Bruzzi, C., Musso, E.: Knowledge-based urban environments and entrepreneurship: inside EU cities. Cities **96** (2020). https://doi.org/10.1016/j.cities.2019.102443

22. Rybka-Iwańska, K., Serradel-Lopez, E.: Smart cities and the search for global talent. In: Visvizi, A., Lytras, M.D. (eds.) Smart Cities: Issues and Challenges: Mapping Political, Social and Economic Risks and Threats, pp. 171–184. Elsevier (2019). ISBN: 9780128166390
23. Dziembała, M.: Smart city as a steering center of the region's sustainable development and competitiveness. In: Visvizi, A., Lytras, M.D. (eds.) Smart cities: issues and challenges: mapping political, social and economic risks and threats, pp.149-169. Elsevier (2019). ISBN: 9780128166390
24. Ferraris, A., Belyaeva, Z., Bresciani, S.: The role of universities in the Smart City innovation: Multistakeholder integration and engagement perspectives. J. Bus. Res. **119**, 163–171 (2020). ISSN: 0148-2963. https://doi.org/10.1016/j.jbusres.2018.12.010
25. Bakıcı, T., Almirall, E., Wareham, J.: A Smart city initiative: the case of barcelona. J. Knowl. Econ. **4**, 135–148 (2013). https://doi.org/10.1007/s13132-012-0084-9
26. Zygiaris, S.: Smart city reference model: assisting planners to conceptualize the building of smart city innovation ecosystems. J. Knowl. Econ. **4**, 217–231 (2013). https://doi.org/10.1007/s13132-012-0089-4
27. Sharifi, A.: A typology of smart city assessment tools and indicator sets. Sustain. Cities Soc. **53**,101936 (2020a). ISSN: 2210-6707. https://doi.org/10.1016/j.scs.2019.101936
28. Sharifi, A.: A global dataset on tools, frameworks, and indicator sets for smart city assessment. Data in Brief. **29**, 105364 (2020b). ISSN: 2352-3409. https://doi.org/10.1016/j.dib.2020.105364
29. Sharifi, A., Kawakubo, S., Milovidova, A.: Urban sustainability assessment tools: toward integrating smart city indicators. In: Yamagata, Y., Yang, P.P.J. (eds.) Urban Systems Design, pp. 345–372. Elsevier (2020). ISBN: 9780128160558. https://doi.org/10.1016/B978-0-12-816055-8.00011-7
30. Sáez, L., Heras-Saizarbitoria, I., Rodríguez-Núñez, E.: Sustainable city rankings, benchmarking and indexes: looking into the black box. Sustain. Cities Soc. **53**, 101938 (2020). ISSN: 2210-6707. https://doi.org/10.1016/j.scs.2019.101938.
31. Wang, M., Zhou, T., Wang,D.: Tracking the evolution processes of smart cities in China by assessing performance and efficiency. Technol. Soc. **63**, 101353 (2020). ISSN: 0160-791X. https://doi.org/10.1016/j.techsoc.2020.101353
32. Dall'O', G., Bruni, E., Panza, A., Sarto, L., Khayatian, F.: Evaluation of cities smartness by means of indicators for small and medium cities and communities: a methodology for Northern Italy. Sustain. Cities Soc. **34**, 193–202 (2017). ISSN: 2210-6707. https://doi.org/10.1016/j.scs.2017.06.021
33. Fox, M.S.: The role of ontologies in publishing and analyzing city indicators. Comput. Environ. Urban Syst. **54**, 266–279 (2015). ISSN: 0198-9715. https://doi.org/10.1016/j.compenvurbsys.2015.09.009
34. Visvizi, A., Perez del Hoyo, R. (eds.): Smart cities and the UN SDGs. Elsevier, New York (2021)
35. Angelidou, M.: Smart city policies: a spatial approach. Cities **41**, 3–11 (2014)
36. Angelidou, M.: Smart cities: a conjuncture of four forces. Cities **47**, 95–106 (2015)
37. Khatoun, R., Zeadally. S.: Smart cities: concepts, architectures, research opportunities. Commun. ACM **59**, 46–57 (2016). https://doi.org/10.1145/2858789
38. Ahvenniemi, H., Huovila, A., Pinto-Seppä, I., Airaksinen, M.: What are the differences between sustainable and smart cities? Cities **60**, 234–245 (2017)
39. Kummitha, P.K.R., Crutzen, N.: How do we understand smart cities? An evolutionary perspective. Cities **67**, 43–52 (2017)
40. European Commission: Digital agenda for Europe: rebooting Europe's economy; directorate-general for communication (European Commission). Brussels, Belgium (2014)
41. ISO 37120: Sustainable cities and communities—indicators for city services and quality of life. ISO: Geneva, Switzerland (2018)
42. Lytras, M.D., Visvizi, A.: Who uses smart city services and what to make of it: toward interdisciplinary smart cities research. Sustainability **2018**, 10 (1998). https://doi.org/10.3390/su10061998

43. Lytras, M.D., Visvizi, A.: Information management as a dual-purpose process in the smart city: collecting, managing and utilizing information. Int. J. Inf. Manage. (2021). https://doi.org/10.1016/j.ijinfomgt.2020.102224
44. Kashef, M., Visvizi, A., Troisi, O.: Smart city as a smart service system: Human-computer interaction and smart city surveillance systems. Comput. Hum. Behav. **124**, 106923 (2021). https://doi.org/10.1016/j.chb.2021.106923
45. Silva, B.N., Khan, M., Han, K.: Towards sustainable smart cities: a review of trends, architectures, components, and open challenges in smart cities. Sustain. Cities Soc. **38**, 697–713 (2018)
46. Li, X., Fong, P.S., Dai, S., Li, Y.: Towards sustainable smart cities: an empirical comparative assessment and development pattern optimization in China. J. Clean. Prod. **215**, 730–743 (2019)
47. Höjer, M., Wangel, J.: Smart sustainable cities: definition and challenges. In: ICT Innovations for Sustainability, pp. 333–349. Springer, Cham (2015)
48. Bibri, S.E., Krogstie, J.: Generating a vision for smart sustainable cities of the future: a scholarly backcasting approach. Eur. J. Futures Res. **7**, 5 (2019). https://doi.org/10.1186/s40309-019-0157-0
49. Etezadzadeh, C.: Smart City–Future City?: Smart city 2.0 as a Livable City and Future Market. Springer (2015)
50. Barile, S., Polese, F.: Smart service systems and viable service systems: applying systems theory to service science. Serv. Sci. **2**(1–2), 21–40 (2010)
51. Maglio, P.P., Lim, C.H.: Innovation and big data in smart service systems. J. Innov. Manag. **4**(1) (2016). https://doi.org/10.24840/2183-0606_004.001_0003
52. Spohrer, J., Maglio, P.P., Bailey, J., Gruhl, D.: Steps toward a science of service systems. Computer **40**(1), 71–77 (2007). https://doi.org/10.1109/MC.2007.33
53. Maglio, P.P., Spohrer, J.: Fundamentals of service science. J. Acad. Mark. Sci. **36**, 18–20 (2008). https://doi.org/10.1007/s11747-007-0058-9
54. Archer, M.: Realist Social Theory: the Morphogenetic Approach. Cambridge University Press, Cambridge (1995)
55. Lombardi, P., Vanolo, A.: Smart city as a mobile technology: critical perspectives on urban development policies. In: Rodríguez-Bolívar, M. (eds.) Transforming City Governments for Successful Smart Cities. Public administration and information technology, vol. 8. Springer, Cham (2015). https://doi.org/10.1007/978-3-319-03167-5_8
56. Cardullo, P., Di Feliciantonio, C., Kitchin, R.: The Right to the Smart City. Emerald Publishing, Bingley, UK (2019)
57. Cardullo, P., Kitchin, R.: Being a 'citizen' in the smart city: up and down the scaffold of smart citizen participation in Dublin, Ireland. GeoJournal **84**, 1–13 (2019). https://doi.org/10.1007/s10708-018-9845-8
58. Black, J.S., van Esch, P.: AI-enabled recruiting: What is it and how should a manager use it? Bus. Horiz. **63**(2), 215–226 (2020)
59. Huang, M.H., Rust, R.T.: Artificial intelligence in service. J. Serv. Res. **21**(2), 155–172 (2018)
60. Paschen, J., Kietzmann, J., Kietzmann, T.C.: Artificial intelligence (AI) and its implications for market knowledge in B2B marketing. J. Bus. Ind. Mark. **34**(7), 1410–2141 (2019)
61. Axinn, W.G., Pearce, L.D.: Mixed Method Data Collection Strategies. Cambridge University Press (2006)
62. Gioia, D., Corley, K., Aimee, H.: Seeking qualitative rigor in inductive research: notes on the Gioia methodology. Organ. Res. Methods **16**(1), 15–31 (2013). https://doi.org/10.1177/1094428112452151
63. Giffinger, R., Fertner, C., Kramar, H., Meijers, E., Pichler-Milanović, N.: Smart cities: ranking of European medium-sized cities (2007). Vienna. Available at http://www.smart-cities.eu/download/smart_cities_final_report.pdf
64. Bibri, S.E., Krogstie, J.: Smart sustainable cities of the future: an extensive interdisciplinary literature review. Sustain. Cities Soc. **31**, 183–212 (2017)
65. Coletta, C., Evans, L., Heaphy, L., Kitchin, R. (eds.): Creating Smart Cities. Routledge, London and New York (2019)

66. Bär, L., Ossewaarde, M., van Gerven, M.: The ideological justifications of the smart City of Hamburg. Cities. **105**, 102811(2020). ISSN: 0264-2751. https://doi.org/10.1016/j.cities.2020.102811
67. Lorquet, A., Pauwels, L.: Interrogating urban projections in audio-visual smart city narratives. Cities. **100**, 102660(2020). ISSN: 0264-2751. https://doi.org/10.1016/j.cities.2020.102660

Livable City: Broadening the Smart City Paradigm, Insights from Saudi Arabia

Abeer Samy Yousef Mohamed

Abstract Quality of life, or Livability, is one of the most important objectives for any community worldwide. Over decades, people have recognized that some features can make places more or less livable. Whereas urbanism is the transformation of rural society into an urban one, the smart city is a paradigm that aims at employing technology to improve, in essence, Livability through sustainable integration of technology, the natural environment, and people's needs. In line with the Saudi Vision 2030 objectives, which aim to take its cities to a level of sustainable urbanization that improves the Quality of Life, the Saudi Green initiative was announced early in 2021 to be the solid foundation for livable cities. This paper examines these developments and, against this backdrop, introduces a new feature to the smart cities paradigm, i.e., Livability.

Keywords Livable city · New paradigm · Smart cities · Saudi Arabia · The Saudi green initiative

1 Introduction

Smartness is not just about incorporating digital interfaces into traditional infrastructure. Technology and data need to be targeted to make better decisions. Cities use smart technologies to improve key Quality of Life indicators by 10–30% [1]. Applying the principles of a smart city with the integration of Livability could be an excellent opportunity.

The COVID-19 pandemic had a major impact on living functions and daily life. Most cities can rely on government support for major emergencies, but this is not always the case in developed countries [2]. Cities need to become more self-sufficient and innovative, involving all stakeholders in crisis response and recovery plans.

A. S. Y. Mohamed (✉)
Effat University, PO Box 34689, Jeddah 21478, Saudi Arabia
e-mail: asmohameddawod@effatuniversity.edu.sa; drabeersamy@hotmail.com

Tanta University, Post Code, Tanta 31111, Egypt

A. Visvizi et al. (eds.), *Research and Innovation Forum 2022*,
Springer Proceedings in Complexity, https://doi.org/10.1007/978-3-031-19560-0_15

1.1 Research Question

Livable communities and resilient cities are buzzwords of the moment. But exactly how could livability be achieved through the smart city concept? The research focuses on this exact question to achieve a responsive city by analyzing smart cities leading model of Saudi Arabia aligned with The Kingdom Green initiatives as creative inspiration.

1.2 Research Hypotheses

- Smart city technology has significant untapped potential to improve the Quality of Life in the city.
- COVID-19 significantly impacts all life aspects; the smart city is one of the important solutions that could help health safety according to the Quality of Life.
- Smart cities are transforming the infrastructure economy, creating space for partnerships and private sector participation.
- The smart city is the key word to achieve Livability within sustainable urbanization and supreme Quality of Life, considering the Kingdom Saudi Arabia model of Smart City that follows the 2030 Vision aligned with the kingdom's Green Initiatives.

1.3 Research Methodology

The research employs a qualitative and a desktop analysis approach to develop a new integrated paradigm of smart cities through theoretical and practical features that follow The Quality of Life enhancement for a livable city application, which assists the city's stakeholders and specialists in the development processes of urban planning to take efficient actions to activate Livability.

2 Smart City

2.1 Smart City Definition and Background

Cities can be defined as "Smart" when investing in people. Social capital and modern transportation and communications Infrastructure promote sustainable economic growth and high Quality of Life standards [3]. While technology is crucial, the ability to apply it consciously and efficiently drives the transformation of cities to smart cities [4].

Table 1 Smart city modeling approaches

Source	Smart city model
Giffinger et al. [6]	Smart cities indicators
Cohen [13]	Smart cities wheel model
IBM [14]	Smart cities nine pillar models

Focusing on the exceptional Quality of Life is undoubtedly considered among vital smart city mainstays to assure the orientation of the higher lifestyle [5]. Giffinger et al. [6] considered "Smart" as an average of prospective acting contemplating the improvement of aware, flexible, transformable, synergistic, individual, self-decisive, and strategic components.

Smart cities will become a central location for information that needs to be managed efficiently and sustainably [7]. Sustainability requires a mechanism to create city dwellers, that is, individuals/citizens and organizations (companies and others) together and enable them to engage innovatively [8].

Smart cities are rooted in creating intelligent infrastructure [9] and connecting people to information and communication technology (ICT). In line with the Sustainable Development Goals (SDGs), SDG11 [10]; growth must respect these three axes [11]:

- Sustainability
- Smartness and
- Inclusion.

Smart cities intersect three focal points – dimensions, process, and results. The way they are built-in, connected, and fueled represents the motor and transportation type [12] (Table 1).

The Common smartness framework focuses on how residents can benefit from smart city initiatives and technologies, measured by time savings, as the indexed model is based on four categories: mobility, health, public safety, and productivity.

2.2 Saudi Arabia 2030 Vision a Path to Livable City

Saudi Arabia's Vision 2030 is a comprehensive and integrated national development project approved by the Council of Ministers on April 25, 2016. This vision is a unique concept of transformative economic and social reform that opens Saudi Arabia to the world and is built around three main themes: a dynamic society, a prosperous economy, and an ambitious country [15].

Many Vision Realization Programs (VPR) B. Lifestyles are created to consistently contribute to economic transformation from an income-based economy to a productive and globally competitive economy, including Quality programs. Launched in 2018, the Quality of Program primarily improves the lives of individuals by promoting the participation of citizens and expatriates in culture and sports [15].

Saudi Arabia's Quality of Life Program aims to enhance the quality of life for individuals and families by creating the environment necessary to develop and support new life choices and lifestyles. These options increase citizen participation, residents, visitor culture, entertainment, sports, tourism, people, and other related activities that enhance the quality of life, thus improving the ranking of Saudi cities worldwide [16].

3 Livable City

A more livable city is the best place to live and provides more resilient and competitive social, economic and environmental benefits; it is also a healthy city that promotes health, well-being, and equity.

3.1 Livability Definition

The key factors for a livable community are [17]:

- Residents' safety, socially connected, and inclusive.
- Environmental compatibility.
- Access to affordable and diverse housing options related to recreation and culture through employment, education, local shops, public spaces and parks, health and community services, public transport, and walking and cycling infrastructure.

Livability is the Quality of Life for people living in cities, regions, and communities—including their physical and mental health. Public transport and health care are key indicators of a city worth living in.

3.2 Livability Rationale Indicators/Index

The Livability Index is a composite score based on aspects of livability, including social infrastructure, walkability, public transport, open public spaces, housing affordability, and local employment [18, 19]. The 13 measures that make up the Livability Index [20] are; street connectivity; housing density; access to community, cultural and recreational destinations; access to child care services; into public schools; access to health services; access to sports and recreation facilities; access to fresh food; access to convenience stores; access to regular public transport; access to a large open public space; low stress related to housing affordability; and local job.

4 Smart Cities: Creativity in Digital Solutions for a More Livable Future

A smart city is a city where resources are used intelligently. effectively, using innovative technology to do the following [21]:

- Cost and energy savings.
- Enhancement of services provided.
- Improve the quality of life.
- Smart cities also offer mobile and network-based services.
- Sms-parking.
- 24-h observation of people with heart disease and other healthy people.
- Remote monitoring of service fleet.
- Electronic school diary.
- What and how all parents can see video surveillance at school for the teaching process of kids and many others.

The smart city methodology will enhance the city's livability, resulting in hi-Quality of Life for all residents. The most valuable achievement in that smart cities selected model could be merged within the existing city urban planning.

5 Saudi Arabia Unique Leading Ship of Smart Cities Development

The Saudi Arabian government is pursuing a fantastic plan to have smart cities as a key function in many specific kingdom cities. The new "smart city" Neom is the most advanced city globally. In the Vision 2030 National Development Plan, the Saudi government has set out its ambition to recognize three Saudi cities among the 100 best cities in the world.

The Quality of Life improves by meeting citizens' needs and demands to ensure that different services are adequately provided. Smart cities can be the next logical and innovative step in creating an environment that helps meet the challenges rather than strengthening a growing urban community.

The Valuable Saudi Green Initiative and Middle East Green Initiative provide a path for the Kingdom to protect the planet. With ambitious goals for the coming decades, national initiatives will improve Quality of Life and the future by increasing reliance on.

Clean energy, offsetting the effects of fossil fuels, and protecting the environment for current and coming generations. Through the Middle East Green Initiative, Saudi Arabia will lead regional efforts to achieve global goals for combating climate change. These green Initiatives are the essential cornerstone for enhancing city Livability within the applications of Smart cities.

5.1 NEOM-Saudi Arabia's World-Leading Smart City

The kingdom of Saudi Arabia pursues economic diversification and modernization programs; NEOM is a major project drawing the universe's attention. The name of NEOM does not hide its ambition either. It is derived from neo, which means "new" in Greek, and mostaqbal, "future" in Arabic. NEOM is one of the key pillars of Saudi Arabia's strategy to diversify its economy [22, 23]. The city focuses basically on nine tracks which are:

- Energy and water
- Mobility
- Biotech
- Food
- Advanced industrialized production
- Media and media production
- Entertaining
- Technical and digital services
- Living (housing, education, healthcare, etc.)

Future technology is the basis of NEOM's development. Focusing on disruptive solutions for transportation from autonomous driving to passenger drones, new food growth and processing methods, and patients' overall well-being. Healthcare, wireless high-speed internet free product "Digital Air", free continuous world-class online education, comprehensive electronic governance with immediate access to urban services, Net zero-carbon house standard building standards, walking and to give just a few examples, such as urban layouts that facilitate cycling, renewable energy is the only source of power. This creates a new way of life that considers human ambitions and perspectives, combining the best future technology with excellent economic prospects. This successful model has been implemented elsewhere in the MENAT region, and NEOM is rapidly becoming the largest zone of its kind. With an average wind speed of 10. 3 m/s [18] and abundant solar resources, NEOM will be a fully renewable energy-powered city.

5.2 NEOM Conceptual Pillars

The concept of NEOM is based on six main pillars [23, 24]:

- People's Orientation: The purpose of the city is to create an "idyllic society" with a comfortable living environment.
- Healthy life and transportation: The city will be built to promote walking and cycling, and advanced technology will be used to create an unprecedented transportation infrastructure [25].
- Automated Services: This will be the first to offer e-Government, a fully automated system for government services.

- Digitization: The Digital Air initiative will provide all citizens with high-speed internet access and online education completely free of charge.
- Sustainability: The city uses only renewable energy, and its building has a "net zero carbon footprint [25].
- Innovative construction: Encourage new technologies and materials to meet future needs.

6 Smart City Paradigm, a New Feature for Smart Cities Through the Application of Livable City

Smart cities are expected to become a new standard for urbanity life worldwide; it could be underpinned mainly by seven technologies that the World Bank believes will disrupt and transform the way cities serve their citizens [22].

- 5G mobile networks are up to 60 times faster than 4G networks.
- Financial technology allows transactions to be conducted without an intermediary such as blockchain, an exchange that currently functions as a guarantor for transactions [22].
- Artificial intelligence will enable smart parking and energy efficiency of buildings.
- A self-driving car will have a significant impact in the coming decades. Studies have shown that the combination of autonomous taxis and high-speed intercity rail systems can reduce the number of vehicles in the city by up to 90% [22, 23].
- Low-cost space exploration and microsatellites are key to powering the 20 billion connected things that research firm Gartner expects to use by 2020 [24].
- Biometrics help raise people worldwide that can prove their identities and re duce fraud, waste, and corruption.
- Drones combined with artificial intelligence can do everything in no time, from parcel delivery to dangerous tasks such as rooftop and tower maintenance checks.

With such a phenomenal rate of population growth, cities face three sustainability challenges:

- Economical in that it allows citizens to develop their economic potential
- Social, opportunity, stability, and safety together affect the Quality of Life
- The environment is created by the city itself or weather or geological events [17].

Smart cities address sustainability and manage congestion, transportation, and energy consumption while improving residents' Quality of Life. A smart city is a new city specially designed and built to maximize the potential of technology. Faster access to healthcare, more integrated transport and less pollution are some benefits [22].

The Smart City Model should be based on a holistic and comparative study of recent success stories, incorporating the new shift in urban planning thought that made the Quality of Life a vital priority of future planning for any city to achieve

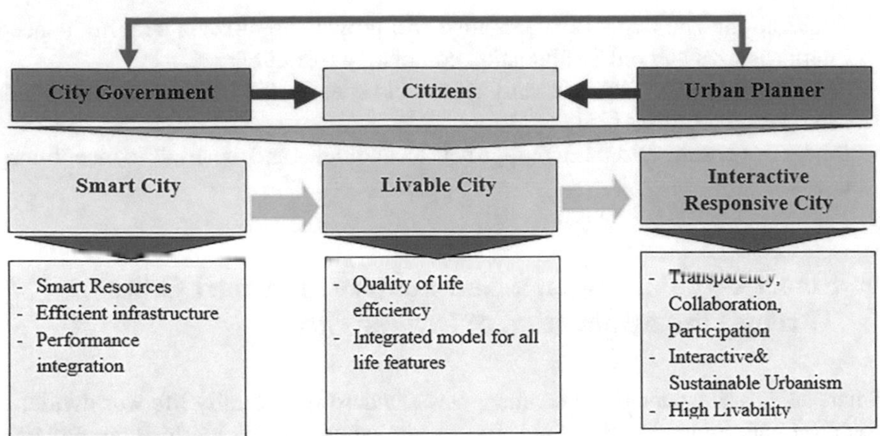

Fig. 1 New paradigm features for smartness, livability, interactive responsive city, *Source* Author

Livability. The Responsive City proposes an integrated paradigm for urban planning, having what is called the integrated paradigm (see Fig. 1).

7 Conclusion

The Quality of Life is the sole of any urban planning development by making Livability the main pillar for any improvement planning of existing cities or new ones. The smart city features with interactive collaboration with citizens will lead to reaching a responsive city.

Smart cities have different models that are proper to be reallocated and adjusted according to the city's nature, including Livability as the main cornerstone for development programming.

After the COVID-19 pandemic, the entire world has been changed after facing many challenges that affected all life aspects, resulting in the importance of The Quality of Life for city residences, which should be the main target of any development planning. Applying the smart city interactive model that engages city citizens within urban planning is the main step toward any success in achieving The Quality of Life that leads to Livability presented in the Interactive Responsive City model (see Fig. 2).

Research Output practical and theoretical implications.

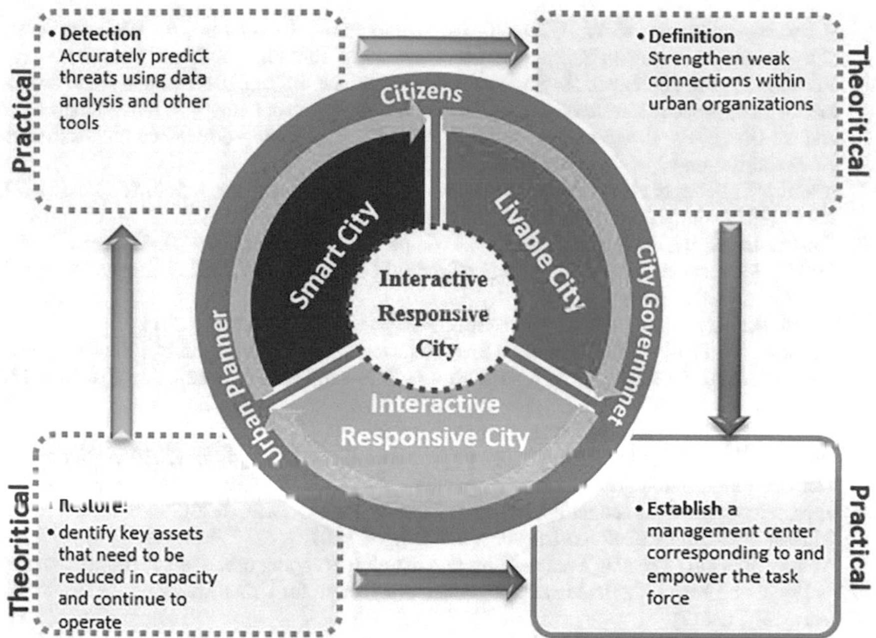

Fig. 2 New smart interactive responsive city urban domains (Theoretical/practical) processes, *Source* Author

References

1. McKinsey Centre for Government.: Smart city solutions: What drives citizen adoption around the globe?, McKinsey & Company. Accessed March 23 2019. https://www.mckinsey.com/indus-tries/public-sector/our-insights/smart-city-solutions-what-drives-citizen-adoption-arou
2. Dipak S., Aithal, P. S.: Smart cities development during and post COVID-19 pandemic—A predictive analysis. Int. J. Manag. Technol. Soc. Sci. (IJMTS). **6**(1), (2021). ISSN 2581–6012. www.srinivaspublication.com
3. Nam, T., Pardo, T. A.: Conceptualizing smart city with dimensions of technology, people, and institutions. In: Proceedings of the 12th annual digital government research conference, pp. 282–291 (2011)
4. Visvizi, A., Troisi, O.: Effective management of the smart city: An outline of a conversation. In: Visvizi, A., Troisi, O. (eds.) Managing smart cities. Springer, Cham, (2022). https://doi.org/10.1007/978-3-030-93585-6_1
5. Geller, A. L.: Smart Growth: A prescription for livable cities. Am. J. Public Health. **93**(9), 1410–1415 (2003)
6. Giffinger, R., Fertner, C., Kramar, H., Meijers, E.: Smart cities: Ranking of European medium-sized cities. Vienna University of Technology (2007)
7. Lytras, M.D., Visvizi, A.: Information management as a dual-purpose process in the smart city: Collecting, managing and utilizing information. Int. J. Inf. Manage. (2021). https://doi.org/10.1016/j.ijinfomgt.2020.102224
8. Kashef, M., Visvizi, A., Troisi, O.: Smart city as a smart service system: Human-computer interaction and smart city surveillance systems. Comput. Hum. Behav. **124**, 106923 (2021). https://doi.org/10.1016/j.chb.2021.106923

9. Batty, M., Axhausen, K. W., Giannotti, F., Pozdnoukhov, A., Bazzani, A., Wachowicz, M., Ou—zounis G., Portugali, Y.: Smart cities of the cuture. Eur. Phys. J. Spec. Top. (2012)
10. Visvizi, A., Pérez-delHoyo, R.: Sustainable Development Goals (SDGs) in the smart city: A tool or an approach? (an Introduction). In: Visvizi, A., Perez del Hoyo, R. (eds.) Smart cities and the UN SDGs. Elsevier (2021), 9780323851510. https://www.elsevier.com/books/smart-cities-and-the-un-sdgs/visvizi/978-0-323-85151-0
11. Kourtit, K., Peter, N.: Smart cities in the innovation age. Innov.: Eur. J. Soc. Sci. Res. **25** (2), 93–95 (2012). https://doi.org/10.1080/13511610.2012.660331
12. Zait, A.: Exploring the role of civilizational competences for smart cities' development. Transforming Government: People, Process and Policy **11**(3), 377–392 (2017). https://doi.org/10.1108/TG-07-2016-0044
13. Cohen, B.: Key components for smart cities. Retrieved from January 15 2014
14. Sderstrm, O., Paasche, T., Klauser, F.: Smart cities as corporate storytelling, city: Analysis of urban trends. Cult., Theory, Policy, Action **18**(3), 307–320 (2014). https://doi.org/10.1080/13604813.2014.906716
15. Vision 2030.: https://vision2030.gov.sa/v2030/overview/
16. Quality of Life Program Delivery Plan.: https://www.vision2030.gov.sa/v2030/vrps/qol/
17. The Economist Intelligence Unit.: Liveanomics: Urban Liveability and Economic Growth, https://eiuperspectives.economist.com/sites/default/files/LON%20-%20IS%20-%20Philips%20liveable%20cities%20Report%2002%20WEB.pdf. (2011)
18. Arundel, J., et al.: Creating liveable cities in Australia: Mapping urban policy implementation and evidence-based National liveability indicators. Centre for Urban Research (CUR), RMIT University, (2017)
19. Higgs, C, et al.: The urban liveability index: Developing a policy-relevant urban liveability composite measure and evaluating associations with transport mode choice. Int. J. Health Geogr., (2019)
20. Lowe, M., et al.: Liveability aspirations and realities: Implementation of urban policies designed to create healthy cities in Australia. Soc. Sci. & Med. **245**, (2020)
21. IESE Business School.: IESE cities in motion index. (2017). https://blog.iese.edu/cities-challenges-and-management/2017/05/25/164/
22. https://www.alj.com
23. https://www.neom.com/en-us
24. https://tomorrow.city/a/neom-saudi-arabia
25. Ali, M.: Multiagent systems applied to smart city—NEOM as a model. J. Eng. Appl. Sci. **7**(1), (2020)

Green Sustainable Urban Systems: The Case of Jeddah

Mady Mohamed, Muhammad ALSurf, and Sanaa AL-Kesmi

Abstract The governments of cities around the world are taking the green city approach into account by turning their cities' weaknesses into opportunities. The Saudi Vision 2030 continues to support the concepts of smart growth with the vision of promoting economic growth and development. At the same time, it ensures that natural resources continue to provide the resources and environmental services that strengthen the foundation of the country. The paper gives an overview of the concepts and principles of green cities and examines the available green sustainable urban tools that may be appropriate for Jeddah City. to help it transform into a green city. The current research adopts the triangulation model of mixed methods that include analytical literature review, questionnaire, and case study analysis. The paper reviews the most common and important Key Performance Indicators (KPIs) of the global Green Rating Systems (GRS) for sustainable cities. The analytical literature review for the tools' official technical manuals and professional exploratory surveys that preceded this research has been critical to confirm the applicability of the suggested system and its KPIs. The sustainable city in UAE has been selected as a case study to be analyzed and employed in developing the suggested SGRS. The results suggested a Saudi Green Rating System (SGRS) that includes eight categories (One required category, 6 mandatory credit categories, and one optional category) with thirty-one KPI, a total weight of 80 points. The suggested GSRS can aid the required transformation in Saudi Cities.

M. Mohamed (✉)
Architecture Department, College of Architecture and design, Effat University, Jeddah 21478, Saudi Arabia
e-mail: Momohamed@effatuniversity.edu.sa

Architecture Department, College of Engineering, Zagazig University, Zagazig 44519, Egypt

M. ALSurf
Regional Manager-Market Dev.-Green Business Certification Inc, Jeddah, Saudi Arabia

S. AL-Kesmi
Master of Science in Urban Design, College of Architecture and Design, Effat University, Jeddah 21478, Saudi Arabia

A. Visvizi et al. (eds.), *Research and Innovation Forum 2022*,
Springer Proceedings in Complexity, https://doi.org/10.1007/978-3-031-19560-0_16

Keywords Key Performance Indicators (KPI) · Green city · Green Rating System (GRS) · Sustainable tools · Jeddah

1 Introduction and Research Background

Cities are the main source of pollution; It causes 60% of carbon dioxide and greenhouse gas emissions. Also, climate change is now challenging cities to reduce their impact and adapt to changing conditions [1]. Hence, the increasing demand for sustainability is leading to rapid changes in policies, laws, and regulations to encourage more sustainable projects [2–4]. The implementation of sustainability could be different for each community, but they share common goals which are a healthy environment, smart growth, and the well-being of the people [5]. The rating systems are used to make projects more sustainable by providing frameworks with a set of criteria that cover several aspects of a project's environmental impact [6]. The Saudi Vision 2030 promotes concepts such as smart economic growth and natural resource sustainability to empower the country [7–9]. The city of Jeddah has a hot and humid climate, and the city's natural boundaries limit the city" growth from the east and the west, where it is surrounded by mountains from the east and the Red Sea from the west, directing urban growth north–south and expanding the built-up area along the coast. During the previous few decades, Jeddah quickly transformed into a modern metropolis, mainly in the north, which is 17 miles from the old city center. However, these rapid developments lack services and infrastructure, which poses challenges for urban planning, such as uncontrolled developments, inadequate urban services, the proliferation of slums, degraded quality of the urban environment, demographic issues, and high energy consumption producing greenhouse gases [10]. Therefore, the study aims to develop a green sustainable urban tool for cities (Saudi Green rating System "SGRS") that is compatible with the Saudi context. The results of the research suggest an SGRS that corresponds to the context of the city of Jeddah, based on the top-rated urban sustainable tools.

2 Green City and Its Principles

The sustainable approaches aim to preserve and improve the urban environment by achieving the three pillars of sustainability, environmental responsibility, social paradigm, and economic prosperity at the urban level [9, 11–14]. Green cities are resilient to environmental disasters and infectious diseases, also promote green behavior such as public transport, and minimize the impact on the ecosystem [15] as well as a foundation for good urban governance [16]. The main goal of Green Cities is the optimal and efficient use of natural resources such as water, energy, and land use [17]. The green planning principles enable residents to live in a livable

Table 1 Sustainable cities' international KPI [19]

International key performance indicators for sustainable cities	Dimension	Indicator
	Economy	Unemployment rate
		Economic growth
	Environment	Green space, mobility, Air quality, Water recycle, water quality, reduce greenhouse gases
	Social	Education, sanitation, health, compact city, housing, quality public space

city and achieve balanced growth, taking into account the environment, economic competitiveness, and equity in equal measure. These principles are [18]:

1. Climate and context
2. Renewable energy for zero emissions
3. Zero waste city
4. Water
5. Landscape, gardens, and biodiversity
6. Sustainable transport and good public space
7. Local and sustainable materials with less embodied energy
8. Density and retrofitting of existing districts
9. Green buildings and districts, using passive design principles
10. Livability, healthy communities, and mixed-use program
11. Local food and short supply chains
12. Cultural heritage, identity, and sense of place
13. Urban governance, leadership, and best practice
14. Education, research, and knowledge
15. Strategies for cities in developing countries.

GRS uses Key Performance Indicators for assessment. The indicators that can identify the problems and pressures also make it possible to diagnose the urban areas that need to be developed through good governance and science-based responses [19]. The Indicator List is an easy-to-use toolkit for cities, guiding how to assess the needs of each city and how to establish baseline objectives and best practices from case studies (Table 1) [19, 20].

3 Research Methodology

The current research adopts the triangulation model of mixed methods that includes an analytical literature review, questionnaire, and case study analysis. Triangulation provides validation through cross-verification from multiple sources to check the consistency of the results [21, 22].

The paper uses the analytical literature review as the first method to review the most common and important Key Performance Indicators (KPIs) of the global Green Rating Systems (GRS) for sustainable cities. The research reviews the tools' official technical manuals in addition to the official homepages of the organizations.

The second method is a professional exploratory survey that preceded this research and has been critical in confirming the applicability of the suggested system and its KPIs using semi-structured closed-ended questionnaires. Likert scale with a range from 1–10 was employed in this survey in addition to the various statistical parameters such as the mean, mode, median, and standard deviation [23]. The questionnaire was distributed through Monkey Survey online tool. The authors posted the survey link on the WhatsApp groups for professionals in Jeddah. The survey reached 95 participants. The total number of completed questionnaires was 72, divided approximately equally in gender which are 38 males with 53% and 34 females with 47%. The majority of the participants were Saudi reaching 83% which represented 60 participants and only 12 non-Saudis presented the remaining percentage which is 17%. The age of the participants ranges from 20 years to above 40 years. In addition, the participants' background includes the participant's educational background as the 53% is the majority that has bachelor's degree, 26% has master's degree, 14% of the participants are doctors in the field, and the lowest percentage is 7% for the diploma degree. The professionality of the participants is divided into several majors which are an urban planner with 8%, urban designer with 10%, project manager with 17%, an architect with 28%, engineer with 22%, and other professions with 15% that presenting the landscape architects and the sustainability consultants. The majority of the participants work in the private sector and only 29% work in the governmental sector. Their years of experience range from 1–10 yrs got the heights percentage and 19% of the participants have 10–15 years also 8% have more than 15 years of experience.

Through the method of the case study analysis, the sustainable city in UAE has been selected as a case study to be analyzed and employed in developing the suggested SGRS. The analysis is conducted using a scoring matrix [24–26] to quantitatively benchmark several sustainable urban tools according to the context of the city of Jeddah to identify the three most sustainable urban tools to be integrated and developed.

4 Results and Discussion

Eleven GRSs have been analyzed and compared to reach a shortlist of three relevant tools for Jeddah City [27]. These are [28]:

1. Reference Framework of Sustainable Cities (RFSC) [29].
2. STAR community rating system (STAR) [5].
3. City Statistics (Urban Audit) [30].
4. Leadership in Energy and Environmental Design (LEED ARCR) [31].

5. City Prosperity Initiative (CPI) [32].
6. Pearl community rating system (Pearl) [33].
7. Global Sustainability Assessment System (GSAS) [34].
8. Climate Positive Development Program (CPDP) [35]
9. Green Building Council of Australia (Green Star) [36].
10. City Performance Tool by Siemens (CyPT) [37].
11. Indian Green Building Council (IGBC) [38].

Table 2 shows the assessment matrix that evaluates the eleven GRS based on the common indicators of the systems, to reach three tools that cover most of the dimensions.

The analysis also evaluates them based on their relevance to Jeddah City context through several criteria such as culture, regional climate, development scale, access to knowledge, indicator dimensions, and financial access.

In developing an SGRS, the three sustainable systems selected are the STAR Community Rating System [5], Indian Green Building Council [17], and Estidama's Pearl Community Rating System [33].

4.1 The Proposed Saudi Green Assessment System for Cities

The proposed system (Saudi Green Rating System, SGRS) is a combination of the three sustainable systems selected, considering mainly the highest rated system, the IGBC, and integrating the other two systems to build a complete system that is compatible with the Saudi urban context. The SGRS combines the main three dimensions of sustainable cities in addition to the governance dimension, education and innovation from the IGBC, and other dimensions such as culture from Pearl and civic engagement from the STAR. The proposed system constructs KPIs under the main categories of land use and built environment, health and well-being, sustainable mobility, water-energy infrastructure management, equity and self-determination, information and communication management, and innovation in urban planning.

5 Validation of the Proposed Saudi Green Rating System (SGRS)

After proposing the system, the research validates it through the other two methods of the triangulation model which include the professional survey and the case study analysis.

Table 2 Sustainable tools assessment, by the Authors

Tool	Economy	Social	Environment	Government	Spatial	Civic	Cultural Education	Innovation	Culture	Regional climate	Development Scale	Availability of Information	Dimensions	Accessibility	Total
									0 Different, 1 Neutral, 2 Typical		0 Different, 1 Typical	1 Accessible, 0 Limited Access	Total Dimension/2	F1 free, 0 required fees	
RFSC	1	1	1	1	1	–	1	–	0	0	1	1	2.5	1	11.5
STAR	1	1	1	–	–	1	1	1	0	1	1	1	3	1	13
Urban Audit	1	1	1	1	–	1	–	–	0	1	1	0	2.5	1	10.5
LEED ARC^R	1	1	1	–	–	–	1	–	0	1	0	0	2	0	**7**
CPI	1	1	1	1	–	1	–	–	1	1	1	0	2.5	1	11.5
Pearl	1	1	1	–	–	–	1	–	2	2	0	1	2	1	12
GSAS	1	1	1	–	1	–	1	–	2	2	0	0	2.5	0	11.5
CPDP	1	1	1	–	–	–	–	–	0	1	1	0	1.5	0	6.5
Green Star	1	1	1	1	–	–	1	1	0	1	0	0	3	0	10
CyPT	1	–	1	–	–	–	–	–	0	1	1	0	1	0	5
IGBC	1	1	1	1	–	–	1	1	1	2	1	1	3	1	15

5.1 Experts' Survey

The questionnaire aims to rate the eight proposed categories and their weighted credits quantitively with one open-ended question asking the experts' recommendations that can enhance the proposed SGRS. The total number of participants that were considered for the analysis was 72. (Appendix 1 presents the questionnaire results of the experts and the professionals in the field).

The majority of the participants voted for all the categories and gave them a high score. They also agreed on most of the proposed KPIs except the following. Under Health and well-being category, they suggested moving the preservation and restoration of water bodies and eco-sensitive zones, the solid waste management, and the regional recycled materials to the water, energy, and infrastructure management category. Also, they recommended combining the public open spaces indicators and considered the health community and safe community goals to be achieved by the category. As for the sustainable mobility category, some of the participants agreed on considering the sustainable mobility plan as a category and recommended combining barrier-free accessibility and access to mass transit facilities. As they believe the weak condition of Jeddah's infrastructure will not support the transportation choices indicator.

As for the water, energy, and infrastructure management category, some of the participants recommended combining the rainwater harvesting and the wastewater treatment and reuse indicators within the water efficiency plan indicator due to the lack of rainwater in Jeddah city and considered the renewable energy part of the efficiency plan indicator. As for the equity and empowerment category, participants recommended combining the local economy and employment opportunities indicators and considered equitable services and access as part of sustainable mobility. Under the Innovation in city planning category, they believe that innovation practices indicators must be optional.

5.2 Case Study Analysis

The research involves an in-depth and up-close examination of a case study through a variety of data resources that are collected (Fig. 1).

The design strategies of the city implemented comprehensive approaches that address the three pillars of sustainability and provide affordable solutions such as density, orientation, and form that promote environmental gains [39]. The city area is approximately 46 hectares and designed to accommodate 2700 residents along with a daily population of 600 [40], this area includes various land uses such as residential, commercial, urban farming, education, health care, leisure, and the innovation center

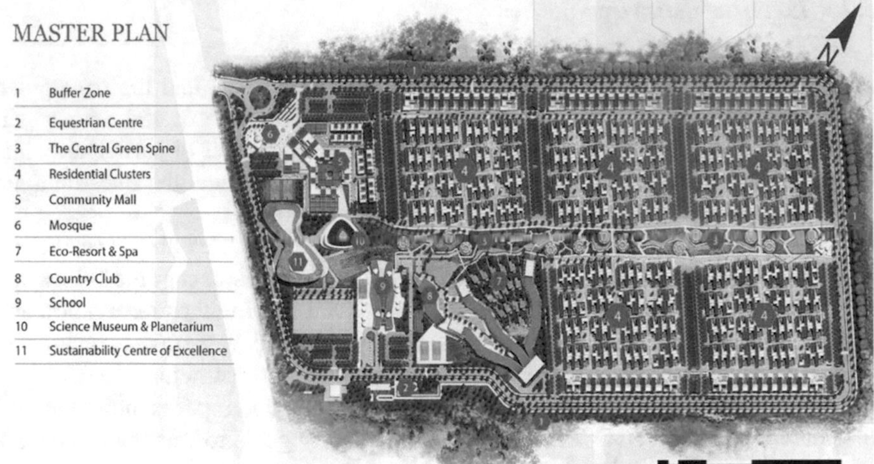

Fig. 1 Master Plan of Dubai Sustainable city—*Source* [39]

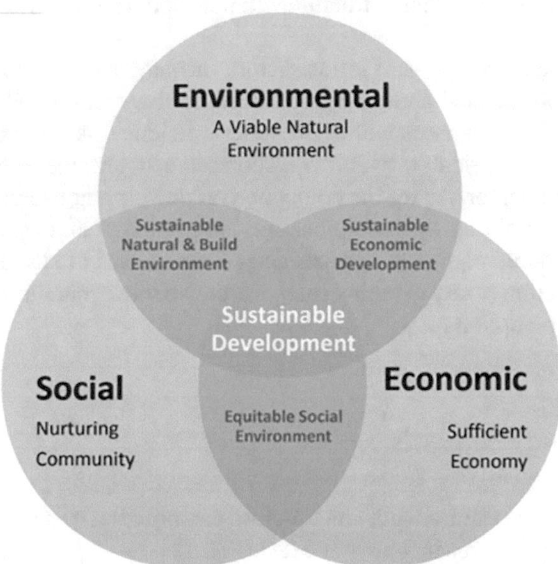

Fig. 2 Sustainable pillars of Dubai sustainable city—*Source* [42]

[41]. The sustainable city of Dubai is the first operational net zero energy city in Dubai [39, 41]. It achieved this through innovation in design, engagement of stakeholders, and future monitoring for sustainable functioning [42] (Fig. 2).

On analyzing the implemented principles of sustainability in the sustainable city, it has been found that.

(1) In The eco vision category, the sustainable city embraced the aspects of urban sustainability intending to fulfill human needs, protect the environment, and fulfill the needs of future generations.
(2) In The land use and built environment category, the city master plan is compressive, creating high-density residential clusters and mixed land uses and implementing green methods to minimize energy consumption. It created a pleasant atmosphere through green fingers that increase connectivity between the units to reduce the UHI phenomenon [39].
(3) In the health and well-being category, the city implemented the public green open spaces and the accessibility by providing different types of open spaces, that connect the city longitudinal by creating urban farming and productive landscape and promoting walkability and sports facilities. The city recycles 85% of the collected materials and organic waste.
(4) In the sustainable mobility category, it limited vehicular access within the city, creating alternative transportation indicator that provides electrical buggies and electrical buses, In addition, it enhances the walkability of the narrow streets and sidewalks throughout the city.
(5) In the water, energy, and infrastructure management category, it is a net-zero energy city that minimizes emissions, by adopting a closed-loop underground system including the transportation routes, and underground waste collection system, and converting wasted energy for heating and cooling the districts. The energy efficiency plan indicator is implemented from the early stages through building orientation and finishing. Also, using renewable energy indicators by locating solar panels on the villa's roof and parking shading system. The water efficiency plan of the city is to reduce 50% of water consumption and 100% of wastewater treatment and reuse indicators by using gray water stored in lakes for landscape irrigation and using bio-filter plants. In addition, collecting rainwater and reusing it to reduce fresh potable water [39, 41].
(6) In the information and communication technology category, It implemented the smart power grid that processes the energy from the solar panels to the city services and buildings, water-smart metering to measure the volume and the flow of the water in the city, also the smart irrigation system that irrigates the landscape and the farm through using smart collector that evaluates the amount of the water and the irrigation timing.
(7) In the equity and empowerment category, it has a vision of achieving sustainability while maintaining societal comforts and environmental stewardship. It involved the owner, key stakeholders, and the local community in the city

project. The sustainable plaza covers the maintenance cost and provides employment opportunities. The sustainable city enhances the resident's level of education by offering educational programs and classes in green living and waste management. The city ensures accessibility to services by all residents as well as maintains the privacy of the residential clusters.

(8) Innovation in the city planning category, the sustainable city created sustainable districts through developing innovation practices. It adopted traditional architectural elements and converted them into an efficient way to reduce energy consumption. In addition, the city established smart units that measure environmental performance and safety.

5.3 Final Proposed SGRS

Table 3 presents the final proposed Saudi Green Rating System (SGRS) (Presented separately in Appendix 2) for cities in an attempt to provide a better understanding of the city rating systems and the adaptation potential in the Saudi context. The updated system was designed based on a deduced system from the analytical literature review, the experts' and professionals' validation, and the case study analysis.

6 Conclusion

This paper presented an overview of eleven known urban sustainability scoring systems to identify the appropriate systems to be developed for Jeddah City conditions. Demonstrate the key performance indicators (KPIs) of the three selected ones, namely the STAR Community Rating System, Indian Green Building Council (IGBC), and Estidama's Pearl Community Rating System. The final proposed SGRS was validated through a survey of 72 professionals in the field and a case study analysis "Dubai Sustainable City" and through case study analysis. The final proposed SGRS includes eight categories (one required category, 6 mandatory credit categories, and one optional category) and thirty-one key performance indicators with a total weight of 80 points including 4 optional points. The output of this research opens the door for the decision-makers in Saudi Arabia to adopt the green rating systems that can aid the required transformation for Saudi Cities.

Table 3 The deduced and final proposed SGRS, by the authors

Modules		Points	Source	Final proposed SGRS	
Eco-vision		Required		Selected	Weight
Required	Eco-vision of the city				Required
Land use planning & built environment		15			15
Credit 1	Compact city planning	4	IGBC	√	4
Credit 2	Affordability housing	4	IGBC	√	4
Credit 3	Green buildings	2	IGBC	√	2
Credit 4	Urban heat island mitigation	3	IGBC	√	3
Credit 5	Infill and redevelopment	2	Star	√	2
Health & well-being		**20**			10
Required	Preservation and restoration of water bodies and eco-sensitive zones	Required	IGBC	Moved	
Credit 1	Public green and open spaces	4	IGBC	Moved	
Credit 2	Accessibility to public green and open spaces	3	IGBC	√	4
Credit 3	Environmental monitoring	2	IGBC	√	2
Credit 4	Solid waste management	4	IGBC	x	
Credit 5	Regional, recycled materials	3	Pearl	Moved	
Credit 6	Community health	2	Star	√	2
Credit 7	Safe communities	2	Star	√	2
Sustainable mobility		**10**			10
Required	Sustainable Mobility plan	Required	IGBC	√	Required
Required	Barrier-free Accessibility	Required	IGBC	√	Required
Credit 1	Access to mass transit facilities	2	IGBC	x	
Credit 2	Pedestrian network	2	IGBC	√	2
Credit 3	Transportation choices	4	Star	√	2
Credit 4	Travel plan	2	Pearl	√	4
Credit 4	Equitable services and access (*Moved from the Equity and Empowerment Category*)	2	STAR	√	2
Water, energy and infrastructure management		25			20
Credit 1	Water efficiency plan	5	IGBC	√	5
Credit 2	Rainwater harvesting	2	IGBC		
	Preservation and restoration of water bodies and eco-sensitive zones *(Moved from the health & well-being category)*			√	2
Credit 3	Waste water treatment and Reuse	5	IGBC	√	2

(continued)

Table 3 (continued)

Modules		Points	Source	Final proposed SGRS	
Eco-vision		Required		Selected	Weight
Required	Eco-vision of the city				Required
Credit 4	Energy efficiency plan	5	IGBC	√	5
Credit 5	Renewable energy	3	IGBC	x	
	Regional, recycled materials *(Moved from the health and wellbeing category)*			√	2
Credit 6	Green infrastructure *(Efficient infrastructure)*	3	Star	√	2
Credit 7	Food system	2	Pearl	√	2
Information and Communication Technology ICT		5			5
Credit 1	ICT applications	5	IGBC	√	5
Equity and empowerment		20			16
Credit 1	Civic engagement	3	Star	√	3
Credit 2	Environmental justice	2	Star	√	2
Credit 3	Civil human rights	2	Star	√	2
Credit 4	Equitable services and access	2	Star	Moved	
Credit 5	Education opportunities	4	Star	√	**4**
Credit 6	Local economy	2	Star	x	
Credit 7	Sustainable awareness	2	Pearl	√	2
Credit 8	Employment Opportunities	3	IGBC	√	3
Innovation in city planning		5			4
Credit 1	Innovation in city planning	2	IGBC	x	
Credit 2	Innovating practices	1	Pearl	√	2
Credit 3	Good governance	2	STAR	√	2
Total		100			80

Appendix 1: Questionnaire Results, by Authors

Classification	Statistical parameters				Scale				Frequency	
Category	Mean	Mode	Median	Standard deviation	1–5	5–6	6–8	9–10	Agree	Disagree
Eco Vision	8.96	10	10	1.614	–	10	10	52		
Land use planning & Built environment	8.99	10	10	1.682	4	–	14	45		
Compact City Planning									68	4
Affordable Housing									67	5
Green Buildings									66	6
Urban Heat Island Mitigation									66	6
Infill and Redevelopments									67	5
Health & Well-being	9.33	10	10	1.3	2	–	10	60		
Preservation and restoration of water bodies and eco-sensitive zones									68	4
Public green and open spaces									68	4
Accessibility to public green and open spaces									68	4
Environmental Monitoring									68	4
Solid waste management									66	6
Regional, recycled materials									66	6
Community Health									66	6
Safe community									67	5
Sustainable Mobility	8.89	10	10	1.896	5	–	13	54		
Sustainable mobility plan									68	4
Barrier-free Accessibility									66	6
Access to mass transit facilities									68	4
Pedestrian network									70	2

(continued)

(continued)

Classification	Statistical parameters				Scale				Frequency	
transportation choices									66	6
Travel plan									69	3
Water, energy, & infrastructure	9.29	10	10	1.486	3	–	8	61		
Water Efficiency Plan									70	2
Rainwater harvesting									67	5
Waste water treatment & reuse									67	5
Energy efficiency plan									68	4
Renewable energy									68	4
Green infrastructure									65	7
Food system									64	8
Information & communication technology	8.79	10	10	1.883	5	–	17	50		
ICT applications									69	3
Equity & empowerment	8.57	10	10	1.934	4	–	15	51		
Civic engagement									68	4
Environmental justice									64	8
Civil human rights									66	6
equitable services and access									62	10
Education opportunities									67	5
Local economy									67	5
Sustainable awareness									66	6
Employment opportunities									69	3
Innovation in city planning	8.89	10	10	1.858	8	–	12	52		

(continued)

(continued)

Classification	Statistical parameters	Scale	Frequency	
Innovation in city planning			70	2
Innovation practices			67	5
Good governance			66	6
Total		72 participants		

Appendix 2: Final Proposed SGRS—Source Authors

No	Category	KPI		Weight
I	Eco vision/required	No	Item	Required
II	Land use planning & Built environment	1	Compact city planning	4
		2	Affordable housing	4
		3	Green buildings	2
		4	Urban heat island mitigation	3
		5	Infill and redevelopments	2
Sub-total				15
III	Health & well-being	6	Availability & accessibility of public open spaces	4
		7	Environmental monitoring	2
		8	Community health	2
		9	Safe community	2
Sub-total				10
IV	Sustainable mobility	10	Sustainable mobility plan	Required
		11	Barrier-free accessibility	Required
		12	Pedestrian network	2
		13	transportation choices	2
		14	Travel plan	4
		15	Equitable services and access	2
Sub-total				10
V	Water, energy, & infrastructure Management	16	Water efficiency plan	5
		17	Energy efficiency plan	5
		18	Efficient infrastructure	2
		19	Preservation and restoration of water bodies and eco-sensitive zones	2
		20	Solid waste management	2
		21	Regional, recycled materials	2
		22	Food system	2
Sub-total				20
VI	Information & communication Technology	23	ICT Applications	5
Sub-total				5
VII	Equity & empowerment	24	Civic engagement	3
		25	Environmental justice	2
		26	Civil human rights	2

(continued)

(continued)

No	Category	KPI		Weight
		27	Education opportunities	4
		28	Sustainable awareness	2
		29	Employment opportunities	3
Sub-total				16
VIII	Innovation in city planning/optional	30	Innovation practices	2
		31	Good governance	2
Sub-total				4
Total credits for the KPIs				**80**

References

1. UN-Habitat.: Urban themes. 2012. Available from https://unhabitat.org/urban-themes/climate-change/
2. Hellström, T.: Dimensions of environmentally sustainable innovation: the structure of eco-innovation concepts. Sustain. Dev. **15**(3), 148–159 (2007)
3. Mohamed, M., et al.: Urban heat island effects on megacities in desert environments using spatial network analysis and remote sensing data: a case study from western Saudi Arabia. J. Remote. Sens., SI Underst. Ur-Ban Syst. Using Remote. Sens. **13**(10), 1941 (2021)
4. Mohamed, M.: Using rating systems as a guide tool for consultancy in maintenance work (LEED as a case study). In: OMAINTEC 2017, the 15th international operation & maintenance conference in the Arab countries, (2017)
5. Communities, S., Council, U. S. G. B.: STAR community rating system. (2007). [cited 2019]. Available from http://www.starcommunities.org/
6. Bernardi, E., et al.: An analysis of the most adopted rating systems for assessing the environmental impact of buildings. Sustain. **9**(7), (2017)
7. Artmann, M., Inostroza, L., Fan, P.: Urban sprawl, compact urban development and green cities. How much do we know, how much do we agree? Ecol. Indic. **96**, 3–9 (2019)
8. Artmann, M., et al.: How smart growth and green infrastructure can mutually support each other—A conceptual framework for compact and green cities. Ecol. Ind. **96**, 10–22 (2019)
9. Richter, B., Behnisch, M.: Integrated evaluation framework for environmental planning in the context of compact green cities. Ecol. Ind. **96**, 38–53 (2019)
10. Mandeli, K. N.: The realities of integrating physical planning and local management into urban development: A case study of Jeddah, Saudi Arabia. Habitat Int. **32**(4), 512-533 (2008)
11. LEGRAND.: Legrand and the sustainable development goals: Integrate the economic, environmental and social components of sustainable development in a balanced way. (2017). Available from http://www.legrand.com/EN/sustainable-development-description_12847.html
12. El Ghorab, H.K., Shalaby, H.A.: Eco and Green cities as new approaches for planning and developing cities in Egypt. Alex. Eng. J. **55**(1), 495–503 (2016)
13. Iwan, S., Thompson, R.G., Kijewska, K.: 2nd International conference green Cities 2016—green logistics for greener cities. Transp. Res. Procedia **16**, 1–3 (2016)
14. Jollands, N.: How to become a green city. (2016). [cited 2019]. Available from https://www.ebrd.com/news/2016/how-to-become-a-green-city.html
15. Maria-Laura, et al.: Green cities—urban planning models of the future. (2014)
16. ADB, A.D.B.: Hue GrEEEn city action plan, p. 60. Vietnam (2015)

17. IGBC.: Green Cities. Indian green building council. (2015). [cited 2018]. Available from https://igbc.in/igbc/
18. Lehmann, S.: Green urbanism: formulating a seriesof holistic principles. SAPIEN.S. **3**(2), (2010)
19. European, C., Directorate-General. E for.: Indicators for sustainable cities. Publications office (2018)
20. Dantes.: Environmental Performance Indicators, EPI. (2006). [cited 2018]. Available from https://dantes.info/Tools&Methods/Environmentalinformation/enviro_info_spi_epi.html
21. Evaluation, B.: Triangulation. (2014) Available from http://www.betterevaluation.org/en/evaluation-options/triangulation
22. Olsen, W.: Triangulation in social research: qualitative and quantitative methods can really be mixed. In: Haralambos, M., Holborn. H (eds.) Developments in sociology causeway. Manchester Research Explorer (2004)
23. Smith, J.: How to find the mean, median, mode, range, and standard deviation. (2018). [cited 2018], Available from https://sciencing.com/median-mode-range-standard-deviation-4599485.html
24. Al-Kesmi, S., Mohammed, M., AlSurf, M.: A proposal for green urban sustainable system. In: Meamaryat internal conference MIC 2018, Effat University, Jeddah, KSA (2018)
25. Mohamed, M., et al.: Sick neighborhood syndromes in hot dry climate. In: The 4th meamary at international conference MIC20, designing for the desert, Effat University, Jeddah, KSA (2020)
26. Fatani, K., Mohamed, M., Al-Khateeb, S.: Survey based sustainable socio-cultural guidelines for neighbourhood design in Jeddah. IOP Conf. Ser.: Earth Environ. Sci. **385**, 012050 (2019)
27. Mohamed, M.: Green Building Rating Systems as Sustainability Assessment Tools: Case Study Analysis, In: Fuentes-Bargues, J. L., Bastante-Ceca, M. J., Hufnagel, L., Mihai, F. C., Iatu, C., (eds.) Sustainability assessment at the 21st century, IntechOpen (2020)
28. Vierra, S.: Green buildings standards and certification systems. Whole building design guide. (2016). [cited 2019], Available from https://www.wbdg.org/resources/green-building-standards-and-certification-systems
29. French Ministry in charge of housing and urban development, a.t.C. Council of European Municipalities and Regions, and Apbisonalaitfos development.: Reference Framework of sustainable cities, RFSC. (2008). [cited 2022 March]
30. Eurostat.: City statistics (Urban audit). (1981). [cited 2022 May]. Available from https://ec.europa.eu/eurostat/web/cities/data/database
31. Council, U. S. G. B.: Leadership in energy and environmental design. (2000). [cited 2019]. Available from https://new.usgbc.org/leed
32. Habitat, U.: City prosperity initiative. (2012). [cited 2022 May]. Available from https://unhabitat.org/programme/city-prosperity-initiative
33. Council, A. D. U. P.: Pearl—BRS—Pearl Building Rating System (PBRS). (2010). [cited 2019]. Available from http://www3.cec.org/islandora-gb/en/islandora/object/greenbuilding%3A101
34. GORD, G. O. F. R. D.: Global Sustainability Assessment System (GSAS). (2007). [cited 2019]. Available from https://www.gord.qa/gsas-trust
35. Group, C. C. C. L.: Climate positive development program, CPDP. London (2009)
36. Africa, G. B. C. S.: Green star tools. (2007). [cited 2019]. Available from https://gbcsa.org.za/certify/green-star-sa/
37. Siemens.: City performance tool by Siemens. (2016). [cited 2022 Feb]. Available from https://new.siemens.com/global/en/products/services/iot-siemens/public-sector/city-performance-tool.html
38. Council, I. G. B.: IGBC Rating Systems. (2001). [cited 2019]. Available from https://igbc.in/igbc/
39. Architecture, B.: Dubai sustainable City. (2013). [cited 2022 May]. Available from https://www.baharash.com/architecture-design/dubai-sustainable-city/

40. Selim , S., El-Bana, N., Taleb, H.: Optimising sustainability at an urban level: a case study of Dubai sustainable city. WIT transactions on ecology and the environment. Sustain. Dev. Plan. **193**, 985–995 (2015)
41. Propsearch.ae.: The sustainable city. (2015) [cited 2022 May]. Available from https://propsearch.ae/dubai/the-sustainable-city
42. NEXUS, S.: The sustainable city—Sustainablity solutions. (2019). [cited 2022 May]. Available from https://www.thesustainablecity.ae/sustainability/

The Need for Green Software in Smart Cities

Laura-Diana Radu

Abstract The evolution of information and communication technologies is the basis of smart cities' progress. They emerge as a potential solution for citizens' problems that live in large urban centres and provide access to various services and products. But all these technologies consume resources and are sources of pollution. Many electronic and electric devices have sensors to perceive the world around them. They collect a large volume of data. In this context, minimizing the effect of these processes and actions on the environment is a big challenge. It requires the involvement of citizens, companies, and the government of smart cities as users or both as users and developers. This paper analyses the necessity of green software development and its characteristics in the context of smart cities evolution.

Keywords Green software · Smart city · Environment · Dashboard · Software evaluation criteria

1 Introduction

The evolution of smart cities is based on information and communication technologies (ICT) in all areas of economic and social life. A wide variety of advanced technologies such as artificial intelligence (AI), edge computing, blockchain, Internet of Things (IoT), often invisible or barely perceptible, have become ubiquitous in citizens' lives [1]. They bring new challenges both for users and for the environment. Regarding the environmental impact, the initiatives are various and consider both hardware and software, whose negative influences are more difficult to perceive. According to Podder et al. [2], "software is the backbone of virtually all the intelligent solutions designed to support the environment". The potential of reducing the negative impact on the environment has various dimensions, such as design for environmental sustainability, power management, data centre design and location, server virtualization, green metrics and assessment tools, renewable resources, and

L.-D. Radu (✉)
Faculty of Economics and Business Administration, Department of Business Information Systems, Alexandru Ioan Cuza University of Iasi, Iasi, Romania
e-mail: glaura@uaic.ro

A. Visvizi et al. (eds.), *Research and Innovation Forum 2022*,
Springer Proceedings in Complexity, https://doi.org/10.1007/978-3-031-19560-0_17

eco-labelling. Green software is part of "green ICT" or "green computing". Its influence on environmental protection has two dimensions mentioned in the literature by the following two concepts green by software and green in software [3]. First refers to the software dedicated exclusively to environmental protection and monitoring and provides efficient resources management for other applications; second refers to minimizing negative influences of software development on the environment and supporting the entire life cycle of sustainable software system engineering [4]. The integration of the environmental concerns into all stages of the software life cycle (development, use, and disposal) is essential [4] and has positive long-term effects. In the context of smart cities evolution that operates based on a large volume of equipment and applications, ICT specialists make significant efforts to reduce resource consumption both for hardware and software development and use. According to Manville et al. [5], "smart cities seek to address public issues via ICT-based solutions based on a multi-stakeholder, municipally based partnership" to improve its dimensions: smart economy, smart environment, smart governance, smart living, smart mobility, and smart people [6]. In smart cities, green software involves the adoption of environmentally friendly ICT to monitor and optimize the use of urban resources to ensure access to the infrastructure [7]. According to Chou et al. [8], these integrate the following three dimensions of environmental sustainability the economics of energy efficiency, the total cost of ownership (including the cost of disposal and recycling), and social imperatives. Regarding software sustainability, one significant aspect refers to technical features that make it adaptable to future needs and changes. In terms of hardware, the current trend is favourable for replacement, but in terms of the software, it is preferable to update it and only in particular cases to develop the application from scratch. Sen et al. [9] introduced the concept of smart software for applications used to operate smart cities. It needs to be energy-efficient and require little hardware to minimize carbon footprint, helping to create a smart, green city.

This paper investigates the need for green software in smart cities. The rest of this paper is organized as follows: Sect. 2 presents the material and methods. The green software criteria followed by an analysis of 10 smart cities dashboards from the viewpoint of environmental impact are presented in Sect. 3. Conclusions are given in Sect. 4.

2 Material and Methods

This study explores the main characteristics of green software in the case of smart cities based on literature. The interest in developing minimally invasive software applications for the environment has increased significantly, especially in smart cities, which require a considerable number of interconnected devices. Each software engineer, developer, tester, and IT administrator must consider that every single line of code may run years from now on zillions of processors, consuming energy and contributing to global climate change [10]. Based on the purpose of our research, presented in the introduction section, we identified the main research questions:

1. What are the evaluation criteria for green software?
2. How green are smart cities' dashboards?

To obtain a sense of the current state of green software concerns, we collected information from a wide variety of sources, including academic publishers (Clarivate Analytics Web of Science and Scopus), professional societies, and online repositories. We used the following keywords ALL = ((“smart city” or “smart cities”) and “green software”) and included only papers in English considered relevant by title, keywords and abstract.

3 Results and Discussion

There are a variety of software applications used in smart cities. Their use has become so natural that it is difficult to separate services from products. Many of the ICT-based services in smart cities are delivered through smartphone apps. Citizens use these devices more and more often due to the benefits they bring to everyday life. Companies and institutions also operate based on a wide variety of interconnected software applications and devices. GPS technologies for improving transportation and traffic flow, big data technologies for health systems, database technologies for education and energy efficiency, pattern recognition software for enhancing security systems, and mobile technologies for involving people in services co-creation or social activities are the most popular ICT applications in smart cities [11, 12]. The problem of environmental impact has naturally arisen. The first concerns were related to the hardware components because the effects were more visible but later extended to the software level. Software design and implementation influence hardware specifications and are responsible for hardware replacement. According to Hilty et al. [13], about 90% of the energy consumed by the hardware is allocated to the application software running on it, according to a study published by GHG Protocol.

3.1 Green Software Criteria

The lack of standards for assessing the compliance of software applications regarding their effects on the environment is an essential issue in the case of green software. A research team coordinated by Professor Stefan Naumann of Environmental Campus Birkenfeld has introduced some sets of criteria for a potential sustainable software certification scheme, known as the “blue angel label for resource—and energy-efficient software” [14]. They mention the following criteria for sustainable by design software: controllability, sufficiency, and power awareness. These reflect the possibility of configuring applications by users according to individual needs, the compatibility of new versions of applications with older hardware, and the control of direct energy consumption by applications [14].

Verdecchia et al. identified several actions that increase the energy efficiency of applications, such as ecology by design, green user experience (UX), blending software and its context, etc. [10]. The integration of concerns for optimizing resource consumption and reducing the negative impact on the environment can be assumed by designers and implemented from the software design. Applications must be compatible with systems and equipment from previous generations, minimizing resources used and creating functions/modules/classes reusable in future applications. The software must run on hardware with an average configuration [15]. The end-user will not have to replace the existing equipment to ensure the operation of the applications. The goal is to maximize the lifespan of the hardware.

As far as possible, the software update is preferred over developing a new one. The focus will be on optimizing the existing ones. The development of cloud solutions decreases energy consumption and allows the end-users to use hardware equipment with limited performance. Smart cities use a wide variety of interconnected applications. They allow citizens access to various public and private services (transportation, health, education, etc.). However, only a part of the resource consumption and CO_2 emissions are quantifiable at the city level. Another significant part is generated by cloud service providers or during the development process. Even under these conditions, end-users could support environmental protection with the help of UX specialists. The latter design options to configure the software according to the individual needs of each user, display warnings about the consumption of resources in the process of use or stop specific unnecessary processes [10].

The existence of software evaluation criteria in terms of environmental influence is even more significant in smart cities, where the number of users and applications is growing very fast. A multitude of applications involves the management of an extensive volume of data, mainly semi-structured or unstructured. Their storage and use require high-performance hardware and high-power consumption. There are various proposals for labelling software applications. In 2015 Kern et al. [16] proposed 39 criteria for the usage phase and 22 for the development phase. They grouped these criteria into the following three categories efficiency, feasibility, and perdurability. Subsequently, in 2020, Deneckère and Rubio [15] proposed 15 criteria for evaluating software applications categorized by life cycle, including the end-of-life phase. Also, in the development phase, they include aspects related to the elaboration of the documentation and the specifications.

Recently, the Linux Foundation, together with Accenture, GitHub, Microsoft, and ThoughtWorks, has initiated the Green Software Foundation to build "an ecosystem of people, standards, tooling, and practices to reduce carbon emissions caused by software development" [17]. Its founders aim to develop and publish green software standards, patterns, and practices, across various domains. Sustainability must become a concern of software development teams, a parameter of applications like security, accessibility, and scalability [18].

3.2 *How Green Are Smart Cities Dashboards?*

Citizens, organizations, and institutions of smart cities are some of the most important software suppliers and users. They have provided a perfect environment for creating digital innovation hubs. Kashef et al. consider a smart city "a laboratory to query several dimensions of the complex human–computer relationship" [19]. Due to the living conditions, smart cities attract talented and intelligent citizens with an essential role in innovation. Companies and local governments collect a wide variety of information and make it available to citizens and entrepreneurs through city dashboards. They present a synthesis of information from various sources. Smart cities' dashboards vary in the diversity and type of data provided. We used the Website Carbon Calculator on 10 cities' dashboards to evaluate the environmental impact. The app estimates how green a website is. It assesses the webpage's carbon footprint based on data transferred during upload. According to its developers [20], the calculation is based on the following indicators: data transfer over the wire, energy intensity of web data, the energy source used by the data centre, and carbon intensity of electricity and website traffic. The results are presented in Table 1.

Two of the ten dashboards—Paris and Vancouver—are clean based on the Website Carbon Calculator evaluation. Paris is cleaner than 69% of web pages tested and Vancouver is cleaner than 65% of them. The result depends on the volume and diversity of the information provided. An assessment of the usefulness of the information presented on these dashboards compared with the negative impact on the

Table 1 Smart cities dashboard (*Source* [20])

Smart city	Website	CO_2/visit (g)	Energy/year (~10.000 visits/month) (kWh)
London	https://data.london.gov.uk/ [21]	4.97	1,254
Chicago	https://data.cityofchicago.org/ [22]	19.08	4,821
Sydney	http://citydashboard.be.unsw.edu.au/ [23]	1.29	325
Edmonton	https://dashboard.edmonton.ca/ [24]	4.68	1,181
Singapore	https://www.singstat.gov.sg/whats-new/visualising-data [25]	6.72	1,698
Helsinki	https://helsinki.dealroom.co/dashboard/f/geo/anyof_Vantaa [26]	1.17	326
Cascais	https://data.cascais.pt/ [27]	2.89	731
Vancouver	https://opendata.vancouver.ca/pages/vandashboard/ [28]	0.59	166
Paris	https://dashboard.paris/pages/home/ [29]	0.50	140
Oslo	https://oslo.dealroom.co/dashboard [30]	1.26	351

environment would be beneficial and could lead to a balance. According to Acar et al. [31], the power consumption of the software is of two types: static—depending on the characteristics of the hardware and dynamic—depending on the source code. Smart development is a component of sustainable development [32] including in the software development field.

Accessing all data through a single platform has benefits too. Citizens will find faster information with fewer resources. Local governments thus become responsible for selecting the best product, considering the impact on the environment. However, to choose the best offers, they need clear criteria for evaluating software applications that could be classified into efficiency classes in terms of the impact on the environment, like hardware products. The diversity of software, the existing regulations in different countries, and the geographical dispersion of resources make it difficult to assess the impact on the environment. Software applications must be evaluated in the development phase and during the use and maintenance phase. Energy consumption must be analysed, together with the number of beneficiaries, the source of energy, and their importance or impact on community members' lives. Labelling software according to its environmental impact can influence end-user decisions. According to Deneckère and Rubio [15], the lack of clear consensus and definitions about sustainability in software engineering is the main obstacle in carrying out these steps.

4 Conclusion

There are various categories of software users. The criteria for assessing applications in terms of environmental impact must consider this variety. For example, developers are interested in labels for libraries or APIs and the standards they need to consider when developing products. In their case, the evaluation criteria must be part of the corporate social responsibility strategy. End users are interested in the features of applications such as operating systems, office suites, browsers, green services, green websites, etc. In smart cities, interconnected applications and devices are ubiquitous. Their number and the amount of information collected and processed are continuously growing. The environmental impact assessment corresponding to their usefulness is needed. To achieve their classification in categories according to the impact on the environment, clear and explicit criteria are required. Large companies in the ICT field have an essential role in the success of such initiatives.

Acknowledgements This work was supported by ERASMUS+ PROGRAMME JEAN MONNET ACTIVITIES, Call for proposals EAC/A02/2019—Jean Monnet Activities, Project number: 620415-EPP-1-2020-1-RO-EPPJMO-MODULE, "European Smart Cities for Sustainable Development (SmartEU)".

References

1. Radu, L.: Disruptive technologies in smart cities: a survey on current trends and challenges. Smart Cities **3**(3), 1022–1038 (2020)
2. Podder, S., Burden, A., Singh, S., Maruca, R.: How green is your software? Harv. Bus. Rev. (2020)
3. Calero, C., Piattini, M.: Introduction to Green in Software Engineering. In: Calero, C., Piattini, M. (eds.) Green in Software Engineering, pp. 3–27. Springer, Heidelberg (2015)
4. Radu, L.: Determinants of green ICT adoption in organizations: a theoretical perspective. Sustainability **8**(8), 731 (2016)
5. Manville, C., Cochrane, G., Cave, J., Millard, J., Pederson, J.K., Thaarup, R.K., Kotterink, B.: Mapping smart cities in the EU. European parliament. In: Directorate-General for Internal Policies. Policy Department: Economic and Scientific Policy A. Publications Office (2014)
6. Camero, A., Alba, E.: Smart city and information technology: a review. Cities **93**, 84–94 (2019)
7. Angelidou, M., Psaltoglou, A., Komninos, N., Kakderi, C., Tsarchopoulos, P., Panori, A.: Enhancing sustainable urban development through smart city applications. J. Sci. Technol. Policy Manag. **9**(2), 146–169 (2018)
8. Chou, D.C., Chou, A.Y.: Awareness of green IT and its value model. Comput. Standards Interf. **34**(5), 447–451 (2012)
9. Sen, M., Dutt, A., Shah, J., Agarwal, S., Nath, A.: Smart software and smart cities: a study on green software and green technology to develop a smart urbanized world. Int. J. Adv. Comput. Res. **2**(6), 373–380 (2012)
10. Verdecchia, R., Lago, P., Ebert, C., De Vries, C.: Green IT and green software. IEEE Softw. **38**(6), 7–15 (2021)
11. Bulu, M.: Upgrading a city via technology. Technol. Forecast. Soc. Chang. **89**, 63–67 (2014)
12. Bifulco, F., Tregua, M., Amitrano, C.C., D'Auria, A.: ICT and sustainability in smart cities management. Int. J. Public Sect. Manag. **29**(2), 132–147 (2016)
13. Hilty, L., Lohmann, W., Behrendt, S., Evers-Wölk, M., Fichter, K., Hintemann, R., Janßen, M., Moser, H., Köhn, M.: Green software: analysis of potentials for optimizing software development and deployment for resource conservation: Establishing and exploiting potentials for environmental protection in information and communication technology (Green IT). UBA TEXTE, 23 (2015)
14. Zimmermann, B.: Green Software: an overlooked factor in the sustainability discourse. https://www.digitalsme.eu/green-software-an-overlooked-factor-in-the-sustainability-discourse/. Accessed 5 December 2021
15. Deneckère, R., Rubio, G.: EcoSoft: proposition of an eco-label for software sustainability. In D.-C.S., Proper, H.A. (Ed.), CAiSE 2020 Workshops, pp. 121–132. Springer (2020)
16. Kern, E., Dick, M., Naumann, S., Filler, A.: Labelling sustainable software products and websites: ideas, approaches, and challenges. In: EnviroInfo/ICT4S (1), pp. 82–91 (2015)
17. Green Software Foundation, https://greensoftware.foundation. Accessed 21 December 2011
18. Lago, P., Koçak, S., Crnkovic, I., Penzenstadler, B.: Framing sustainability as a property of software quality. Commun. ACM **58**(10), 70–78 (2015)
19. Kashef, M., Visvizi, A., Troisi, O.: Smart city as a smart service system: human-computer interaction and smart city surveillance systems. Comput. Hum. Behav. **124**, 106923 (2021)
20. Website Carbon Calculator: Website Carbon Calculator. https://www.websitecarbon.com/. Accessed 21 February 2022
21. London Datastore Logo: https://data.london.gov.uk/. Accessed 8 December 2021
22. Chicago Data Portal: https://data.cityofchicago.org/. Accessed 12 November 2021
23. Sydney City Dashboard: http://citydashboard.be.unsw.edu.au/.Accessed 12 January 2021
24. City of Edmonton. Citizen Dashboard: https://dashboard.edmonton.ca/. Accessed 26 January 2021
25. Singapore Dashboard: https://www.singstat.gov.sg/whats-new/visualising-data. Accessed 20 November 2021

26. NewCoHelsinki: https://helsinki.dealroom.co/dashboard/f/geo/anyof_Vantaa. Accessed 20 October 2021
27. Cascais Data: https://data.cascais.pt/. Accessed 15 November 2021
28. City of Vancouver.Open Data Portal: https://opendata.vancouver.ca/pages/vandashboard/. Accessed 8 December 2021
29. Living in Paris. Public service quality in numbers https://dashboard.paris/pages/home/. Accessed 18 November 2021
30. Oslo Dashboard: https://oslo.dealroom.co/dashboard. Accessed 22 November 2021
31. Acar, H., Alptekin, G., Gelas, J.P., Ghodous, P.: The impact of source code in software on power consumption. Int. J. Electron. Bus. Manag., 42–52 (2016)
32. Gerli, P., Navio Marco, J., Whalley, J.: What makes a smart village smart? A review of the literature. Transforming Government: People, Process and Policy (2022)

Smart Mobility in Smart City: A Critical Review of the Emergence of the Concept. Focus on Saudi Arabia

Aroob Khashoggi and Mohammed F. M. Mohammed

Abstract Today, half of the world's population lives in urban areas where information and communication technology (ICT) are natural catalysts for innovations. Progressing urbanization requires improved urban planning and management to make urban spaces more inclusive, safe, resilient, and sustainable, as outlined in SDG11. The concept of smart city offers a way of reconciling the diversity of challenges that the urban space is exposed to today, whereby ICT is considered a tool in this respect. Arguably, smart cities create the opportunity to integrate advances in ICT in the fabric of the city, thus creatively and innovatively provide improved quality of life. The domains of mobility, economy, environment, etc. are just a few examples. The objective of this paper is to focus on the domain of mobility in the smart city and to examine it from a historical perspective, i.e. starting by introducing the concept of smart mobility and its role in creating smart cities, identifying the related technologies and challenges, exploring the different types of smart mobility, and outlining international initiatives. The application of smart mobility in the Saudi Arabia context is highlighted.

Keywords Smart cities · Smart mobility · Smart communities · Smart cities dimensions · Social interaction · Quality of life · SDG 11

1 Introduction

Rapid urbanization and high demands on the quality of life created a need for enhancements in all areas of infrastructure within cities to make urban areas more inclusive, safe, resilient, and sustainable. Therefore, the idea of smart cities has emerged to cope with resources and limited spaces worldwide by highly advanced technology. Scholars specified these dimensions into six main categories: smart economy, smart

A. Khashoggi · M. F. M. Mohammed (✉)
Effat University, Jeddah, Saudi Arabia
e-mail: mfekry@effatuniversity.edu.sa

A. Khashoggi
e-mail: ankhashoqji@effat.edu.sa

A. Visvizi et al. (eds.), *Research and Innovation Forum 2022*,
Springer Proceedings in Complexity, https://doi.org/10.1007/978-3-031-19560-0_18

people, smart governance, smart environment, smart living, and smart mobility [1]. While extant literature exists that elaborates on each of these components of the smart city, a little less has been written about these domains from a historical perspective, i.e. in a manner that would seek to trace the emergence and the evolution of the concept. By focusing on smart mobility, this paper seeks to do just that. The argument in this paper is structured as follows. First, the concept of smart mobility is introduced.

Smart mobility has the potential to move from a critical problem to a possible source of improvement and transformation for modern cities. Urban mobility planning in the contemporary world must begin considering the need to promote safer, integrated, and, above all, sustainable means of transport [2]. The research utilizes a qualitative investigative methodology to comprehensively investigate the research topic and provide a complementary and supplementary exploration by collecting, reviewing, and analyzing the literature on smart mobility's characteristics, concepts, models, and challenges.

2 Smart Mobility

One of the most difficult topics to face in urban areas is mobility, as it involves both economic and environmental aspects that need higher technologies and virtuous people behaviors [2]. Smart mobility plays a crucial role in the ecosystems of complex smart cities as it is an important factor to allow cities to transform into intelligent cities [3]. Smart mobility can be identified as a general term to describe technologies related to transportation in urban areas. It represents a new transportation approach, including creating a sustainable system [4]. Benevolo et al. [5] has identified many advantages of smart mobility and categorized the main objectives into six points illustrated below (see Fig. 1). Smart mobility impacts the transportation systems in cities, leading to an impact on the environment. Consequently, it has become an increasingly present theme in sustainability agendas. The concept of smart mobility has evolved mainly from merging the digital revolution with the transport industry. Thus, new technologies have increased transport network efficiency [6].

3 Smart Mobility Technologies

Smart mobility is largely permeated by ICT, used in both backward and forward applications to support the optimization of traffic fluxes and collect citizens' opinions about livability in cities or the quality of local public transport services [5]. Within the latter domain, researchers found evidence that traffic is important to allow flexible movement within cities, but traffic time negatively impacts individual well-being. Prior studies also show a strong positive relationship between traffic time and city life satisfaction. Traffic has a measurable negative effect on the life quality of citizens,

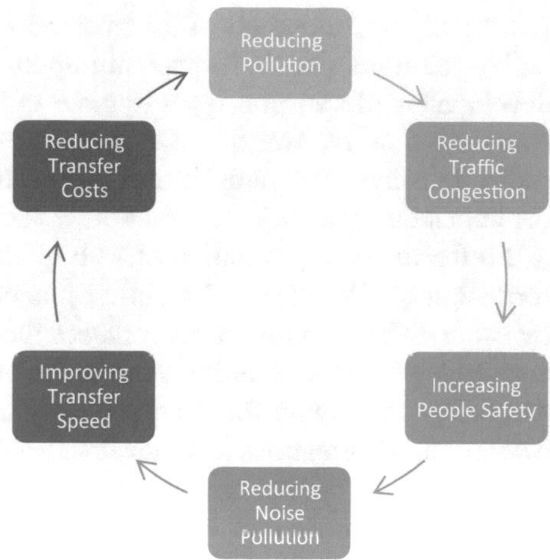

Fig. 1 Smart Mobility's main objectives [5], Developed by the author

and commute time plays a significant role in cities and entire countries. Therefore, the optimization and improvement of transport, mobility, and logistics in urban areas is a crucial goal in realizing the potential of smart cities. The authors stress that mobility is critical for regions' and cities' intellectual development and growth. The huge problem of traffic congestion requires the development of more sophisticated transportation concepts. Apart from these impacts on citizens' quality of life, traffic leads to air pollution that negatively impacts respiratory health. As one possible solution, several nations are deploying electric vehicles and charging stations to build green transportation networks [7, 8].

The technology of information and communication (ICTs) and digital technologies have improved the smart city concept. Government agencies have begun to invest in ICT systems, such as applications for individual use, cameras in monitoring systems, and sensors, aiming at positive changes in citizens' daily lives [5]. In this context, the use of ICTs in urban mobility has played a prominent role in promoting more sustainable and efficient transport, as in the case of interconnected services and technologies that allow car-sharing, bike-sharing, ride-sharing, buses, and trains [9]. In combination with the need for more flexible mobility and low CO_2 emissions, these technologies have resulted in the dissemination of these new initiatives in the field of mobility, thus making such mobility more intelligent [6].

4 Smart Mobility History

The history of urban transport knew many changes caused by diverse people's travel choices that gave birth to three types of cities. Walking Cities was the first idea proposed by British architect Ron Herron in 1964. Ron characterized the walkable city as tight streets with a large population density, and different mixed land uses.

After that the Automobile City emerged after the second world war. A city type built upon transport modes had an important changeover due to the technological development of transport types to move quickly to any destination. Then in the late twentieth century, Transit Cities emerged to develop public transport to integrate trains' railways, and trams' routes to spread the cities while reducing the density of the population [10, 11].

Furthermore, the evolution of metropolitan areas worldwide resulted in a transformation of lifestyles and mobility practices; this made individuals move freely by using different transportation modes increasingly for diversified reasons. A new term called "Urban Mobility" emerged to solve traffic problems such as congestion and delays to account for the complexity of mobility practices and the strong link between transformations and movements. Then Smart mobility came to attention in the early 1980s by having electric cars solve air pollution problems in larger cities [12].

Moreover, in 1990, California introduced its first zero-emission regulations by creating new projects for improving electric vehicles. Zero-emission vehicles became a new term applied. However, in this context, the focus should be on the sustainable development conversation. Since the late eighties of the last century, a new focus began on developing alternative resources of energy and technologies in transportation sector. This focus has become a major pillar of the current environmental and alternative energy sector, and it has even come to lead this sector. Issues related to the alternative energy development in transportation sector have been integrated into the much wider scope of sustainable development [11, 13].

5 Smart Mobility Challenges

Nowadays, many mobility initiatives influence how people move around their city and get their daily needs as transportation is one of the essential daily routines. As smart mobility grows daily, multiple organizations are starting to invest more. Transportation improvement has created an enthusiasm that can be applied to its width because the main beneficiaries are the consumers [5]. However, urban scholars and urban planners are working on expanding Smart Mobility to improve the goals of city policies, reduce traffic congestion, and encourage greater productivity. There are many ways to explore Smart mobility concepts where new projects could change the daily routine of people moving around the city [3]. Many challenges can occur in this process. Information security, payment security, data privacy, and heterogeneous networks are classical challenges common in any information system and will not disappear in urban mobility systems. In smart mobility's ecosystem, legal challenges involve different players co-ordinating together efficiently, e.g., financial companies, private transportation providers, city administration, and users. Another challenge smart mobility will deal with is the privacy of both the citizens and the companies [11].

On the other hand, social challenges are critical as smart mobility can create job loss for many. Similar to the logic that states that robots will replace human workers, the introduction of self-driving vehicles is causing truck and taxi drivers to worry as people will not require drivers anymore. Moreover, smart mobility needs smart people who can efficiently handle advanced complex technologies as not all users are equally comfortable using new mobility services [12].

6 Smart Mobility: The Case of Saudi Arabia

Each generation brings innovations that can lead to 'smarter' urban transportation results. These innovations could include direct technological improvements, such as the internal combustion engine (ICE), or they could integrate new approaches to manage the transportation sector, which opens the door to new business models in the world of transportation. Additionally, these advances could enhance mobility stakeholders beyond the traditional public transportation operators, private vehicle owners, and producers. Transport innovations encourage more active, public, and shared transportation, telecommuting, and better use of existing road capacity to address important concerns such as vehicular emissions, transportation cost, travel time, congestion, safety, accessibility, and social equity. Butler et al. developed sustainability evaluation criteria to measure smart mobility systems. Butler et al. identified five main types of smart mobility: Intelligent transport systems, Demand responsive transport, autonomous vehicles, shared mobility, and electric mobility [3].

Smart mobility has started in implementation phases in various cities worldwide such as Singapore, UAE, Netherlands, France, USA, and many more [11]. Singapore aims to become the world's first true Smart Nation by utilizing advanced technologies to the fullest while improving the quality of life, strengthening businesses, and building stronger and bigger opportunities. Singapore is creating smart neighborhoods, such as the Tengah neighborhood, built upon smart planning tools with advanced technologies with smart initiatives to be implemented across the city [13]. Some of these initiatives are presented in Table 1.

The initiatives for implementing the Smart Mobility concept ranged from applying the system in new planned cities to the integration of the system within existing cities. The Netherlands, as an existing classical city, aimed to become one of the leading smart cities worldwide. The Netherlands started implementing the smart mobility concept in 2013, which already led to empowering the developments within the city. The Nether- lands national government has promoted the Netherlands as a frontrunner in smart mobility. Several governmental programs have been initiated to have more efficient and effective approaches to transport issues than traditional measures [15]. The Netherlands initiatives range from traffic information services to multimodal travel planners and from communicating automated vehicles to car-sharing services [16]. The Amsterdam government initiated a project called *AmSmarterdam* to improve the city with advanced technologies such as smart traffic management

Table 1 Smart mobility initiatives [14]

Project description	Country
Intelligent junctions: Smoothing the traffic flows and reducing door-to-door travel times by interactive communication between cars, lorries, public transport vehicles, bicycles, and emergency services with traffic lights and other signals and sensors	The Netherlands
Joint deployment: Motorists receive detailed information on road works over a secure WiFi connection. Cars serve as mobile traffic information sources and pass these data onto traffic control centers. Autonomous vehicles on public roads, following a fixed route, the so-called, WEpods (without a steering wheel or pedals)	The Netherlands
Travel between Ede-Wageningen railway station and the campus	The Netherlands
FREILOT project: Communication between the traffic lights at the busy arterial road and the lorries enabled the system to provide the drivers with speed advice, displaying the remaining time to green	The Netherlands
Connected Traffic Management Center (TMC) and Connected Fleets; Travel Time Reliability as a City Service for Connected Freight; and Safer Pedestrian Crossings for Connected Citizens	Denver, USA
"Smart Spine" corridors in Pittsburgh to improve the connections between isolated neighborhoods and major employment, education, and healthcare centers	Pittsburg, USA

to have a smarter flow, which manages and monitors sensors across Amsterdam to report traffic flow and parking availability. Amsterdam started with a pilot project for the smart parking platform that reduced the average required time to find a parking spot by 43% while reducing the parking cost to help drivers make smarter choices by revealing the cheapest options within a given area [15].

Another example of cities starting to implement smart mobility projects in Croatian cities as the country started different shared mobility projects such as public bike sharing, various ride-sharing apps, and apps for optimizing the usage of on-street and car parks by having smart parking and smart card in public transport [11]. Also, in the US, Denver created different initiatives such as the Connect Traffic Management Center (TMC) and Connected Fleets; Travel Time Reliability as a City Service for Connected Freight; and Safer Pedestrian Crossings for Connected Citizens. Another project in the U.S. called "Smart Spine" corridors in Pittsburgh improved the connections between isolated neighborhoods and major employment, education, and healthcare centers [6].

The smart city concept and smart city six dimensions are deeply rooted in the Saudi 2030 vision as it is clear that Saudi Arabia strives to transform its cities into smart ones, such as Riyadh, Dammam, and Jeddah cities or build new smart cities from scratch such as Neom, that includes the line city, and Oxagon. Saudi Arabia has a clear vision and commitment, and it prepared the required resources to achieve the 2030 vision [17].

For Neom as an example of smart city application Saudi Arabia; they are planning on building the first smart futuristic megacity in Neom development; that introduces a

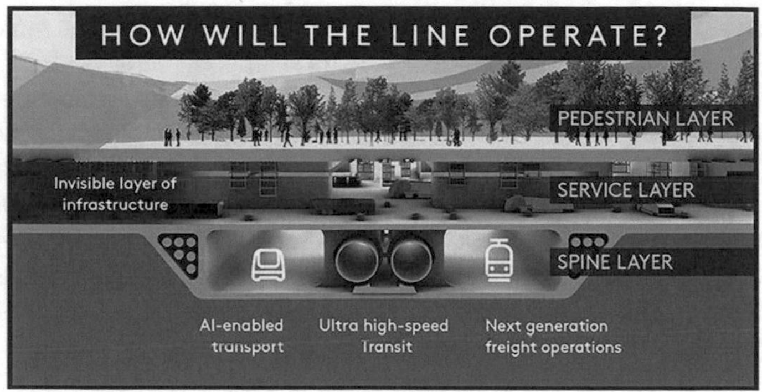

Fig. 2 The line city smart mobility layers. *Source* Neom Official Website, 2022 [18]

new model of urbanization and sustainability. The city is under construction and will be completed by 2030 [17]. Neom aims to provide a unique lifestyle to its residents by giving future technology in transportation, growing and processing food, healthcare, the Internet of things, and digital air as they build a city called The Line built on smart mobility (see Fig. 2). The line city is a 170-km linear city that creates a belt of communities connected without cars or roads, making the city completely free of cars while giving the residents direct access to nature and providing all of their daily needs within walking distance of five minutes [18].

Another example of smart mobility in Saudi Arabia started with SAPTCO governmental company as it started a smart mobility initiative called "Rekab," Which focuses on shared mobility at affordable prices through the use of smartphone applications. The initiative started in late 2021 in limited spaces in the main cities of Saudi Arabia" Riyadh, Jeddah, and Dammam. Saudi Arabia is gradually moving toward smart mobility; for example, Madinah established the first electrical charging unit to charge electric vehicles in March 2022. The main lesson learned of each case study are mentioned in Table 1 which shows a summary of a few examples of smart mobility initiatives [17].

7 Conclusion

Smart cities can be implemented by building a smart city from scratch using smart technologies and smart planning, such as the Tengah neighborhood in Singapore or the Line City in Saudi Arabia, which is built mainly on smart mobility. Alternatively, smart cities can also be implemented by transforming the current cities into smarter cities by plugging-in smart initiatives such as smart traffic systems, parking systems, electric vehicles, and smart shared mobility systems.

The concepts of smart mobility provide a broader horizon for implementing the concept of smart cities, intuitively with the creation of new cities, and in a more effective and flexible way to develop and modernize existing cities and increase their efficiency to cope with the era of digital transformation.

Urban Mobility in the cities has become one of the biggest problems for local municipalities as they started to create sustainable and environmentally acceptable solutions for urban mobility because of the increasing number of private cars, road accidents, congested roads in the traffic network, less public space for people, and economic stagnation. Smart Mobility deals with various past and real-time data; with the help of various advanced technology solutions in each field of transport and traffic science, the possibilities for implementing technology into the transport sector are increasing.

References

1. Albino, V., Berardi, U., Dangelico, R.: Smart cities: definitions, dimensions, and performance. J. Urban Technol. **22**(1), 3–21 (2015). https://doi.org/10.1080/10630732.2014.942092
2. Montes, J.: A historical view of smart cities: definitions, features and tipping points. SSRN Electron. J. (2020). https://doi.org/10.2139/ssrn.3637617
3. Butler, L., Yigitcanlar, T., Paz, A.: How can smart mobility innovations alleviate transportation disadvantages? Assembling a conceptual framework through a systematic review. Appl. Sci. **10**(18), 6306 (2020). https://doi.org/10.3390/app10186306
4. Carrasco-Sáez, J., Butter, M.C., Badilla-Quintana, M.: The new pyramid of needs for the digital citizen: a transition towards smart human cities. Sustainability **9**(12), 2258 (2017). https://doi.org/10.3390/su9122258
5. Benevolo, C., Dameri, R.P., D'Auria, B.: Smart mobility in smart city. Lecture Notes in Information Systems and Organisation, 13–28 (2015). https://doi.org/10.1007/978-3-319-23784-8_2
6. Munhoz, P.A., da Costa Dias, F., Kowal Chinelli, C., Azevedo Guedes, A.L., Neves dos Santos, J.A., da Silveira e Silva, W., Pereira Soares, C.A.: Smart mobility: the main drivers for increasing the intelligence of Urban mobility. Sustainability **12**(24), 10675 (2020).https://doi.org/10.3390/su122410675
7. Karger, E., Jagals, M., Ahlemann, F.: Blockchain for smart mobility—literature review and future research agenda. Sustainability **13**(23), 13268 (2021). https://doi.org/10.3390/su132313268
8. Troisi, O., Kashef, M., Visvizi, A.: Managing safety and security in the smart city: Covid-19, emergencies and smart surveillance. In: Visvizi, A., Troisi, O. (eds.) Managing Smart Cities. Springer, Cham (2022). https://doi.org/10.1007/978-3-030-93585-6_5
9. Lytras, M.D., Visvizi, A.: Information management as a dual-purpose process in the smart city: collecting, managing and utilizing information. Int. J. Inf. Manage. (2021). https://doi.org/10.1016/j.ijinfomgt.2020.102224
10. Arroub, A., Zahi, B., Sabir, E., Sadik, M.: A literature review on smart cities: paradigms, opportunities, and open problems. 2016 International Conference on Wireless Networks and Mobile Communications (WINCOM) (2016). https://doi.org/10.1109/win-com.2016.7777211
11. Brčić, D., Slavulj, M., Šojat, D., Jurak, J.: The role of smart mobility in smart cities. Road Rail Infrastr. V (2018).https://doi.org/10.5592/co/cetra.2018.812
12. Kashef, M., Visvizi, A., Troisi, O.: Smart city as a smart service system: human-computer interaction and smart city surveillance systems. Comput. Hum. Behav. **124**, 106923 (2021). https://doi.org/10.1016/j.chb.2021.106923
13. Singapore's Ministry of Foreign Affairs, 2018, Towards a sustainable and resilient Singapore

14. Khashoggi, A.N.: Measuring Community Perception of Smart Mobility Systems in Saudi Future Cities, M.Sc. thesis of urban design. Effat University, Jeddah, KSA (2022)
15. Manders, T., Klaassen, E.: Unpacking the smart mobility concept in the Dutch context based on a text mining approach. Sustainability **11**(23), 6583 (2019). https://doi.org/10.3390/su11236583
16. Manders, T., Cox, R., Wieczorek, A., Verbong, G.: The ultimate smart mobility combination for sustainable transport? A case study on shared electric automated mobility initiatives in the Netherlands. Transp. Res. Interdis. Perspect. **5**, 100129 (2020). https://doi.org/10.1016/j.trip.2020.100129
17. Doheim, R.M., Farag, A.A., Badawi, S.: Smart city vision and practices across the Kingdom of Saudi Arabia—a review. Smart Cities Issues Challenges, 309–332 (2019). https://doi.org/10.1016/b978-0-12-816639-0.00017-x
18. The line. NEOM. (n.d.). Retrieved April 5, 2022, from https://www.neom.com/en-us/re-gions/whatistheline

How About Value Chain in Smart Cities? Addressing Urban Business Model Innovation to Circularity

Francesca Loia, Vincenzo Basile, Nancy Capobianco, and Roberto Vona

Abstract In the last years, the concept of "smart city" has been the subject of increasing attention in urban planning and governance due to the sustainability challenges and great technological advancements. Although the relevant literature highlights those smart cities can create a fertile environment to drive innovation from a technological, managerial and organizational, and policy point of view how smart cities function, in terms of their value chain and in light of sustainability and circularity is still an open question. Based on these considerations, this work aims to investigate how smart cities deliver value by combining environmental, societal, and financial priorities to re-imagine their core business models and shift the boundaries of urban competition in light of circular economy principles. To reach this goal, the tool Sustainable Business Model Innovation Canvas has been applied in the urban context with the aim to create and leverage an environmental and societal surplus in light of circularity in the smart city framework. Preliminary results shed light on new vistas on the future and new challenges of smart cities particularly relevant for facing the current complexity of the economy, social, and environmental changes.

Keywords Smart city · Sustainability · Circularity · Sustainable business model · Canvas · Innovation · Value chain

1 Introduction

The concept of a "smart city" has been the subject of increasing attention in urban planning and governance due to the sustainability challenges and great technological advancements [5, 12, 17, 25]. Cities and metropolitan areas account for about 60% of Gross Domestic Product, but they are also responsible for 70% of CO_2 emissions and more than 60% of resource use. As part of this transition towards sustainability, the development of smart cities includes creating circular cities with a focus on the well-being of the entire ecosystem.

F. Loia (✉) · V. Basile · N. Capobianco · R. Vona
Department of Economics, Management and Institutions, University of Naples 'Federico II', Naples, Italy
e-mail: francesca.loia@unina.it

A. Visvizi et al. (eds.), *Research and Innovation Forum 2022*,
Springer Proceedings in Complexity, https://doi.org/10.1007/978-3-031-19560-0_19

Although the relevant literature highlights that smart cities can create a fertile environment to drive innovation, how smart cities function, in terms of their value chain, in order to support sustainability and circularity is still an open question. Based on these considerations, this work aims to investigate how smart cities deliver value by combining environmental, societal, and financial priorities to re-imagine their core business models and shift the boundaries of urban competition in favor of circular approaches. This paper, thus, tries to answer the following research question: "How do smart cities shape their business models for combining environmental, societal, and financial priorities in favor of circular approaches?". To reach this goal, the tool Sustainable Business Model Innovation Canvas has been applied in the urban context.

This paper is structured as follows: Sect. 2 frames smart cities as a driver of innovation for sustainability and circularity; Sect. 3 introduces the methodological approach based on an application of the Sustainable Business Model Innovation Canvas; Sect. 4 regards the results; Sect. 5 lists main implications and it also draws conclusions and directions for future research.

2 Literature Review

The concept of "smart city" has been deeply investigated in urban planning and governance [10, 17, 26, 27]. As broadly discusses by the consolidated literature, a smart city can be defined as a conceptual framework that combines key components such as technology (infrastructures of hardware and software), people (creativity, diversity, and education), and institutions (governance and policy) [17].

The ICT infrastructure is related to numerous emerging and streamline technologies, such as the Internet of Things (or IoT) and Big Data, together with cognitive computing, advanced analytics and business intelligence, 5G networks, anticipatory and context-aware computing and advanced distributed data warehouse platforms, which serve as enablers of sophisticated smart city services [9, 12, 14, 20, 28]. The technological evolution makes it possible to gather and analyse data across machines, enabling faster, more flexible, and more efficient processes to produce higher-quality urban services [12, 13, 15, 28].

Smart cities have also emerged as a possible solution to sustainability problems deriving from rapid urbanization, climate change, and population growth [4, 24]. Environmentally oriented definitions, thanks to digital technologies based on platforms and applications which enable the analysis, monitoring, and optimization of urban physical systems (energy, water, waste, transportation, mobility, consumption, and others), can be used to improve the use of resources and decrease emissions. For the right monitoring of these assets, the key tool is the sensor network and the Internet of Things technologies, able to connect the data from sensors to the virtual context [1, 3]. In line with the concept of CE, defined by Pietro-Sandoval et al. [21], more attention is devoted to the implementation of circular principles to decouple

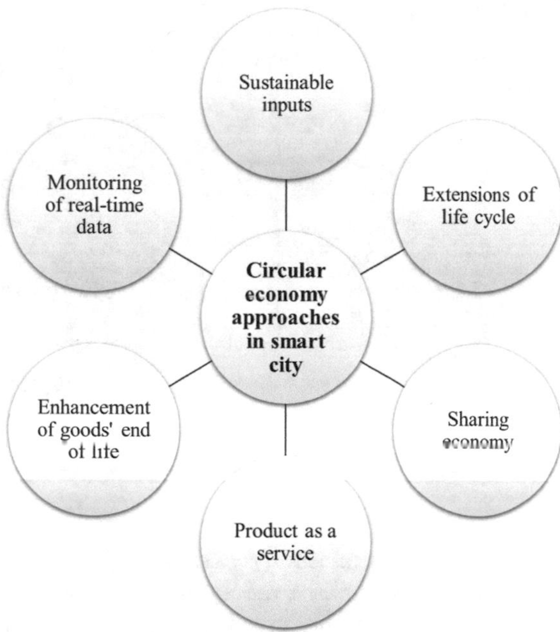

Fig. 1 Circular Economy approaches in smart city. *Source* Authors' elaboration

economic growth in cities from primary resource extraction through reuse, repair, and recycling [16].

In fact, CE principles can determine an urban revolution; they sustain the adoption of sustainable inputs (renewable sources, reuse, recycling), an extension of the useful life of assets and products, adoption of sharing economy initiatives, product as a service, enhancement of the end of life of goods (re-cycling, reuse, upcycling), and monitoring of real-time data through platforms (as shown in Fig. 1). In particular, recent debate has highlighted the need to explore the relation of smart, sustainable, and circular cities more systematically, focusing on practical applications—such as the Business Model tool—that could enable a deeper understanding of the considered domain [28].

3 Methodology

With the aim to describe how smart cities function, in terms of their value chain, for contributing to sustainability and circularity, the Business Models' (BMs) [19] have been investigated. Several business model frameworks have been proposed in the literature, as highlighted by [11]. However, rising environmental evolutions have led to the development of Business Model Innovation (BMI) which is defined as a novel way of creating, delivering, and capturing value, which is achieved through a change of one or multiple components in the business model [18].

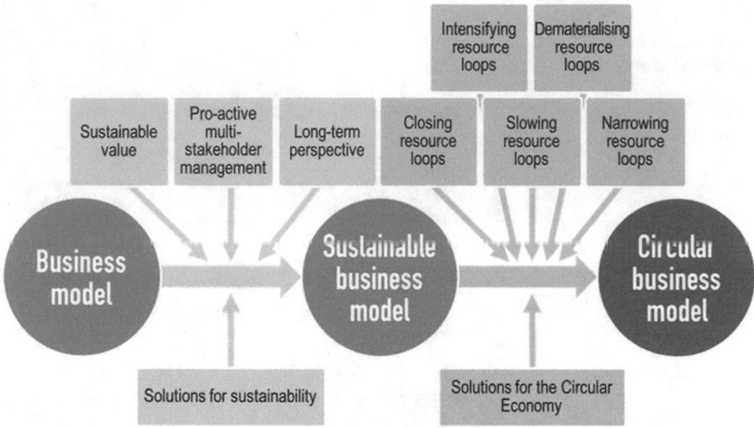

Fig. 2 Sustainable and circular business models. *Source* Geissdoerfer et al. [7]

As CE and sustainability have gained greater attention from governments, industry, and academia, a variety of BMI approaches have been proposed to suit the CE and sustainability principles [2, 7], as shown in Fig. 2. In this context, the alternative concept of the sustainable business model (SBM) has been broadly investigated due to its potential to bring a competitive advantage to organizations by boosting the conventional business models with the aim to meet the sustainable development while maintaining productivity and profitability [23]. Circular business models (CBM), similar to sustainable business models, aim to create, capture, and deliver value to improve resource efficiency by extending the lifespan of products and parts that thereby realizes environmental, social, and economic benefits [22].

In the smart city context, designing business models is essential to systematically assess the most relevant aspects of investment propositions stimulating and fostering communication and engagement of relevant stakeholders who could potentially liaise the replication on large-scale of smart cities solutions. Some studies on sustainable business models for smart city have been carried out by applying the tool of Business Model Canvas [6, 8], however further research efforts are needed for following the principles of the CE.

4 Results and Discussion

In order to answer to the research question, Sustainable Business Model for smart city in light of CE practices has been proposed [22] (Fig. 3).

In the case of the SBM, the ***key partners*** include all the target users, involved in the network of the smart city ecosystem, for which the value in terms of CE is created and whose needs are addressed through a smart city project. Network can

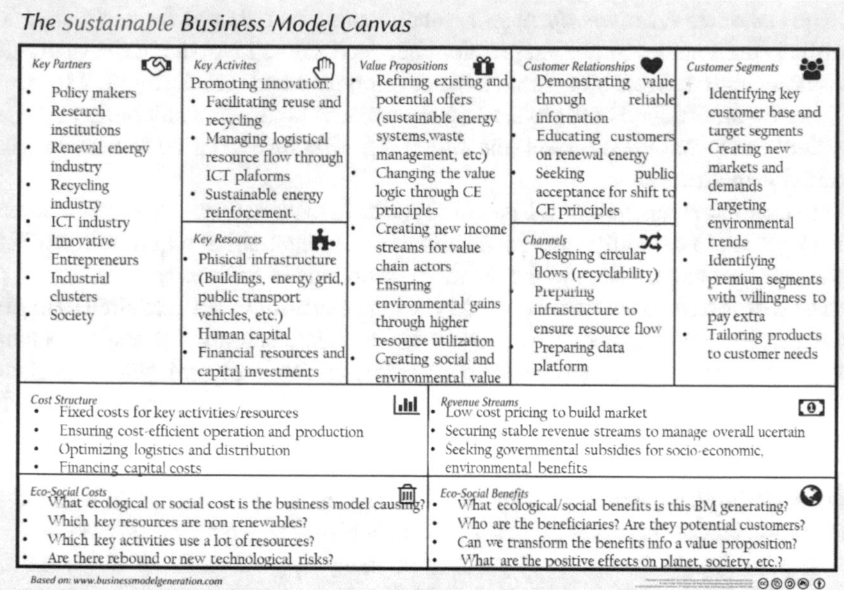

Fig. 3 Sustainable Business Model for smart city in light of circularity. *Source* Authors' elaboration from Giourka et al. [8], Reim et al. [22] and Basile et al. [2]

include community, business and research organizations, decision-making bodies (government), and nonprofit organizations [8, 22].

The ***key activities*** regard the management and delivery of activities by actors involved in the smart city context. The focus is mainly on activities essential to transition to CE aimed at promoting more innovation based on the reuse and recycling approach, resource flow through ICT platforms, and sustainable energy reinforcement.

The ***key resources*** refer to physical assets (which include buildings, energy grid, public transportation vehicles and systems, wireless networks, waste management systems, etc.), human assets (which can consist of the political will toward circularity within smart cities, the creation of a culture of innovation and sustainability between citizens), intellectual assets (which include patents and legal, regulatory frameworks aimed to support circularity and sustainability initiatives), and financial assets (which include access to capital and subsidies and tax incentives to realize smart solutions) [8, 22].

The ***value proposition*** consists of smart city initiatives in terms of CE principles that address the critical need to rapidly reduce emissions and curb climate change. From the perspective of citizens, smart city projects can reduce energy bills and improve air quality. Different value propositions can be derived from the various areas of smart cities, such as mobility, safety, health, energy, water, waste, economic development, housing, and community engagement [8, 22].

The ***customer relationship*** regards relationships that should be established for enabling continuous co-creation practices between actors involved in the smart city ecosystem aimed at reaching sustainable development and circularity [8, 22].

The ***channels*** regard both physical (design of circular flows, availability of infrastructure) and technological asset (the digital infrastructure required for enabling the smart city applications).

The ***customers*** regard the high potential and need to work on creating new markets. It is important to carefully consider customer needs and market requirements and to adapt products and services to the potential customers in smart city.

The ***cost structure*** is related to the key resources and activities required to implement smart city solutions such as fixed costs, i.e., salaries, rents, IoT sensor installation process and services, network/grid infrastructure, technology installations, land, or costs that vary based on incentives or rewards provided to end-users. The SC-BMC helps to identify public and private funding and to facilitate collaboration between multiple actors. Economies of scale can also provide a cost advantage in the case of large-scale implementation of solutions, causing the average cost of investment to decrease as output increases (e.g., sensors recognizing free parking spaces).

CE initiatives in the smart city can offer many opportunities for new ***revenue streams*** for companies. Seeking and obtaining government grants is an important activity to increase the revenue stream. However, companies must demonstrate that CE-related products and services are profitable. Upgrading infrastructure, and technologies and opening up public city data can create new value chains and opportunities in a sustainable way. For example, revenues for private actors can come from user fees (i.e., end users pay for units of energy from renewable sources), or from loans/rents/leases (i.e., granting an investor trade excess energy produced by the PV systems on buildings for a fixed period in return for a fee).

Ecosocial costs refer to the urban operations which have the potential to impact on the environment. In this case, ecosocial costs may be related to bio-physical measures such as CO_2 e emissions, ecosystem impacts, natural resource depletion, human health, political perspectives, non-developmental areas, and pollution.

Ecosocial benefits, on the other hand, allow for better understanding where the organisation's biggest environmental impacts lie within the BM and provide insights for where the organization may focus on creating environmentally-oriented innovations.

5 Conclusions

Nowadays, cities need to provide for the majority of the world's citizens while rapidly decreasing environmental impact and taking up the challenge of circularity.

This work aims to, by applying the managerial tool of the Sustainable Business Model, highlight the smart city potentialities related to circular economy practices. As a result, the value proposition that emerged consists of smart city initiatives based on circular economy principles which respond to the need of improving resource efficiency, reducing emissions, and containing climate change for realizing environmental, social, and economic benefits. Smart cities, therefore, have the opportunity to drive circular economies by encouraging the use of renewable energy, energy saving, sustainable consumption and production, sustainable transportation, conservation of natural resources, and sustainable waste management [21]. Such policies—useful for building resilience in the urban context—are also promoted by Sustainable Development Goal 11 of the 2030 Agenda.

However, the work provides initial understandings that pave the way for additional studies for a deeper understanding of the issue. In this sense, the partial nature of the research does not allow us to generalize, although the insights from this preliminary exploratory study may offer a useful stimulus for future qualitative studies.

References

1. Arasteh, H., Hosseinnezhad, V., Loia, V., Tommasetti, A., Troisi, O., Shafie-khah, M., Siano, P.: Iot-based smart cities: a survey. In: 2016 IEEE 16th International Conference on Environment and Electrical Engineering (EEEIC), pp. 1–6. IEEE (2016)
2. Basile, V., Capobianco, N., Vona, R.: The usefulness of sustainable business models: analysis from oil and gas industry. Corp. Soc. Responsib. Environ. Manag. **28**(6), 1801–1821 (2021)
3. Chui, K.T., Lytras, M.D., Visvizi, A.: Energy sustainability in smart cities: artificial intelligence, smart monitoring, and optimization of energy consumption. Energies **11**(11), 2869 (2018)
4. Ciasullo, M.V., Troisi, O., Grimaldi, M., Leone, D.: Multi-level governance for sustainable innovation in smart communities: an ecosystems approach. Int. Entrep. Manag. J. **16**(4), 1167–1195 (2020)
5. Dameri, R.P.: Smart city definition, goals and performance. In: Smart City Implementation, pp. 1–22. Springer, Cham (2017)
6. Díaz-Díaz, R., Muñoz, L., Pérez-González, D.: Business model analysis of public services operating in the smart city ecosystem: The case of SmartSantander. Futur. Gener. Comput. Syst. **76**, 198–214 (2017)
7. Geissdoerfer, M., Vladimirova, D., Evans, S.: Sustainable business model innovation: a review. J. Clean. Prod. **198**, 401–416 (2018)
8. Giourka, P., Sanders, M.W., Angelakoglou, K., Pramangioulis, D., Nikolopoulos, N., Rakopoulos, D., Tzovaras, D.: The smart city business model canvas—a smart city business modeling framework and practical tool. Energies **12**(24), 4798 (2019)
9. Kashef, M., Visvizi, A., Troisi, O.: Smart city as a smart service system: human-computer interaction and smart city surveillance systems. Comput. Hum. Behav. **124**, 106923 (2021)
10. Komninos, N., Schaffers, H., Pallot, M.: Developing a policy roadmap for smart cities and the future internet. In: eChallenges e-2011 Conference Proceedings, IIMC International Information Management Corporation, pp. 1–8. IMC International Information Management Corporation (2011)
11. Krumeich, J., Burkhart, T., Werth, D., Loos, P.: Towards a component-based description of business models: a state-of-the-art analysis (2012)
12. Lytras, M.D., Visvizi, A.: Who uses smart city services and what to make of it: toward interdisciplinary smart cities research. Sustainability **10**(6), 1998 (2018)
13. Lytras, M.D., Visvizi, A., Sarirete, A.: Clustering smart city services: perceptions, expectations, responses. Sustainability **11**(6), 1669 (2019)
14. Lytras, M.D., Visvizi, A., Jussila, J.: Social media mining for smart cities and smart villages research. Soft. Comput. **24**(15), 10983–10987 (2020)

15. Lytras, M.D., Visvizi, A., Chopdar, P.K., Sarirete, A., Alhalabi, W.: Information management in smart cities: turning end users' views into multi-item scale development, validation, and policy-making recommendations. Int. J. Inf. Manage. **56**, 102146 (2021)
16. MacArthur, E.: Towards the circular economy. J. Ind. Ecol. **2**(1), 23–44 (2013)
17. Nam, T., Pardo, T.A.: Conceptualizing smart city with dimensions of technology, people, and institutions. In: Proceedings of the 12th Annual International Digital Government Research Conference: Digital Government Innovation in Challenging Times, pp. 282–291 (2011)
18. Osterwalder, A., Pigneur, Y.: Business Model Generation: a Handbook for Visionaries, Game Changers, and Challengers, Vol. 1. Wiley (2010)
19. Osterwalder, A., Pigneur, Y., Tucci, C.L.: Clarifying business models: origins, present, and future of the concept. Commun. Assoc. Inf. Syst. **16**(1), 1 (2005)
20. Polese, F., Troisi, O., Grimaldi, M., Loia, F.: Reinterpreting governance in smart cities: an ecosystem-based view. In: Smart Cities and the un SDGs, pp. 71–89. Elsevier (2021)
21. Prieto-Sandoval, V., Jaca, C., Ormazabal, M.: Towards a consensus on the circular economy. J. Clean. Prod. **179**, 605–615 (2018)
22. Reim, W., Parida, V., Sjödin, D.R.: Circular business models for the bio-economy: a review and new directions for future research. Sustainability **11**(9), 2558 (2019)
23. Schaltegger, S., Lüdeke-Freund, F., Hansen, E.G.: Business models for sustainability: a co-evolutionary analysis of sustainable entrepreneurship, innovation, and transformation. Organ. Environ. **29**(3), 264–289 (2016)
24. Toli, A.M., Murtagh, N.: The concept of sustainability in smart city definitions. Front. Built Environ. **6**, 77 (2020)
25. Troisi, O., Fenza, G., Grimaldi, M., Loia, F.: Covid-19 sentiments in smart cities: the role of technology anxiety before and during the pandemic. Comput. Hum. Behav. **126**, 106986 (2022)
26. Visvizi, A., Lytras, M.D.: Rescaling and refocusing smart cities research: From mega cities to smart villages. J. Sci. Technol. Policy Manag. (2018)
27. Visvizi, A., Lytras, M.D.: Sustainable smart cities and smart villages research: rethinking security, safety, well-being, and happiness. Sustainability **12**(1), 215 (2019)
28. Visvizi, A., Lytras, M.D., Damiani, E., Mathkour, H.: Policy making for smart cities: Innovation and social inclusive economic growth for sustainability. J. Sci. Technol. Policy Manag. (2018).

The Role of Social Platform in the Constitution of Smart Cities: A Systematic Literature Review

Ciro Clemente De Falco and Emilia Romeo

Abstract The diffusion of social networks has made it possible to utilize human users as sensors to inspect the city environment and human activities. The increasing availability of data and the possibility of exploiting them with analytics has sparked an interesting discussion: how social media can provide valuable information for smart city planners? Such insights, related to economic activities, sustainable city design, environmental impacts, and responses to climate change, may contribute to the improvement in the quality of human life within a smart city? Thus, this paper aims to perform a Systematic Literature Review regarding the data collected and analyzed from social media and their use within smart cities, examining the extent to which this particular data source is considered in the current literature and how it is characterized in terms of benefits and challenges. The insights derived from the critical review of the literature point out gaps, which may inform future research and theory development in this area and support practitioners' decision-making on the reconfiguration of smart cities.

Keywords Social media · Smart city · Twitter · Literature review

1 Introduction

In the last twenty years, the "smart city" concept has received growing attention in urban planning and governance [1–3], becoming one of the main cornerstones of development for a contemporary society capable of strategically solving (or mitigating) problems normally generated by rapid urbanization. As discussed in the literature, a smart city can be defined as complex sets of technology, people, and institutions [3]. Townsend defines the smart city as where ICTs are combined with infrastructure,

C. C. De Falco
University of Naples Federico II, Naples, Italy
e-mail: ciroclemente.defalco@unina.it

C. C. De Falco · E. Romeo (✉)
University of Salerno, Fisciano, Salerno, Italy
e-mail: eromeo@unisa.it

A. Visvizi et al. (eds.), *Research and Innovation Forum 2022*,
Springer Proceedings in Complexity, https://doi.org/10.1007/978-3-031-19560-0_20

architecture, everyday objects, and even bodies to solve social, economic, and environmental problems [4]. Giffinger and Gudrun [5] then identified a starting point model for a smart city that lists its six constituent dimensions: 1. Smart mobility 2. Smart economy 3. Smart environment 4. Smart governance 5. Smart people 6. Smart living. Accordingly, the transformation of urban centres into smart cities and better management relies on the virtuous synergy between all these dimensions, on efficient networks and infrastructure and direct participation of citizens through new digital channels. As part of city governing processes, it is critical to monitor and critically analyse human activities for strategic planning for better decision making and to meet emerging challenges in city governance [6]. Not surprisingly, different have been developed to follow cities' evolution in the different dimensions that characterize a smart city. For example, standard 37,120 from the International Organization for Standardization (ISO) proposes 19 subject areas (see Fig. 3) to assess urban sustainability. The authors of ISO 37120 argue that cities will achieve several benefits collecting data regarding the listed indicators, such as more effective governance, better service provision, and better decision-making processes [7]. Moreover, the Smart City as a urban management concept implies using ICT and big data [8]. Accordingly, different stakeholders, such as city planners or governing authorities, could leverage the massive increase in human data generation for intelligent processing of the information flows. Social media, web platforms and smartphone apps, works as data sources, recording every aspect of life and producing heterogeneous data on a large scale. This proliferation offers new opportunities to monitor and analyse information about human activity within the smart cities. For instance, Mora and colleagues [9] studied social media as an example of the evolution of citizen communication habits to generate insights for developing citizen-centred smart services and policies. Despite demonstrating that new analytical tools could assist smart city governance, no study so far has provided a comprehensive description of the areas in which online human-generated data through social media could be leveraged for smart cities management [10]. Therefore, in line with the need to manage critical issues in the administration of smart cities through governance and citizens' inclusion [11, 12], this study intends to conduct an exploration of the literature regarding the use of data extracted from human activities on social media using the ISO 37120 standard as a useful component of a multi-method urban analysis [13] to understand the dimensions positively impacted from the use of social media in smart cities.

Thus, the paper aims to answer the following research questions:

RQ1 What is the role of social media in constructing a smart city?

RQ2 What are the domains in which data extracted from social media are used in Smart cities?

RQ3 How does these data impact the domains identified by the ISO 37120 standard?

2 Methodology

A Systematic Literature Review (SLR) based on content analysis is performed to answer research questions. Drafted from Tranfield, Easterby-Smith and their colleagues [14, 15] the steps undertaken to conduct the SLR were the following: (1) planning the review, (2) identifying and evaluating a sample of relevant studies, (3) extracting and synthesizing data, (4) reporting, (5) utilizing the findings (Fig. 1). To cover the issue, Scopus and Web of Science have been used for the paper research according to their extensiveness and relevance. We analysed only scientific papers to obtain the most relevant literature in databases. With the keywords related to the smart city and social media we defined this search string: ((“smart city” OR “smart cities”) AND (“social media” OR “social network” OR “Weibo” OR “Facebook” OR “Twitter” OR “Tiktok” OR “Instagram” OR “Reddit” OR “WeChat” or “Whatsup”)). The search produced 517 articles (746 articles, of which 229 were duplicates). Finally, to answer the research questions, we read the abstracts/articles to understand the use of social media, and we used keyword network analysis to understand the application areas of social media. Then, the several uses found in the literature have been categorized according to ISO 37120 indicators to bring out the best practices to be implemented for better and sustainable management of smart cities.

Phase	Description	Activities	Output
1	Planning the review	Forming a review panel and mapping the field of study	Authors searched Scopus and Web of Science, for the most relevant papers regarding the theme. So, conference proceedings, book chapters and white papers were excluded
2	Identifying & Evaluating Studies	Conducting a systematic search	To identify the relevant sources, authors use the following search string: ((“smart city” OR “smart cities”) AND (“social media” OR “social network” OR “weibo” OR “facebook” OR “twitter” OR “tiktok” OR “instagram” OR “reddit” OR “wechat” OR “whatsapp”))
3	Extracting & Synthesising Data	Conducting data extraction	Studies were abstracted, coded, and assembled to address the research questions and to highlight the areas in which the data extracted from social media are used to build a smart city
		Conducting data synthesis	Authors have conducted the synthesis of the studies searched in the following way: a) keyword network analysis to understand the application areas of social media; b) read the abstracts/articles to understand the different kind of use made of social media
4	Reporting	Reporting the findings	Thanks to the analysis result it is possible: 1) to understand the application areas of social media 2) to understand the different uses made of social media
5	Utilising the Findings	Informing research and practice	The authors discuss the usefulness of social media in building smart cities

Fig. 1 Literature review phases

3 Findings

3.1 The Role of Social Media in Building Smart Cities

The review shows that social media have multiple roles in constructing smart cities (RQ1). These roles are intertwined with the broader transformations that are affecting our contemporary society, such as datafication. Datafication [16] refers to the process by which subjects, objects, and practices are transformed into digital data. The increasing mediatization and digitalization of societal discourses and social interaction are accelerating the datafication process making social media a valuable data source where our digital traces are stored. This characteristic makes social media a source of helpful information for constructing a smart city [17]. Multiple systems have been developed to collect and analyze user contents to produce valuable information for the public and the private sector. It should be noted that social media are not only a source of textual material but also a source of geolocalized data. In fact, user-Generated Content (UGC) often has a geographic reference embedded, which transforms social media into a big warehouse of spatial data [18]. The analyzed papers show how this data can be helpful in ordinary and extraordinary urban land management. Furthermore, social media can be considered relational infrastructures that allows different types of interaction at several levels of interaction. Interaction between human actors and between human actors and institutions is very useful, for example, to increase trust and thus foster the development of public or private services. In addition, social networks bring together (a) human actors and (b) non-human actors. These two other levels of interaction follow the development of algorithms, and the last one is the result of the research related to the topic of the Social Internet of things (SIoT). The SIoT is defined as an Internet of things where things can establish social relationships with other objects autonomously with respect to humans [19].

So, in the construction of smart cities, social media can play an important role both as a source of data and as a relational infrastructure. Concerning the management of the smart city, while, on the one hand, social media can promote the transparency of institutions and their democratic nature, on the other hand, they are also powerful means of control, in fact, in some articles, the central theme is privacy and how it can be violated. Then, the analysis focused on a) the subjects that can use the potential of social media, and b) which are the most used. However, the potential of social media can be exploited by public institutions, private companies, and citizens for different purposes; while the most widely used social media in the West seems to be Twitter. This is not by chance, however, the result of Twitter's data management policies. Twitter, unlike other platforms, allows much easier access to data. It is no coincidence that Twitter is also a source of data for social scientists because of the possibility of accessing large quantities of information, even free of charge.

3.2 Main Domains of Social Media Data

To answer the RQ2, a network analysis useful to identify specific research areas [20] through the analysis of co-occurrences of the articles' keywords was carried out.

The co-occurrence analysis result has been organized and displayed in a reticular form (Fig. 1), on which we run the cluster analysis to identify keywords groups in the network. That groups constitute the fundamental topics of the research field under examination [21] regarding smart cities and their construction through the analysis enabled by social media.

The software used for keyword processing was VOSviewer [22].

In the first cluster, studies are collected regarding the use of geolocated data extracted from social media that provides to city planners insights to make decisions regarding effective smart city development. Rossi and colleagues [23] studied how to model geographical information for the next destination prediction in human mobility. This kind of analysis finds several applications in smart cities scenarios, from optimizing the efficiency of electronic dispatching systems to predicting and reducing traffic jams. Saldana and colleagues [24] elaborated an approach to forecast possible future traffic events in the city, analyzing crowd-sensed data collected on Twitter from a geospatial perspective. The study of Li and colleagues [25] demonstrates the usefulness of understanding citizen activities based on their geolocated activity storylines.

The second cluster collects work that focuses on the impact of social media on citizen participation. From the study of Mangialardi and colleagues [26] emerged that social media facilitates value creation, strengthens the connections between culture and local development, and enables participatory governance of archaeological heritage.

Prandi and collegues [27] find that travel planners and mobile applications related to cultural heritage can play an interesting role in developing smart cities when they are integrated, engaging the user in touristic and entertainment activities, making them a source of cultural resources. Vikas [28] studied the use of social media in natural disasters, deducing that for effective public participation in governance through social media, proactive government intervention is necessary. Hojda and Martins [29] stated that social media increase social participation and urban resilience within smart cities, improve local management, minimize urban problems, save lives during emergencies, and contribute to a better quality of life. Social media can deliver new and interactive contact channels between the government and society, contributing directly to social empowerment in the city's operational management. In their study, citizens were able to use these communication channels to dialogue with the city logistic managers, be well informed about emergencies, and report urban problems. Also, Díaz-Díaz and Pérez-González [30] proved that virtual social media are effective tools for civil society, as they can set the political agenda and influence the framing of political discourse; however, they should not be considered the main channel for citizen participation (Fig. 2).

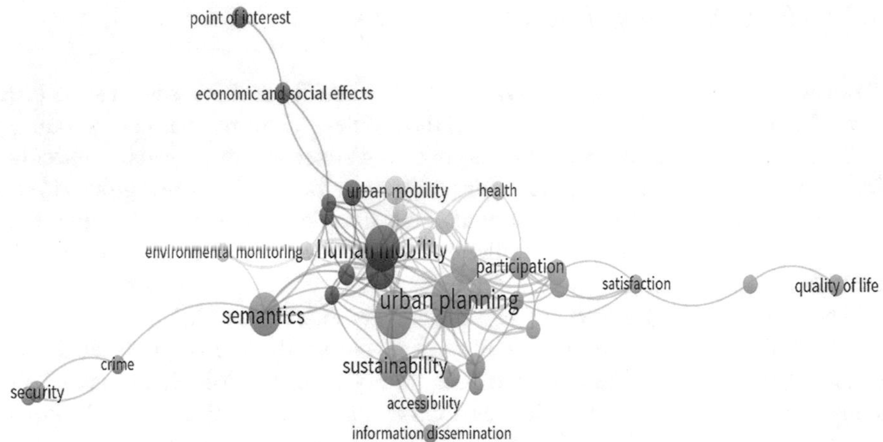

Fig. 2 Keyword's network

This cluster regards the uses of social media data in the tourism sector and their impact on the city's management.

Encalada and colleagues [31] identified, thanks to data collected from social networks, the most popular places from tourists' point of view. Their work showed how this data could allow decision-makers to imagine new ways of planning and managing cities towards a sustainable 'smart' future. Femenia and colleagues [32] showed how tourists' roles and experiences mediated by social media are decisive in smart destinations. The authors stated that destination management organisations, in their policies and actions, should consider the privacy issues derived from the mobile devices' use. Mora and colleagues [9], instead, in the context of city management, investigated the possibilities offered by social networks in sport. This research describes a systematic method to find the most popular areas for sports and to improve, then, the decision-making process of urban planning.

In the last cluster, it is possible to find environmental management works, from the air quality to urban monitoring. Candelieri and colleagues [33] studied how social media data—integrated into a collaborative decision support network—aims to support water and air quality management and assess the impact on health. On the same line, Terroso e Cecilia [34] demonstrated how social media data are used to optimize the taxis' journeys to improve road traffic pollution, which is one of the main factors affecting urban air quality. Di Dio and colleagues [35] stated that social media platforms became effective tools for engaging citizens in environmental conscious mobility habits. Then, Thakuriah and colleagues [36] highlighted the importance of purposefully designed multi-construct and multi-instrument data collection approaches to respond to complex urban challenges: their work highlights the need to link social media analyses to policy frameworks to translate data into impactful decision-making.

Themes	Social media and smart cities research
Economics	
Education	
Energy	
Environment and climate change	
Finance	
Governance	x
Health	x
Housing	
Population and social conditions	
Recreation	x
Solid Waste	
Safety	x
Sport and Culture	x
Telecommunications	x
Transportation	x
Urban/local agriculture and food security	
Urban Planning	x
Wastewater	
Water	x

Fig. 3 ISO 37120 core and sample research

3.3 *Social Media Impact Within Smart Cities*

To address the RQ3 and to highlight the areas in which social media can have a positive impact within the smart city context, we considered both the results of the content analysis (RQ1) and the results of the analysis of co-occurrences (RQ2) to enumerate the areas less explored in the literature regarding the topic in question. From the association of the themes extracted from the articles in the sample and the indicators that make up the ISO 37120 standard, it emerges that the themes that have been studied the most concern the following indicators: Transportation and Urban Planning, followed by Recreation, Governance, Sport and Culture and Health (Fig. 3). Accordingly, it is possible to consider these resultant dimensions the most impacted by social media use.

4 Discussion and Conclusion

The SLR highlighted the fields of smart cities on which the social media data impacted and how the literature investigated this phenomenon.

The content analysis results suggested a series of theoretical-conceptual results through the keywords' co-occurrence analysis (Fig. 2), and the collected data on themes studied and dimensions impacted.

Regarding the RQ1, the findings have shown that (1) social media act as a database of both valuable textual data and geolocalized ones in the construction of the smart city; and (2) the social infrastructure that they represent is able to connect not only people but also human and non-human actors. However, it is important to pay attention to the dark side of social media, which can be used as a powerful means of control. Harmonizing and controlling this aspect of social media should be a subject for future reflection and research.

From the thematic analysis (RQ2) 4 clusters emerged: the first deal with the general use of geolocated data extracted from social networks; the second collects works on citizen participation; the third regards the concepts of smart tourism and the consequent urban planning resulting from it, while the fourth deal with the environmental monitoring applied in different contexts.

Then, thanks to the comparison between the core themes of the ISO 37120 and the themes most discussed in the literature (RQ3), it was possible to find the areas in which the literature has to focus both to improve the knowledge about this interesting topic and to better qualified the ways to manage and build smart cities in the future.

This paper provides a synthesis of the current conceptual and empirical literature on the use of social media in the management of smart cities. The first theoretical contribution of this study pertains to identifying the current state of the art of social media impacts and the way researchers have investigated these issues. Another implication of the review lies in harmonizing the existing knowledge in four macro-areas that underline the main stressed topics in literature. Moreover, the cluster analysis shows that little attention has been given to the use of social media for the social sustainability of smart cities.

Furthermore, the use of the ISO to categorize the areas of impact of social media data in improving the management of smart cities is a departure point to map the gaps present in the literature and then to develop a research agenda. Regarding the practical implications, this study could help policymakers to adopt good practices found in literature to better manage the smart cities.

For example, citizens' participation could potentially increase if policy makers use social media platforms to raise citizens' awareness about various issues. It is no coincidence that during the Covid-19 pandemic some institutional communications took place on social media, and on social media public figures were engaged by institutions to increase awareness and thus resilience.

Moreover, regarding the resilience concept, social media monitoring systems could help people in emergencies or disasters. However, in this work, the social participation remains an unanswered question because there weren't case studies aimed at increasing citizen cohesion and participation in social activities in smart cities. However, this could be a starting point for further investigation to under-stand why, nowadays, smart city development seems do not have a positive im-pact on this dimension.

However, the paper shows some limitations. For example, the results need to be updated with works that will be published in the near future and it will be interesting to understand if the trends highlighted in this article could radically changes or not.

References

1. Nam, T., Pardo, T.A.: Smart city as urban innovation: Focusing on management, policy, and context. In: Proceedings of the 5th International Conference on Theory and Practice of Electronic Governance, pp. 185–194, ACM Digital Library (2011)
2. Visvizi, A., Lytras M.: Smart Cities: Issues and Challenges: Mapping Political, Social And Economic Risks and Threats. Elsevier-US: New York, NY, USA (2019).https://doi.org/10.1016/C2018-0-00336-9
3. Visvizi, A., Lytras, M.D., Damiani, E., Mathkour, H.: Policy making for smart cities: Innovation and social inclusive economic growth for sustainability. J. Sci. Technol. Policy Manag. **2**(9), 126–133 (2018)
4. Townsend, A.M.: Smart Cities: Big Data. Civic Hackers and the Quest for a New Utopia. WW Norton, New York (2013)
5. Giffinger, R., Gudrun, H.: Smart cities ranking: an effective instrument for the positioning of the cities. ACE: Archit. City Environ. **IV**(12), 7–25 (2010)
6. De Guimarães, J.C.F., Severo, E.A., Júnior, L.A.F., Da Costa, W.P.L.B., Salmoria, F.T.: Governance and quality of life in smart cities: Towards sustainable development goals. J. Clean. Prod. **253**, 119926 (2020)
7. McCarney, P.: Building high calibre city data. Econ. Dev. J. **16**(2), 7–17 (2017)
8. Visvizi, A., Troisi, O.: Effective Management of the Smart City: an outline of a conversation. In: Visvizi, A., Troisi, O. (eds.) Managing Smart Cities. Springer, Cham (2022). https://doi.org/10.1007/978-3-030-93585-6_1
9. Mora, H., Pérez-delHoyo, R., Paredes-Pérez, J.F., Mollá-Sirvent, R.A.: Analysis of social networking service data for smart urban planning. Sustainability **10**(12), 4732 (2018)
10. Li, Z., Zhu, S., Hong, H., Li, Y., El Saddik, A.: City digital pulse: a cloud based heterogeneous data analysis platform. Multimedia Tools Appl. **76**(8), 10893–10916 (2017)
11. Kitchin, R.: Making sense of smart cities: addressing present shortcomings. Camb. J. Reg. Econ. Soc. **8**(1), 131–136 (2015)
12. Castelnovo, W., Misuraca, G., Savoldelli, A.: Smart cities governance: The need for a holistic approach to assessing urban participatory policy making. Soc. Sci. Comput. Rev. **34**(6), 724–739 (2016)
13. Mair, S., Jones, A., Ward, J., Christie, I., Druckman, A., Lyon, F.A.: Critical review of the role of indicators in implementing the sustainable development goals. In: Filho, W.L. (eds.) Handbook of Sustainability Science and Research, pp 41–56. Springer Nature, Cham (2018)
14. Easterby-Smith, M., Thorpe, R., Jackson, P.: Management Research. SAGE Publications, London (2012)
15. Tranfield, D., Denyer, D., Smart, P.: Towards a methodology for developing evidenceinformed management knowledge by means of systematic review. Br. J. Manag. **14**(3), 207–222 (2003)
16. Southerton C.: Datafication. In: Schintler L., McNeely C. (eds.) Encyclopedia of Big Data. Springer, Cham (2020)
17. Lytras, M.D., Visvizi, A., Chopdar, P.K., Sarirete, A., Alhalabi, W.: Information Management in Smart Cities: Turning end users' views into multi-item scale development, validation, and policy-making recommendations. Int. J. Inf. Manage. **56**, 102–146 (2021)
18. Elwood, S., Goodchild, M.F., Sui, D.Z.: Researching volunteered geographic information: Spatial data, geographic research, and new social practice. Ann. Assoc. Am. Geogr. **102**, 571–590 (2012)
19. Atzori, L., Iera, A., Morabito, G., Nitti, M.: The social internet of things (siot)–when social networks meet the internet of things: Concept, architecture and network characterization. Comput. Netw. **56**(16), 3594–3608 (2012)
20. López-Fernández, M.C., Serrano-Bedia, A.M., Pérez-Pérez, M.: Entrepreneurship and family firm research: a bibliometric analysis of an emerging field. J. Small Bus. Manage. **54**(2), 622–639 (2016). https://doi.org/10.1111/jsbm.12161

21. Manesh, M.F., Pellegrini, M.M., Marzi, G., Dabic, M.: Knowledge management in the fourth industrial revolution: mapping the literature and scoping future avenues. IEEE Trans. Eng. Manage. **68**(1), 289–300 (2020)
22. Van Eck, N.J., Waltman, L.: Citation-based clustering of publications using CitNetExplorer and VOSviewer. Scientometrics **111**(2), 1053–1070 (2017)
23. Rossi, A., Barlacchi, G., Bianchini, M., Lepri, B.: Modelling taxi drivers' behaviour for the next destination prediction. IEEE Trans. Intell. Transp. Syst. **21**(7), 2980–29891 (2019)
24. Saldana-Perez, M., Torres-Ruiz, M., Moreno-Ibarra, M.: Geospatial modeling of road traffic using a semi-supervised regression algorithm. IEEE Access **7**, 177376–177386 (2019)
25. Miah, S.J., Vu, H.Q., Alahakoon, D.: A social media analytics perspective for human-oriented smart city planning and management. J. Am. Soc. Inf. Sci. **73**(1), 119–135 (2022)
26. Mangialardi, G., Corallo, A., Esposito, M., Fortunato, L., Monastero, A., Schina, L.: An integrated and networked approach for the cultural heritage lifecycle management. ENCATC J. Cult. Manage. Policy **6**(1), 80–95 (2016)
27. Prandi, C., Melis, A., Prandini, M., Delnevo, G., Monti, L., Mirri, S., Salomoni, P.: Gamifying cultural experiences across the urban environment. Multimedia Tools Appl. **78**(3), 3341–3364 (2019)
28. Vikas: ICT and disaster management: a study of the social media use in 2015 Chennai city floods. Int. J. Public Administr. the Digital Age **4**(3), 29–41 (2017)
29. Hojda, A., Martins, P.R.: Communication as a Tool for Expanding Social Participation: The Case of the Rio Operations Center. In: Azeiteiro, U.M., Akerman, M., Leal Filho, W., Setti, A.F.F., Brandli, L.L. (eds.) Lifelong Learning and Education in Healthy and Sustainable Cities, pp. 521–537. Springer, Cham (2018)
30. Díaz-Díaz, R., Pérez-González, D.: Implementation of social media concepts for e-government: case study of a social media tool for value co-creation and citizen participation. J. Organ. End User Comput. (JOEUC) **28**(3), 104–121 (2016)
31. Encalada, L., Boavida-Portugal, I., Cardoso Ferreira, C., Rocha, J.: Identifying tourist places of interest based on digital imprints: towards a sustainable smart city. Sustainability **9**(12), 2317 (2017)
32. Femenia-Serra, F., Perles-Ribes, J.F., Ivars-Baidal, J.A.: Smart destinations and tech-savvy millennial tourists: hype versus reality. Tourism Rev. **74**(1), 63–81 (2019)
33. Candelieri, A., Archetti, F., Giordani, I., Arosio, G., Sormani, R.: Smart cities management by integrating sensors, models and user generated contents. WIT Trans. Ecol. Environ. **179**, 719–730 (2013)
34. Terroso-Saenz, F., Muñoz, A., Cecilia, J.M.: QUADRIVEN: a framework for qualitative taxi demand prediction based on time-variant online social network data analysis. Sensors **19**(22), 4882 (2019)
35. Di Dio, S., La Gennusa, M., Peri, G., Rizzo, G., Vinci, I.: Involving people in the building up of smart and sustainable cities: how to influence commuters' behaviors through a mobile app game. Sustain. Cities Soc. **42**, 325–336 (2018)
36. Thakuriah, P.V., Sila-Nowicka, K., Hong, J., Boididou, C., Osborne, M., Lido, C., McHugh, A.: Integrated multimedia city data (iMCD): a composite survey and sensing approach to understanding urban living and mobility. Comput. Environ. Urban Syst. **80**, 101427 (2020)

Smart Cities at Risk: Tech Breakthrough or Social Control. Chinese Case Study

Marina S. Reshetnikova, Galina A. Vasilieva, and Ivan A. Mikhaylov

Abstract Around the world, cities have become the hub of shaping the post-Covid world by taking inclusive measures in combining economic recovery with environmental sustainability. Many cities have become more active in using various smart city tools due to the crucial role of digitalization. In China, the trend towards the introduction of digital technologies in the processes of urban management is becoming more noticeable than in any other part of the world. The purpose of this article is to analyze and identify trends in the development of smart cities in China, as well as to assess the relationship between technological and human aspects. This article is an empirical study that was carried out to analyze and evaluate the model of urban development in China. The authors focused on cluster development and determined the geographical distribution of Chinese smart cities. The authors concluded that smart cities in China have their own characteristics, which can bring great benefits to citizens and income growth. But the development is uneven, there is a certain imbalance that does not correspond to the original list of intellectual infrastructures and social requirements. The study confirms that technological aspects are more developed than infrastructure, which slows down the sustainable development of smart cities in China. Of course, this opens great opportunities for effective city management and decision-making, but, on the other hand, such efforts to monitor all the actions of city-dwellers raise many concerns for the psychological safety and privacy.

Keywords Smart city · China · Cluster · Covid-19

1 Introduction

Today, the urban population grows from year to year. This trend is a test for cities sustainability and requires development and comprehensive optimization of the city's infrastructure. Innovative solutions have long become an integral part of our daily life.

M. S. Reshetnikova (✉) · G. A. Vasilieva · I. A. Mikhaylov
RUDN University, Miklukho-Maklaya, 6, Moscow, Russian Federation
e-mail: Reshetnikova-ms@rudn.ru

A. Visvizi et al. (eds.), *Research and Innovation Forum 2022*,
Springer Proceedings in Complexity, https://doi.org/10.1007/978-3-031-19560-0_21

Cities tend to create a single digital urban ecosystems and are increasingly resorting to the use of technologies such as the Internet of Things, artificial intelligence (AI), blockchain. During the Covid-19 pandemic, the urgency of this issue has become especially evident, when a variety of smart city tools have helped to control and contain the spread of infection [1–3]. All this suggests that at present, the digitalization of an increasing number of cities is simply inevitable. Therefore, in countries around the world, the number of smart cities is growing, and due to the growing number of innovative solutions, their development models are constantly evolving [4]. Today, China is the undisputed leader in smart cities solutions, using monitoring, big data, and artificial intelligence technologies to simplify the management of cities and public spaces. The government of the country has introduced a national strategy in the field of smart cities and directs public resources to support their growth, which confirms the importance of developing this area. If the things keep going it might happen that almost half of all smart cities in the world soon will be concentrated in China, as today there are 800 pilot smart city programs in the country that are in the planning or implementation stage [5, 6].

With the development of smart cities concept, questions about its ambiguous impact on the lives of citizens are increasingly being raised, since many technologies perform the functions of monitoring and controlling society, which it does not always even know about, thereby penetrating people's private lives [7].

Although in certain cases this can have a positive effect, as in the case of countering the spread of Covid-19, when many countries, primarily China, used these technologies to bring unscrupulous citizens who did not comply with the quarantine to account.

The purpose of the article is to analyze the trends of smart cities in China and identify development patterns, as well as try to answer the question what is a "smart city", progress or a modern tool for controlling society?

2 Literature Review

There have been many definitions of the smart city, but it is typically based on different technological innovations to improve the lives of city-dwellers. The concept of a smart city aims to ensure sustainable growth in the quality of life of citizens [8], to establish a favorable business ecosystem [9], to enable centralized, continuous, and most importantly transparent urban management [10], as well as to improve the efficiency of public spending [11]. That is, a smart city is primarily a city for a people, where all residents participate in its management with the help of AI, digital technologies that have a cross-cutting property in all areas.

While the definition of a smart city is still a matter of debate two dimensions are transversal to multiple definitions. In the literature, this is expressed by the citizen-led or private-led [12] as well as technology-driven [13, 14]. The issue of assessing the relationship between the main dimensions of smart cities also occupies one of the key places in scientific discussions.

Notwithstanding the persisting disagreements around the definition of a smart city, the authors will use the two dimensions to analyze both the different types and conceptions of smart cities and the multiple aspects in which they reinforce old and introduce new ethical problems.

From one hand fast-developing technologies such as AI and Big Data draw a future image of sustainable cities and give huge opportunities for efficient urban management and decision making [15–17]. In contrast the excessive effort to monitor any activity within cities raise multiple concerns about the psychological security and privacy of citizens [7]. Several researchers address the importance of ensuring privacy and security in the smart city context [18, 19], but most of them focused on cyber-security and privacy-related issues related to ICT infrastructures [20].

Beyond the risks to ICT infrastructure and applications, major privacy and psychological concerns arise from the underlying risks of data storage and usage [21]. The problem of the malicious use of AI has not been yet addressed and adequately discussed. Nevertheless, AI's impact extends well beyond individual consumer choices. It is starting to transform basic patterns of urban management, by providing governments and private companies with unprecedented capabilities to monitor citizens and shape their choices.

3 Methods

This research was conducted in the context of smart cities in China. The selection of the cities as the study context is justified as follows: (a) cities included in 19 clusters of smart cities (13th Five Year Plan for Economic and Social Development); (b) the cities are preliminarily completed the smart concept and practices. The research used the following secondary sources to gather and analyze data: (a) online resources were adopted to obtain data from the China Statistical Yearbook [22] and the Yearbook of China Information Industry [23]; (b) the National Economy and Society Developed Statistical Bulletin [24], and the official website of the municipal government. It should be noted that Hong Kong and Taipei were not included in the sample due to methodological features.

We suppose that there are two pillars that shape smart city initiatives and elucidate the impact of governance, communities, economy, natural environment, and infrastructure [25]. The Structures pillar referring to the existing infrastructure of the cities, and the Technology pillar describing the technological provisions and services available to the inhabitants. Each pillar is evaluated over five key areas: health and safety, mobility, activities, opportunities, and governance.

In this study to reduce the dimensionality of large data sets the authors used PCA (Principal Component Analysis) to assess the relationship between the components of smart cities in China by two factors, namely technological and intellectual infrastructure (Tables 1 and 2).

Table 1 Factors for assessing the structures development of smart cities

Pillar	Explanation factors	Operating indices	Unit
Structures	Health and safety	Health expenditure	10^4 yuan
		Air quality standard achievement	%
	Mobility	Passengers in a vehicle	10^4 person-times
		Public transportation vehicles	Number per 10^4 persons
		Network penetration	%
	Activities	Postal and telecommunication services	10^4 yuan
		Science and technology expenditure	10^4 yuan
	Opportunities	High-tech industrial output value	10^4 yuan
		Enterprises in hi-tech industry development zone	number
	Government	Maintaining and building cities expenditure	10^4 yuan

Sources China Statistical Yearbook, China City Statistics Yearbook, Yearbook of China Information Industry

Table 2 Factors for assessing the technologies development of smart cities

Pillar	Explanation factors	Operating indices	Unit
Technologies	Health and safety	Electronic public services categories	Number
	Mobility	Modal split of passenger transport	10^4 person-inland
	Activities	Number of websites hosted	Sites per 10^3 population
	Opportunities	Labor employment in information transmission computer services and software	%
	Government	Cloud and big data service platforms	10^4 yuan

Sources China Statistical Yearbook, China City Statistics Yearbook, Yearbook of China Information Industry

4 Results

The results show that the state of infrastructure and technology in the selected cities are not congruent. If we talk about the existing infrastructure, then we can note a pronounced leader in all five areas—Zhuhai (Fig. 1). The most developed factor is activities, which is expressed in the presence of a sufficient number of green areas, cultural and entertainment facilities. The least developed is mobility, which is characterized by such indicators as the presence of traffic jams, the state of public transport infrastructure. It is also worth noting that the biggest problems in this area are observed in Beijing, Shanghai, and Guangzhou, which are the most significant smart cities in their clusters.

Assessments of technological aspects demonstrate high level of development for all areas in these cities. But we can still see a noticeable shift in the direction of activities, as in the previous chart (Fig. 2). Thus, it allows us to conclude that the assessment of the existing urban infrastructure is lower than the technological aspects in the five areas considered. This means that the authorities of the country are focusing on digitalization and automation of urban life processes rather than infrastructure aspect.

The presence of general trends in the development both in the existing urban infrastructure and in technologies for most cities, which indicates the implementation of a unified policy in the field of smart cities and a problem of mobility.

The Chinese government is focusing on the cluster development of smart cities. In the 13th Five Year Plan for Economic and Social Development (13th Five Year Plan), 19 clusters of super-cities were identified, which will account for about 80% of China's entire economy.

It is no coincidence that the government plans and promotes the cluster development of cities, thus it supports urbanization in the country, the growth rate of which, in turn, directly correlates with the economic growth rate [26]. Clustering stimulates the growth of productivity and investment activity, as the growth in the number of enterprises leads to the growth of the labor market, and the higher the income of the citizens, as well as higher consumption. On the positive side, the clusters combine developed cities and less developed provinces (thus Jing-Jin-Ji City Cluster includes the developed Beijing and Tianjin, as well as the less developed Hebei province),

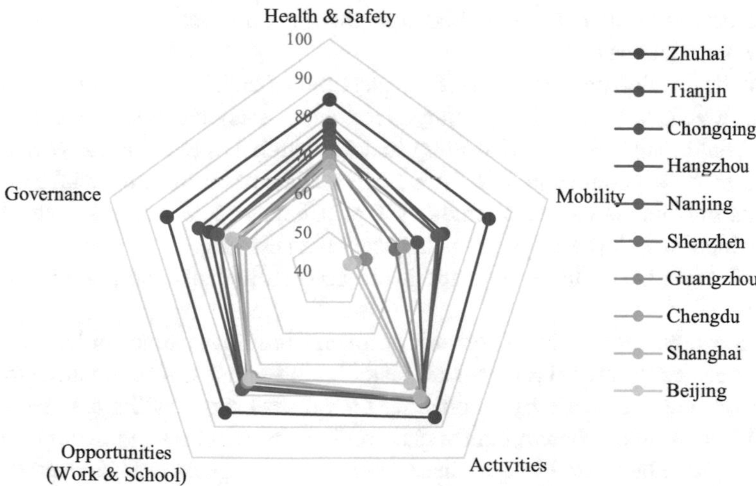

Fig. 1 Assessing the Existing Urban Infrastructure of China's Smart Cities. *Source* compiled by the authors. Data collected from: China Statistical Yearbook, China City Statistics Yearbook, Yearbook of China Information Industry

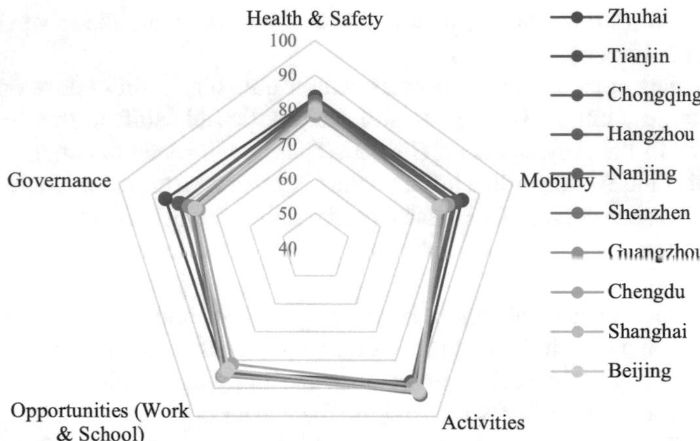

Fig. 2 Assessing the Existing Urban Technologies of China's Smart Cities. *Source* compiled by the authors. Data collected from: China Statistical Yearbook, China City Statistics Yearbook, Yearbook of China Information Industry

which leads to a redistribution of resources. Due to this, the less developed cities of the provinces, which are at the first stages of industrialization, can move up the value chain and abandon the heavy industry that pollutes the environment, and the developed centers also move up the chain and concentrate their activities on innovative industries.

Government priorities are aligned towards 3 main clusters: Jing-Jin-Ji City Cluster, Yangze River Delta City Cluster, Greater Bay Area City Cluster. It is they that, according to the plan, should become the most innovative and competitive at the international level.

With the development of smart cities, questions of their ambiguous impact on the lives of citizens are increasingly being raised, since many technologies perform the functions of monitoring and controlling society, which it does not always even know about, thereby penetrating people's private lives. One of the controversial smart city solutions in China is the social credit system (China's social credit system, SoCS, SCS), which over the past few years has entered a new stage of its development and standardization [27]. This system applies to both individuals and public or private companies.

SoCS can be attributed to a network of diverse initiatives aimed at building trust in Chinese society. Its goal is to make it easier for individuals and companies to make various business decisions by automating the process and creating a single record tracked in real time, where a high social credit score will be an indication that the other party can be trusted in a business context. These goals are achieved through several main mechanisms: data collection and exchange; punishments, sanctions, and rewards; control of "black" (punishing) and "red" (encouraging) lists.

Today, China is one of the most intensive video surveillance countries in the world, the net number of cameras in the country has already reached 8 million. Camcorders are everywhere in China, and their capabilities are increasing every year (Fig. 3). China leads in the number of cameras, overtaking the US, Japan, and European countries. Among all other countries, China also leads in the ratio of the number of cameras per 1,000 people. It is also worth noting that out of the 20 largest cities in the world in terms of the number of CCTV cameras, 15 are in China.

It is not a secret for the citizens of Chinese cities that their social activity is monitored by the government. The authorities have strongly denied that such technologies can be used for "mind control", stating that active video surveillance is part of the smart cities system, which allows them to track and control traffic, as well as respond faster to emergencies. So, for example, surveillance cameras track traffic violations by displaying photos of those who crossed the road in the wrong place. According to the police, such measures reduce the crime rate in the country. Also, there is a slight correlation between the number of cameras and the level of crime: where there are more cameras, crime is lower, but it is important to remember that only one indicator is involved in such statistics, without considering other parameters, which cannot claim absolute reliability. Today, technology has stepped far, and the Covid-19 epidemic has prompted its use more openly. Nowadays, the Chinese authorities do not hide the active use of cameras and biometrics in smart cities development. In addition, they now have a legal basis and a "good" purpose for their use.

So, for example, now there is a practice of installing cameras over the doors of an infected person. Also, at the moment, even if you are healthy and wearing a mask, CCTV cameras are able to recognize your face, informing the authorities about your location. On the one hand, Chinese people agree that a system in which you can easily be told that you were in the same space as an infected person is effective enough and useful in dealing with an emergency, helping to reduce the incidence;

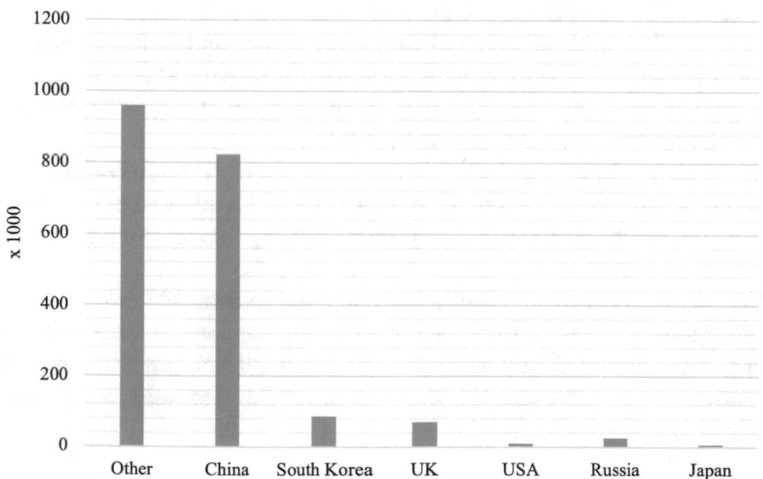

Fig. 3 Surveillance camera statistics by countries (2021). *Source* [28] Bischoff, P.: Surveillance camera statistics: which cities have the most CCTV cameras? Comparitech, https://www.comparitech.com/vpn-privacy/theworlds-most-surveilled-cities/, last accessed 2021/07/22

however, on the other hand, it is inhumane and violates the personal boundaries of the inhabitants.

Summing up, it is worth noting that smart city technological instrument is very ambiguous, on the one hand, the active use of AI, blockchain and IoT has a lot of proven advantages and practical benefits, on the other hand, it is an encroachment on the privacy and mental health of citizens. As for the social credit system, SoCS remains an opaque system due to the huge volume and decentralized storage of information [29, 30]. But Chinese society cannot get away from it due to cultural and historical characteristics. It also represents Xi Jinping's aspirations for a more modern and efficient government, and its development will remain a high priority for years to come. All this allows us to conclude that the line between a smart city, as part of world progress and as a tool for controlling society, is very thin in modern conditions.

Nevertheless, the Chinese authorities are not going to stop, planning to use the possibilities of AI more and more intensively every year, this is what causes the most concern among the citizens of the country.

5 Discussion

Protection of personal psychological security and overall privacy is of extreme importance to individuals. Protecting the privacy of citizens is an important issue in the context of AI solutions proliferation. Such technologies have only recently become widespread and already today they are in our cars, houses, and personal belongings. They are also an important part of facilitating our daily lives through advanced technology in decision making and privacy.

In such conditions, the question of the safety of data and its non-falling into the hands of unauthorized persons for malicious use is even more acute. Of course, there are a huge number of positive aspects of using AI in smart city data collection, but it remains important who has access to this data.

By highlighting the risks and problems of using AI in this area, it becomes possible to point out a wide range of problems that need to be addressed. But it remains unknown whether these issues are being addressed in practice within the framework of smart city projects.

In the conditions of existence of a huge amount of available information, cases of inability to separate truth from fake information become possible. Since during the analysis the collected data array may be biased. There are also concerns that when data analysis is outsourced to corporations, policy decisions may be driven by corporate interests.

6 Conclusion

Both as a solution to traditional urban problems and as new opportunities for future development, smart cities come with their own risks and challenges. As a solution to traditional urban problems, smart cities are often seen to improve the efficiency of urban governance. This approach conceptualizes smart cities mainly in terms of technology and their optimization potential, which, as discussed, can exacerbate the complexity of urban life and its multifaceted challenges. The widespread and excessive introduction of technological solutions can carry the danger and risks of both exacerbating pre-existing problems generated by old solutions and creating new ones primarily related to malicious use of AI.

Technologies based on motion detection, predictive analytics, drones, and other autonomous devices are often introduced in the name of public safety and improving the quality of urban traditional services, as well as optimizing socio-economic costs. However, they may pose a threat and have negative impact to the privacy of citizens, and psychological security. Growing role of private companies shifts from public accountability to privately led projects deserve consistent scrutiny and rectification.

References

1. Inn, T.L.: Smart city technologies take on Covid-19. World Health, 841 (2020)
2. Allam, Z., Jones, D.S.: On the coronavirus (Covid-19) outbreak and the smart city network: universal data sharing standards coupled with artificial intelligence (AI) to benefit urban health monitoring and management. Healthcare **8**(1), 46 (2020)
3. Troisi, O., Fenza, G., Grimaldi, M., Loia, F.: Covid-19 sentiments in smart cities: the role of technology anxiety before and during the pandemic. Comput. Hum. Behav. **126**, 106986 (2022)
4. Byun, J., Kim, S., Sa, J., Kim, S., Shin, Y.T., Kim, J.B.: Smart city implementation models based on IoT technology. Adv. Sci. Technol. Lett. **129**(41), 209–212 (2016)
5. Hao, L., Lei, X., Yan, Z., ChunLi, Y.: The application and implementation research of smart city in China. In: 2012 International Conference on System Science and Engineering (ICSSE), pp. 288–292. IEEE (2012)
6. Shen, L., Huang, Z., Wong, S.W., Liao, S., Lou, Y.: A holistic evaluation of smart city performance in the context of China. J. Clean. Prod. **200**, 667–679 (2018)
7. Ismagilova, E., Hughes, L., Rana, N.P., Dwivedi, Y.K.: Security, privacy, and risks within smart cities: Literature review and development of a smart city interaction framework. Inf. Syst. Front., 1–22. (2020)
8. Toli, A.M., Murtagh, N.: The concept of sustainability in smart city definitions. Front. Built Environ. **6**(77) (2020)
9. Appio, F.P., Lima, M., Paroutis, S.: Understanding smart cities: innovation ecosystems, technological advancements, and societal challenges. Technol. Forecast. Soc. Chang. **142**, 1–14 (2019)
10. Visvizi, A., Troisi, O.: Managing Smart Cities Sustainability and Resilience Through Effective Management, Springer Cham (2022)
11. França, R.P., Monteiro, A.C.B., Arthur, R., Iano, Y.: Smart cities ecosystem in the modern digital age: an introduction. In Data-Driven Mining, Learning and Analytics for Secured Smart Cities, pp. 49–70. Springer, Cham (2021)
12. Cohen, B., Munoz, P.: Sharing cities and sustainable consumption and production: towards an integrated framework. J. Clean. Prod. **134**, 87–97 (2016)

13. Echebarria, C., Barrutia, J.M., Aguado-Moralejo, I.: The smart city journey: a systematic review and future research agenda. Innov. Europ. J. Soc. Sci. Res. **34**, 159–201 (2020)
14. Kummitha, R.K.R., Crutzen, N.: How do we understand smart cities? An evolutionary perspective. Cities **67**, 43–52 (2017)
15. Kitchin, R.: Big data and human geography: Oopportunities, challenges, and risks. Dialogues Human Geogr. **3**(3), 262–267 (2013)
16. Kitchin, R.: The real-time city? Big data and smart urbanism. GeoJournal **79**(1), 1–14 (2014)
17. Kashef, M., Visvizi, A., Troisi, O.: Smart city as a smart service system: Human-computer interaction and smart city surveillance systems. Comput. Human Behav. **124**, 106923 (2021)
18. Atlam, H.F., Wills, G.B.: IoT security, privacy, safety and ethics. In: Digital Twin Technologies and Smart Cities, pp. 123–149. Springer, Cham (2020)
19. Martin, T., Karopoulos, G., Hernández-Ramos, J.L., Kambourakis, G., Nai Fovino, I.: Demystifying COVID-19 digital contact tracing: a survey on frameworks and mobile apps. Wirel. Commun. Mobile Comput. (2020)
20. McDonald, N., Forte A.: Powerful privacy norms in social network discourse. Proc. ACM Human-Comput. Interac., 1–27 (2021)
21. Webster, C.W.R., Löfgren, K.: The value of big data in government: the case of 'smart cities'. Big Data Soc. **7**(1) (2020)
22. National Bureau of Statistics of China: China Statistical Yearbook. China Statistical Press, Beijing (2020). http://www.stats.gov.cn/tjsj/ndsj/2020/indexeh.htm. Accessed 21 January 2022
23. Yearbook Editorial Board: Yearbook of China Information Industry. Publishing House of Electronics Industry, Beijing (2016)
24. 2020 Statistical Bulletin on the Economic and Social Development, http://www.zj.gov.cn/art/2020/8/12/art_1229216136_54388053.html. Accessed 21 January 2022
25. Chourabi, H.: Understanding smart cities: an integrative framework. 45th Hawaii International Conference on System Sciences, pp. 2289–2297 (2012)
26. Lu, D., Tian, Y., Liu, V.Y., Zhang, Y.: The performance of the smart cities in China—a comparative study by means of self-organizing maps and social networks analysis. Sustainability **7**(6), 7604–7621 (2015)
27. Liang, F., Das, V., Kostyuk, N., Hussain, M.M.: Constructing a data-driven society: China's social credit system as a state surveillance infrastructure. Policy Internet **10**(4), 415–453 (2018)
28. Bischoff, P.: Surveillance camera statistics: which cities have the most CCTV cameras? Comparitech, Homepage, https://www.comparitech.com/vpn-privacy/theworlds-most-surveilled-cities/. Accessed 22 July 2021
29. Wong, K.L.X., Dobson, A.S.: We're just data: exploring China's social credit system in relation to digital platform ratings cultures in Westernised democracies. Global Media China **4**(2), 220–232 (2019)
30. Blomberg, M.V.: The social credit system and China's rule of law. In: Social Credit Rating, pp. 111–137. Springer Gabler, Wiesbaden (2020)

COVID-19 and Smart City in the Context of Tourism: A Bibliometric Analysis Using VOSviewer Software

Mirko Perano, Claudio Del Regno, Marco Pellicano, and Gian Luca Casali

Abstract Since the beginning of the pandemic (at the end of 2019), papers have been published in journals firstly around the topic of medical area (such as clinical futures and complications, non-pharmaceutical interventions, treatment, virology, etc.), then around other topics (financial, economic, managerial and tourist). Covid-19 has generated impacts that have propagated from the medical field to the depths of human and relational behavior, modifying thought and action even in consumption. Many territories have reacted with specific policies of social limitations, others have devoted attention to technology to monitor the pandemic. This paper focuses on deepening the relationship between the Covid-19 and the Smart cities in the context of tourism. A bibliometric analysis and word concepts (word co-occurrences matrix) has been used to answer the research questions. A VOSviewer software has been used to process 4818 journal papers extracted from WoS (Web of Science) database between November 2019 and March 2022.

Keywords Covid-19 · Smart cities · Tourism · Technologies · Bibliometric analysis

M. Perano (✉) · C. Del Regno · M. Pellicano
University of Salerno, Fisciano, SA, Italy
e-mail: mperano@unisa.it

C. Del Regno
e-mail: cldelregno@unisa.it

M. Pellicano
e-mail: pellicano@unisa.it

G. L. Casali
Queensland University of Technology, Brisbane, QLD, Australia
e-mail: luca.casali@qut.edu.au

A. Visvizi et al. (eds.), *Research and Innovation Forum 2022*,
Springer Proceedings in Complexity, https://doi.org/10.1007/978-3-031-19560-0_22

1 Introduction

The Covid-19 has been affecting the entire world since the last two years generating negative impacts on territories both from the economic and social side. It can be assumed that from the environmental side, the Covid-19 has positively impact on the global reduction of pollution emissions ascribable to the lockdown policies imposed by the different territories to manage the pandemic [1]. By using a stakeholder' lens [2], one of the most important impacts of the Covid-19 on the territories is related to the value decreased due to the dramatic reduction of the relations among peoples-peoples, peoples-firms, firms-public entities, people-public entities. These lasts play roles of facilitating the value co-creation [3, 4] process in specific territories with the main aim to "generate economic growth and development" [5]. By using a relational lens [6–10], territories have suffered the value disruption directly linked with the reduction of the number of interactions among people-organizations-public entities, in specific territories and in different sectors such as tourism, manufacturing, food, culture, etc.

Considering policies geared toward the use of technology for pandemic control, an interesting aspect to investigates is direct to the smart cities. The concept of smart cities is associated with the possibility of effective use of ICT-enhanced tools in city and urban space [11]. These last "uses digital technologies to enhance performance and well-being, reduce costs and resource consumption, and engage more effectively and actively with its citizens" [12]. Technology a creates opportunity to enhance this well-being by constantly improving the quality of the extant technology in these cities. Technology, through the right application of flexible skills and knowledge, allow the attainment of innovation [13]. The use of technology has already been thought to contribute in solving problems in the context of the health care [14], mobility [15], security, energy consumption, crime [16] and research and development promotion [17]. This picture of smart cities requires that thorough consideration is given to smart services and applications capable of supporting that ecosystem of needs and functions [18]. By using a Fuzzy Formal Concept Analysis, Troisi et al. [19] investigate citizens attitudes in European cities providing insights useful to managers to identify the potential negative effects of smart technologies. Wang et al. [20] propose multi-dimensional technologies to enhance the trust in the shopping activities guaranteeing the social distancing to protect people from the effect of Covid-19.

Each of these technological aspects are linked to the smartness with the main aim to contribute to the improvement of the citizenship' well-being. Therefore, the main aim of this paper is to provide a bibliometric analysis of research on Covid-19 and Smart-city in the context of Tourism and also answering the following research question:

RQ1: What are the main implications of Covid-19 for smart cities?

2 Methodology

Bibliometrics is widely used for the quantitative research assessment of academic output [21]. Bibliometrics analysis uses bibliometric networks to map the relationships between citations, co-citations, co-authorship and co-occurring keywords to map author, institutional, word and thematic relationships [22]. Keyword co-occurrence networks map keywords (noun phrases) extracted from publication titles, abstracts or author-supplied keywords. The resultant keyword bibliometric network consists of nodes and edges. Nodes represent the selected keywords; edges represent the relationships between pairs of nodes–between the keywords [23]. Nodes and edges can also be used to represent citation and co-citation relations. Bibliometric networks can be weighted with the edge weight representing the extent of the relationships between any two nodes. Nodes that are more connected or related have higher edge weights [24].

The co-occurrences of two keywords are the number of publications in which both keywords occur together in the title, abstract, or keyword list. A term or keyword map is a two-dimensional map in which the distance between two terms can be interpreted as indicating the relatedness of the terms. In general, the smaller the distance between two terms, the stronger the terms are related to each other [25]. The relatedness of terms is determined based on co-occurrences in documents [23].

The application of bibliometrics analysis using a comprehensive keyword bibliometric network was implemented to examine the relationships between Covid-19 and smart cities in the context of tourism.

3 Analysis and Findings

3.1 Bibliometric Data

A WoS (Clarivate) bibliometric database search was undertaken across all fields using the search string: [("Covid-19" OR "SARS" OR "Smart city" OR "smart cities") AND ("land management" OR "policy makers" OR "tourism")].

Document selection was limited to the English language published journal articles, early access, proceeding papers, review articles, and meeting abstracts. Four thousand eight hundred and eighteen articles were selected. These records were exported as plain text files. A limitation of WoS is that data downloads in plain text format are limited to 500 records. Multiple extracts were done with subsequent file editing occurred to remove header and footer lines. These separate text files were then merged for bibliometric mapping.

3.2 Items for Analysis

According to Börner et al. [26], journals, papers, authors, and words are most commonly selected as the type of item to analyze in bibliometric analyses. Each type of item provides a different visualization of a field of science and results in a different analysis. In the present study, we choose to analyze word concepts (word co-occurrences) to best answer our research questions. A bibliometric map showing the associations between words in this scientific domain was run using VOSviewer software.

The construction of a bibliometric map is a process that consists of three steps. In the first step, a similarity matrix is calculated based on a co-occurrence matrix. In the second step, a map is constructed by applying a VOS mapping technique to the similarity matrix. In the third step, the map is translated, rotated, and reflected.

The VOS mapping technique requires a similarity matrix as input. A similarity matrix is obtained from a co-occurrence matrix by normalizing the latter matrix by correcting the matrix for differences in the total number of item occurrences. The most popular similarity measures for normalizing co-occurrence data are the cosine and the Jaccard index. However, VOSviewer uses an association strength measure [23].

3.3 Publications by Author

From the publications extracted for this research more than 15,000 authors have been identified as being taking part in publishing at least 1 article. Thus, out of such a large number of authors the top most cited Authors are Professor Hall C. Michael currently at the University of Canterbury, Professor Stefan Gossling at University of Linnaeus and Lund University and Professor Daniel Scott University Research Chair at University of Waterloo and Vice Chancellor Research Chair at University of Surrey (see Table 1). An interesting result is that in within the top 20 Authors by citation they differ quite dramatically in term of number of publications ranging from 1 to 28 which is the maximum number of publications published by an author (alone or in collaboration with other authors).

In the top author it is interesting but at the same time a bit expected to see that he has been publishing with many authors from different clusters (different colors) support the multi-disciplinary nature of this research area (Fig. 1).

3.4 Publications by Journal

Analyzing the 4818 articles collected for this research emerges that they are all scatter around more than 900 journals suggesting that the main topic of this research is highly multidisciplinary in nature. Out of 924 journals 654 (50.32%) published only one

Table 1 Publications by Author citations

N	Author	Documents	Citations
1	Hall, C.M	13	1388
2	Gossling, S	6	1322
3	Scott, D	3	1285
4	Castaldelli-Maia, J.M	2	1077
5	O'Higgins, M	1	1060
6	Torales, J	1	1060
7	Ventriglio, A	1	1060
8	Wang, Y	28	958
9	De Marco, A	2	921
10	Cagliano, A.C	1	917
11	Mangano, G	1	917
12	Neirotti, P	1	917
13	Scorrano, F	1	917
14	Gordon, C.J	4	906
15	Gotte, M	4	906
16	Tchesnokov, E.P	4	906
17	Liu, X	22	839
18	Singh, B	3	820
19	Li, S	9	766
20	Ward, J	3	677

article on the main research topic in withing the researched time, 188 (20.3%) two journals, 86 (4.44%) 3 journals, 41 (3.03%) 4 journals and 144 five and more journals (Table 2). Thus, as in similar studies [27] it is arguable that only those journals with 5 and more articles can be considered better sources in the area of Covid-19 & Smart-cities in the context of tourism. Secondly, the top 20 journals only account for 26% of the overall published articles on this topic (Table 3). Thirdly, in within the first top five journal three are from a tourism background.

Another discovery is that only few journals started actively publishing on this topic in 2019 which includes Tourism Management, Cities, Sensors, IEEE access and few others (Fig. 3). However, the majority of the other journals have started publishing on this topic only between the end of 2020 and the beginning of 2021 (Fig. 2).

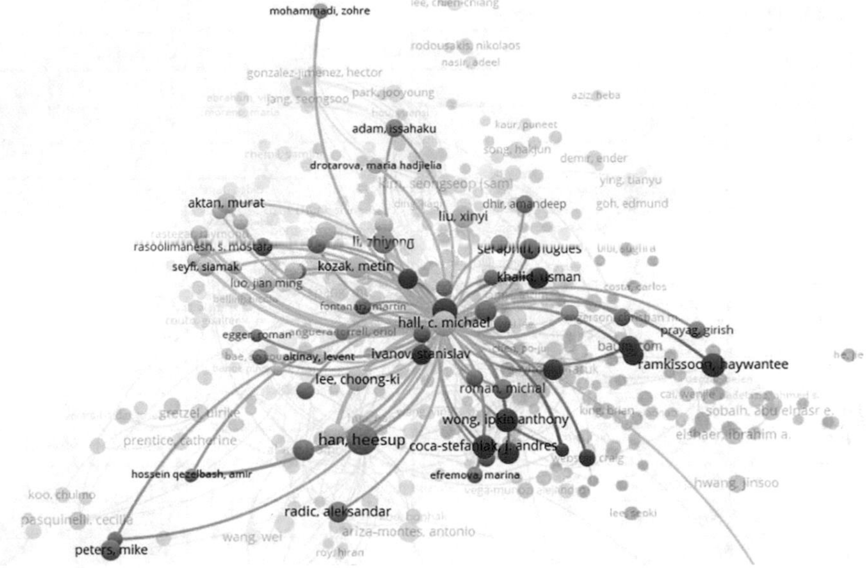

Fig. 1 Professor Michael C. Hall co-authors map

Table 2 Summary of productivity of Journals (2019–2022)

Production volume by journal	Journals	% Of 924
1 published article	465	50.32
2 published articles	188	20.3
3 published articles	86	4.44
4 published articles	41	3.03
5 published articles or more	144	12.55
Total	924	100

3.5 Publications by Country

Covid-19 and Smart- City is a global research topic involving more than 90 counties worldwide. Table 4 lists the top 15 countries that have produced the largest number of academic papers in the area of research discussed in this paper and accounting for an overall 98.8% of all publications. USA is the country that has produced the largest number of publications in between the researched timeframe with 892 papers followed by People of China with 785 and England with 523. Poland and France are correspondently at the 14th and 15th place with 123 and 115 publications each. An important discovery is that 45.5 of the overall publications on the topic have been produced by the top three countries.

Table 3 Publication by Journal

Ranking	Title	N. of Articles	%
1	Sustainability	319	6.6
2	Current issues in tourism	140	2.9
3	Int. Journal of Environment. Res. and Public Health	131	2.7
4	International Journal of Hospitality Management	71	1.5
5	Plos One	55	1.1
6	Journal of Sustainable Tourism	45	0.9
7	Worldwide Hospitality and Tourism Themes	44	0.9
8	International Journal of Contemporary Hospitality Management	43	0.9
9	Journal of Hospitality and Tourism Management	43	0.9
10	Tourism Management	43	0.9
11	Frontiers in Psychology	38	0.8
12	Tourism Geographies	38	0.8
13	Tourism Economics	37	0.8
14	Anatolia-International Journal of Tourism and Hospitality Research	35	0.7
15	Tourism Recreation Research	34	0.7
16	Annals of Tourism Research	30	0.6
17	Journal of Travel Research	30	0.6
18	Frontiers in Public Health	29	0.6
19	Journal of Medical Internet Research	29	0.6
20	International Journal of Tourism Cities	28	0.6

By reviewing the map in Fig. 4 emerges that the three top countries in the list are actually all part of their independent clusters with USA leading the light green cluster, England the green and People of China the turquoise one.

4 Discussions and Conclusions

This paper aims to clarify the state of the art about the relationship between Covid-19 and Smart city in the context of tourism by using a bibliometric analysis.

Table 1 provide insights about the most cited Author from the dataset extracted and analyzed. A list of 20 authors has been extracted and from the analysis emerges how Authors differs in terms of relationship between number of publications (from 2 to 28) and citations (from 677 to 1388). Authors with a high citation index with only 1 publication on the topic here investigated are Professors O'Higgins, M., Torales, J. and Ventriglio, A. with 1060 citations till 2022 March. Authors with a high number

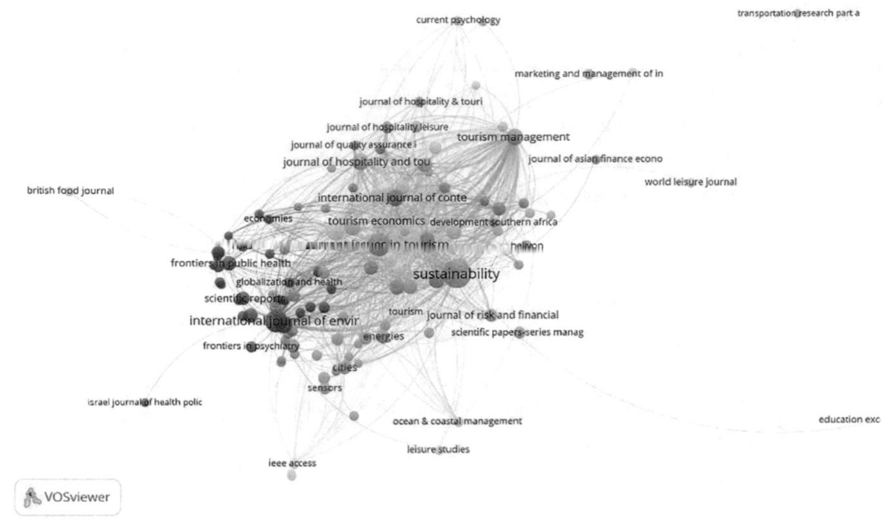

Fig. 2 Map of journals with 5 publication or more

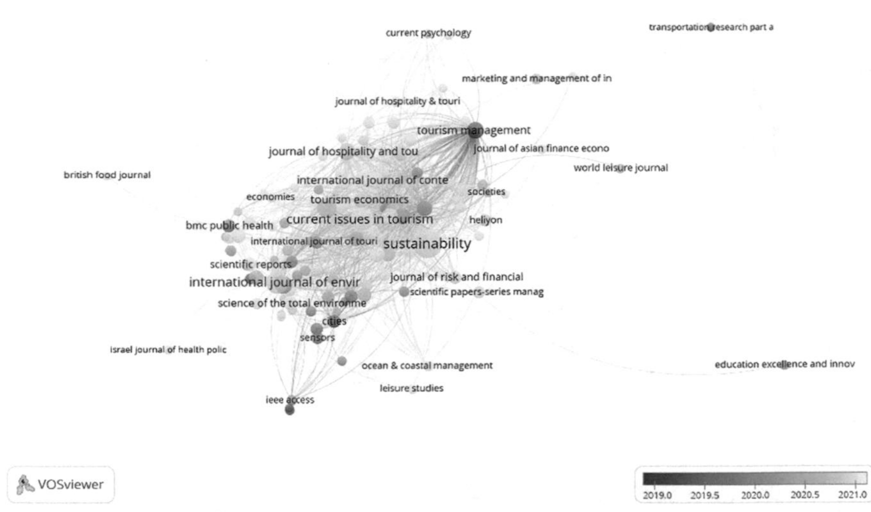

Fig. 3 Map of journals with 5 publication or more by time

of publications are Professor Wang, Y. with 28 papers, Professor Liu, X. with 22 papers and Professor Hall, C.M. with 13 papers. Authors with a high ratio (number of publications/citations) are Professors O'Higgins, M., Torales, J. and Ventriglio, A. with 1060 citations per 1 paper and Professor Scott. D. with 428.33 per paper. About the results of 28 Journal papers published by Professor Micheal C. Hall, a co-authors map has been realized (Fig. 1).

Table 4 Number of publications in co-authorship by country (2019–2022)

R	Co-authorship by country	Number	% of 4818
1	USA	892	18.5
2	People of China	785	16.3
3	England	523	10.8
4	Australia	385	7.9
5	Italy	339	7.0
6	Spain	331	6.9
7	India	268	5.6
8	South Korea	201	4.2
9	Canada	194	4.0
10	Germany	183	3.8
11	Portugal	161	3.3
12	Turkey	149	3.0
13	South Africa	126	2.6
14	Poland	123	2.5
15	France	115	2.4

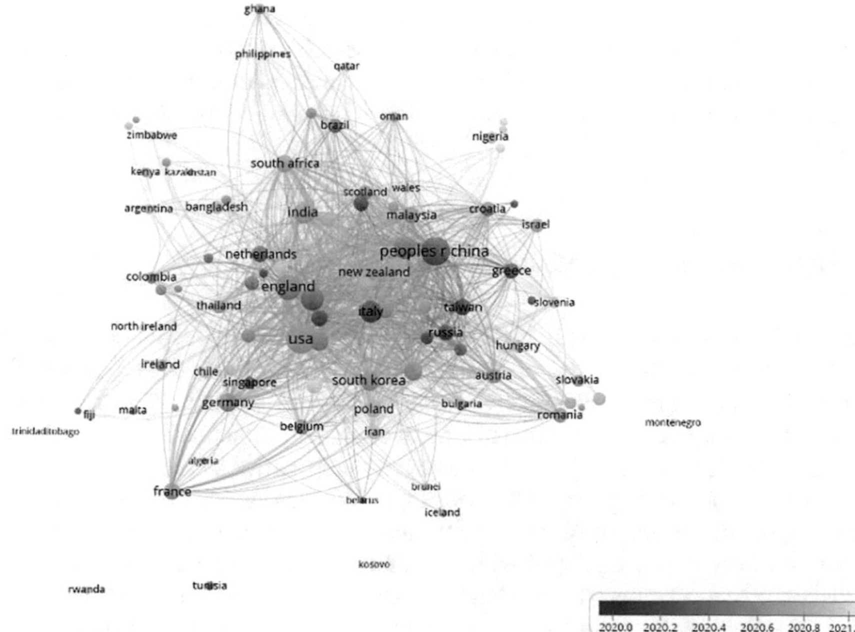

Fig. 4 Map of co-authorship by Country

Table 2 describes the summary of the productivity of the Journals. From November 2019 and 2022 March, a total of 924 Journals published the amount of the paper detected in the database used in this research and 144 Journals published 5 or more articles (with a ratio of 12.55% on 924). Figure 2 show a map of Journals with 5 publications or more and Fig. 3 represents a map of Journals with 5 publication or more by time.

From Table 3 it is detected than more than 1200 pieces of research about the topic under investigation, have been published as articles in 20 journals. The journal with higher ratio (6.6%) is Sustainability Journal, then Current Issue on Tourism (2.9%) and International Journal of Environmental Research and Public Health (2.7%). 12 out of 20 of the Journals deal with a tourism topic in the title but, in general, it can be stated that papers investigate the relationship between Covid-19 and Smart cities in the context of tourism have a highly multi-disciplinary approach.

Table 4 describes the number of publications in co-authorship by country from 2019 to 2022. 15 countries have been detected: USA with a ratio of 18.5% and a number of publications in co-authorship of 892 is the first country (light green cluster in Fig. 4); People of China with a ratio of 16.3% with a number of publications in co-authorship of 785 is the second one (turquoise cluster in Fig. 4); England with a ratio of 10.8% with a number of publications in co-authorship of 523 (green cluster in Fig. 4) and the last is the France with a ratio of 2.4% with a number of publications in co-authorship of 115. Figure 4 suggests that the top research countries on this topic they have created their own main research hubs by collaborating with particular countries rather than working with each other. Therefore, such behaviors could be observed from a strategic view as three potential competitors rather than collaborators.

5 Implications, Limitations and Future Research

The main implications consist in highlighting the research on the topic of Covid-19 in relation to that of Smart City in the context of tourism. This can be considered, until now, a unique bibliometric analysis characterized by this aim. This paper also contributes for the oncourse research in term of awareness about the extant literature, the main authors, co-authors, journals and country more active in this research topic.

Also, the number of papers (4818) on the topic under research seem to suggest that technology could generate positive impact in managing and containing the pandemic effects. Territory highly equipped with technology such as the Smart Cities could guarantee more wellbeing in term of health. Technology can provide opportunities to reduce the effects of the pandemic by matching public policy and land management. From the analysis emerges that the main implication of the Covid-19 in the smart cities are the aspect related to the technology management and the rules related to this area in the smart cities. In this direction, public organizations can benefit from this study with a higher awareness of their territory equipped with a technology and about the opportunity to define specific policies to leverage technology to improve not only the well-being [11–16], but also the health aspects.

This paper is not free from limitation. Firstly, the main criteria used by VOSviewer is based on the citations index of the papers. Therefore, papers that currently have not met the minimum threshold level of citations to be accepted by the software to be analyzed, they can potentially be included in future research. Another limitation can be seen in technical decisions such as the choose of the English language and to select journal articles, early access, proceeding papers, review articles, and meeting abstracts excluding other potential sources. Also, only the WoS has been used in this bibliometric work. Finally, that chosen methodology is limited to provide descriptive analysis about the topic under this research rather than much more in depth of results and implications.

References

1. Bashir, M.F., Ma, B., Komal, B., Bashir, M.A., Tan, D., Bashir, M.: Correlation between climate indicators and COVID-19 pandemic in New York, USA. Sci. Total Environ. **728**, 138835 (2020)
2. Freeman, R.E.: Strategic Management: A Stakeholder Approach. Cambridge University Press, Cambridge (2010)
3. Vargo, S.L., Lusch, R.F.: Evolving to a new dominant logic for marketing. J. Mark. **68**(1), 1–17 (2004)
4. Vargo, S.L., Lusch, R.F.: Service-dominant logic: continuing the evolution. J. Acad. Manag. Sci. **36**(1), 1–10 (2008)
5. Perano, M., Abbate, T., La Rocca, E.T., Casali, G.L.: Cittaslow & fast-growing SMEs: evidence from Europe. Land Use Policy **82**, 195–203 (2019)
6. Pellicano, M.: Occasionalità o stabilità nelle relazioni di impresa alla ricerca della vitalità sistemica. Esperienze d'impresa **8**(2) (2000)
7. Pellicano, M.: L'impresa Relazionale. Giappichelli, Torino (2017)
8. Pellicano, M., Ciasullo, M.V.: La visione Relazionale D'impresa. Giappichelli, Torino (2010)
9. Ciasullo, M.V., Troisi, O.: La visione relazionale dell'impresa. In: Un Approccio Per la Strategic Governance. Giappichelli, Torino (2017)
10. Perano, M., Cerrato, R.: IL Bilancio Sociale tra Pianificazione Strategica e Co-creazione di Valore. Il caso dell'Università di Salerno. Giappichelli, Torino (2017)
11. Visvizi, A., Lytras, M.: Smart Cities: Issues and Challenges: Mapping Political, Social and Economic Risks and Threats. Elsevier, Amsterdam (2019)
12. Snow, C.C., Håkonsson, D.D., Obel, B.: A smart city is a collaborative community: lessons from smart Aarhus. California Manag. Rev. **29**(1), 92–108 (2016)
13. Troisi, O., Visvizi, A., Grimaldi, M.: The different shades of innovation emergence in smart service systems: the case of Italian cluster for aerospace technology. J. Bus. Indus. Market. (2021). https://doi.org/10.1108/JBIM-02-2020-0091
14. Bhunia, S.S., Dhar, S.K., Mukherjee, N.: A fuzzy approach for provisioning intelligent health-care system in smart city. In: 2014 IEEE 10th International Conference on Wireless and Mobile Computing, Networking and Communications, pp. 187–193. IEEE, Larnaca (2022)
15. Farid, A.M., Alshareef, M., Badhesha, P.S., Boccaletti, C., Cacho, N.A.A., Carlier, C.I., Corriveau, A., Khayal, I., Liner, B., Martins, J.S.B., Rahimi, F., Rossett, R., Schoonenberg, W.C.H., Stillwell, A., Wang, Y.: Smart city drivers and challenges in urban-mobility, health-care, and interdependent infrastructure systems. IEEE Potent. **40**(1), 11–16 (2020)
16. Truntsevsky, Y.V., Lukiny, I.I., Sumachev, A.V., Kopytova, A.V.: A smart city is a safe city: the current status of street crime and its victim prevention using a digital application. In: MATEC Web of Conferences, vol. 170, p. 01067. EDP Sciences, Lel Ulis (2018)

17. Visvizi, A., Troisi, O.: Effective management of the smart city: an outline of a conversation. In: Visvizi, A., Troisi, O. (eds.) Managing Smart Cities, pp. 1–10. Springer, Cham (2022)
18. Lytras, M.D., Visvizi, A., Chopdar, P.K., Sarirete, A., Alhalabi, W.: Information management in smart cities: turning end users' views into multi-item scale development, validation, and policy-making recommendations. Int. J. Inf. Manage. **56**, 102146 (2021)
19. Troisi, O., Fenza, G., Grimaldi, M., Loia, F.: Covid-19 sentiments in smart cities: the role of technology anxiety before and during the pandemic. Comput. Hum. Behav. **126**, 106986 (2022)
20. Wang, X., Wong, Y.D., Chen, T., Yuen, K.F.: Adoption of shopper-facing technologies under social distancing: a conceptualisation and an interplay between task-technology fit and technology trust. Comput. Hum. Behav. **124**, 106900 (2021)
21. Álvarez, C., Urbano, D., Amorós, J.E.: GEM research: achievements and challenges. Small Bus. Econ. **42**(3), 445–465 (2013)
22. Cobo, M.J., López-Herrera, A.G., Herrera-Viedma, E., Herrera, F.: An approach for detecting, quantifying, and visualizing the evolution of a research field: a practical application to the Fuzzy sets theory field. J. Informet. **5**(1), 146–166 (2011)
23. Van Eck, N.J., Waltman, L., Van den Berg, J., Kaymak, U.: Visualizing the computational intelligence field. IEEE Comput. Intell. Mag. **1**(4), 6–10 (2006)
24. Chen, C.: CiteSpace II: detecting and visualizing emerging trends and transient patterns in scientific literature. J. Am. Soc. Inform. Sci. Technol. **57**(3), 359–377 (2006)
25. Van Eck, N.J., Waltman, L.: Software survey: VOSviewer, a computer program for bibliometric mapping. Scientometrics **84**(2), 523–538 (2010)
26. Börner, K., Chen, C., Boyack, K.W.: Visualizing knowledge domains. Ann. Rev. Inf. Sci. Technol. **37**(1), 179–255 (2003)
27. Cavalcante, W.Q.D.F., Coelho, A., Bairrada, C.M.: Sustainability and tourism marketing: a bibliometric analysis of publications between 1997 and 2020 using vosviewer software. Sustainability **13**(9), 4987 (2021)

Augmented Reality (AR) for Urban Cultural Heritage Interpretation: A User Experience Evaluation

Sema Refae, Tarek Ragab, and Haitham Samir

Abstract Heritage interpretation plays a vital role in understanding and perceiving heritage values. This research investigates the role of Augmented Reality (AR) in the heritage interpretation process through conducting a pilot AR study in historic Jeddah by simulating the old images of existing buildings through an interactive augmentation in an instantaneous mobile. The study measures the users' experience in five aspects: Informative, Interactive & enjoyable, Realistic & intuitive, Convenience, and Memorable, which have been addressed in the literature to describe AR's influence on visitors' experience in cultural heritage sites. The study verifies the links between heritage sites' interpretation and visitor experience. It also proposes five factors leading to an enhanced visitor experience through the employment of AR in cultural heritage sites. To verify the viability of the AR model in the interpretation of cultural heritage values and the overall visitor experience, a survey was conducted among a group of local visitors and foreign tourists. The survey's analysis helps trace the reactions and responses to using AR interpretation in a specific cultural context like Saudi Arabia. It confirms the acceptance of the visitors and tourists of the AR interpretation model in simulating lost physical urban memory and confirms that implementing AR in cultural heritage interpretation contributes positively to the overall experience and interaction of the heritage site visitors.

Keywords Urban heritage · Heritage interpretation · Augmented reality · Visitor experience

---

S. Refae
Architecture Department, Dar AlHekma University, Jeddah, Saudi Arabia
e-mail: srefae@dah.edu.sa

T. Ragab · H. Samir (✉)
College of Architecture and Design, Effat University, Jeddah, Saudi Arabia
e-mail: Haitham.samir@ngu.edu.eg

T. Ragab
e-mail: tragab@effatuniversity.edu.sa

H. Samir
School of Engineering, New Giza University, Cairo, Egypt

A. Visvizi et al. (eds.), *Research and Innovation Forum 2022*,
Springer Proceedings in Complexity, https://doi.org/10.1007/978-3-031-19560-0_23

1 Introduction

There is a substantial emerging concern for cultural heritage preservation in Saudi Arabia, accompanied by ongoing efforts to enable cultural heritage sites for tourism and visitation. To support these efforts, new techniques are being developed to overcome the hinders posed by the lack of finance and human resources or even by the extended time needed for traditional heritage preservation techniques. Such techniques shall contribute to the capacity of the concerned stakeholders to manage heritage locations which eventually entices the cultural tourism industry. Within this paradigm, this research argues that incorporating augmented reality (AR) tools and applications shall enhance the physical interaction with historic sites through a quality interpretation of the values associated with cultural sites. This research raises an important question that has been attempted to answer through the research development: "How and to what extent does the utilization of AR contribute to improve the heritage sites visiting experiences through the interpretation of these sites' values?".

To answer this question, and due to the scope of this paper, the research adopts a set of previously defined aspects of visitor-perception interpreted values, which are used as indicators for visitors' experiences and perception quality. These indicators are informativity, interactivity & enjoyability, reality & intuition, Convenience, and, finally, memorability. Then, the research lays a theoretical background that includes revising the literature to review the potential employment of AR in values interpretation and users' perceptions. This was followed by developing a simulation AR model to measure a set of the five indicators of the visitor experience. In building the simulation model interface, the Appypie platform is used to simulate the old images of the existing buildings and locations in the historic Jeddah through interactive augmentation in an instantaneous mobile application. To verify the viability of the model in interpretation the cultural heritage values and the overall visitor experience, a survey was conducted among a group of local visitors and foreign tourists to measure and evaluate its implications on the users' perception, emotions, attachment, and exploration of the concerned cultural heritage sites.

2 The Role of Values Interpretation in Cultural Heritage Sites

Interpretation incorporates the various means of representing heritage to people [1]. In 1985, Cohen distinguished between information and interpretation [2]. Beck and Cable [3] defined interpretation as "an informational and inspirational process designed to enhance the understanding, appreciation, and protection of our cultural and natural legacy." Interpretation refers to different activities that enhance the awareness of heritage sites that drew the principles of the ICOMOS Charter for the Interpretation and Presentation of Cultural Heritage Site [4]. Expert knowledge has indicated that having a proper interpretive plan is necessary to ensure the heritage site's long-term success [5].

3 The Potential of Using AR for Heritage Sites Visiting Experience

The user experience can be improved by new ways of site presentation and interpretation [6]. AR allows for additional cultural heritage interaction, exploration, and interpretation possibilities. AR gives visitors the freedom to explore and personalize their visit [7].

3.1 The Information Transferred by AR

Aspects related to information are essential variables in the tourists' experience. Tourists use informative features that support their visits. They guide their decisions, shape their gaze, determine their routes, and even influence the interpretation they make of places. Donaire and Galí argued that informative features are a central instrument for mapping tourists' itineraries. Therefore, it is not unexpected to find that the relationship between the models of itineraries and the informative variables should be highly significant [8].

3.2 The Immersive Experience of AR as a Form of User-Site Interaction

AR can enable novel interaction between visitors and the authentic setting in which the building, site, or urban environment is encountered for heritage sites. AR interaction allows for new experiences with various forms of cultural heritage [9]. On the other hand, the current technology in smartphones and tablets is recommended for improving interactivity with the cultural heritage sites and, in turn, improving how individual visitors navigate these cultural sites [10].

Jung et al. confirmed that the aesthetics of AR have a significant influence on noticing enjoyment. Aesthetic designing of AR implementation is essential to assure positive perception and lead to positive emotion and behavioral intentions [11].

Table 1 Factors that affect the utilization of AR as a tool to enhance visitors' experience in heritage sites

Interpretive principle			
Interpretive	Experience	Factors lead to a good experience	Informative (Awareness)
			Interactive & enjoyable (Experience)
			Convenience (Acceptance)
			Realistic & intuitive
			Memorable (Memory & Identity)

3.3 *Active Engagement with the Memory*

AR has been identified as one-way visitor memory can be augmented or improved, especially when helping find customized personal experiences. This encourages these visitors to revisit the site. Through more advanced AR, visitors form a new understanding and identity and develop a different perspective or meaning for the heritage site [9] (Table 1).

4 The Pilot Study

The study investigates visitors' preferences and feedback on using AR and its potential in enhancing the user visitation experience and its implications on their perception, emotions, attachment, and exploration of cultural heritage sites. The investigation was followed by a survey of the visitors' experience quality. The analysis of the results is used to measure the previously mentioned five aspects relevant to visitors' experience and the overall value of the application.

4.1 *Model Development, Limitations, and Site Selection Criteria*

Three limitations affected developing the AR application: technical implications, time, and site. There was an evident technical problem in building the AR marker from image reality. The time was also limited in producing a professional 3d model and developing the model to be more convenient to use. The limitation of the site selection options, due to limited data and documents available for the old urban context of the current historical sites, was also evident through the study development.

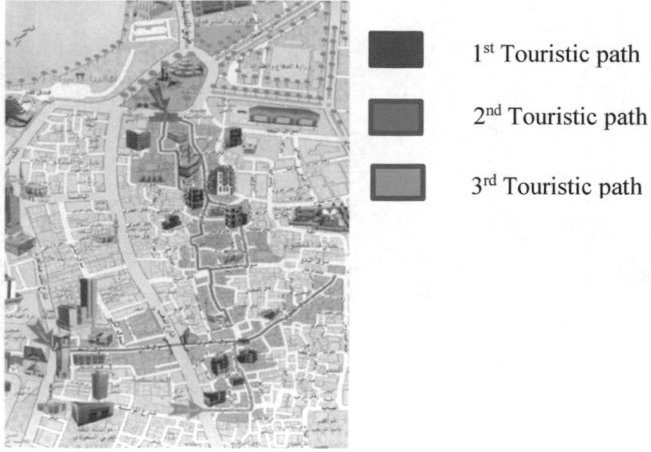

Fig. 1 Jeddah historical touristic path (SCTA, 2012)

The researchers investigated the optimum locations (touristic routes) to implement the AR simulation model. Three existing touristic paths designated by the Saudi Commission for Tourism and National Heritage (SCTH), as shown in (Fig. 1), were selected. The three routes are located under the UNESCO protection ordinance. Following a close observation, a particular path was selected that shows valuable urban elements and has always been a vibrant location.

4.2 AR Simulation Tour Design and Location

The AR simulation tour will pass through nine stops in the second touristic path. The path will allow users to view nine AR views created from nine markers. The views include 2D old photographs and audio videos (Fig. 2).

4.3 The Pilot Study Methodology and Technique

An experimental AR model was developed to measure the five factors that lead to a good visiting experience and evaluate the AR interpretation capacity. The production of the AR model started by selecting a technology platform considering that the application can work with both Android and iOS systems. The app generates two types of AR: old photographs and videos (Fig. 3).

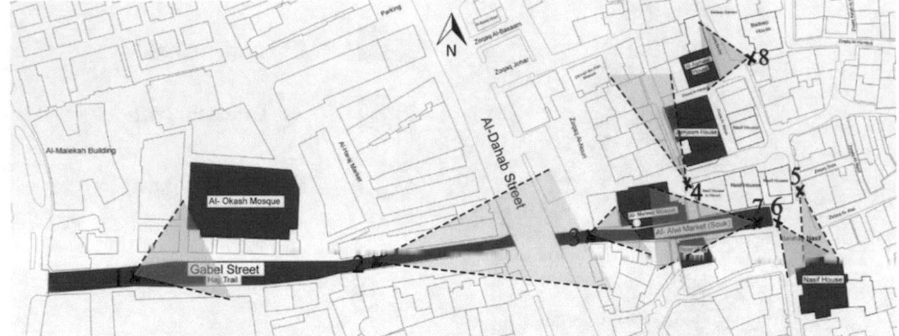

Fig. 2 The AR path simulation (by authors)

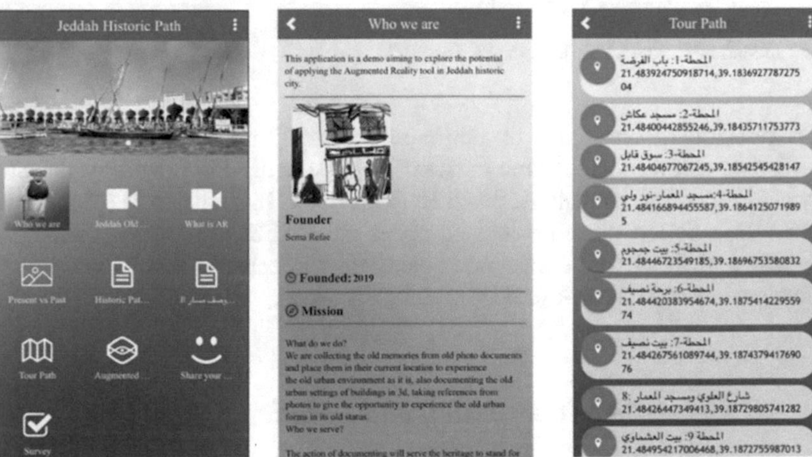

Fig. 3 Mobile app design and features (by authors)

4.3.1 Building the AR Model Interface

The Appypie Builder platform and Vuforia platform are used for building the AR application. It has an AR feature in its beta version. The platform allowed to create a mobile app without utilizing coding skills. Building a demo app began with naming the app "Jeddah historical path" (JHP). The nine features shown in Fig. 4 were added. In addition to a "Survey" feature to the application that includes a designed survey for the visitors to respond to after the tour ends. Figures 4, 5, and 6 indicate the process of building the images and videos in JHP.

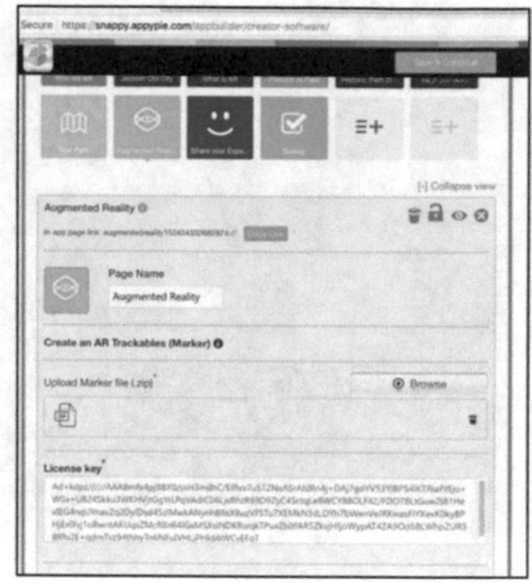

Fig. 4 The app builder face, adding key license and uploading the marker file from Vuforia, (by authors)

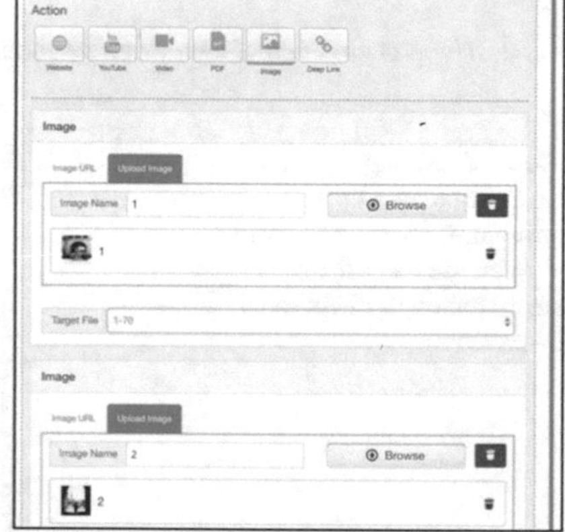

Fig. 5 Selecting a file from the target file database and uploading an image as an AR to attach it with the marker (by authors)

4.3.2 Evaluation Methodology

A survey was conducted among groups of local visitors and tourists to investigate the designed AR interpretation viability. Thirty-three participants were able to experience the AR app in different times and days from 23rd to 28th of April 2019 (Fig. 7).

5 Data Analysis

The first part of the questionnaire explored how the participants are aware of and familiar with AR as a new technology tool. The results show that (64) percent of the participants are aware or have heard about augmented reality. These groups are mostly ranged from 21 to 40 years old. However, (76) percent of them had never experienced AR tools before.

Fig. 6 The process of creating a video by combining the audio with photo (by authors)

Fig. 7 The image marker was printed on a board throughout the tour to avoid not viewing the AR maker during the experiment tour (by authors)

5.1 Informative

The results indicated that (55) percent of the participants had partially increased their awareness regarding the heritage and culture of Jeddah's historic site, and for (45) percent, the experience had increased their awareness a lot. No answers were recorded for negative responses or neutral ones. Also, the results indicated that (94) percent were not disturbed by using the AR experience. Only (3) percent were partially disturbed, and (3) percent were disturbed a lot. It indicates that AR does not disturb awareness. Accordingly, the experience shows a significant role of AR in increasing information and knowledge & expanding visitors' awareness (Figs. 8 and 9).

Did the AR experience increase your understanding and awareness of the heritage and culture of the historical site?

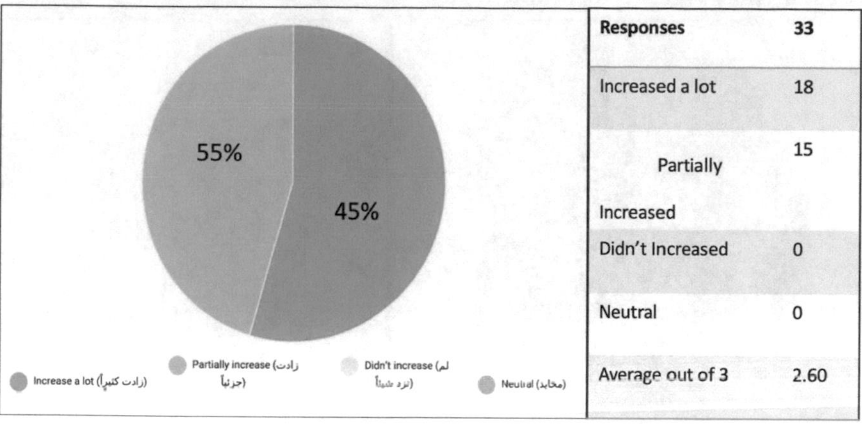

Responses	33
Increased a lot	18
Partially Increased	15
Didn't Increased	0
Neutral	0
Average out of 3	2.60

Fig. 8 Survey result—AR increased the awareness of heritage site (the authors)

Did the AR disturb your understanding and awareness of the historical site?

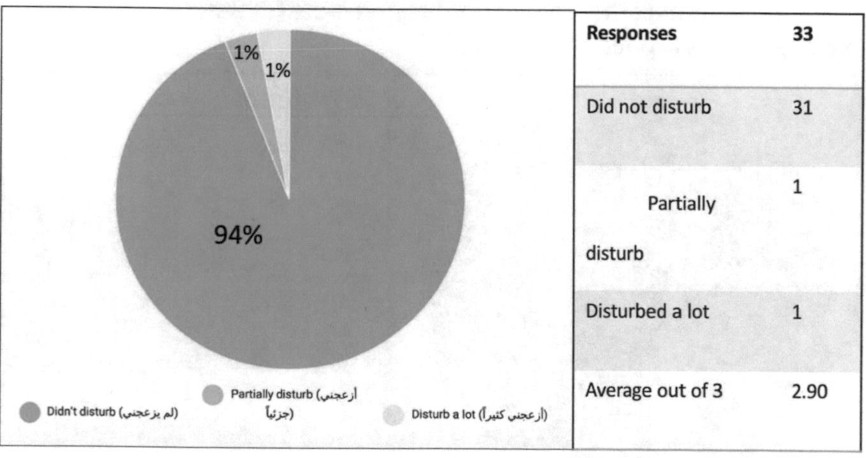

Responses	33
Did not disturb	31
Partially disturb	1
Disturbed a lot	1
Average out of 3	2.90

Fig. 9 Survey result—AR disturb the understanding & awareness (the authors)

5.2 *Interactive and Enjoyable*

All participants felt that the AR interpretation experience is unique and exciting. The data indicated that (100) percent of participants agreed on the role of AR application in increasing the entertainment and enjoyment of their tour. This result confirms the high impact of AR on the interactive, enjoyable factor added to the visitors of historical sites (Figs. 10 and 11).

Do you feel that the experience provides unique and exciting information from the past?

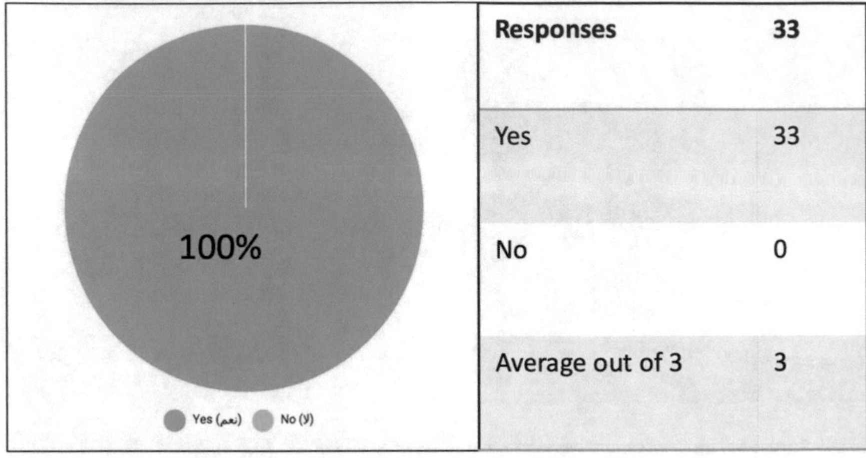

Fig. 10 Experience survey results (by authors)

Do you think that the AR application increases the entertainment aspect of your tour?

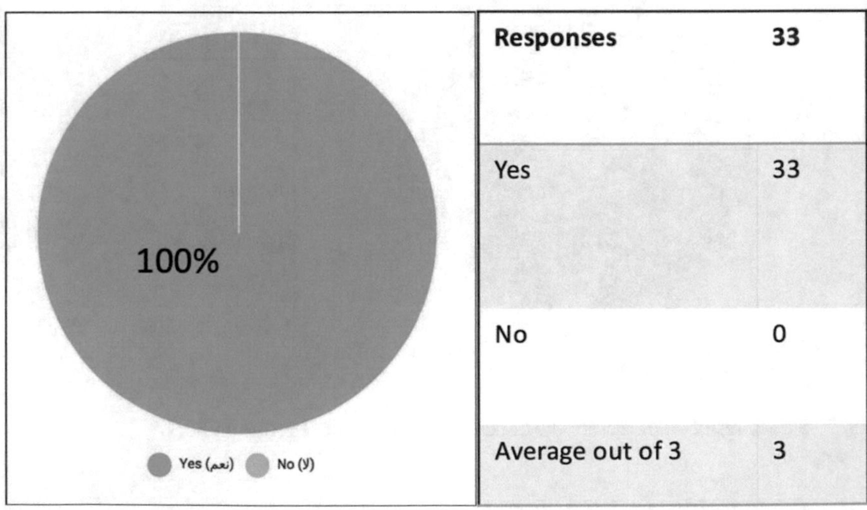

Fig. 11 Entertainment benefit survey results (by authors)

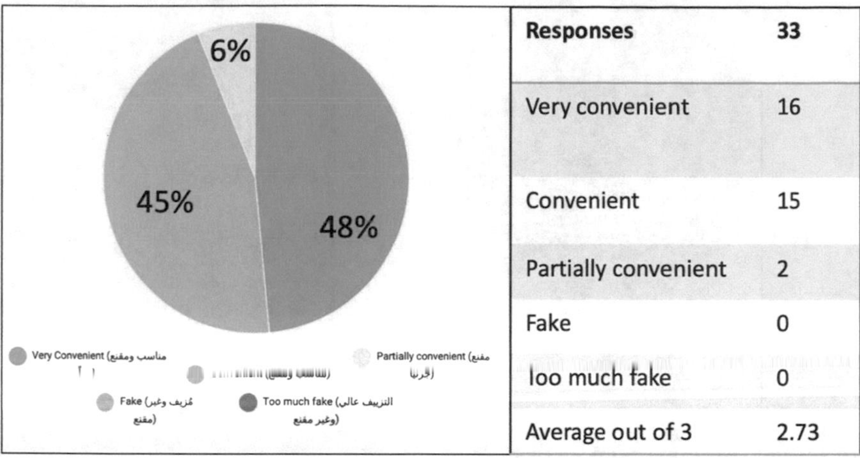

Responses	33
Very convenient	16
Convenient	15
Partially convenient	2
Fake	0
Too much fake	0
Average out of 3	2.73

Fig. 12 Acceptance survey result (by authors)

5.3 *Realistic and Intuitive*

The data indicated that (48) percent of the participants felt that the production of ARin historic Jeddah is very convenient for them, and (45) percent of them felt that it is convenient for them. Only (6) percent of the participants felt that it is partially convenient. None of the participants felt that the augmentation of the application is fake or too fake (Figs. 12 and 13).

The data also indicated that participants' perception of the aesthetics and beauty of the past from AR recorded (73) percent of the participants had a high level of feelings, and (27) percent had moderate feelings. No participant had low feelings or no feelings at all. It means that the AR enhances the feelings of beauty and aesthetics of the lost memories. These results confirm that the production of AR through the 2D photo is convenient.

5.4 *Convenience*

The data indicated that (94) percent of the participants accepted exploring the place with AR technology and a tour guide, and only (6) percent of the participants were unsure. None of the participants disagreed with using the augmented reality tool and preferred only the tour guide for information, which indicates high acceptance of the AR interpretation. The result confirms that the AR interpretation app is convenient to use (Fig. 14).

Did you feel the aesthetics and beauty of the past with augmented reality?

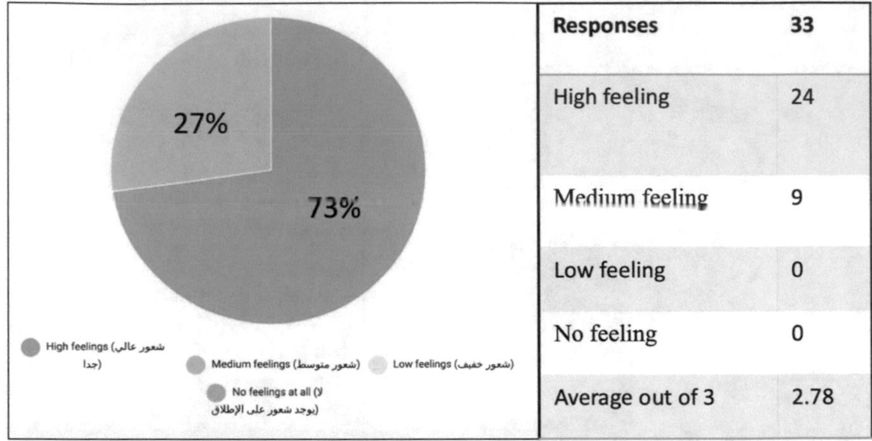

Responses	33
High feeling	24
Medium feeling	9
Low feeling	0
No feeling	0
Average out of 3	2.78

Fig. 13 Memory survey result (by authors)

Do you agree to explore the place with the augmented reality technology along with the presence of a tour guide or would you prefer to listen to only the tour guide?

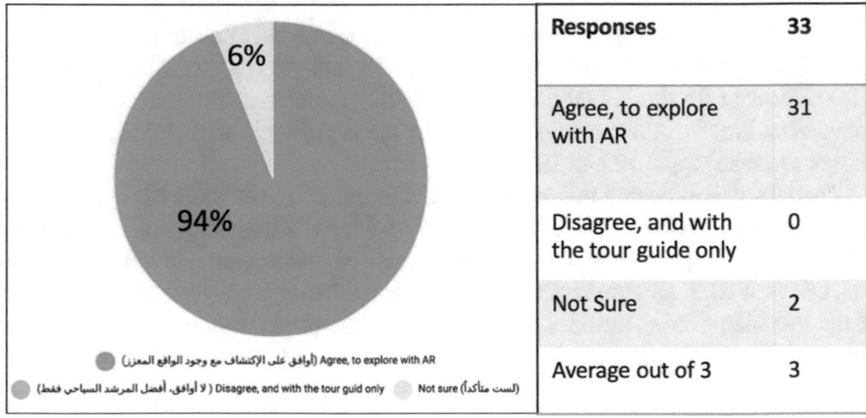

Responses	33
Agree, to explore with AR	31
Disagree, and with the tour guide only	0
Not Sure	2
Average out of 3	3

Fig. 14 Acceptance survey result–exploring the heritage site with AR (by authors)

5.5 *Memorable*

The data indicated (67) percent had simulated feelings of emotional attachment to the place in the past, and (21) percent had moderate simulated feelings of emotional attachment. Moreover, (12) percent presented a low profile of emotional feeling attachment to the past. From the responses, the researcher can assess that the weak feelings could associate with the young and non-Saudi participants who do not have memories or relations to the local culture (Figs. 15 and 16).

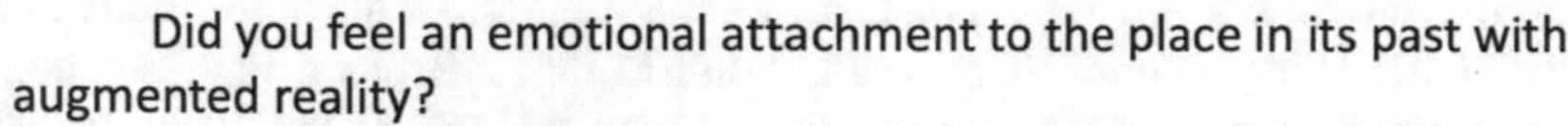

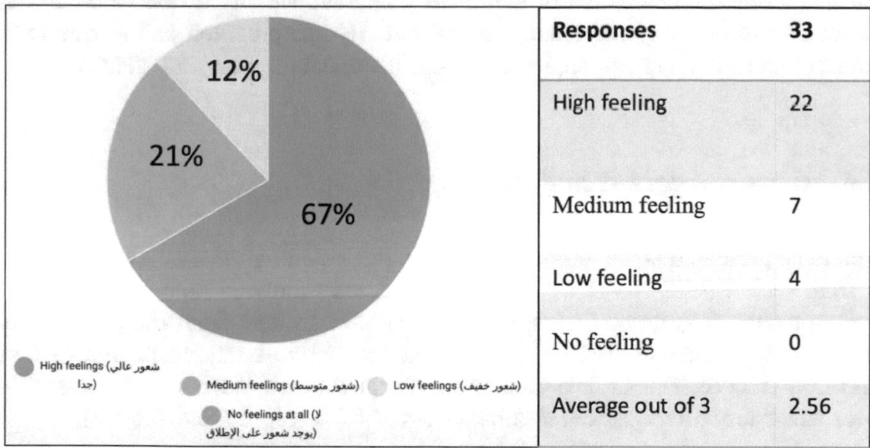

Responses	33
High feeling	22
Medium feeling	7
Low feeling	4
No feeling	0
Average out of 3	2.56

Fig. 15 Memory survey result—AR & feeling of emotional attachment to the place in the past (by authors)

Did the experience increase your awareness of the authenticity and identity of the Hijazi architecture and its special urban context?

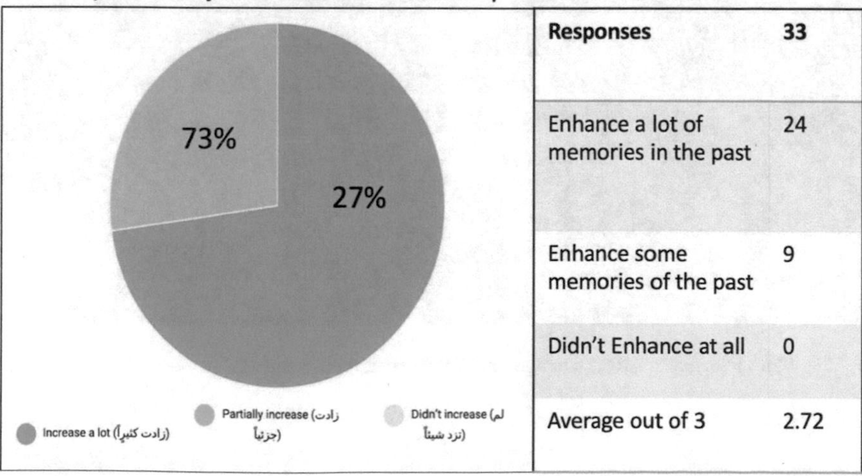

Responses	33
Enhance a lot of memories in the past	24
Enhance some memories of the past	9
Didn't Enhance at all	0
Average out of 3	2.72

Fig. 16 Identity survey result (by authors)

The data indicated that (73) percent of the participants felt that the experience increased their awareness of the authenticity and identity, of the site, and (27) percent of the participants thought it had partially increased their awareness. No participant thought that the experience did not increase their awareness of the authenticity and identity of the Hijazi context. The results confirm that AR has a high potential for increasing the perception of the identity and authenticity of the place. It was concluded from the questionnaire results that AR interpretation had a significant impact on visitors' perception, including old historical memory and identity.

5.6 Results Analysis and Findings

The table presented below provides the outcome, counting the total average out of 3 from each aspect. The results are all considered positively high from the analysis outcome. It is noted that there are remarkable results from the analysis for the questions related to the following aspects: Interactive & enjoyable, informative, and Convenience. The findings from the analysis of the results confirmed a general acceptance for utilizing AR in historical sites of KSA (Fig. 17 and Table 2).

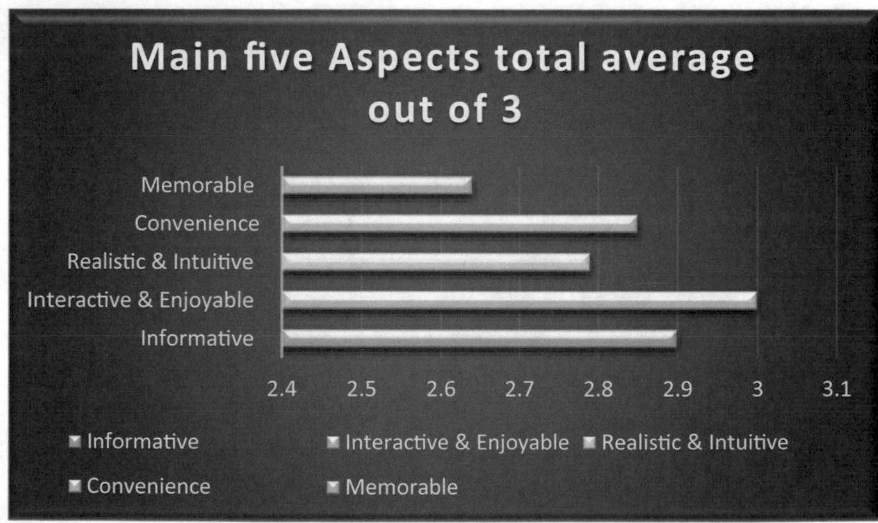

Fig. 17 Main Five aspects total average out of three charts, (by authors)

Table 2 Result analysis outcome measuring the five aspects & sub aspects out of three (by authors)

Main five aspects	Sub-aspects	Average out of 3	Total aspect average out of 3
Informative	Awareness	2.60	**2.90**
	Understanding	2.90	
Interactive & enjoyable	Excitement	3	**3**
	Enjoyable	3	
Realistic & intuitive	Realistic	2.84	2.79
	Aesthetic	2.73	
Convenience	Convenience	3	2.85
	Easy to use & handy	2.69	
Memorable	Memory	2.56	2.64
	Identity	2.72	

6 Discussion and Conclusions

The experimental AR model used in Jeddah's historic site aims to investigate visitors' reactions and responses to the use of AR tool for values interpretation. The study measured five aspects (1) Informativity, (2) Interactivity & enjoyability, (3) Reality & initiative, (4) Convenience, (5) Memorability. These aspects are extracted from the literature on how AR interpretation affects visitors' experience and interaction and changes cultural heritage perception. The analysis of the survey results and findings is primarily positive and helped trace the reactions and responses to using AR interpretation in a specific culture like Saudi Arabia and confirmed the acceptance of the local visitors and tourists of the AR interpretation model in evoking the lost physical urban memory. The results showed that the impact of the memorable aspect is the lowest, and the less feeling and attachment with the place was from the users who are not familiar with the old urban context or the young generation. However, the researchers observed the positive user's behavior during the tour that endorses high social value and a unique approach to identity aspects through the different reactions to the lost urban memory between the old and young generations.

The study confirms that implementing AR in cultural heritage interpretation will improve the overall experience and the five aspects leading to an excellent experience due to the essential impact of the AR interpretation model in enhancing the user's enjoyment and perception, and understanding of the cultural heritage.

References

1. Park, HY.: Heritage Tourism. London: Routledge, (2013).
2. Cohen, E.: The tourist guide: the origins, structure and dynamics of a role. Ann. Tour. Res. **1**(12), 5–29 (1985)

3. Beck, L., & Cable, T. T.: Interpretation for the 21st century: fifteen guiding principles for interpreting nature and culture. Sagamore Pub. Llc. (2002)
4. Icomos.org.: The ICOMOS charter for the interpretation and presentation of cultural heritage sites. http://icip.icomos.org/downloads/ICOMOS_Interpretation_Charter_ENG_04_10_08.pdf. Last accessed 13 Oct 2021
5. Robertson, I.J.: 2016 Introduction: heritage from below. In: Heritage from below, pp. 15–42. Routledge
6. Styliani, S., et al.: Virtual museums, a survey and some issues for consideration. J. Cult. Herit. **4**(10), 520–528 (2009)
7. Higgett, N., Chen, Y., Tatham, E.: A user experience evaluation of the use of augmented and virtual reality in visualising and interpreting Roman leicester 210AD. J. Cult. Herit. **4**(10), 520–528 (2009)
8. Donaire, J.A., Gali, N.: Modeling tourist itineraries in heritage cities Routes around the Old District of Girona. Agustin Santana Talavera, COMITÉ EDITORIAL DIRECTOR **3**(6), 435–449
9. Trisciuoglio, M., Yu, W.: Towards a "Digital CultHeriScape"–the changing of the cultural landscape's perception. http://citeseerx.ist.psu.edu/viewdoc/download?doi=10.1.1.659.9916&rep=rep1&type=pdf. Accessed 13 April 2021
10. Nieto, J.E., et al.: Management of built heritage via HBIM project: a case of study of flooring and tiling. Universitat Politècnica de València Virt. Archaeol. Rev. **14**(7), 1–12 (2016)
11. Jung, H.T., et al.: Cross-cultural differences in adopting mobile ARat cultural heritage tourism sites. Int. J. Contemp. Hosp. Manag. **3**(30), 621–1645 (2018)

Protecting Video Streaming for Automatic Accident Notification in Smart Cities

Pablo Piñol, Pablo Garrido, Miguel Martinez-Rach, and Manuel Perez-Malumbres

Abstract Video streaming over VANETs is a very challenging task due both to the network characteristics (packet-loss-prone networks), and to the video streaming requirements (high bandwidth), and so, solutions need to be provided in order to keep the received video over a minimum quality threshold. The specific case of use studied applies video streaming over VANETs to Automatic Accident Notification in Smart Cities. When a vehicle suffers an accident it sends an automatic notification to the emergency services and also to the surrounding vehicles in order to warn them (this notification can also be sent by a vehicle whose driver witnesses the accident). Then, a short video is recorded and sent to the emergency services. This video can be very useful to evaluate the seriousness of the situation. In this work, a simulation framework is implemented and several alternatives to improve the quality of the video received are proposed and evaluated, so it can be as much useful as possible, maintaining the quality of the reconstructed video over a minimum acceptable threshold.

Keywords Automatic accident notification · Smart cities · Video streaming · VANET

1 Introduction

1.1 Smart Cities and Connected Vehicles

In the definition of "smart city" by the IEEE Smart Cities Initiative [1], you can find that smart cities include smart mobility as one of its goals. One of the smart mobility fields of work is related to Connected and Autonomous Vehicles (CAVs). A connected vehicle is a vehicle which, making use of its communication capabilities, may interact with other vehicles (V2V—Vehicle-to-Vehicle), with the existing

P. Piñol (✉) · P. Garrido · M. Martinez-Rach · M. Perez-Malumbres
Universidad Miguel Hernandez de Elche, 03202 Elche, Spain
e-mail: pablop@umh.es

A. Visvizi et al. (eds.), *Research and Innovation Forum 2022*,
Springer Proceedings in Complexity, https://doi.org/10.1007/978-3-031-19560-0_24

infrastructure (V2I—Vehicle-to-Infrastructure) and/or with any other device or entity (V2X—Vehicle-to-Everything), which may include pedestrians, cyclists, etc.

This connectivity allows the formation of Vehicular Ad-hoc Networks (VANETs). In VANETs, vehicles can have a high relative speed between them. These networks have a highly changing topology, because vehicles may take different routes. We can have sparse networks (missing of network links) and also congested networks (packet saturation). Therefore, VANETs are packet-loss-prone networks. One of the main benefits of VANETs is Safety. We can find useful applications such as lane change assistance, forward collision warning, electronic emergency brake light, automatic accident notification, etc.

1.2 Video Streaming for Automatic Accident Notification

In the European Union, by means of regulation 2015/758 of the European Parliament [2], it is mandatory to implement the eCall (emergency Call) system in every new vehicle since 31 March 2018. In this work we analyze video streaming over VANETs in a specific application: Automatic Accident Notification. When a vehicle suffers an accident, two tasks are carried out. First, a warning message is sent to the emergency services (112/911) with useful information as the location of the accident, the type of accident, and the number of vehicles and people involved. This information is also spread to the surrounding vehicles in order to warn them. Then, a short video is recorded and sent. It can be useful for policemen in order to coordinate the intervention, for the fire brigade to plan the necessary procedures (extrication of people, extinguishing a fire, etc.), and for the medical emergency services in order to evaluate the seriousness of the situation (number and gravity of injured people, medical aerial transportation required or not, etc.). This 360° video footage of the surroundings of the vehicle can be useful to evaluate the environment of the accident and to discern if there are any other victims, like cyclists or pedestrians. A video footage of the inside of the vehicle can also be taken in case that occupants cannot answer an automatic eCall.

Sending video over VANETs is a very challenging task. As previously mentioned, VANETs characteristics make them packet-loss-prone networks, and, moreover, video streaming has high demanding requirements (video sequences need a high bandwidth to be transmitted). The combination of these two handicaps originates the need of proposals in order to keep the received video over a minimum quality threshold. In this work we address the technical part of transmitting a video footage in case of accident. We focus on applying different techniques to guarantee the quality of the video received by the emergency services.

1.3 Related Work

The development and management of smart cities does not only depend on governments and organizations, but also, as stated in [3], on citizens. This is true because an unavoidable interaction between humans and technologies arises [4, 5]. In emergency calls this interaction occurs mostly in a shocking manner, but it may ameliorate with the transmission of video. In [6], the authors carried out a study of the advantages of making a video call instead of a phone call, in case of a contingency. Video can provide a better overview of current hazards and help in situations where people cannot describe the situation properly (state of panic, stress, deaf persons, unconscious, …). In [7], the authors explore the experience of people making emergency calls. In the callers' opinion, video transmission may make available contextual information, get over communication problems and lessen the uncertainty of the situation. In [8], the authors work in the design of future and advanced emergency call systems, where the video and pictures transmission is an important part of the new contributions.

2 Simulation Framework

A simulation framework has been implemented and several alternatives have been proposed to protect the quality of the video received by the emergency services, so it can be as much useful as possible.

2.1 Simulators

To set up and carry out the experiments we have combined the use of OMNeT++ [9] (a simulation library and framework), Veins [10] (VEhicles In Network Simulation—an open source framework for running vehicular network simulations), and SUMO [11] (Simulation of Urban Mobility—an open source traffic simulation package), together with an own development named VDSF-VN (Video Delivery Simulation Framework over Vehicular Networks) [12]. With the use of these packages and other tools (specified in Sect. 2.2), we have created a scenario for the simulations. In this scenario a vehicle suffers an accident and begins the transmission of a video sequence towards the nearest RSU (Road Side Unit). This RSU, which is part of the infrastructure, is in charge of delivering the video footage to the emergency services in a timely manner.

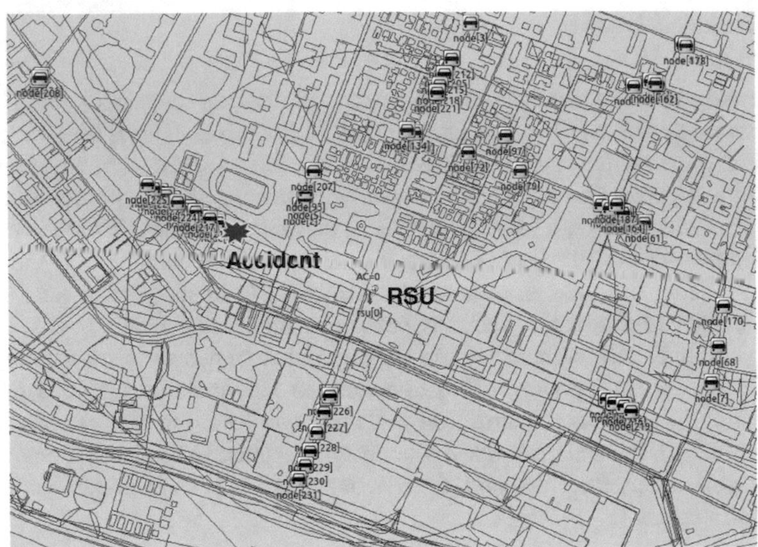

Fig. 1 Close-up of the implemented scenario (OMNeT++ simulator view). The RSU and the vehicle which has suffered an accident are tagged

2.2 VDSF-VN Framework

For the creation of the scenario and the traffic flows we have used GatcomSUMO tool [13], which is part of VDSF-VN [14]. With this tool we have downloaded a real map from OpenStreetMap (OSM) [15] and converted it into the format that SUMO can manage. We have also created several vehicle flows inside the selected region and we have placed in the map the RSU which will receive the video transmission from the crashed vehicle.

We have selected a square region of 2000 × 2000 m of the city of Honolulu in the Hawaii Islands. In Fig. 1 you can see a close-up of the scenario created. In the center of the image you can see the RSU and, inside its coverage range, the vehicle which has suffered an accident.

2.3 Video Encoding

For the encoding of the video footage we have used HEVC (High-Efficiency Video Coding) standard [16]. We have chosen a well-known video sequence, included in the Common Test Conditions [17] for the evaluation of HEVC encoder, named "Race Horses", with a resolution of 832 × 480 pixels and a frame rate of 30 frames per second.

Table 1 Bitrate (Mbps) and PSNR (dB) for the encoded video sequence (Race Horses)

Encoding mode	QP	Tiles/frame	Bitrate (Mbps)	PSNR (dB)
AI	31	1 til/frm	5.80	36.14
AI	31	2 til/frm	5.82	36.14
AI	31	4 til/frm	5.87	36.14
AI	31	8 til/frm	5.91	36.14
I7P	28	1 til/frm	3.23	36.29
I7P	28	2 til/frm	3.24	36.28
I7P	28	4 til/frm	3.27	36.28
I7P	28	8 til/frm	3.28	36.27

We have evaluated two different encoding methods to see which one of them can provide a better resilience. The sequence has been encoded in AI (All Intra) and I7P modes. In AI mode, all the frames are encoded by using intra-picture prediction exclusively, so every frame of the sequence is encoded independently. In I7P mode, the first frame of every 8 is encoded as an intra frame, and the 7 remaining frames are P (predicted) frames. For each encoding mode, we have tested 4 different layouts (number of tiles per frame). Tiles are independently encoded (and decoded) rectangular regions of a frame. Some robustness can be achieved because of the independency from the rest of the tiles. We have used GatcomVideo tool [14] to encode the video, aiming at a compressed video quality of 36 dB (PSNR—Peak Signal-to-Noise Ratio).

In Table 1, the bitrate and PSNR of the encoded video streams are shown. We need a value of 31 for the Quantization Parameter (QP) for AI mode and a value of 28 for I7P mode. I7P mode is way more efficient as it only requires a bitrate around 3.25 Mbps to achieve the same quality than AI mode, which needs a bandwidth around 5.85 Mbps to get a similar quality.

3 Experiments and Results

3.1 Background Network Traffic Conditions

In this scenario, in addition to "ideal conditions" (no background network traffic), three different rates of traffic have been injected. These conditions simulate light, medium, and heavy traffic loads. We have injected a payload of 512 bytes (4096 bits) from 10 sources (nearby vehicles in the radius of coverage) at the following packets-per-second (pps) rates: 0 pps, 25 pps, 50 pps, and 75 pps, which generate the corresponding background traffic rates: 0 Mbps, 1 Mbps, 2 Mbps, and 3 Mbps. In the following subsections we will present the results under these network conditions.

3.2 Encoding Modes and Number of Tiles Per Frame

First of all, we have compared the performance regarding packet losses. I7P mode is more efficient in terms of compression and, therefore, the bandwidth required is lower than for AI mode. This provides an intrinsic protection because fewer collisions may arise. In Fig. 2 it can be seen that, in ideal conditions and with a low network load, both encoding methods provide very similar packet loss rates. Instead, for medium background traffic load (50 pps), with AI mode, around a 3% more of packet losses occur than with I7P mode, and for high background traffic load (75 pps), this difference increases up to around 5%.

Regarding the different layouts evaluated, in Fig. 2 we can also observe that when we use 4 or 8 tiles/frame to encode the video sequence, we always get lower levels of packet loss than when 1 or 2 tiles/frame are used, for both encoding modes.

But when we reconstructed the videos and measured their quality, we realized that results did not exactly correlate with packet losses. In Fig. 3, the quality of the reconstructed videos is shown. The red dashed line delimits the upper threshold of the video quality. This upper bound is 36.29 dB which is the maximum quality that the decompressed video can achieve if none of the packets get lost (see Table 1). The green dashed line indicates the minimum acceptable quality threshold, under which the video is not considered useful for the emergency services (27 dB). As it can be seen, AI mode, even suffering from higher packet loss rates than I7P mode, shows a greater robustness and keeps reconstructed video over the minimum threshold with low back ground traffic (25 pps). Whereas, I7P mode only keeps inside the bearable range for the 8 tiles/frame layout for that network load. For both encoding modes, this layout (8 tiles/frame) has always the best performance, followed by the layout with 4 tiles/frame. In AI mode, these two layouts remain inside the acceptable margins even with medium traffic conditions (50 pps).

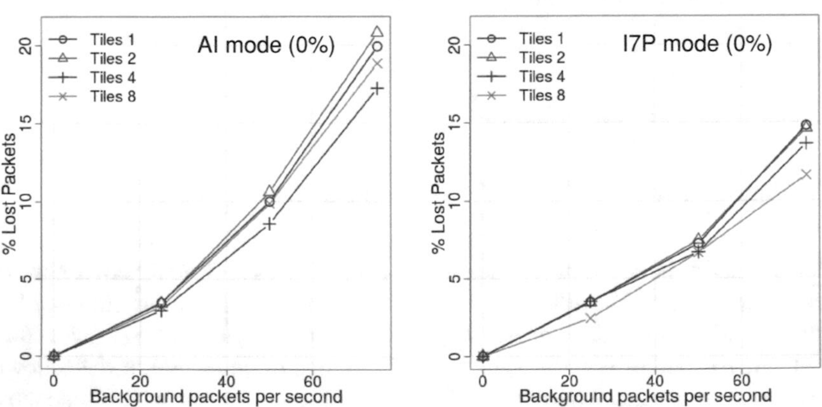

Fig. 2 Percentage of lost packets at different background rates (0, 25, 50, 75pps) with 0% of priority packets. Left: AI mode. Right: I7P mode

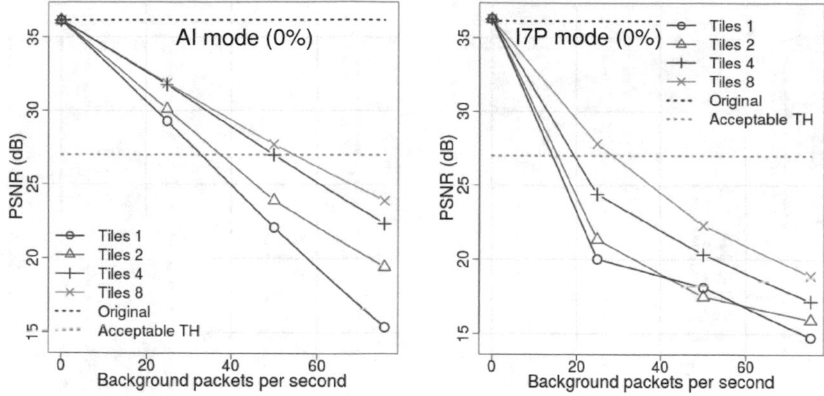

Fig. 3 PSNR (dB) of the reconstructed video sequence at different background rates (0 pps, 25pps, 50pps, 75pps) assigning a high priority to 0% of video packets. Left: AI mode. Right: I7P mode

3.3 Protection of Video Packets

In order to add an extrinsic protection, we have applied the prioritization of video packets. This is based on the Quality of Service (QoS) mechanism included in VANETs. Several queues are implemented and each one of them has a different priority. In order not to collapse other applications, we have assigned a higher priority to 75% of video packets (instead of 100%). In Fig. 4, the outcome of the use of QoS priorities can be seen. Now, AI mode shows a robust behavior in all the spectrum of network loads, even for heavy values of background traffic. As it can be seen, I7P mode, even with the "help" of a higher priority (and needing nearly half of the bandwitdh than AI mode), does not reach the minimum acceptable quality for heavy traffic loads (75 pps).

3.4 Different Compression Rates

We have also tested the best combination of the evaluated mechanisms (AI encoding mode, 4 or 8 tiles/frame, and 75% of packets prioritized) with other levels of compression, obtaining video sequences with higher (and lower) PSNR values corresponding to higher (and lower) bitrates. For soft compression where the initial quality value is high, the wireless channel becomes saturated and this fact complicates the good reception of the transmitted video. On "the other side of the coin", by applying a higher level of compression (which implies lower packet loss rates), the generated video bitstreams have lower PSNR values which, afterwards, entail lower quality of the reconstructed videos. So, we can conclude that the best choice is to use the lowest level of compression possible which does not saturate the communications channel.

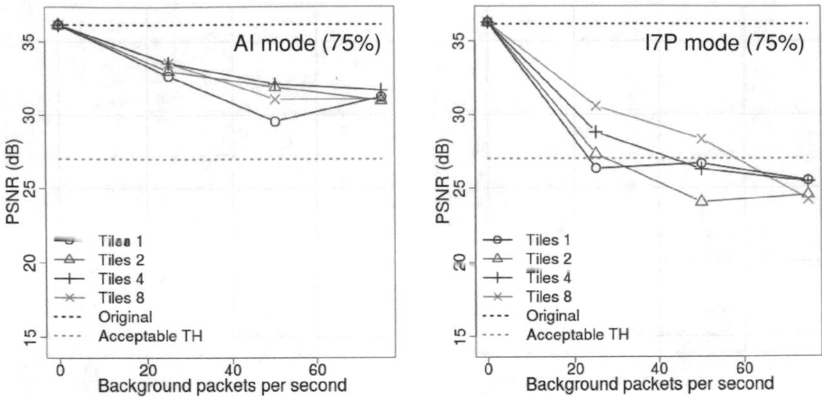

Fig. 4 PSNR (dB) of the reconstructed video sequence at different background rates (0 pps, 25pps, 50pps, 75pps) assigning a high priority to 75% of video packets. Left: AI mode. Right: I7P mode

4 Conclusions and Future Work

4.1 Conclusions

Several works emphasize the benefits of video transmission in emergency situations. These benefits can help to better analyze the environment of an accident and to give a quicker and better response to incidents. In this work we have evaluated the feasibility of emergency video transmission over VANETs. The best choice to guarantee that video streams keep over a minimum quality threshold combines the selection of two intrinsic methods (AI mode combined with 8 or 4 tiles/frame layouts) and an extrinsic method, based on assigning emergency video packets a higher priority in order to preserve most of them from packet losses. Also the encoding of the video streams at compression levels which do not saturate the wireless channel while keeping a good initial level of quality has shown to be the most appropriate decision.

4.2 Future Work

Future directions of our research include the evaluation of multi-hop emergency video transmission in zones with no nearby infrastructure available and/or in shadow areas where the poor infrastructure coverage complicates the communication. Another future line of research focuses in the design and evaluation of adaptive mechanisms which may adjust the level of protection to the specific network conditions at the moment after the accident has happened.

Funding: This research was supported by the Valencian Ministry of Innovation, Universities, Science and Digital Society under Grant GV/2021/152, and by the Spanish Ministry of Science, Innovation and Universities and the Research State Agency under Grant RTI2018-098156-B-C54, cofinanced by FEDER funds (MCIU / AEI / FEDER, UE).

References

1. IEEE Smart Cities Initiative Homepage. https://smartcities.ieee.org. Last accessed 30 Mar 2022
2. Regulation (EU) 2015/758 of the European Parliament and of the Council of 29 April 2015 concerning type-approval requirements for the deployment of the eCall in-vehicle system based on the 112 service and amending Directive 2007/46/EC (2015)
3. Visvizi, A., Troisi, O.: Effective Management of the Smart City: An Outline of a Conversation. In: Visvizi, A., Troisi, O. (eds) Managing Smart Cities. Springer, Cham (2022)
4. Zait, A.: Exploring the role of civilizational competences for smart cities' development. Transf. Govern. People Proc. Policy **11**(3), 377–392 (2017)
5. Kashef, M., Visvizi, A., Troisi, O.: Smart city as a smart service system: Human-computer interaction and smart city surveillance systems. Comp Hum Behav **124** (2021)
6. Neustaedter, C., Jones, B., O'Hara, K., Sellen, A.: The benefits and challenges of video calling for emergency situations. In: Proceedings of the 2018 ACM Conference on Human Factors in Computing Systems (CHI), pp. 1–14. ACM, Montreal (2018)
7. Singhal, S., Neustaedter, C.: Caller needs and reactions to 9–1–1 video calling for emergencies. In: Proceedings of the 2018 Designing Interactive Systems Conference (DIS), pp. 985–997. ACM, New York (2018)
8. Dash, P., Neustaedter, C., Jones, B., Yip, C.: The design and evaluation of emergency call taking user interfaces for next generation 9-1-1. Front. Hum. Dynam. **4**, 2673–2726 (2022)
9. OpenSim Ltd.: OMNeT++ Discrete Event Simulator. http://www.omnetpp.org. Accessed 30 Mar 2022
10. Sommer, C., German, R., Dressler, F.: Bidirectionally coupled network and road traffic simulation for improved IVC analysis. IEEE Trans. Mobile Comput. (TMC) **10**(1), 3–15 (2011)
11. Lopez, P.A., Behrisch, M., Bieker-Walz, L., Erdmann, J., Flötteröd, Y.P., Hilbrich, R., Lücken, L., Rummel, J., Wagner, P., Wiessner, E.: Microscopic traffic simulation using SUMO. In: The 21st IEEE International Conference on Intelligent Transportation Systems (ITSC), pp. 2575–2582. Maui, USA (2018)
12. Garrido, P.P., Malumbres, M.P., Piñol, P., Lopez, O.: Source coding options to improve HEVC video streaming in vehicular networks. Sensors **18**(9), 1–15 (2018)
13. Garrido, P.P., Malumbres, M.P., Piñol, P.: GatcomSUMO: a graphical tool for VANET simulations using SUMO and OMNeT++. In: Proceedings of the 2017 SUMO User Conference, pp. 113–133. Berlin, Germany (2017)
14. Garrido, P.P., Malumbres, M.P., Piñol, P., López, O.: A simulation tool for evaluating video streaming architectures in vehicular network scenarios. Electronics **9**(11), 1–15 (2020)
15. OpenStreetMap. http://www.openstreetmap.org. Accessed 30 Mar 2022
16. Joint Collaborative Team on Video Coding (JCT-VC): High Efficiency Video Coding (HEVC). In: Technical Report Recommendation ITU-T H.265|ISO/IEC 23008–2. Geneva, Switzerland (2018)
17. Bossen, F.: Common test conditions and software reference configurations. In: Technical Report JCTVC-L1100. Geneva, Switzerland (2013)

Creating a Smart Transit Option for a Car-Dependent City: The Case of Jeddah City

Nada T. Bakri and Asmaa Ibrahim

Abstract The city of Jeddah faces an increase in motorization and car dependency in transportation systems, resulting in increased traffic congestion and more negative effects such as an increase in pollution and car accidents. Nowadays, the citizens are not in favour of using public transportation services due to the lack of several factors, which include but are not limited to mal-planning in the land use distribution causing long journeys from one destination to another, quality of hygiene, safety, destination and location, affordability, and other factors that this research aspires to determine and highlight. The interpretation of "smart city" should not be limited to using advanced technology in transportation but should also include the smartness in adopting different transit options to satisfy the different social needs of the different community sectors. The execution methodology of this research is based on exploring the current understanding of Transit-Oriented Developments (TOD) that contributed to bettering the transportation issues that occurred in car-dependent cities. Following that, an empirical case study of Jeddah city will be conducted to divide the city into homogeneous zones supported by a qualitative data gathering of different stakeholder's opinion that aims to gain a better understanding of Jeddah's condition with public transportation modes up to this date. The city of Jeddah is experiencing a surge of car dependency, given the increasing traffic congestion, which causes the city to suffer from increased pollution and vehicle accidents.

Keywords Transit-oriented · TOD · Walkability · Car-dependency

N. T. Bakri (✉) · A. Ibrahim
Effat University, Jeddah City, KSA, Saudi Arabia
e-mail: NadaTBakri@gmail.com

A. Ibrahim
e-mail: asibrahim@effatuniversity.edu.sa

A. Visvizi et al. (eds.), *Research and Innovation Forum 2022*,
Springer Proceedings in Complexity, https://doi.org/10.1007/978-3-031-19560-0_25

1 Research Introduction

Urban planners and decision-makers face a difficult challenge with cities that grow rapidly. Cities in Saudi Arabia have experienced rapid growth and urban sprawl in the last five decades, with Jeddah city leading this event [1]. Jeddah City has a rich functional dimension that includes a movement system that passes long distances from north to south. However, it faces a problem with limited urban transportation and the lack of public transit amenities and infrastructure that supports walkability

Jeddah City has become a car-dependent city since the travel pattern has changed dramatically due to the linear urban sprawl in the last four decades. The city expansion resulted in only 7% of the citizens using public vehicles [2]. Jeddah nowadays has only four public transportation systems provided by the government & private companies:

SAPTCO: Saudi Arabian Public Transportation company.

Coaster Buses: Operated by Individuals.

Taxi Service or "Ojrah."

Uber and Careem services.

When the sprawl started to come into action in the 80s, as Al-Joufie mentioned, the city started expanding dramatically. Even the main city attractions and locations started varying from north to south (Fig. 1). The citizens of Jeddah city do not favour using public transportation services due to the lack of several factors. These factors include but are not limited to mal-planning in the land use distribution causing long journeys from one destination to another, quality of hygiene, safety, destination and location, affordability, and other factors. Therefore, this research investigates this challenge to promote walkability and the dependence on transit options.

Jeddah is experiencing a surge of car dependency, given the increasing traffic congestion, which causes the city to suffer from increased pollution and vehicle accidents. This research aims to understand the reasons that discourage citizens of Jeddah city from using public transportations. It proposes a framework incorporating different stakeholders attempting to motivate private car users to use transit-oriented services and options for walkability if relevant.

Based on the problem statement of this study, the objectives of this research incorporate the following:

- Investigating the main factors that motivate citizens to use public transportation services through a literature review and an empirical case study in Jeddah. These factors will be later compared with the socio-economical characteristics of Jeddah city to ensure its correspondence with the needs and desires of the users.
- Determining the factors that promote the dependence on walkability and transit options by analyzing different international and national case studies that have previously formulated plans for achieving this objective. This objective is essential to understand the practical design approaches and learn from them to be considered

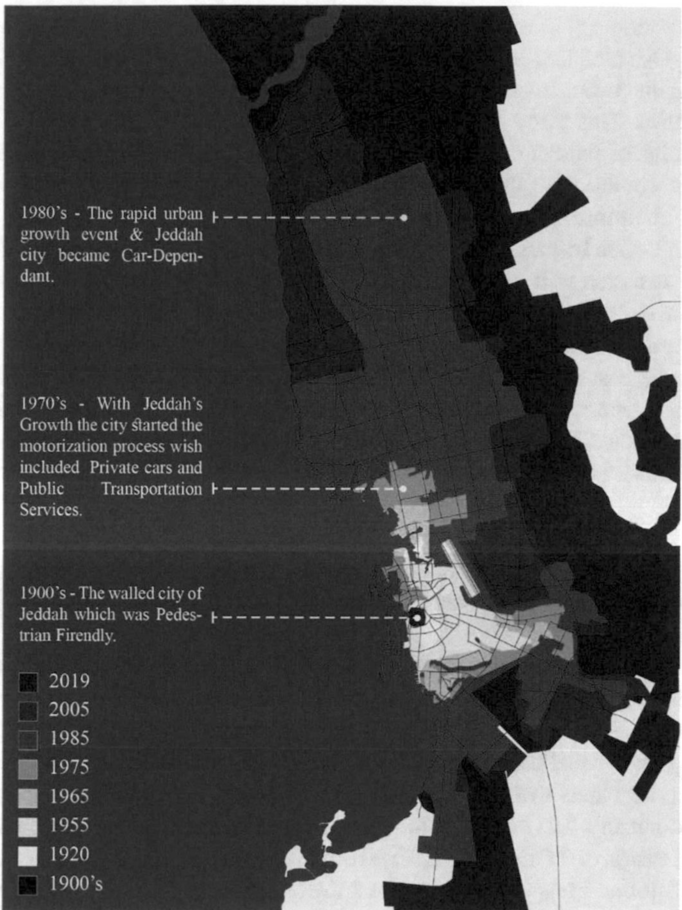

Fig. 1 A map of the urban growth of Jeddah city between the 1900's–2019

in the case of Jeddah city. Moreover, and finally, the study will analyze these factors concerning the preference of the citizens of Jeddah city. The case studies will be selected based on criteria identifying similarities with Jeddah City. The empirical part of this study shall explore the unique aspects of Jeddah City and deduce the specific factors for promoting walkability related to Jeddah City in particular.

- Analyzing how citizens select their mode of mobility inside the different parts of Jeddah.

This research is limited to 3 municipalities in Jeddah: Northern Ubhur, Southern Ubhor, and New Jeddah. This limitation is due to the lack of time within this research process and the lack of accessibility to participants from other municipalities. Also, the limitations of this study include not being able to reach stakeholders representing the local public transportation companies of the public and private sectors.

The methodology of this research is based on exploring the current understanding of Transit-Oriented Developments (TOD) that contribute to bettering the transportation issues that occur in car-dependent cities—also determining the impacts of TOD on such cities. This study will ensure a comprehensive understanding of TODs and what the city of Jeddah could benefit from. Later, this study aspires to investigate the current condition of the transportation modes and citizens of Jeddah city up to this date. Therefore, this research will consider using an evidence-based approach through an action research methodology that includes three levels of investigation. First, this research will explore through a qualitative methodology the concept of TOD by investigating its applications and approaches to implementation. Next, the study will conduct an analytical methodology to explore two transit service projects in the middle east to learn about the implementation of TOD in cities with similar conditions to Jeddah city. Following that, an empirical case study of Jeddah city will be conducted through a mixed-approach methodology to divide the city into homogeneous zones supported by a quantitative analysis of a distributed survey that aims to gain a better understanding of Jeddah's citizen's socioeconomic characters and their satisfaction level of public transportation modes up to this data. Finally, the study will use a deductive methodology to analyze all findings and recommend a framework for the future implementations of TOD in the proposed areas in Jeddah city, if applicable.

The study is divided into six chapters. The first chapter outlines this study's research background, importance, goals, and objectives and the methods utilized to complete it. The following chapter explores the general understanding of TOD using an inductive methodology. This chapter also investigates the principles of TOD and the requirements that users of such projects need while comparing them to the case of Jeddah city. The third chapter will analyze different case studies related to the successful usage of TOD in cities; The fourth chapter explores the historical development of public transportation modes introduced in Jeddah city, highlighting the change in the transportation system that different authors documented. This exploration will aid in initiating a framework of design that will help the citizens of Jeddah with resilient travel modes that reflect their culture and respond to their socioeconomic needs. The fifth chapter presents the empirical case study in Jeddah city. It investigates the city's physical and socioeconomic character. It divides it into categorized homogeneous zones to have a cross-sectional analysis and ensure the representation of all socioeconomic sectors. Later, the study seeks to determine the satisfaction of Jeddah's citizens by surveying to analyze the requirements that would motivate the citizens to use different transportation modes instead of relying on private cars. Finally, the last chapter will conclude the study by compiling the different results of the study chapters, recommending a framework for TOD implementation in different areas of the study, and proposing future research areas (Fig. 2).

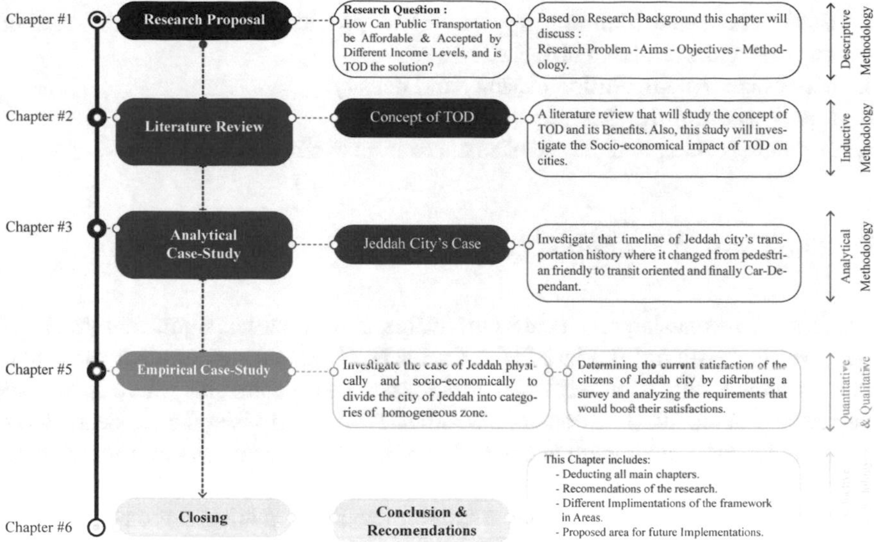

Fig. 2 The structure flow diagram of the study

2 Literature Review: The Concept of Transit-Oriented Developments

In highly urbanized cities such as Jeddah city of Saudi Arabia, a large percentage of the population depends on private cars for daily travel. This study aims to investigate the concept of TOD and discover the factors that helped motivate the citizens toward using different transit-oriented solutions. The study selected the TOD concept for its efficiency and ability to offer diverse mobility and transportation options.

Transit-Oriented Development (TOD) is a concept that has been developed as a cure for car-dependent cities. Chan, Nakamura & Imura (2016) described TOD as a radical concept that focuses on an alternative method of travel that is more efficient than vehicles. Therefore, TOD leads to compact, walkable, mixed-use neighbourhoods based on high-quality transport systems.

C40 cities advise planners to use the eight principles of TOD that were stated by the Institute of Transportation and Development Policy (ITDP) that would lead to efficient use of the TOD concept [3], and these principles were:

1. **Promote walkability** by offering the users of the neighbourhood the required infrastructures and facilities, such as paved sidewalks with different usable zones.
2. Provide the users with the opportunity to **choose non-motorized transportation modes** such as cycling.
3. Develop a street and **path network that is dense** and well-planned to serve the user's daily travel destinations.
4. Plan developments to be **located nearby high-quality transportation services**.

5. Used **mixed-use planning** approaches to provide users with their required services within walking distance.
6. **Maximize transportation** capacity and density.
7. Develop **commuter-friendly** areas.
8. Regulate parking and the use of roads to **improve mobility**.

3 The Case of Jeddah City

Jeddah is the second-largest city in Saudi Arabia, and its size has significantly affected the travel-ling pattern [4]. Since 2007, the usage of private cars has increased up to 93% in Jeddah city. This fact urges this research to understand where Jedda city has reached in the 2020s to conclude whether TOD could assist the car-dependency situation. Jeddah is an important city with a rich history, and its strategic location has been glorified through the words of many historians. The city of Jeddah has a history dating back over 3000 years when it was utilized as a fishing settlement. During that time, a tribe known as the 'Quda'ah' moved into the region, and legend shows that the city got its name from the Qudah tribe. The numbers account for more than 96 per cent of all daily travel, and many of Jeddah's roadways are congested. The UN-Habitat conducted a study in 2019 to examine the functionality of Jeddah city's road network. Forty-eight percent of the population, or two million individuals, had access to the city centre within a 15-min drive. In addition, three essential city districts were subjected to a pedestrian accessibility investigation. This analysis shows that the culture of walkability is attached to the concept of sports activity and exercising (Fig. 3).

4 The Empirical Case Study of Jeddah City

The study aims to analyze four main homogeneous zones of Jeddah city to understand better the different categories of locations and citizens that this study will face. The empirical case study identifies its primary data from interviews and distributed questionnaires. Later, the study will continue to compare the primary data collected with the secondary data gathered previously to design a framework that resembles a vision that will create an efficient and desirable solution for the problem of car dependency. This research phase will look at four primary homogenous zones in Jeddah to further study the different areas and inhabitants that this research will encounter. The region of Aubhor in northern Jeddah and the area of new Jeddah in the centre zone of Jeddah city will be the only homogenous zones that this study will investigate due to the limitation of time and resources (Fig. 4). This analysis aims to study the zona's road network, street pattern, and socioeconomic character.

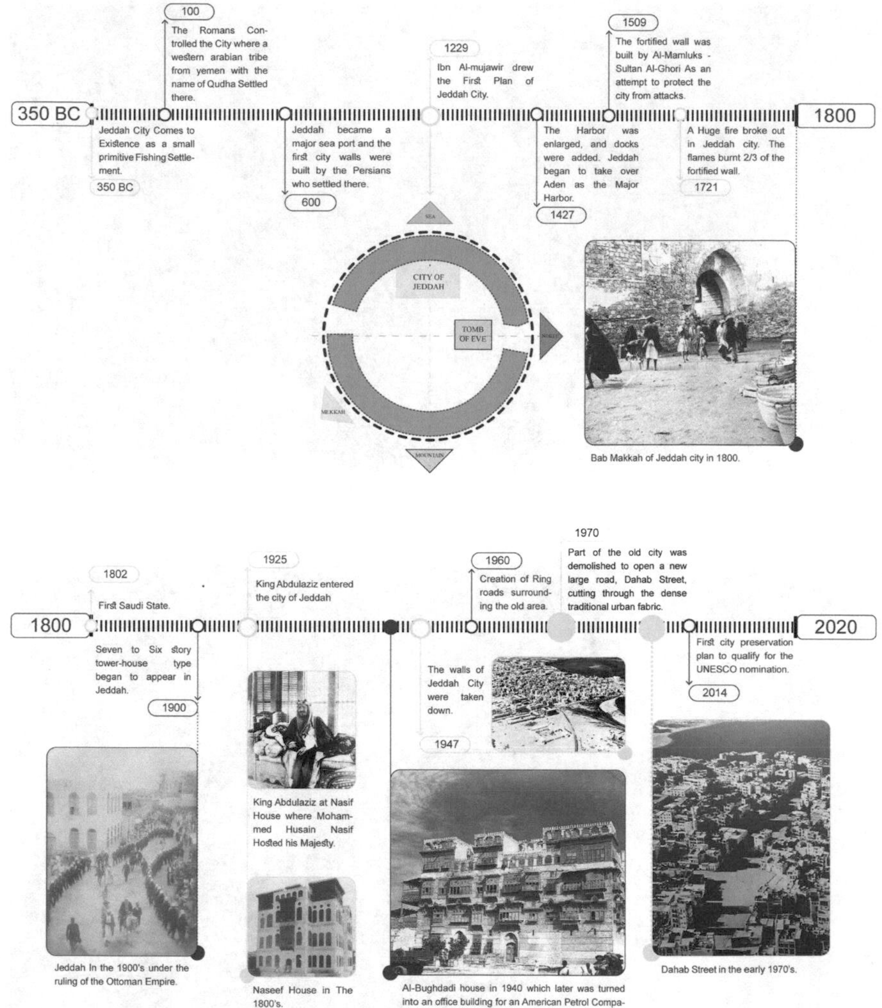

Fig. 3 The Morphological Changes of Jeddah city between 1970–2007

4.1 Category#1: The Zone of Aubhor, Northen Jeddah City

This category comprises 25 districts under two major municipalities of Jeddah - Ubhur and Northen Ubhur Municipality. However, only 6 of those districts are homogeneous regarding social, physical & economic aspects. The six districts are North Ubhur, Al-Safari, Al-Shara'a, Al-Yaqoot, South Ubhur, and Al-Asalah.

Fig. 4 The Selected Homogeneous Zones of Jeddah City (Category 1,2)

Category.1 is significant for three prominent landmarks as the zone's main attraction and destination. As shown in Fig. 5, The first prominent landmark (1) is Jeddah's economic city which is expected to be completed in 2025. Also, the first landmark includes Jeddah's Tower, which is expected to compete with the international towers of the world in its height. The second landmark on map (2) resembles one of the Al-Haramian train stations. Finally, the third landmark on map (3) is the King Abdulaziz International Airport of Jeddah city. These landmarks resemble the prominent landmarks and attractions of category.1. The network of categories.1 roads, as shown in Fig. 6, is focused on two main highways of Jeddah city. These highways are Al-Madina rd. and King Abdulaziz rd. which are the leading used rd. to reach most zone destinations, as shown in Fig. 13. As shown in Fig. 7, the street pattern in category.1 is mainly fragmented parallel. The streets in this zone have minimal curves following the major arterial roads, which eases the driving experience between districts and neighbourhoods (Figs. 8, 9, 10, 11 and 12).

The population in category.1 varies between 1 and 28 p/ha. Compared with the other categories this study discusses, this category includes the least population density since it expanded in the late 2000s. The zone of category.1 includes excellent housing conditions in high-socioeconomic-status districts (Level 1), with villas designed for one or two families being the norm. These areas often have a low population density and two or three automobile lanes on their roadways. South Aubhor has low socioeconomic level districts with low housing and living conditions (Level 3). These are the only regions where poor non-Saudis can afford to live due to the low cost of housing. Central and southern Jeddah City areas have dense populations, and power and water are frequently unavailable. As a result, many save water in tanks to use when they need it. The drainage system is insufficient, and surface water just needs a tiny quantity of rain to arrive [5].

4.2 Category#2: The Zone of New Jeddah, Central Jeddah City

Category.2 is the largest category in this analysis. It includes two significant municipalities of Jeddah city: Ubhor and New Jeddah. This category includes ten districts that will be discussed: Al-Basateen, Al-Shate'e, Al-Mohammadeya, Al-Murjan, Al-Nae'em, Al-Nahdha, Al-Khaldeyah Al-Rawdhah, Al-Zahrah, and Al-Salamah District. Category.2 was selected due to its location's significance near the King Abdulaziz Airport (1), as shown in Fig. 8. Also, it includes one of Jeddah city's primary touristic locations, the Waterfront of Jeddah city. This category includes one of the largest central commercial areas of Jeddah city, where most commercial and office buildings are located. The network of category.2 roads, as shown in Fig. 9, are also focused on the two main highways of Jeddah city. **Al-Madina rd.** and **King Abdulaziz rd.** are the most often used highways to access most zone locations. Also, Another major street that tourists and entertainment facility users

Fig. 5 Category#1 of the empirical casestudy of Jeddah city_landmarks locations

Fig. 6 Category#1 of the empirical casestudy of Jeddah city_network analysis

Fig. 7 Category#1 of the empirical casestudy of Jeddah city_street pattern

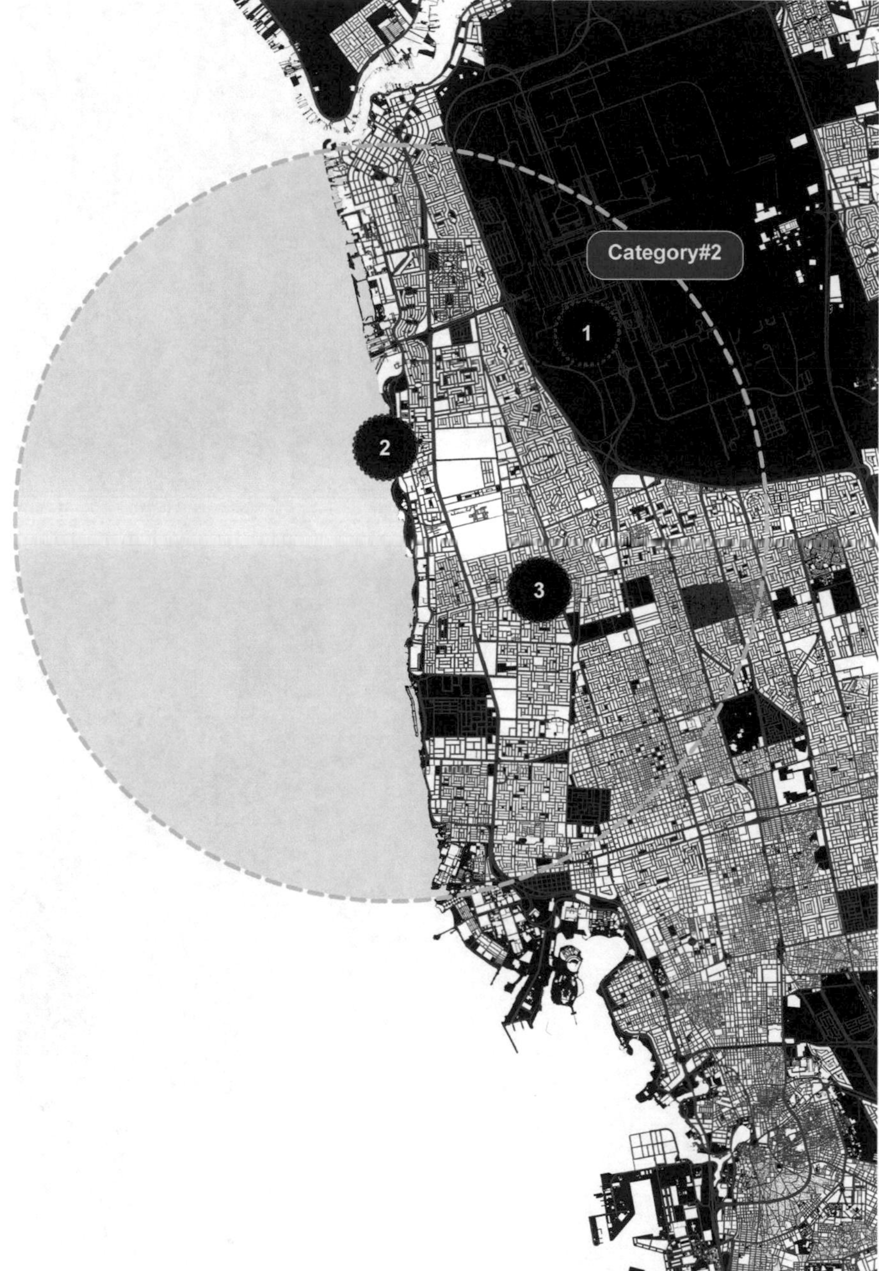

Fig. 8 Category#2 of the empirical casestudy of Jeddah city_landmarks locations

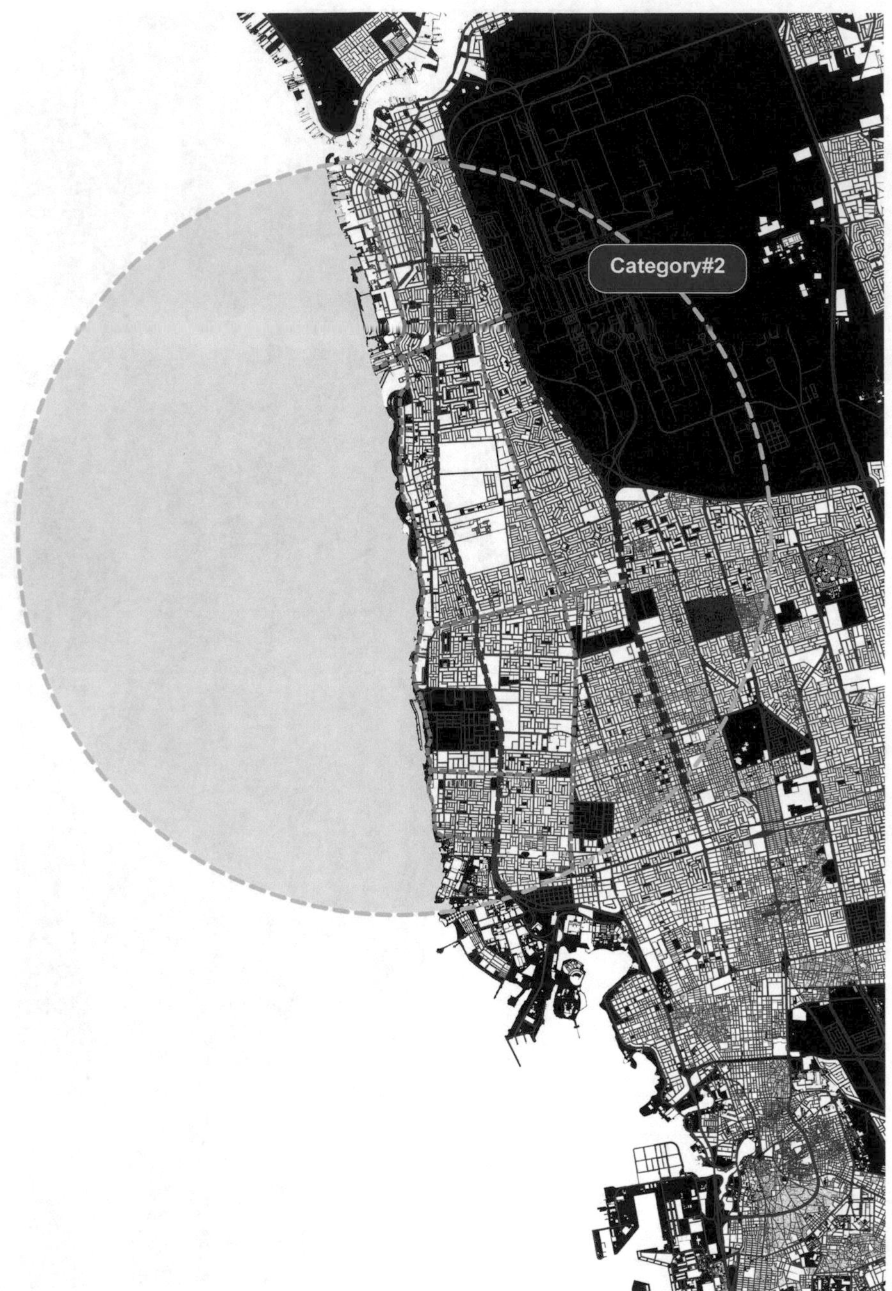

Fig. 9 Category#2 of the empirical casestudy of Jeddah city__network analysis

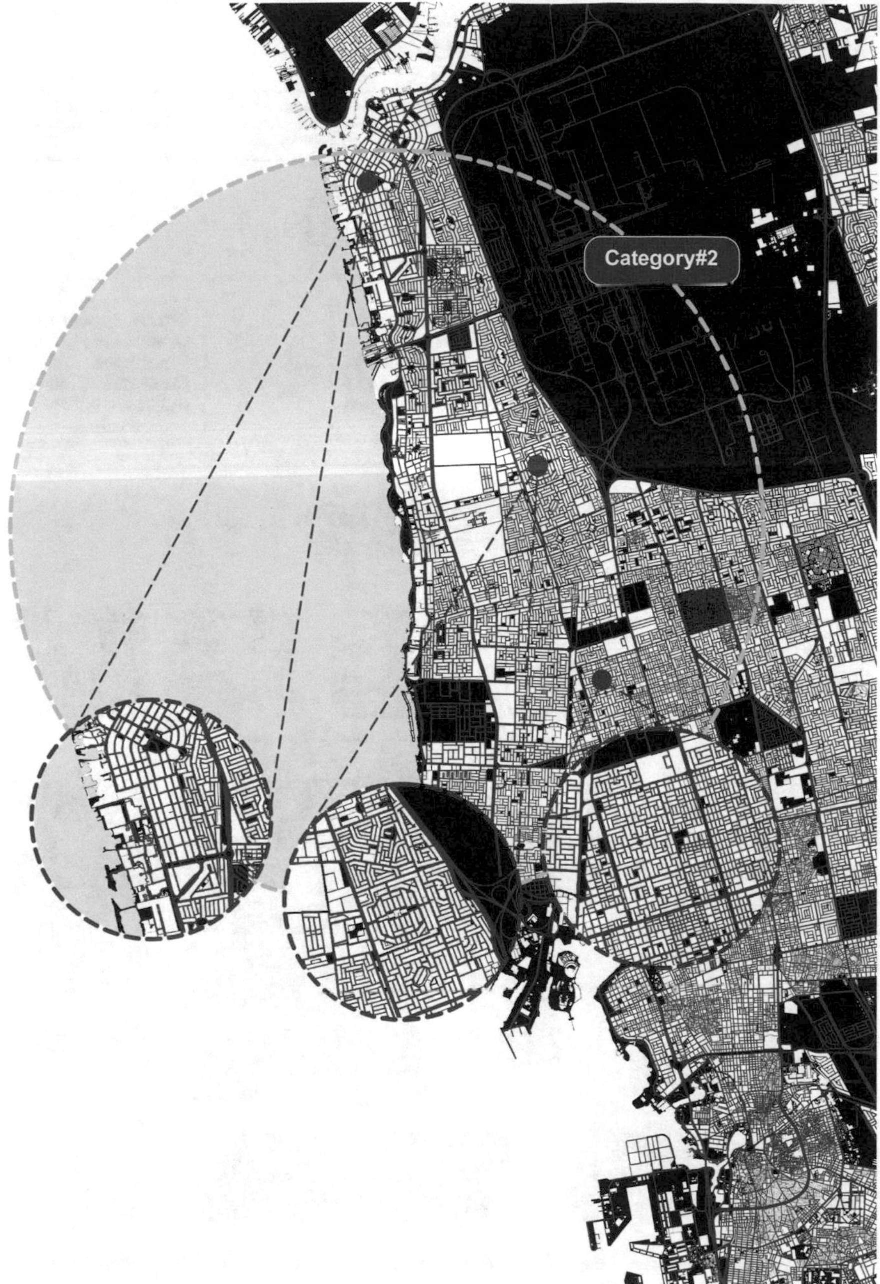

Fig. 10 Category#2 of the empirical casestudy of Jeddah city__street pattern

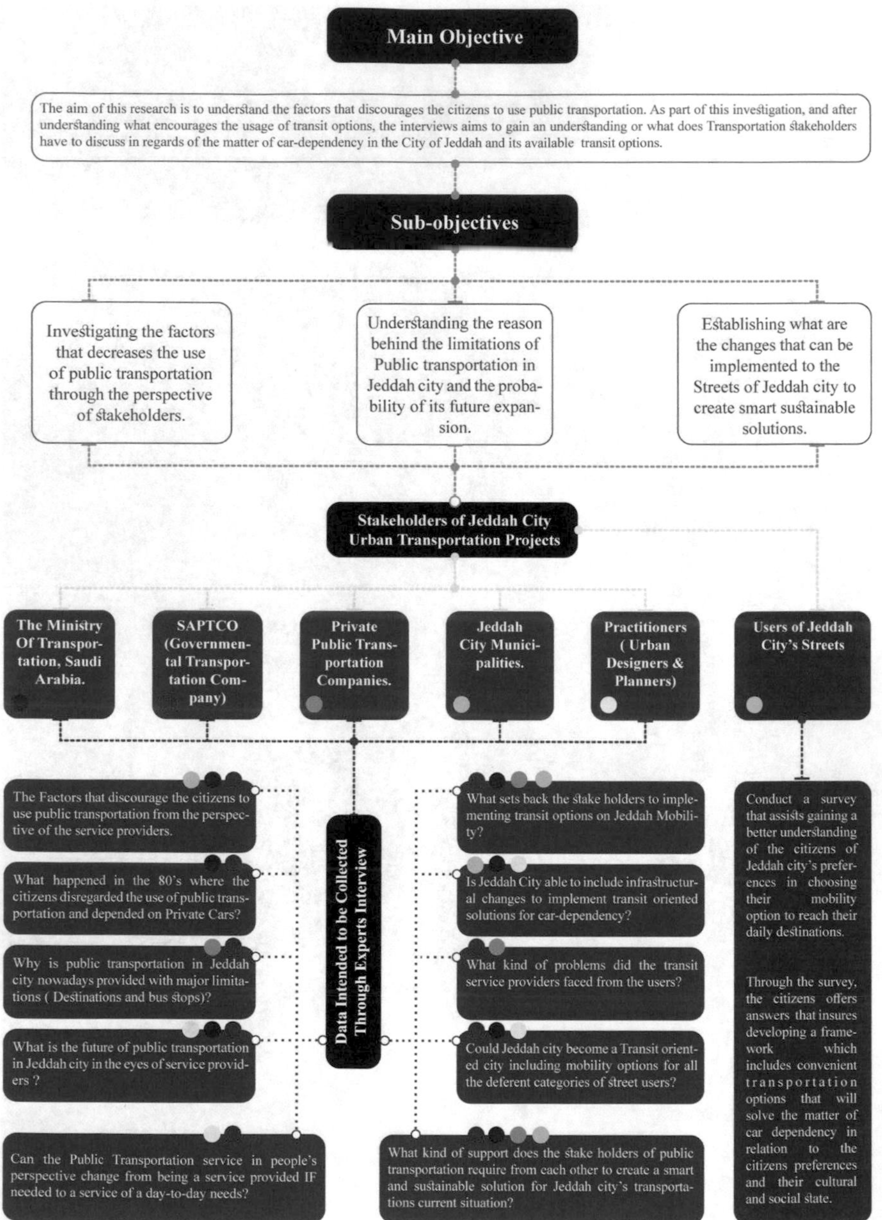

Fig. 11 Category#2 the mindmap of the urban transportation stakeholders in Jeddah City (Red circles identify the interviews and questionnaire conducted)

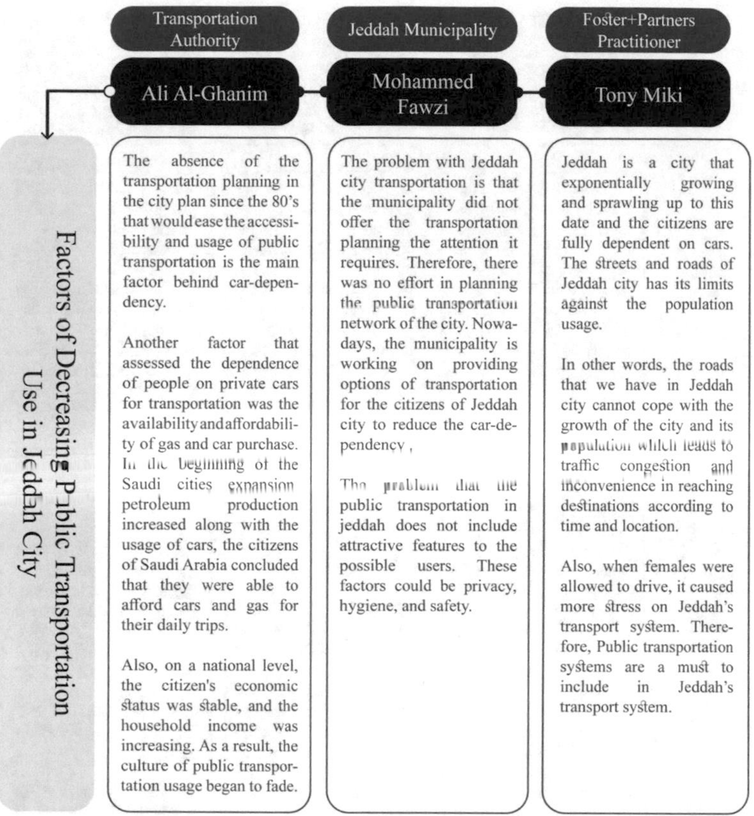

Fig. 12 The perspective of stakeholders on the factors of decreasing public transportation use in Jeddah City

mainly use is Al-Corniche rd. which overlooks the seashore of Jeddah city. Other main streets are included in category.2 are **Hira st., Prince Sultan st., Sari st. and Prince Mohammed bin Abdulaziz st.** The street design in category.2 is mostly fragmented parallel, as illustrated in Fig. 10. The roadways in this zone feature few turns and follow main arterial roads and highways, making travelling between districts and neighbourhoods easier. Housing conditions in high-socioeconomic status neighbourhoods (Level 1) are outstanding [5]. The population density in category.2 fluctuates between 1 and 43 p/ha. Most of these areas are densely populated and primarily Saudi, with more than 6–8 individuals living in a 4–5 bedroom flat.

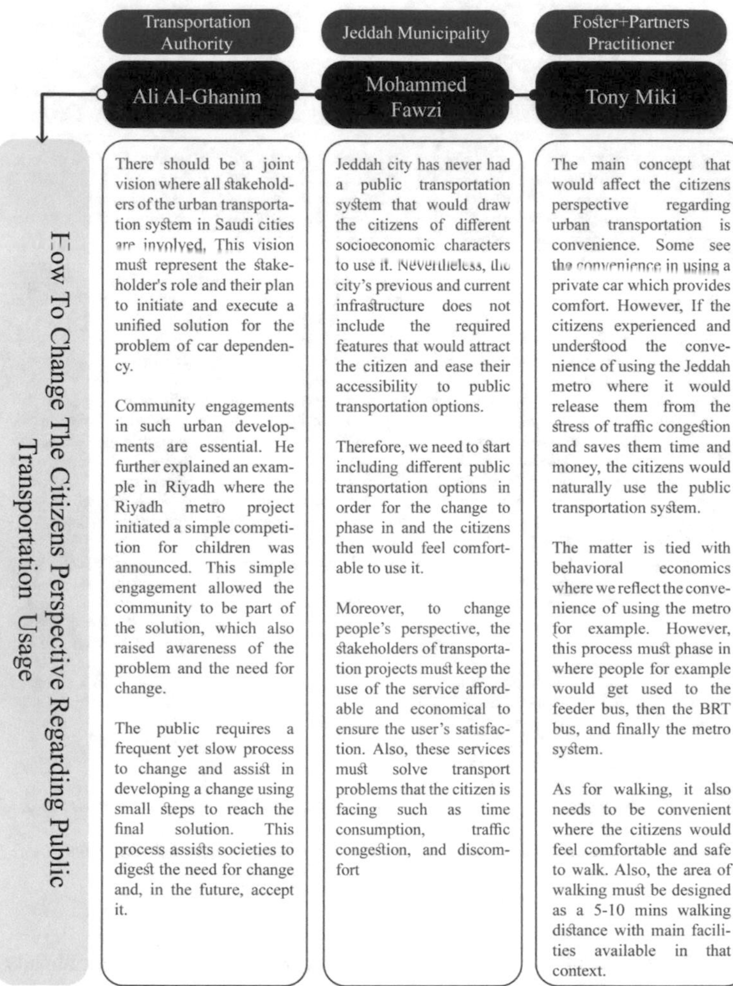

Fig. 13 The stakeholders' perspective on how to change the citizen's perspective regarding public transportation use

4.3 Interviews of Primary Data Collection

The research conducts semi-structured interviews with five main stakeholders of Jeddah's urban transportation model. As shown in Fig. 11, the main aim of this study is to discover what variables deter individuals from using public transportation. Moreover, the figure identifies the five main stakeholders of Jeddah's urban transportation system. The interviews are conducted with four stakeholders, excluding the users of Jeddah streets, where a survey is distributed to reflect their opinion and preferences.

Interviews were conducted with members of the Jeddah Urban Transportation Development Authority (TDA) to understand the current situation of transportation through the eyes of different members involved in the transportation planning process. The questions were based on the themes of the chapter on urban transport and were designed to examine the city's transport sector's social, cultural and economic aspects. The Interview shows that there are many different opinions and perspectives when discussing the phenomena of car dependency in Saudi Arabia. The first factor discussed was the neglect of transportation planning during the expansion of Jeddah city. The second factor that the authority representative mentioned was the rise of the oil and gas industry and the affordability of cars. This factor disregards any transportation planning that includes public transportation services. Jeddah Municipality believes that the streets we have in Jeddah city cannot cope with the city's continuous expansion and development.

On the other hand, A practitioner of Foster + Partners believes that it would be better to start the change slowly and frequently, and the citizens will adapt. The transportation authority saw the solution in creating a joint vision that all stakeholders agree on and engaging the community. Many solutions require major and minor changes in the city to solve the problem of car dependency. The authority of transportation and the municipality of Jeddah saw that making infrastructural changes is a huge challenge that requires significant funding. In Contrast, the practitioner of Foster + partners argues that the solution is that the government should provide incentivization, and stakeholders should develop guidelines and requirements of what the streets need to become transit-oriented.

Many factors caused Jeddah city to become car-dependent. First is the city's rapid growth in a short period compared to other cities and the neglect of public transportation planning. Second, the affordability of gas and car purchases when the urban sprawl started, and third, the fact that the streets were not and cannot cope with growth. The citizens have their perspective oriented to use private cars, but this can be changed if a plan is set to start the change. Therefore, the solution must also attract citizens and investors to fund such a project. The ideal change is to make Jeddah city a network of different TOD projects where each project is connected to the other yet serves the area and is location efficiently (Fig. 14).

4.4 *Study Guidelines Towards Creating TOD Networks in Jeddah City*

By deducing the primary and secondary data collected, the study has concluded major guidelines resulting from a correlation between the three themes of Jeddah's socioeconomic, cultural, and physical environments study and the variable of walkability, non-motorized, and public transportation. Accordingly, six diagrams of urban transportation guidelines have been created to identify the role of the project's future investors, the project advertisement, and urban project designers. The first thing

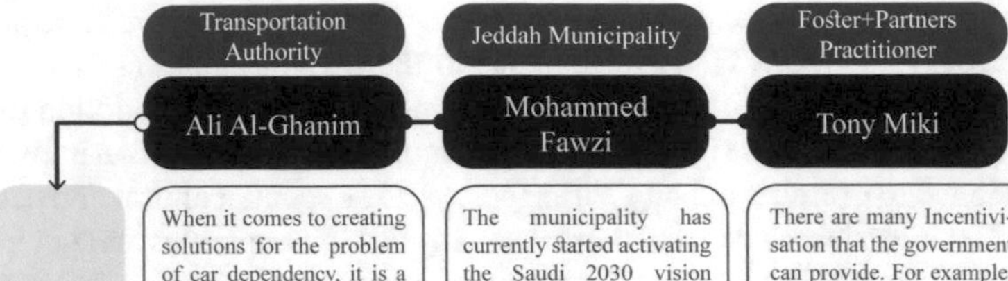

What Kind of Change Could Jeddah City Encounter Regarding Urban Transportation

When it comes to creating solutions for the problem of car dependency, it is a challenge to make infrastructural changes. The first challenge that must be considered is funding which it is difficult for the government's budget to cope with the city's growth. Considering the current municipal regulations of street design in Saudi Arabia, the government will not be able to fund additional costs required to design streets that support TOD's concept.

Therefore, the solution must be economical, considering what the government can afford and what the private sector can fund. The problem is that specialists study the perfect international standard solutions and require to implement them in cities of Saudi Arabia while disregarding the government and private sector's budget. He further explains that every location must have its solution due to the difference in culture, society, and other aspects of the city.

The municipality has currently started activating the Saudi 2030 vision within it's plans of Jeddah city roads and transportation planning. The municipality plan includes increasing the number of parking lots in many locations of the city.

Moreover, the municipality's direction is to increase the requirements that supports pedestrian and cyclist's activities on the streets of Jeddah city. Also, The municipality is including bus lay-by's in bus stops to ensure the use's comfort. However, the public transportation projects require investors and fundings as they require a large budget.

By using the 2030 Vision of Saudi Arabia in humanizing Saudi cities, The municipality is turning its direction towards creating regulations and policies that are pedestrian friendly and supporting public transportation systems to make infrastructural changes and ease the citizen's transport experience in many cities of Saudi Arabia and on top of them Jeddah city.

There are many Incentivisation that the government can provide. For example, the city will need to by some lands to implement the stations for the metro system where TOD can take place. Also, developers will get interested in the lands surrounding that station where commercial and residential buildings can be located and the station there turns into a cash machine where it brings in more developers.

Jeddah city is able to include infrastructural changes to develop the existing areas to become walkable. .

Also, the city needs to develop a guideline and requirements of what the streets need to like. However, the key would be "where would you want to make an area walkable". The city of Jeddah needs to become a network of TOD areas. The solution would be is that we could have stations located on a road where the areas in between them are walkable so it becomes a cluster of walkable areas.

Fig. 14 The stakeholders' perspective on what kind of change could Jeddah city encounter regarding urban transportation

stakeholders must remember is that Jeddah city cannot become one holistic transit-oriented development. Nevertheless, the city is not able to become entirely walkable. Therefore, the city must become a network of other transit-oriented developments that include walkable zones between the neighbourhood streets to reach transit stops and stations. These guidelines are described in the following Figs. 15, 16, 17, 18, 19, 20.

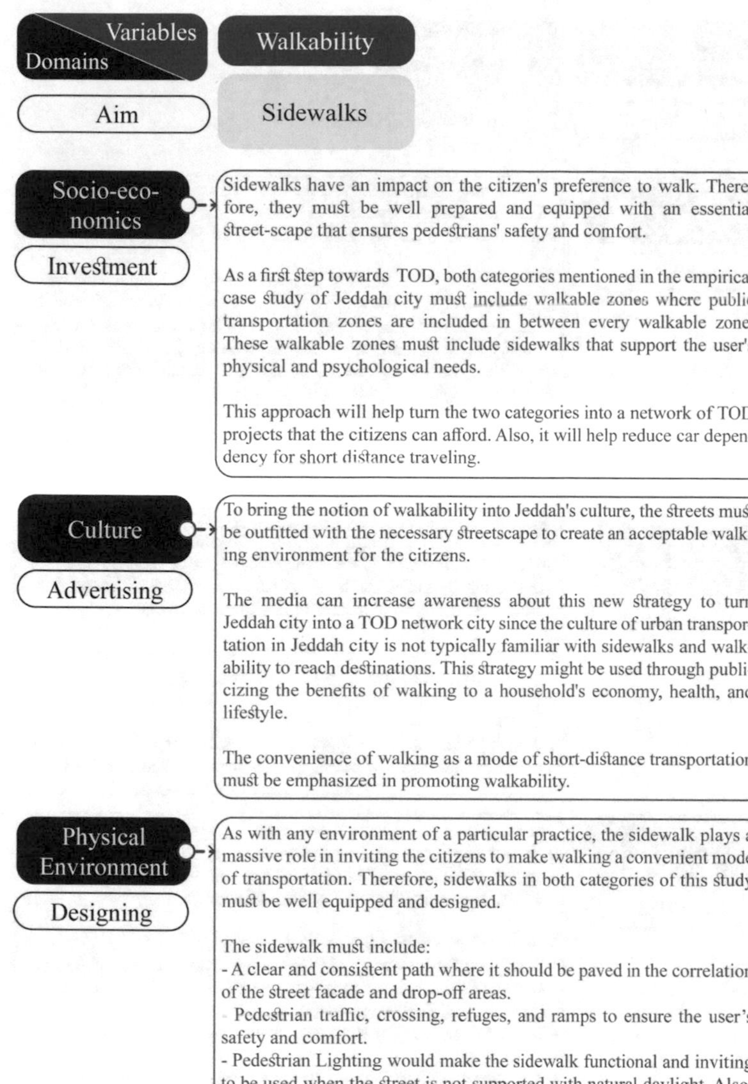

Fig. 15 The study guidelines of Walkability in Jeddah city_Sidewalks

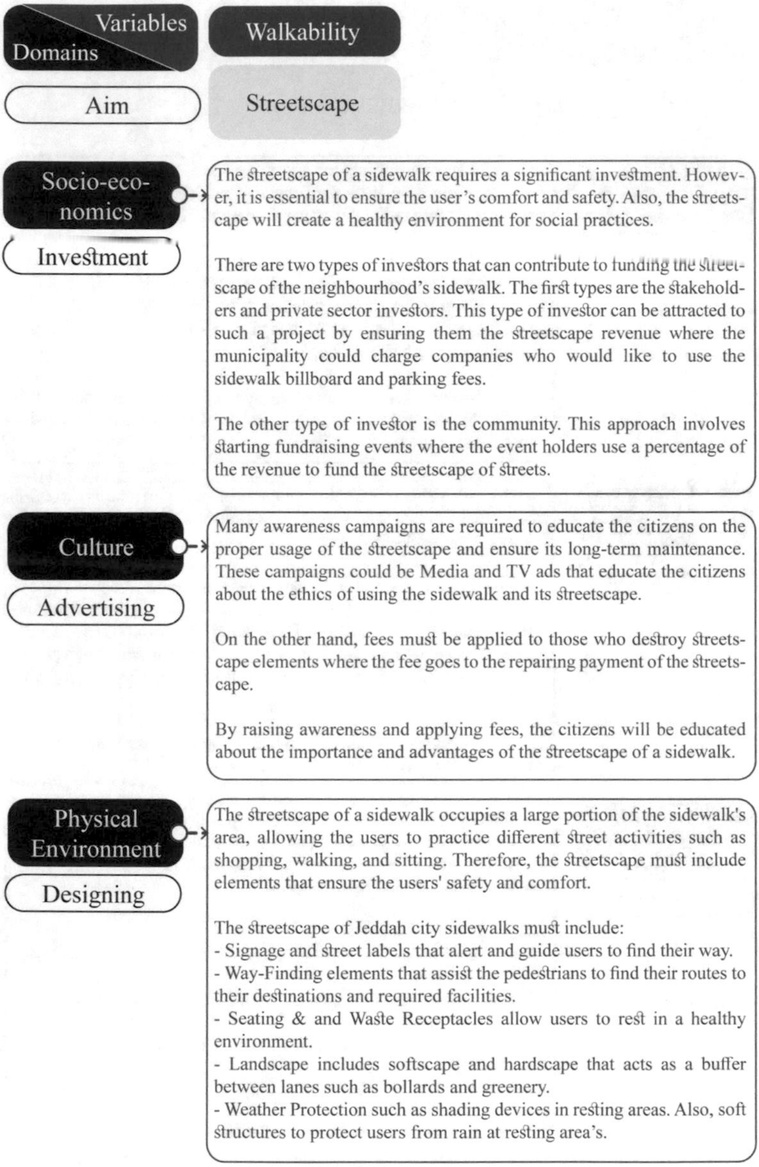

Variables / Domains	Walkability
Aim	Streetscape
Socio-economics Investment	The streetscape of a sidewalk requires a significant investment. However, it is essential to ensure the user's comfort and safety. Also, the streetscape will create a healthy environment for social practices. There are two types of investors that can contribute to funding the streetscape of the neighbourhood's sidewalk. The first types are the stakeholders and private sector investors. This type of investor can be attracted to such a project by ensuring them the streetscape revenue where the municipality could charge companies who would like to use the sidewalk billboard and parking fees. The other type of investor is the community. This approach involves starting fundraising events where the event holders use a percentage of the revenue to fund the streetscape of streets.
Culture Advertising	Many awareness campaigns are required to educate the citizens on the proper usage of the streetscape and ensure its long-term maintenance. These campaigns could be Media and TV ads that educate the citizens about the ethics of using the sidewalk and its streetscape. On the other hand, fees must be applied to those who destroy streetscape elements where the fee goes to the repairing payment of the streetscape. By raising awareness and applying fees, the citizens will be educated about the importance and advantages of the streetscape of a sidewalk.
Physical Environment Designing	The streetscape of a sidewalk occupies a large portion of the sidewalk's area, allowing the users to practice different street activities such as shopping, walking, and sitting. Therefore, the streetscape must include elements that ensure the users' safety and comfort. The streetscape of Jeddah city sidewalks must include: - Signage and street labels that alert and guide users to find their way. - Way-Finding elements that assist the pedestrians to find their routes to their destinations and required facilities. - Seating & and Waste Receptacles allow users to rest in a healthy environment. - Landscape includes softscape and hardscape that acts as a buffer between lanes such as bollards and greenery. - Weather Protection such as shading devices in resting areas. Also, soft structures to protect users from rain at resting area's.

Fig. 16 The study guidelines of Walkability in Jeddah city_Streetscape

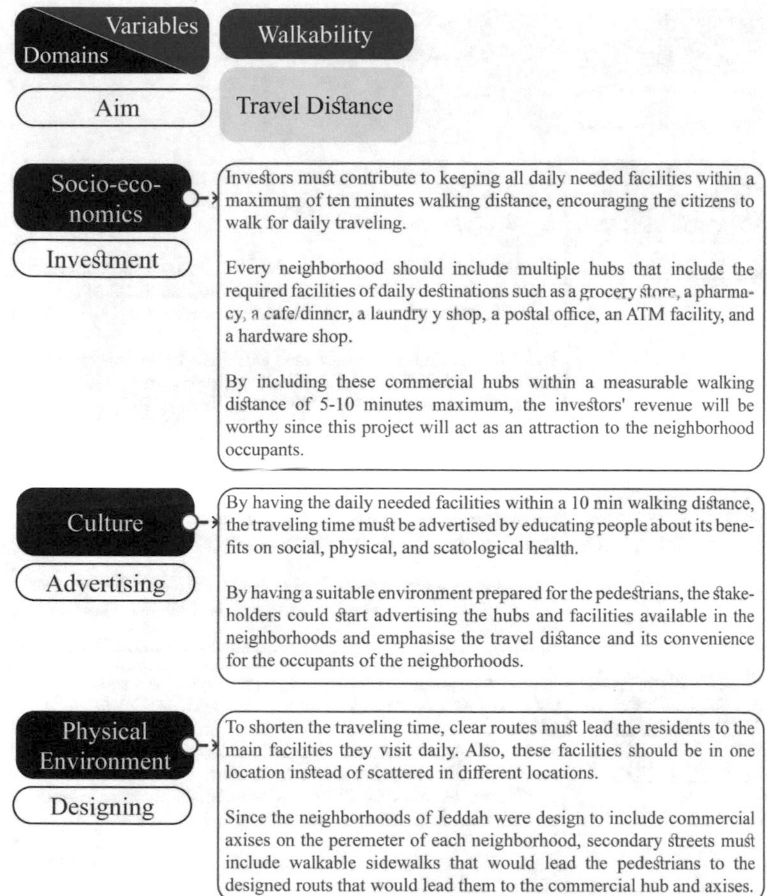

Fig. 17 The study guidelines of Walkability in Jeddah city_Travel Distance

5 Conclusion

Cities that are quickly growing present a significant task for urban planners and decision-makers. Jeddah City has a diverse functional component that comprises a long-distance movement system from north to south. This study tackles the issue of car dependency by studying the principles of TOD and how to create guidelines that represent a holistic vision toward an iconic urban transportations system in Jeddah city. As a result of interviewing three of the major stakeholders of the urban transportation sector in Jeddah city, the city requires a holistic vision that would define the stakeholder's role to achieve a network of TOD in Jeddha city's plan. This network of TOD is essential to create walkable zones that pedestrians can use to reach

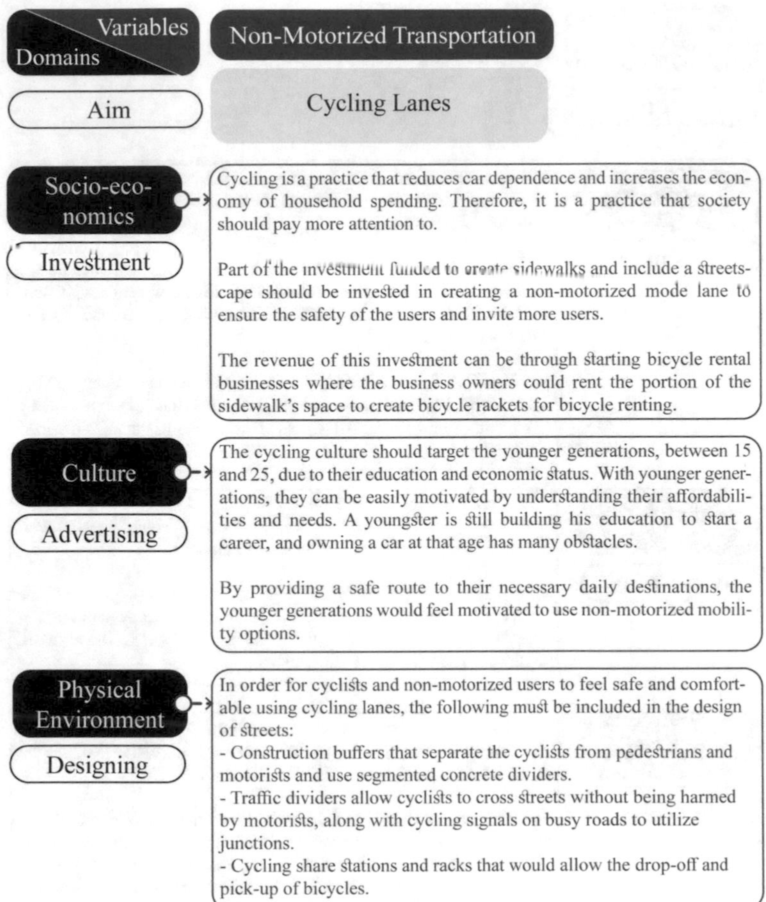

Fig. 18 The study guidelines of Non-motorized Transportation in Jeddah city_Cycling Lanes

public transit stations or stops. The guidelines discussed in this study are the essential requirements that will ensure revenue to the investors funding this project, the introduction of public transportation into the urban transportation culture of Jeddah city, and, most importantly, motivating citizens of Jeddah city to practice walkability and use public transportation. These guidelines must be taken into consideration by the transportation stakeholders to start solving the problem of car dependency and high traffic junctions on the streets of Jeddah city. By that, the city of Jeddah will endorse a resilient and intelligent urban transportation system that will affect many generations and citizens with different socioeconomic levels as the city will become humanized and easily navigated.

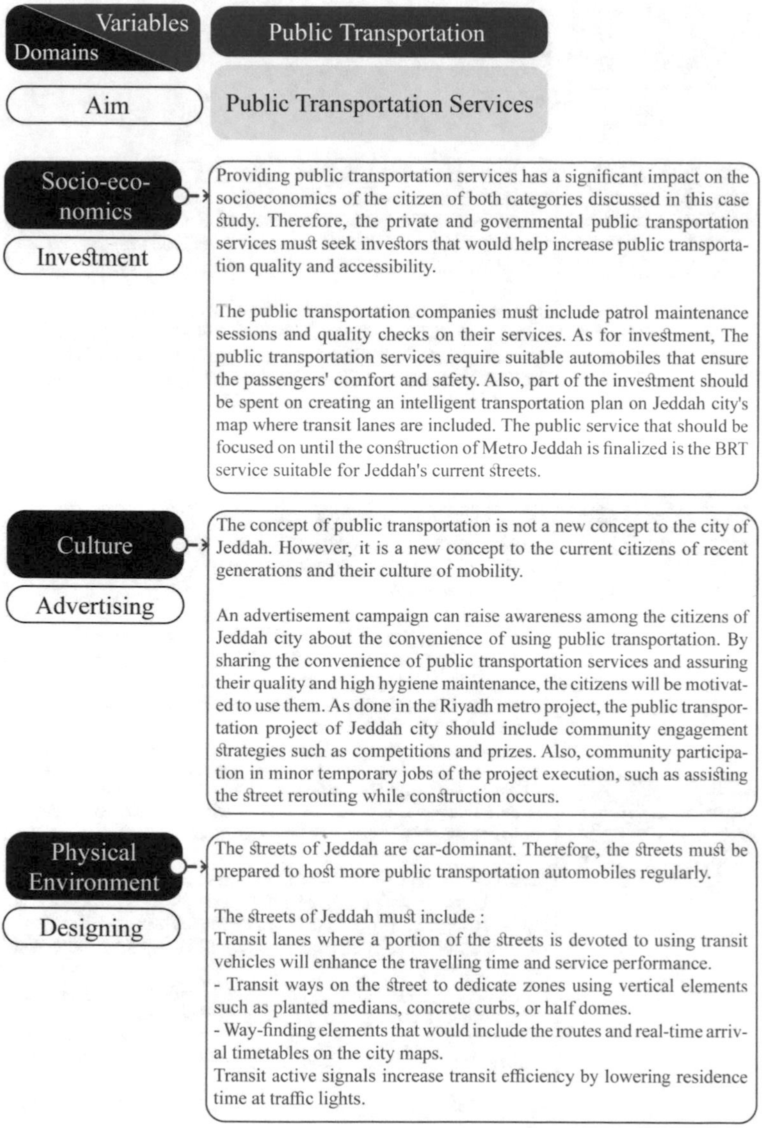

Variables / Domains	Public Transportation
Aim	Public Transportation Services
Socio-economics — Investment	Providing public transportation services has a significant impact on the socioeconomics of the citizen of both categories discussed in this case study. Therefore, the private and governmental public transportation services must seek investors that would help increase public transportation quality and accessibility. The public transportation companies must include patrol maintenance sessions and quality checks on their services. As for investment, The public transportation services require suitable automobiles that ensure the passengers' comfort and safety. Also, part of the investment should be spent on creating an intelligent transportation plan on Jeddah city's map where transit lanes are included. The public service that should be focused on until the construction of Metro Jeddah is finalized is the BRT service suitable for Jeddah's current streets.
Culture — Advertising	The concept of public transportation is not a new concept to the city of Jeddah. However, it is a new concept to the current citizens of recent generations and their culture of mobility. An advertisement campaign can raise awareness among the citizens of Jeddah city about the convenience of using public transportation. By sharing the convenience of public transportation services and assuring their quality and high hygiene maintenance, the citizens will be motivated to use them. As done in the Riyadh metro project, the public transportation project of Jeddah city should include community engagement strategies such as competitions and prizes. Also, community participation in minor temporary jobs of the project execution, such as assisting the street rerouting while construction occurs.
Physical Environment — Designing	The streets of Jeddah are car-dominant. Therefore, the streets must be prepared to host more public transportation automobiles regularly. The streets of Jeddah must include : Transit lanes where a portion of the streets is devoted to using transit vehicles will enhance the travelling time and service performance. - Transit ways on the street to dedicate zones using vertical elements such as planted medians, concrete curbs, or half domes. - Way-finding elements that would include the routes and real-time arrival timetables on the city maps. Transit active signals increase transit efficiency by lowering residence time at traffic lights.

Fig. 19 The study guidelines of Public Transportation in Jeddah city_Public Transportation Services

Variables / Domains	Public Transportation
Aim	Transit Stations/Stops
Socio-economics — Investment	The transit station has an impact on the preferences of transportation mode selection. Also, the location of these transit stops affects the citizens' willingness to use public transportation. Therefore, investors play a significant role in funding the government and the transportation service companies with comfortable and safe stops for citizens' use. By having these stops located in every neighbourhood of the two categories discussed in this study, the public transportation service availability and use will increase. To ensure a contentious revenue to the investors of the transit stops and stations, these stops could include advertisement boards rented to companies and private businesses. Also, stations could include facilities such as mini markets and small cafes to be rented to serve the users of the public transportation services.
Culture — Advertising	The culture of urban transportation in Jeddah city is not familiar with transit stations/stops. Also, the user's time must be occupied as they wait for their ride to arrive. Therefore, the station/stop is located in a strategic location 10 minutes away from their residence yet near shops or a commercial axis. The maintenance, quality, and hygiene of these stations/stops are essential to motivate the citizens to become public transportation users. Hence, these stops require a regular patrol that would ensure the high hygiene stability of the stops and require a fee from any users responsible for destroying such facilities.
Physical Environment — Designing	The transit stops require a portion of the sidewalk in different locations. However, they play a vital role in motivating citizens' use of public transportation at different socio-economic levels. These stations require the following elements on the sidewalk : - Transit stations with a clearly defined space that is sheltered and include seating for the user is to rest as they wait for their ride. - An accessible boarding area permits the user with wheelchairs to load into service vehicles. - Ticket/card membership vending machines allow users to buy or charge their membership cards for public transportation. - Waste bins to keep the stop clean. - Signage and signals to alert and inform the user are about loading areas. - Transit stations of Larger buildings on broader roadways or medians,

Fig. 20 The study guidelines of Public Transportation in Jeddah city_Tranit Stations/Stops

References

1. Author, F.: Article title. Journal 2(5), 99–110 (2016)
2. Al-joufie, M.: Exploring the determinants of public transport system planning in car-dependent cities. Proc. Soc. Behav. Sci. **216**(2016), 536 (2015)
3. Al-joufie, M.: User's satisfaction level of transportation system quality in Jeddah city, Saudi Arabia. ARPN J. Eng. Appl. Sci. **11**(5), 1 (2016)
4. Al-joufie, M.: The impact assessment of increasing population density on Jeddah road transportation using spatial-temporal analysis. Sustainability **13**(1455), 2 (2020)
5. C40 Cities: Transit Oriented Development. Good Practice Guide, 6–9(2016)

Smart Urban Planning and Videogames: The Value of Virtualization as Prognosis of Urban Scenes

Raquel Pérez-delHoyo, Laura Ferrando-Martínez, and Fernando Tafalla-Porto

Abstract Urban planning and architecture have always played a fundamental role in videogames, directly or indirectly. Together with the characters and the historical plot of the game, they have made it possible to create attractive virtual worlds in which to enjoy experiences that go beyond fiction. Today, we are at a turning point. New technologies applied to the development of videogames provide a great opportunity to rethink new smart ways of planning and managing the city that are more participatory and inclusive in the post-Covid world. Within this context, the main objective of this paper is to explore and highlight the experimental possibilities that virtual space can offer to urban planning and architecture of a smart city. In this sense, it is proposed the integration in the urban planning process of a methodology that seeks to speculate on the possibilities that virtual models can offer, merging virtual space and real space in a collaborative work. As a result, the existence of a symbiotic relationship between reality and virtuality is confirmed, making possible new resilient ways of understanding space, as well as testing new strategies, systems and complex social, political and/or economic organizations in virtual urban environments very close to reality.

Keywords Smart cities · Urban planning · Videogames · Virtual city

1 Introduction

Throughout history, games have indirectly played a fundamental role in the development of human beings. From an early age, games offer human beings a direct experience with the material world, allowing socialization between individuals. Games facilitate learning and provide the acquisition of skills and knowledge about the environment. Through games, human beings have the opportunity to learn about colors, shapes, materials, textures and even the space around us.

R. Pérez-delHoyo (✉) · L. Ferrando-Martínez · F. Tafalla-Porto
Department of Building Sciences and Urbanism, University of Alicante, Campus de San Vicente del Raspeig, 03690 Alicante, Spain
e-mail: perezdelhoyo@ua.es

A. Visvizi et al. (eds.), *Research and Innovation Forum 2022*,
Springer Proceedings in Complexity, https://doi.org/10.1007/978-3-031-19560-0_26

It is easy to find in memory childhood games made up of pieces of various shapes that allow the creation of different configurations of empty and full spaces, experiment on the possible compositions of spaces, or test the stability of the elements in height. Therefore, it can be stated that the first approach of human beings to architecture, unconsciously or innocently, is experienced when playing with the elements to discover the opportunities of space [1].

Human beings live in the physical space of architecture, although it cannot be ignored that, today, this physical space coexists with the virtual world thanks to the important development of new technologies [2]. The vertiginous development of the virtual space makes it necessary to reflect in depth on the opportunities that new technologies can offer society in order to continue evolving [3, 4]. In this evolution of virtual space, the development of videogames has played a fundamental role. In the process of creating videogames, the representations of architecture and the city, as well as the different ways of inhabiting and interacting with them, have been experimented with and reflected upon. Thus, games have created stories that allow the player to have a direct experience with the virtual space [5].

In this context, the main objective of this research is to reflect on the range of possibilities that the duality of virtual and physical space offers for designing the city. To this end, the paper addresses the following research questions: Can the experimentation of virtual space and videogames offer new reflections and ideas to improve the current processes of management and planning of cities? Is it possible to consider virtual space as a tool for the architectural and urban design of smart cities?

2 Concepts and Relationships Videogame-Architecture

The part that occupies an object, the capacity of a place and the extension that contains matter are some of the definitions that we can use to refer to physical space. Architectural space refers to the place whose production is the object of architecture. But how is virtual space defined? Virtual space shares with physical and/or architectural space the premise that both are spatial production. Therefore, they could be considered as two sides of the same coin [6]. Although it may be thought that there should be a conceptual limitation between the two spaces, there is really no need for a delimitation that would lead us to differentiate the basis or nature of both spaces [1].

From this hypothesis, certain similarities can be evidenced. Physical and/or architectural space and virtual space share concepts that are essential to understand how they function. In the first place, the influence of architecture on videogames can be found, for example, in such an essential point as the use of the same terms or concepts to identify each of these spaces.

Arcade games are a good example of this idea. This is a videogame genre that emerged to encompass a series of games originally conceived as arcade machines. Among these games, which became very popular in the 1980s, were the *mythical*

Pac-Man (1980) or *Space Invaders* (1978), among others. The point is that these arcade machines, which were called Arcade machines, were located in arcades or Arcade salons, from which they received their name. These arcades, in turn, received their name from the Penny Arcades, which were the first machines used in the United States and England to play with coins. Likewise, the Penny Arcades received their name from the place they occupied, which was the arcade portico. Thus, it can be seen how, over time, the architectural concept of the space originally occupied by the first gaming machines has been used to identify both the new videogame machines and the new places they occupy and, finally, also to identify the videogames themselves.

Secondly, the close relationship between architecture and videogames can also be found in the ways in which perspective and forms of representation are used. It is clear that perspective has played an essential role in the evolution of both architectural representation and the development of videogames.

Videogame developers commonly use tools and forms of representation that can be considered purely architectural, such as two-dimensional representation by means of plans, elevations and/or sections, or three-dimensional representation with the use of different types of perspective.

Starting from the basis of two-dimensional representation, examples can be cited where the development of the game's story is done directly through the plan view of the virtual space, as is the case of the videogame *Hotline Miami* (2012). Regarding the use of elevation representation, a good example is the legendary game *Donkey Kong* (1981). Another clear example of the use of section as a technique for defining space is the game *This War of Mine* (2014), which was created exclusively from the sections of the buildings and streets that make up the scenario where the game's story takes place.

On the other hand, in terms of three-dimensional representation, in the well-known cities of *Pokémon, Zelda* or *Final Fantasy* videogames, it is possible to observe similarities. It is common in most Japanese role-playing games to use the Egyptian perspective, which is an architectural three-dimensional representation technique. Other three-dimensional representations, for example, are axonometric and conic perspectives, which can be observed in games such as *Diablo III* (2012) or *Counter Strike: Condition Zero* (2004), respectively.

The truth is that videogames have brought millions of people into contact with architecture, even leading them to design buildings or build cities without professional training in architecture. On many occasions, videogames have also inspired a whole generation of architects. In this regard, mobile games have been an important influence. Creative games such as *Minecraft* or *Roblox*, which allow building elements of cubic dimensions with blocks of materials as far as the imagination can reach, are a good example. Likewise, in the field of the city, mobile games such as *My City-Entertainment Tycoon, Springfield, Build Away!, Smurfs' Village, Pixel Plex, SimCity BuildIt* or *Megapolis,* have been very influential, with the placement of buildings, parks, services or roads to create the best city and meet their needs. Thus, on many occasions videogames may also have inspired a whole generation of architects.

Therefore, it can be stated that there is a direct relationship between architecture and videogames that human beings assume from very early stages and in an indirect way. The development of videogames has always been accompanied by the means of representation of architecture and forms of spatial representation [6]. The point is, how has this relationship between architecture and videogames evolved since the emergence of the virtual world?

3 Evolution of Videogames in Relation to Architecture

Since their origin, videogames have had a close relationship with architecture, mainly in the way of understanding and visualizing cities (see Fig. 1). It may be thought that the cities represented in videogames are utopian and invented, however, videogames offer users the opportunity to get to know existing cities or ancient cities that have already disappeared. Over time, cities have become more and more present in videogames and have become more precisely defined [1, 5], as can be seen in some of the earliest games such as *Grand Theft Auto I* (1997) and *Grand Theft Auto II* (1999). These were fully two-dimensional games that offered the opportunity to virtually traverse a scenario with characteristics similar to the real world, without the need to behave in a totally "legal" way. Therefore, these games offered users a vision and perspective of the city totally different from the one they were used to experiencing.

With the emergence of three-dimensional worlds that add the concept of depth to representation, new opportunities opened up. The development of games such as *Super Mario 64* (1996) set standards both in the representation of three-dimensional environments and in the way of traversing them. From this moment on, the variety of proposals multiplied with the aim of breaking with the logical representation of interactive environments in videogames. Thus, the possibility of activating changing elements in virtual worlds gradually emerged, as well as combinations of different perspectives and points of view that, in short, resemble the individual visual perception of the human being in the physical city.

It is at the present time when all these reflections acquire greater value. Society is at a turning point where virtual space can be considered, for the first time, as surpassing physical space [5]. Many videogame developers have worked to achieve a reliable image of the city. The trend in the design and development of videogames has been the use of the open world, allowing the player to freely traverse a large space, as in the case of *World of Warcraft* (2004). On the other hand, the reproduction of the image of the city has always been accompanied by the representation of the social situations that take place in it. As a result, videogames are increasingly able to offer the player a more personal experience with virtual space. A good example of this type of game is the videogame *Cyberpunk 2077* (2020).

There is also a specific category of videogames that have gone a step further by allowing the player to design his own scenarios and/or virtual worlds. This category of games is of special interest in this research, since it is directly related to the concept

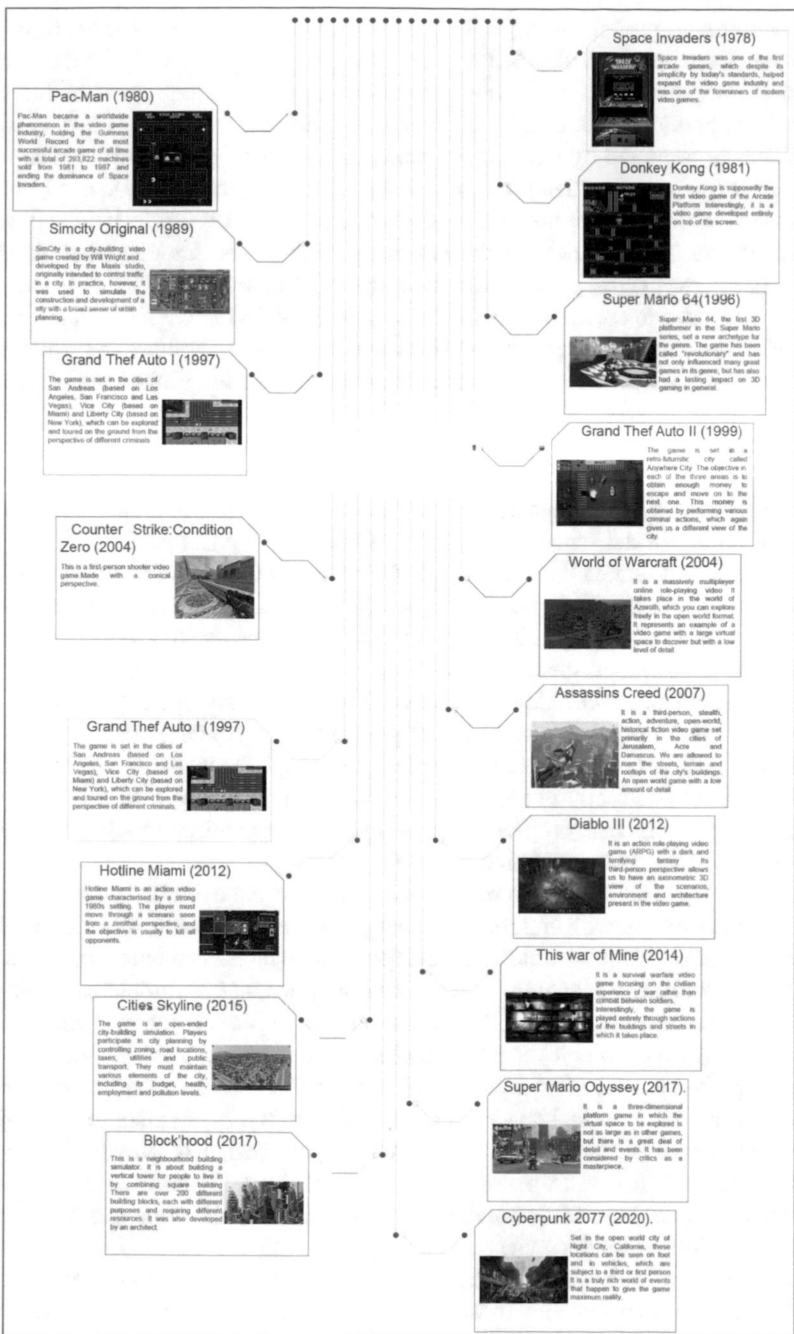

Fig. 1 Evolution of videogames in relation to architecture. Timeline. *Source* Own preparation

of smart citizen participation [7, 8]. These games offer users the opportunity to make decisions in the configuration of the space and design of a city. In this regard, videogames can also be useful and valuable tools for learning about architecture and urban planning. For example, the videogame *Simcity Original* (1989), in which the city is represented by cavalier projections, is already used in some schools of architecture to teach students the principles of urban planning [9]. Also, a very significant example is the popular videogame *Minecraft* (2009), considered a tool for design and architecture by both developers and players. Another example is the game *Block'hood* (2017), which was developed by an architect. In this videogame the player is free to create his own neighborhood within the city. The game allows for high-rise construction and offers users the opportunity to both make architectural design decisions and manage various resources, without the limitations of the real world. A step further has been taken by the videogame *Cities Skyline* (2015), in which users must design their own complete city, deciding the location of the different residential or commercial neighborhoods, urban densities, street layouts, as well as the management of resources and waste, which is a great opportunity to approach urban planning for the game users and, therefore, for the citizens.

Thus, it can be stated that, at present, the relationship between architecture and videogames has advanced to a level of development that goes beyond the mere image or representation of the city [10]. The contribution between both physical and virtual worlds has transcended mere visualization. Architecture and urban planning can now be fed back to the new developments of videogames and the opportunities they offer [5]. This generates a wide range of possibilities that should be the subject of research and experimentation in real contexts. In short, the methods and techniques that videogames currently use to tell and develop the different stories of the game are, in a way, promoting progress towards new, more smart and inclusive ways of designing, planning and managing the city.

From this conviction as a starting point, and once the evolution of videogames and their relationship with architectural design has been analyzed, it is time to think about the future and the specific tools that these videogames can put at the service of architecture and urban planning. There is no doubt that, at first, architecture directly influenced the development of videogames, but now it is evident how virtual spaces and videogame technologies are influencing the creation of new architecture and cities [10]. The creation of a "digital twin" with which to experience shapes, spaces, textures or colors in virtual space, at a time prior to the materialization of objects in physical space is already an essential part of the design of new buildings and cities [5, 11, 12]. It is not unknown that many of the software used by architects and urban planners for architectural and urban design use the same graphic engines and/or capabilities of videogames.

The latest developments in videogames are focused on the creation of much more complex virtual environments, in which it is possible to recreate the functioning of social organisms or economic systems. Trial and error mechanisms or the development of experimental virtual models are just a few examples of the multiple tools that videogames offer city planners to explore the creation of real worlds. These tools make it possible to test and verify the evolution of different possible alternatives in the face of future problems, or to test the effects of various adverse situations, as in the case of the recent pandemic situation. Today it is already possible to imagine the possibility of building a virtual model of a neighborhood or a city. On a larger scale, it would be possible to create a country, i.e., a virtual state with infrastructures, cities and digital inhabitants, in which governments, companies and citizens could experiment with the elements to be innovated and make decisions of all kinds, both decisions affecting the physical elements that make up the city, as well as decisions that determine strategic policies in relation to the management of resources to be preserved or exploited, energy consumption, pollution levels or any other element relevant to the design and management of the city.

In summary, the creation of a virtual model can be the basis for testing future proposals that should be tested prior to their implementation in the design and planning process of cities, in order to choose the most optimal and effective alternative. The use of the virtual model therefore offers society an experimental scenario for the implementation of future strategies that allow for smarter and more inclusive city planning and management. There is also an important component of trial and error that allows, based on virtual models, the creation of strategies adaptable to the rapid changes of today's world and society.

4 Conclusions

From the research conducted, it can be deduced that videogames have been linked to architecture since its beginnings. First, architecture inspired the virtual space of videogames and, today, videogame technology allows architecture to design real physical space. From this point of view and looking to the future, the rapid evolution of videogames could, in the not too distant future, allow the creation of complex virtual models of a city, becoming true experimental laboratories to test the impact of any urban, economic or social transformation to be implemented in the real city. In this way, thanks to videogame technology, a test prior to the definitive implementation of changes in the physical space of our cities is potentially possible as part of the process of smarter urban planning.

Today, the pandemic situation has reopened the debate and invites reflection on the development of cities. It has become evident that urban planning and management mechanisms must be reconverted into agile processes that can be quickly adjusted to changing situations in short periods of time, ensuring that the transformations and changes in the configuration of the city do not, at any time, harm the functions and interactions that take place in it. Therefore, agile tools for experimentation in the urban space are necessary to allow the testing of design and planning strategies, as well as city management plans. In this regard, videogame technology takes on a special value with the use of virtual models.

Urban laboratories already exist in several cities around the world, such as Shanghai or Singapore (created by 51world: https://youtu.be/kCS0AyWVS1s) or recently Wellington (developed by BuildMedia: https://youtu.be/NraF12qN4gs). Twin cities in virtual space serve as laboratories to understand the rules that govern real-world cities and plan their future. The simulation of thousands of parameters that affect the operation of cities–transportation systems, traffic, energy, temperature, pollution, people, consumption–is becoming increasingly perfect, making it possible to predict changes in the dynamics of the city and how these changes affect the well-being of its inhabitants and the economy. However, there is still a need to better understand how to channel citizen talent into initiatives to co-create the city. In this sense, videogames, which are increasingly perfect simulation models that duplicate physical reality and allow user interaction, as if it were an advanced *SimCity*, for example, represent a great opportunity for smart cities.

A final consideration to highlight is the fact that, from a didactic point of view, videogames are helping to understand the functioning and complexity of urban planning in schools of architecture and, therefore, to make the knowledge of the city accessible not only to students but to all citizens in general. Therefore, today videogame technology, beyond being pointed out in many contexts as a distraction, is contributing in a specific way to bringing citizens closer to their city, which may mean, in the medium term, an unprecedented boost to the concept of public participation in citizens' awareness.

References

1. Díez Fornes, C.: Arquitectura y videojuegos. Creación o disolución de barreras. Doctoral dissertation, Universitat Politècnica de València (2021)
2. Pérez-delHoyo, R., Mora, H.: Knowledge society technologies for smart cities development. In: Smart Cities: Issues and Challenges, pp. 185–198. Elsevier (2019)
3. Varela-Guzmán, E., Mora, H., Visvizi, A.: Exploring the differentiating characteristics between the smart city and the smart society models. In: The International Research & Innovation Forum, pp. 313–319. Springer, Cham (2021)
4. Bellard, M.: From architecture to videogames. In: Architects After Architecture, pp. 228–234. Routledge (2020)
5. Díaz Vázquez, P.: Arquitectura y videojuegos: relaciones. Doctoral dissertation, Universidade da Coruña (2018)
6. Hidrobo, M.: Ciudad y Videojuegos. Arquitasa. https://arquitasa.com/ciudad-y-videojuegos-ii-mario-hidrobo. Accessed 04 Jan 2022
7. Bello-Maldonado, G.D.: Un videojuego que promueve la participación ciudadana interactiva en las reformas urbanas. Virtu@lmente **7**(2) (2019)
8. Khan, Z., Ludlow, D., Loibl, W., Soomro, K.: ICT enabled participatory urban planning and policy development: The UrbanAPI project. Transf. Govern. People Proc. Policy **8**(2) (2014)
9. Bruscato, U., García Alvarado, R., Ruiz Tagle, J.: Ciudades virtuales-experiencias múltiples. In: XII Congreso Iberoamericano de Gráfica Digital SiGraDi. Cuba (2008)
10. Pearson, L.: Worlds that are given: how architecture speaks through videogames. Thresholds **46**, 264–275 (2018)
11. Visvizi, A., Troisi, O.: Effective management of the smart city: an outline of a conversation. In: Managing Smart Cities, pp. 1–10. Springer, Cham (2022)
12. Kashef, M., Visvizi, A., Troisi, O.: Smart city as a smart service system: human-computer interaction and smart city surveillance systems. Comput. Hum. Behav. **124**, 106923 (2021)

AI-Based Remoted Sensing Model for Sustainable Landcover Mapping and Monitoring in Smart City Context

Asamaporn Sitthi and Saeed-Ul Hassan

Abstract In recent years, numerous attempts have been documented in the smart city context to make cities and human settlements more inclusive, safe, resilient, and sustainable by combining the power of ICT tools with AI/Machine Learning backed remote sensing technologies. Using remote sensing technologies, this study aims to enhance methodologies for mapping and monitoring changes in terrestrial Landcover resources in Thailand's Dong Phayayen-Khao Yai National Park. The goal is to investigate and develop a remote sensing technique for classifying terrestrial Landcover by compensating for topographic effects. Changes were detected using the Landsat 5-TM and Landsat 8 OLI satellites, and deviations from solar and terrain were rectified before the satellite imagery was identified using a Random Forest classifier. It improves efficiency in identifying terrestrial forest regions by combining high-level numerical modelling data (Digital Elevation Model: DEM) with it. The results showed that in the Khao Yai National Park area, the extraction of terrestrial Landcover areas using Long-term Landsat satellite photos performed significantly, with an accuracy of 82.05 percent. The goal of this study is to leverage the power of AI to make the best use of a wide range of terrestrial forest resources. This includes the significance of conducting a comprehensive evaluation of legislation governing the management of terrestrial forest resources.

Keywords Landcover · Land use · Smart city · Remote sensing · Artificial intelligence · Machine learning · Sustainable development

A. Sitthi (✉)
Department of Geography, Faculty of Social Science, Srinakharinwirot University, 114 Sukhumvit 23, Wattana District, Bangkok 10110, Thailand
e-mail: asamaporn@g.swu.ac.th

S.-U. Hassan
Department of Computing and Mathematics, Management Manchester Metropolitan University, Manchester, UK
e-mail: s.ul-hassan@mmu.ac.uk

A. Visvizi et al. (eds.), *Research and Innovation Forum 2022*,
Springer Proceedings in Complexity, https://doi.org/10.1007/978-3-031-19560-0_27

1 Introduction

Land use planning is the process by which a centralised entity governs land usage [1]. As a result, it is commonly done to achieve more palatable social and environmental effects as well as more cost-effective resource utilisation [2]. It is significant to research the proportions and criteria for diverse land use kinds, but it leads to the establishment of natural balance and sustainable Land use.

Changes in Land use and Landcover affect a highly productive ecosystem that helps to mitigate the effects of climate change [3]. A growing awareness of Land use and Landcover change has resulted in several measures to preserve ecosystem function. Several Landcover conservation programmes have been carried out worldwide, including forest ecosystems and small-scale agriculture [4]. Long-term changes in Land use and Landcover could considerably impact the surrounding environment. As a result, current data on Land use and Landcover size, distribution, and changes in nearby Land use and Landcover are critical for future ecosystem management and decision-making [5].

According to detailed research, the harmful by-products from intensive Land use and Landcover continue to damage nearby geographical areas. National, regional, and global scales are typically used to collect quantitative data [6]. The type and manner of production, the volume and intensity of management activities, and the physiological location of Land use and Landcover are all interrelated aspects that must be determined. However, there are no established standards for mapping and monitoring environmental changes, making it difficult to assess the long-term effects of such actions on ecosystems [7].

Dong Yen–Khao Yai Forest, which includes Khao Yai National Park, Thap Lan National Park, and Khao Yai National Park, is regarded as one of Thailand's most important forest areas. Dong Yai Wildlife Sanctuary, Pang Sida National Park, Ta Phraya National Park, and Pang Sida National Park. It is a biodiverse environment in terms of both vegetation and wildlife. The terrain is challenging to navigate, and the mountains are steep. It spans six provinces, including Nakhon Ratchasima, Saraburi, Na-khon Nayok, Prachin Buri, Sa Kaeo, and Buriram, and has a total area of 3,874,863 rai, making it a rich source of flora and fauna. However, natural forest areas are being converted to regions with different Land use types, such as agricultural lands. The ecosystem's balance is affected by community habitat.

Approaches based on satellite remote sensing have proven to be time and cost-effective for monitoring ecosystems [8, 9]. Satellite imaging has been widely employed in a variety of Land use and Landcover research, such as mapping areas and identifying subtypes. Several studies summarised several remote sensing sensors (optical, RADAR, and aerial image) in tropical areas for mapping and monitoring changes. Despite its many benefits, aerial photography, high-resolution satellite imagery, and RADAR images have significant limitations that limit their use for interannual changes. The fundamental limitations are the availability of continuous temporal resolution, high-resolution imaging covering a small area, complicated data rectification and processing, and the related high cost.

In this study, we examine the capacity of Landsat satellite imagery to retrieve and map Land use and Landcover variables in Thailand's Dong Phayayen-Khao Yai Forest Complex and the influence on nearby mangrove areas. We analyse the Land use Landcover changes of the surroundings to understand the long-term impact on Land use and Landcover change and to interpret conversion rates near the ecosystem to evaluate the short-term change dynamics. This research can improve the Land cover quality in the region by strengthening sustainable forest management [10], including climate change mitigation and adaptive responses to sustainable and smart city development plans [11, 12].

2 Study Area and Dataset

Pang Sida National Park, Khao Yai National Park, Ta Phraya National Park, Thap Lan National Park and Dong Yai Wildlife Sanctuary are part of the Khao Yai Forest Complex, which runs from Khao Yai to the Cambodian border. The region stretches from the Sankamphang mountain range's western end to the Nakhon Ratchasima Province plateau's southwestern border (see Fig. 1). The majority of this site is located in Nakhon Ratchasima, although it also includes the provinces of Saraburi, Prachinburi, and Nakhon Nayok in Thailand. The terrain in the study area is mountainous and complex. Because the region has a diverse variety of forest types and the Royal Thai Forest Department possesses historical field survey data and extant field inventories, it might be a future validation site.

This study used satellite imagery data from the Land Processed Distributed Active Archive Center (LP DAAC), including Landsat TM-5 and OLI-8 satellite images, to prepare terrestrial data (see Table 1). Between 1999 and 2015, terrestrial forest areas were represented by 30*30 square metres with minimal cloud cover. The Fig. 2 depicts a satellite image of the study area.

3 Methodology

This paper used optical remote sensing data to conduct a long-term spatial and temporal analysis of Landcover change from 1999 to 2015. To create an algorithm for tracking Land use and land cover, it is necessary to study relevant information such as spectral signatures in order to understand the specific properties and reflections of Landcover types that differ from others. The accuracy of the Landcover data obtained from satellite imagery will then be evaluated (Accuracy Assessment), along with the field inventory data, as shown in Fig. 3.

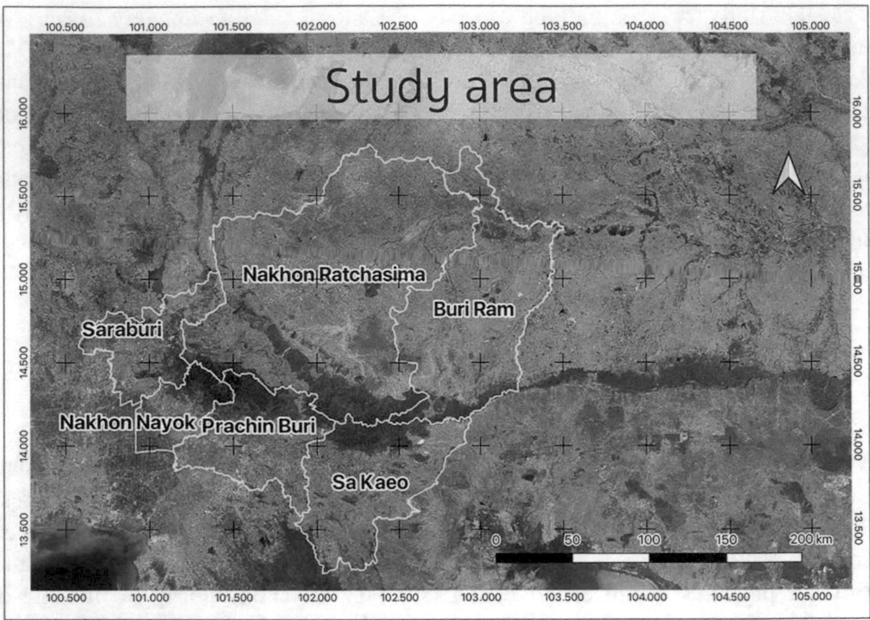

Fig. 1 Map of study area using in this research

Table 1 Landsat 5 TM and landsat 8 OLI specifications and path/row

Number	Date of acquisition	Satellite/sensor	Reference system/Path/Row
1	21/11/1999	Landsat 5 TM	WRS–2 /127/50
2	20/01/2015	Landsat 8 OLI	WRS–2 /128/49
3	21/11/1999	Landsat 5 TM	WRS–2 /128/50
4	20/01/2015	Landsat 8 OLI	WRS–2 /128/51
5	21/11/1999	Landsat 5 TM	WRS–2 /129/49

3.1 Spectral Signature Analysis

The spectral signature analysis will be used to classify that area. The indicator traits associated with Landcover reflectance must be investigated. The spectral signature is based on data from optical satellite imagery to identify and compare the reflectance of different objects. Depending on the composition, they have similar reflectance values, which influences Landcover reflectance.

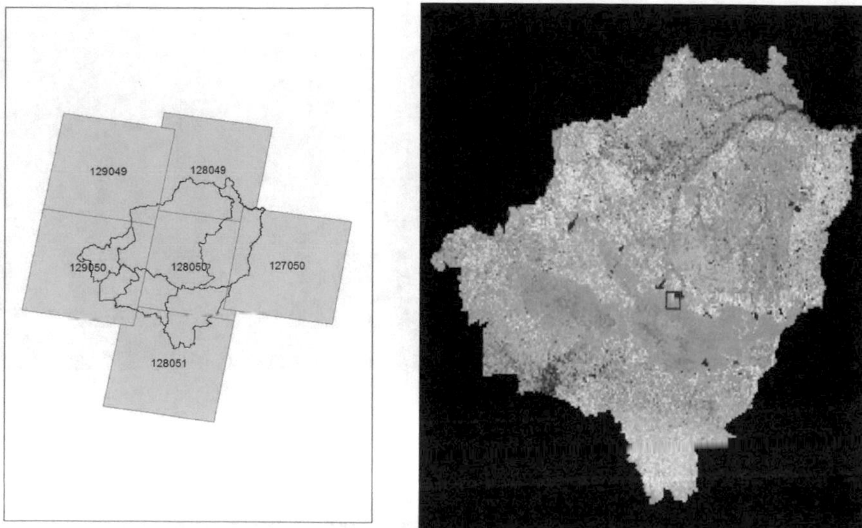

Fig. 2 Satellite imagery covers the study area

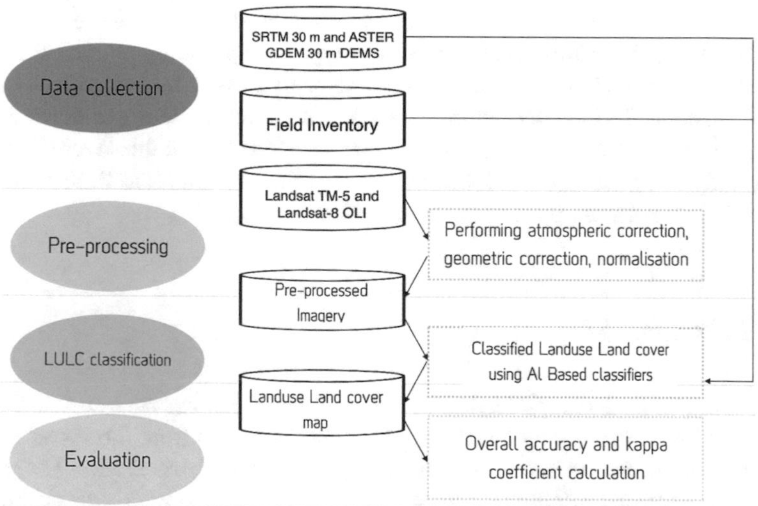

Fig. 3 Methodological flow of the research work

3.2 AI-Based Landcover Classification

Identifying the causes of Landcover change is a critical issue. From 1999 to 2015, we developed an Al-based approach to identify the distribution of Landcover altered by human actions and natural phenomena. The Random Forest classifier techniques can

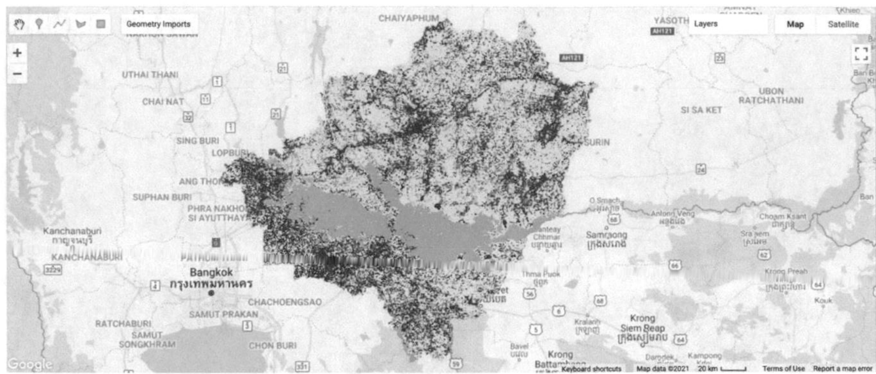

Fig. 4 AI-based land cover classification

provide information about the change's drivers. Random forest classifier (RF) is a type of machine learning model created from various decision trees. Random Forest increases the number of trees to multiple trees, resulting in increased work efficiency. More specifically, the Random Forest model is a popular machine learning model.

The Random Forest classifier can be used in remote sensing to classify Landcover types. Prior to processing, satellite imagery is analysed, and terrain aberrations are corrected. The images are classified into four types using Random Forest classifiers: forest, agriculture, water, urban area, and others. To classify the test pixels, training data from Thailand's Land Development Department was randomly selected with the distinguishing features of each Landcover type. Al-based techniques were developed in this study using Google Earth Engine and the Raster toolbox in R, as well as the open code GIS software package Quantum GIS (see Fig. 4).

3.3 Validation Site

After classification is complete, accuracy is evaluated. Further we evaluate the classification results' validity by comparing the types obtained from the visual classification to those surveyed in the field. Note that if such comparisons can be made for all types, it generates additional data in the form of an error matrix to evaluate the validity of each type and the overall classification result to help report evaluating the accuracy of the results obtained. Also note that validation site surveys used stratified random sampling techniques to determine Landcover type. The validation from such surveys has been used effectively in the study. The validity assessment results used in this study are based on classification results with inventory data references. To assess overall accuracy, data references were divided into 70 percent training data and 30 percent testing data, as shown in Fig. 5.

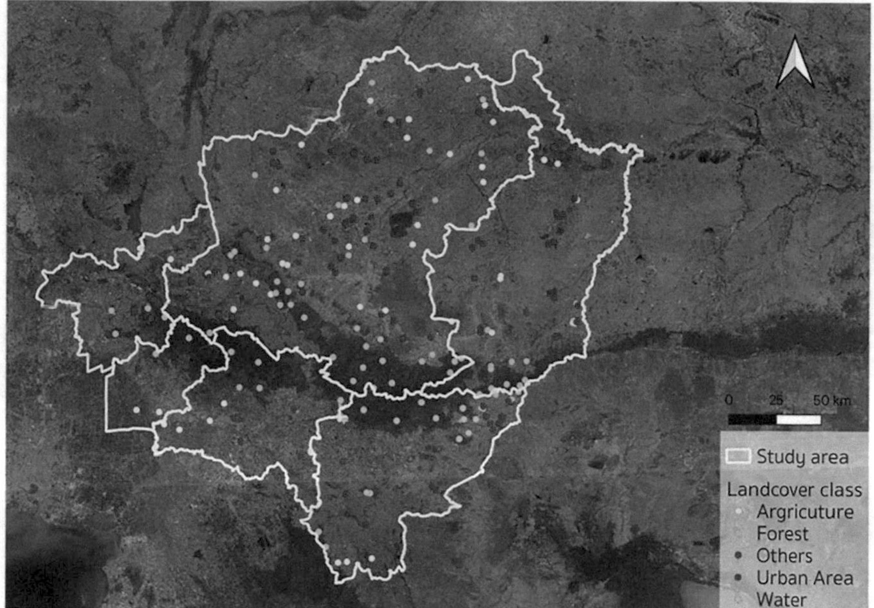

Fig. 5 Sampling data for land cover validation

4 Result and Discussion

This section presents the collection of the findings following the objective of the study to develop an algorithm for mapping Landcover areas using an AI-based classifier. The results can be broken down into the following subsection.

4.1 Landcover Map from 1999 and 2015 Landsat Images

The study area's Landcover distribution was estimated using an AI-based method based on an RF classifier. Figure 6 depicts Landcover maps from 1999 to 2015. The total area in each Landcover class of the year is shown in square kilometres and as a percentage of the distribution. According to our interpretation, the majority of the areas in 1999 were covered by agricultural and forest land. The remaining area is made up of water, urban areas, and bare soil. On the other hand, urban areas increased the distribution of total area in 2015.

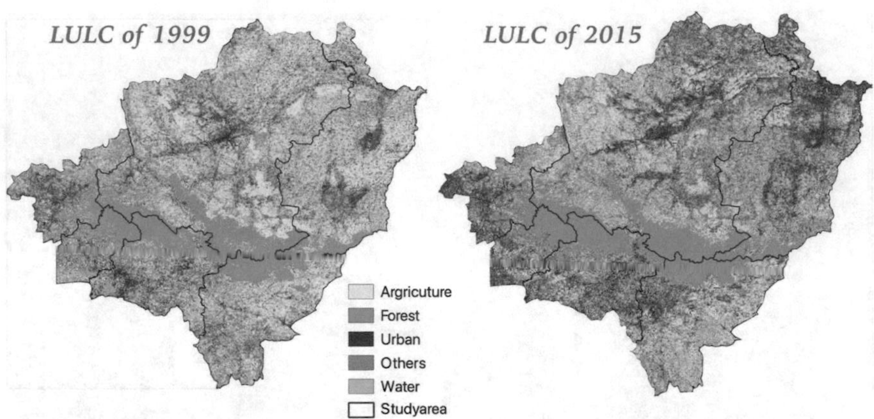

Fig. 6 Landcover map and areas of class distribution in 1999 and 2015

4.2 Accuracy Assessment

The 2015 Landcover mapping was validated using ground control points from a field survey. A confusion matrix was used to determine how many members of a predicted class were correctly classified. Many measurements have been used to improve the error matrix's interpretation. The Kappa coefficient is the percentage of agreement between the classifier's observed and predated classes [5].

Table 2 displays the overall accuracy and kappa statistics for evaluating classification performance, which are 82.05 percent and 0.75, respectively. In the case of individual Landcover types, the majority of all classes have producer accuracy greater than 60%. The urban, water, and forest classes have been precisely classified. Because there were no field survey points available in 2005, the accuracy of the Landcover was not assessed.

Table 2 Landcover accuracy assessment

Landcover class	Producers accuracy (%)	Users accuracy (%)
Agricultural	71.32	74.53
Forest	92.15	95.98
Urban areas	81.25	86.26
Others	69.41	57.84
Water bodies	93.33	89.98

4.3 Landcover Change from 1999 to 2015

The change in Landcover has been observed over a 15-year period. The loss, gain, and net change of each Landcover area distribution are shown in Fig. 7 and Table 3 using a change detection matrix. Table 3 displays a change detection matrix with Landcover classes from 1999 images in columns and Landcover classes from 2015 images in rows. Based on this matrix, the results show that agriculture and forest have increased (positively) in area from 1999 to 2015, while other classes have decreased (negative change). In terms of changing the specific Landcover class. Since 1999, we've noticed that forest and agricultural areas have grown in size, while urban (build-up) areas have shrunk. These changes are primarily the result of government policy aimed at preserving forest areas for sustainable development and smart cities.

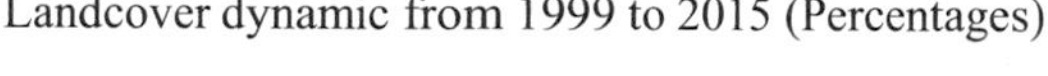

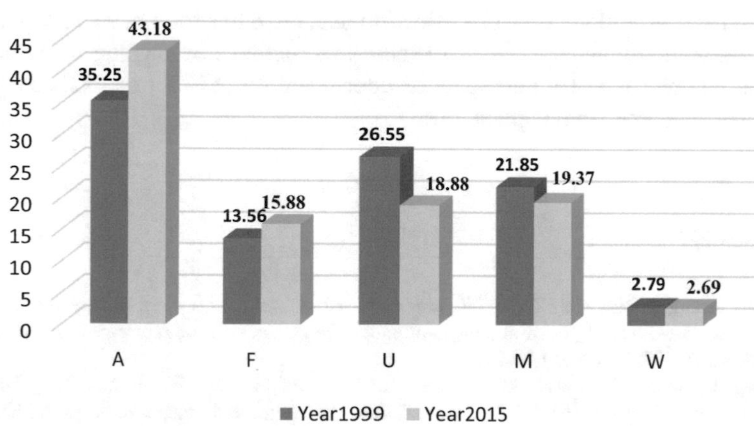

Fig. 7 Landcover dynamic from 1999 to 2015

Table 3 Landcover dynamic from 1999 to 2015 (Percentages)

Landcover class	Agriculture	Forest	Urban	Others	Water
Agriculture	44.40	8.65	32.32	41.45	21.38
Forest	2.69	70.17	2.17	3.79	4.15
Urban	26.08	7.24	44.79	25.16	30.09
Others	24.89	12.37	18.03	27.26	16.94
Water	1.94	1.57	2.69	2.34	27.44
Class total	100.00	100.00	100.00	100.00	100.00
Image difference	+7.93	+2.32	−7.67	−2.48	−0.10

5 Concluding Remarks

In this study, we use AI to classify satellite imagery data alongside topographical elevation model data such as SRTM DEM and ASTER GDEM DEM. We analyse Landsat 5 TM and 8 OLI satellite imagery with a spatial resolution of 30 * 30 sq.m. However, the classifier performs better in the forest and agricultural classes. Landsat satellite images were extracted to classify land cover. AI-based techniques based on RF classifier were used to differentiate between each Landcover class, including water, others, urban area, forest, and agricultural.

The overall classification accuracy is 80.76 percent when validated using ground survey points. These are tracked as changes in detection over 15-year time periods in the study area and forest complex. The agricultural and forest industries have been found to have the highest rate of change. The forest area is rapidly expanding in terms of area. As a result of government policy, water area is decreasing. Incorporating advanced techniques like object-based and other machine learning techniques, on the other hand, would improve classification accuracy. In addition, we also plan use geospatial natural language [13, 14] processing methods by tapping the power of machine learning [15, 16] and advance deep learning [17–20] methods for more comprehensive geospatial applications.

References

1. Li, H., Zhang, J., Zhang, S., Zhang, W., Zhang, S., Yu, P., Song, Z.: A framework to assess spatio-temporal variations of potential non-point source pollution risk for future land-use planning. Ecol. Indic. **137**(1), 108751 (2022)
2. Wang, J., Bretz, M., Dewan, M.A., Delavar, M.A.: Machine learning in modelling land-use and land cover-change (LULCC): current status, challenges and prospects. Sci. Total Environ. **31**, 153559 (2022)
3. Rimal, B., Sharma, R., Kunwar, R., Keshtkar, H., Stork, N.E., Rijal, S., Rahman, S.A., Baral, H.: Effects of land use and land cover change on ecosystem services in the Koshi River Basin, Eastern Nepal. Ecosyst. Serv. **38**(1), 100963 (2019)
4. Vittek, M., Brink, A., Donnay, F., Simonetti, D., Desclée, B.: Landcover change monitoring using landsat MSS/TM satellite image data over West Africa between 1975 and 1990. Remote Sens. **6**(1), 658–676 (2014)
5. Forkuo, E.K., Frimpong, A.: Analysis of forest cover change detection. Int. J. Remote Sens. Appl. **2**(4), 82–92 (2012)
6. Navalgund, R.R., Jayaraman, V., Roy, P.S.: Remote sensing applications: an overview. Curr. Sci. **25**, 1747–66 (2007)
7. Xu, Z., Wang, Y.: Radar satellite image time series analysis for high-resolution mapping of man-made forest change in Chongming Eco-Island. Remote Sens. **12**(20), 3438 (2020)
8. Chauvenet, A.L., Reise, J., Kümpel, N.F., Pettorelli, N.: Satellite-Based Remote Sensing for Measuring the Earth's Natural Capital and Ecosystem Services. Zoological Society of London, Institute of Zoology (2014)
9. Cihlar, J.: Landcover mapping of large areas from satellites: status and research priorities. Int. J. Remote Sens. **21**(6–7), 1093–114 (2000)

10. Dorren, L.K., Maier, B., Seijmonsbergen, A.C.: Improved Landsat-based forest mapping in steep mountainous terrain using object-based classification. For. Ecol. Manage. **183**(1–3), 31–46 (2003)
11. Albanese, F., Caprioli, M., Tarantino, E.: Using machine learning algorithms for Landcover classification of Ikonos imagery. In: Proceedings of the XXII rd. ACI/ICA International Cartographic Conference (2005)
12. Alqurashi, A., Kumar, L.: Investigating the use of remote sensing and GIS techniques to detect land use and land cover change: a review. Adv. Remote Sens. 193–204 (2013)
13. Iqbal, S., Hassan, S.U., Aljohani, N.R., Alelyani, S., Nawaz, R., Bornmann, L.: A decade of in-text citation analysis based on natural language processing and machine learning techniques: an overview of empirical studies. Scientometrics **126**(8), 6551–6599 (2021)
14. Hassan, S.U., Imran, M., Iqbal, S., Aljohani, N.R., Nawaz, R.: Deep context of citations using machine-learning models in scholarly full-text articles. Scientometrics **117**(3), 1645–1662 (2018)
15. Hassan, S.U., Saleem, A., Soroya, S.H., Safder, I., Iqbal, S., Jamil, S., Bukhari, F., Aljohani, N.R., Nawaz, R.: Sentiment analysis of tweets through Altmetrics: a machine learning approach. J. Inf. Sci. **47**(6), 712–726 (2021)
16. Safder, I., Mahmood, Z., Sarwar, R., Hassan, S.U., Zaman, F., Nawab, R.M., Bukhari, F., Abbasi, R.A., Alelyani, S., Aljohani, N.R., Nawaz, R.: Sentiment analysis for Urdu online reviews using deep learning models. Expert. Syst. **28**, e12751 (2021)
17. Rahi, S., Safder, I., Iqbal, S., Hassan, S.U., Reid, I., Nawaz, R.: Citation classification using natural language processing and machine learning models. In: International conference on smart Information communication Technologies (pp. 357–365). Springer, Cham (2019)
18. Hassan, S.U., Shabbir, M., Iqbal, S., Said, A., Kamiran, F., Nawaz, R., Saif, U.: Leveraging deep learning and SNA approaches for smart city policing in the developing world. Int. J. Inform. Manag. **56**, 102045 (2021)
19. Mahmood, Z., Safder, I., Nawab, R.M., Bukhari, F., Nawaz, R., Alfakeeh, A.S., Aljohani, N.R., Hassan, S.U.: Deep sentiments in roman urdu text using recurrent convolutional neural network model. Inform. Proc. Manag. **57**(4), 102233 (2020)
20. Safder, I., Hassan, S.U., Visvizi, A., Noraset, T., Nawaz, R., Tuarob, S.: Deep learning-based extraction of algorithmic metadata in full-text scholarly documents. Inform. Proc. Manag. **57**(6), 102269 (2020)

Education and Technology Enhanced Learning: The Post-Covid "New Normal"

Higher Education in the Face of Educational Paradigm Shifts–From Face-to-Face to Distance Learning

Janusz Gierszewski, Andrzej Pieczywok, and Wojciech Pietrzyński

Abstract Higher education has recently undergone changes in the didactics of education caused mainly by the COVID-19 pandemic. The vast majority of universities have changed the form of education from stationary to remote. Academic teachers and students have faced serious challenges in fulfilling their tasks when face-to-face communication is very difficult or impossible. The aim of the article was to find out students' opinions on the effectiveness of remote learning in the context of a change in the paradigm of education. In order to more fully understand the issue of the effectiveness of remote learning caused by the COVID-19 pandemic, a diagnostic survey method was used and the tool was a survey questionnaire. Responses from respondents were collected using Google Forms electronic form. The survey revealed that students rated the effectiveness of remote learning moderately. In their opinion, remote learning as a permanent form of education raises controversies and concerns. These concerns are due to the stability of the internet connection during remote classes both on the part of the lecturers and the students themselves and the moderate psychological support received from the university. The study was conducted in Poland during the COVID-19 pandemic, an emergency situation. Therefore, the authors do not usurp the right to construct conclusions from the research on a general level in normal conditions of student education.

Keywords Higher education · Learning theories · Learning paradigm · Remote education · Pandemic

---

J. Gierszewski (✉) · W. Pietrzyński
Pomeranian University in Slupsk, 76-200, Slupsk, Poland
e-mail: janusz.gierszewski@apsl.edu.pl

W. Pietrzyński
e-mail: wojciech.piestrzynski@apsl.edu.pl

A. Pieczywok
Kazimierz Wielki University, 85-064, Bydgoszcz, Poland
e-mail: a.pieczywok@wp.pl

A. Visvizi et al. (eds.), *Research and Innovation Forum 2022*,
Springer Proceedings in Complexity, https://doi.org/10.1007/978-3-031-19560-0_28

1 Introduction

The pandemic situation in terms of teaching in higher education has fundamentally changed the paradigm of education, as remote learning has gone from being a complementary form to being the primary one. In view of the pandemic situation caused by COVID-19, there was a rather significant change in the form of higher education–a shift from face-to-face teaching to distance learning. This situation caused many changes in higher education institutions. Classes started to take place in the form of teleconferences–via various communicators such as Zoom, Skype or Microsoft Teams–or homework assignments to be completed on your own. Many teachers and students were unable to use this form of teaching effectively. There were technical problems (limitations in the supply of Internet or too slow connection) and economic problems–the high cost of purchasing equipment and a professional platform, including the training of lec-turers. As a result, the lecturers could not properly control the quality of knowledge transfer as they were not able to respond to individual questions and needs of students [1].

Given the timing of the pandemic, there has also been a reorganisation of education among students in security-related courses.

The need to devote more time to the development of e-learning materials also proved problematic for students. This situation was further exacerbated by the lack of motivation and self-discipline among students to prepare for classes themselves. In addition, the inability to organise practical classes remotely became apparent. The challenge during the pandemic became the organisation of online examination sessions and the implementation of work placements. The COVID-19 pandemic and above all its media representations and the outbreak of collective hysteria also had an impact on stress and anxiety levels among students. Fatigue, burnout, monotony of the day leading to boredom are just some of these effects.

Therefore, the authors of this study are interested in the opinion of students on the effectiveness of remote education in the situation of a COVID-19 pandemic threat [2].

Moreover, the conclusions formulated in this way will contribute to the evaluation of the applied educational paradigm in higher education. They will also indicate the value of exemplary thinking and action of teachers.

In contemporary analyses of the content of education, it is worth considering the assumptions of multiparadigmality [3]. It is Thomas Kuhn's theory, revealing the variability of learning, that may prove useful in undertaking such a task. In the approach to education, in thinking about learning and teaching, it is worth revealing the diversity of approaches. Change should become a key issue in considering the process of education, and thus the principles and content of teaching.

To a large extent, the above considerations can provide an excellent platform for contemporary interpretations and research explorations of learning paradigms in remote education. They can also enrich diverse paradigms of the educational process. Recognising the wealth of potential hidden in information technologies in the form of means facilitating the elimination of the disadvantages of face-to-face education.

2 Research Methodology

The subject of the research was the analysis of remote education in the perception of students of security studies. The aim of the research was to find out students' opinions on the effectiveness of remote education in the context of education paradigms. The formulated main research problem as follows: How do students (of security studies) evaluate the effectiveness of remote studying in the area of acquired knowledge and shaped skills in the context of objectivity and subjectivity (corresponding teaching philosophy and learning philosophy)?

In addition to the main problem, specific problems were formulated:

Q.1 How would students rate the stability of the connection during remote classes?
Q.2 To what extent do students receive technical support from the university during remote learning?
Q.3 To what extent do students rate the psychological support during remote learning?
Q.4 How effective do students think distance learning is?
Q.5 How do students perceive remote learning as a permanent form of education?

The following research hypotheses were also adopted:

H. 1. Stability of connection during remote classes on the part of lecturers was high and moderate on the part of students.
H. 2. The university's technical support to students was very poor.
H. 3. The psychological support received from the university, students rated moderate.
H. 4. Students rate the effectiveness of distance learning moderately.
H. 5. Students perceive remote learning as a permanent form of education moderately positively.

In order to more fully understand the issue of the effectiveness of remote learning caused by the COVID-19 pandemic, a diagnostic survey method was used and the tool was a survey questionnaire. The questionnaire was developed by the research team of the Pomeranian Academy in Słupsk, Poland. Thanks to the introduction of the Likert scale, a possibility of graded evaluation of the phenomenon of obligatory distance learning was created for the respondents. Responses from respondents were collected using Google Forms electronic form. The collected data were processed using Microsoft Office package, descriptive statistics spreadsheets, ANOVA analysis of variance, Tukey post hoc test. The research was conducted on students of Security Studies at the Pomeranian Academy in Słupsk in the Republic of Poland (mainly from the northern region of the country). The criterion adopted was year of attendance (first, second and third year), as it was recognised that there may be significant differences in the perception of distance and in-person learning due to the previous experience of in-person learning experienced by older students.

3 Essence, Concepts and Teaching Principles of Remote Education

Generally speaking, remote learning (distance education) is a part of didactics and has many characteristics that characterise it. It is a systematically implemented didactic project that includes: methodical preparation, presentation of learning material and control of the learning process, as well as support for learners, most often without the direct participation of the teacher. The project also includes new media, which by their planned use influence changes in learners.

When trying to provide a definition of distance education we come across many problems with different approaches to this category in the literature. The variety that occurs results both from the desire to accurately translate English terms, as well as is a manifestation of the desire to include in one category diverse forms of education united by the idea of increasing the availability of education to the greatest number of people.

It seems that a comprehensive problem of distance education, considering its main idea and changing technical conditions of its implementation was presented by Mirosław Kubiak. He points out that it is a method of conducting the teaching process in conditions when teachers and students are distant from each other and are not in the same place, using to transmit information, in addition to traditional ways of communication, also modern, very modern telecommunications technologies transmitting: voice, video, computer data and printed materials [4].

Distance learners are physically separated from the institution that delivers a particular course. In addition, the contract between the parties involved in the learning process requires that the teaching includes: the checking of learning outcomes, the provision of instructions for learning, the preparation of the learner for examinations and their conduct by the institutions organising the learning process. All this should be achieved through individual and group communication in the physical absence of the teacher [5].

The phenomenon of change is therefore not new in education, but it is of particular importance in distance learning because of the role that technology plays in it. The emergence of new communication tools causes an avalanche of changes in the form of new methods and forms of education, as well as the organisation of learning. Theory therefore has an important role to play in determining the mechanisms and processes that occur in the learning environment [6].

In distance education we observe many theoretical concepts. Desmond Keegan, in his work The Foundations of Distance Education [7] published in 1986, distinguishes three historical approaches: the theory of industrialization of teaching by Peters [8] in 1983; the theory of independence and autonomy by Wedemeyer [9] and Moore [10]; the theory of interaction and communication formulated by: Baath [11], Sewart [12], Daniel and Marquis [13] and Holmberg [14].

To the mentioned concepts it is worth adding another one, developed by Perraton [15], which is a synthesis of the mentioned theories of distance education and is a kind of education. The principles of multimedia learning are based on instructional design and aim to minimise extraneous cognitive load and manage internal load. In instructional design, Richard E. Mayer and Roxana Moreno made partial use of the model developed by Baddeley and Hitch model of working memory and the resulting important conclusion about the parallel flow of information: visual and auditory.

Analysing the development of online learning, there is a clear trend towards the creation of learning principles. This has been fostered by emerging cognitive theories of education, which have paved the way for multifaceted explorations. One result is the emergence of an outline of principles for multimedia-supported online learning, presented by Clark and Mayer, among others.

Many lecturers and students claim that remote classes are less effective than so-called "contact classes". There are also many enthusiasts who believe that technological innovations are superior to "traditional" teaching. Certainly, the development of classes using remote learning has started a revolution in teaching. The conducted verification of learning outcomes does not always show a significant difference. Hence, it is worthwhile to find out students' opinions on remote learning.

4 Selected Theories of Education in a Digital World

Theories of distance learning include many approaches and views resulting from the adopted theoretical basis, such as humanistic or cognitive. The authors, due to a certain limitation of the framework of the article and the multiplicity of theories, decided on a simplified division of distance learning theories, which includes a historical approach.

Four fundamental theories were selected which have influenced the development of educational didactics today. In the group of concepts referred to as the historical approach, the following theories were presented: industrialisation of teaching, independence and autonomy, interaction and communication, and synthesis of historical theories.

One of the most relevant theoretical concepts of the twentieth century in distance education was based on the model of industrial production of distance education. Its creator, Peters, assumed that distance education could be analysed by comparison with industrial production of goods.

Peters characterises distance education as a method of transmitting knowledge, skills and attitudes that is rationalised by the use of division of labour and organisational principles and the extensive use of technical media. These means should be used to reproduce high quality teaching materials that enable instruction to be given to learners at the same time and in any place.

The use of well-defined learning objectives, as in the instructional design of the system, allowed the translation of the principles governing teaching–learning into concrete solutions to enhance the effectiveness and efficiency of distance education.

One of the pioneers of distance education, Wedemeyer [16], formulated the theory of learner autonomy (freedom in learning). It assumes the necessity of the occurrence of autonomy in learning. The aforementioned conditions of learning are today the basis of a modern online educational system. It should be emphasized that Wedemeyer's way of thinking about distance education was in line with the principles of humanism postulated by didacticians and the proposals of andragogists.

Moore [17], who developed the concept of transactional distance, is also worth citing at this point. It combines both Otto Peters' view of distance learning as a highly organised mechanical system and Charles A. Wedemeyer, who emphasised the learner and his interactive relationship with the teacher.

One of the founders of the communicative interaction theory is Börje Holmberg. His theory assumes that distance learning is an interaction similar to a conversation between a learner and a teacher. Fundamental to this is the concept of so-called didactic conversation, which refers to both real and simulated conversation. In order to facilitate this, Holmberg has developed a didactic guide.

He based his theory on correspondence communication between teacher and learner and the industrial organisation of the learning process. He assumed that learning can only occur when: the learner is active, the focal point of education is the learning process and communication, learning takes place in a planned manner under the supervision of the educational organisation.

Holmberg believes that the basis of distance learning is the interaction between teacher and learner. Central to this are motivation and empathy, as well as autonomy and communication. The simulated interactions that occur in a substantive conversation can develop thinking, create conditions for different views and approaches to clash and enable solutions to be sought to existing problems.

A characteristic feature of the communicative interaction theory is the absence of direct teacher supervision of the class and the teacher's absence during the learning process. Simulating a conversation (conversation) involves learners working with text contained in instructional materials.

The concept of synthesis of existing theories has been proposed by Hilary Perraton, who bases her idea on the combination of different concepts by choosing 5 covers the ways of distance learning, another 4 refer to the need to intensify the dialogue and the last 5 deal with making it a method.

Many studies compare the effectiveness of traditional "contact" learning with that of new technologies, the most recent of which are differentiated forms of digital distance learning. Therefore, in the opinion of the authors of this article, it is assumed that in the level of acquired knowledge there are no significant differences in the effectiveness of learning in the framework of "contact" education with distance learning using electronic media. Thanks to the conducted research, we have feedback on remote learning in security-related fields of study.

It is well known that highly rated classes result in greater learning efficiency. However, this correlation depends on a number of variables. We assumed that it depends on, among other things, the stability of the internet connection, technical and psychological support from the university.

5 Analysis of Test Results

The results of our own research and verification of the working hypotheses are presented below (Tables 1 and 2).

Verification of H. 1. Stability of connection during remote classes on the part of lecturers was high and moderate on the part of students.

The hypothesis was completely unconfirmed. Students rated their own internet connection stability low (M = 2.47), with first-year students rating it lowest (M = 2.00). At the same time, second and third year students rated slightly higher (M = 2.84; M = 2.47). Despite this, the evaluation of first-year students is significantly different from that of second- and third-year students ($p < 0.01$). An even greater inconsistency between assumptions and reality was found in the evaluation of the stability of the Internet connection on the part of the lecturers, (low rating; M = 2.21; M-I = 1.84; M-II = 2.63; M-III = 2.34). In the pair of internal treatments, significant differences were found between the first and second and third years of study ($p < 0.01$) (Table 3).

Verification H. 2. The university's technical support to students was very poor.

Working hypothesis 2 was not confirmed. The students experienced moderate technical support from the university during remote learning. They rated the above support moderately (M = 3.76; M-I = 4.13; M-II = 3.40; M-III = 3.56), where significant differences were found within the study group between the first year and the second and third years ($p < 0.01$) (Table 4).

Verification of H. 3. The psychological support received from the university, students rated moderate.

Hypothesis 3 was confirmed. Students rated psychological support moderately (M = 3.56; M-I = 3.91; M-II = 3.10; M = III = 3.52). The internal group measurement allows us to conclude that there are statistically significant differences only between the first and second year of study ($p < 0.01$) (Table 5).

H. 4. Students rate the effectiveness of distance learning moderately.

Hypothesis 4 was fully confirmed. This is evidenced by the following results (M = 3.67; M-I = 4.13; M = II = 3.19; M-III = 3.49). At the same time, significant differences were found between the assessment of the effectiveness of distance learning made by first-year students and second- and third-year students ($p < 0.01$) (Table 6).

H. 5 Students perceive remote learning as a permanent form of education moderately positively.

Hypothesis 5 was partially confirmed. Only first-year students perceived remote learning in a moderately positive way (M-I = 3.54) The others rated it low (M-II = 2.75; 2.77; M = 3.09). Hence, significant differences were found within the study group between the first year and the others ($p < 0.01$).

Table 1 Descriptive statistics, results of analysis of variance, Tukey's post hoc test for the variable stability of connection during remote activities by students

Group	N	Min.	Max.	M	SD
First year of studies	81	1	5	2.00	0.84
Second year of studies	57	1	5	2.84	0.88
Third year of studies	53	1	5	2.79	0.96
Total	191	1	5	2.49	0.97
--------------------	--------------------	--------------------	--------------------	--------------------	--------------------
Source	SS	df	MS	Test F	p
Between groups	31.29	2	15.65	19.83	$p < 0.01$
Within groups	148.29	188	0.79		
Total	179.59	190			
--------------------	--------------------	--------------------	--------------------	--------------------	--------------------
Comparative pairs	Tukey HSD Q statistics	Tukey HSD p-values Q	Materiality level	--------------------	--------------------
1st year versus 2nd year	7.76	0.001	**$p < 0.01$	--------------------	--------------------
1st year versus 3rd year	7.14	0.001	**$p < 0.01$	--------------------	--------------------
2nd year versus 3rd year	0.41	0.899	Irrelevant	--------------------	--------------------

Source own elaboration

Table 2 Descriptive statistics, results of analysis of variance, Tukey post hoc test for the variable stability of connection during remote classes by lecturers

Group	N	Min.	Max.	M	SD
First year of studies	81	1	5	1.84	0.84
Second year of studies	57	1	5	2.63	0.83
Third year of studies	53	1	5	2.34	0.91
Total	191	1	5	2.21	0.92
--------------------	--------------------	--------------------	--------------------	--------------------	--------------------
Source	SS	df	MS	Test F	p
Between groups	22.13	2	11.06	14.85	$p < 0.05$
Within groups	144.06	188	0.74		
Total	162.19	190			
--------------------	--------------------	--------------------	--------------------	--------------------	--------------------
Comparative pairs	Tukey HSD Q statistics	Tukey HSD p-values Q	Materiality level	--------------------	--------------------
1st year versus 2nd year	7.50	0.001	**$p < 0.01$	--------------------	--------------------
1st year versus 3rd year	4.63	0.003	**$p < 0.01$	--------------------	--------------------
2nd year versus 3rd year	2.50	0.118	Irrelevant	--------------------	--------------------

Source own elaboration

Table 3 Descriptive statistics, results of analysis of variance, Tukey post hoc test for the variable technical support of students from the university

Group	N	Min.	Max.	M	SD
First year of studies	81	1	5	4.13	0.85
Second year of studies	57	1	5	3.40	0.84
Third year of studies	53	1	5	3.56	1.06
Total	191	1	5	3.76	0.96
--------------------	--------------------	--------------------	--------------------	--------------------	--------------------
Source	SS	df	MS	Test F	p
Between groups	20.68	2	10.33	12.44	$p < 0.05$
Within groups	156.24	188	0.83		
Total	176.92	190			
--------------------	--------------------	--------------------	--------------------	--------------------	--------------------
Comparative pairs	Tukey HSD Q statistics	Tukey HSD p-values Q	Materiality level	--------------------	--------------------
1st year versus 2nd year	6.57	0.001	**$p < 0.001$	--------------------	--------------------
1st year versus 3rd year	5.00	0.001	**$p < 0.001$	--------------------	--------------------
2nd year versus 3rd year	1.32	0.610	Irrelevant	--------------------	--------------------

Source own elaboration

Table 4 Descriptive statistics, results of analysis of variance, Tukey post hoc test for the variable psychological support of students by the university

Group	N	Min.	Max.	M	SD
First year of studies	81	1	5	3.91	0.94
Second year of studies	57	1	5	3.10	1.01
Third year of studies	53	1	5	3.52	0.99
Total	191	1	5	3.56	1.02
--------------------	--------------------	--------------------	--------------------	--------------------	--------------------
Source	SS	df	MS	Test F	p
Between groups	21.96	2	10.98	11.53	$p < 0.05$
Within groups	178.97	188	0.95		
Total	200.93	190			
--------------------	--------------------	--------------------	--------------------	--------------------	--------------------
Comparative pairs	Tukey HSD Q statistics	Tukey HSD p-values Q	Materiality level	--------------------	--------------------
1st year versus 2nd year	6.77	0.001	**$p < 0.01$	--------------------	--------------------
1st year versus 3rd year	3.16	0.068	Irrelevant	--------------------	--------------------
2nd year versus 3rd year	3.21	0.062	Irrelevant	--------------------	--------------------

Source own elaboration

Table 5 Descriptive statistics, results of analysis of variance, Tukey post hoc test for the variable students' evaluation of the effectiveness of distance education

Group	N	Min.	Max.	M	SD
First year of studies	81	1	5	4.13	0.81
Second year of studies	57	1	5	3.19	1.21
Third year of studies	53	1	5	3.49	0.99
Total	191	1	5	3.67	1.07
--------------------	--------------------	--------------------	--------------------	--------------------	--------------------
Source	SS	df	MS	Test F	p
Between groups	32.24	2	16.12	16.15	$p < 0.05$
Within groups	187.63	188	0.99		
Total	219.87	190			
--------------------	--------------------	--------------------	--------------------	--------------------	--------------------
Comparative pairs	Tukey HSD Q statistics	Tukey HSD p-values Q	Materiality level	--------------------	--------------------
1st year versus 2nd year	7.72	0.001	**$p < 0.01$	--------------------	--------------------
1st year versus 3rd year	5.17	0.001	**$p < 0.01$	--------------------	--------------------
2nd year versus 3rd year	2.20	0.265	Irrelevant	--------------------	--------------------

Source own elaboration

Table 6 Descriptive statistics, results of analysis of variance, Tukey post hoc test for the variable perception of remote education students as a permanent form of education

Group	N	Min.	Max.	M	SD
First year of studies	81	1	5	3.54	1.12
Second year of studies	57	1	5	2.75	1.35
Third year of studies	53	1	5	2.77	1.20
Total	191	1	5	3.09	1.27
--------------------	--------------------	--------------------	--------------------	--------------------	--------------------
Source	SS	df	MS	Test F	p
Between groups	28.36	2	14.18	9.59	$p = 0.0001$
Within groups	277.94	188	1.48		
Total	306.30	190			
--------------------	--------------------	--------------------	--------------------	--------------------	--------------------
Comparative pairs	Tukey HSD Q statistics	Tukey HSD p-values Q	Materiality level	--------------------	--------------------
1st year versus 2nd year	5.30	0.001	**$p < 0.01$	--------------------	--------------------
1st year versus 3rd year	5.06	0.001	**$p < 0.01$	--------------------	--------------------
2nd year versus 3rd year	0.12	0.899	Irrelevant	--------------------	--------------------

Source own elaboration

6 Conclusions

Synchronous teaching, called “remote”, consists of students and lecturers attending classes at the same time, but outside the lecture hall, and their contact is usually in the form of videoconferencing.

However, the most useful in evaluating the effectiveness of distance learning seems to be the constructivist theory, which takes a more creative approach. The student is treated in this paradigm as an active and independent participant, who, using the information and experiences he has previously acquired, deepens his knowledge. The most important role in this type of teaching is played by programs enabling real-time conversations (e.g. Microsoft Teams, Meet, Zoom, Google Class-room or ClickMeeting).

The assumed assumption that there are no significant differences in the level of acquired knowledge in the effectiveness of "contact" learning with distance learning using electronic media has turned out to be not entirely true due to several regularities. In remote learning, one should keep in mind:

1. Internet connection stability.
2. Technical support from the university.

Thus, a significant shortcoming of remote education is the quality of equipment and internet connections of students and lecturers. In order to compare the effects of the learning process, it should be assumed that the basic methods of education must be the same with these variants (hypothesis 1 and 2). In other words, if the same graphical and verbal elements occur in “contact” learning, they must occur in the same shape in “remote” learning and without technical interference. This is because the learning process is a psychological (hypothesis 3) component of active learning based on the constructivist paradigm, occurring independent of the medium that is used. Therefore, it is important for universities to reflect on the choice of modes of communicating with students, i.e., which one can be used to enhance effective learning. It would also be helpful to indicate the choice of subjects (recommended by academic teachers and students), which are possible to implement remotely without “losing” the quality of the verification of learning outcomes (subjects of theoretical and practical character).

What used to be considered different and separate in the learning process is now very similar (hypothesis 4) Despite appearances, distance classes are not new. The study found that there is no noticeable difference between “face-to-face” and “remote” classes.

Time will tell (hypothesis no. 5) when the differences between remote and “contact” learning will become blurred (we assume that soon). It is certainly necessary to increase the teaching offer for students related to remote working.

References

1. Gierszewski, J., Pieczywok, A.: The challenges of studying during the SARS-Cov-2 pandemic. Sec. Dimens. Int. Natl. Stud. **34**(34), 22–44 (2020). https://doi.org/10.5604/01.3001.0014.5600
2. Daniela, L., Visvizi, A.: Remote Learning in Times of Pandemic: Issues, Implications and Best Practice, 1st edn. Routledge. https://doi.org/10.4324/9781003167594 (2021)
3. Johnson, A.F., Roberto, K.J., Rauhaus, B.M.: Policies, politics and pandemics: course delivery method for US higher education institutions amid COVID-19. Transf. Govern. People Proc. Policy **15**(2), 291–303 (2021). https://doi.org/10.1108/TG-07-2020-0158
4. Gierszewski, J., Pieczywok, A., Piestrzyński, W.: Studying during the COVID-19 pandemic: personal and technical aspects. In: Proceedings of the 37th International Business Information Management Association (IBIMA 2021), pp 11777–11784, ISBN: 978-0-9998551-6-4. Cordoba (2021)
5. Visvizi, A., Lytras, M.D., Sarirete, A.: Emerging technologies and higher education: management and administration in focus. Emerald Publishing Limited, Management and Administration of Higher Education Institutions at Times of Change (2019)
6. Klus-Stańska, D,; Paradygmaty Dydaktyki [Didactic Paradigms] PWN, Warsaw (2018)
7. Kubiak, M.J.: Virtual Education. MIKOM, Warsaw (2000)
8. Peters, O.: Distance teaching and industrial production: A comparative interpretation in outline. In: Sewart, D., Keegan, D., Holmberg, B. (eds.) Distance Education: International Perspectives, pp. 95–113. Croom Helm, London (1983)
9. Wedemeyer, C.A.: Independent study. In: Knowles, A.S. (ed.)The International Encyclopedia of Higher Education. Northeastern University, Boston, MA (1977)
10. Moore, M.G.: Toward a theory of independent learning and teaching. J. High. Educ. 44, 66–69 (1973)
11. Baath, J.A.: Distance students' learning–empirical findings and theoretical deliberations. Distan. Educ. 1, 6–27 (1982)
12. Staff Development Needs in Distance Education and Campus-Based Education: Are They So Different? In: Sewart, D. (ed.), Croom Helm, London (1987)
13. Daniel, J.S., Marquis, C.: Interaction and independence: Getting the mixture right. Teach. Distan. 15, 25–44 (1979)
14. Holmberg, B.: Theory and Practice of Distance Education. Routledge, London, New York (1989)
15. Perraton, H.: Open and Distance Learning in the Developing World, 2nd edn. Routledge, Taylor & Francis Group, London, New York (2006)
16. Wedemeyer, Ch.A.: Learning at the Back Door: Reflections on Non-Traditional Learning in the Lifespan, Madison. University of Wisconsin, WI) (1981)
17. Moore, M.G., Kearsley, G.: Distance Education: A Systems View, Belmont. Wadsworth Publishing Company, CA) (1996)
18. Mayer, R.E.: Multimedia Learning. Cambridge University Press, New York (2001)
19. Mayer, R.E., Moreno, R.: A Cognitive Theory of Multimedia Learning: Implications for Design Principles. (1998)
20. Clark, R.C., Nguyen, F., Sweller, J.: Efficiency in Learning: Evidence-Based Guidelines to Manage Cognitive Load. John Wiley, Somerset (NJ) (2011)
21. Holmberg, B.: The sphere of distance-education theory revisited. (2022). https://eric.ed.gov/?id=ED386578

Generic Competences in Higher Education After Covid-19 Pandemic

Raquel Ferreras-Garcia, Jordi Sales-Zaguirre, and Enric Serradell-López

Abstract This paper aims to analyse whether acquisition of students' generic competences has been influenced by Covid-19 pandemic in an online higher education environment. In order to analyse these effects we present an empirical study to know the influence of Covid-19 pandemic in generic competences of business and economics students. A quantitative analysis is conducted in a sample of 221 students from the online bachelor's degree programme on Business and Economics at the Universitat Oberta de Catalunya. The analyses show that the achievement of most part of competences has been affected by the pandemic, but in different levels. Moreover, the analyses also show which competences are affected in higher levels and which one in lower levels. Moreover the study analyses the possible differences depending on whether the students were at the beginning of the studies or at the end. The key finding is the effect of Covid-19 pandemic on generic competences, which creates an opportunity to undertake actions to address the lowest achieved and to generate strategies in order to adapt to this new situation.

Keywords Covid-19 pandemic · Generic competences · Higher education

1 Introduction

SARS-CoV-2 has caused a global pandemic of the Covid-19 disease, which has generated an unprecedented crisis in all socioeconomic areas, including education. At the end of March 2020, the governments of the majority of European countries decided to lock down the population and countries closed their borders.

R. Ferreras-Garcia (✉) · E. Serradell-López
Universitat Oberta de Catalunya, Barcelona, Spain
e-mail: rferreras@uoc.edu

E. Serradell-López
e-mail: eserradell@uoc.edu

J. Sales-Zaguirre
Institut Químic de Sarrià, Universitat Ramon Llull, Barcelona, Spain
e-mail: jordi.sales@iqs.url.edu

A. Visvizi et al. (eds.), *Research and Innovation Forum 2022*,
Springer Proceedings in Complexity, https://doi.org/10.1007/978-3-031-19560-0_29

Around the world, face-to-face universities had to make the transition from using classic teaching techniques to implementing remote emergency teaching in just a few days, with virtual platforms suddenly becoming the focus of interaction. The online universities had a clear advantage in this exceptional situation since their procedures linked to teaching and learning were already online [1].

Despite this supposed advantage (students from online universities were supposedly already fully adapted to an entirely remote teaching and learning system), we do not have enough information on whether students' academic performance from online universities was affected by the pandemic.

Throughout the crisis, researchers have sought to better understand the effects the pandemic has had on students. Various works have contributed by analysing the effect of virtualisation on face-to-face university students, leaving aside the effect the pandemic may have had on online university students' learning. Notably, the analyses made until now have focused on students' academic performance basically in terms of qualifications. However, and despite the crucial role competences have in the learning process [2] and in the creation of links in the labour market [3], there are as yet no analyses on the effect the pandemic may have had on competences.

Thus, it is not clear how the pandemic has influenced students' academic performance in terms of competences of those who were studying in an online university, so more research is needed in order to fill this gap.

Bearing this in mind, the aim is to establish what impact the Covid-19 pandemic has had on the development of generic competences.

This issue is expressed in the following research question:

RQ1: What is the impact of the pandemic on competency development?

Moreover, it could be interesting to know if the course of study of students has some kind of influence.

This issue is expressed in the following research question:

RQ2: Are there any differences in the impact of the pandemic on competency development between the first year students and the last year students?

2 Background

The Covid-19 pandemic has had a huge impact at all socioeconomic levels and education has been no exception. The new reality, with the declaration of the state of emergency and the lockdown of the population, forced all face-to-face universities to change to online teaching, thus affecting most of the university population. Nonetheless, this virtual learning environment is not new; it has existed for some years and it is implemented by different universities.

The online universities and the universities that were leaning towards virtualisation before the pandemic responded well, adapting quickly to the sudden change, while the universities with little or no experience with virtualisation faced difficulties to migrate to digital platforms and to carry on the learning process [4]. Research into this sudden paradigm shift and the effects on students' academic performance is an aspect in process of study.

The Covid-19 pandemic hugely affected those at the heart of the university model, the students. They had to cope with many factors that influenced their learning experience [5].

There is a growing interest in understanding and investigating the impact of the Covid-19 pandemic on higher education students. The elevated number of works in this area are varied in the topics they cover and the results obtained (for example, [6–11]), most of them focused on the effect of the emergency e-learning situation on students' academic performance from face-to-face universities.

The implementation of EHEA put competences as a fundamental part of the teaching–learning process of European universities. Competences provide reference points for teachers in terms of results and learning levels, while also facilitating easy comparison between disciplines [12]. In the Tuning project [2], competences represent 'a dynamic combination of attributes, abilities and attitudes. Fostering these competences is the object of educational programmes. Among these competences we find generic competences. Generic competences are the ones that are interdisciplinary in nature and common to all degree programs. These competences have been widely analysed in studies prior to the pandemic (for example, [13–17]), but to our knowledge no previous work has explicitly analysed the effect the pandemic has had on students' generic competences.

Clearly, the changes due to the Covid-19 pandemic may have affected students' acquisition of competences. However, we have no knowledge about this possible effect, so more research is needed in order to answer this question and improve our knowledge of the impact that the pandemic might have had on the development of competences, especially of those students from online universities.

3 Methodology

3.1 Measurement of Variables

The sample consisted of students in their first and last course at the Faculty of Economics and Business from the online university Universitat Oberta de Catalunya during the first semester of the 2020/21 academic year.

A questionnaire was designed to obtain the perception of the effect of the pandemic on the development of students' generic competences. The students completed it in January 2021, so that students' opinions had been collected after some months of pandemic. In total the questionnaire was completed by 221 students.

The first part of the questionnaire included questions designed to collect information on the characteristics of the sample, such as gender, age and course. The

Table 1 Generic competences

G1	Learning and responsibility
G2	Identifying and appropriately using the information
G3	Critical analysis
G4	Application of knowledge to practice
G5	Taking decisions in uncertain situations
G6	Adapting to new situations
G7	Team work and collaboration
G8	Negotiation
G9	Creativity
G10	Entrepreneurship
G11	Innovation
G12	Leadership
G13	Oral and written communication
G14	Communication in a third language
G15	Change management
G16	Using ICT and adapting to new technological environments
G17	Time management in uncertainty
G18	A commitment to ethics
G19	Selecting, analysing and taking decisions based on the information
G20	Planning and organisation
G21	Initiative and working autonomously

second part correspond to the evaluation of the effect the pandemic had had on their acquisition of the generic competences. The competences were selected from the items included in the white paper of the Bachelor's Degree in Economics and Business Administration [18]. These items were used in previous research focused on the level of generic and specific competences of students enrolled on the Economics and Business Administration and Management degree programmes for the purpose of measuring the effect of the different learning tools, such as business simulators and business plans ([14, 19–21]). The questions referring to competences were designed using an ordinal 5-point Likert scale where 1 meant "Very low degree of effect" and 5 meant "Very high degree of effect". The generic competences are described in Table 1.

3.2 Data Analysis

The statistics software SPSS version 25 was used to carry out the quantitative analysis. On analysing the distribution of the variables, the Kolmogorov–Smirnov test led us to reject the null hypothesis at a significance level of $\alpha = 0.05$ and to conclude that

Table 2 Distribution by course

Course	Frequency	Percentage
First year	93	42.1
Last year	128	57.9
Total	221	100

the generic competences variables did not follow a normal distribution. To answer the research questions, a descriptive calculation analysis of the mean for the different competences was carried out, moreover a Mann–Whitney test was carried out on the difference between the means.

Cronbach's alpha coefficient was calculated to check the reliability of the measurement scale [22]. The results show an alpha coefficient of 0.979 for the sample, which means it is highly reliable [23].

The distribution of the students by gender shows that 54.3% were men and 45.7% were women. The average age of the students was 35.18 years with a range between 18 and 62 years.

Table 2 shows the frequencies and percentages for the 'course' variable.

4 Results

Table 3 shows the means and standard deviations of the generic competences variables. The results show that the students perceive that the pandemic has affected the development of their generic competences, with values above 2.5 points in all cases.

The competences most affected by the pandemic were those related to time management (G17, G20), the use of ICT (G16), team work (G7), taking decisions (G5), change management (G15) and autonomous work (G21), all of them with values above 3 points. Meanwhile, the competences they considered least affected by the pandemic were those related to communication (G13 and G14), with values of 2.65 and 2.52 points.

The means of the competences were calculated for students of first and last year to ascertain whether there are course differences in competency acquisition. The results show (Table 4) that the effect of pandemic over competency development is higher for first year students than for last year students.

In addition, competences with a higher affectation for first year students were those related to adapting to new situations (G6), time management (G17), team work (G7) and the use of ICT (G16). Meanwhile for last year students the higher effect was over adapting to new situations (G6) and time management (G17). It should be noted that for first year students most of competences were evaluated over 3 points, while last year students just evaluated two competences over 3 points.

Table 3 Means of the generic competences

	Minimum	Maximum	Mean	Standard deviation
G1	1	5	2.92	1.322
G2	1	5	2.79	1.322
G3	1	5	2.74	1.302
G4	1	5	2.86	1.258
G5	1	5	3.07	1.311
G6	1	5	3.30	1.369
G7	1	5	3.09	1.451
G8	1	5	2.82	1.360
G9	1	5	2.93	1.360
G10	1	5	2.99	1.378
G11	1	5	2.90	1.365
G12	1	5	2.94	1.445
G13	1	5	2.65	1.389
G14	1	5	2.52	1.316
G15	1	5	3.03	1.329
G16	1	5	3.11	1.407
G17	1	5	3.26	1.315
G18	1	5	2.82	1.469
G19	1	5	2.87	1.349
G20	1	5	3.10	1.409
G21	1	5	3.01	1.466

Table 4 Means of the competences by course

Course	G1	G2	G3	G4	G5	G6	G7	G8	G9	G10	G11
First	3.12	2.97	2.90	2.95	3.25	3.48	3.30	3.00	3.09	3.25	3.12
Last	2.77	2.66	2.62	2.80	2.94	3.16	2.93	2.69	2.81	2.80	2.74

Course	G12	G13	G14	G15	G16	G17	G18	G19	G20	G21
First	3.12	2.82	2.72	3.14	3.27	3.43	3.10	3.09	3.25	3.17
Last	2.80	2.52	2.38	2.95	2.99	3.13	2.62	2.72	2.98	2.90

Thus, it may be said that first year students were more affected by the pandemic when we refer to competency development and therefore the course is a variable that affects students' competency acquisition.

A Mann–Whitney U test comparing the means was conducted to analyse possible differences by course (Table 5). At a significance level of $\alpha = 0.05$, the results

Table 5 Significant statistical tests of the competences for the course variable

	G1	G3	G5	G7	G8	G10
Mann–Whitney U test	5120.000	5197.000	5153.000	5091.000	5192.000	4875.500
Wilcoxon W test	13,376.000	13,453.000	13,409.000	13,347.000	13,448.000	13,131.500
Z	−1.818	−1.648	−1.744	−1.875	−1.659	−2.344
Asymptotic significance (2-sided)	0.069*	0.099*	0.081*	0.061*	0.097*	0.019**

	G11	G14	G17	G18	G19
Mann–Whitney U test	5033.500	5074.500	5188.500	4889.000	5049.500
Wilcoxon W test	13,289.500	13,330.500	13,444.500	13,145.000	13,305.500
Z	−2.001	−1.923	−1.669	−2.325	−1.969
Asymptotic significance (2-sided)	0.045**	0.054*	0.095*	0.020**	0.049**

Significance codes: p-value 0.05 '**'; p-value 0.1 '*'

show significant differences between first and last year students for generic competences related to entrepreneurship (G10), innovation (G11), ethics (G18) and information (G19). At a significance level of $\alpha = 0.1$, learning and responsibility (G1), critical analysis (G3), taking decisions (G5), team work (G7), negotiation (G8), communication (G14) and time management (G17) also showed significant differences.

5 Conclusions

The main objective of education is to provide industry with highly qualified graduates equipped with appropriate competences. In that sense, programmes of study should prepare future graduates and give them the competences and training required to access the labour market with sufficient grounding for their professional development.

The aim of this study is to explore the effect the Covid-19 pandemic has had on the development of students' generic competences. To do this, an analysis was carried out in the context of an online university, taking into account whether the course is a key factor in the affectation of competences among university students.

The analysis contributes as innovative research, providing knowledge about the effect of the pandemic on the development of students' generic competences, which has not been previously investigated.

This work has provided interesting results on the effect of the pandemic on students' competences. The results indicate that the pandemic has affected the development of students' generic competence. Furthermore, the course influences this development, with differences observed between first year course students and last year course students.

The study provides universities and researchers with relevant information on how the pandemic has affected students. The universities will be able to take measures in line with these results to improve the competences affected by the pandemic, as well as adapt better their teaching should they be faced with a future emergency situation.

References

1. Cranfield, D.J., Tick, A., Venter, I.M., Blignaut, R.J., Renaud, K.: Higher education students' perceptions of online learning during Covid-19: a comparative study. Educ. Sci. **11**(8), 403 (2021)
2. González, J., Wagenaar, R.: Tuning Educational Structures in Europe. University of Deusto, Bilbao (2003)
3. Sánchez-Rebull, M.V., Campa-Planas, F., Hernández-Lara, A.B.: Dolceta, educación online para los consumidores: módulo de alfabetización financiera en España. Profesional de la Información **20**(6), 682–688 (2011)
4. Tang, Y.M., Chen, P.C., Law, K.M., Wu, C.H., Lau, Y.Y., Guan, J., He, D., Ho, G.T.S.: Comparative analysis of Student's live online learning readiness during the coronavirus (Covid-19) pandemic in the higher education sector. Comput. Educ. **168**, 104211 (2021)
5. Grant, K., Gedeon, S.: The impact of Covid-19 on university teaching. In: Remenyi, D., Grant, K., Singh, S. (eds.) The University of the Future: Responding to Covid-19, p. 161. Academic Bookshop, Reading, UK (2020)
6. Lemay, D.J., Doleck, T.: Online learning communities in the Covid-19 pandemic: Social learning network analysis of twitter during the shutdown. Int. J. Learn. Anal. Artif. Intell. Educ. **2**(1), 85–100 (2020)
7. Adnan, M., Anwar, K.: Online learning amid the Covid-19 pandemic: students' perspectives. Online Submiss. **2**(1), 45–51 (2020)
8. Iglesias-Pradas, S., Hernández-García, Á., Chaparro-Peláez, J., Prieto, J.L.: Emergency remote teaching and students' academic performance in higher education during the Covid-19 pandemic: a case study. Comput. Hum. Behav. **119**, 106713 (2021)
9. El Said, G.R.: How did the Covid-19 pandemic affect higher education learning experience? An empirical investigation of learners' academic performance at a university in a developing country. Adv. Hum.-Comput. Interact. **2021**, 1–10 (2021)
10. Prat, J., Llorens, A., Salvador, F., Alier, M., Amo, D.: A methodology to study the university's online teaching activity from virtual platform indicators: the effect of the Covid-19 pandemic at Universitat Politècnica de Catalunya. Sustainability **13**(9), 5177 (2021)
11. Alarabiat, A., Hujran, O., Soares, D., Tarhini, A.: Examining students' continuous use of online learning in the post-Covid-19 era: an application of the process virtualization theory. Inf. Technol. People (2021)
12. López-Bonilla, J.M., López-Bonilla, L.M.: Holistic competence approach in tourism higher education: an exploratory study in Spain. Curr. Issue Tour. **17**(4), 312–326 (2014)
13. Chan, C.K., Fong, E.T., Luk, L.Y., Ho, R.: A review of literature on challenges in the development and implementation of generic competencies in higher education curriculum. Int. J. Educ. Dev. **57**, 1–10 (2017)

14. Ferreras-Garcia, R., Sales-Zaguirre, J., Serradell-López, E.: Developing entrepreneurial competencies in higher education: a structural model approach. Educ.+ Train. **63**(5), 720–743 (2021)
15. Okolie, U.C., Igwe, P.A., Nwosu, H.E., Eneje, B.C., Mlanga, S.: Enhancing graduate employability: Why do higher education institutions have problems with teaching generic skills? Policy Futures Educ. **18**(2), 294–313 (2020)
16. Suleman, F.: The employability skills of higher education graduates: insights into conceptual frameworks and methodological options. High. Educ. **76**(2), 263–278 (2018)
17. Williams, R.: National higher education policy and the development of generic skills. J. High. Educ. Policy Manag. **41**(4), 404–415 (2019)
18. ANECA: Libro Blanco. Título de grado en economía y empresa. Agencia Nacional de Evaluación de la Calidad y Acreditación, Madrid (2005)
19. Ferreras-Garcia, R., Hernández-Lara, A.B., Serradell-López, E.: Entrepreneurial competences in a higher education business plan course. Educ.+ Train. **61**(7/8), 850–869 (2019)
20. Fitó-Bertran, À., Hernández-Lara, A.B., Serradell-López, E.: The effect of competences on learning results an educational experience with a business simulator. Comput. Hum. Behav. **51**, 910–914 (2015)
21. Hernández-Lara, A.B., Serradell-López, E., Fitó-Bertran, À.: Do business games foster skills? A cross-cultural study from learners' views. Intang. Cap. **14**(2), 315–331 (2018)
22. Cronbach, L.J.: Studies of acquiescence as a factor in the true-false test. J. Educ. Psychol. **33**(6), 401 (1942)
23. De Vellis, R.F., Dancer, L.S.: Scale development: theory and Applications. J. Educ. Meas. **31**(1), 79–82 (1991)

Universities and Digital Skills' Development in Colombia

Maddalena della Volpe, Alexandra Jaramillo-Gutiérrez, and Andrés Henao-Rosero

Abstract The development of human capital is a driver for economic growth. Future educational systems, based on digital platforms, can make us more resilient when confronting unforeseen crises such as the Covid-19 pandemic, which gave a sudden boost to digital transformation in society, in order to access services which had to be digitalized for external circumstances of isolation, highlighting gaps and the need to adapt to a new context. A great barrier is represented by the lack or insufficiency of digital skills in different age groups, which underlines the fundamental importance of training to improve quick access to these skills, and adequately face one's daily activities without running the risk of being expelled from the labor market. It is up to universities to deliver digital skills, to assess skills demand from job market to address right policies. In this work, we explore digital needs in Colombia and consider the alarming lack of skills in this sector, with its serious implications on social life. We administered a questionnaire to students of the Catholic University of Pereira (Colombia) to verify whether the high educational level they are involved in corresponds to an adequate development of digital skills. The study highlighted a gender gap in terms of perception by the students, especially related to advanced digital skills, such as the creation of new digital content, knowledge of programming languages and management of given tools.

Keywords Digital skills · Education · Colombia · Digital divide · Gender

M. della Volpe (✉) · A. Jaramillo-Gutiérrez
University of Salerno, 84084 Salerno, Italy
e-mail: mdellavolpe@unisa.it

A. Jaramillo-Gutiérrez
e-mail: ajaramillogutierrez@unisa.it

A. Henao-Rosero
Catholic University of Pereira, 660002 Pereira, Colombia

A. Visvizi et al. (eds.), *Research and Innovation Forum 2022*,
Springer Proceedings in Complexity, https://doi.org/10.1007/978-3-031-19560-0_30

1 Introduction

The development of human capital is a driver for economic growth. Digital platforms can play an important role in the future of education if there is equal access to economic opportunities [1]. Covid-19 illustrated the limit of our educational systems. It boosted digital transformation in society, highlighting gaps and the need to adapt to a new context, underlining the fundamental importance of training to improve quick access to these skills, adequately face one's daily activities and not running the risk of being expelled from the labor market [2]. It is up to universities to improve training of digital skills, as well as assessing the demand for skills in the labor market to address the right policies [3]. Now more than ever, with the adoption and integration of advanced digital technologies such as Artificial Intelligence, Internet of Things, Big Data, we are moving from a hyper-connected world to a digitized world in the economic and social dimension [4].

In this work, we explore the digital context of Colombia and consider the lack of skills in this sector, with its serious implications on social life. In the first section, we explore the implications of digitization in the current context. In the second one, we present data on digital access in Colombia. In the third section, we explore gender gap as a characteristic of the digital divide considering the integration of women within STEM disciplines. In the fourth section, we describe the research methodology, based on the administration of a questionnaire to the students of the Catholic University of Pereira (Colombia) to verify whether their high educational level corresponds to an adequate development of digital skills, a necessary condition to enter the labor market. Finally, we reflect on the results that emerged to elaborate our final considerations.

2 Digital Skills

In recent years, there has been an increase in the number and sophistication of digital technologies, becoming more necessary to carry out daily activities. At the educational level, the offer of online learning, virtual education, distance learning or through MOOCs is multiplying: digitization allows students to exploit new opportunities [4]. Digital technology and new business models have brought disruptive changes in production, since digitization, digital culture and the way people interact with technologies are a key necessity for their healthy growth and for their re-qualification in line with future employment [1, 5, 6]. In the labor market, analytical, digital skills and leadership are transversal in all sectors and are strongly integrated with digital technology, highlighting the real trends that will impact future jobs [3, 7]. Only with the acquisition of these skills can a digital divide be avoided. Therefore, it becomes essential that the population is digitally qualified [8–10].

Digital skills are defined by ITU [11] as technological skills acquired in whole or in part before entering the labor market and used to interact with technology, to design, create, use tools and solutions in different sectors. Furthermore, they generate different digital users and professional profiles. ITU divides these into basic, intermediate and advanced digital skills. Basic skills are fundamental for carrying out basic tasks, which include hardware (use of printer, keyboard), software (word processing, file management) and basic online operations (e-mail, research, online questionnaires). Intermediate skills allow the production and use of digital content, facilitating entry into the labor market. Finally, the advanced ones are those needed by ICT specialists, experts in science, technology, engineering and mathematics (STEM) and professionals in the field of social studies, mobility, analytics and cloud (SMAC).

With digitization, a digital divide arises, understood as the different levels that exist between those who have access and can make use of Information and Communication Technologies (ICT) and, more specifically, the differences in access and use of internet-based digital services. This difference is accentuated by inequalities such as literacy, costs, geographic areas, socioeconomic status, age and gender [12]. In developing countries, this difference becomes even more evident when one takes into account obstacles such as difficulties in accessing devices and the Internet, the relevance of content, access capacities or the shutdown of the Internet by some governments.

3 Digital Access in Colombia

In 2020, only 67% of the population in the Latin American and Caribbean (LAC) area has access to Internet, a value far below the average of 88.5% in North America and 82.5% in Europe. In countries like Colombia, only 65% of people are users, with the highest participation among people aged 15–24 (84.7%) and between 25 and 74 (62%). Women represent the majority of Internet users (66% versus 64% of men), revealing that there is almost gender parity in terms of Internet use.

However, in this area the development of digital skills constitutes a great challenge, because less than 25% of the LAC population and only 5% of Colombians have advanced digital skills; less than 30% of the LAC population and 22% of Colombians have intermediate skills; less than 40% in the LAC area and 31% in Colombia have basic skills. Furthermore, only 37% of households have a computer at home and only 52% of households have access to the Internet at home [4].

In Colombia, the government operates in the framework of digitization thanks to the National Development Plan 2018–2022 *Pact for Colombia, Pact for equity*. The *digital future belongs to all*, both planning tools for digital development, where the challenge is increasing access and use of ICT, empowering citizens, connecting households and the development of the digital society and industry 4.0 [13]. Several initiatives have been proposed to support learning in ICT as digital citizenship, with the aim of working on digital skills training through online programs.

As a result of the *Computers for Education program*, the country has obtained a large number of computers per student, but only 2 thirds of these are connected to the

Internet. Colombia has a lower average of students with a broadband connection than the one recorded in OECD countries. Throughout the country, the Internet connection service is used more in urban areas (66.6%) than in rural areas (23.9%). However, its quality is perceived negatively [12].

To clarify the socio-economic differentiation of social strata in Colombia, it should be noted that, to define the level of belonging to a stratum, the government refers to the classification of properties per dwelling in the country, where residents with greater economic capacity belong to the highest strata. Costs for public services fall proportionately with the social level considered [14]. The main point is that 33% of the houses belong to the population with the lowest economic income and only 1.2% of the houses belong to the highest strata: these data immediately highlight the accentuated gap between social strata that exists in the country. Furthermore, the percentage of the population over the age of 5 using the Internet is 69.8% (78% in urban areas and 43.1% in rural areas): there are therefore significant gaps in the use of the Internet by areas [12].

4 The Gender Gap as a Feature of the Digital Divide

The gender gap has been affected by the impact of the pandemic, due to the slowdown in progress made in various economies and industries in terms of gender. In 2021, the gender gap widened: the sectors most affected by closures such as retail or health, and those that suffered from rapid digitization, were those where women worked the most.

Furthermore, it should be added that the pressures generated by home care fall to a greater extent on women: in fact, they are employed 3 times more than men in the care of family members and in housework [15, 16].

Globally, the percentage of the gender gap already closed is 68% (as of 2021), a negative value compared to the previous year (−0.6%). More years (135.6) are now needed to bridge this gap in the world. Compared to the political empowerment index, only 22% of the gap was filled, recording a lower value than the other indices.

In terms of education, Colombia's efforts could focus on integrating women into STEM disciplines: only 13.76% of women graduates come from studies related to these disciplines, compared to 35.12% of men. Only 2% of women attended degree courses in computer science, communication and technology (compared to 8.44% of men) [17].

Women need to play a more decisive role in the digital economy to ensure their participation in the Industry 4.0 and not widen the gender gap. Being more involved in STEM disciplines can have a strong impact and help other women be more present in AI-related sectors, machine learning, deep learning and language processing, all competences strongly projected towards the jobs of the future [18].

5 Methodology

The goal of our research was to identify the digital skills acquired by undergraduate students during their studies and detect their readability for the job market. This process was also informed by the literature review on digital access and gender gap in Colombia. We expected to find further cues of how this gap is being reflected on a high educational level and whether there are indications on mediating factors for future adjustments. We expect that the higher the skills considered (from basic to advanced), the higher the difference in perceived self-competence (to the detriment of women).

Between October and November 2021, we administered an online questionnaire with multiple choice questions to 96 students enrolled at the Catholic University of Pereira. Criteria for selection was attending the last two years to courses in business administration, international trade and marketing, belonging to the Faculty of Economic and Administrative Sciences. Their ages were between 20 and 25 years old. This selection made it possible to investigate the level of digital skills acquired by students close to graduation and therefore near their entry into the job market.

In our research, the perceptions of young women reveal that they mostly consider their digital skills to be poor or bad, and to a much greater extent than men. This creates a gender gap to the detriment of female students, who rarely perceive their skills at an excellent level and are more often at a good or sufficient level.

The questionnaire used for the survey consisted of 30 items, based on a Likert scale from 1 to 5, where 1 corresponds to very bad and 5 to excellent. It investigated the perception of each student in relation to their level of knowledge and the use of some digital tools at a theoretical and practical level. The analysis used the indicators proposed by ITU [19] and the European Commission [20] to establish the level of digital skills possessed by students (basic, intermediate, advanced digital skills). These indicators have been modified and enriched to better adapt to the purposes of the research. Taking the aforementioned levels of digitization as a reference, the study proceeds with a descriptive analysis of the data collected.

6 Results Analysis

In our survey, we found a greater participation of female students in terms of response to the questionnaire (53.13% against 46.88% of males). The most numerous age group of students is from 20 to 21 (54.32%), followed by students under the age of 20 (19.75%), and those between 22 and 23 (16.05%). Finally, a minority is represented by those between 24 and 25 (9.88%).

Digital skills have become a necessity at work and academic level: in our research we found that digital skills were acquired by most students on their own (56.25%), only in the second instance thanks to school or university courses (33.33%). To a lesser extent they learned from friends (7.29%), family members (2.08%) or other sources (1.04%). In relation to gender, we found that males have acquired digital skills on their own to a greater extent than females (71.11% against 43.14%), while the latter attend school or university courses to a greater extent (43, 14% compared to 22.22% of males).

Furthermore, it appears that male students believe that their distance learning skills are excellent (42.71%) or good (39.58%) to a greater extent. No student thinks they are bad, suggesting that distance learning achieved in the years 2020 and 2021 was carried out by students without major problems. In relation to gender, women believe they possess adequate skills to a greater extent than men (88.24% between excellent and good against 75.56% of males).

In the next paragraph, we conduct the analysis of the perception of skills, according to basic, intermediate or advanced level. We point out that, after having reported the values relating to the different levels of skills, we also analyze the results that express the perception of their abilities, segmenting them by gender (Table 1).

Table 1 Students skills level by gender

Skill		Excellent	Good	Sufficient	Bad	Very bad
Basic skills						
Copy or move a file or folder	Female	86,27%	7,84%	5,88%		
	Males	91,11%	6,67%	2,22%		
Sending emails with attachments	Female	94,12%	3,92%	1,96%		
	Males	93,33%	6,67%	0%		
Select the appropriate tool, device or service for certain tasks	Female	33,33%	27,45%	29,41%	5,88%	3,92%
	Males	28,89%	53,33%	13,33%	4,44%	0,00%
Intermediate skills						
Use basic arithmetic formulas in a spreadsheet	Female	54,90%	31,37%	9,80%	3,92%	0%
	Males	53,33%	24,44%	20,00%	0%	2,22%
Knowledge about the use of cloud services, such as Google Drive, Dropbox and One Drive for file sharing	Female	43,14%	25,49%	25,49%	3,92%	1,96%
	Males	53,33%	31,11%	13%	2,22%	0%
Ability to use online learning tools to improve their skills through tutorials or online courses	Female	50,98%	31,37%	15,69%	1,96%	0,00%
	Males	46,67%	37,78%	11,11%	2,22%	2,22%
Advanced skills						
The ability to write a computer program, using a specialized programming language	Female	1,96%	3,92%	21,57%	19,61%	52,94%
	Males	2,22%	6,67%	13,33%	35,56%	42,22%
Use of data tools (databases, data mining, analysis software) which manage and organize complex information to make decisions and solve problems	Female	17,65%	23,53%	33,33%	15,69%	9,80%
	Males	17,78%	24,44%	37,78%	15,56%	4,44%

6.1 Basic Skills

In relation to basic skills, male students believe that their abilities to copy or move a file or folder are excellent (88.54%), and at the gender level they perceive this competence as excellent in greater proportion (91.11%) compared to women (86.27%): a gap that aligns with DANE [12], where 81.6% of Colombian males have been found to possess this basic competence compared to 78.9% of females. The results of our survey also show that, in most cases, university students perceive their digital skills as excellent or good compared to that of Colombians in general.

In sending emails with attachments, students consider their skills to be excellent to a greater extent (93.75%) and no one perceives themselves as poor or bad: these values are well above the national national figures (80,7%) of citizens believe they have this competence) [12]. Furthermore, men consider their skills excellent (93.33%) and good (6.67%), while women perceive themselves as excellent (94.12%), good (3.92%) and sometimes sufficient (1,96%).

Regarding the ability to select the appropriate tool, device or service for certain tasks such as, for example, a professional video call, the highest percentage of students believe that their competence is sufficient (39.58%) and excellent (31.25%), while few consider it bad (2.08%).

Concluding the analysis of basic skills, we can detect a gender difference in their perception, even if we are referring to an initial level of skills. These data are well above the values presented by Colombian citizens, where only 26.4% of the population over 7 have basic ICT skills [12].

6.2 Intermediate Skills

Beyond the basic level of digital skills, people could be better prepared to face digitization, which poses new challenges every day. In the intermediate level, for example, we have the ability to use basic arithmetic formulas in a spreadsheet. Students consider their mastery of this skill to be excellent (54.17%) and good (28.13%), while few consider it bad (1.04%). Women have a better perception (86.27% between excellent and good) than men (77.78%).

In relation to their knowledge about the use of cloud services, such as Google Drive, Dropbox and One Drive for file sharing, students consider their skills to be excellent (47.92%), with a gender gap greater than 10% in favor of men (53.33% against 43.14% of women), and it is the latter who consider their skills as poor or very bad (3.92% poor and 1.96% bad).

Regarding the ability to use online learning tools to improve their skills through tutorials or online courses, students have a more positive perception: 48.96% believe that their competence is excellent and 34.38% deems it good. In a greater percentage, men consider this ability to be at a very bad level (2.22% compared to no female).

6.3 Advanced Skills

A discouraging panorama emerges from the analysis of advanced skills. Comparing the data obtained by the students with the Colombian national ones we note that they are quite aligned. In Colombia, only 4.69% of people over the age of 5 who use a computer know any advanced programs, most of them are men (15%) rather than women (9%) [12].

Among the students interviewed, the ability to write a computer program, using a specialized programming language, is quite rare. Only 2.08% believe that their skills are excellent and 5.21% consider them good. The highest percentage of students believe they have poor competence (47.92%). The panorama, then, becomes even more difficult for women, who in a greater proportion consider their skills poor, with a gap of more than 10% (52.94% of women against 42.22% of men). In the use of data tools such as databases, data mining, analysis software, which manage and organize complex information to make decisions and solve problems, most students believe that their skills are sufficient (35.42%) and a large percentage who are at a low level (15.63%).

The results reveal the urgency of intervening on the training level for young people, as there is an evident lack of skills at both intermediate and advanced levels. The role that the university must play in digitalization education is essential.

7 Conclusions

Although Colombia shows a low gender gap in education, recording a successful result compared to other countries, global participation of women in the digital labor market remains low. In our research, the perceptions of young women reveal that they mostly consider their digital skills to be poor or bad, and to a much greater extent than men. This confirms and furthers a gender gap to the detriment of female students, who rarely perceive their skills at an excellent level and are more often at a good or sufficient level.

In the adoption of digital skills, which must indisputably involve an entire network of actors who support and motivate digitization, the government, for its part, implements and motivates students' autonomous digital learning through free training programs.

Young people entering the labor market must have the skills necessary to be part of a digital economy without running the risk of being excluded due to a lack of skills. As highlighted during the pandemic, digitalization and connectivity have been key players in the development of business activities, including the expansion of new forms of work and the generation of new businesses [19]. One of the most evident challenges that young people have to face is the development of advanced digital skills which, as shown by our research, strongly involves young people about to graduate in disciplines related to economic and administrative sciences. Digital skills are, in fact, the key to entering the world of work. A worrying void was found in our analysis with respect to the use of data tools to manage and organize information that supports and improves decision-making and organizational problem solving.

The analysis of the answers given by students, conducted in our research, can represent a further push for the design and implementation of complementary courses related to digital skills for university students, so as to promote and motivate learning and the development of digital skills of which students, to date, show little mastery. Finally, we also believe government and international programs and funding should be continued and expanded.

References

1. WEF: Shaping an Equitable, Inclusive and Sustainable Recovery: Acting Now for a Better Future (2021)
2. CEPAL: Tecnologías digitales para un nuevo futuro. Santiago (2021)
3. Rivoir, A., Morales, M.J.: Políticas digitales educativas en América Latina frente a la pandemia de covid. UNESCO, Argentina (2021)
4. ITU: Digital Development Deshboard. https://www.itu.int/en/ITU-D/Statistics/Dashboards/Pages/Digital-Development.aspx. Accessed 10 Dec 2021
5. Visvizi, A., Daniela, L., Chen, C.W.: Beyond the ICT- and sustainability hypes: a case for quality education. Comput. Hum. Behav. **107** (Jun 2020). https://doi.org/10.1016/j.chb.2020.106304
6. Visvizi, A., Lytras, M., Daniela, L. (eds.): The Future of Innovation and Technology in Education: Policies and Practices for Teaching and Learning Excellence. Emerald Publishing, (2018). ISBN: 9781787565562
7. 7WEF: The Future of Jobs Report 2020. Geneva (2020)
8. Strack, R., Kaufman, E., Kotsis, A., Sigelman, M., Restuccia, D., Taska, B.: What's trending in jobs and skills. Burning Glass Technologies (2019)
9. Iivari, N., Sharma, S., Ventä-Olkkonen, L.: Digital transformation of everyday life- How COVID-19 pandemic transformed the basic education of the young generation and why information management research should care? Int. J. Inf. Manag. **55**, 102183 (2020). https://doi.org/10.1016/j.ijinfomgt.2020.102183
10. Winarsih, Indriastuti, M., Fuad, K.: Impact of Covid-19 on digital transformation and sustainability in small and medium enterprises (SMEs): a conceptual framework. In: Barolli, L., Poniszewska-Maranda, A., Enokido, T. (eds.) Complex, Intelligent and Software Intensive Systems. CISIS 2020. Advances in Intelligent Systems and Computing, vol. 1194. Springer, Cham (2021). https://doi.org/10.1007/978-3-030-50454-0_48
11. ITU: Digital Skills Assessment Guidebook. Genova (2020)
12. DANE: Encuesta de Calidad de Vida. Bogotá, Colombia. https://www.dane.gov.co/index.php/estadisticas-por-tema/tecnologia-e-innovacion/tecnologias-de-la-informacion-y-las-comunicaciones-tic/indicadores-basicos-de-tic-en-hogares. Accessed 12 Jan 2022
13. MINTIC: Plan TIC 2018–2022: El Futuro Digital es de Todos. Colombia (2018)
14. DANE: Ley 142 de 1994. https://www.dane.gov.co/files/dig/ley142_1994.pdf. Accessed 12 Dec 2021
15. OECD: Latin American Economic Outlook 2020: Digital Transformation for Building Back Better. Paris (2020)
16. Berniell, I., Berniell, L., de la Mata, D., Edo, M., Marchionni, M.: Gender gaps in labor informality: the motherhood effect. J. Dev. Econ. **150**, p102599 (2021). https://doi.org/10.1016/j.jdeveco.2020.102599

17. WEF: Global Gander Gap Report 2021. Geneva (2021)
18. UNESCO: Science Report: The Race Against Time for Smarter Development. Schneegans, S., Straza, T., Lewis, J. (eds). UNESCO Publishing, Paris (2021)
19. ITU: ICT indicators for the SDG monitoring framework. Technical information sheets prepared. https://www.itu.int/en/ITU-D/Statistics/Documents/intlcoop/sdgs/ITU-ICT-technical-information-sheets-for-the-SDG-indicators-Sept2015.pdf. Accessed 12 Dec 2021
20. European Commission: Digital Economy and Society Index. https://ec.europa.eu/digital-single-market/en/digital-economy-and-society-index-desi. Accessed 12 Dec 2021
21. OECD: Bridging the Rural Digital Divide. OECD Digital Economy Papers (2018)

Addressing the Digital Divide in Online Education: Lessons to Be Drawn from Online Negotiation

Christian Pauletto

Abstract During the Covid-19 pandemic, the digital divide was often identified as an obstacle to online teaching, and inequalities in education were sometimes observed. This challenge could be addressed by making online teaching more efficient through the application of good practices. Given that there are similarities between diplomatic negotiating practice and teaching practice, it is worth examining whether good practices developed in online diplomatic negotiation are transferable to online teaching. The aim of this paper is to examine good practices in online negotiation and explore how such good online practices could be transferred to teaching. The research is informed by the author's first-hand experience from practice in both international diplomatic negotiation and academic teaching. The paper identifies a number of practices from online negotiation that are equally effective in online teaching. It focuses on communicational aspects, mutual understanding (monitoring and optimisation of understanding), motivation to listen, attention, active participation, and non-verbal communication. Some unresolved challenges of online teaching remain though, which are not addressed in this paper.

Keywords Academic teaching · Digital diplomacy · Digital divide · Diplomacy · Higher education · International negotiation · Online teaching

1 Introduction

> Teaching by its very nature is a complex social, personal and cognitive process that relies on effective communication and relationships between educators and students, and as such it is an emotional experience for both.
>
> Naylor and Nyanjom [11]

The Covid-19 pandemic and the ensuing increased adoption of online education has challenged education policies worldwide and particularly in poor countries. From

C. Pauletto (✉)
International University in Geneva, Geneva, Switzerland
e-mail: cpauletto@iun.ch

A. Visvizi et al. (eds.), *Research and Innovation Forum 2022*,
Springer Proceedings in Complexity, https://doi.org/10.1007/978-3-031-19560-0_31

a public policy perspective, decision-makers had to tackle two problems at once. First, the existing level of "digital divide" and second, the existing level of "educational divide" (education poverty / learning poverty) (see Sect. 2). These two types of the divide persist in the North–South context, between geographical sub-divisions of a country, and between strata of the society. This makes it necessary to enquire how best to improve the migration of *ex cathedra* lectures onto online and hybrid modules.

The aim of this paper is to explore whether experiences from online international diplomatic negotiation are transferable to online teaching in Higher Education Institutions (HEIs). A case is made that such a comparative perspective would not only spur cross-fertilisation. It would also permit shifting the spotlight to specific aspects of teaching, namely the communicational and interpersonal dimensions (monitoring and optimisation of understanding, motivation to listen, attention, active participation, non-verbal communication). This matters because in an innovation or transformation process, adoption of a new technology and adaptation of the communicational technique go hand in hand [12, 20].

Besides, foregrounding the communicational aspect of good practices is interesting as it shifts the focus away from technological refinements such as a multiplication of new and more sophisticated digital tools. This matches well the challenges characterised by a digital and an educational divide.

Some unresolved challenges of online teaching remain though [12], which are not addressed in this paper.

2 The Digital Versus Educational Divide

A question that has arisen during the pandemic was whether the migration to online teaching might increase the degree of "educational divide". The idea was that because of the prevailing "digital divide", disadvantaged groups could not keep pace with the requirements of online learning. Besides scholarly literature, international organisations were well-placed to explore those two types of the divide, and a succinct review of their publications suggests that their views are converging. For once, it is amazing how international bodies with different focus and missions reached an almost perfect consensus on the matter.

Starting with the view of the UN agency in charge of digital issues helps grasping the main interrelationship. The ITU [6, p. 2] found that the COVID-19 pandemic had highlighted many global inequalities, including the digital divides between and within countries. The ITU also found that countries with strong digital ecosystems are more resilient to shocks and emergencies and have better responded to the pandemic (*cf*. also ITU [7]; Lai and Widmar [9]).

Unsurprisingly, what applied at macro (cross-sectoral) level also obtained at sector-specific levels, as highlighted by the UN body in charge of the youth: UNICEF & IHD [16, p. xix] found that the impact of the pandemic had been disastrous in the sphere of education, with school closures and the digital divide exposing the fault lines between the haves and the have-nots. A main reason invoked by UNICEF &

IHD [16, p. xxvii] to explain such impact was that due to school closures resulting from the pandemic, online classes became the main avenue for learning. The UNICEF concluded that the digital divide affected the respondent families adversely, because many of them could not afford devices such as smartphones and lacked digital literacy and access to adequate Internet connectivity.

Similar views were shared by the UN body in charge of education, science and culture. UNESCO [15, p. 3] established a linkage between the pandemic and the educational divide, to the extent that the varied impact of COVID-19 on enrolment across countries of different income levels demonstrated how the pandemic exacerbated education inequality. In terms of response, UNESCO [14, p. 17] warned that online education should not lead to the exclusion of the disadvantaged ones: "two very distinct situations are often confused. It is one thing to employ digital tools in teachers' pedagogical work with students", but "in digital societies it is unthinkable to exclude from school the possibilities of access to knowledge and communication that are increasingly seen as a necessary component of daily life—that we correctly speak of 'digital divides' shows how important internet access and device connectivity have become".

But those views didn't remain confined to specialised agencies. The top political level concurred. The UN Secretary General declared that the crisis brought a deeper understanding of the digital divide and related equity gaps, which required urgent attention [17, p. 12]. Concerned that distance learning in high income countries covers about 80–85%, while this drops to less than 50% in low-income countries, he explained that this shortfall can largely be attributed to the digital divide, with the disadvantaged having limited access to basic household services such as technology infrastructure (ibid.).

What about economic international institutions? In the case at hand, no divergence could be detected from that side either. The World Bank [23, p. 5] too was alarmed that the most disadvantaged children and youth had the worst access to schooling, highest dropout rates, and the largest learning deficits. The organisation even predicted that these adverse impacts may reverberate for a long time, as lower human capital in the current student cohort—concentrated among the most disadvantaged—would perpetuate the vicious cycle of poverty and inequality.

Academia supported those views. Blackman [3, p. 10] considers that another challenge the pandemic posed to the equity and inclusion agenda is that emergency home schooling is likely to disproportionately affect children from lower-income households, with higher rates of education loss. Azevedo et al. [2, p. 29] expect that "the world is poised to face a substantial setback to the goal of halving the number of learning poor and will be unable to meet the goal by 2030 unless drastic remedial action is taken", and anticipate that about 43 percent of children will still be learning poor in 2030 (*cf*. also Azevedo [1]). This reverberates at national scale, for example when Bokayev et al. [4, p. 287] conclude, in the case of Kazakhstan's transition to online learning, that vulnerable groups, disadvantaged rural regions and remote localities should be at the forefront of policy measures.

If online teaching is there to stay, then efforts to reduce the educational divide ought to be strengthened (*cf*. Pauletto [13]). But for those efforts to be successful, the challenge of the digital divide should be managed. Disseminating and implementing non-technical good practices to facilitate online teaching would contribute to such endeavours. Informed by the practice of online international negotiation, the next sections attempt to focus on such good practices.

This paper is structured in several sections dedicated to selected good practices. Instead of starting with a stand-alone section on diplomatic negotiating practice, each topical section starts with one or more paragraphs presenting the rationale and use of a particular good practice in online negotiation, and on that basis proceeds with the transferability to HEIs. More elaborated considerations on the characteristics and rationale of online diplomatic negotiation are expounded in Pauletto [12].

The author believes that the good practices described in this article may contribute to a successful online teaching. The survey of students' satisfaction for his teaching tenure actually improved in 2020 (online) and 2021 (hybrid) compared to 2019.

3 Focused Speech

The impact most often observed or perceived in online negotiations is a higher risk of misunderstanding compared to face-to-face conversation. Therefore, when migrating online, the adjustments to undertake in the communication methods would aim at minimising misunderstandings.

Just like online negotiators, professors should adapt their speech to an online audience and get used to formulate their messages in a more focused and straightforward manner. The main risk in an online class is in fact a very simple one: to get misinterpreted.

4 Thorough Preparation

Probably, the manner in which an online session starts may be another reason why misunderstandings are more likely to occur. In an in-person event, participants would always have some time in the corridors or immediately before the event to informally exchange a few words. Seasoned negotiators would confirm that this exchange would typically pertain to the aim, the expectations, the objectives or the focus of the ensuing session. Online, the process is too tight to allow for informal preliminary chatting among diplomats.

In a classroom too, the abrupt way of starting interaction between instructor and students when they are online potentially has the same impact. To mitigate the risks just described, one major lesson from online negotiation is to devote more time for preparing a session. To be effective, an online negotiator needs to prepare his online session very carefully and fine-tune every aspect. The negotiator needs to be prepared to spell out clearly at the session's outset what the purpose, the objective and modalities of the session are and make sure that all participants share the same understanding.

This good practice seems to be transferable one-to-one to a teaching environment. The Covid-19 experience quickly proved that to avoid a failed lecture the complete lecture must be more carefully prepared, planned and fine-tuned. The plan and the rollout need to be presented to students at the start of the class or module. The teacher must be prepared to start by explaining what the purpose of the class or module is, what are the teaching modalities, and what are the expectations. Efforts must be put into ensuring that students will be able to make sense of what happens during the session and understand why it is organised in the way it is.

5 Seamlessness

Though computer technology is well performing, it sometimes also is relatively heavy and cumbersome. For example, switching from one functionality or conferencing configuration to another may take much more time than corresponding negotiating-floor procedures. When a process is very interactive, such as a multilateral conference, this becomes an issue. The session should be planned in such a way as to minimise switching back and forth between functionalities or speakers. A seamless process permits to minimise the loss of participants' attention and prevents wasting precious time.

Especially in multilateral conferences, some ways of circumventing the rigidity of technology have proved effective. One way is to record speakers' interventions and have them ready to be switched on from the conference's control room when the floor becomes theirs.

Similar challenges occur in teaching, especially in interactive classes. However, the use of video-recorded material is less easily transferable from a negotiating into a teaching context. Having the teacher's lecture recorded on video and then imparted to the students would radically modify the teaching modalities, and alter them from synchronous to asynchronous. However, there may be specific opportunities, such as the presentation that students use to do in traditional classes. For that specific purpose, an option is for students to consign beforehand the electronic file containing their video presentation.

6 Paper-Based Self-preparation

Keeping concentrated during long hours (or days) of diplomatic negotiation is a challenge for every negotiator, and more so if there are long sequences of speech. In traditional negotiating settings, negotiators tend to prefer explaining their reasons and arguments *viva voce* before submitting written material to the other party. A useful practice when migrating a negotiation online is to rely less on oral in-session explanations and more on submitting written material to the other parties ahead of the session, so to say "for self-study".

In online teaching too, students' attention and focus are a major challenge. Just as in online negotiation, specific means to address this should be implemented. One such means is the flipped classroom. The concept of flipped classroom is not new, but it gained attention during the pandemic. One reason is that asking the students to engage more in (paper-based) self-study allows to alleviate the challenge of online engagement.

7 Satisfaction and Marks of Appreciation

Togetherness, informal moments, and non-verbal means of communication provide ample ways of sharing satisfaction among participants. In online negotiations, their absence needs to be compensated. In the author's experience, a good negotiating practice is to redouble the explicit signs of appreciation. Statements such as "I appreciate it", or "We are making good progress", or "We are on the right track" are often enough to compensate the lack of physical (non-verbal) interactions serving that purpose.

As in negotiation, marks of appreciation gain added weight in online teaching. Appreciation pertains to several aspects of teaching, e.g. the course of the session in general, the class collectively, or individual students. The professor has to never miss an opportunity of expressing verbally and explicitly his appreciation. This may include appreciation of how fast the session is progressing, how well the taught material was then implemented in group works, or how relevant a student's remark was.

8 Managing Rigidities

In negotiation a lot goes on "in the corridors". Similarly in university teaching, especially in smaller institutions. Typically, the educator would approach a few students during breaks and ask whether everything was clear, whether the pace of teaching is right, and so on. In the absence of such informal opportunities in virtual meetings, the instructor will be less able to adapt to the circumstances of the ongoing session.

The challenge is double because neither the negotiator nor the teacher should depart from the core rule of preparing in detail a seamless session. It thus becomes difficult to change the modalities when resuming an online session after a break, even assuming that an "online break" too could constitute an opportunity for informal feedback.

Tight planning may bring negotiators in another predicament. In the negotiating setting, a major loss that occurs in virtual sessions is in terms of on-the-spot feedback. Similarly in teaching. Because everything is so well-prepared and fine-tuned, students do not dare anymore to interfere. This reduces the quality of ongoing feedback both ways, students to faculty and faculty to students. This needs to be offset by an increase use of ex-post feedback tools, such as short quizzes or self-tests.

A more serious challenge is in terms of negotiator's, respectively educator's, adaptability. If some parameters happen to change just before or during the class compared to what was anticipated during the preparation, it is less easy to adjust, simply because too many things have been fine-tuned. The advice "not to depart" from the modalities shared with the students needs to be kept.

More generally (and beyond the difficulty of undertaking short-term or on-the-spot adjustments), setting the online course procedures and functionalities in great detail and communicating them in equally great detail to students may have a downside. The downside is that the instructors feel compelled to carry over the same procedures and functionalities all along the teaching programme, in order to avoid misunderstanding, and to save time and money. This reduces the variety in teaching method, introduces rigidity and standardisation. Exactly the same way as negotiators often observe in online negotiations. Thus, the organisation of online classes must be approached with a long-run perspective. This requires an anticipation effort.

9 Conclusion

Informed by first-hand practical experience in both fields, this paper has investigated whether online international negotiating practice displays features that are transferable to online teaching practice. The research concludes that in both practices, changes in methods are indispensable when migrating online. As much as one would not negotiate in the same manner online, one would not teach in the same way online. The need for adaptation and adjustment thus goes well beyond the technical dimension. The major change is that in the online context, the preparation of a session, whether for negotiation or teaching, must be more fine-tuned. Besides, more care must be lent to precisely describing the objectives, approach and entire rollout of the session. The professor, respectively the negotiator, should explicitly explain what the modalities of the session are. Any departures from the agreed modalities, even minute ones, shall be explicitly laid out. Preparation must therefore adopt a long-term perspective.

Actively managing the motivation and participation of the audience is a challenge common to online negotiation and teaching. Best practices such as increasing the amount of written material shared ahead of a negotiating session are transferable to teaching. Submission of pre-recorded video material seems trickier, but is advisable for specific purposes. Alternative ways of sharing marks of appreciation and satisfaction are easily transferable from negotiation to teaching, provided the protagonists are firmly convinced of the merits thereof. Last but not least, focused and straightforward speech becomes essential to avoiding misinterpretation.

References

1. Azevedo, J.P.: Learning Poverty: Measures and Simulations. Policy Research Working Paper, 9446. World Bank (2020)
2. Azevedo, J.P., Hasan, A., Goldemberg, D., Geven, K., Iqbal, S.A.: Simulating the Potential Impacts of COVID-19 School Closures on Schooling and Learning Outcomes: A Set of Global Estimates. The World Bank Research Observer (2021). https://doi.org/10.1093/wbro/lkab003
3. Blackman, S.: The impact of Covid-19 on education equity: a view from Barbados and Jamaica. Prospects, 1 15 (2021). https://doi.org/10.1007/s11125-021-09568-4
4. Bokayev, B., Torebekova, Z., Abdykalikova, M., Davletbayeva, Z.: Exposing policy gaps-the experience of Kazakhstan in implementing distance learning during the COVID-19 pandemic. Transform. Gov. People Process Policy **15**(2), 275–290 (2021)
5. Daniela, L., Visvizi, A. (eds.): Remote Learning in Times of Pandemic. Routledge, London New York (2022). ISBN:9780367765705
6. ITU. Giga: Empowering communities in Asia and the Pacific through school connectivity. Geneva (2021).https://www.itu.int/dms_pub/itu-d/opb/tnd/D-TND-03-2021-PDF-E.pdf
7. ITU. Pandemic in the Internet age: From second wave to new normal, recovery, adaptation and resilience. Geneva (2021).https://www.itu.int/hub/publication/D-PREF-EF.PANDEMIC_01-2021/
8. Johnson, A.F., Roberto, K.: Policies, politics and pandemics-course delivery method for US higher educational institutions amid COVID-19. Transform. Gov. People Process Policy **15**(2), 291–303 (2021)
9. Lai, J., Widmar, N.: Revisiting the digital divide in the COVID-19 era. Appl. Econ. Perspect. Policy **43**(1), 458–464 (2020). https://doi.org/10.1002/aepp.13104
10. Miller, M.: Minds Online: Teaching Effectively with Technology, 279 pp. Harvard University Press, Cambridge, MA (2014). ISBN:978-0674368248
11. Naylor, D., Nyanjom, J.: Educators' emotions involved in the transition to online teaching in higher education. High. Educ. Res. Dev. **40**(6), 1236–1250 (2021). https://doi.org/10.1080/07294360.2020.1811645
12. Pauletto, C.: Online teaching in HEIs vs. online international negotiation: analogies and transferable good practices. In: Visvizi, A., Lytras, M.D., Jamal Al-Lail, H. (eds.) Moving Higher Education Beyond Covid-19: Innovative and Technology-Enhanced Approaches to Teaching and Learning. Emerald Publishing, Bingley (2023) (forthcoming)
13. Pauletto, C. : Numérisation de l'éducation et Société 4.0/Digital Learning and Society 4.0. In: Paul, E., Demierre, P. (eds.) Smart à tout prix ? Défis de la numérisation au temps de la Covid-19./Smart at Any Cost? The Challenges of Digitalisation in the Time of Covid-19. Fondation Jean Monnet pour l'Europe (FJME), Lausanne, Collection débats et documents, no. 26, pp. 16–20 (2022)
14. UNESCO. Education in a post-COVID world: Nine ideas for public action. UNESCO, International Commission on the Futures of Education (2020)
15. UNESCO. COVID-19: reopening and reimagining universities, survey on higher education through the UNESCO National Commissions–UNESCO Global Survey (2021).https://en.unesco.org/news/new-unesco-global-survey-reveals-impact-covid-19-higher-education
16. UNICEF IHD. Assessing Impact of the COVID-19 Pandemic on the Socio-economic Situation of Vulnerable Populations through Community-based Monitoring. New Delhi (2021)
17. UNSG. Policy Brief: Education During COVID-19 and Beyond. New York (2020). https://www.un.org/development/desa/dspd/wp-content/uploads/sites/22/2020/08/sg_policy_brief_covid-19_and_education_august_2020.pdf
18. Visvizi, A., Lytras, M., Daniela, L. (eds.): The Future of Innovation and Technology in Education: Policies and Practices for Teaching and Learning Excellence. Emerald, Bingley (2019). ISBN:978-1-78756-556-2
19. Visvizi, A., Jussila, J., Lytras, M., Ijäs, M.: Tweeting and mining OECD-related microcontent in the post-truth era: a cloudbased app'. Comput. Hum. Behav. **107**(June), 105958 (2020). https://doi.org/10.1016/j.chb.2019.03.022

20. Visvizi, A., Daniela, L., Chen, C.-W.: Beyond the ICT-and sustainability hypes: a case for quality education. Comput. Hum. Behav. **107**(June), 106304 (2020). https://doi.org/10.1016/j.chb.2020.106304
21. Visvizi, A., Field, M., Pachocka, M. (eds.): Teaching the EU: Fostering Knowledge Understanding in the Brexit Age, 204 pp. Emerald Publishing, Bingley (2021). ISBN:978-1-80043-274-1
22. World Bank: Ending Learning Poverty: What Will It Take?. Washington, DC (2019). https://open-knowledge.worldbank.org/handle/10986/32553
23. World Bank: The Covid-19 Pandemic: Shocks to Education and Policy Responses. Washington, DC (2020). https://www.worldbank.org/en/topic/education/publication/the-covid19-pandemic-shocks-to-education-and-policy-responses

Preparing High School Teachers for Teaching STEM Online: Exploration in Senegal

Christelle Scharff, Samedi Heng, Ndeye Massata Ndiaye, and Babacar Diop

Abstract The Covid-19 pandemic has impacted more than 1.1 billion (68%) learners worldwide. Schools were partially or fully closed depending on the health restrictions implemented by each government. Online learning was the obvious solution for ensuring education continuity. In developing countries, it was particularly difficult for institutions and instructors to transition to online teaching. Institutions had no infrastructures and practices in place; instructors did not have the required material in digital form and were not trained. The majority of students did not have computers and Internet access. In some cases, electricity was an additional obstacle. Instructors require training not only on mastering tools and practices but also on how to design their courses to adapt to the online format. In particular, they need to be exposed to strategies for interacting with and engaging students. We present our findings related to the obstacles encountered by STEM (Science, Technology, Engineering, Mathematics) instructors to transition to online teaching in Senegal. This study is based on a 6-week training for 79 STEM high school instructors. We also explore the potential and shortcomings of leveraging mobile technology for online teaching in this context.

C. Scharff
Seidenberg School of Computer Science and Information Systems, Pace University, New York, USA
e-mail: cscharff@pace.edu

S. Heng (✉)
HEC Liège, Université de Liège, Liège, Belgium
e-mail: samedi.heng@uliege.be

N. M. Ndiaye
Pôle Science, Technologie et Numérique, Université Virtuelle du Sénégal, Dakar, Senegal
e-mail: ndeyemassata.ndiaye@uvs.edu.sn

B. Diop
mJangale, Coding School, Thiès, Senegal
e-mail: babacar@mjangale.com

A. Visvizi et al. (eds.), *Research and Innovation Forum 2022*,
Springer Proceedings in Complexity, https://doi.org/10.1007/978-3-031-19560-0_32

Keywords Covid-19 · Education · Developing countries · Online teaching · Senegal · STEM

1 Introduction

The Covid-19 pandemic has spread worldwide and has affected every aspect of our daily life. Many restrictions such as lockdown, social distancing, etc. were adopted by governments to limit virus transmission. All sectors had to adapt and be creative to continue their activities during this abnormal period. Working from home was imposed on all non-essential sectors. Schools and universities were fully or partially closed. More than 1.1 billion learners worldwide were impacted [7, 10]. In developed countries, online learning was the obvious solution for ensuring education continuity; however, it was impossible for the majority of developing countries to implement. These countries were facing numerous barriers preventing them from transitioning to online teaching. The most remarkable obstacles were access to technology and the Internet.

STEM refers to Science, Technology, Engineering, and Mathematics. Many countries, developing and developed alike, consider STEM as a means to develop and as a sector where jobs will continue to grow. STEM aims at preparing a competitive workforce with skills going beyond technical and scientific ones, such as problem-solving, critical thinking, resilience, creativity, communication, and digital literacy [5]. In 2013, the government of Senegal decided to reorient higher education to focus on STEM, for example offering STEM university programs for students with humanities high school degrees [3]. This was also to react to a long literature-focused tradition of studies greatly influenced by Senegalese authors such as Leopold Sedar Senghor and Mariama Ba.

Adopting technologies for teaching STEM in the classroom remains challenging for teachers and requires preparation and practice [2]. Successful integration of technology in teaching depends greatly on familiarity with tools, but also adequate understanding of its advantages and impact in the classroom. With the Covid-19 pandemic, teachers faced additional challenges. They had to move to online teaching completely without warning. There is a difference between using technology to teach and teaching online. Teachers needed to adapt and change the format of their courses as well as their pedagogy. To teach online, teachers need to master a number of digital tools, e.g., for recording videos, teaching synchronously online, communicating with students, posting material in an eLearning platform, grading students' work, etc. to offer effective quality courses [8, 9].

The Covid-19 pandemic has impacted negatively everyone. Nevertheless, on the positive side, it is considered as a driver for many digital transformations, including in education. It provided teachers with the opportunities to rethink and reengineer the way they work, teach, and interact with students.

This research aims at understanding online teaching of STEM in developing countries and how it could be improved. It uses a capacity-building initiative of online teaching of STEM offered to 79 STEM teachers in Senegal as a basis.

This research is exploratory and focuses on the following research questions in the context of Senegal:

1. What are the *obstacles* to STEM online teaching?
2. What are the *tools/practices* that could jumpstart STEM online teaching?
3. What is teachers' *opinion* about STEM online teaching?
4. What are the *factors for enabling* STEM online teaching?

The remaining of this paper is structured as follows. Section 2 provides the research background. Section 3 exposes the research methodology. Section 4 shows the results. Finally, we discuss the results and conclude the paper in Sect. 5.

2 Research Background

In 2013, the government of Senegal decided to reorient higher education by emphasizing STEM and the use of ICT in teaching and learning. This strategy led to the creation of the Virtual University of Senegal (UVS) [3]. UVS offers online programs for more than 60,000 students. The number of high school students in science still remains low compared to humanities and social sciences, with a ratio of 5 times, according to the statistics provided by the government in 2021 [6]. Senegal was impacted by the pandemic and schools and universities were closed from March 16th, 2020 to June 2nd, 2020. Most schools and universities did not offer any courses during this period, disrupting the academic calendar. Being an online university, UVS was the least impacted university. To the best of our knowledge, there is no published study on the education landscape during that period in Senegal. However, there is a plethora of studies conducted on this topic worldwide [1, 8, 9]. For example, Romano [8] made a survey of more than 400 primary and secondary school teachers in Italy about the Covid-19 period. This study looked at the emotional state of teachers during the pandemic, revealing high emotional impact on mindset and behavior due to the lockdown. In addition, it presented the difficulties, barriers, and obstacles encountered when teaching online. The difficulties presented were "*lack of teachers' familiarization with digital devices*" (22.41%), "*lack of specific and technical training on online and distance learning*" (28.45%), and "*temporary and not systematic actions of training provided by different institutions*" (44.25%). The barriers were "*unstable or absent Wi-Fi connection*" (16.09%), "*teachers' low skills in technology-based instructional design*" (25%), and "*unpredictability of the pandemic evolution*" (48.85%). Finally, the obstacles were "*no equal distribution of technological devices among students*" (86.49%), "*teachers' loneliness and poor cooperation in the professional community*" (89.65%), "*sporadic spots and pills of technical learning, lack of compliance with families of the students*" (38.50%), and "*no cooperation with colleagues for curriculum design*" (65.90%). The authors also identified three elements that facilitate online learning: "*cooperation among teachers, school administrators, instructional designers*" (76.44%), "*systematic, bottom-up and situated training about technical artifacts and digital learning*" (56.61%), and "*cooperation and compliance with families*" (48.85%). Sedaghatjou et al. [9] reported the challenges faced by 110 international STEM faculty members during Covid-19 when they switch to the online environment; participants from developing countries are not representative (<20%). The challenges reported were "*synchronized meeting*" (23%), "*assessment*" (22.4%), "*projects*" (13.8%), "*evaluations*" (13.3%), "*deconstructing the course and syllabus*" (10.2%), "*synchronized sessions*" (4.5%), "*other*" (4.6%) and "*update reading*" (3.1%). The authors also mentioned that students' engagement, participation, and motivation were difficult for teachers to evaluate when teaching STEM online. The authors used the Khan's framework for e-learning which addresses online issues, strategies, and practices for flexible, and open and distance learning environments [4]. It considers that instructors' development needs to consider the following areas: Pedagogical, Technological, Interface Design, Evaluation, Management, Resource Support, Ethical, and Institutional. Sedaghatjou et al. [9] reveals that "*online evaluation*" and "*pedagogy*" were two dimensions particularly disrupted, followed by "*pedagogical challenges*", and "*adaptation to the new technology*".

3 Research Methodology and Data Collection

In order to answer our research questions, we studied an online training for STEM high school teachers offered post pandemic in Senegal (December 2021–March 2022). The pandemic revealed needs of such training but teachers training in online education is still not widely offered. The training we proposed was composed of two phases. The first phase proposed a 6-week online training focusing on online course instructional design (scenarization), digital tools for teaching online (video conferencing, online assessment, discussion boards, videos), teaching STEM online (simulation tools, projects, labs), collaborative tools for instructors' and students' teamwork, online pedagogy (breakout rooms, polls, flipped classroom) and students' motivation and engagement in online environment. The course also discussed the use of mobile devices and platforms to teach online and the use of mobile web sites versus mobile apps. The course was administered synchronously on Zoom. Numerous tools were presented and used such as Zoom for video-conferencing (with breakout rooms, polls, and recording features), Khahoot and Google Docs for quiz design, Google Presentation and Jamboard for collaborative work, Discord for discussions, GeoGebra and PhET for STEM simulations, and YouTube for online videos. In the second phase, STEM instructors formed teams to create an online activity for their students and presented their implemented projects to their peers. Activities ranged from organizing a Zoom session on a specific topic followed by a Khahoot or Google Forms quiz, creating a YouTube video covering a course module followed by a Khahoot, Google Forms quiz or questions/answers session on WhatsApp. The details of the training is available at: http://stem4senegal.com.

3.1 Data Collection

We adopted an exploratory research approach to answer our research questions. It allows us to explore the questions with more flexibility and have more insight and understanding for future investigations. We designed entry and exit questionnaires with closed and open-ended questions for collecting both quantitative and qualitative data. Specifically, the entry questionnaire was about understanding the context of the teachers and their technology and online readiness. The exit questionnaire collected information about teachers' perception and obstacles and opportunities of online teaching. The surveys were administered using a Google Form. The research was IRB-approved. Quantitative data were exploited in Excel. We followed the grounded theory approach for analyzing open questions and presentations' contents.

3.2 Participants

79 participants (9 women and 70 men) were selected from 12 regions of Senegal (out of 14) from 70 different institutions. They were recruited through a call for participation posted on social media and shared with teachers' organizations. The selection of the participants was based on their answers to motivation questions, their gender and their location. The ratio of women participants reflects the lack of women instructors in the STEM disciplines. Participants were teaching math, physics, and life and earth sciences. They taught courses in middle and high schools.

4 Results

In this section, we present the results derived from our survey to answer the research questions. 51 participants answered the entry survey and 49 the exit survey.

Table 1 Questions and answers related to schools' infrastructure

Questions	Yes (%)	No (%)
Q1: Do you have electricity in your school?	**84.31**	15.69
Q2: Do you have a computer room in your school?	45.09	**54.90**
Q3: Do you use your school's computer room?	19.60	**80.40**
Q4: Do you have a projector in your classroom?	29.41	**70.59**
Q5: Do you have problems accessing the Internet in the area/region where you teach?	37.25	**62.74**
Q6: Do you have Internet access in the school where you teach?	**50.09**	49.01

4.1 Obstacles to STEM Online Teaching

Schools' Infrastructures. Table 1 shows that most schools are not well-equipped, lacking computers, computer labs and projectors. Electricity still remains a problem since only 84.30% of schools have electricity. It is interesting to notice that more than half of the schools have computer rooms but only 19.60% of the teachers use computer rooms for teaching. The questionnaire included an open question on "*Why don't instructors use the school's computers?*". Most answers were '*lack of computers in the computer rooms*', '*unstable electricity*', '*unstable Internet connection*', '*no projector*' and '*the available computers are old*'. In addition, the answers mentioned that computers were reserved for administration and IT instructors. In addition, Internet is still a major constraint for schools. Only half of the schools have access to Internet limiting activities that teachers and students can perform online.

Instructors' Skills and Readiness for Online STEM Teaching. When it comes to online teaching, computer skills are required. On a Likert scale of 5, the average of the answers to the question "*How familiar are you with using a computer (word processing, Internet, etc.)?*" was 4.08. It means that most teachers were familiar with computers. 68.63% of them use some software in class or to teach their students in their preparations. However, only 13.73% of them reported making their students use software in class. Nearly half of them used videos in their teaching. All teachers have at least a smartphone but only 62.75% have a computer and only 23.53% of them have Internet at home through Wifi or similar. This supports their answers on the question about how they access the Internet: 96% of them access the Internet via their mobile phone and 5.88% of them access the Internet via cable internet at home or from their neighbors. Regarding eLearning, the majority of them (62.75%) are aware of how eLearning is facilitated through platforms. The platforms mentioned the most are Google Classroom, Virtual University of Senegal, Coursera, Zoom, Edx, and FUN-MOOC (France Université Numérique). We also asked teachers what comes first to their mind when talking about online education. The most common five answers to this question were "*Internet*", "*Interesting*", "*Zoom*", "*UVS*", and "*Computers*". A significant number of teachers (70.58%) do not feel ready for teaching online (Table 2).

Table 2 Questions and answers related to instructors' skills and readiness

Questions	Yes (%)	No (%)
Q7: Do you own a computer?	**62.75**	37.25
Q8: Do you use videos to teach?	45.10	**54.90**
Q9: Do you use software in your courses or to teach your students?	**68.63**	31.37
Q10: Do you ask your students to use software in your class?	13.73	**86.27**
Q11: Do you own a smartphone?	**100**	0
Q12: Do you have Internet at home?	23.53	**76.47**
Q13: Do you know any e-Learning/distance learning platforms?	**62.75**	37.25
Q14: Do you think you are prepared to teach online?	29.41	**70.58**

Table 3 Questions and answers related to students' skills and readiness

Questions	Yes (%)	No (%)
Q15: Do you think students are prepared for online learning?	19.60	**81.39**
Q16: Do students face difficulties accessing a smartphone?	**81.39**	19.60
Q17: Do students face difficulties accessing a computer?	**98.04**	1.96
Q18: Do students face difficulties in accessing Internet?	**94.12**	5.88
Q19: Do you think that students can afford to access the Internet regularly to take a course?	37.25	**62.75**

Students' Skills and Readiness for Learning STEM Online. The results presented in Table 3 are the answers from teachers' perspectives. We argue that these results are relevant as teachers are well aware of the situation of their students. According to them, students are not ready for online learning. 19.60% of them reported that students were ready. The main obstacles are that their students have difficulty accessing computers (98.04%), smartphones (81.39%), and Internet (94.12%). They also think that 37.25% of their students can afford to access the Internet regularly to take the course.

Obstacles to STEM Online Teaching in Senegal. Table 4 exposes the obstacles raised by teachers to pursue online teaching. The most mentioned obstacle is the Internet (51.61%). Both students and teachers do not have a good Internet connection at home as well as at school (unstable and slow). In addition, Internet is expensive. The second obstacle is related to hardware infrastructure (35.94%); it refers to computers, projectors, smartphones, and computer rooms. The third obstacle is mastering tools (22.58%). Teachers need training on using the wide range of tools that are used in teaching online, from video conferencing and communication tools to E-learning platforms. The vast majority of the teachers in the training never used Zoom before. Teachers also mentioned that it is difficult to follow the work of their students online and to gauge their understanding of the material, engagement and motivation (13.36%). Teachers do not envision online pedagogy and setup time as major obstacles to delivering online courses. Internet and infrastructures are put

Table 4 Obstacles related to online teaching

Obstacles	Occurrences (%)
Internet	51.61
Hardware infrastructures	35.94
Mastering tools	22.58
Gauging students' understanding, engagement and motivation	13.36
Pedagogy for online teaching	5.99
School restrictions	4.60
Electricity	4.60
Setup time	2.76

Table 5 Obstacles related to online teaching

Obstacles	Occurrences (%)
YouTube	65.32
Kahoot	65.32
Course website	53.22
Google Form	50.00
Google presentation/PowerPoint	43.14
Geogebra	41.53
Moodle	34.27

as the first and main obstacles to tackle. Some schools forbid the use of smartphones and computers in the classroom, preventing teachers from creating bring-your-own-device activities in the classroom (4.60%).

4.2 Jumpstarting STEM Online Teaching

The exit survey included the question: "*Which tools would you like to learn for transitioning to online teaching?*" The most mentioned tools are presented in Table 5. YouTube is the most cited tool (65.32%) and a course website was most cited than a complete platform such as Moodle (53.22 vs. 34.27%). Some other tools were also cited, with low occurrences, such as Padlet, STEM simulation tools, Google Classroom, ZipGrade, Latex, Python, Excel, Microsoft Teams, and Maple.

Table 6 Teachers, STEM and online education

Questions	Yes (%)	No (%)
Q20: Do you know what the term STEM means?	35.29	**64.70**
Q21: Do you think we can ignore online education in the future?	5.88	**94.11**

4.3 Teachers' Opinion about STEM Online Teaching

Only 35.29% of the teachers knew the meaning of the term STEM (STIM in French) before the training. They are in majority convinced that online teaching cannot be ignored (94.11%). When we asked "*What is specific to teaching STEM online?*", they answered '*using software or simulation software*' and '*new pedagogy for teaching: synchronous, asynchronous, student-centered learning, flip learning*'.

In the exit survey, teachers were asked to describe how they would apply what they have learned in the training in their courses, either if they decide to pursue online teaching on their own, during a situation similar to Covid-19 when online teaching is the only option, or to augment their face-to-face classes with online components. It is noteworthy to mention that teachers do not think that neither synchronous nor asynchronous online courses are a good option for their students (10.6% and 23.4% respectively). 74.5% of them think that a combination of both, synchronous and asynchronous components, are best. This is logical considering the intermittent access to Internet of teachers and students. Teachers proposed scenarios for teaching online. Most of them aim at creating a course website, teaching with Zoom, recording classes and posting videos on YouTube, using simulation tools, and doing formative assessment with quizzes. One shared scenario is: "I plan to organize sessions with Zoom to give my lessons live. I will use PhET [for physics simulations]. After each lesson, a Kahoot [quiz] will be sent to students to check their understanding and I will use Google Forms for evaluations". Teachers also mentioned that they would use simulation software such as PhET and GeoGebra, and Kahoot and Google Forms in their face-to-face classes while teaching or as homework assignments (Table 6).

4.4 Factors for Enabling STEM Online Teaching

We identified three main factors that can accelerate teaching STEM online, as well as the introduction of online STEM activities in and outside of the classroom: (1) leadership at the school level; (2) government investment; and (3) mobile technology.

Schools' Role in Enabling STEM Online Teaching. The analysis of the answers to the question "*In your opinion, what is the role of schools in encouraging online teaching and learning?*" resulted in three main points: (1) develop schools' infrastructure; (2) provide adequate teachers' training; and (3) raise awareness of STEM and online teaching to students and parents. The following quotes illustrate the answers: "First, stop banning students from using smartphones, tablets and PCs. Promote the creation of computer rooms in schools. Introduce introductory computer courses and encourage teachers to follow students through exercises and even online courses on weekends and holidays. Encourage the creation and animation of ICT clubs" and "It will first be necessary to modify the internal regulations which prohibit the use of the smartphone. Then rehabilitate the computer rooms and finally introduce students and some teachers to the basics of computer science. Provide logistical resources and support teacher training on online teaching".

Government's Role for Enabling STEM Online Teaching. Teachers answered the question: "*In your opinion, what is the role of the government in supporting public middle and highschools in terms of online teaching and learning?*" They think that the government should provide means to schools to set up computer rooms, and provide Internet access and digital tools to be used in schools. Government should organize training for teachers at scale in using digital tools and encourage them to teach online. The following quote illustrates this aspect: "Train staff in digital education, equip schools with Internet, computers, projector, servers and digital or IT tools. Promote the creation and production of digital educational resources ...".

Mobile Technology. We observed that most teachers used mobile phones to attend the training. We asked them about the option of using a mobile phone for teaching: "*Do you think it is possible to teach online with a mobile phone?*" The answer was 3.98 on the 5-Likert scale. Teaching online with a mobile phone could be a good option for developing countries since both students and teachers have limited access to computers and the majority of them have a smartphone. However, teaching with a smartphone has its limitations. Teachers highlight that they have low- or middle-end Android smartphones in general. Physical limitations of smartphones include screen size, storage and battery life. Software-wise, the mobile app version of a software/platform version has often less features compared to the web-based one. In Zoom itself, it is not possible to record a meeting or create breakout rooms on a phone. Switching between screens is not convenient. Editing a Google Presentation is harder than on a computer, switching between editing and reading modes and making the process of importing pictures very cumbersome. Teachers also mention phone notifications as distracting during online teaching or online learning.

5 Discussion and Conclusion

This paper is an attempt to understand the obstacles and enablers of STEM online education in Senegal. Some of these results can be transferred to other developed and developing countries. The study was carried out, post pandemic, in a STEM online teaching training for 79 STEM teachers representing the diversity of the 12 regions (out of 14) of Senegal. It showed that we need to focus, first, on the obstacles and enablers of online education in general before focusing on STEM online education. Access to Internet is the main obstacle for teachers to create online activities and courses. Students are not ready to learn online as they lack digital skills, cannot afford current Internet pricing, and may not have access to quality smartphones. While teachers have access to smartphones, preparing and delivering courses from a phone comes with numerous limitations, including a lack of features with respect to the computer version. Instructors feel strongly that online learning is the future and that hybrid courses would be beneficial to their students. They also need more training to be ready for online or hybrid teaching.

In Senegal, where a large number of students are oriented to the Virtual University of Senegal (UVS) without previous exposure to online learning, high school teachers have a crucial role to play in providing them with first experiences in the online world. School and government leaderships are the enablers of online education and there is a need for consequent investment in technology in the classroom, as well as for extracurricular activities, but also in changing the policies (authorizing personal phones and computers in the classrooms) and equipping teachers with the required material to be productive.

This study has wider implications for resilience in education. There is a need of middle and high school teachers training at scale to prepare them to teach online either completely or partly. While higher education institutions have infrastructures in place and trained instructors, middle and high schools are still not ready for unforeseen situations such as Covid-19. This research presents a scenario of online teaching that can be used by organizations interested in providing training to teachers in developing counties. It illustrates, in particular, the needs of careful planning and user-centered design of such initiatives to take the context of teachers and students into account, and the limitations of mobile devices and applications to engage teach and learn online.

Acknowledgements The STEM online training project implementation was supported by a US Citizen Diplomacy Action Fund (CDAF) grant.

References

1. Daniela, L., Visvizi, A.: Remote Learning in Times of Pandemic: Issues. Routledge, Implications and Best Practice (2021)
2. El-Deghaidy, H., Mansour, N.: Science teachers' perceptions of stem education: possibilities and challenges. Int. J. Learn. Teach. **1**(1), 51–54 (2015)
3. Gaglo, K., Seck, M., Ouya, S., Mendy, G.: Proposal for an online practical work platform for improving the teaching of stem. In: 2018 20th International Conference on Advanced Communication Technology (ICACT), pp. 738–743. IEEE (2018)
4. Khan, B.H.: Introduction to e-learning. Int. Handb. e-learn. **1**, 1–40 (2015)
5. Morrison, J.: Ties stem education monograph series, attributes of stem education. National research council (2010). Preparing teachers: Building evidence for sound policy (2006)
6. Office du BAC website. http://officedubac.sn. Accessed Apr 2022
7. Pokhrel, S., Chhetri, R.: A literature review on impact of Covid-19 pandemic on teaching and learning. High. Educ. Fut. **8**(1), 133–141 (2021)

8. Romano, A.: How to cultivate personal learning and professional growth in times of disruption, resistance and collective transformation? In: The International Research & Innovation Forum, pp. 165–178. Springer (2021)
9. Sedaghatjou, M., Hughes, J., Liu, M., Ferrara, F., Howard, J., Mammana, M.F.: Teaching stem online at the tertiary level during the Covid-19 pandemic. Int. J. Math. Educ. Sci. Technol. 1–17 (2021)
10. UNESCO. https://en.unesco.org/covid19. Accessed April 2022

Predicting Academic Performance of Students from the Assessment Submission in Virtual Learning Environment

Hajra Waheed, Ifra Nisar, Mehr-un-Nisa Khalid, Ali Shahid, Naif Radi Aljohani, Saeed-Ul Hassan, and Raheel Nawaz

Abstract The growth of online learning platforms has transformed the educational world, and the research community is now more interested than ever in discovering optimal ways to exploit online education repositories, including students' interactions with such platforms. Early prediction of student academic success has been a prevalent study issue, and multiple studies have articulated several measures and predictors for such analysis. In this work, we hand-engineer numerous factors relevant to a student's first evaluation in a course using data obtained from the Open University in the United Kingdom. A series of typical machine learning methods runs an array of trials on two feature sets, incorporating demographics with interactions and assessment data and ignoring demographics. The outcomes are examined to find key characteristics and determinants of student success. Such research is expected to aid in developing appropriate educational policies for an active layer of student support and intervention.

Keywords Student academic performance · Student assessment · At-risk students · Machine learning · Virtual learning environment

1 Introduction

With the vehemently wide-spread increase in the online learning platforms and virtual learning environments (VLEs), the accumulated educational data repositories

---

H. Waheed (✉) · I. Nisar · M.-N. Khalid · A. Shahid
Information Technology University, 346-B, Ferozepur Road, Lahore, Pakistan
e-mail: hajra.waheed@itu.edu.pk

N. R. Aljohani
King Abdulaziz University, Jeddah, Saudi Arabia

S.-U. Hassan
Manchester Metropolitan University, Manchester M15 6BH, United Kingdom

R. Nawaz
Staffordshire University, Staffordshire ST4 2DE, United Kingdom
e-mail: raheel.nawaz@staffs.ac.uk

A. Visvizi et al. (eds.), *Research and Innovation Forum 2022*,
Springer Proceedings in Complexity, https://doi.org/10.1007/978-3-031-19560-0_33

provide sufficient prospects to exploit and tap learner's behaviour, by addressing and understanding various dimensions of issues affiliated with educational stakeholders. With the emergence of such educational interactional data several research communities have surfaced, exhibiting an interest in analysing learner's behaviour, interpreting their patterns and extracting meaningful insights from this explorative examination [1]. Overall, these predictive analyses can facilitate educational communities in formulating educational policies and strategies to improve learners' behaviour and optimize their environment. Engagement with the online learning platforms generates digital footprints, that can be tapped to ascertain important predictors of student success [2]. This has yielded to the emergence of many disciplines and the inclusion of numerous terms, associated with the exploration of the educational data, such as academic analytics, predictive analytics and learning analytics [3].

The learning analytics research paradigm emphasizes on interpreting the learner behaviour to have an in-depth understanding of their performance such as the required actions associated with performance, assembling methods to have an enhanced understanding of their behaviour, that yield towards an optimal environment [4]. One such method traces student behaviour to early predict the potential grades in an-going course through an application of Markov models [5, 6]. Moreover, such implementations assist in the formation of a transformed optimistic educational setting that supports institutes in maintaining resource allocation procedures, ultimately facilitating them towards forming a positive reputation [7, 8]. In the recent past, online educational systems have emerged as a rising phenomenon, tapping and understanding the behaviour of students, analysing student success that can assist to form policies for student retention [9–12], ultimately optimizing the pedagogical methods and environment [13, 14]. However, evidence on the in-course assessments and their association with the academic performances of students, are limited. Assessments are an important attribute reflecting students' industrious nature and consequently their grade is impacted by the performances they demonstrate in such assessments. More specifically, first assessment in a course is given considerable emphasis to highlight the significance of such activities and their impact on the academic performances of students. In line with this, our study has the following objectives, leveraging student assessment performance to predict their academic performances in a course.

- To start with, we use the Open University Learning Analytics Dataset (OULAD) to engineer new features, pertaining to student activity in the VLE, reflecting their interactions with the online platform on an in-depth level.
- Moreover, leveraging the power of first assessment (A1) in a course, we predict the academic performances of students using their interactions and engagement with the VLE till A1, by leveraging various machine learning models.
- Finally, a multitude of feature sets are applied to early predict the academic performances of students, that is by using students' interaction data with the VLE and by merging demographic information with student engagement data.

The rest of the paper has been organized as follows: Sect. 2 presents literature review. Section 2 presents related work followed by data and methodology in Sect. 3. Then, Sect. 4 presents results and finally conclusions are presented in Sect. 5.

2 Review on Student Academic Performance and Assessment

The paradigm of learning analytics extensively emphasizes on assessing students' learning patterns, their grades and providing targeted support to the learners in need of an intervention. To get state-of-the-art results, it is necessary that you have an accessible data that provides information about students [15]. As this field is one of the emerging fields, so researchers are doing valuable work in this area by leveraging machine learning and deep learning techniques. Different studies have utilized different types of data including high school, university, online learning environments or distance learning university data. The learning analytics research community focuses on student retention through the prediction of their academic performance and by identifying the behaviour of dropout students.

Okubo and Takayoshi [16] deployed a deep learning-based approach to find out the final grade of the students. They trained Recurrent Neural Network (RNN) of 937 students attending 6 courses to identify the important learning activities necessary to achieve a specific grade. These learning activities including in Active Learner Point (ALP) like attendance, quizzes, report etc. contributing to obtain a certain final grade, inferred from the weights of the obtained RNN. Francis and Babu [17] deployed a hybrid approach for the performance prediction of students in higher education. Their hybrid technique based on classification and clustering models produces far better results than standalone clustering and classification techniques, in terms of accuracy for students' academic performance. Different attributes of instructors are identified that are necessary to engage the students, including the instructor's presentation in the video. This study also provides practical solution for instructors of MOOCs.

Qiu [18] analysed factors that affect students' engagement in the course. Their results reveal an important concept of social/peer influence, that is if a student's friend has a certificate, then the tendency to get the certificate in the student himself also increases. They also observed that the tendency to question more or perform good does not corelate with a student earning a certificate. Their 'learning effectiveness' model outperforms alternative methods in this field. Shahiri and Husain [19] provide an overview on Educational Data Mining (EDM) techniques to find out the students' final grade and the important features including external assessment, psychometric factors, CGPA and extra-curricular activities, that play a vital role in students' performance prediction. A number of studies revealed that different classification techniques like Decision Tree, Neural Network are commonly used for student performance prediction [20–22].

Table 1 Features computed from students clickstream data

Feature type	Feature
Assessment	Score in A1, submission status of A1, number of days student submitted A1 before deadline, number of days student submitted A1 after deadline
Overall clicks	Student total clicks, Student total clicks before course starts, student total clicks after course started, student total clicks before A1 submission, total clicks on A1 submission date
Logged clicks	No. of days student logged in on VLE, first day student was active on VLE, last day student was active on VLE before A1, first day student was active on VLE after course started, no of days student logged in VLE till A1 submission deadline
Clicks for each activity	Clicks before A1 deadline on {dataplus, forumng, glossary, oucollaborate, oucontent, resources, subpages, homepage, URL}, Clicks on A1 deadline on {dataplus, forumng, glossary, oucollaborate, oucontent, resources, subpages, homepage, URL}

3 Data and Methodology

This study uses OULAD dataset, containing data from the courses presented in Open University, UK from sessions 2013–14 [23]. The data set contains both demographic data and click stream interactions of students with the VLE. Assessment results and interaction logs of 32,593 students with the VLE in 22 courses is present in the dataset [24, 25]. A detail of the features engineered is provided in Table 1.

After the feature engineering process, conventional machine learning algorithms were deployed to predict the academic performances of students from the newly generated features, including assessment and demographics related information. An array of two feature-sets were deployed: (a) feature-set consisting of engagement interactions and assessment related information including the newly generated features, (b) feature-set including engagement interactions, assessment information and demographics. The deployed machine learning algorithms include logistic regression (LR), decision tree (DT), random forest (RF), XGB boosting algorithm. The details of the experiments conducted for each model are provided in the next section.

4 Experiments and Discussion

This section describes the experiments conducted to analyse the objectives of this study. Various deployed machine learning models were tuned to find the optimal parameters, details are provided in this section.

4.1 Early Prediction of Students Academic Performance

We predict student academic performance in a course and all courses combined based on A1 scores and students' engagement information till the submission of A1. For course wise and overall performance of students in all courses combined, four different experiments were performed to measure the prediction performance of different machine learning algorithms based on different predictors (features) that is (a) all selected features, (b) assessments + clickstream features. Table 2 provides an overview of experiments and student records. The machine learning models used in this analysis and their results with two different set of features are presented in Figure 1a, b. Results show that models based on all attributes and models which are trained with assessments + clickstream attained almost same results. Also, LR, XGB and RF show better performance in terms of AUC (0.88 AUC) than DT which shows 0.87 AUC. In terms of Recall LR shows better performance (0.81 recall).

Table 2 Experiments for student academic performance

Analysis	Features	Training set	Testing set
All courses	All attributes (demographics, assessments, clickstream)	16,690	5,564
	Assessments + clickstream	16,690	5,564
Course FFF	All attributes (demographics, assessments, clickstream)	4,773	1,139
	Assessments + clickstream	4,773	1,139

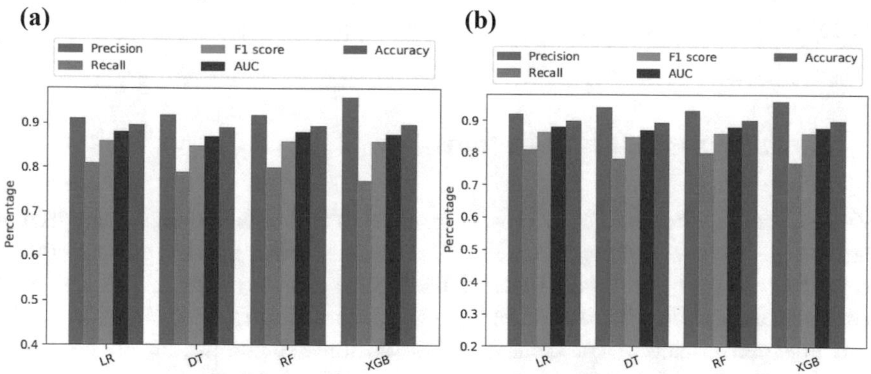

Fig. 1 **a** Academic performance classification without demographics. **b** Academic performance classification with all attributes

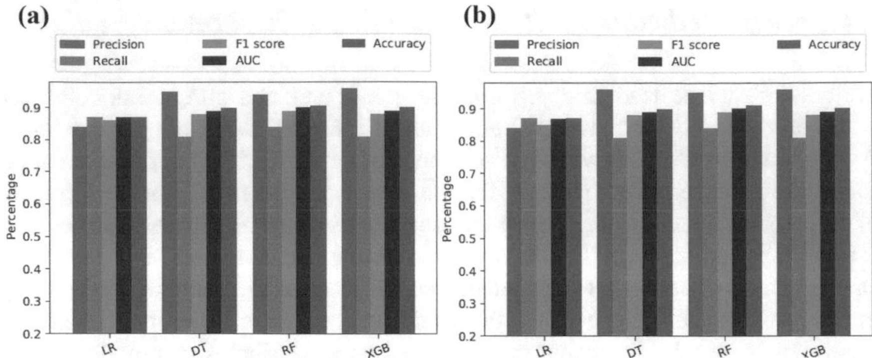

Fig. 2 **a** Course-level academic performance classification without demographics. **b** Course-level academic performance classification with all attributes

4.2 Course Level Student Academic Performance

For course level analysis, 2014B presentation of course FFF (the course title has been anonymised in the data), was used for testing and all other presentations of this course for the years 2013 and 2014 were used for training. The results of machine learning models for course wise analysis using assessments + clickstream and with all attributes are presented in this subsection. According to Fig. 2a, b models based on all attributes and models which are trained with assessments + clickstream attained almost same results. However, RF proves to be a better predictor compared to other deployed algorithms with an AUC of 0.9 for both predictor sets.

5 Conclusions and Implications

Most of the previous studies acknowledge that student data should be analysed at early stage of course to support timely interventions, which provide greater support to teachers to early recognize those students who are at-risk of failure and need support in their studies. The study deploys two different feature sets for predicting the academic performances of students (a) all the attributes including engagement data, assessment data and demographics (b) excluding the demographics data. Results reveal that in the first experiment, where all courses are combined as one course, all the deployed machine learning algorithms perform with a negligible difference for both feature-sets and RF performs best amongst all models. For the second experiment, where one particular course was chosen to predict the performances of students, and test train sets consisted of this one particular course, a slight difference can be observed between the results of the two feature-sets. Feature-set including demographics has a slightly better performance compared to the other feature-set, thus highlighting the significance of demographics in the prediction of academic performances.

The existing literature asserts the prediction power to significantly improve with the progression of the course. However, a key indicator is how early the academic performance can be predicted to support and assist the learning analytics community for better prediction models. This study employs data using A1 as an important activity, where interaction patterns of each student are accumulated till A1, their A1 submission and A1 score is countered to analyse the improvement in the prediction models. Such analysis will facilitate the educational community towards the well-being of the students in need of support and help, formulate policies and committee structures to improve the student body by providing recommendations and counselling methods. In future, Also, in addition to quantitative data, we plan to incorporate open ended questionnaires from the participants to apply advance natural language processing [26, 27] techniques and deep learning methods [28–30] to better understand learning behaviour of the student and their sentiments [31, 32] from textual data.

References

1. Baneres, D., Rodriguez-Gonzalez, M.E., Serra, M.: An early feedback prediction system for learners at-risk within a first-year higher education course. IEEE Trans. Learn. Technol. **12**(2), 249–263 (2019). https://doi.org/10.1109/TLT.2019.2912167
2. Visvizi, A., Lytras, M.D., Sarirete, A.: By means of conclusion: ICT at the service of higher education in a transforming world. In: Visvizi, A., Lytras, M.D., Sarirete, A. (eds.) Management and Administration of Higher Education Institutions in Times of Change, Bingley, UK (2019)
3. Nawaz, R., Sun, Q., Shardlow, M., Kontonatsios, G., Aljohani, N.R., Visvizi, A., Hassan, S.U.: Leveraging AI and machine learning for national student survey: actionable insights from textual feedback to enhance quality of teaching and learning in UK's higher education. Appl. Sci. **12**(1), 514 (2022)
4. Sicilia, M., Visvizi, A.: Blockchain and OECD data repositories: opportunities and policy-making implications. Library Hi Tech **37**(1), 30–42 (2019)
5. Hu, Q., Rangwala, H.: Course-specific Markovian models for grade prediction. In: Pacific-Asia Conference on Knowledge Discovery and Data Mining, pp. 29–41 (2018)
6. Polyzou, A., Karypis, G.: Grade prediction with course and student specific models. In: Pacific-Asia Conference on Knowledge Discovery and Data Mining, pp. 89–101 (2016)
7. Baker, R.S., Inventado, P.S.: Educational data mining and learning analytics. In: Learning Analytics, pp. 61–75. Springer (2014)
8. Daniel, B.K.: Big data in higher education: the big picture. In: Big Data and Learning Analytics in Higher Education, pp. 19–28. Springer (2017)
9. Rienties, B., Boroowa, A., Cross, S., Kubiak, C., Mayles, K., Murphy, S.: Analytics4Action evaluation framework: a review of evidence-based learning analytics interventions at the Open University UK. J. Interact. Media Educ. **2016**(1) (2016)
10. Waheed, H., Hassan, S.-U., Aljohani, N.R., Wasif, M.: A bibliometric perspective of learning analytics research landscape. BIT **37**(10–11), 941–957 (2018)
11. Wasif, M., Waheed, H., Aljohani, N.R., Hassan, S.-U.: Understanding student learning behavior and predicting their performance. In: Cognitive computing in technology-enhanced learning, pp. 1–28. IGI Global (2019)
12. Brdesee, H.S., Alsaggaf, W., Aljohani, N., Hassan, S.-U.: Predictive model using a machine learning approach for enhancing the retention rate of students at-risk. Int. J. Semant. Web Inform. Syst. (IJSWIS) **18**(1), 1–21 (2022)

13. Li, J., Wong, Y., Kankanhalli, M.S.: Multi-stream deep learning framework for automated presentation assessment. In: 2016 IEEE International Symposium on Multimedia (ISM), pp. 222–225 (2016)
14. Hassan, S.-U., Waheed, H., Aljohani, N.R., Ali, M., Ventura, S., Herrera, F.: Virtual learning environment to predict withdrawal by leveraging deep learning. Int. J. Intell. Syst. **34**(8), 1935–1952 (2019)
15. Kavitha, M., Raj, D.: Educational data mining and learning analytics-educational assistance for teaching and learning. Int. J. Comput. Organ. Trends **41**(1), 21–25 (2017)
16. Okubo, F., Yamashita, T., Shimada, A., Konomi, S. (2017) Students' performance prediction using data of multiple courses by recurrent neural network. Proc. ICCE2017, 439–444 (2017)
17. Francis, B.K., Babu, S.S.: Predicting academic performance of students using a hybrid data mining approach. J. Med. Syst. **43**(6), 162 (2019)
18. Qiu, J., et al.: Modeling and predicting learning behavior in MOOCs. In: Proceedings of the Ninth ACM International Conference on Web Search and Data Mining, pp. 93–102 (2016)
19. Shahiri, A.M., Husain, W.: A review on predicting student's performance using data mining techniques. Procedia Comput. Sci. **72**, 414–422 (2015)
20. Okubo, F., Yamashita, T., Shimada, A., Ogata, H.: A neural network approach for students' performance prediction. In: Proceedings of the Seventh International Learning Analytics & Knowledge Conference, pp. 598–599 (2017)
21. Hlosta, M., Zdrahal, Z., Zendulka, J.: Ouroboros: early identification of at-risk students without models based on legacy data. In: Proceedings of Seventh International Learning Analytics & Knowledge Conference, Vancouver, British Columbia, Canada, pp. 6–15 (2017)
22. Azcona, D., Hsiao, I.-H., Smeaton, A.F.: Detecting students-at-risk in computer programming classes with learning analytics from students' digital footprints. User Model. User-Adapt. Interact. **29**(4), 759–788 (2019)
23. Kuzilek, J., Hlosta, M., Zdrahal, Z.: Open university learning analytics dataset. Sci. Data **4**, 170171 (2017)
24. Waheed, H., Hassan, S.-U., Aljohani, N.R., Hardman, J., Alelyani, S., Nawaz, R.: Predicting academic performance of students from VLE big data using deep learning models. Comput. Hum. Behav. **104**, 106189 (2020)
25. Waheed, H., Anas, M., Hassan, S.U., Aljohani, N.R., Alelyani, S., Edifor, E.E., Nawaz, R.: Balancing sequential data to predict students at-risk using adversarial networks. Comput. Electr. Eng. **93**, 107274 (2021)
26. Rahi, S., Safder, I., Iqbal, S., Hassan, S.U., Reid, I., Nawaz, R.: Citation classification using natural language processing and machine learning models. In: International Conference on Smart Information Communication Technologies, pp. 357–365. Springer, Cham (2019)
27. Iqbal, S., Hassan, S.U., Aljohani, N.R., Alelyani, S., Nawaz, R., Bornmann, L.: A decade of in-text citation analysis based on natural language processing and machine learning techniques: an overview of empirical studies. Scientometrics **126**(8), 6551–6599 (2021)
28. Hassan, S.U., Imran, M., Iqbal, S., Aljohani, N.R., Nawaz, R.: Deep context of citations using machine-learning models in scholarly full-text articles. Scientometrics. **117**(3), 1645–1662 (2018)
29. Hassan, S.U., Saleem, A., Soroya, S.H., Safder, I., Iqbal, S., Jamil, S., Bukhari, F., Aljohani, N.R., Nawaz, R.: Sentiment analysis of tweets through Altmetrics: A machine learning approach. J. Inform. Sci. **47**(6), 712–726 (2021)
30. Hassan, S.U., Shabbir, M., Iqbal, S., Said, A., Kamiran, F., Nawaz, R., Saif, U.: Leveraging deep learning and SNA approaches for smart city policing in the developing world. Int. J. Inform. Manag. 1;**56**, 102045 (2021)
31. Safder, I., Mahmood, Z., Sarwar, R., Hassan, S.U., Zaman, F., Nawab, R.M., Bukhari, F., Abbasi, R.A., Alelyani, S., Aljohani, N.R., Nawaz, R.: Sentiment analysis for Urdu online reviews using deep learning models. Exp. Syst. **28**, e12751 (2021)
32. Mahmood, Z., Safder, I., Nawab, R.M., Bukhari, F., Nawaz, R., Alfakeeh, A.S., Aljohani, N.R., Hassan, S.U.: Deep sentiments in roman Urdu text using recurrent convolutional neural network model. Inform. Process. Manag. 1;**57**(4), 102233 (2020)

Covid-19 and Quasi-Covid-19 Pandemic Vicariant Innovations in the Educational Context: Comparative Cases

Xhimi Hysa and Vilma Çekani

Abstract The COVID-19 pandemic has been a disruptive event in many industries while providing innovation opportunities as well. The transition to remote working in the educational context, on one side, has found universities unprepared, and on the other side has tested their creativity. The aim of this study is to explore emerging innovations and new dynamic capabilities aroused at universities during the pandemic and the quasi- Covid-19 period. In this study, the term "vicariant innovation" is used to express the way innovations themselves have been subject to frequent changes while stimulating more dynamic capabilities for institutions' management, professors, and students. The methodology used in this study is a qualitative one by conducting semi-structured in-depth interviews. The sample consists of 12 students per university, respectively: EPOKA University in Albania and the University of Salerno in Italy. Findings show that the Covid-19 vicariant innovation changed students' habits of studying and although the universities are from 2 different countries the students' thoughts towards this phenomenon are almost the same. Furthermore, vicariant innovations are seen as the future of the education system. The originality of this work consists of the methodology used to explain the vicariance of innovation under the umbrella of educational context during the quasi- Covid-19 period.

Keywords Vicariance innovation · Education system · EPOKA University · University of Salerno · Educational platforms

X. Hysa
POLIS University, Street Bylis 12, Highway Tiranë-Durrës, Km 5, Kashar, SH2, 1051 Tirana, Albania
e-mail: xhimi_hysa@universitetipoli.edu.al

V. Çekani (✉)
University of Salerno, Street Giovanni Paolo II, 132, 84084 Fisciano, Salerno, Italy
e-mail: vcekani@unisa.it

A. Visvizi et al. (eds.), *Research and Innovation Forum 2022*,
Springer Proceedings in Complexity, https://doi.org/10.1007/978-3-031-19560-0_34

1 Introduction

The Covid-19 pandemic affected the global education systems just as it affected the global economy and the national healthcare systems. The quarantine and movement restrictions interrupted the traditional way of learning which led to distance learning. The latter, roots back in the 1800 and 1900s, was introduced through parcel post, radio, television, and online [1]. Allen and Seaman [2] consider online education as distance education in which technology is used as means of communication, and at least 80% of the course is taught online. Furthermore, the challenge faced by education organizations about remote learning was the fact that the course contents would be 100% online and that the instructors would have needed to prepare the same organizational structure for e-learning as it would have been in the presence [3]. Furthermore, academicians underline the importance of vicariant solutions in the education system, which main conducted studies stay under the umbrella of inclusive education as it is a pure example of the significance of vicariance [4]. Berthoz [5] defined vicariance as a substitution, which implies using different means to solve a problem that will bring the same results. In this regard, vicariance and the Covid-19 pandemic go hand to hand, given that contingency solutions emerged by providing similar educational services as in face-to-face learning. Thus, this study comes with a precise scope: to explore emerging innovations and new dynamic capabilities aroused at universities during the Covid-19 pandemic and the post-Covid-19 period. The objectives of this research study are:

1. To explore quasi-Covid-19 pandemic vicariant solutions in the education system.
2. To understand how vicariant innovations coped with Covid-19 consequences in the education system.

This study uses a qualitative methodology. First, a brief literature review is conducted to present and show the conceptual relationship between vicariance, Covid-19, and the education systems. Then, in-depth interviews have been conducted to gather data in two universities: the University of Salerno (Italy) and EPOKA University (Albania). The originality of this work consists in giving a panorama of how vicariant solutions had an impact on students during the Covid-19 pandemic period.

2 Literature Review

This section's aim is to give a brief overview of the vicariant innovation and its relationship with the education system.

2.1 Theoretical Background

Vicariance is used for the first time in the 1950s and has its genesis in biology defined as involving varieties, species, communities, or the like that have evolved out of effective contact with one another even though they live in similar habitats but are separated [6–8]. As its definition implies, vicariance is found in different fields of study, due to the variety and evolution. Meanwhile, the etymological origin of the word in Latin means "a substitute, change, interchange or that takes place of a person" [9]. In psychology, it is used to describe the differences within society, starting from groups, individuals, ethnic groups, and genders in terms of what different solutions they give to solve the same problem [10, 11]. Likewise, the French psychologist Alain Berthoz argues that vicariance incorporates processes that are part of different scientific domains and can be defined as a substitute of a mechanism/process by another mechanism/process to bring the same results [12]. In addition, he distinguishes two types of vicariance: functional vicariance (substituting a mechanism/process that can lead to the same results by using different means), vicariance of use (the ability of individuals to use the same tools in different ways, depending on their objectives/limits so that an object can be used for different purpose and tasks). On the other hand, Prigogine and Stengers [13] consider it a creative deviation from the previous extant path.

2.2 Vicariant Innovation and Education

Vicariance gives a significant contribution to the "information variety attribution" to a person, team, or organization, with functional or transformational attributions [14]. In this regard, vicariance is a dynamic capability to survive in complex environments [15–17]. Scholars often define it analogically as it deals with re-combining, resetting, and re-shaping [18, 19], by breaking the rules, inventing, and problem creating [20]. In addition, vicariance introduces "the need for a culture of variety" that is connected and intervenes with the definition of innovation itself [21, 22], whereas it also merges its role of diversity creator. In this vein, vicariance identifies four processes: substitution, replacement, innovation, and resilience. Therefore, to execute these processes a behavior analysis that leads to a similar function is important to be conducted to know which to substitute, and then to use this replacement to create new strategies/tools to perform the same tasks. Scholars like Simone et al. [15, 23], state that individuals and organizations can execute different vicariant processes to adapt to everyday situations, and this happens for several reasons starting from their past experiences, trust/distrust, triggers/stimuli that they get from external factors. Furthermore, taking into consideration the aforementioned reasons it can be underlined that vicariance gives a significant contribution to management and decision making, both on theoretical and practical levels. According to Berthoz [6], vicariance represents an adaptive and learning tool, which in education it is incorporated with the methods that the academic staff finds appropriate, and which should be unique and original and can be adapted to the educational system in different circumstances such as Covid-19 itself. These vicariant solutions come because of their past experiences by taking into consideration how much they know their students [24–26]. In the education system, vicariance allows adapting a dynamic and flexible teaching approach that permits solving different situations that are represented by a variety of solutions that mediate relationships between places, knowledge, instructors, and students [27]. Covid-19 pandemic emergency, constrained academics to find different means to conduct the teaching–learning process triggering their creativity and resilience [28].

3 Methodology

This study relies on the explorative design through comparative case studies (a comparative analysis between two universities: the University of Salerno in Italy, and EPOKA University in Albania). The reason why these two universities are chosen is that each author is part of this university, which makes the data more reliable. The data gathering process follows the example of the study conducted by Flavin [29] related to students' perception of ongoing disruptive innovations in education. Given the nature of in-depth interviews, and because of Covid-19 restrictions, it was seen as more reasonable to conduct the interviews online. Also, there are advantages because given that it was conducted online, the participants had less pressure and could express their opinion truthfully (the interview was anonymous as well). The number of participants is 24, which indicates 12 interviewees per university, following Dworkin's guidelines for qualitative studies sample size [30, 31]. The sample of this study uses a non-probability sampling approach and includes students from different departments, to give a panorama of how the Covid-19 pandemic affected students in different fields of study. The age gap varies from 20 to 27 years old. The following Tables 1 and 2 provide detailed information about the sample, which are part of the first 2 questions of the survey. In addition, Fig. 1 will show the questions of the survey.

4 Findings

In both universities, overall, students share similar thoughts even though in the case of EPOKA University, they did not only use the Zoom, Moodle, and Microsoft teams but used also Google Meets or Google classroom. Graphs 1 and 2 show which platforms were used mostly in both universities:

In the third question only 1 student from the University of Salerno, stated that:

> Often bugs and gives problems of connection.

Table 1 University of Salerno students background (*Source* Authors' elaboration)

Department	Program	Students	Degree
Department of Management and Innovation System	Big Data Management	4	Ph.D
Department of Business Science and Innovation system	Business Innovation and Informatics	1	M.Sc
Department of Physics	Physics	1	Ph.D
Department of Industrial Engineering	Chemical Engineering	2	Ph.D
Department of Industrial Engineering	Management Engineering	1	M.Sc
Department of Economic Sciences and Statistics	Business Economics	3	M.Sc

Table 2 EPOKA University students' background (*Source* Authors' elaboration)

Department	Program	Students	Degree
Department of Business Administration	Business Administration	4	B.Sc
Department of Business Administration	Business Informatics	2	B.Sc
Department of Economics	Economics	1	M.Sc
Department of Architecture	Architecture	1	B. Arch
Department of Business Administration	Business Administration	3	M.Sc
Department of Banking and Finance	Banking and Finance	1	B.Sc

Semi-structured in-depth interview

1. In which university are you studying?
- EPOKA University
- University of Salerno
2. Which is the department and the program you are studying?
3. Can you describe the educational platforms used during Covid-19?
4. What is your opinion regarding the usage of these platforms?
5. Were you familiar with them before?
-Yes
*No
6. Which did you use most and why do you prefer it? (if you knew any)
7. Did e-learning change the routine of your academic routine in the quasi-post Covid-19 phase?
8. Were you skeptical, towards this educational change and why?
9. How the hybrid teaching is affecting the educational system?
10. Can you specify its pros and cons?

Fig.1 Semi-structured in-depth interview questions (*Source* Authors' elaboration)

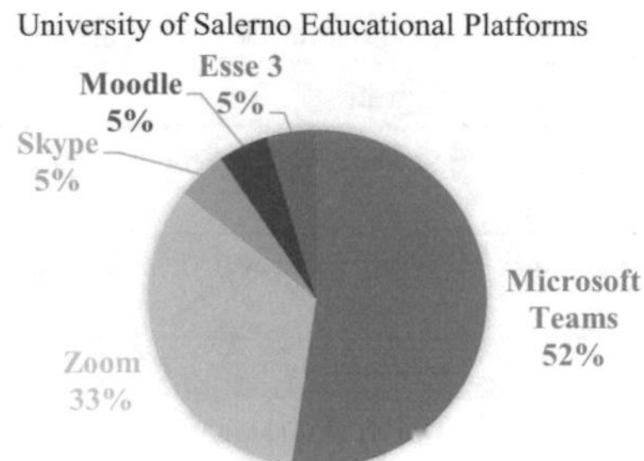

Graph 1 University of Salerno educational platforms during Covid-19 (*Source* Authors' elaboration)

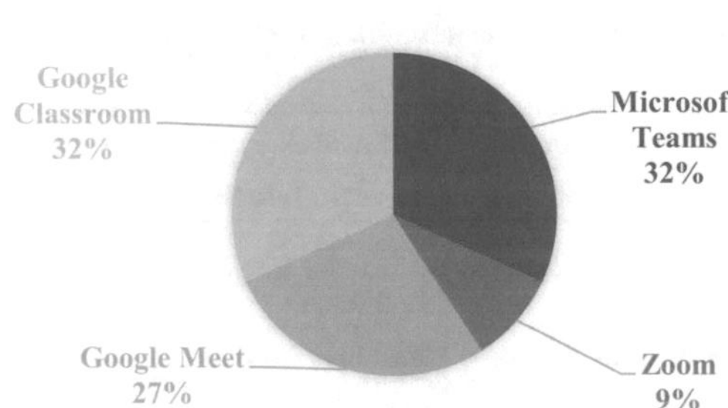

Graph 2 EPOKA University educational platforms during Covid-19 (*Source* Authors' elaboration)

Meanwhile, the other students stated that they found the educational platforms easy to use, helpful, good substitution, and necessary during Covid-19.

On the other hand, 2 students from the University of EPOKA, stated that:

> The platforms that were used during the pandemic were not effective. This software should be more interesting to help the lecturers and students' interaction.

> Drawbacks were mostly due to a lack of connection by one party or the other.

While other students found them effective, beneficial in HyFlex learning, and easy to use.

In the fourth question, Graph 3 represents how familiar students were with these platforms.

In addition to this graph, only 4 students from 12 knew different educational platforms used during the Covid-19, while 7 said they did not know, and 1 stated that *"I wasn't familiar with all of them except for Skype"*). On the other hand, 6 students confirmed that they knew these platforms, while 5 did not know before, meanwhile, 1 student stated that *"slightly familiar, only had used them on a couple of occasions"*.

Were you familiar with them before?
24 responses

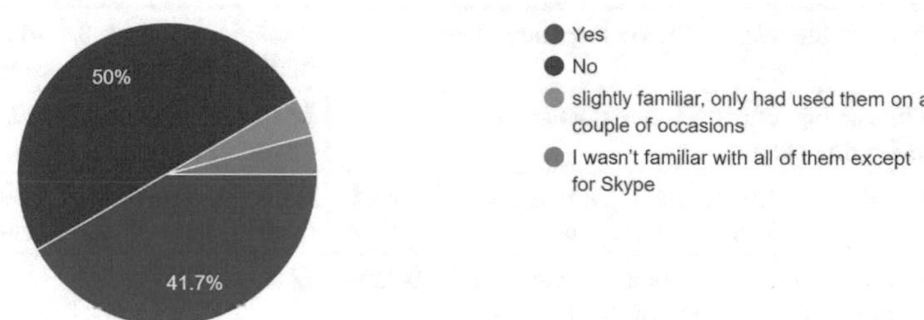

Graph 3 Familiarity of the 24 students with the educational platforms (*Source* Google form graph)

On the fifth question, which was about their favorite educational platform, students from the University of Salerno, were divided between Microsoft Teams and Zoom:

> Zoom, easier to use and deploy. It connects easily across desktops, laptops, and mobile devices. It has breakouts rooms for students to work in short groups.
>
> Teams, easy to use and create a meeting.

On the other hand, students from EPOKA University mostly affirmed that: *"In my opinion, Microsoft Teams is the best option, because it's user-friendly, increases focus on work, and team productivity, and we can easily access materials or respond to chats/conversations."* While other students stated that:

> *"Google classroom. It is a good way to keep everything in one place and easily accessible by any place if you know your email and password."*

On the sixth question, students were asked how e-learning changed their academic routine, most of the students from the University of Salerno, stated that their studying routine changed, but some did not. Doctoral students were the ones who felt more comfortable taking courses and doing research at home rather than in presence. Students of EPOKA University affirmed that e-learning changed a lot of their usual learning routine as well their daily habits, such as sleeping more and being lazy.

On the seventh question, students were asked if they were skeptical towards this change in the educational system, whereas in the University of Salerno students stated that:

> Yes, I was. I thought it was going to be less effective.
>
> No, as I said, whether it is physical or e-learning it is just a guide, I still have to study the topics and learn them.

Students at EPOKA University generally replied similarly as students from the University of Salerno:

> I was at the beginning because it was a new change.
>
> I was skeptical because of many technical problems that could happen and also the participation of the students in the lesson.

Table 3 A summary of students' feedback for hybrid teaching (*Source* Authors' elaboration)

Pros	Cons
You can attend your lesson from anywhere	Sometimes they make you lazy and you don't want to attend lessons
Timesaving, effortless, making use of technological advances	Lower concentration level, less interaction, lower motivation to study
Flexibility in the program that may meet the needs of some students who work part-time	A challenging task for lecturers to deliver courses with a split audience
More flexible; easy to be conducted; students have more free time to work during studies	Less productive
Some students may feel more motivated as they have more freedom in the learning method	On the other hand, students may not feel motivated to learn as they are not being "observed" by professors
Students aren't obliged to follow courses only in classes	Technical difficulties may appear that can make learning hard

On the eighth question, students were asked about their opinion related to hybrid teaching. Students at the University of Salerno mostly stated that:

> I think it's giving people the chance to correctly balance educational and personal life.

> I think it affects badly because students have more difficulty focusing on the lessons.

Students from EPOKA University generally stated that:

> Hybrid learning is helping us break the barriers of a classroom, due to the technological innovations. It is also enabling the educational system with new ways of learning, communicating, and working.

> In general, I think it is making the students a bit lazy, especially after they got used to e-learning. But some other students may find it more productive than the in-presence learning.

Table 3 will show a summary of the pro and cons of hybrid teaching based on the answers that 24 students gave.

5 Discussion and Conclusion

This study gives a significant contribution to the vicariance innovation in the education system because variance innovation is used mostly as a solution to aid students with specific needs (disabilities). The study specifies how Covid-19 triggers vicariance in an education system where different solutions are offered for students. In this way a different perspective of vicariance innovation, by proposing solutions also to students without disabilities. The results show similarities in students' perceptions at both universities. Even though the modalities and tools used for the teaching process differ from one university to another, still students express similar thoughts regarding their educational journey during the Covid-19 period. Findings show that the University of Salerno focused mostly on educational platforms such as Microsoft Teams and Zoom, while EPOKA University is fairly distributed among Google Classroom, Google Meet, and Microsoft Teams (using Zoom only sporadically). Students studying for Ph.D programs shared different opinions regarding online teaching and hybrid one. Most of them felt comfortable because of the nature of their study (research) while students with bachelor's or master's found it more difficult the e-learning process. On the other side, professors had to adapt and create a new digital learning atmosphere to provide the same service as it would have been in presence. These efforts and consequent innovations assume in this study the shape of what we call "vicariant innovation", a typical emergent innovation of different shades [32]. In addition, adapting and finding new solutions helps both parties for good results. As is shown in the section above, students had problems in the beginning but later it was seen as an effective methodology. Not everybody had the necessary tools to follow the courses, a reason why many preferred the hybrid teaching education as a midway. The findings suggest that students find it hard at a certain point to adapt to the new change after 2 years of online learning because to most of them this period changed their daily routine and in one way or another made them lazier. It is important to emphasize that online teaching is not new, it was used before but not as an obligatory methodology as it was and as it currently is transformed into a hybrid one. Education institutions due to their nature and importance in the societies changes come with big responsibilities, that is why this experimental period will show results in the future of how effective it is and that it should move its traditional way of teaching, even though this may not apply to every academic sector.

References

1. Kentnor, H.E.: Distance education and the evolution of online learning in the United States. Curr. Teach. Dial. 17 (1), pp. 21–34 (2015)
2. Allen, I.E., Seaman, J.: Sizing the opportunity: the quality and extent of online education in the United States, 2002 and 2003. Sloan Consortium (NJ1) (2003)
3. Linda, D., Visvizi, A. (eds.): Remote Learning in Times of Pandemic: Issues, Implications and Best Practice. Routledge, (2021)
4. Agrillo, F., Di Gennaro, D.C., Sibilio, M.: The concept of vicariance in the teaching-learning process: a possible didactic tool for promoting school inclusion. In: Proceedings of ICERI 2016 Conference, pp. 2244–2250 (2016)
5. Berthoz, A.: La vicarianza. Il cervello emulatore di mondi. Codice, Torino (2015)
6. Croizat, L.: Manual of Phytogeography: An Account of Plant-Dispersal Throughout the World. Springer (2013)
7. Visvizi, A., Lytras, M.D., Aljohani, N.R. (eds.): Research and Innovation Forum 2020: Disruptive Technologies in Times of Change. Springer Nature (2021)
8. Nelson, G.: From Candolle to Croizat: comments on the history of biogeography. J. Hist. Biol. **11**(2), 269–305 (1978)
9. Online etymology dictionary. https://www.etymonline.com/search?q=vicar
10. Reuchlin, M.: Vicarious processes and interindividual differences. J. Psychol. **2**, 133–145(1978)
11. Lautrey, J.: Universel et différentiel en psychologie: Symposium de l'APSLF (Aix en Provence, 1993). FeniXX (1995)

12. Berthoz: ALa vicariance: le cerveau créateur de mondes. Odile Jacob (2013)
13. Prigogine, I., Stengers, I.: Order out of chaos: Man's new dialogue with nature. New Science Library, Boulder, CO (1954)
14. Simone, C., Arcuri, M., La Sala, A.: Be vicarious: the challenge for project management in the service economy. In: Toulon-Verona Conference 'Excellence in Services' (2017)
15. Visvizi, A., Troisi, O., Grimaldi, M., Loia, F.: Think human, act digital: activating data-driven orientation in innovative start-ups. Eur. J. Innov. Manag. **25**(6), 452–478 (2022)
16. Di Nauta, P., Simone, C., Sarno, D.: Investigating the determinants of resilience: new insights from the vicariance. "Workshop di Organizzazione Aziendale—WOA 2018. The resilient Organization: Design, Change, and Innovation in the Globalized Economy", Track 3, pp 1–9 (2018)
17. Troisi, O., Fenza, G., Grimaldi, M., Loia, F.: Covid-19 sentiments in smart cities: the role of technology anxiety before and during the pandemic. Comput. Hum. Behav. **126**, 106986 (2022)
18. Quattrociocchi, B., Calabrese, M., Hysa, X., Wankowicz, E.: Technology and innovation for networks. J. Organ. Transf. Soc. Change **14**(1), 4–20 (2017)
19. Quattrociocchi, B., Simone, C., Calabrese, M., Iandolo, F., Fulco, I.: Innovation between redundancy and vicariance: the rising need for a culture of variety. In: Cybernetics and Systems, pp. 232–237. Routledge (2018)
20. Visvizi, A., Lytras, M.D., Daniela, L.: (Re) defining smart education: towards dynamic education and information systems for innovation networks. In: Enhancing Knowledge Discovery and Innovation in the Digital Era, pp. 1–12. IGI Global (2018)
21. Barile, S., Espejo, R., Perko, I., Saviano, M. (eds.): Cybernetics and Systems: Social and Business Decisions. Routledge (2018)
22. Troisi, O., Visvizi, A., Grimaldi, M.: The different shades of innovation emergence in smart service systems: the case of Italian cluster for aerospace technology. J. Bus. Indus. Mark. (2021)
23. Loia, V., Maione, G., Tommasetti, A., Torre, C., Troisi, O., Botti, A.: Toward smart value co-education. In: Smart Education and e-Learning, pp. 61–71. Springer, Cham (2016)
24. Zollo, I., Sibilio, M.: Possible applications of creative thinking within a simplex didactics perspective. Athens J. Educ. **3**(1), 67–84 (2016)
25. Altet, M.: Plural analysis of a teaching-learning sequence. CREN notebooks. CRDP Pays-de-la-Loire, Nantes
26. Dworkin, S.L.: Sample size policy for qualitative studies using in-depth interviews. Arch. Sex. Behav. **41**(6), 1319–1320 (1999) (2012)
27. Cassibba, R., Ferrarello, D., Mammana, M.F., Musso, P., Pennisi, M., Taranto, E.: Teaching mathematics at distance: a challenge for universities. Educ. Sci. **11**(1), 1 (2020)
28. Visvizi, A., Lytras, M. D., Daniela, L. (eds.): The Future of Innovation and Technology in Education: Policies and Practices for Teaching and Learning Excellence. Emerald Group Publishing (2018)
29. Flavin, M.: A disruptive innovation perspective on students' opinions of online assessment. Res. Learn. Technol. **29** (2021)
30. Troisi, O., Visvizi, A., Grimaldi, M.: The different shades of innovation emergence in smart service systems: the case of Italian cluster for aerospace technology. J. Bus. Indus. Mark. (2021)
31. Bokayev, B., Torebekova, Z., Abdykalikova, M., Davletbayeva, Z.: Exposing policy gaps: the experience of Kazakhstan in implementing distance learning during the COVID-19 pandemic. Transf. Govern.: People Process Policy **15**(2), 275–290 (2021)
32. Johnson, A.F., Roberto, K.J., Rauhaus, B.M.: Policies, politics and pandemics: course delivery method for US higher educational institutions amid COVID-19. Transf. Govern.: People Process Policy **15**(2), 291–303 (2021)

More Than Virtual Reality: Tools, Methods and Approaches to Hotel Employee Training

Jan Fiala, Jan Hán, Jan Husák, Karel Chadt, Štěpán Chalupa, Jiřina Jenčková, Martin Kotek, Michal Kotek, Martina Perutková, Tomáš Průcha, Lucie Rohlíková, Jakub Stejskal, and Anna Visvizi

Abstract In this paper, the use of virtual reality (VR) in the education of future hotel and tourism industry employees is explored. Through interdisciplinary co-operation of researchers and practitioners in the field of technology, hospitality, and pedagogy, a 3D training module simulating the environment of a hotel reception was developed. As a part of the project, the key elements of communication between/among hotel staff and various types of guests were identified, and variant scenarios of communication situations were built. The development of scenarios followed the principles of effective simulation, including realism (immersion), drama (involvement), and challenge (motivation). The methodology of training communication skills of students using VR was compiled against the backdrop of the leading sources existing in the literature and practice. The latter embraces the principle of the "action review cycle", which contains (1) a description of expectations from the simulation, (2) a description of experience, and (3) an analysis of differences between expectations and actual experience. In this study, the basic action review cycle was extended by "reflection",

J. Fiala · T. Průcha · L. Rohlíková (✉)
Department of Computer Science and Educational Technology, University of West, Bohemia, Univerzitni 8, 301 00 Plzen, Czech Republic
e-mail: lrohlik@kvd.zcu.cz

J. Hán · K. Chadt · Š. Chalupa
The Institute of Hospitality Management and Economics, Svídnická 506/1, 181 00 Praha 8 Prague, Czech Republic

J. Husák · M. Kotek · M. Kotek · M. Perutková · J. Stejskal
Center for Digitalization and Educational Technologies (CEDET), Czech Institute of Informatics, Robotics, and Cybernetics (CIIRC), Czech Technical University in Prague (CTU), Jugoslavskych Partyzanu 1580/3, 160 00 Praha 6 Dejvice, Czech Republic

J. Jenčková
Perfect Hotel Concept, Bártlova 35/10, Horní Počernice, 193 00 Praha Prague, Czech Republic

A. Visvizi (✉)
SGH Warsaw School of Economics, Al. Niepodległości 162, 02-554 Warsaw, Poland
e-mail: avisvi@sgh.waw.pl

Effat University, Jeddah 21551, Saudi Arabia

A. Visvizi et al. (eds.), *Research and Innovation Forum 2022*,
Springer Proceedings in Complexity, https://doi.org/10.1007/978-3-031-19560-0_35

i.e. quality of communication and professional accuracy. The VR tool elaborated on this paper remains a pilot and so more research and testing are needed to make it marketable.

Keywords Virtual Reality (VR) · Students · Future hotel employees training · Higher education · Communication skills

1 Introduction

"Virtual Hotel" is an application designed for training, and especially for the acquisition and development of communication skills by students, future hotel employees. The app was developed during the Covid-19 pandemic and was designed to alleviate the concerns and risks associated with the spread of Covid-19. Practice proves, however, that the VR-enhanced tool thus developed has a great practical purchase beyond the pandemic, and so will be usable in the years to come. The essence of the "Virtual Hotel" is that in the context of VR, the student wears a head-mounted display (HMD) and, so immerses him/herself in a virtual hotel environment. Here, the critical communication situations, mimicking real life scenarios are simulated. The necessity to develop and enhance communication skills, and thus also public appearances, is key for the value creation in the hospitality industry. The training thus provided through the "Virtual Hotel" helps students to break down the so-called "speaking/speech anxiety", i.e. the fear of public speech [1, 2]. Another important part of the development of communication skills is the practice of social communication. Indeed, literature [3] suggests that, the most effective way to develop this set of skills is through role-playing, where students play out situations similar to those they may encounter in the real world. As in the practice of public speaking, they then have their performance evaluated by an expert, ideally also by classmates, a self-assessment on the basis of a video recording follows. This immediate and comprehensive feedback helped to limit the use of filler words and superfluous terms [4] in communication. A specific area of development of students' communication skills is the practice of active behavior. It consists mainly in training the following: how to paraphrase messages, express emotions, summarize information, clarify, and ask questions. Such training again, ideally, takes place under the guidance and subsequent provision of feedback and reflection.

In the VR environment, the user has the opportunity to move freely and interact with active objects, such as a telephone or computer. A methodology was created for effective training, according to which teachers and students will use the application. In this sense, the methodology is an irreplaceable application manual of the created model of the hotel section, processed by virtual reality technology. The methodology is based on a theoretical framework for the implementation of educational training in a virtual environment, including inspiration from other projects and initiatives in the field of virtual reality. The objective was to find similar projects in the available literature that use virtual reality in education, in higher education, or directly develop communicative skills using virtual reality. The methodology also includes a description of the created virtual hotel, its functionalities, and basic types of training implementation. This methodology was created on the basis of theoretical study of professional literature and practical experience of project implementation team members from previous projects related to virtual reality - especially from experience with training future teachers in "virtual classroom", where it was verified in practice, and continuously improved with a three-phase training model [5].

2 The Research Model: Immersive Technologies and Communication Skills

An interesting source from the point of view of inspiration for the use of immersion technologies for the training of communication skills is the article "Designing Virtual Gaming Simulations" [6]. The platform presented in this article uses branched-out scenarios and pre-recorded video sequences to train the paramedics that the participant goes through in a virtual gaming environment. The whole process should then be cyclical [7]. The simulation team in this case consisted of professional nurses, instructors, web developers, audiovisual experts, standardized patients, and support staff. The whole team met regularly to design the simulation and its goal. Even before the start of the development itself, it was necessary to clearly identify the learning objectives (according to the relevant curricular document), from which the scenario design, branching nodes, and the overall appearance of the product were based. Outputs must not be strictly "tailored" to one institution, as they would then make it impossible for other students outside the test environment to use the simulation.

From the point of view of simulation preparation, the main activity was the creation, revision, and testing of the scenario. In this phase, the characters with which the participant will interact during the simulation are also designed and divided. These can be played by real people, in this case other students and family members. A test read follows, during which the script is finalized. Furthermore, the scenario is forwarded for the preparation of audiovisual recordings for the story of the simulation, due to the very specific implementation, in this particular case, I will include in this phase generally the preparation of the simulation in terms of graphics—in a virtual environment it is modelling objects and environments.

Before including the simulation in the learning process, it is important to test the simulation with the creative team, and then the preparation of the students for the simulation can begin. Before the simulation itself, the student will gain essential information about the performance:

- description of learning objectives,
- description of the simulation story,
- instruction on the movement of simulations,
- familiarization with the technology used.

Subsequently, the student enters the simulation and immerses himself in the story. The results show that there is a suitable presence of other people, namely a team consisting of, for example, didactics, moderator, and technician, ready to solve technical problems and take care of the simulation record and, last but not least, the researcher conducting the survey [8]. The initial data collection also begins at this stage.

In a post-simulation research [9], the authors examined 3 types of reflection: (a) self-debriefing, (b) facilitator-led in-person debriefing, and (c) evaluation authority in a virtual environment (facilitator-led virtual debrief). The research shows the benefits of providing reflection to others (especially classmates), but also emphasizes the need for self-reflection, which should precede a group debate, so that students first think about their own performance and try to analyze it themselves. The researcher performs another data collection here.

The authors describing the Simul@b project list three factors that ensure the effectiveness of simulations in a real environment: (a) realism, (b) drama, and (c) challenge. By combining these factors, it is possible to create effective and learning-supporting simulations even in a virtual environment [10].

3 Methodology

Based on the study of professional materials and previous experience, a model of the virtual hotel and a methodology of three-phase communication skills training were created. The model was gradually improved and refined in 2021, on the basis of feedback from 40 testers from among students and teachers of the University of Hospitality Management and employees of co-operating hotels/experts from practice.

3.1 Design-Based Research

We chose design-based research to create and test an immersive training system in virtual reality. It is a method that was proposed by Collins [12] and Braun [13], and which has become more widespread, especially in connection with the work of Cobb [14]. Design-based research includes 4 phases [15] and these phases were followed in the development of the model "Virtual Hotel". The phases include (see Fig. 1): analysis of practical problems by researchers and practitioners; development of a solution with a theoretical framework; evaluation and testing of solutions in practice; and documentation and reflection on the production of "design principles".

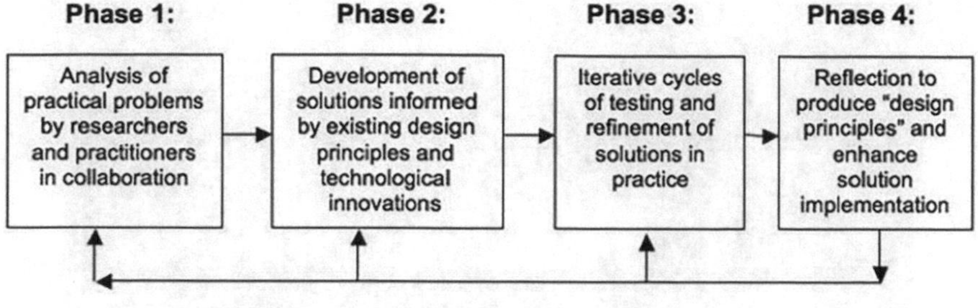

Fig. 1 Design-based research according to Reeves (2000) [15]

3.2 Description of the Educational Tool: The Virtual Hotel

Initially, to develop the model of the virtual hotel, photogrammetry and Neos VR were employed. The idea was that Neos VR will facilitate the process of connecting multiple users. Eventually, the use of the Neos VR platform was dropped. Instead, the Unity game engine employed. As a result, the model hotel reception uses Steam VR. This ensures compatibility with most currently used HMDs. To provide a comprehensive experience, the model hotel reception was extended to an entire front office (see Fig. 2 for an insight). The so created model contains many objects that the trainee can actively engage with, e.g. a computer, property management system, door keys etc. The objective here is to increase the realism of the simulation. Specific predefined scenarios are then linked to these objects and their use. An initial scenario was created for the initial testing, in which the student had to make a change in the reservation by telephone (see Fig. 3).

3.3 Three-Phase Training of Students in the Virtual Hotel

A methodology was created for training in the virtual hotel—a document that specifies the individual stages of training, and deals with important aspects of the development of communication skills of the target group in an immersive environment. The methodology is designed to directly correspond to the current possibilities and limits of the created virtual hotel model, and lead to the effective development of communicative skills of students in the hotel industry.

Fig. 2 Visualization of the hotel reception

1. Initial familiarisation with the virtual hotel environment Self-reflection, discussion, familiarisation with the realisation of the 2nd phase.
2. Virtual training (immersion into specific situations) Self-reflection, feedback, discussion, analysis of the recording.
3. Virtual training (repeated specific situations) Self-reflection, feedback, discussion, analysis of the recording.

Fig. 3 Three-phase training model

4 Results and Discussion

For the purpose of testing the usability of the "Virtual Hotel", including the quality of the immersion in the VR environment, 40 testers, aged 21–56, were engaged. Based on the evaluation of their personal experience and the feedback they provided, a case can be made that the model "Virtual Hotel" provides a very realistic form of immersion. In fact, many testers outlined how surprised they were by the realism of the scene and the naturality of the ambience. Importantly, the quality of the experience evaluated by the testers was aided by the quality of the 3D. The latter eliminated the problem of cyber sickness. Indeed, none of the individuals testing the app had to quit training due to cyber sickness.

For trained persons, but also for associate students, the feedback provided by the teacher, as well as the discussion that develops over individual communication situations, is of great importance. It is excellent if people with direct experience in the hotel industry also take part in the training. It is only necessary to select experienced mentors for this role who will not be too strict, so as not to discourage students or contest them even in the case of "childish" mistakes. It is advisable to work with practitioners who have experience in training new staff in hotels.

With this type of training, it is not possible to evaluate the trained persons too strictly; there is always a need for a large degree of empathy and helpfulness on the part of the evaluators, and an effort to provide very nice and fundamentally positive feedback.

The advantage of simulations in a VR-enhanced environment consists of the fact that a given situation can be repeated. Also, prospective participants/trainees can join the exercise from anywhere. Considering the lack of resources, increasing numbers of students, and public health concerns, factor which frequently make internships, as a form of training, impossible, the VR "Virtual Hotel" model offers a way of navigating the challenge [6]. As a result, irrespective of the existing constraints, students still get the opportunity of as-close-to-real-as-possible training.

Telling and involving the student in the story is important for the player's interest and motivation in performing the requirements in game. This can be performed by creating a scenario. It has already been mentioned above that stories can be divided into explicit and implicit—strictly given and looser, but both types must lead the player to the goal, and thereby meet the learning objectives.

5 Conclusion and Future Work

The objective of this paper was to elaborate on the value added of a VR-enhanced "Virtual Hotel" application. The latter was developed as a means of reconciling three challenges higher education, and especially hospitality management programs face today, including shrinking resources, rising number of students, and public safety concerns. Based on recognized, three-tiered methodology, focused on the development of communication skills, the model thus developed was tested by 40 individuals. The results prove that the mode of training and the skills thus emulated on prospective hotel employees meet the objectives and the needs of the hospitality industry at large. Through the VR-enabled deep immersion in the hotel realm, the "Virtual Hotel" app offers a viable addition to traditional forms of training employed in the hotel industry. The value of the VR-enhanced training tool will increase in the future. This is due to the fact that the current generation of university students is particularly prone to use VR-enhanced tools. More research is needed to enhance the "Virtual Hotel" in view of equipping it with additional virtual spaces and related situational scenarios. As the model expands, it is equally important to train the instructors to employ this VR-enhanced tool as effectively as possible. A system of rewards and fines needs to be implemented in the application too, both to increase the motivation of the trainees and to increase the stress demands that come along this profession.

Acknowledgements This content of this article reflects the outcomes of research conducted with the financial support of the project of the Technology Agency of the Czech Republic No. TL04000153, “Application of virtual reality tools in the training of communication skills of hotel staff in order to alleviate concerns and risks associated with the spread of Covid 19”.

References

1. Frisby, B.N., Kaufmann, R., Vallade, J.I., Frey, T.K., Martin, J.C.: Using virtual reality for speech rehearsals: an innovative instructor approach to enhance student public speaking efficacy. Virt. Real. **32**, 21 (2020)
2. Menzel, K.E., Carrell, L.J.: The relationship between preparation and performance in public speaking. Commun. Educ. **43**, 17–26 (1994). https://doi.org/10.1080/03634529409378958
3. Metusalem, R., Belenky, D.M., DiCerbo, K.: Skills for Today: What We Know about Teaching and Assessing Communication. Pearson, London (2017)
4. Hazel, M., McMahon, C., Schmidt, N.: Immediate feedback: a means of reducing distracting filler words during public speeches. **23**, 29 (2011)
5. Duffek, V., Fiala, J., Hořejší, P., Mentlík, P., Polcar, J., Průcha, T., Rohlíková, L.: Pre-service teachers’ immersive experience in virtual classroom. In: Visvizi, A., Lytras, M.D., Aljohani, N.R. (eds.) Research and Innovation Forum 2020, pp. 155–170. Springer International Publishing, Cham (2021)
6. Verkuyl, M., Lapum, J.L., St-Amant, O., Hughes, M., Romaniuk, D., Mastrilli, P.: Designing virtual gaming simulations. Clin. Simul. Nurs. **32**, 8–12 (2019). https://doi.org/10.1016/j.ecns.2019.03.008
7. Dieker, L.A., Rodriguez, J.A., Lignugaris/Kraft, B., Hynes, M.C., Hughes, C.E.: The potential of simulated environments in teacher education: current and future possibilities. Teach. Educ. Spec. Educ. **37**, 21–33 (2014). https://doi.org/10.1177/0888406413512683
8. Badilla Quintana, M.G., Vera Sagredo, A., Lytras, M.D.: Pre-service teachers’ skills and perceptions about the use of virtual learning environments to improve teaching and learning. Behav. Inform. Technol. **36**, 575–588 (2017). https://doi.org/10.1080/0144929X.2016.1266388
9. Verkuyl, M., Lapum, J.L., Hughes, M., McCulloch, T., Liu, L., Mastrilli, P., Romani-uk, D., Betts, L.: Virtual gaming simulation: exploring self-debriefing, virtual de-briefing, and in-person debriefing. Clin. Simul. Nurs. **20**, 7–14 (2018). https://doi.org/10.1016/j.ecns.2018.04.006
10. González Martínez, J., Camacho Martí, M., Gisbert Cervera, M.: Inside a 3D simulation: realism, dramatism and challenge in the development of students’ teacher digital competence. AJET (2019). https://doi.org/10.14742/ajet.3885
11. Collins, A.: Toward a design science of education. Technical Report No. 1 (1990)
12. Brown, A.L.: Design experiments: theoretical and methodological challenges in creating complex interventions in classroom settings. J. Learn. Sci. **2**, 141–178 (1992). https://doi.org/10.1207/s15327809jls0202_2

13. Cobb, P., Confrey, J., diSessa, A., Lehrer, R., Schauble, L.: Design experiments in educational research. Educ. Res. **32**, 9–13 (2003). https://doi.org/10.3102/0013189X032001009
14. Reeves, T.: Design research from a technology perspective. Educ. Des. Res. 52–66 (2006)
15. Muccini, H.: Detecting implied scenarios analyzing non-local branching choices. In: Pezzè, M. (ed.) Fundamental Approaches to Software Engineering, pp. 372–386. Springer, Berlin Heidelberg, Berlin, Heidelberg (2003)

Holocaust Education in Digital Media

Marek Bodziany and Justyna Matkowska

Abstract The article aims to analyze the digital forms of education about Sinti and Roma persecution during World War II, its challenges. Digital media has increasingly become a valuable tool used by educators, academic lecturers in Holocaust, Genocide studies education, including the Sinti and Roma. The paper discusses various digital commemoration forms, digital mapping discovering historical sites online, virtual reality, its application in Holocaust education, digital storytelling, testimonials archives, digital archives, social media. Digital media, its application in Holocaust education might be challenging for educators. However, innovative technologies create a range of various educational tools, practices, diverse approaches, which might significantly improve the process of learning about the Holocaust, Genocide. Digital media has vast potential in Holocaust education. Using this framework allows for identifying, discussing digital media in education about the Sinti and Roma Genocide.

Keywords Digital media · Holocaust · Education

1 Introduction

Holocaust education in digital media in the case of Sinti and Roma persecution is becoming a topic of discussion among scholars, educators, activists, and Roma youth. The young generation of Sinti and Roma is becoming increasingly aware and interested in deepening their knowledge and promoting Romani history, including the Nazi persecution of Sinti and Roma during WWII.

M. Bodziany (✉)
The General Kosciuszko Military University of Land Forces, 51-151, Wroclaw, Poland
e-mail: marek.bodziany@awl.edu.pl

J. Matkowska
Department of History, University at Albany, State University of New York, 1400 Washington Avenue, Albany, NY 12222, USA
e-mail: jmatkowska@albany.edu

Centre for Migration Studies, Adam Mickiewicz University of Poznan, Poznan, Poland

A. Visvizi et al. (eds.), *Research and Innovation Forum 2022*,
Springer Proceedings in Complexity, https://doi.org/10.1007/978-3-031-19560-0_36

In the case of the research on the Sinti and Roma Genocide, it should be emphasized that it is challenging to clarify the exact number of victims and percentage of the European Romani community who was murdered during WWII [1]. Estimated and "symbolic figure" of Roma and Sinti victims of the Nazi Genocide is half-million [2]. Sinti and Roma Genocide during the Nazis period is often referred as the "Forgotten Genocide."

Taking that into account, the goal of many Roma scholars, educators, and activists, including young members of Roma groups, is to educate the public about Sinti and Roma's tragic experience and persecution under Nazi oppression during WWII. The trends of using digital media in Holocaust education turned out to be a crucial tool in teaching about the Sinti and Roma Genocide and promoting the tragic fate of Romani people among Roma and non-Roma society.

In recent years, digital media has become a key tool in Holocaust and Genocide studies education, including the education on Nazi persecution of Sinti and Roma. The process of learning about the Holocaust has improved thanks to innovative technologies, various educational tools, and public access to resources and practices. The access to diverse approaches and digital media tools uncovered the vast potential of Holocaust education.

Through an increasing number of educational platforms and programs, institutions can fulfill their educational role and, at the same time, test innovative technology in order to future improvement of recipients' experience, especially for young participants (including Sinti and Roma). Holocaust education is no longer necessary and not the most non-exciting material to learn for the young audience, but it has an engaging and explorative character. The reason Holocaust education enjoys great popularity is that the process of learning is now more individualized, interesting, accessible, engaging, and encouraging.

2 Methodology

This study identifies the impact of the usage of digital media to educate and raise awareness of the Sinti and Roma Holocaust. The main premise for undertaking this subject of research was the assumption that there is a significant deficit of knowledge about the martyrdom of the ethnic groups studied and that digital media may become a tool to strengthen social awareness on this subject. Therefore, a research problem has been specified, the essence of which focuses on the question of the impact of digital media on education in the field of the Roma and Sinti Holocaust. For the purposes of the research, we adopted a qualitative method to analyse the findings. Therefore, research was based on the analysis of sources (literature and secondary sources) as well as analysis of the content contained in digital media.

It should be emphasized that the presented paper dedicated to the education of Sinti and Roma Holocaust is not an exhaustive study because Holocaust also applies to other groups such as Jews, Jehovah's Witnesses, black people, Slavs, Homosexuals, disabled people, political prisoners, Russian POW's, etc.

3 Holocaust Education in Digital Media

Currently, Holocaust museums, as well as research and educational institutions are using the potential of digital technologies in order to pursue their cultural and educational tasks. Institutions create diverse media materials to communicate memorial and educational content. Social media content and user-friendly technology reach a broad range of people and help to access historical and scientific knowledge, raise awareness of Genocide, and increase Holocaust knowledge [3]. On the other hand, the Internet also plays a negative role in Holocaust distortion and trivialization [4].

Digital practices of Holocaust education are a constantly evolving phenomenon "involving relationships between computational logics, programs and software, interfaces, users, and wider traditions of media representation, museological, archival and memory practices, and emerging characteristics of digital cultures" [5]. Digital media has impacted the way of thinking about the Holocaust, including education, research, commemoration, virtual reality, and digital Holocaust memory. Researchers, educators, and lecturers often become producers of digital memory and new educational platforms based on modern approaches, methodologies, and technologies [5].

3.1 Digital Commemoration

Digital commemorative practices offer expanded visibility and new technical capabilities, and at the same time, they raise the issues of long-standing concepts of commemoration. The transfer from in-person commemoration to social media platforms caused by the Covid-19 pandemic created new, modern approaches to commemoration, which adopted the culture of commenting and sharing in order to commemorate the Holocaust from a distance. Digital commemoration cannot replace or be fully compared to a personal encounter with the Holocaust site or survivor. However, virtual memorials, virtual tours, videos, and photos have refined Holocaust commemoration.

The various forms of digital commemoration created a commemorative sphere that is based on platforms such as YouTube, Instagram, Facebook, and Twitter. The content of particular activities is interrelated and linked by the usage of hashtags. Because of the pandemic, museums and memorials established a new way to use the digital media for commemorational goals and virtual tours in order to "reduce distance and strive for a sense of presence at the sites of Holocaust commemoration in times of social distancing" [6].

In the case of virtual commemoration of Sinti and Roma victims of the Nazi Genocide, we should emphasize that the main events take place on May 16th and August 2nd. In 2015 European Parliament declared August 2nd as Roma Genocide Remembrance Day [7], which is commemorated annually and has an international character. Its symbolism is associated with the remembrance of the liquidation of *Zigeunerlager* (Romani family camp) in Auschwitz-Birkenau, which took place on August 2nd,

1944 [8]. During the Covid-19 pandemic, official August 2nd commemoration events have been organized and promoted on digital media platforms.

To meet the expectation, in 2020 [9] and 2021 [10], the "Virtual Commemoration Ceremony" has been organized by Central Council of German Sinti and Roma. There are also Sinti and Roma commemoration website [11] which contains a range of resources, statements, testimonies, videos, and various materials. Another important platform is the dedicated website *2August.eu* [12], which contains materials such as links to local Commemorations across Europe, a *Virtual Name-Reading Ceremony*, and statements of Young Roma and non-Roma who commemorate Sinti and Roma Genocide and stand against racism.

During the Covid-19 pandemic, museums, memorials, and educational centers had to cope with the challenges of restrictions and the development of new practices [13]. The digital commemoration has provided new standards for modern and engaging Holocaust education, which is accessible to everyone from all over the world and decreases the physical distance. Thanks to the pandemic, digital activities of museums and memorials moved beyond the existing commemorative practices of social media and transferred into the environment with reduced distance and no limits in access and presence at the commemoration of Holocaust places. In the post-Covid-19 era, Zoom and digital video platforms might develop and provide new and more interactive forms of commemoration.

3.2 Digital Mapping and Discovering Historical Sites Online

Digital, interactive maps of historical sites and places of memory across Europe are a particularly valuable tool, which provides a topography of the terrain and a comprehensive illustration of history. Those online platforms are useful for discovering sites that commemorate sites of Genocide and resistance, including the Sinti and Roma Genocide and resistance.

The most recent platform was created by Yahad In-Unum, which is an international organization raising the consciousness of the sites of Jewish and Roma mass executions by Nazi killing units in Eastern Europe. The interactive map of "Holocaust by bullets" contains more than 1900 documented execution sites as well as over 1100 sites available for consultation [14].

Another great interactive source dedicated only to Sinti and Roma is the map of sites of resistance, which is useful for discovering European places of Romani Resistance acts and events. The map is part of the project undertaken by the European Roma Institute for Arts and Culture [15]. There are also a few other maps, such as a list of memorial sites of Sinti and Roma [16], or Map of Stolpersteine –Gunter Demnig, which commemorates the victims of National Socialism by installing commemorative brass plaques on the pavement in front of their last address of choice [17].

Digital mapping and active exploration of historical places and of memory with the usage of virtual and augmented reality technology accessible online represent an especially useful, widely available, and effective pedagogical tool in Holocaust History and cultural education. The learning method of active, virtual exploration has been proved more effective than education based on exhibition. Also, the method based on combining historical knowledge with a spatialized environment is much more effective than using a spatial presentation alone. In conclusion, the education-based digital mapping and discovery of historical sites online are significantly improving the process of learning through individual, interactive, and engaging exploration of historical sites [18].

3.3 Virtual Reality and Its Application in the Holocaust Education

Most recent, modern digital tools allowed broad and public access to 3D models of historical sites. Virtual reality can offer numerous opportunities for education and commemoration of the Holocaust and historical memorial sites. The concept behind the virtual reality projects is to reconstruct the sites and learn about their history through digitized documents and photos. The recent project of 3D reconstruction of the Bergen-Belsen concentration camp [19] or virtual tour of Memorial and Museum Auschwitz-Birkenau [20]. Those platforms focus on Holocaust from a broader educational perspective. So far, has not emerged virtual reality platforms dedicated only to Sinti and Roma persecution.

It can be argued that augmented reality has better educational value than virtual reality because of the physical immersion within the environment and stimulation of certain senses, such as being able to smell the environment and touch. In addition, the museum tour includes a live tour guide, who can provide background and answer asked questions compared to the programmed guide of virtual reality, which has limited and pre-recorded narrative and answers [21].

Virtual reality in education on the Holocaust has been proven a great learning method. On the other hand, the usage of technology to visualize Holocaust images carries the risks of potential ethical challenges and implications related to exposing users—including young students—to dramatic images and traumatic stories and events imagery [22]. It should be emphasized that in the case of Romani culture, the traumatic experience and abuse are the topic of cultural taboo, so the Romani students would not take part in this specific method of education, which includes the topics of taboo.

3.4 Testimonial Archives and Digital Storytelling of the Narrative of the Holocaust

Testimonial archives and digital storytelling of the narrative of the Holocaust are highly effective methods of teaching about WWII and the Holocaust [23]. From the end of World War II to 2010 were collected approximately 100,000 Holocaust survivors' testimonies [24]. Testimonies of Sinti and Roma represent the small number of testimonies; e.g., among over 55,000 testimonies collected by the USC Shoah Foundation, there are only 406 Sinti and Roma survivors' accounts. In addition, the opportunities for live Holocaust survivor testimonies are decreasing year by year.

One of the biggest sources of knowledge and most comprehensive archive of National Socialism victims and survivors is the international center on Nazi persecution—Arolsen Archive (formerly called International Tracking System). The collection contains information and data about over 17.5 million victims of Nazi persecution, including Sinti and Roma [25].

Another popular platform is IWitness—the free educational platform of the USC Shoah Foundation, which contains stories of survivors and witnesses of Genocide [26]. In addition, Project IWalks [27] provides access to USC Shoah Foundation's Visual History Archives (VHA) video testimonies of Holocaust survivors and victims and local history walks in order to teach about racism and democracy. The goal is to present the regional stories and allow students to use digital technology while learning about their community.

In the case of testimonials and archives is important to mention the project RomArchive, particularly the pilot section "Voices of the Victims, "which focuses on Sinti and Roma victims. It is an archival project focused on the collection of narratives from the victim's perspective [28]. Another important web-based collection of video and audio educational materials about the History of Sinti and Roma Genocide and persecution during the Nazi era is portal *Tajsa.eu,* which is produced in 12 European regions [29].

Digital archives and testimonies have unlimited potential in Holocaust education. Testimonies and stories of survivors represent important part of the Holocaust narrative and there are available at major and local archives. Holocaust education resources include virtual interactive testimonies, documentary films, and popular culture.

4 Conclusions

In the Covid-19 pandemic and lockdown, digital media has become a space for memory, commemoration, and education about the Holocaust, including the Sinti and Roma Genocide. Technology has provided innovative solutions to meet the current expectations as well as distance and safety challenges related to the Covid-19 pandemic. Lockdown contributed to the popularization of digital technology in

Holocaust education and increased the use and functions of the Internet in this matter. Thanks to diverse educational techniques and tools, Holocaust education and commemoration are accessible to recipients from all over the world. Virtual commemoration and meetings, as well as Holocaust history lessons and research, are publicly available on many platforms and accessible to people of all ages. Technology has also positively impacted the collaboration and performance among scientists and educators, including the research on the Sinti and Roma Genocide during WWII, which is still underrepresented in Genocide studies discourse.

The rapid development of interactive approaches to museums and memorials—caused among other by the Covid-19 pandemic—play an important educational role. Digital media established "freedom" to visit without the time and place limits so that online visitors' experience is individual and personal. People who have already visited the sites of the Holocaust are eager to expand their knowledge, and for those who have never been and seen it, virtual visiting can encourage and inspire them to deepen their knowledge and broaden their horizons.

Current development and implementation of high-quality, engaging programs and curricula for primary, secondary, and university students, as well as the general public offered through the use of various digital media platforms, has a positive impact on communication and participation in Holocaust education. Despite the historical topic, virtual access also made it easier to reach young people and engaged them in the process of learning about the History of the Holocaust in an attractive and access-free way. It should also be noted that for Romani participants, the Holocaust education is at the same time a Historical and cultural lesson for Sinti and Roma and a unique opportunity to gain specific knowledge of the persecution of ancestors.

In general, Holocaust education based on digital media tools usage scientifically increased the knowledge about Sinti and Roma persecution during WWII. Moreover, the increased presence of Sinti and Roma Genocide discourse in Holocaust education might turn out to be a positive aspect of lockdown and the popularization of digital technology caused by the Covid-19 pandemic.

Over the last decades, digital media has continued the process of raising awareness, educating, and informing the major society (including the young generation) about the Sinti and Roma Genocide during the Nazi period. In addition, digital media impact the growing research and knowledge production of the Sinti and Roma Genocide. Digital media helps to promote the Romani people's historical experience in Nazi-occupied Europe and Nazi concentration and extermination camps from World War II.

The positive impact of digitization is that participants can e-visit and get to know and Holocaust site and take part in commemoration without leaving home, incurring costs, or standing in lines for a ticket. It is also essential that thanks to digital media, online participants are not deprived or excluded from having contact with the survivor, appropriately commemorating victims as well as being able to see the exhibit, testimony, photographic images, and artifacts. Similar to in-person lectures, digital media Holocaust education provokes students and participants to reflect on WWII Genocide and its aftermath. The goal of digital media education for virtual and in-person visitors and participants is that the long-term memories such as images, videos, and testimonies will remain in the consciences of society, and Holocaust, including the Sinti and Roma persecution, will never be forgotten.

References

1. Kapralski, S.: Roma Holocaust: the end of silence. In: Mirga-Kruszelnicka, A., Acuña E.C., Trojański, P. (eds.) Education for remembrance of the Roma Genocide. Scholarship, commemoration, and the role of youth, p. 43. Libron, Cracow (2015)
2. Kenrick, D.: The genocide of the gypsies: what we now know and what we still don't know. In: Schulze, R. (eds.) The Holocaust in History and Memory, vol. 3, p. 28 (2010)
3. Manca, S.: Digital memory in the post-witness era: how holocaust museums use social media as new memory ecologies. Information **12**(1), 31, 2–4 (2021)
4. Bauer, Y.: Creating a "Usable" past: on holocaust denial and distortion. Israel J. For. Aff. **14**, 209–227 (2020)
5. Walden, V.G.: Defining the digital in digital holocaust memory, education and research. In: Walden, V.G. (eds.) Digital Holocaust Memory, Education and Research. Palgrave Macmillan, Cham (2021)
6. Ebbrecht-Hartmann, T.: Commemorating from a distance: the digital transformation of holocaust memory in times of COVID-19. Media Cult. Soc. **43**(6), 1095–1112, 1109 (2021)
7. Press Releases: MEPs urge end to Roma discrimination and recognition of Roma Genocide Day. https://www.europarl.europa.eu/news/en/press-room/20150414IPR41851/meps-urge-end-to-roma-discrimination-and-recognition-of-roma-genocide-day. Accessed 5 Apr 2022
8. Sinti and Roma (Gypsies) in Auschwitz. http://auschwitz.org/en/history/categories-of-prisoners/sinti-and-roma-gypsies-in-auschwitz/. Accessed 4 Apr 2022
9. Virtual Commemoration Ceremony 2020. European Holocaust Memorial Day for Sinti and Roma. https://www.youtube.com/watch?v=1-8__KSyaYs. Accessed 6 Apr 2022; Preview of commemoration, The European Holocaust Memorial Day for Sinti and Roma 2020. https://www.youtube.com/watch?v=VTlkekHwGwc. Accessed 6 Apr 2022
10. Virtual Commemoration Ceremony 2021. European Holocaust Memorial Day for Sinti and Roma. https://www.youtube.com/watch?v=cPRPfLkPPUs&t=4s. Accessed 6 Apr 2022
11. Sinti and Roma Commemoration website. https://www.roma-sinti-holocaust-memorial-day.eu/. Accessed 6 Apr 2022
12. Dikh He Na Bister. https://2august.eu/. Accessed 6 Apr 2022
13. Bodziany, M., Matkowska, J.: Digitization of material culture resources and its impact on poles' participation in "Cyberculture" during the Covid-19 pandemic. In: Visvizi, A., Troisi, O., Saeedi, K. (eds.) Research and Innovation Forum 2021. Springer, Cham (2021)
14. Holocaust by bullets. https://yahadmap.org/#map/. Accessed 6 Apr 2022
15. Interactive map of sites of Romani Resistance. https://eriac.org/roma-resistance-map/roma-resistance-map. Accessed 6 Apr 2022
16. Memorial sites of Sinti and Roma. https://verortungen.de/gedenkorte/. Accessed 6 Apr 2022
17. Map of Stolpersteine–Gunter Demnig. https://stolpersteinmap.de/#13/53.5686/9.9841. Accessed 6 Apr 2022
18. Blancas, M., Wierenga, S., Ribbens, K., Rieffe, C., Knoch, H., Billib, S., Verschure, P.: Active learning in digital heritage: introducing geo-localisation, VR and AR at holocaust historical sites. In: Walden, V.G. (ed.) Digital Holocaust Memory, Education and Research, pp. 145–176. Palgrave Macmillan, Cham (2021)
19. Memory in a digital age: a virtual reconstruction of Bergen-Belsen, Wiener Holocaust Library in London. https://wienerholocaustlibrary.org/exhibition/memory-in-a-digital-age-a-virtual-reconstruction-of-bergen-belsen/. Accessed 6 Apr 2022

20. Auschwitz-Birkenau–virtual tour. https://panorama.auschwitz.org/tour1,en.html. Accessed 6 Apr 2022
21. Challenor, J., Minhua, M.A.: Review of augmented reality applications for history education and heritage visualisation. Multimodal Technol. Interact. **3**(2), 39 (2019)
22. Gardner, E.: Does a VR Auschwitz simulation cross an ethical line? https://www.alphr.com/life-culture/1007241/does-a-vr-auschwitz-simulation-cross-an-ethical-line. Accessed 3 Apr 2022
23. Marcus, A.S., Maor, R., McGregor, I.M., Mills, G., Schweber, S., Stoddard, J., Hicks, D.: Holocaust education in transition from live to virtual survivor testimony: pedagogical and ethical dilemmas. Holocaust Studies. J. Cult. Hist. **28**(3), 279–301 (2021)
24. Greenspan, H.: Survivors' accounts. In: Hayes, P., Roth, J.K. (eds.) The Oxford handbook of holocaust studies, pp. 414–427. Oxford University Press, Oxford (2010)
25. Arolsen Archive. https://arolsen-archives.org/en/. Accessed 8 Apr 2022
26. Portal IWitness. https://iwitness.usc.edu/sfi/. Accessed 8 Apr 2022
27. Portal IWalks. https://www.pamatnik-terezin.cz/iwalks. Accessed 8 Apr 2022
28. RomArchive. https://www.romarchive.eu/en/voices-of-the-victims/voices-victims-infotext/. Accessed 8 Apr 2022
29. Web-based collection of video/audio podcasts and educational resources of Sinti and Roma Genocide. tajsa.eu. Accessed 8 Apr 2022

Reinforcement Learning Through Gamification and Open Online Resources in Elementary School

María de los Ángeles Tárraga Sánchez, María del Mar Ballesteros García, Héctor Migallón, and Otoniel López Granado

Abstract We are now in the post-confinement phase, a confinement decreed by the authorities in many countries due to COVID-19, which has led to the suspension of face-to-face teaching at all educational stages. Educational institutions and teachers have worked together to provide online teaching to enable students to achieve the required competencies at each educational level during the period of confinement. In this paper we show an online education proposal focused on preschool education that began to be developed in the period of confinement, for students between 3 and 5 years old. The adaptation and successful use of this proposal for the post-confinement period is presented, in which the return to face-to-face education has taken place. This proposal is a significant part of a curriculum design conceived as a game-based project in which the part related to computer games is presented here. It is important to note that this proposal focuses on generations that are already digital natives, so it is necessary to pay attention to both the content and the design of the proposed computer games, as well as to maintain the motivation of the students. The methodology presented allows for live resources that evolve along with the trainees.

Keywords Gamification · Open resources · 3–5 years-old · Online learning

M. de los Ángeles Tárraga Sánchez · M. del Mar Ballesteros García
CEIP Bec de L'Águila, San Vicente del Raspeig, E-03690, Alicante, Spain
e-mail: ma.tarragasanchez@edu.gva.es

M. del Mar Ballesteros García
e-mail: mm.ballesterosgarc@edu.gva.es

H. Migallón (✉) · O. López Granado
Department of Computer Engineering, Miguel Hernández University, 03202 ElcheAlicante, Spain
e-mail: hmigallon@umh.es

O. López Granado
e-mail: otoniel@umh.es

A. Visvizi et al. (eds.), *Research and Innovation Forum 2022*,
Springer Proceedings in Complexity, https://doi.org/10.1007/978-3-031-19560-0_37

1 Introduction

Distance teaching is not a new concept, although it is still a very current one. For example, the UNED, Spain's national university of distance education (http://www.uned.es), was created in 1972. In the same year, Moore [1] defines distance education as the type of instructional method in which teaching practices occur separate from the learning process, so that communications between teacher and student must take place by means of printed, electronic, mechanical, or other techniques. In [2], authors, endorsed by UNESCO (United Nations Educational, Scientific, and Cultural Organization), establish that distance educational system must facilitate the participation of all those who want to learn without imposing traditional admission requirements and without obtaining an academic degree or any other certificate, and the system must be able to overcome the distance between teaching staff and students, using this distance as a positive element for the development of learning autonomy. In [3], O. Peters defines distance education as a method of imparting knowledge, skills and attitudes which requires the division and organization of the work to be done, and should make intensive use of technical resources, especially for the reproduction of high-quality teaching material, which can instruct large groups of students at the same time in remote areas. This method is defined as an industrial form of teaching and learning.

On the other hand, the concept of e-learning, directly related to distance education or online learning, refers to the method of content dissemination and rapid learning using the resources provided by information technology and the Internet. In some sectors, such as secondary and higher education, business, and non-formal training, it refers to the flexible delivery of content and programs over the Internet that focus on supporting communities of practice. This article is based on the successful work designed ad-hoc to be able to continue with the learning process, in the period of compulsory confinement, of children from 3 to 5 years old. This work continues to be useful in normal circumstances, both inside and outside the classroom. The tools used, which are fundamental to the success of the work developed and their availability to the global educational community, will also be discussed. The proposal presented has high degree of dynamism and great adaptability. It must be considered that we are discussing interactive multimedia resources developed by early childhood education teachers.

2 Objectives

It is evident that long hours of continuous study without breaks lead inexorably to the fact that learners detest learning. One way to avoid this consequence, especially at preschool age, is to use games, which on the one hand allows the learner to have fun but, on the other hand, if the game is properly designed, it allows to continue learning along the route designed by the teacher. This technique could be considered as the creation of effective teaching and learning environments to enhance the traditional approach. Our proposal is based on the design of computer games that permit an easy adaptation to effectively break the group dynamics without interrupting the learning process, and on the promotion of the use of these games outside the classroom to consolidate knowledge in a funny way. Game-based learning, although not infallible, is one of the recommended approaches to enhance interest and motivation to learn in general [4], and especially for children of preschool age [5, 6]. Usually game-based learning in a real environment involves children in a game world, which allows them to interact with the learning material while motivating them to improve their knowledge and skills through competitive activities with rules, objectives, feedback, interactions, and results [7]. These objectives remain unchanged in our proposal, considering that the learning material is the Information and Communication Technology (ICT) resources available both in the classroom and at home. Although our proposal can be used with mobile devices, such as smart phones or tablets, it is not considered a primary target. It is true that there are hundreds, or thousands of educational proposals based on computer games, but it is also true that the educational community as a whole is beginning to question the content, relevance and suitability of these proposals [8]. In fact, it is known that we cannot isolate children from technology, but we must make sure that it does not harm them [9] because they are digital natives who live surrounded by this technology and apps [10]. Our proposal allows us to cover two relevant goals, on the one hand to control the content consumed by children and on the other hand to ensure the relevance of the content. It is important to note that these contents have been designed and specially adapted to promote the learning path designed by the teacher, in some cases at a group level and in others at an individual level.

It is very important to bear in mind that many applications that include the concept of educational in their title do not respond to the developmental needs of the age group they are aimed [8]. Several studies [8, 11] suggest lines of research for the design of resources to be used in early childhood education environments, which is in line with our proposal, since the design of resources is carried out by the teacher of the learner who will use them.

We should not ask ourselves whether technology also belongs in early childhood education, but rather how we can efficiently leverage it for the education of young children [12]. In this context, educators, developers, and designers must ensure that resources intended for young children have a solid theoretical basis and follow high quality standards to effectively contribute to the developmental progress of young children. In this context, our proposal offers a perfect collaboration between these three actors, since they all converge in the teacher, who performs the tasks of development and design in addition to his or her own task as an educator.

Many studies over the past decades have shown that learner-centered design addresses the need for learner engagement [13, 14], and this is not unrelated to preschool education [15]. Our proposal aims to design resources that also meet this objective. In fact, it is an easily achievable objective, since the learners can take part in the design of the resources, either in the design stage or in the subsequent adaptation stage. This makes it easier for all members of the group to feel protagonists and therefore linked and motivated to their use.

3 Platforms and Tools Analysis

The proposal presented in this article is part of a curricular model that has been under development for years, and part of this curricular model is a game-based model, including the proposal and other additional resources. The original model required an adaptation, due to the confinement, which involved the development of many resources such as computer games. This adaptation obviously cannot be done with all the resources used in face-to-face teaching, but it could be done with some of the resources, and the result is a very powerful and useful tool. The original set of resources developed can be considered, and were developed with that concept, as learning objects. While it is admittedly true that there is no clear definition of learning objects [16], since learning objects vary both in terms of size and scope, content, design, and technical implementation, and therefore pinpointing the essence of a learning object is not an easy task. One thing that most descriptions have in common is that they focus on how learning objects are created, used, and stored, rather than what they look like. One perspective, which fits in with our proposal, is that which compares learning objects to LEGOTM building blocks, i.e., learning objects are small units that can be fitted together and organized in many ways to produce personalized learning experiences [17]. As mentioned above, the proposal presented is composed of a set of learning objects, or more specifically a set of Virtual Learning Objects (VLO). The VLOs become as facilitators of learning processes in different levels of education, used for sharing and reusing content in any area of knowledge. In fact, it is one of the most widely used solutions to achieve reusability, accessibility, durability, and interoperability of educational resources [18].

As previously said, the COVID-19 pandemic caused educators, from preschool to university level, to unexpectedly move their teaching practice to online environments. The challenges to be solved in this online education are well documented: feelings of disconnection or the difficulty of responding to individual needs, for example. These challenges, suddenly supervened by the pandemic, occurred for all educators and learners, who were suddenly immersed in this environment and were likely unfamiliar with its possibilities.

As the work done during the pandemic by the educators in this proposal progresses, important issues related to MOOCs emerge, even though the work pills are designed primarily as games. It should be noted that teachers must be designers, developers, and facilitators of learner-centered, face-to-face educational experiences, and that they become designers and developers who see MOOCs as a way to foster similar learning experiences in an online environment. Other than the MOOC model, there are many models and frameworks that aim to provide teaching or learning to individuals who are distant, but also to those with individual limitations to access knowledge (spatial, temporal, technological, psychosocial, and socio-economic limitations). As a result, several online learning systems have been developed. Among them we can mention Adaptive Hypermedia Systems (AHS), Learning Management Systems (LMS), Virtual Learning Environments (VLE), Knowledge Management Systems (KMS), and very recently MOOCs and Learning Experience Platforms (LXP). None of these models fits the needs of our proposal because our proposal combines several features of game-based learning and MOOCs, therefore we named our proposal as MOOG2L (Massive Online Open Games to Learn). In addition, none of the online systems considered meet the requirements for teachers to be designers, developers, and facilitators of learner-centered instruction at the same time. Perhaps the type of system most like our needs would be the LXP, but with the handicap of lacking tools that allow the design and development of resources without the need for skills in any environment or programming language. Then, the selected tool that combines all the capabilities required by our project is Genially (https://genial.ly/), which cannot be catalogued in any of the online systems listed before but allows us to have a model similar to the MOOC and LMS models, and at the same time allows us to be developers of the computer games that compose our proposal. It is therefore the basic tool to be able to develop our proposal, and above all it is the tool that allows us to adapt all our proposals to the individual and group needs of the learners.

The chosen tool is important because it overcomes the technological impediments that usually prevent the design of resources as desired by teachers. In addition, once designed, it allows to analyze the performance, mainly in face-to-face teaching, and to improve them, by the teachers themselves, without high development costs.

It is necessary to emphasize that it is a part of a game-based project that makes use of other technologies and advanced teaching resources. For example, includes resources designed with Makey-Makey (https://makeymakey.com/#), which is a simple device implementing the Human Interface Device (HID) protocol easily connected to any computer using a USB port, where no additional software or drivers are required, but it requires computational programming skills using Scratch. Games have also been developed using educational robots, in particular the "Bee-bot" (https://www.terrapinlogo.com/bee-bot-family.html), which does not require special programming kills, but the catalog of games to be developed is much more limited. Many other non-interactive resources have been designed using Canva (https://www.canva.com/), and some others.

4 Proposal

As mentioned, we define our proposal as a MOOG2L (Massive Online Open Games to Learn) system, designed, developed, and offered by teachers using Genially to their group of students between 3 to 5 years old. The relevance of our proposal is based on two main aspects, the first one is that the set of resources developed (two of them are the ones presented here) can be freely used and adapted, and the second one is that the process of development and/or adaptation of these computer games does not require programming knowledge. Aspects prior to development, such as the conceptualization or design of these resources are part of the study and planning of the complete curricular design, so they will not be analyzed here because they are beyond the scope of this article.

As said, in this paper just two resources of all available ones will be presented. The examples presented show different learning objectives, different levels of interactivity, different levels of playability and different levels of reusability. It is clear that the starting point of any resource is the learning objective. The first resource, shown in Fig. 1 is based on the well-known game of the goose or the snake and the ladder, where the main objective is the learning of vocabulary. Figure 1 shows the most complete version in terms of number of objectives worked on, which translates into less interactivity. Before explaining the different developments of this game, that is, the different adaptations that are developed according to the age of the students, the learning objectives and the characteristics that define the group, we will slightly explain the functioning of the game as shown in Fig. 1 (version adapted to four-year students), which is available at https://view.genial.ly/6227bc5b58b95f0019c9ae1b/game-joc-de-labecedari. As mentioned above, it is a group game; the basic instructions are explained on the screen itself. The game is designed for a specific group of learners and has been modified to suit the age of the students. In the example shown the group is identified, from the age of 3, with the tiger, as the group is currently 5 years old (last year of the cycle) the tiger that appears in Fig. 1 is already an adult tiger. In our way of working, the tiger is not a pet, and the group identifies themself with this animal, as already mentioned, from 3 to 5 years old and this choice is made by consensus. Each group identifies themself with a different animal, which implies that this resource is adapted for each single group. Adaptation to other groups can be performed quickly. As can be seen, in this version all the letters of the alphabet are included; the player must say a word that begins with the letter of the square in which it has fallen. Different resources decide the complexity and allow or disallow a word containing that letter instead of requiring that it begins with that letter, as well as decide which letters have a higher frequency of occurrence. All these adaptations are quick to implement. This game is introduced in the classroom, using the interactive whiteboard, with the teacher as a catalyst, and mothers, fathers and/or guardians are informed so that the game can be continued at home. Thus, a family member acts as the master of the game. In one of the versions provided to families, some letters are replaced by photos of the students in the group, which allows families to identify all members of the group. After developing a base resource, it is necessary to be able to make adaptations that motivate the learner to continue using the resource and continue the learning process both inside and outside the classroom. It is necessary to bear in mind that the alphabet of the Valencian language includes the compound letter "ny", as it can be seen in Fig. 1.

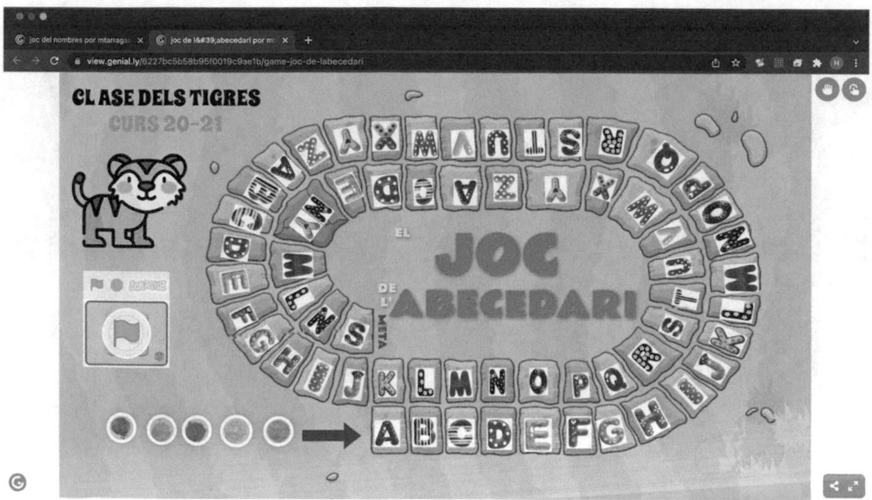

Fig. 1 Resource based on the game of the goose, for vocabulary learning

The second resource presented is an example of a game that includes an objective related to language, discovering new words, but it is also designed as a basic tool for the project developed with the objective of learning what insects are, knowing some of them, and their primordial role in nature. We have created a version of the game that contains screens of the games developed for 3-, 4- and 5-year-old students, available at https://view.genial.ly/6232219fa4a8c800181e4e79/learning-experience-challenges-el-insectes. The game is structured as a flip screen game. The first screen, shown in Fig. 2, is the presentation of the game. The next screen requires ordering the letters of a word, and if the word is composed correctly, by dragging and dropping, they can watch a short video about the corresponding insect and then move on to the next screen. Once the ordering of each word has been completed, it must be checked in order to move forward or retry. As the game is designed as a flip screen game, the game can be switched from one screen to another, thus breaking a possible demotivating routine that discourages the use of the resource. In the second part of the game, the learning objective is changed so that new insects and its nouns are learned.

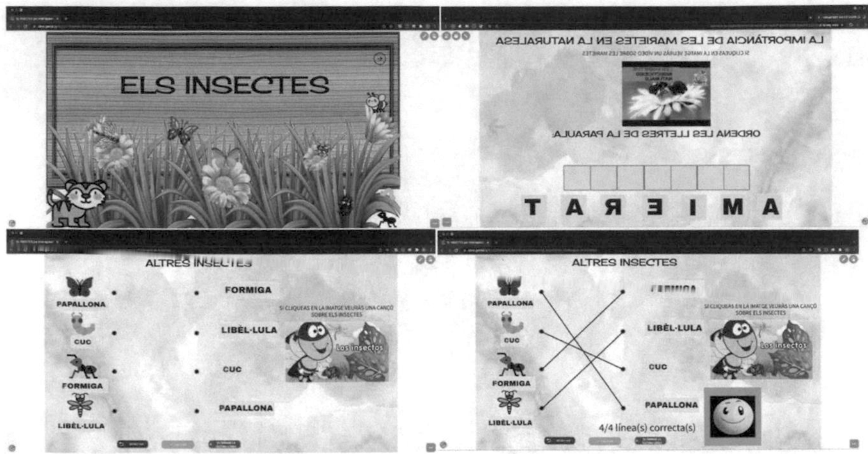

Fig. 2 Insects game

References

1. Moore, M.G.: Learner autonomy: the second dimension of independent learning. In: Convergence, pp. 76–88 (1972)
2. MacKenzie, N., Postgate, R., Scupham, J.: Open Learning: Systems and Problems of Post-secondary Education. UNESCO (1976)
3. Peters, O.: Open Learning: Systems and Problems of Post-secondary Education. Croom Helm Ltd., Beckenham, Kent (1983)
4. Chen, C.H., Law, V.: Scaffolding individual and collaborative game-based learning in learning performance and intrinsic motivation. Comput. Hum. Behav. **55**, 1201–1212 (2016). https://doi.org/10.1016/j.chb.2015.03.010
5. Cojocariu, V.M., Boghian, I.: Teaching the relevance of game-based learning to preschool and primary teachers. Procedia Soc. Behav. Sci. **142**, 640–646. https://doi.org/10.1016/j.sbspro.2014.07.679. The Fourth International Conference on Adult Education, Romania (2014)
6. Erhel, S., Jamet, E.: Digital game-based learning: Impact of instructions and feedback on motivation and learning effectiveness. Comput. Educ. **67**, 156–167 (2013). https://doi.org/10.1016/j.compedu.2013.02.019
7. Kim, B., Park, H., Baek, Y.: Not just fun, but serious strategies: using meta-cognitive strategies in game-based learning. Comput. Educ. **52**, 800–810 (2009). https://doi.org/10.1016/j.compedu.2008.12.004
8. Papadakis, S., Kalogiannakis, M., Zaranis, N.: Educational apps from the android google play for greek preschoolers: a systematic review. Comput. Educ. **116**, 139–160 (2018). https://doi.org/10.1016/j.compedu.2017.09.007
9. Ebbeck, M., Yim, H.Y.B., Chan, Y., Goh, M.: Singaporean parents' views of their young children's access and use of technological devices. Early Childhood Educ. J. **44**, 127–134 (2016)
10. Guernsey, L., Levine, M., Chiong, C., Severns, M.: Pioneering literacy in the digital wild west: empowering parents and educators. Campaign for Grade-Level Reading, Washington, DC (2012)
11. Crescenzi, L., Jewitt, C., Price, S.: The role of touch in preschool children's learning using ipad versus paper interaction. Austr. J. Lang. Literacy **37**, 86–95 (2014)
12. Shapiro, J.: The New Childhood: Raising Kids to Thrive in a Digitally Connected World. Hachette, UK (2019)

13. Norman, D.A., Spohrer, J.C.: Learner-centered education. Commun. ACM **39**, 24–27 (1996)
14. Watson, S.L., Reigeluth, C.M.: The learner-centered paradigm of education. Educ. Technol. 42–48 (2008)
15. Yenawine, P.: Visual thinking strategies for preschool: using art to enhance literacy and social skills. ERIC (2018)
16. Smith, R.S.: Guidelines for Authors of Learning Objects. New Media Consortium (2004)
17. Hodgins, H.W.: Into the future a vision paper. In: Commission on Technology and Adult Learning (2000)
18. Cedeño, D.M.C., Sentí, V.E., Rodríguez, J.P.F.: Technological environment for the creation of learning objects to support the teaching-learning process of cuban universities. Revista Cubana de Información en Ciencias de la Salud (ACIMED) **23**, 116–129 (2012)

Minecraft as a Tool to Enhance Engagement in Higher Education

Salem AlJanah, Pin Shen Teh, Jin Yee Tay, Opeoluwa Aiyenitaju, and Raheel Nawaz

Abstract The popularity of online teaching has increased in the last decade, especially in the last two years, where many institutions were forced to close their campuses due to Covid-19 restrictions. Although online teaching may provide an easy alternative to on-campus teaching, it often introduces a number of challenges. One of the major challenges is the lack of student engagement. Teachers typically use student engagement as an indicator to identify student strengths and weaknesses to tailor their teaching and delivery methods to meet student needs. Hence, the success of the learning process often relies on student engagement. Educational games are typically used in schools, especially during early years, to enhance student engagement. However, they are rarely used in universities as the games are typically designed for children. Recently, a number of advanced educational video games, such as Microsoft Minecraft Education Edition (MC:EE), have been released. Unlike traditional educational games, designed for children, these games can be used in universities. To provide insight into how educational video games can be used in a university setting, we ran an experiment on a group of university students to teach them fundamental programming concepts using Python programming language. In

S. AlJanah (✉)
College of Computer and Information Sciences, Imam Mohammad Ibn Saud Islamic University (IMSIU), Riyadh, Saudi Arabia
e-mail: ssaljanah@imamu.edu.sa

P. S. Teh · O. Aiyenitaju · R. Nawaz
Department of Operations, Technology, Events and Hospitality Management, Manchester Metropolitan University, Manchester, UK
e-mail: P.Teh@mmu.ac.uk

R. Nawaz
e-mail: R.Nawaz@mmu.ac.uk

J. Y. Tay
Department of Computing and Mathematics, Manchester Metropolitan University, Manchester, UK
e-mail: J.Tay@mmu.ac.uk

A. Visvizi et al. (eds.), *Research and Innovation Forum 2022*,
Springer Proceedings in Complexity, https://doi.org/10.1007/978-3-031-19560-0_38

this experiment, MC:EE was used as the main delivery tool. The feedback received after the experiment were mainly positive. Based on the feedback and experimental results, educational video games can enhance student engagement during online lessons.

Keywords Video games · Online teaching · Innovative education · Minecraft education

1 Introduction

Online teaching is fast becoming a viable alternative to face-to-face teaching. Even more so since the emergence of Covid-19 pandemic [1], where some institutions have been forced to move teaching off-campus [2]. However, the use of online teaching introduces a lack of student engagement. Without an adequate level of engagement, it might be difficult to identify student strengths and weaknesses or modify the teaching and delivery methods to meet their needs. Hence, an inadequate level of engagement typically leads to poor student performance [3].

Student engagement refers to the level of interest, attention, and effort shown inside the class [4]. In online classes, the level of student engagement is often lower than that of on-campus classes [5]. This might be because it is easier for students to get bored or feel isolated [6]. To address this issue, Martin and Bolliger suggest considering learner-to-content, learner-to-instructor, and learner-to-learner engagement in the design and delivery of online courses [7]. One way to achieve this is to use unconventional delivery methods, e.g., educational video games.

This paper explores the effect of using a video game, Microsoft Minecraft Education Edition (MC:EE) [8], on student engagement during online lessons. In detail, the rest of the paper is organized as follows. Section 2 discusses the use of games in education. Section 3 describes the experiment methodology. Sections 4 and 5 describe the experiment results and result analyses, respectively. Section 6 presents a set of guidelines for using video games in education. Section 7 concludes the paper.

2 Related Work

Educational games have been widely used in different education settings. Liu and Chen [9] conducted a study on a group of elementary school students to evaluate the effectiveness of using a card game, Conveyance Go, to assist students to acquire certain knowledge and skills. A group of 18 fifth-grade students was used in the study. The study reported that using the game had enhanced the students' knowledge and attitudes towards learning new skills.

Knautz et al. [10] carried out an experiment to evaluate the effectiveness of game-based learning using an interactive text adventure game. A group of 91 students undertaking an information literacy course was used in this experiment. The evaluation results showed that 75% of the students were engaged in the game, which improved their overall performance and grade point average (GPA). The results also showed that the failure rate was decreased by 13% after using the game.

Mathrani et al. [11] used a game called LightBot [12] to teach programming fundamentals. The experiment involved two different groups of students. The first group contained 20 students who had yet to commence the programming course, whereas the second group contained 24 students who completed the course. The experimental results showed that the use of the game improved student perception and engagement in the two groups of students.

Liao et al. [13] carried out a year-long study on a group of elementary school students to assess the use of game-based learning to improve student engagement and performance in writing. A group of 245 third-grade students was used in the study. The study results showed that the student interest, engagement, and performance improved after using the tool.

Felszeghy et al. [14] conducted an online competition using Kahoot [15], a game-based quizzing tool, on a group of first-year university students to evaluate the use of game-based learning. They found that using the tool has improved the students' overall performance and engagement.

In a more recent study conducted by Dorfner et al. [16], a custom Bingo game was used to deliver a biomedical engineering course. A group of 38 students was used in the study. Zoom [17] and Blackboard [18] were used to facilitate communications during the online class. The study results showed that 68% of the students were more engaged and 72% were more motivated and focused when the game was used during the class. As some students were not engaged in the experiment, the authors believe that games may not be able to replace pedagogy, but they can be used as additional delivery methods to improve the student learning experience.

Video games are more widely used in schools, rather than in post-18 education settings. This observation has motivated us to set a goal to integrate video games into teaching and learning in a higher education setting. We have decided to investigate the feasibility of this idea by running a small-scale trial lesson, whereby students will learn how to write simple computer programs using the Python programming language.

3 Methodology

After exploring video game options such as MC:EE [8], Scratch [19], CodeMonkey [20], and Blockly Games [21], we chose MC:EE because of its integrated Python programming language interface. Furthermore, MC:EE licensing requires only an Office 365 account credential which all our staff and students already have. Finally, Minecraft is a very popular game among students [22].

Our study involved ten students from the Department of Operations, Technology, Events and Hospitality Management (OTEHM) at Manchester Metropolitan University studying in the BSc (Hons) Business Technology programme. The participants

Fig. 1 Screenshots of students working collaboratively in MC:EE

were between the ages 18–21. Three were first-year students and seven were second-year students. Among the participants, there were six males and four females. All participants were recruited on a voluntary basis, i.e., without receiving any monetary benefits.

Prior to the class, the required tools and installation instructions were emailed to the participants. All participants were able to install the tools with the given instructions. The participants were then asked to attend a one-hour online class. The class was delivered in a virtual campus created using MC:EE environment. At the beginning of the class, students were given an overview of the session tasks. The class was semi-supervised, whereby the students were given the freedom to work on the tasks without being interrupted by the teaching staff. However, they could reach out to tutors if they needed assistance. Figure 1 displays four screenshots to show the students' active engagement in the virtual campus.

Five programming tasks with increasing difficulty were given to the students. The instructions to complete each task were written on a noticeboard as shown in Fig. 2. A sample output was displayed next to each noticeboard as a reference. In the example shown in Fig. 2, students were asked to build a structure of two blocks

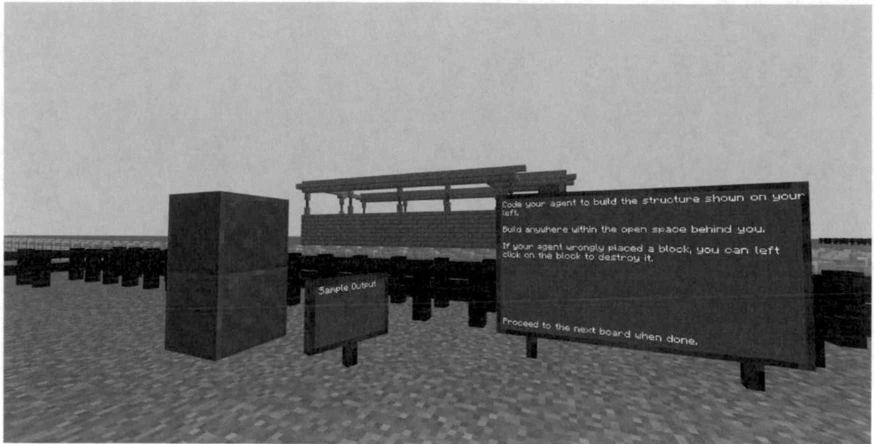

Fig. 2 A sample task in the session

Fig. 3 Completing a task using block code

stacked vertically on top of each other. The learning objective of this task was to understand and apply a fundamental programming concept called sequence control structure.

The students could recreate the sample by giving instructions to their agent. An agent is an in-game robot which can be programmed to build a structure. The instructions can be given through a Code Builder interface. The interface provides three methods of input: (i) Block code, (ii) JavaScript code, and (iii) Python code. In this experiment, the students were asked to use either block code (see Fig. 3) or Python code (see Fig. 4). Students with no or minimal programming background used block code, whereas students with some programming experience used Python code.

Fig. 4 Completing a task using Python code

Questions

1. Do you have experience playing Minecraft prior to the session?
2. How much did you enjoy the session?
3. Do you agree that MC:EE can be used as an effective educational tool for teaching and learning?
4. How challenging were the tasks given in the session?
5. Do you have comments or suggestions?

Fig. 5 Feedback form

After completing a task, the students proceed to the next task area to complete another task. This is repeated until all tasks are completed or until the one-hour session has finished. After the session, the participants were asked to complete a feedback form to evaluate the use of MC:EE to improve teaching and learning. The form contained five questions as presented in Fig. 5.

4 Results and Discussion

The feedback form consisted of four quantitative questions and one qualitative question.

Question-1: Do you have experience playing Minecraft prior to the session?

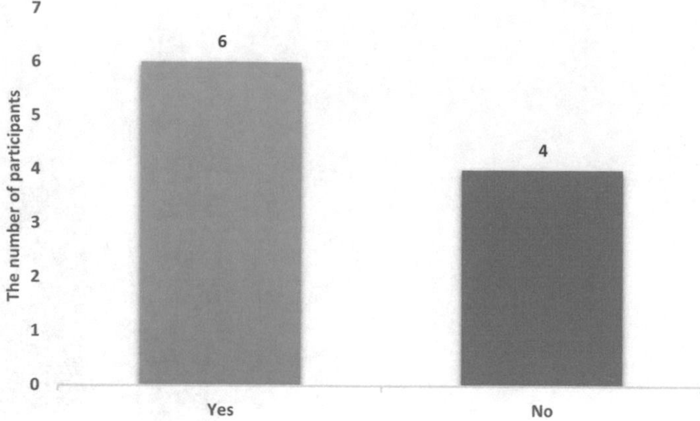

Fig. 6 Prior Minecraft experience

4.1 *Prior Experience*

This question aimed to understand whether existing knowledge in Minecraft affects the overall learning experience of the students within this experiment. Figure 6 shows that six participants have prior experience in playing Minecraft before the session, whereas four participants have never played the game before. This observation shows that the results obtained from analysing the remaining questions are likely to apply to a wider population who may not have prior experience in Minecraft.

4.2 *Level of Satisfaction*

This question aimed to examine the level of satisfaction of the participants after using MC:EE. Figure 7 shows that all students are either satisfied or very satisfied with the session. This indicates that MC:EE is an effective tool to increase the level of excitement and satisfaction inside the class.

Question-2: How much did you enjoy the session?

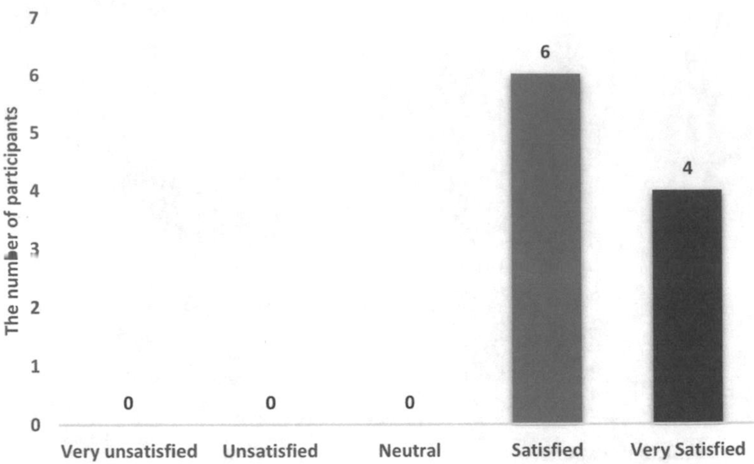

Fig. 7 MC:EE session satisfaction

4.3 MC:EE for Teaching and Learning

This question aimed to examine whether the participants believe that MC:EE could be used as an effective educational tool for teaching and learning. Figure 8 shows that apart from one student who was neutral, all students either agreed or strongly agreed that MC:EE could be used as a delivery method. This gives us a strong indication that MC:EE can be used as an effective educational tool in teaching and learning.

Question-3: Do you agree that MC:EE can be used as an effective educational tool for teaching and learning?

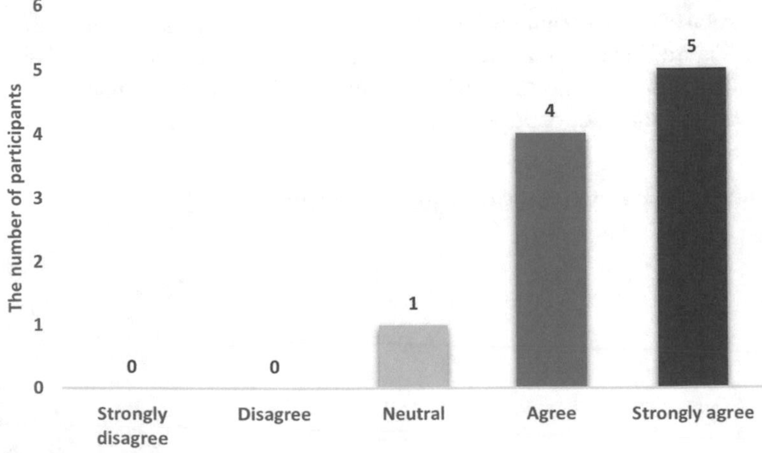

Fig. 8 MC:EE effectiveness as an educational tool for teaching and learning

4.4 *Tasks Difficulty Level*

This question aimed to examine whether MC:EE could be used as an effective educational tool to cater to different groups of students based on their level of understanding of a topic. Figure 9 shows that one participant considered the tasks unchallenging, seven were neutral, and two considered them challenging. This result shows that the difficulty level of the task within the session can be increased or decreased to cater for different groups of students depending on their level of understanding of the topic.

Question-4: How challenging were the tasks given in the session?

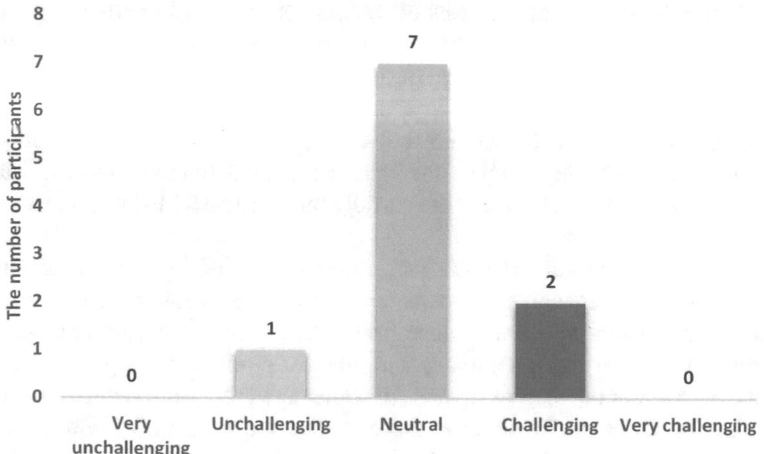

Fig. 9 Tasks difficulty level

4.5 *Qualitative Feedback*

This question aimed to capture any other insights not captured by the quantitative questions.

Question-5: Do you have comments or suggestions?

- "It was interactive"
- ". . . the game element of it was great."
- ". . . can work well with as a group, in a competitive sense."
- "High level of interaction between players and the game."
- ". . . playing a game while learning which is a big plus in my books."
- "I like being able to move around, seeing other people in the space . . ."
- "This was a well thought out activity which was engaging and gradually built my skills . . ."

The feedback show that MC:EE is a highly interactive tool that added an extra layer of social interaction among learners.

5 Analysis

The participants provided largely positive feedback. Figure 7 shows that all participants enjoyed the session. Furthermore, as can be seen from Fig. 8, all participants (except one) agree that MC:EE could be used as an effective educational tool for teaching and learning. Hence, the use of MC:EE can improve the learning experience. Figure 9 shows that the tasks in the session were not challenging enough. This is probably because some participants may have prior knowledge of Python from a previous unit, they have taken during their first year of study. This is not an immediate concern because the main purpose of the lesson was to see how students would react towards using the MC:EE tool and not so much on the tasks themselves. Furthermore, the difficulty level of the tasks can be easily customised in MC:EE. The qualitative feedback show that the participants found MC:EE social, interactive and engaging.

The feedback and analysis show that educational video games could enhance the level of student engagement considerably. While the feedback show promising results, it is premature to make a generalised conclusion. We recommend conducting further studies involving (i) a larger number of participants, e.g., more than 25 students, (ii) a wider range of units, and (iii) other age groups are required to support the findings of this study. To facilitate these studies, we are working with colleagues across the department to implement the following plan in the coming semesters. For (i), we are planning to integrate the trial described in this study into a workshop activity for the entire cohort (approximately 100 students) of first-year BSc (Hons) Business Technology students. This should demonstrate whether the positive feedback obtained in this study scale to a larger number of participants. For (ii), we are working closely with colleagues teaching in the BA (Hons) Events Management and Master of Business Administration programme to use MC:EE in some of their workshop activities. This should demonstrate whether MC:EE can also be used effectively for other (non-coding) units. For (iii), we are working with colleagues teaching in the master's degree apprenticeship for experienced working professionals to integrate MC:EE into one of the theory-based units workshop activities. This should demonstrate whether MC:EE is useful among working professionals.

6 Further Discussions

To successfully use video games in different disciplines or professional areas, we suggest applying the five-step process introduced by Huang and Soman [23]. The process is as follows.

1. **Understand the education context and target audience**.
   What is the background around education? Who is your learner?
2. **Define your learning goals**.
   What are the learning objectives?
3. **Structure the learning experience**.
   What are the learning stages/milestones?
4. **Identify resources**.
   What are the resources needed to use video games?
5. **Apply gamification**.
   Which gamification approach should be applied? e.g., personal or social?

In addition to applying the steps, we suggest conducting periodic reflections to identify and address issues that may come up during the process.

7 Conclusion

In this paper, we conducted an experiment to evaluate the use of MC:EE to improve student engagement during online lessons. Based on the experimental results, we recommend the use of educational video games to attract student attention and engagement, especially when the learning material is delivered in an intensive format such as block teaching. The use of such methods could encourage students to be more engaged and motivate them to expand their knowledge beyond what is taught in class.

Even though the experiment has shown promising results, the number of participants in the experiment is rather small. Therefore, these results may not apply to a large group of participants, or when additional variables are introduced (e.g., units other than coding). Hence, we plan to conduct further experiments with a large number of participants across a wide range of units to validate the results.

References

1. Szczepański, P., Pacer, M.: Impact of the Covid-19 pandemic on education: a case from a military academy. Int. Res. Innov. Forum, 179–192 (2021)
2. Johnson, A.F., Roberto, K.J., Rauhaus, B.M.: Policies, politics and pandemics: course delivery method for US higher educational institutions amid Covid-19. Transform. Govern.: People Process Policy **15**(2), 291–303. 2021. https://doi.org/10.1108/TG-07-2020-0158
3. Singh, A., Rocke, S., Pooransingh, A., Ramlal, C.J.: Improving student engagement in teaching electric machines through blended learning. IEEE Trans. Educ. **62**(4), 297–304 (2019). https://doi.org/10.1109/TE.2019.2918097
4. Bond, M., Buntins, K., Bedenlier, S., Zawacki-Richter, O., Kerres, M.: Mapping research in student engagement and educational technology in higher education: a systematic evidence Map. Int. J. Educ. Technol. High. Educ. **17**(2), 1–30 (2020). https://doi.org/10.1186/s41239-019-0176-8

5. Kim, D., Lee, Y., Leite, W.L., Huggins-Manley, A.C.: Exploring student and teacher usage patterns associated with student attrition in an open educational resource-supported online learning platform. Comput. Educ. **156**, 1–14 (2020). https://doi.org/10.1016/j.compedu.2020.103961
6. Kebritchi, M, Lipschuetz, A, Santiague, L.: Issues and challenges for teaching successful online courses in higher education: a literature review. J. Educ. Technol. Syst. **46**(1), 4–29 (2017). https://doi.org/10.1177/0047239516661713
7. Martin, F., Bolliger, D.U.: Engagement matters: student perceptions on the importance of engagement strategies in the online learning environment. Online Learn. J. **22**(1), 205–222 (2018). https://doi.org/10.24059/olj.v22i1.1092
8. Minecraft Education Edition: Minecraft Official Site. https://education.minecraft.net/en-us. Accessed 20 Mar 2022
9. Liu, E.Z.F., Chen, P.K.: The effect of game-based learning on students' learning performance in science learning—a case of "Conveyance Go". Procedia-Soc. Behav. Sci. **103**, 1044–1051 (2013). https://doi.org/10.1016/j.sbspro.2013.10.430
10. Knautz, K., Wintermeyer, A., Orszullok, L., Soubusta, S.: From know that to know how—providing new learning strategies for information literacy instruction. In: European Conference on Information Literacy, pp. 417–426 (2014). https://doi.org/10.1007/978-3-319-14136-7_44
11. Mathrani, A., Christian, S., Ponder-Sutton, A.: PlayIT: game based learning approach for teaching programming concepts. J. Educ. Technol. Soc. **19**(2), 5–17 (2016)
12. LightBot Homepage. https://lightbot.com/. Accessed 3 Apr 2022
13. Liao, C.C., Chang, W.C., Chan, T.W.: The effects of participation, performance, and interest in a game-based writing environment. J. Comput. Assist. Learn. **34**(3), 211–222 (2018). https://doi.org/10.1111/jcal.12233
14. Felszeghy, S., Pasonen-Seppänen, S., Koskela, A., Nieminen, P., Härkönen, K., Paldanius, K., Gabbouj, S., Ketola, K., Hiltunen, M., Lundin, M., Haapaniemi, T., Sointu, E., Bauman, E., Gilbert, G., Morton, D., Mahonen, A.: Using online game-based platforms to improve student performance and engagement in histology teaching. BMC Med. Educ. **19**(1), 1–11 (2019). https://doi.org/10.1186/s12909-019-1701-0
15. Kahoot!|Learning Games|Make Learning Awesome! https://kahoot.com/. Accessed 3 Apr 2022
16. Dorfner, N., Zakerzadeh, R.: Academic games as a form of increasing student engagement in remote teaching. Biomed. Eng. Educ. **1**(2), 335–343 (2021). https://doi.org/10.1007/s43683-021-00048-x
17. Zoom Homepage. https://zoom.us/. Accessed 3 Apr 2022
18. Blackboard Learn: An Advanced LMS. https://www.blackboard.com/. Accessed 3 Apr 2022
19. Scratch Homepage. https://scratch.mit.edu/. Accessed 20 Mar 2022
20. CodeMonkey Homepage. https://www.codemonkey.com/. Accessed 20 Mar 2022
21. Blockly Games Homepage. https://blockly.games/. Accessed 20 Mar 2022
22. Ming, G.: The use of minecraft education edition as a gamification approach in teaching and learning mathematics among year five students. In: Proceedings: International Invention, Innovative & Creative (InIIC) Conference, vol. 4, no. 2, pp. 44–48 (2020)
23. Huang, W., Soman, D.: A practitioner's guide to gamification of education. Research report series: behavioural economics in action. Rotman School of Management, University of Toronto, 1–29 (2013)

Pre-service Teacher Training in an Immersive Environment

Václav Duffek, Jan Fiala, Petr Hořejší, Pavel Mentlík, Tomáš Průcha, Lucie Rohlíková, and Miroslav Zíka

Abstract In close co-operation with training experts using virtual reality and teachers of the faculty preparing teachers, a training module called the "Virtual Classroom" was created. Users from the ranks of future teachers put on the head display and find themselves in an immersive classroom environment. There is common school equipment that is normally found in the classroom, and there are avatars—pupils or avatars—parents. The avatars are controlled by the teachers, or students who assist the teachers, and respond to avatars in specific situations given by framework-prepared scenarios. Avatars communicate with the future teacher, and also implement several events that have a visual representation (e.g. reporting, interrupting—talking together, etc.). The future teacher always has a specific assignment—what to tell the students or parents, and at the same time, he/she is forced to improvise in various partial communication situations initiated verbally, or by visual action by didactics and avatars. The paper presents the gradual development of the virtual classroom, as the model was improved during testing, and also presents the results of pilot studies. The paper also describes a three-phase model of training, which proved successful in working with a virtual classroom during pilot studies.

Keywords Virtual reality · Higher education · Pre-service teachers · Teacher training · Curriculum · Didactics

---

V. Duffek · P. Mentlík
Centre of Biology, Geoscience and Environmental Education, University of West Bohemia, Univerzitni 8, Pilsen 301 00, Czech Republic

J. Fiala · T. Průcha · L. Rohlíková (✉) · M. Zíka
Department of Computer Science and Educational Technology, University of West Bohemia, Univerzitni 8, Pilsen 301 00, Czech Republic
e-mail: lrohlik@kvd.zcu.cz

P. Hořejší (✉)
Department of Industrial Engineering and Management, University of West Bohemia, Univerzitni 8, Pilsen 301 00, Czech Republic
e-mail: tucnak@kpv.zcu.cz

A. Visvizi et al. (eds.), *Research and Innovation Forum 2022*,
Springer Proceedings in Complexity, https://doi.org/10.1007/978-3-031-19560-0_39

1 Introduction

The integration of technologies including virtual environments into education in various fields has become common practice in some places around the world [1, 2]. This is evidenced by the growing number of studies, publications, and experiments that are in the application level on the activity of students built in virtual reality [3]. The most frequently mentioned applications are in medical fields, in the military sector, in aviation, but also in fields such as stonework [4], and for the purposes of this work, especially in pedagogy. Recently, the possibilities of using virtual reality in the university environment have also been mapped in some detail, but the authors state that the design of specific applications is not based on any commonly established learning theories [5]. A basic idea of the possibilities of using virtual reality in teacher education is provided by a survey study by Billingsley et al. [6]. The results of this study confirm that immersive environments can help future teachers to improve their specific classroom skills and bring more realistic learning opportunities to teacher education programmes, without the risk of negatively impacting real pupils in schools [7]. This work is based on the assumption that the potential of virtual reality can, with the help of properly planned teaching activities, contribute to streamlining and improving the teaching processes of future teachers, by becoming a didactic-technical means of stimulating the development of their communicative competencies. Work in already verified and described research from the areas of the above-mentioned fields seeks inspiration for the design of such activities. The presumption of the importance of virtual reality for the future of education is also supported by the authors of the Horizon Report expert study from 2021 [8].

We can see also many studies on metaverse which is the promising new approach in the area of virtual reality [9, 10]

The initial idea for determining the research problem is to consider the role of virtual reality in the development of teachers' communicative competencies. Several authors have already shown that repeating not only communication, but also other skills, can help students to improve and develop their skills [11]. The work is based on the assumption that specific, repeated activity of future teachers in a virtual environment, based on simulating communication with their students' parents, based on real pedagogical situations, can help them develop their communicative competencies in this particular context [3]. The research problem therefore deals with the study of the possibilities of using the virtual environment in the preparation of future teachers, and revealing its limitations in specific teaching activities. Therefore, the work seeks to contribute to the development of pedagogy and the description of the specific use of the virtual environment as another possible technical didactic tool ready for integration into the educational process.

2 Methodology

A team of didactics and technicians has been working to create a virtual classroom since 2019. The initial virtual model is gradually being improved, tested, and re-improved. We are basing this on the methodology of design-based research [12–15].

The actual virtual training then takes place according to a three-phase model [16]. The aim of training in a virtual hotel is to prepare target groups not only for various situations that the teacher experiences in practice. We are currently working with the possibility of teaching in a virtual classroom (e.g. the teacher explains a critical point of the curriculum to the pupils) or simulates the teacher's performance at a meeting with parents.

The virtual classroom application has one room model. To ensure the greatest possible degree of authenticity, this model is based on a real basis, which was a primary school classroom in Rokycany. The basic parameters of the room for which demands were made include its proportions, basic equipment, and various student works, which provide authenticity to the virtual space, and help students towards a greater degree of immersion. The model was created in the Blender modelling environment within the Bachelor's thesis of a FPE UWB student in Pilsen [17]. Subsequently, it was supplemented by freely available models, such as benches and chairs, and models created using the SketchUp 2017 programme.

However, the most important models that the respondent spontaneously pays attention to are the created characters. Within the research and planned research, there are two types of characters available—models corresponding to the proportions of adults, and models of 6th to 8th graders. An alternative to Blender and Maya 3D modelling programmes—Adobe Fuse CC graphics software—was chosen to create them. It is a specialized product designed for 3D game developers. Within four sections (three in the programme, the fourth in the associated Mixamo service) it is possible to create any character, including clothes and skeletons (so-called rigu), which is necessary for the application of movements in the form of animations.

In the autumn of 2021, a pilot research project was conducted to find out the possibilities offered by the immersive environment of the Virtual Classroom for practicing communication skills with beginning teachers. Five volunteers from the Faculty of Education of the University of West Bohemia in Pilsen took part in the research. The whole research was guided by the author's pre-planned methodology for practicing communicative skills, although during the research, the author acquired several ideas for improvement (Fig. 1).

3 Results

All of the participants in the research agreed that, in its current form, the virtual classroom environment offers a high degree of immersion. The class model is inspired by a real primary school class, and its graphic design, including a number of details,

Fig. 1 Performance and evaluation of student performance

such as the sink and didactic posters on the walls, enhances the feeling of presence in space. However, the students were stressed during the activities by the class lighting model, and the sophisticated gloomy environment visible through the windows of the virtual space. Among the often mentioned shortcomings was the absence of the possibility of screening and controlling the presentation, which the student would prepare for class meetings and show it to parents, all of the participants showed interest in this possibility.

Control and movement in the virtual space did not cause a problem for any of the participants. It was confirmed that it was appropriate to physically isolate the immersed student in an empty part of the room so that he would not be afraid of hitting another subject. A single chair served as a point of reference for all of the students, which, however, rather limited the students' movement as a result of trying to instinctively stick to it. The other students in the role of parents watched from behind the screen. The most limiting factor turned out to be the need to connect the headset to a running Virtual Classroom application with a computer using cabling, due to the hardware complexity of the software. The students testified that they subconsciously tried to avoid the cabling, and not risk disconnecting it.

Although the research consisted of four phases (see above), all of the participants in the training confirmed the functionality of the three-phase model, i.e. the first meeting in the spirit of both the technology and the training plan, teaching scenarios and their placement in the pedagogical context, and two more. The meeting rehearses one situation with the observation of students' progress. The fourth meeting only confirmed that practicing the same situation for the third time was unnecessary for all involved, and it was more appropriate to practice a new situation, which can be considered as the beginning of the next test cycle. Changing the scripts is therefore more than desirable, always after the second performance on the same topic.

The participants appreciated the importance and benefits of immediate feedback after the performance, the subsequent discussion at the end of the meeting, and subsequent individual evaluation by experts, but did not show interest in monitoring their own performance of the record due to self-evaluation—most often due to embarrassment and unnecessary waste of time. Therefore, it turned out that video recording is more important for the research team.

4 Discussion

The virtual classroom is a technical teaching tool that has been developed to support the education of future teachers. The tool has already gone through several test rounds in various contexts [15, 18]. The described research was aimed at revealing its potential for training the non-didactic skills of teachers. In its current state, it offers a very realistically modelled environment, which in conjunction with a suitably chosen learning scenario provides its users with a high degree of immersion, and thereby serves as a unique tool for developing their professional competencies.

The pilot study proved to focus on the development of scenarios for virtual classroom training on three factors, ensuring the effectiveness of simulations in a real environment: (a) realism, (b) drama, (c) challenge [19]. And the assumption that a combination of these factors can be used to create effective and learning-friendly simulations in a virtual environment has been confirmed [20].

This method of teaching, i.e. dramatization in the form of role-playing in a virtual environment, combined with a large number of students and lack of opportunities for their involvement in practical training, amplified by the consequences of the Covid-19 virus epidemic, offers safe and simple forms of short activities, the opportunity to satisfy large numbers at minimum cost (excluding simulation development costs) [21]. Training in a virtual environment provides students with the opportunity to practice and develop specific skills in standardized activities, so that the experience of them is the same for everyone, which students would not encounter in real world practice. In addition, the virtual environment is also a safe environment, without the risk that misconduct will have real consequences.

Specific pilot testing of the use in communicative competence training revealed a very wide didactic potential of this tool, in accordance with the results described by Spencer et al. [3] and Lorenzo [22].

The individual performances also showed the optimal training time for one meeting. Training in the virtual classroom is more intensive, and ten minutes of training can replace 30–60 min of training in a real world environment [23]. It was assumed that the duration of one presentation should be exactly ten minutes, which is the length of training of future teachers which has been proven by other authors [24]. However, with the increasing quality of performances, the performance time also increased. Therefore, due to the nature of the topics, an optimal time of up to fifteen minutes for one performance will be expected for further research.

5 Conclusion and Future Work

The results show that simulation in a virtual environment is an effective method in teacher education. The only limitations at the moment stem from the technical condition of the tool being developed. The set didactic teaching methodology and testing methodology proved to be suitable, therefore they will be kept for further research, with only minor modifications. The technical condition of the virtual class will be subject to wider changes that will respect the existing findings, and therefore contribute to unlocking the further potential of the tool. The openness of our solution offer the opportunity to use our application in different languages and contexts. The impact of the solution is in the innovative way how to integrate virtual reality to didactic training of future teachers.

The implemented pilot study showed that further development of the virtual classroom model is needed to create a broader library of interactions and tools that could be used in the virtual classroom. It is also very important to work on the interface for the involvement of associate students, so that the individual avatars are individually controllable by the training participants who speak for the avatars—so far, the virtual classroom is controlled by only one person sitting at the computer for all of the virtual avatars.

Acknowledgements This publication of the paper was supported by University of West Bohemia in the project SGS-2022-033 New Approaches in the Computer Science Teacher Preparation.

References

1. Visvizi, A., Daniela, L., Chen, C.W.: Beyond the ICT- and sustainability hypes: a case for quality education, Comput. Hum. Behav. **107** (2020). https://doi.org/10.1016/j.chb.2020.106304
2. Visvizi, A., Lytras, M.D., Sarirete, A.: Emerging technologies and higher education: management and administration in focus. In: Visvizi, A., Lytras, M.D., Sarirete, A. (eds.) Management and Administration of Higher Education Institutions in Times of Change. Emerald Publishing, Bingley, UK. ISBN: 9781789736281, https://books.emeraldinsight.com/page/detail/Management-and-Administration-of-Higher-Education-Institutions-in-Times-of-Change/?K=9781789736281
3. Spencer, S., Drescher, T., Sears, J., Scruggs, A.F., Schreffler, J.: Comparing the efficacy of virtual simulation to traditional classroom role-play. J. Educ. Comput. Res. **57**, 1772–1785 (2019). https://doi.org/10.1177/0735633119855613
4. Makransky, G., Borre-Gude, S., Mayer, R.E.: Motivational and cognitive benefits of training in immersive virtual reality based on multiple assessments. J. Comput. Assist. Learn. **35**, 691–707 (2019). https://doi.org/10.1111/jcal.12375
5. Radianti, J., Majchrzak, T.A., Fromm, J., Wohlgenannt, I.: A systematic review of immersive virtual reality applications for higher education: design elements, lessons learned, and research agenda. Comput. Educ. **147**, 103778 (2020). https://doi.org/10.1016/j.compedu.2019.103778
6. Billingsley, G., Smith, S., Smith, S., Meritt, J.: A systematic literature review of using immersive virtual reality technology in teacher education. J. Interact. Learn. Res. **30**, 65–90 (2019)
7. Kaufman, D., Ireland, A.: Enhancing Teacher Education with Simulations. TechTrends: Linking Research and Practice to Improve Learning, vol. 60, pp. 260–267 (2016)

8. Pelletier, K., Brown, M., Brooks, D.C., McCormack, M., Reeves, J., Arbino, N., Bozkurt, A., Crawford, S., Czerniewicz, L., Gibson, R., Linder, K., Mason, J., Mondelli, V.: 2021 EDUCAUSE Horizon Report Teaching and Learning Edition. EDU (2021)
9. Suzuki, S., et al.: Virtual experiments in metaverse and their applications to collaborative projects: the framework and its significance. Procedia Comput. Sci. **176**, 2125–2132 (2020)
10. Siyaev, A., Jo, G.: Towards aircraft maintenance metaverse using speech interactions with virtual objects in mixed reality. Sensors **21**(6) (2021). https://doi.org/10.3390/s21062066
11. Frisby, B.N., Kaufmann, R., Vallade, J.I., Frey, T.K., Martin, J.C.: Using virtual reality for speech rehearsals: an innovative instructor approach to enhance student public speaking efficacy. Virt. Real. **32**, 21 (2020)
12. Collins, A.: Toward a design science of education. Technical Report No. 1. (1990)
13. Brown, A.L.: Design experiments: theoretical and methodological challenges in creating complex interventions in classroom settings. J. Learn. Sci. **2**, 141–178 (1992). https://doi.org/10.1207/s15327809jls0202_2
14. Cobb, P., Confrey, J., diSessa, A., Lehrer, R., Schauble, L.: Design experiments in educational research. Educ. Res. **32**, 9–13 (2003). https://doi.org/10.3102/0013189X032001009
15. Reeves, T.: Design research from a technology perspective. Educ. Des. Res. 52–66 (2006)
16. Duffek, V., Fiala, J., Hořejší, P., Mentlík, P., Polcar, J., Průcha, T., Rohlíková, L.: Pre-service teachers' immersive experience in virtual classroom. In: Visvizi, A., Lytras, M.D., Aljohani, N.R. (eds.) Research and Innovation Forum 2020, pp. 155–170. Springer International Publishing, Cham (2021)
17. Vachovec, L.: Tvorba knihovny modelů pro prostředí tréninku studentů učitelství ve virtuální realitě. Creation of a library of models for an environment for training of students of education in virtual reality (2021)
18. Duffek, V., Horejsi, P., Mentlik, P., Polcar, J., Prucha, T., Rohlikova, L.: Pre-service teacher training in the virtual classroom: pilot study. In: Beseda, J., Rohlikova, L., Duffek, V. (eds.) E-Learning: Unlocking the Gate to Education Around the Globe, pp. 201–210. Center Higher Education Studies, Praha 5 (2019)
19. González Martínez, J., Camacho Martí, M., Gisbert Cervera, M.: Inside a 3D simulation: realism, dramatism and challenge in the development of students' teacher digital competence. AJET (2019). https://doi.org/10.14742/ajet.3885
20. Tyler-Wood, T., Estes, M., Christensen, R., Knezek, G., Gibson, D.: SimSchool: an opportunity for using serious gaming for training teachers in rural areas. Rural Spec. Educ. Quart. **34**, 17–20 (2015). https://doi.org/10.1177/875687051503400304
21. Verkuyl, M., Lapum, J.L., St-Amant, O., Hughes, M., Romaniuk, D., Mastrilli, P.: Designing virtual gaming simulations. Clin. Simul. Nurs. **32**, 8–12 (2019). https://doi.org/10.1016/j.ecns.2019.03.008
22. Lorenzo, C.-M.: Teacher's skill improvement by role-play and simulations on collaborative educational virtual worlds. J. Educ. Comput. Res. **50**, 347–378 (2014). https://doi.org/10.2190/EC.50.3.d
23. Dieker, L.A., Hughes, C.E., Hynes, M.C., Straub, C.: Using simulated virtual environments to improve teacher performance. School-Univ. Partners. **10**, 62–81 (2017)
24. Straub, C., Dieker, L., Hughes, C., Hynes, M.: Proceedings from TeachLivE 2014: 2nd National TLE TeachLivE Conference. Presented at the January 1 (2014)

Business Sector Response to Covid-19: Strategic Recovery

Antifragility in Innovative Start-Ups: Resources, Relationships, People

Vincenzo Corvello, Serafina Montefresco, and Saverino Verteramo

Abstract Contexts of uncertainty, such as the pandemic, cause exogenous shocks for different players in the global economic system. Some actors, however, react by turning crises into opportunities: a property called antifragility. This study has the goal to identify antecedents of antifragility in innovative start-ups. The paper presents the results of a survey conducted on Italian innovative start-ups during the Covid-19 crisis to investigate the links between the antifragile reaction and factors as intangible capital, availability of uncommitted tangible resources (or slack), technologies and absorptive capacity.

Keywords Innovative start-up · Antifragility · Black swan · Intellectual capital · Absorptive capacity

1 Introduction

Innovative start-ups, that represent the future of the economic activity of a nation [1, 2] are also one of the most vulnerable economic actors facing the typical problems of smallness and newness [3].

Covid-19 has created upheavals in many fields and also at the business level. In late 2020, less than one in five Italian firms reported to have been spared from the crisis [4].

Such conditions pose serious challenges to the possibility of start-ups to survive.

The concepts of resilience and antifragility both deal with the ability to react to shocks. They are sometimes confused but they present important differences: in particular, resilient organizations resist shocks antifragile ones get better [5].

For start-ups, antifragility represents an opportunity to improve their competitive position during crises.

We hypothesized that intellectual capital [6], the availability of slack resources [7] and absorptive capacity of an organization [8] are antecedents of antifragility.

V. Corvello (✉) · S. Montefresco · S. Verteramo
Department of Mechanical, Energy and Management Engineering,
University of Calabria, Rende, Italy
e-mail: vincenzo.corvello@unical.it

A. Visvizi et al. (eds.), *Research and Innovation Forum 2022*,
Springer Proceedings in Complexity, https://doi.org/10.1007/978-3-031-19560-0_40

Through a statistical survey we test this hypotheses on a sample of Italian innovative start-ups, during the Covid-19 emergency.

2 Theoretical Background

The reaction at the disruptive events is not the same for every organization but strongly depends on their intrinsic characteristics and the exogenous event with which they fall into. This defines the spectrum of the main possible reactions: fragility, resilience, and antifragility.

Resilience allows to lead the system to pre-crisis situation and normal operational regimes. It requires both the ability to reflect and learn and the ability to change [9, 10].

Being antifragile, instead, means being able to achieve a new condition, significantly better than the previous one, by exploiting the opportunities offered by the crisis [11, 12].

A start-up is a temporary organization designed to search for a repeatable and scalable business model [13]. Over the years, start-ups have received strong attention from governments because they not only lead technological development but also trigger the boost of economic and social growth [14].

Even if in some cases they prove to be fragile, in others their specific characteristics enable them to be better prepared to cope with a crisis. Being innovative, indeed, is a precondition of being resilient and possibly antifragile [15, 16].

Several studies suggest that the difference between survival and failure of a new firm, especially if high-tech [17] lies in its *intellectual capital*, divided in three distinctive elements: *human*, *organizational* (or *structural*) and *relational capital* [18, 19].

Based on the discussion above, this study proposes the hypothesis summarized in Fig. 1.

H1: *intellectual capital (in its three components) has a positive impact on antifragility in innovative start-ups.*

Innovative start-ups are organizations, creative and dynamic, that generate knowledge and depend on the innovative combination of their internal knowledge [20] with those of external partners [21]. The ability to recognize the value of new external information, assimilate it and apply it for commercial purposes identifies the *absorptive capacity* of the firm [8, 22].

We hypothesize that.

H2: *absorptive capacity has a positive link with antifragility in innovative start-ups.*

It is impossible to generalize which are the most suitable resources to face a crisis, because this is closely dependent on the characteristics of the event, of the start-up

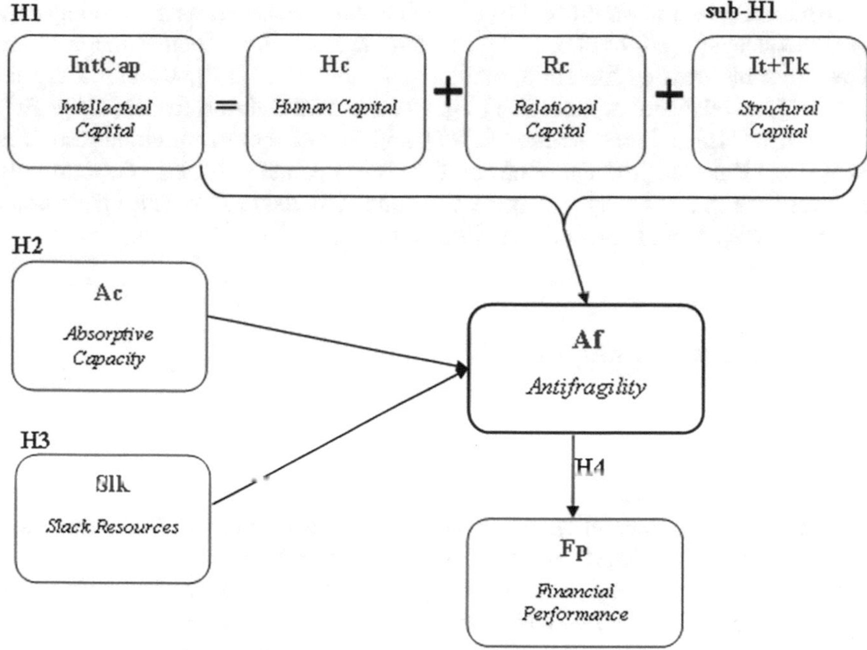

Fig. 1 Research hypotheses. Own elaboration

and the strategic objectives it intends to reach. Availability of slack resources allows the entrepreneurial team to be more flexible in face of a crisis [23, 24].

H3: *the availability of uncommitted resources (or slack) has a positive impact on the antifragility of the innovative start-ups.*

Besides we formulate the following hypothesis linking antifragility to the perception of an improved financial performance.

H4: *antifragility has a positive impact on the overall financial performance of the start-up.*

The four hypotheses are summarized in Fig. 1.

3 Methodology

This study employs a questionnaire survey approach to collect data for testing the validity of the research hypotheses. The random sample is extracted from the specific "Register" of Italian innovative start-ups [25].

To measure the firm's antifragility, a "proactive resilience" measure developed by Jia and colleagues [26] has been adapted. The measures for the other factors were taken from the existing literature: intellectual capital (IC) [27]; structural capital (SC) [27] and relational capital (RC) [28]. The measure of absorptive capacity (AC) derived from Hsu and Fang's research [27], and the one for uncommitted resources (SLK) from Essuman [29]. Question based on seven-point Likert scale (1 = strongly in disagreement; 7 = strongly in agreement) and dichotomous (Yes/No) were used, to avoid ambiguity and obtain a quantitative measure.

3.1 Data Analysis and Results

The anonymous replies to the questionnaire, correctly completed, were 144. The main characteristics of the final sample are reported in Table 1. IBM SPSS software was used to analyse data.

To compute the reliability and validity of all questionnaire items we used Cronbach's Alpha (Table 2) with lower limit of at least 0.70.

The first column of the Spearman's matrix shows the existence of positive correlation between antifragility variable and other independent variables (see Table 3).

Linear regression was used to test the research hypotheses (Tables 4 and 5).

Several models (single and multiple) were generated with the linear regression method to check the relationships between the independent variables and dependent variable and to find the most influential predictor for an antifragile reaction in innovative start-ups.

The estimate accuracy was evaluated with the determination coefficient R^2. However, this coefficient has been corrected (R^2 adjusted), to takes into account the number of independent variables and avoid over-fitting problems. An important assumption of goodness of the regression method is the absence of linear dependence between independent variables. For this, the analysis of collinearity was performed with calculation of two indices: Tolerance (Ti) and Variance Inflation Factor (VIF). Very low values of Ti (<0.25) and high values of VIF, are indicative of the possible existence of collinearity. The values of the Ti and VIF associated with the predictors show a range which fall within acceptable limits and implying no need for concern about multicollinearity.

Intellectual capital is a unique asset for the firm so the three components have been replaced with the single independent variable IntCap. This confirmed our research hypotheses H1, H2 and H3.

Finally, the following linear regression model allowed the verification and confirmation of the last hypothesis proposed (H4).

Results have been summarized in Fig. 2.

Table 1 Main characteristics of the sample. Own elaboration

	N	%	≤19 Emp.	>19 Emp.
Macro-activity industrial sector				
Agriculture, forestry and fishing	2	1.39	2	0
Other service activities	*45*	31.25	32	13
Others	*15*	10.42	13	2
Accommodation and catering services	*1*	0.69	0	1
Financial and insurance	*3*	2.08	2	1
Real estate	*1*	0.69	0	1
Manufacturing	*27*	18.75	15	12
Professional, scientific and technical	*11*	7.64	9	2
Wholesale and retail trade	*5*	3.47	5	0
Construction	*1*	0.69	1	0
Supply of electricity, gas, steam and air-condition	*3*	2.08	2	1
Education	*1*	0.69	0	1
Health and social care	*9*	6.25	7	2
Information and communication	*18*	12.50	12	6
Transport and storage	2	1.39	2	0
Total	**144**			
Enterprise size (number of employees)				
0–4	30	20.83		
5–9	38	26.39		
10–19	34	23.61		
20–49	25	17.36		
50–249	16	11.11		
Least 250	1	0.9		
Total	**144**			

4 Discussion of Results

The concept of antifragility is still little known and studied in the organizational sphere and even less in start-ups. Therefore, this research seeks to expand the current knowledge [10] and understand the possible antecedents of antifragility in the case of innovative start-ups. Intellectual capital, absorptive capacity, and the availability of uncommitted resources have been considered as possible antecedents.

The analysis of data provided by the sample of the 144 innovative start-ups from Italy has allowed to fully support the four hypotheses formulated.

The effect of the independent variables on antifragility is positive and statistically significant.

Table 2 Reliability and validity. Own elaboration

	Variables	Scale	Item	Cronbach's α
Dep. variable	***Af*** *(Antifragility)*	Likert 1–7	[AF1, AF2, AF3, AF4]	0.902
Indep. variables	***Hc*** *(Human Capital)*	Likert 1–7	[HC1, HC2, HC3]	0.808
	Rc *(Relational Capital)*	Likert 1–7	[RC1, RC2, RC3, RC4]	0.797
	Sc *(Structural Capital)*:			
	• ***It*** *(Information System)*	Likert 1–7	[IT1, IT2, IT3, IT4, IT5]	0.905
	• ***Tk*** *(4.0 Technology)*	Dichotomous	[TK01, TK02, TK03, TK04, TK05, TK06, TK07, TK08, TK09, TK10]	–
	Slk *(Availability of resources)*	Likert 1–7	[SLK1, SLK2, SLK3, SLK4, SLK5]	0.759
	Ac *(Absorptive capacity)*	Likert 1–7	[AC1, AC2, ~~AC3~~]	0.736
	Fp *(Financial performance)*	Likert 1–7	[FP1, FP2, FP3, FP4, FP5]	0.943

The results confirm the importance of intellectual capital for the antifragile reaction of a start-up. Human capital, for instance, can be the cornerstone of antifragility, especially in the case of young innovative start-ups that are based on the capabilities of the entrepreneurial team rather than on fixed capital or economies of scale. However, the results show that IC is no sufficient. It is essential to develop both good absorptive capacity and availability of uncommitted resources. Of all combinations tested, the most representative models are obtained by considering intellectual capital as a single intangible asset of the firm.

Moreover, if antifragility allows to take advantage of new opportunities and obtain more benefits, as the theory argues, then this must have positive repercussions on financial performance. This was indeed confirmed by the linear regression model which supports H4.

The results of the study are important from the point of view of innovative start-ups. Our findings are in line with other studies demonstrating that intellectual capital is one of the most important and critical resources for the organization (e.g. [30–33])) and confirm the importance of slack resources (e.g. [7]) and absorptive capacity (e.g. [8, 34, 35]) not only for the purposes of innovation, growth, success and resilience, as supported by the aforementioned research, but are also for organizational antifragility and, through antifragility, for the future performance of the firm [10].

Table 3 Spearman's correlation matrix. Own elaboration

Spearman's correlation matrix

		Af	*Hc*	*Rc*	*It*	*Tk*	*Slk*	*Ac*	*Fp*
Af	Correlation coefficient	**1.000**							
	Sig. (2-tailed)								
Hc	Correlation coefficient	0.365^{**}	**1.000**						
	Sig. (2-tailed)	0.00							
Rc	Correlation coefficient	0.353^{**}	0.461^{**}	**1.000**					
	Sig. (2-tailed)	0.000	0.000						
It	Correlation coefficient	0.413^{**}	0.432^{**}	0.335^{**}	**1.000**				
	Sig. (2-tailed)	0.000	0.000	0.000					
Tk	Correlation coefficient	0.197^{*}	0.113	0.121	0.290^{**}	**1.000**			
	Sig. (2-tailed)	0.018	0.177	0.150	0.000				
Slk	Correlation coefficient	0.437^{**}	0.216^{**}	0.441^{**}	0.330^{**}	0.116	**1.000**		
	Sig. (2-tailed)	0.000	0.009	0.000	0.000	0.167			
Ac	Correlation coefficient	0.494^{**}	0.403^{**}	0.380^{**}	0.305^{**}	0.066	0.436^{**}	**1.000**	
	Sig. (2-tailed)	0.000	0.000	0.000	0.000	0.434	0.000		
Fp	Correlation coefficient	0.720^{**}	0.318^{**}	0.373^{**}	0.413^{**}	0.170^{*}	0.582^{**}	0.429^{**}	**1.000**
	Sig. (2-tailed)	0.000	0.000	0.000	0.000	0,041	0.000	0.000	

**Correlation is significant at the 0.01 level (2-tailed)
*Correlation is significant at the 0.05 level (2-tailed)

5 Conclusions and Recommendations

The current literature lacks references to the concept of antifragility in start-ups, and lacks evidence on the antecedents behind antifragility. This research aims to help fill this gap. Our survey revealed a positive overall perception of the concept

Table 4 Linear regression models (with unique variable *IntCap*). Own elaboration

Linear regression models					
Dependent variable	**Af**				
Model		**M12**	**M13**		
Independent variables		*IntCap*	*IntCap*	*Ac*	*Slk*
Summary of the model					
R^2		28.39%		39.48%	
R^2 adjusted		**27.89%**		**38.18%**	
Std. error est		1.088		1.007	
SSR		66.66		92.678	
SSE		168.11		142.092	
SST		234.77		234.77	
F		56.307		30.438	
Sig.		0		0	
Linear regression line coef					
Intercept β0		1.061		0.024	
β std. coef		**0.533****	**0.333****	**0.258****	**0.191***
t-student		7.504	4.336	3.258	2.452
Sig.		0	0	0.001	0.015
Collinearity					
Tolerance Ti		1	0.734	0.687	0.714
VIF		1	1.362	1.456	1.4

Regression model is significant at the *0.05, **0.01 level

of antifragility and of its positive relationship with the perceived performance of the firm.

The research has some limits. Some aspects were not considered (e.g., founding team characteristics such as leadership, experience, gender and age, training, and education) that could impact on the results. The sample represents a small part of the total of innovative Italian start-ups; it is focused only on a single type of companies (innovative start-ups), a single country (Italy) and a single crisis (Covid-19); the sample sectoral clusters were not balanced and prevented targeted investigations. In addition, the measure adopted for antifragility has been inferred from one of resilience [26] that may not fully grasp the purpose: a more specific measure would be more suitable to this end. These limitations could be a starting point for further future research that should also investigate the long-term effects of antifragility on the performance of start-ups.

Table 5 Regression antifragility—financial performance. Own elaboration

Linear regression models	
Dependent variable	**Fp**
Independent variables	**Af**
Summary of the model	
R^2	53.62%
R^2 adjusted	**53.30%**
Std. error est	0.936
SSR	143,895
SSE	124,451
SST	268,346
F	164,185
Sig.	0.000
Linear regression line coef	
Intercept β0	0.513
β std. coef	**0.783****
t-student	12,813
Sig.	0.000
Collinearity	
Tolerance Ti	1.000
VIF	1.000

Regression model is significant at the *0.05, **0.01 level

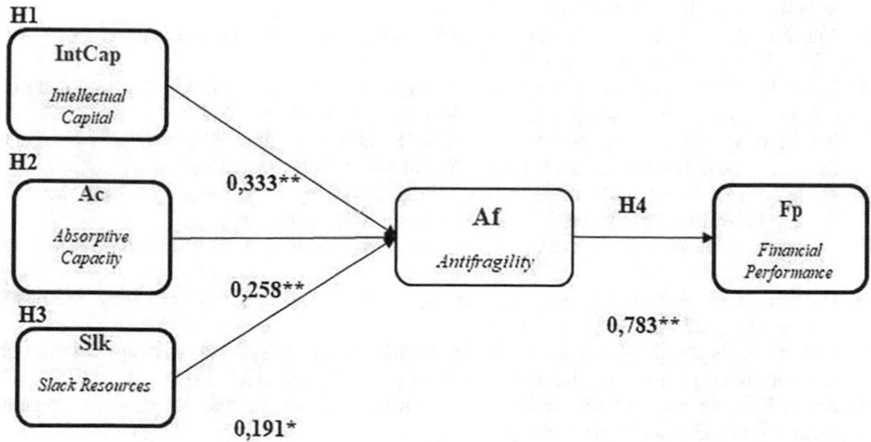

Fig. 2 Research model result. Own elaboration

References

1. Weiblen, T., Chesbrough, H.W.: Engaging with start-ups to enhance corporate innovation. Calif. Manage. Rev. **57**(2), 66–90 (2015)
2. De Groote, J.K., Backmann, J.: Initiating open innovation collaborations between incumbents and start-ups: how can David and Goliath get along? Int. J. Innov. Manag. **24**(2), 1–33 (2020)
3. Stinchcombe, A.L.: Organizations and social structure. Handb. Organ. **44**, 142–193 (1965)
4. ISTAT: Rapporto sulla competitività dei settori produttivi - Edizione 2021, Roma: Istituto Nazionale di Statistica (2021)
5. Taleb, N.N.: Antifragile: Things That Gain from Disorder. Random House Publishing Group, UK (2012)
6. Mahmood, T., Mubarik, M.S.: Balancing innovation and exploitation in the fourth industrial revolution: role of intellectual capital and technology absorptive capacity. Technol. Forec. Soc. Chang. **160**(120248), 1–9 (2020)
7. Leuridan, G., Demil, B.: Exploring the dynamics of slack in extreme contexts: a practice-based view. Human Relat. (2021).
8. Cohen, W.M., Levinthal, D.A.: Absorptive capacity: a new perspective on learning and innovation. Adm. Sci. Q. **35**(1), 128–152 (1990)
9. Duchek, S.: Organizational resilience: a capability-based conceptualization. Bus. Res. **13**, 215–246 (2020)
10. Ramezani, J., Camarinha-Matos, L.M.: Approaches for resilience and antifragility in collaborative business ecosystems. Technol. Forecast. Soc. Change, 151 (2020)
11. De Bruijn, H., Größler, A., Videira, N.: Antifragility as a design criterion for modelling dynamic systems. Syst. Res. Behav. Sci. **37**(1), 23–37 (2020)
12. Ruiz-Martin, C., López-Paredes, A., Wainer, G.: What we know and do not know about organizational resilience. Int. J. Prod. Manag. Eng. **6**(1), 11–28 (2018)
13. Blank, S.G., Dorf, B.: The Start-Up Owner's Manual: The Step-by-Step Guide for Building a Great Company. I ed. s.l.: K&S Ranch, Inc. (2012)
14. Skawińska, E., Zalewski, R.I.: Success factors of start-ups in the eu—a comparative study. Sustainability **12**(19:8200), 1–28 (2020)
15. Kuckertz, A., et al.: Start-ups in times of crisis—a rapid response to the COVID-19 pandemic. J. Bus. Venturing Insights **13**, e00169 (2020)
16. Linnenluecke, M.K.: Resilience in Business and management research: a review of influential publications and a research agenda. Int. J. Manag. Rev. Iss. **19**, 4–30 (2017)
17. Khalique, M., Abdul Nassir bin Shaari, J., Hassan Md. Isa, A.: Intellectual capital in small and medium enterprises in Pakistan. J. Intell. Capital **16**(1), 224–238 (2015)
18. Edvinsson, L., Malone, M.: Intellectual Capital. "Realizing Your Company's True Value by Finding Its Hidden Brainpower". Harper Business, New York (1997)
19. Bontis, N.: Intellectual capital: an exploratory study that develops measures and models. Manag. Decis. **36**(2), 63–76 (1998)
20. Desouza, K.C., Awazu, Y.: Knowledge management at SMEs: five peculiarities. J. Knowl. Manag. **10**(1), 32–43 (2006)
21. Presutti, M., Boari, C., Majocchi, A.: The Importance of proximity for the start-ups' knowledge acquisition and exploitation. J. Small Bus. Manage. **49**(3), 361–389 (2011)
22. Zahra, S.A., George, G.: Absorptive capacity: a review, reconceptualization, and extension. Acad. Manag. Rev. **27**(2), 185–203 (2002)
23. Kowalkiewicz, M., Safrudin, N., Schulze, B.: The Business Consequences of a Digitally Transformed Economy. Shaping the Digital Enterprise, pp. 29–67 (2017)
24. Oswald, G., Kleinemeier, M.: Shaping the Digital Enterprise—Trends and Use Cases in Digital Innovation and Transformation. Springer Nature, Germania (2017)
25. MiSE & InfoCamere. Homepage https://start-up.registroimprese.it. Accessed 7 June 2021
26. Jia, X., Chowdhury, M., Prayag, G., Chowdhury, M.M.H.: The role of social capital on proactive and reactive resilience of organizations post-disaster. Int. J. Disas. Risk Reduct. **48**(101614) (2020)

27. Hsu, Y.-H., Fang, W.: Intellectual capital and new product development performance: the mediating role of organizational learning capability. Technol. Forecast. Soc. Chang. **76**(5), 664–677 (2009)
28. Ojha, D., Salimath, M., D'Souza, D.: Disaster immunity and performance of service firms: the influence of market acuity and supply network partnering. Int. J. Prod. Econ. **147**, 385–397 (2014)
29. Essuman, D., Boso, N., Annan, J.: Operational resilience, disruption, and efficiency: Conceptual and empirical analyses. Int. J. Prod. Econ. **229**(107762) (2020)
30. Kianto, A., Sáenz, J., Aramburu, N.: Knowledge-based human resource management practices, intellectual capital and innovation. J. Bus. Res. **81**, 11–20 (2017)
31. Hormiga, E., Batista-Canino, R.M., Sánchez-Medina, A.: The role of intellectual capital in the success of new ventures. Int. Entrep. Manag. J. **7**, 71–92 (2011)
32. Alberti, F.G., Ferrario, S., Pizzurno, E.: Resilience: resources and strategies of SMEs in a new theoretical framework. Int. J. Learn. Intellect. Cap. **15**(2), 165–188 (2018)
33. Aldianto, L., et al.: Toward a business resilience framework for start-ups. Sustainability **13**(6), 1–19 (2021)
34. Carvalho, C.E., Rossetto, C.R., Piekas, A.A.S.: Innovativeness in Brazilian start-ups: the effect of the absorptive capacity and environmental dynamism. Int. J. Innov. Learn. **20**(1), 1–17 (2021)
35. Gölgeci, I., Kuivalainen, O.: Does social capital matter for SCR? The role of absorptive capacity and marketing-supply chain management alignment. Ind. Mark. Manage. **84**, 63–74 (2020)

Changing the Rules of the Game: The Role of Antifragility in the Survival of Innovative Start-Ups

Valentina Cucino, Antonio Botti, Ricky Celenta, and Rico Baldegger

Abstract During Covid-19, some innovative start-ups not only absorbed the shock but also improved afterwards through the development of capabilities related to antifragility. Drawing on the antifragility literature we have selected a set of internal capabilities that influence the survival of start-ups. Then, we applied qualitative benchmarking of the fuzzy-set to examine which interactions between the aforementioned internal capabilities affect the survival of innovative start-ups. We used a unique dataset of 37 innovative start-ups that survived in Italy after the Covid-19 lockdown. Our results suggest the interaction between some specific capabilities (*creativity*, *flexibility* and *collaboration*) are antecedents of antifragility. The interaction between these capabilities enables start-ups to survive during a crisis.

Keywords Innovative start-ups · fsQCA · Antifragility · Crisis · Firm performance

1 Introduction

Businesses are continually challenged by unforeseen disruptive events such as Covid-19 [1–3]. Indeed, in times of crisis, when companies suffer the loss of profits, customers and key employees, it is very difficult for them to survive and achieve financial results. Furthermore, the loss of profits and customers becomes more relevant when it comes to innovative small businesses. Indeed, due to their small size,

---

V. Cucino (✉)
Scuola Superiore Sant'Anna, Pisa, Italy
e-mail: valentina.cucino@santannapisa.it

A. Botti · R. Celenta
Università degli Studi di Salerno, Salerno, Italy

A. Botti
Ipag Business School, Parigi, France

R. Baldegger
School of Management Fribourg, Fribourg, Switzerland

A. Visvizi et al. (eds.), *Research and Innovation Forum 2022*,
Springer Proceedings in Complexity, https://doi.org/10.1007/978-3-031-19560-0_41

their investment in R&D, lack of experience and high costs, innovative start-ups are the ones most affected by these changes [1].

However, in some turbulent contexts (e.g. Covid-19), some innovative start-ups have managed to survive without suffering a loss of profit. These start-ups managed to absorb the shocks and get better afterward through the development of capabilities related to antifragility [4, 5]. Antifragility helps organizations to overcome problems that companies cannot foresee but that can have serious consequences [4]. Moreover, antifragility can help companies to overcome change by becoming as adaptive, reactive and flexible as possible [2].

However, few studies have investigated which capabilities related to antifragility are linked to the survival of innovative start-ups during a crisis. In this study, we explore the relationship between antifragility and business survival using fuzzy-set qualitative benchmarking (fsQCA) with key variable related to business survival. More concretely, drawing on the anti-frailty literature [2] we have selected a set of internal capacities that influence the survival of start-ups. Then, we applied qualitative benchmarking of the fuzzy set [6, 7] to examine which interactions between the aforementioned internal capabilities affect the survival of innovative start-ups. We used a unique dataset of innovative start-ups that survived in Italy after the Covid-19 lockdown. Our results suggest the interaction between some specific capabilities (*creativity*, *flexibility* and *collaboration*) amplifies the effect of antifragility on the survival of start-ups. Building on these findings, we offer some theoretical contributions to the innovation management literature, including future research questions, and provide some managerial implications.

2 Antecedents of Antifragility in Start-Ups

Although there is no correct definition of antifragile, we discuss antifragility when systems absorb shock and improve after shock [4, 5, 8–10]. In particular, Taleb first discusses the concept, distinguishing it from the concept of resilience [11]. Thus, in this study we follow Taleb's definition [4], which identifies antifragility as the capability to prosper in a volatile environment and with unexpected shocks. More concretely, while resilient organizations are able to resist without changing once it has recovered from the shock, antifragile organizations are able to benefit from the shock, learning from their mistakes and improving themselves [5, 11, 12]. In other words, an antifragile organization is not only able to survive the shock, but is able to actively strengthen itself. Thus, through learning by doing, the antifragile organizations go beyond the concept of resilience, because they aim at continuous improvement through new perspectives of generativity [2].

However, most scholars have focused on resilience in organizations, leaving limited research on antifragility in organizations. In particular, there are few studies on antifragility studies in SMEs or start-ups [10] and it is not yet clear what are the abilities that lead to its development of antifragility [2, 13]. Several capabilities—internal and external to SMEs—contribute to the development of antifragility [14]. For example, the study by Leuridan and Demil, discuss how unused internal resources, are able to make companies thrive during a crisis [15]. However, it is not certain what are the antecedents linked to antifragility. To fill this gap, our study addresses the following research question: What is the combination of capabilities that enable start-ups to overcome a crisis? In the following, we will develop the most discussed capabilities in the literature that enable antifragility.

2.1 Capabilities to Enhance Antifragility

In line with existing research [2, 16] we identify four capabilities capable of enabling antifragility: (a) "creativity" is understood as the ability to imagine new solutions for identifying opportunities in difficult contexts, which can help companies adapt quickly to external shocks [5, 17, 18]; (b) "flexibility" to change is understood as the ability to quickly change inputs or the way outputs are delivered to ensure that changes caused by the interrupted event can be flexibly managed by the organization, which can help companies adapt quickly to external shocks [18, 19]; (c) "simplicity" is understood as the ability to translate complex elements into more easily resolved elements, which can help companies adapt quickly to external shocks [4, 20]. Finally, (d) "collaboration" is understood as the ability to create lasting bonds with the various players in the ecosystem [21, 22]. According to these authors, such capabilities can support and ensure efficient and effective business readiness, response, and recovery against disruption.

3 Methodology

We used fsQCA to test the relationship between the four capabilites that improve antifragility and an ROI increase to occur in the start-ups. We have adopted fsQCA for two reasons. Firstly, fsQCA is widespread in management research [23, 24] because it enables the identification of multiple patterns leading to the outcome. In addition, fsQCA is suitable for identifying an attribute configuration, as it allows us to assess how different causes combine to influence relevant outcomes [6] such as the survival of start-ups during a crisis.

3.1 Data Sources

In November 2021, we developed an ad hoc questionnaire. In part one we collected data on the enterprise, in part two we collected data on capabilities, and in part three we collected data on performance. Before sending the questionnaire, we contacted five start-ups in order to obtain useful feedback for our research. Subsequently, we revised the questionnaire and presented it, via email, to 600 Italian start-ups, active before January 2020. In April 2022, through Google Form, we received 37 questionnaires (Table 1).

Table 1 Descriptive statistics of the sample

ROI change in 2021 compared to 2019	Number of start-ups	%
Between −30 and −6%	6	16.2
Between −5 and −1%	0	0
Between −1 and 0%	5	13.5
Between 0 and +1	7	18.9
Between +1% and +5%	5	13.6
Between +6% and +30%	14	37.8
Survival	Number of start-ups	%
Yes	37	100
Creativity	Number of start-ups	%
Capability present before the Covid-19 pandemic	27	73
Capability developed during or after the Covid-19 pandemic	9	24.3
Absent	1	2.7
Flexibility	Number of start-ups	%
Capability present before the Covid-19 pandemic	24	64.9
Capability developed during or after the Covid-19 pandemic	13	35.1
Absent	0	0
Simplicity	Number of start-ups	%
Capability present before the Covid-19 pandemic	29	78.4
Capability developed during or after the Covid-19 pandemic	5	13.5
Absent	3	8.1
Collaboration	Number of start-ups	%
Capability present before the Covid-19 pandemic	21	56.8
Capability developed during or after the Covid-19 pandemic	8	21.6
Absent	8	21.6

3.2 Calibration

Through calibration, we operationalized data collected as scores that allow defining whether or not the condition belongs to a predefined group [25]. We carried out the transformation of data into scores by using the direct calibration proposed by Ragin [25] in Table 2.

The second step involves the construction of the truth table. Each row of the truth table represents one of the logically possible combinations of conditions, including cases for which there is no empirical evidence [7]. Through the third step we had to reduce the number of rows in the truth table considering the frequency threshold and the consistency threshold. Following Ragin [25], we applied a frequency threshold of 1 and a coherence threshold of 0.8, respectively. Finally, we used the QuineM-cCluskey algorithm (used in the fsQCA 3.0 software package) to logically minimize the declarations of sufficiency, simplify the complexity, and obtain a more parsimonious response [6] (Table 3).

Rows are labeled as follows: 1, membership in the set, 0 non-membership in the set, 13 rows are not displayed in the truth table as they contain no empirical evidence.

Table 2 Calibration

Outcome/condition	Description	Calibration
ROI variation	Categorical variable specifying the change in ROI in 2021 compared to 2019	Between −30 and −2% → 0 Between −2 and +2% → 0.5 Between +2 and +30% → 1
Creativity (Cr)	Categorical variable specifying whether the firm had this capability	Had it before Covid-19 → 1 Developed it → 0.5 Does not have it → 0
Flexibility (F)	Categorical variable specifying whether the firm had this capability	Had it before Covid-19 → 1 Developed it → 0.5 Does not have it → 0
Simplicity (S)	Categorical variable specifying whether the firm had this capability	Had it before Covid-19 → 1 Developed it → 0.5 Does not have it → 0
Collaboration (Co)	Categorical variable specifying whether the firm had this capability	Had it before Covid-19 → 1 Developed it → 0.5 Does not have it → 0

Table 3 Truth table

Conditions (Cr, F, S, Co)						
Rows	Cr	F	S	Co	Cases	ROI increase
1	1	1	0	1	6	1
2	1	1	1	0	6	0
3	1	1	1	1	5	0

Table 4 Analysis of the necessary conditions

Outcome: ROI increase	Consistency	Coverage
Creativity	0.891304	0.650794
~Creativity	0.108696	0.454545
Flexibility	0.891304	0.672131
~Flexibility	0.108696	0.384615
Simplicity	0.695652	0.592593
~Simplicity	0.304348	0.700000
Collaboration	0.695652	0.640000
~Collaboration	0.304348	0.583333

3.3 Analysis of Necessary Conditions

Through the analysis of necessary conditions, it is determined whether certain of the analyzed causal conditions examined can be considered necessary conditions for an ROI increase to occur in the start-ups. For this analysis, one with a consistency greater than 0.9 is identified as a necessary condition [7]. Consistency analyses how well empirical evidence supports the link between the configuration and the outcome [25]. Considering that the consistency for each condition does not exceed the threshold of 0.9, we can state that our analysis shows that none of the conditions alone is necessary to determine the increase of the ROI (Table 4).

3.4 Analysis of Sufficient Conditions

Sufficient condition analysis identifies all conditions that are sufficient for the outcome to occur. Consistent with Kraus et al. [26] we consider a frequency threshold of 1.0 and a threshold of 0.80. The fsQCA method allows us to analyze the combinations of conditions that lead to the occurrence of the ROI increase, our outcome of interest.

Table 5 shows how the use of the intermediate solution allows us to identify a configuration sufficient for increasing ROI: the combination of the conditions of creativity, flexibility and collaboration. In line with existing research [23], for our analysis we use an "intermediate solution". The intermediate solution is a balance between complexity and parsimony because the intermediate solution is a superset of the most complex solution and a subset of the most parsimonious [25].

Table 5 Intermediate solutions

	Raw coverage	Unique coverage	Consistency
Creativity*Flexibility*Collaboration	0.304348	0.304348	0.875
Solution coverage: 0.304348			
Solution consistency: 0.96			

4 Discussion and Conclusion

During Covid-19, some innovative start-ups not only absorbed the shock but also improved later through the development of capabilities related to antifragility [4, 7]. Although this ability can be distinguished into internal and external company [27] it is not clear what the antecedents of antifragility are. In this article we address the following research question: What is the combination of capabilities that allows start-ups to overcome a crisis? We provide a motivation to guide scholars to focus on the antecedents of antifragility in start-ups.

Our paper has two important theoretical contributions. First, our study enriches the literature on antifragility [2, 5, 8] by showing how an optimal combination of capabilities related to antifragility is the coexistence of (a) *creative* capabilities, understood as the skills that allow the company to imagine new solutions in difficult contexts (b) *collaboration* capabilities, understood as the ability to collaborate with actors and create lasting bonds within the ecosystem and finally (c) capabilities to change or *flexibility*, understood as the ability to quickly change business plans. More concretely, the innovative start-ups that did not suffer loss of turnover during the pandemic linked to Covid-19 are those start-ups prone to change, capable of imagining creative solutions and establishing lasting bonds with stakeholders. Second, our study enriches existing research in the field of innovation management [28–30] by arguing that even start-ups and not just SMEs or large companies manage to survive during crises.

Finally, our study has managerial implications for entrepreneurs and managers. In particular, our study suggests to entrepreneurs and managers to create an organizational climate within the company that is open to collaboration, creativity and flexibility [31, 32]. In this way, employees will be in favor of proposing innovative ideas to overcome crises [23].

However, although several studies have shown the relevance of the Covid-19 phenomenon in Italy [33, 34], our study considers only one country. Future research should consider a larger sample and different countries.

References

1. Chesbrough, H.: To recover faster from Covid-19, open up: managerial implications from an open innovation perspective. Ind. Mark. Manage. **88**, 410–413 (2020)
2. Ramezani, J., Camarinha-Matos, L.M.: Approaches for Resilience and Antifragility in Collaborative Business Ecosystems. Technological Forecasting and Social Change, p. 151. Elsevier (2020)
3. Mariacarmela, Passarelli Giuseppe, Bongiorno Valentina, Cucino Alfio, Cariola (2023) Adopting new technologies during the crisis: An empirical analysis of agricultural sector. Technological Forecasting and Social Change 186122106-S0040162522006278 122106. https://doi.org/10.1016/j.techfore.2022.122106
4. Taleb, N.N.: Antifragile: Things That Gain from Disorder. Random House Inc., New York (2012)
5. Hespanhol, L.: More than smart, beyond resilient: networking communities for antifragile cities. In: Proceedings of the 8th International Conference on Communities and Technologies, pp. 105–114. Troyes, France (2017)
6. Ragin, C.C., Sonnett, J.: Between complexity and parsimony: limited diversity, counterfactual cases, and comparative analysis. In: Vergleichen in der Politikwissenschaft (eds.) VS Verlag für Sozialwissenschaften, pp. 180–197 (2005)
7. Schneider, C.Q., Wagemann, C.: Set-theoretic methods for the social sciences: a guide to qualitative comparative analysis. Cambridge University Press, Cambridge (2012)
8. Blečić, I., Cecchini, A.: Antifragile planning. Plan Theory **19**(2), 172–192 (2019)
9. Conz, E., Magnani, G.: A dynamic perspective on the resilience of FRMS: a systematic literature review and a framework for future research. Eur. Manage. J. **38**, 400–412 (2020)
10. Corvello, V., Verteramo, S.: Antifragility in small and medium service enterprises. In: Excellence in Services University of Salerno 24th International Conference Proceedings, Salerno (2021)
11. Russo, D., Ciancarini, P.: Towards antifragile software architectures. Procedia Comput. Sci. **109**, 929–934 (2017)
12. de Bruijn, H., Größler, A., Videira, N.: Antifragility as a design criterion for modeling dynamic systems. Syst. Res. Behav. Sci. **37**(1), 23–37 (2020)
13. Chroust, G., Aumayr, G.: Resilience 2.0: computer-aided disaster management. J. Syst. Sci. Syst. Eng. **26**, 321–335 (2017)
14. Giménez-Fernández, T., Luque, D., Shanks, D.R., Vadillo, M.A.: Probabilistic cuing of visual search: neither implicit nor inflexible. J. Exp. Psychol. Hum. Percept. Perform. **46**(10), 1222–1234 (2020)
15. Leuridan, G., Benoît, D.: Exploring the dynamics of slack in extreme contexts: a practice-based view. Human Relat., April 2021.
16. Corvello, V., Verteramo, S. Nocella, I., Ammirato, S.: Thrive during a crisis: the role of digital technologies in fostering antifragility in small and medium-sized enterprises. J. Amb. Intell. Human. Comput. (2022)
17. Gotham, K.F., Campanella, R.: Toward a research agenda on transformative resilience: challenges and opportunities for post-trauma urban ecosystems. Crit. Plan **17**, 9–23 (2010)
18. Troisi, O., Grimaldi, M.: Guest editorial: data-driven orientation and open innovation: the role of resilience in the (co-)development of social changes. Transf. Govern.: People Process Policy **16**(2), 165–171 (2022)
19. Carvalho, H., Cruz-Machado, V., Tavares, J.G.: A mapping framework for assessing supply chain resilience. Int. J. Logist Syst. Manag. **12**(3), 354–373 (2012)
20. Gorgeon: Anti-fragile information systems. Proceedings of the ICIS 2015. In: Proceedings 36th International Conference on Information Systems, pp. 1–19 Fort Worth, TX, USA (2015)
21. Berkes, F.: Environmental Governance for the Anthropocene? Social-Ecological Systems, Resilience, and Collaborative Learning. Sustainability, **9**(1232) (2017)
22. Graça, P., Camarinha-Matos, L.M.: Performance indicators for collaborative business ecosystems—literature review and trends. Technol. Forecast. Soc. Chang. **116**, 237–255 (2017)

23. Cucino, V., Del Sarto, N., Di Minin, A., Piccaluga, A.: Empowered or engaged employees? A fuzzy set analysis on knowledge transfer professionals. J. Knowl. Manag. **25**(5), 1081–1104 (2021)
24. Giulio, Ferrigno Giovanni Battista, Dagnino Nadia, Di Paola (2021) R&D alliance partner attributes and innovation performance: a fuzzy set qualitative comparative analysis. *Journal of Business & Industrial Marketing 36*(13) 54–65. https://doi.org/10.1108/JBIM-07-2020-0314
25. Ragin, C.C.: Redesigning Social Inquiry: Fuzzy Sets and Beyond. University of Chicago Press, Chicago (2008)
26. Kraus, S., Ribeiro-Soriano, D., Schüssler, M.: Fuzzy-set qualitative comparative analysis (fsQCA) in entrepreneurship and innovation research–the rise of a method. Int. Entrep. Manag. J. **14**(1), 15–33 (2018)
27. Gimenez-Fernandez, E.M., Sandulli, F.D., Bogers, M.: Unpacking liabilities of newness and smallness in innovative start-ups: investigating the differences in innovation performance between new and older small FRMS. Res. Policy **49**(10), 40–49 (2020)
28. Ferrigno, G., Cucino, V.: Innovating and transforming during COVID-19: insights from Italian firms. R&D Manag. **51**(4), 325–338 (2021)
29. Cucino, V., Ferrigno, G., Piccaluga, A.: Recognizing opportunities during the crisis: a longitudinal analysis of Italian SMEs during Covid-19 crisis. In: Castaldo, S., Mocciaro Li Destri, A., Penco, L., Ugolini, M.. (eds.) Leveraging intersections in management theory and practice, SIMA Conference. vol. 1 pp. 37–41. Fondazione CUEIM (2021)
30. Corvello, V., Grimaldi, M., Rippa, P.: Start-ups and open innovation. Eur. J. Innov. Manag. **20**(1), 2–3 (2017)
31. Visvizi, A., Troisi, O., Grimaldi, M., Loia, F.: Think human, act digital: activating data-driven orientation in innovative start-ups. Eur. J. Innov. Manag. **25**(6), 452–478 (2022)
32. Reshetnikova, M.S.: Future China: AI leader in 2030?. In: Visvizi, A., Troisi, O., Saeedi, K. (eds.) Research and Innovation Forum 2021. RIIFORUM 2021. Springer Proceedings in Complexity. Springer, Cham (2021)
33. Troisi, O., Visvizi, A., Grimaldi, M.: The different shades of innovation emergence in smart service systems: the case of Italian cluster for aerospace technology. J. Bus. Indus. Mark., vol. ahead-of-print No. ahead-of-print (2021)
34. Polese, F., Botti, A., Monda, A.: Value co-creation and data-driven orientation: reflections on restaurant management practices during COVID-19 in Italy. Transf. Govern.: People Process Policy **16**(2), 172–184 (2022)

Digital Readiness and Resilience of Digitally Servitized Firms: A Business Model Innovation Perspective

Maria Vincenza Ciasullo, Raffaella Montera, and Miriana Ferrara

Abstract The paper aims to provide an overview of the state-of- the-art interplay between digital readiness (DR), business model innovation (BMI), and organizational resilience (OR), by analysing how digital technology-enabled BMI influences the achievement of OR at a different intensity of DR. A qualitative study was conducted on three digitally servitized manufacturers of different industries. Research findings demonstrate that a heterogenous DR's intensity of the firms implies that different BMI's components are affected, entailing the achievement of a distinct form of OR. The investigation serves as pilot study which deeps the link between BMI and OR at the light of the role played by digital technologies. The investment in factory-integrated level of DR innovates all BM's components by increasing the possibility to reach a transformative OR.

Keywords Digital readiness · Business model innovation · Organizational resilience · Digital servitization

M. V. Ciasullo · R. Montera (✉) · M. Ferrara
Department of Management and Innovation Systems, University of Salerno, 84084 Fisciano, Italy
e-mail: rmontera@unisa.it

M. V. Ciasullo
e-mail: mciasullo@unisa.it

M. Ferrara
e-mail: mferrara@unisa.it

M. V. Ciasullo
Faculty of Business, Design and Arts, Swinburne University of Technology, Kuching, Malaysia

Department of Management, University of Isfahan, Isfahan, Iran

R. Montera
Department of Economics and Business, University of Florence, 50127 Florence, Italy

A. Visvizi et al. (eds.), *Research and Innovation Forum 2022*,
Springer Proceedings in Complexity, https://doi.org/10.1007/978-3-031-19560-0_42

1 Introduction

Business environment is characterized by an increasing uncertainty due to factors as the climate change, political instability, and economic crises [1]. Therefore, organizations are called to innovate their business models (BMs) that holistically respond and adapt to the new and emerging market needs, by trying to achieve organizational resilience (OR). In manufacturing setting, huge opportunities for business model innovation (BMI) are provided by digital servitization (DS) referring to the adoption of digital technology to provide new value-creating and revenue-generating ways [2]. Despite this represents an extremely urgent issue, scholars have not deeply analysed the relation between BMI and OR yet [3]. In fact, these concepts have been studied sporadically, and never pointing out the link between them [4]. Moreover, understanding which BMs' components are influenced by digital technologies and how organizations handle them regarding OR is strategically important for dealing with complexity [5, 6]. In this vein, the organizations' digital readiness (DR) makes the difference and requires further empirical investigations because various degrees of DR can be observed in practice [7]. Thus, the paper aims to to provide an overview of the state-of- the-art interplay between DR, BMI and OR in the context of digitally servitized firms by answering how digital technology-enabled BMI influences the achievement of OR at a different intensity of DR.

This paper improves the awareness of how reach OR in a BMI perspective, but mostly contributes to highlight the role of digital technologies in organizations to face the complexity. Particularly, the paper enriches DR literature, by analyzing the link between DR and BMI.

The paper is structured as follows: Sect. 2 highlights the main scientific contributions on DS, DR, OR and BMI; Sect. 3 focuses on the method adopted; Sect. 4 presents the main findings; Sect. 5 discusses the results and provides some conclusions.

2 Literature Review

2.1 DS, DR, and BMI

DS is an increasingly prevalent trend in industry [8], for which the literature is still in an emergent phase [9]. It allows the industrial firms' adaptation to current technological developments by offering digitally enabled advanced services [9, 10]. Their new service offerings are made possible by deploying a broad range of digital technologies [11]. Indeed, it is also defined as the degree of utilization of digital to support and/or act as a substitute for physical goods [12]. Accordingly, DS is linked to "the transformation in processes, capabilities, and offerings within industrial firms and their associate ecosystems to progressively create, deliver, and capture increased service value arising from a broad range of enabling digital technologies" [13].

DR is defined as the willingness to switch and adopt digital technology to respond quickly to business environment instability, catch up new opportunities, and achieve objectives faster [14]. Many models, with distinct approaches and domains, have been developed to assess the DR [15, 16], as [17] highlight in their overview of 16 different models. This reveals the scientific community's effort of overcoming the binary concept of DR towards a tiered conceptualization [7].

In this paper, the digitalization levels proposed by [18] for servitized manufacturing firms are adopted to conceptualize the DR. In this logic, the higher is the level of digitalization, the higher is the DR [14]. In particular, DR could be low, digital and factory-integrated. Respectively, when firms present a low degree of digitalization, DR only regards the support for creating new customers database or the implementation of Customer Relationship Management software. Moreover, when firms present a moderate degree of digitalization (named digital level), so that they implement digital tools towards apps or cloud computing, DR regards the ability for manufacturers to deliver distinct service offerings, by adding value to the service solution. Finally, the level of digitalization into firms could be high, so that they combine synergistically new technologies, by interconnecting products, solutions, and processes. In this case, DR is factory-integrated, because it allows to achieve higher value for both customers and the firms' internal processes.

Various levels of DR could differently affect BM. Thus, we recall the concept of BMI as effective form of innovation [19], that allows firms to introduce new ways to organize the business and quickly respond to environmental changes [20]. BMI includes both creating value for customers and users and capturing value for the firm [21]. This study considers the value creation, value delivery, and value capture as the main components of BMI. The levels of DR, which are influenced by the degree of digitalization, could affect in various ways the three components. Regarding value creation, the introduction of new technologies, such as Internet of Things, could combine different offerings with unique opportunities [22], by creating novel offering configurations and by understanding better user needs [23]. Regarding value delivery, digitalization allows firms to acquire and apply new capabilities in a more systematic way, revising operational processes and activities for global delivery and revising roles and responsibilities in the industrial ecosystems [5]. Finally, digitalization can affect value capture mechanisms both on cost reduction through the improvement of internal processes, and revenue increasing through the addition of new and increased revenue streams [5].

2.2 *OR and BMI*

OR is defined as the ability to face a difficulty or a turbulence, and the ability to "absorb" turbulence and make it part of the system to respond to it [4]. In management field, resilience means systems' ability to recover from sudden shocks, even changing the system [24, 25]. In the case of complex adaptive systems, such as industrial ecosystems, resilience is related to reactivity, adaptability, and transformability capabilities, which leads to the notion of reactive, adaptive, and transformative resilience [26]. Reactive resilience is the capacity of a system to rebound from trauma and recover [27] and to return to its previous state after experiencing unexpectable situations [28], by regaining pre-adversity position quickly and effectively. The adaptive resilience regards the capacity to face adversity and recovering from the crisis through a series of adaptive interventions [28] that could bring to a new equilibrium condition [27]. Transformative resilience regards not only resistance to shock and conservation of existing structures, but it is also deals with the systems' ability to reorganize, reconfigure, restructure, in response to disruptions [27]. It is the systems' capability to recognise and face a risk adversity, and to grow thanks to the turbulences. Transformative resilience means to intercept risks and to process them in a structured and systematic way, so that the organisation manages the turbulence and becomes able to take advantages from difficult moments.

BMI is tightly linked to resilience, because both imply actions to rapidly respond to turbulences [29], both stress the readiness to adapt to the changing environment [4]. Indeed, BMI presents, by itself, an adaptable organizational behavior [30]. In fact, through BMI, organizations innovate their BM to cope with external influences to ensure survival [31]. From this point of view, BMI can be a tool to build OR [4].

3 Method

A multiple case study was adopted as an appropriate research design to study novel and contemporary phenomena. We chose to build a theoretical sampling, by selecting cases that would express how manufacturers innovate BMs [32]. Therefore, the sample is composed of three Italian manufacturing firms which operate into B2C. They already promoted initiatives to realize DS by implementing digital technologies, like Internet of Things and Cloud Computing, and digital tools for data analysis to get closer to consumer needs. To maintain the confidentiality, the firms are named Alpha, Beta and Gamma.

We collect both primary and secondary data, as the adoption of multiple sources, and we used data triangulation to obtain results reliable. We used secondary data mainly to deeply clarify the context of our analysis, so as to be more able to correctly interpret the specific contributions of the interviews. As secondary data, we use presentations, white papers, reports, and documents produced by companies involved in the research. Regarding primary data, we conducted a qualitative study using semi-structured individual interviews that are considered more powerful than other types of interviews for qualitative research because they allow to acquire in depth information and evidence from interviewees while considering the focus of the study [33]. The participants were managers directly involved in DS and BMI processes thanks to their leader roles. Particularly, we interviewed 6 key respondents (i.e., general manager, chief information officer, R&D managers). They were asked open-ended questions, supported by an interview guide focused on the main topics of BM and DS, also including a company profile. For example, "How the implementation of new technologies has affected business model?" or "How the implementation of new technologies influenced the way your organization used to get in touch with customers?" were some questions. Each interviews lasted approximately 50 min and were held through virtual meetings platforms (i.e., Zoom). To analyze the data, we used the transcriptions of recorded interviews. We use the cross-case analysis to compare the companies' BMs and determine similarities and/or differences [34], also in terms of achievement of OR.

4 Findings

Alpha is a large firm that operates in fashion industry. This firm has invested particularly on Cloud Computing and Internet of Things. The general manager stated: "Our first goal is to satisfy the customer shop experience, for this reason we focus on value creation". Indeed, both technologies converge into the creation of new value, particularly through the provision of an app. Thus, customers can customize their shop experience, by selecting preferences, wish-lists, so that the app will propose items linked to their desires. Technologies' usage allows the firm to always be aware of customers preferences, by gaining the possibility to adapt its value offerings over the time.

Beta is a large firm that operates in beverage industry. The main technologies adopted are Big Data and Analytics and Blockchain. The R&D manager affirmed: "We wanted to assure the best experience possible to our customers, that is why we committed in considering their desires". Therefore, the firm uses intensively tools such as social media, where customers write comments and reactions and where firms often require customers participation to surveys. These data are analysed to provide detailed visibility into customer behavior patterns, so that the firm becomes able to better meet customer needs. Moreover, blockchain strengthened value delivery, because it allows the firm to monitor beverage safety and traceability, by avoiding that substandard or counterfeit product reaches consumers. Besides, this technologies' usage permits the firm to better analyze and monitor both customer needs and how products are delivered.

Gamma refers to a large firm that operates in automotive industry. It deals with the development, manufacturing, and selling of motor vehicles. The chief information officer interviewed affirmed: "Our customers are mostly busy men and women in business, for this reason we provide new technological options that make them save time and be aware of vehicle needs". Indeed, the firm has integrated systems of Artificial Intelligence (AI) to make vehicles connected: in this way, AI software can identify and communicate the need for vehicles maintenance and provide notifications in advance. Moreover, a manager declares "we would like to communicate the most as possible driving experience to our customers", in fact they introduced a virtual reality room, in which every customer can immerse into vehicle driving experience. Then, using technologies into an integrated and connected way, AI is used to perform tasks like creating schedules and managing efficiently workflow, by reducing costs and improving effectiveness.

5 Discussion and Conclusion

Research findings demonstrate that a heterogenous DR's intensity of the digitally servitized firms implies that different BMI's components are affected, entailing the achievement of a distinct form of OR (Fig. 1). This is in line with [35] who highlight the need to consider digital technologies as the levers/means rather than goal/end of innovation development. In the same direction, [36] recognize the important role that sophisticated Information and Communication Technologies (ICTs) play in managing of organizational continuity in the face of critical events.

Alpha shows a low DR because digital technologies, such as Digital Computing and Internet of Things, are used to increase customer satisfaction and value for customers. Since the digitalization impacts only the BMI's component of value creation, the organizational response to environmental turbulence is limited. Thus, we argue that the firm faces turbulence with a reactive OR that consists in avoiding the most as possible negative effects and regaining previous equilibrium.

Beta shows a digital DR because Big Data and Analytics and Blockchain implemented into the company, are used on one hand, to improve the customers' experience and to meet better their needs; on the other hand, to make products delivery safer. Consequently, digital technologies impact BM in value creation and value delivery. Since the use of digital technologies is wider, they are managed to enable the firm to intervene with changes that bring to a new equilibrium and thereby towards an adaptive OR.

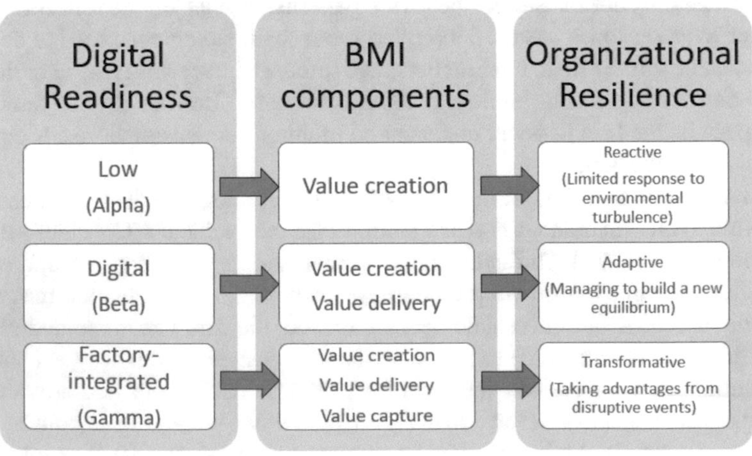

Fig. 1 Relations between DR, BMI components and OR

The adoption of digital technologies in Gamma is interconnected between products and processes. This means that Gamma has pursued an integrated-factory DR that allows to reorganize and reconfigure rapidly the entire organizational structure in case of exogenous changes, also significantly innovating all BM's components. Thus, ICTs drivers, combined dynamically, release a constant innovative tension [37] by ensuring a transformative OR. Therefore, the company has reached the possibility to take advantages from disruptive changes.

In conclusion, this investigation serves as pilot study which, comparing the interplay between DR, BMI and OR in the context of three digitally servitized firms, reveals that innovation on BMs is strictly linked to how much organizations are ready to digitalization. The higher is DR, the higher is its impact on BMI. But mostly, we observe that BMI could influence organizational resilience, depending on which components contribute to reach it. As theoretical-managerial implications, some studies [4] are confirmed: firms should take into the account the relation between BMI and OR to better face turbulences. In addition, firms should invest in building a high DR, to increase the possibility to reach transformative resilience. Lastly, digitally servitized firms have to manage technology by stressing not only technical perspectives, but also socio-technical ones.

Anyway, there are some limitations. First, the study has been conducted only on Italian firms; thus, it would be interesting to analyse what happen in other territorial contexts. Second, the research focused on three firms; so, taking in consideration a higher number of case studies–diversified by company size, manufacturing subsectors, business strategies, and digitalization intensity–would increase the generalization of results.

References

1. Chroust, G., Aumayr, G.X.: Resilience 2.0: computer-aided disaster management. J. Syst. Sci. Syst. Eng. **26**(3), 321–335 (2017)
2. Svahn, F., Mathiassen, L., Lindgren, R.: Embracing digital innovation in incumbent firms: how Volvo cars managed competing concerns. MIS Q. **41**(1), 239–253 (2017)
3. Schaffer, N., Pérez, P., Weking, J.: How business model innovation fosters organizational resilience during COVID-19. In: Conference: Americas Conference on Information Systems (2021)
4. Buliga, O., Scheiner, C.W., Voigt, K.I.: Business model innovation and organizational resilience: towards an integrated conceptual framework. J. Bus. Econ. **86**, 647–670 (2016)
5. Rachinger, M., Rauter, R., Muller, C., Vorraber, W., Schirgi, E.: Digitalization and its influence on business model innovation. J. Manuf. Technol. Manag. **30**(8), 1143–1160 (2019)
6. Granig, P., Hilgarter, K.: Organisational resilience: a qualitative study about how organisations handle trends and their effects on business models from experts' views. Int. J. Innov. Sci. **12**, 525–544 (2020)
7. Lassnig, M., Müller, J. M., Klieber, K., Zeisler, A., Schirl, M.: A digital readiness check for the evaluation of supply chain aspects and company size for Industry 4.0. J. Manuf. Technol. Manag. **33**(9) (2021)

8. Kohtamaki, M., Parida, V., Patel, P.C., Gebauer, H.X: The relationship between digitalization and servitization: the role of servitization in capturing the financial potential of digitalization. Technol. Forecast. Soc. Change **151**, 119804 (2021)
9. Sjödin, D., Parida, V., Kohtamaki, M., Wincent, J.: An agile co-creation process for digital servitization: a micro-service innovation approach. J. Bus. Res. **112**, 478–491 (2020)
10. Ciasullo, M. V., Montera, R: Digital servitization in BtoB manufacturing systems: combining theory and practice for competitiveness enhancement, 1st ed. FrancoAngeli, Milano (2021)
11. Sklyar, A., Kowalkowski, C., Tronvoll, B., Sorhammar, D.: Organizing for digital servitization: a service ecosystem perspective. J. Bus. Res. **104**, 450–460 (2019)
12. Holmström, J., Partanen, J.: Digital manufacturing-driven transformations of service supply chains for complex products. Supply Chain Manag. An Int. J. **19**(4), 421–430 (2014)
13. Sjödin, D., Parida, V., Jovanovic, M., Visjnic, I.: Value creation and value capture alignment in business model innovation: a process view on outcome-based business models. J. Prod. Innov. Manag. **37**(2), 158–183 (2020)
14. Nasution, R., Qodariah, A.: Windasari: the evaluation of digital readiness concept: existing models and future directions. The Asian J. Technol. Manag. **11**(2), 94–117 (2018)
15. Chonsawat, N., Sopadang, A.: Defining SMEs' 4.0 readiness indicators. Appl. Sci. **10**(24), 8998 (2020)
16. Sriram, R.M., Vinodh, S.: Analysis of readiness factors for Industry 4.0 implementation in SMEs using COPRAS. Int. J. Qual. Reliab. Manag. **38**(5), 1178–1192 (2020)
17. Mittal, S., Khan, M.A., Purohit, J.K., Menon, K., Romero, D., Wuest, T.: A smart manufacturing adoption framework for SMEs. Int. J. Prod. Res. **58**(5), 1555–1573 (2020)
18. Frank, A.G., Mendes, G.H., Ayala, N.F., Ghezzi, A.: Servitization and Industry 4.0 convergence in the digital transformation of product firms: a business model innovation perspective. Technol. Forecast. Soc. Change **141**, 341–351 (2019)
19. Chesbrough, H.: Business model innovation. It's not just about technology anymore. Strateg. Leadersh. **35**(6), 12–17 (2007)
20. Kastalli, I.V., van Looy, B.: Servitization. Disentangling the impact of service business model innovation on manufacturing firm performance. J. Oper. Manag. **31**(4), 169–180 (2013)
21. Björkdahl, J., Holmén, M.: Business model innovation–the challenges ahead. Int. J. Prod. Dev. **18**(3/4), 213–225 (2013)
22. Cenamor, J., Sjödin, D.R., Parida, V.: Adopting a platform approach in servitization: leveraging the value of digitalization. Int. J. Prod. Econ. **192**, 54–65 (2017)
23. Parida, V., Sjödin, D., Reim, W.: Reviewing literature on digitalization, business model innovation, and sustainable industry: past achievements and future promises. Sustainability **11**(2), 391 (2019)
24. Chroust, G., Kepler, J., Finlayson, D.: Anticipation and systems thinking: a key to resilient systems. In: Proceedings of the 60th Annual Meeting. ISSS. 1, pp. 1–12 (2016)
25. Ciasullo, M.V., Montera, R., Douglas, A.: Building SMEs' resilience in times of uncertainty: the role of big data analytics capability and co-innovation. Transform. Gov. People, Process Policy **16**(2), (2022)
26. Dahlberg, R.: Resilience and complexity: conjoining the discourses of two contested concepts. J. Curr. Cult. Res. **7**, 541–557 (2015)
27. Ramezani, J., Camarinha-Matos L.M.: Approaches for resilience and antifragility in collaborative business ecosystems. Technol. Forecast. Soc. Change **151** (2020)
28. Dahles, H., Susilowati, T.P.: Business resilience in times of growth and crisis. Ann. Tour. Res. **51**, 34–50 (2015)
29. Johnson, M.W.: Seizing the White Space—Business Model Innovation for Growth and Renewal. Harvard Business Press, Boston (2010)
30. Abraham, S.: Will business model innovation replace strategic analysis? Strateg. Leadersh. **41**(2), 31–38 (2013)
31. Foss, N.J., Saebi, T.: Fifteen years of research on business model innovation: How far have we come, and where should we go? J. Manag. **43**(1), 200–227 (2017)

32. Eisenhardt, K.M., Graebner, M.E.: Theory building from cases: opportunities and challenges. Acad. Manag. J. **50**(1), 25–32 (2007)
33. Ruslin, R., Mashuri, S., Rasak, M.S.A., Alhabsyi, F., Syam, H.: Semi-structured interview: a methodological reflection on the development of a qualitative research instrument in educational studies. IOSR J. Res. Method Educ. (IOSR-JRME) **12**(1), 22–29 (2022)
34. Yin, R.K.: Case Study Research: Design and Methods, 2d edn. Sage Publications, London (1994)
35. Visvizi, A., Troisi, O., Grimaldi, M., Loia, F.: Think human, act digital: activating data-driven orientation in innovative start-ups. Eur. J. Innov. Manag. **25**(6) (2021)
36. Visvizi, A., Lytras, M.D., Aljohani, N.R.: Disruptive technologies in times of change. In Conference: Research and Innovation Forum 2020, Springer Nature (2021)
37. Troisi, O., Visvizi, A., Grimaldi, M.: The different shades of innovation emergence in smart service systems: the case of Italian cluster for aerospace technology. J. Bus. Ind. Market. (2021)

How Startups Attained Resilience During Covid-19 Pandemic Through Pivoting: A Case Study

Francesco Polese, Carlo Alessandro Sirianni, and Gianluca Maria Guazzo

Abstract Covid-19 has restrained the managerial and innovative capacities of startuppers, forcing them to downsize their businesses or close them down altogether. Most resilient startups only, those able to manage changes, have been able to endure. The present study aims to understand how startup founders make their decisions in adverse contexts, such as the one during and after Covid-19. Therefore, following an in-depth study of relevant literature, this paper seeks to investigate the dynamics that lead startuppers to opt for a "pivot" of their business and, consequently, what are the main difficulties encountered in this process. A qualitative survey was conducted, following the Yin methodology for a case study, where data were collected through interview. The study aims to demonstrate what are the dynamics that drive a startup to change its business model and what characteristics the company must have in order for this change to produce positive performance. The study lays the foundations for future research on the topic of resilience and pivoting in the startup environment, representing a guideline for scholars and managers to ensure that the change in strategy and corporate mission guarantees the survival of the startup.

Keywords Resilience · Startup · Pivoting

1 Introduction

Covid-19 pandemic has imposed restrictions on contemporary organizations, which have dealt with unexpected events generating chaos and complexity. This has influenced the operation of businesses, limiting their activity and consequently affecting

F. Polese · C. A. Sirianni · G. M. Guazzo (✉)
Department of Business Science – Management and Innovation System, University of Salerno, 84084 Fisciano, SA, Italy
e-mail: giguazzo@unisa.it

F. Polese
e-mail: fpolese@unisa.it

C. A. Sirianni
e-mail: carlo.sirianni@unisa.it

A. Visvizi et al. (eds.), *Research and Innovation Forum 2022*,
Springer Proceedings in Complexity, https://doi.org/10.1007/978-3-031-19560-0_43

their performance. The negative impact of the pandemic has affected all economic sectors, especially SMEs and startups worldwide, which, by their nature, are more vulnerable to these conditions.

Startups can be defined as a subset of SMEs that are no more than three to five years old; they have fewer resources than other types of businesses and are continually subject to risk due to the high uncertainty in which they operate, caused by market competitiveness, business models, innovative products and/or services that can undermine their existence or ensure their daily activities [1].

Hence, there is the need for startups to adopt business systems that allow them to cope with unexpected situations characterized by a high degree of uncertainty, in order to survive or even seize opportunities to grow and prosper in a volatile and uncertain environment. In order to achieve this goal, startups are often required to change their strategy along the way, to pivot in order to show resilience. According to Polese et al., pivoting can be viewed as a restructuring of the business idea, acting as a lever that can support the transition to a less turbulent phase [2, 3]. This necessity has led scholars and managers to pay special attention to the concept of resilience [4, 5]. Vogus and Sutcliffe [6], in fact, define organizational resilience as the ability of the organization to manage difficult conditions in order to ensure business continuity. However, studies on organizational resilience focus mainly on large or established SMEs, while empirical research on the resilience of startups is still rather limited.

This study aims to investigate how pivoting is perceived in the context of startups in terms of a resilient approach to cope with uncertain and disruptive situations. To this end, after a careful review of the literature on the concept of resilience and on the aspects that define a resilient business, a case study is conducted, comparing whether what has been learned theoretically also has an empirical value. Therefore, the CEO of the Italian startup Daleca was interviewed to give responses and hopefully to fill the gap in the current literature regarding the resilience in startups.

This chapter is divided into five sections: 1. Introduction, 2. Literature review, 3. Methodology, 4. Case study, 5. Discussions and Conclusions.

Results made it possible to observe how pivoting is perceived, what are the aspects to be considered for its management and advance the first steps towards the definition of the sub-dimensions of the construct, offering managerial cues on the strengthening of certain capacities aimed at stimulating resilient behaviors.

2 Literature Review

In the difficult scenario caused by Covid-19 pandemic, many companies have been requested to assume a resilient mindset aimed at the survival of the organizational reality. Business resilience allows organizations to adapt quickly to disruptions, seeking to preserve the assets of the enterprise, ensuring its continuity [7, 8]. Doing business entails implementing strategic choices without, however, having knowledge of the value of other possible solutions; this involves having to operate under conditions of high uncertainty as well as achieving unwanted results or not always in line with expectations [9]. Smaller firms and startups are more able than others to cope with adverse events because of their greater flexibility, adaptability, and innovative capacity [10]. Indeed, innovative and adaptive capabilities play a key role in post-crisis recovery [11–14].

According to Dahles and Susilowati [15], the concept of resilience can be explained through three different perspectives. Scott and Laws [16] consider resilience as a return to a previous state, defined as "normality". The second approach views resilience as the ability to recover after a crisis through bailout orders, restoration of damaged infrastructure, and rebuilding markets. Finally, resilience is seen as a way to produce different conditions following a crisis. In the latter case, the business model changes drastically and in an unplanned and uncontrolled way, allowing the firm to experiment with new operating methods and new partners, serve new markets, deliver different products, respond to new consumer needs, generate new knowledge leading management to better cope with crisis situations, and to create value [17, 18].

The existing literature presents a theoretical framework on the drivers that allow a business to be resilient, namely dynamic capabilities, technological capabilities, agile leadership, knowledge sets and the ambidexterity of innovation; however, this conceptualization does not take into consideration some aspects that characterize startups, for example the scarce tangible and intangible resources possessed, as well as the life cycle in which they are.

Every startup, in fact, goes through six phases (i.e., bootstrap, seed, early stage, early growth, exit), each of which defines activities, objectives and the growth process of the startup itself. In the context of startups, pivoting could be considered a resilient solution, when a change in the conduct of business is needed. Therefore, the objective of this work is to identify which aspects facilitate a change of strategy for a startup, determining its resilience to crisis situations.

2.1 Pivoting

Pivoting is a key aspect of the Lean Startup methodology, according to which it is possible to concretely validate a business idea through three steps: (1) transforming the idea into a product, (2) measuring its effects, (3) learning from the results obtained. This model is called the Build-Measure-Learn (BML) model. This approach allows to test the business model and verify its validity, in this way the management of the startup may decide whether to opt to pivot or persevere in the strategy pursued [19]. This is a strategic reorientation aimed at a reallocation of resources, activities, attention and a correct implementation of this methodology may result in the success of the startup in situations of high uncertainty [20, 21].

Ries presents ten different pivoting strategies that can guarantee a paradigm shift in the business model of a startup; these are represented in Table 1.

The literature on pivoting turns out to be rather restrained and poorly debated, making its complete conceptualization difficult. In this regard, Hampel et al. [22]

Table 1 Types of pivot

Types of pivot	Description
Zoom-in	One product feature becomes the whole product
Zoom-out	Functionality previously considered the entire product now becomes a simple feature of a larger product
Customer segment	It becomes necessary when you serve the wrong type of customers
Customer need	The target audience manifests a need worth solving, but it is not the one identified by the startup
Platform	Moving from a support platform to an application or vice versa
Business architecture	Occurs when the startup decides to change the architecture of its business, focusing for example on a niche market rather than the mass market
Value capture	Change the way the startup generates value, in monetary terms
Engine of growth	Startup changes its growth strategy to increase profitability
Channel	Change the way the startup reaches out to its customers
Technology	Startup adopts completely different technology to offer the same product/service

argue that pivoting can be conceptualized depending on the management objective; in this case, pivoting is seen as a change, namely, when the intention to change or modify something (usually the business model) arises. Besides, according to Crilly et al. [23], the pivot can be seen as a path that allows the startup to survive and grow when available resources are limited; the pivot can also be seen as a response to the failure of previously made assumptions, making the startup unable to create value [24, 25].

Ghezzi [26], on the other hand, argues that pivoting can be considered as a series of steps in which a process is extruded, consisting of trial and error. Finally, the pivoting strategy could be seen as a state of waiting, in which startups find themselves, a quiescent phase where the concepts and the business model are not well defined yet but remain rather fluid, so as to understand what is the best approach to reach the market [27, 28].

3 Methodology

The present study adopts a qualitative, inductive approach to understand which aspects facilitate pivoting and how they determine business resilience. For this reason, the case study methodology consists of the use of interview, as suggested by Yin [29]. To this end, a 90-min semi-structured telephone interview with open-ended questions was conducted, recorded and subsequently transcribed for higher quality results and further analysis.

The interviewee, as founder and CEO of the startup, has intervened in all decision-making processes of the startup, from its inception to the present; therefore, he has full knowledge of the steps that have characterized the change of strategy in the conduct of the business.

4 Case Study: Daleca

Daleca offers an online service of professional speed dating through a software as a service (saas) developed in-house, where both profit and non-profit organizations can arrange networking events to let people get to know each other. Each video call takes place on the proprietary app and randomly connects two people attending the company's event for about 15 min. Initially, the startup, which was founded in late 2019, offered an offline speed dating service that allowed people to meet their potential soulmate; later, the Daleca team created an MVP (minimum-viable product) trying to replicate the experience offered in the past through an app for smartphones. However, despite encouraging early results, the startup decided to opt for a pivot during its seed phase. The main reason for this change in strategy was due to the role of the CEO and co-founder and his desire to offer a service that would have had a more significant impact, especially from a social point of view.

With regard to this transition, the major difficulties encountered, according to the startup founder, were to align the rest of the team with the new mission and vision; in fact, the CEO explained: "*The moment the motivation that drove you to start and pursue an entrepreneurial project changes, the goals and objectives of your idea change, you find yourself having to understand whether the people with whom you started are still interested in being part of the project, sharing your values and intentions that lead you to overturn the initial idea*". At the same time, there were difficulties at the skills level, in fact: "*If the values should be aligned, it is not necessarily that the skills, such as knowledge of the market, are too. If you know the first market you were referring to, it is not certain that you will also know the second one, and this involves a whole series of new investments needed to enter this new market*".

These issues were crucial, according to the interviewee, so that two partners with whom he had embarked on the entrepreneurial journey left the project. In this regard, the Covid-19 played an important role. Indeed, the CEO admitted: "*A startup is a group of people who join forces to follow a vision, but with the pandemic they cannot meet in person and this inevitably leads to a lack of motivation because everyone is forced to work on his/her own. Remote work is fine for corporations, where people are driven by economic interest, by a salary; whereas in a startup, motivation is key, the synergies that are created within the team are fundamental. Covid-19 did not allow us to celebrate a result, share emotions and sensations of working together and, obviously, a simple video call cannot give the feeling of belonging to a team. In this scenario, a startup would no longer be considered as a particularly suitable model in times of crisis as far as team management and cohesion are concerned*".

Table 2 Summary of pivot in Daleca

Pivot type	Triggering factor	Startup pivot phase
Customer segment	Role of CEO and co-founder	Seed

Resilience, that is the ability to withstand the shocks and difficulties mentioned above, which occurred thanks to pivoting, is to be found, according to the CEO, in the leadership attitude possessed, understood as "*The ability to know how to involve the right people, create aggregation and spread a corporate culture that looks at maximizing personal satisfaction, allowing each member of the team to realize themselves, thanks to their awareness of creating something useful, spreading and making people understand the same values, especially if, with the pivoting, the team has lost participants that no longer believed in that vision*". Moreover, another aspect that positively influenced the pivoting and determined its realization was the feedback received; in fact: "*It is the feedback, the continuous relationship with the mentors of your startup and potential customers that drives the pivoting and then makes it effective*". As the CEO states: "*It is not money that drives the world, it is relationships, which is why we decided to pivot. Also, in terms of resilience, having a mindset focused on creating value, not just monetary value, leads to the realization that you can leverage intangible assets, such as relationships, that allow you to overcome even those periods when your startup fails to generate revenue*" (Table 2).

5 Discussions and Conclusions

Making startups involves operating in a climate of uncertainty, forcing the founders to find solutions capable of ensuring the survival of the company and thus resisting difficulties, through continuous experimentation of ideas and activities.

Daleca has experienced a customer segment pivot to be not largely attributed to a lack of interest from the target customers initially identified, since the CEO preferred to serve a new type of user. In addition, the construction of the team was a delicate step in the success of the pivot, because, as emerged during the case study, there could be a misalignment between the original team and the new mission that the startup intends to pursue. Regarding this issue, startups find themselves facing major changes in team composition. According to Bajwa et al. [30], this type of pivot can be called 'team pivot' (to be added to the categorization proposed by Ries in Table 1). This type of pivot could involve both members with a key role (such as co-founders), as with Daleca, and members with less important roles.

The study also seeks to examine the resilient response of a startup to the difficulties generated by the pandemic. In the literature there are no studies that empirically demonstrate the dynamics at the basis of resilience; however, even in this case, the role of the CEO was fundamental. From the interview it emerged, in fact, the central role of the leadership capacity, to the detriment of the role of other equally important capacities (dynamic, technological, cognitive, and innovative) but that have affected in a lesser way the success of pivoting and the resilience of the startup in the face of difficulties.

The reason is probably to be found in the phase in which pivoting for Daleca occurred, in the seed stage. This phase is characterized by a series of evaluations, regarding the work of the team, the development of prototypes, the entry into the market and the evaluation of support mechanisms such as accelerators and incubators. For these reasons, the seed stage is very hectic and uncertain [31]. On the other hand, all past studies examined to conduct this research focused on startups that had already been launched and established, where the resilience was determined more by aspects concerning the innovative capacity of the organizations. In addition, another aspect to consider in successful pivoting is relationships. The structural limitations of startups and their reduced resources push them to create stronger ties with the different actors they interact with, in order to overcome internal limitations and co-create value [32–34].

Despite the difficulties that have emerged, today Daleca is a startup in the middle of its early stage with a solid base of clients who have already chosen the service for their events, thus starting to attract the attention of external funders for obtaining capital; this has been possible thanks to pivoting.

This paper shows the main difficulties that a startup experiences during the pivot and which aspects are able to decree its success, determining a resilient response to these issues. The success of this pivot is due to the leadership skills of the CEO, understood as both the ability to involve team members and relational skills.

However, due to the limited sample size, generalization of the results to determine a comprehensive framework is not possible. In this regard, further quantitative studies are needed to better tackle the main difficulties that a change of strategy entails within a startup and which aspects make it able to cope with such changes.

References

1. Simeone, C.L.: Business resilience: Reframing healthcare risk management. J. Healthc. Risk Manag. **35**, 31–37 (2015)
2. Mele, C., Pels, J., Polese, F.: A brief review of systems theories and their managerial applications. Serv. Sci. **2**(1–2), 126–135 (2010)
3. Barile, S., Pels, J., Polese, F., Saviano, M.: An introduction to the viable systems approach and its contribution to marketing. J. Bus. Mark. Manag. **5**(2), 54–78 (2012)
4. Troisi, O., Grimaldi, M.: Managing global epidemic through resilience: a prism for resilient smart cities. In: The International Research & Innovation Forum, pp. 333–341. Springer, Cham (2021, April)
5. Troisi, O., Grimaldi, M.: Guest editorial: Data-driven orientation and open innovation: the role of resilience in the (co-)development of social changes. Transform. Gov. People Process Policy **16**(2), 165–171 (2022)
6. Vogus, T.J., Sutcliffe, K.M.: Organizational resilience: towards a theory and research agenda. In:2007 IEEE International Conference on Systems, Man and Cybernetics, pp. 3418–3422. IEEE (2007, October)

7. Gans, J.S., Stern, S., Wu, J.: Foundations of entrepreneurial strategy. Strateg. Manag. J. **40**(5), 736–756 (2019)
8. Packard, M.D., Clark, B.B., Klein, P.G.: Uncertainty types and transitions in the entrepreneurial process. Organ. Sci. **28**(5), 840–856 (2017)
9. Engle, N.L.: Adaptive capacity and its assessment. Glob. Environ. Chang. **21**, 647–656 (2011)
10. Visvizi, A., Troisi, O., Grimaldi, M., Loia, F.: Think human, act digital: activating data-driven orientation in innovative start-ups. Eur. J. Innov. Manag. (2021)
11. Troisi, O., Visvizi, A., Grimaldi, M.: The different shades of innovation emergence in smart service systems: the case of Italian cluster for aerospace technology. J. Bus. Ind. Market. (2021)
12. Polese, F., Botti, A., Grimaldi, M., Monda, A., Vesci, M.: Social innovation in smart tourism ecosystems: how technology and institutions shape sustainable value co-creation. Sustainability **10**(1), 140 (2018)
13. Polese, F., Botti, A., Monda, A.: Value co-creation and data-driven orientation: reflections on restaurant management practices during COVID-19 in Italy. Transform. Gov. People Process Policy **16**(2), 172–184 (2022)
14. Visvizi, A., Lytras, M.D., Damiani, E., Mathkour, H.: Policy making for smart cities: innovation and social inclusive economic growth for sustainability. J. Sci. Technol. Policy Manag. (2018)
15. Dahles, H., Susilowati, T.P.: Business resilience in times of growth and crisis. Ann. Tour. Res. **51**, 34–50 (2015)
16. Scott, N., Laws, E.: Tourism crises and disasters: enhancing understanding of system effects. J. Travel Tour. Mark. **19**, 149–158 (2006)
17. Aldianto, L., Anggadwita, G., Permatasari, A., Mirzanti, I.R., Williamson, I.O.: Toward a business resilience framework for startups. Sustainability **13**(6), 3132 (2021)
18. Vargo, S.L., Peters, L., Kjellberg, H., et al.: Emergence in marketing: an institutional and ecosystem framework. J. Acad. Mark. Sci. (2022)
19. Ries, E.: The Lean Startup: How Today's Entrepreneurs Use Continuous Innovation to Create Radically Successful Businesses. Crown Business (2011)
20. Kirtley, J., & O'Mahony, S.: What is a pivot? Explaining when and how entrepreneurial firms decide to make strategic change and pivot. Strateg. Manag. J. (2020)
21. Felin, T., Gambardella, A., Stern, S., Zenger, T.: Lean startup and the business model: experimentation revisited. In: Forthcoming in Long Range Planning (Open Access) (2019)
22. Hampel, C.E., Tracey, P., Weber, K.: The art of the pivot: how new ventures manage identification relationships with stakeholders as they change direction. Acad. Manag. J. **63**(2), 440–471 (2020)
23. Crilly, N.: Fixation' and 'the pivot': balancing persistence with flexibility in design and entrepreneurship. Int. J. Des. Creat. Innov. **6**(1/2), 52–65 (2018)
24. Conway, T., Hemphill, T.: Growth hacking as an approach to producing growth amongst UK technology start-ups: an evaluation". J. Res. Mark. Entrep. **21**(2), 163–179 (2019)
25. Troisi, O., Maione, G., Grimaldi, M., Loia, F.: Growth hacking: Insights on data-driven decision-making from three firms. Ind. Mark. Manag. **90**, 538–557 (2020)
26. Ghezzi, A.: Digital startups and the adoption and implementation of lean startup approaches: effectuation, bricolage and opportunity creation in practice. Technol. Forecast. Soc. Chang. **146**, 945–960 (2019)
27. Bahrami, H., Evans, S.: Super-flexibility for real-time adaptation: perspectives from Silicon Valley. Calif. Manag. Rev. **53**(3), 21–39 (2011)
28. McDonald, R.M., Eisenhardt, K.M.: Parallel play: startups, nascent markets, and effective business-model design. Adm. Sci. Q. **65**(2), 483–523 (2020)
29. Yin, R.: Case Study Research: Design and Methods. SAGE Publications (2003)
30. Bajwa, S.S., Wang, X., Duc, A.N., Abrahamsson, P.: How do software startups pivot? Empirical results from a multiple case study. In: International Conference of Software Business, pp. 169–176. Springer, Cham (2016, June)
31. Salamzadeh, A., Kawamorita Kesim, H.: Startup companies: life cycle and challenges. In: 4th International Conference on Employment, Education and Entrepreneurship (EEE), Belgrade, Serbia (2015)

32. Marcon, A., Ribeiro, J.L.D.: How do startups manage external resources in innovation ecosystems? A resource perspective of startups' lifecycle. Technol. Forecast. Soc. Chang. **171**, 120965 (2021)
33. Polese, F., Payne, A., Frow, P., Sarno, D., Nenonen, S.: Emergence and phase transitions in service ecosystems. J. Bus. Res. **127**, 25–34 (2021)
34. Kashef, M., Visvizi, A., Troisi, O.: Smart city as a smart service system: human-computer interaction and smart city surveillance systems. Comput. Hum. Behav. **124**, 106923 (2021)

Investigating the Role of Dynamic Capabilities and Organizational Design in Improving Decision-Making Processes in Data-Intensive Environments

Hadi Karami, Sofiane Tebboune, Diane Hart, and Raheel Nawaz

Abstract Business environments are getting increasingly dynamic and data-intensive because of the emerging technologies and advances in data science, and information and communication technologies, which require enterprises to make regular and quick decisions to cope with the changes. This paper explores how big data influences decision-making processes and, consequently, organizational design in turbulent business environments. This study uses a qualitative approach (multiple case-study) by applying interviews to gain rich and illuminating data from organizations that use large data sets as a source of information based in the UK. In total, 12 participants from 9 organizations were chosen for the interviews who had a deep understanding of organizational and information-processing mechanisms, such as CEOs (chief executive officers), data analysts, data consultants, CIOs (chief information officers) and middle managers. This study contributes to decision-making theory by providing new insights about dynamic decision making in the context of big data and a better understanding of organizational strategies (either developing new dynamic capabilities or reconfiguring the current ones) for working with and leveraging value from big data. In addition, for the practical aspect, it contributes to guiding decision-makers in evaluating their organizations in terms of required capabilities and processes to become better enabled to reap value from big data.

Keywords Big data · Decision making · Dynamic capabilities · Organizational design

---

H. Karami (✉) · S. Tebboune · D. Hart · R. Nawaz
Faculty of Business and Law, Department of Operations, Technology, Events and Hospitality Management, Manchester Metropolitan University, Manchester, UK
e-mail: H.karami@mmu.ac.uk

S. Tebboune
e-mail: S.tebboune@mmu.ac.uk

D. Hart
e-mail: D.hart@mmu.ac.uk

R. Nawaz
e-mail: R.nawaz@mmu.ac.uk

A. Visvizi et al. (eds.), *Research and Innovation Forum 2022*,
Springer Proceedings in Complexity, https://doi.org/10.1007/978-3-031-19560-0_44

1 Introduction

Business environments are becoming increasingly complex and fast-changing due to various factors, such as Industry 4.0 technologies and advances in data science, and information and communication technologies [1]. "Such advances in digital technologies offer ripe opportunities for firms with superior dynamic capabilities to gain an advantage by deploying them faster and smarter than their rivals" [2:70]. At the same time, they bring about challenges as well. Despite the increasing amount of data collected and stored by companies, many are still looking for ways to improve their competitive advantage [3].

Due to the increasing number of data processing tasks and the complexity of the process, many organizations struggle to make effective decisions. Analyzing increasingly massive data sets is beyond the capacity of traditional databases. This is why they must adopt agile and flexible methods to process their data. Therefore this study seeks to shed light on some of the changes and capabilities required to take advantage of big data and improve decision-making processes.

2 Literature

Advanced Information technologies have the potential to transform the way organizations operate. Their characteristics, such as providing faster and cheaper communication, greater control over participation, and the ability to store and retrieve vast amounts of data, are expected to significantly impact the structure and design of organizations [4]. One of these phenomena is Big Data (BD), which in simple terms, could be defined as extensive data sets characterized by various features such as volume (exponentially growing), variety (text, image, signal, audio, video etc.) and velocity (timeliness in the acquisition and utilizing) [5]. As it has significant implications for organizations, it is vital to study its impact on organizations.

Investigating the impact of Big Data on organizations requires understanding the interplay between three key elements, including technology (computational power), analysis (identification of patterns in data), and mythology (the cultural aspect of BD referring to believing in the potential of BD in providing insights and intelligence by tapping on advance AI and Machine Learning methods) [6]. For example, the quality of data is only one aspect of the equation when it comes to determining the value of big data. Culture also plays a significant role in determining the success of big data initiatives for policy making [7]. In other words, leveraging business analytics to take advantage of data to make sense of decisions requires great changes. Those changes will not happen without changing the organizational infrastructures such as leadership, planning, culture, and structure. Therefore this study aims to shed light on some of those key aspects required to make the most of big data.

According to the bibliometric literature review conducted by Rialti et al. [8], dynamic capabilities is the primary theoretical approach that scholars have used to investigate the effects of big data on organizations. This is because, in increasingly dynamic environments, only organizations will stay ahead that can sense opportunities, seize them more effectively, and support the organizational transformation by redesigning and external shaping, and dynamic capabilities would enable this [2, 9–11].

In terms of technical aspects, Rialti et al. [8] argue that new data analysis tools based on artificial intelligence are required to handle such massive data sets. Accordingly, "Collaborations, knowledge exchange and big data analytics, which can be facilitated by the effective use of technology, heavily determine big data decision-making capabilities" [12:4]. It seems that relying on legacy systems and IT or investing merely in the big data technologies may not be enough to effectively reap value from big data to improve organizational decision making. In this sense, Mikalef and Krogstie [13] believe the importance of IT governance has been studied broadly. However, there is little effort in researching and establishing big data governance, which is defined as an ability of a firm to organize its resources to maximize the value and insight that could be generated from information [13]. Therefore, it is crucial to investigate the impacts of Big Data on organizational design such as structure, culture, and IT infrastructures, and the required capabilities in order to reap value from Big Data.

The decision-making process is referred to as a mental process that involves coming up with a solution that is ideal for a given problem. This is a central component of a company's management structure [14]. Data-driven decisions have gained significant attention due to the values they can provide throughout the businesses and in all tiers of organizations. For example, data-driven mindsets can hugely benefit marketing strategies [15]. In this sense, organizations need to create their big data analytics [16] to provide their decision making processes with valuable information. In doing so, the integration between various elements of organizations is imperative. For example, organizations can improve their operations and develop new knowledge through the ability to integrate the various digital and human components [17]. In this regard, it is imperative to study how decision processes are influenced by big data, and which are components are required to enhance those dynamic decision-making processes.

3 Methodology

This study aims to investigate how Big Data influences decision-making processes and, consequently, organizational design. In other words, how organizations make sense of big data by changing their decision-making processes and developing new capabilities.

Investigating the influence of such data sets on organizations calls for focusing on a few specific areas such as knowledge management, capabilities necessary to deal with big data, organizational routines and processes, and data itself as a source [8]. For this study, a mono qualitative method has been chosen based on the interpretivism philosophical stance and the research question's essence. This study seeks to study the decision-making processes (as a particular management activity), and how managers design their organizations by developing dynamic capabilities to keep pace with fast-changing and data-intensive environments.

A multiple case study approach has been chosen for this research to gain an in-depth understanding of the topic concerned. The cases of this study include organizations that are using big data as a primary source of information for decision-making based in the UK. Qualitative data was collected through semi-structured interviews for this study. In total, 12 participants from 9 organizations were chosen for the interviews who had a deep understanding of organizational and information-processing mechanisms, such as CEOs, data analysts, data consultants, CIOs, and middle managers.

The grounded theory method was used to analyze the qualitative data and interpret the participants' perspectives. The grounded theory method is believed to be very effective in context-based and process-oriented studies [18]. The grounded theory method was conceptualized in social sciences as a way of generating theories that are based on empirical data. It aims to explore the current state of knowledge instead of prescribing solutions. This method helps researchers analyze the data until they reach theoretical saturation [19]. The coding approach is employed to analyze the data by means of attaching conceptual labels to the data. Emerged open codes developed into subcategories and those subcategories eventually shaped the main themes of the study. Then the relationship between the constructs was analyzed by the researchers to build theories. It is important to note that the constant comparison between data and literature has been made to help the researchers to better conceptualize the theories. The dynamic capabilities approach was used as a lens during the data analysis to guide the coding stage. For example, concepts concerning informal Dialogue (meso-level) were labelled and coded as interpersonal communication. Therefore, the constant shift between literature and data has helped the researchers to remain focused on the subject being investigated.

4 Results

The three main research themes that emerged from the data analysis process were decision making, dynamic capabilities, and organizational design.

4.1 Decision Making

Several subcategories associated with decision-making, including normative decision-making, intuitive decision-making, and team-processes, were identified by the participants. The findings show that data analysis plays a vital role in supporting decisions in terms of providing managers with real-time insights. This is relevant in all tiers of decision making, including operational, tactical, and strategic decision making. However, participants believed that it is not always easy to analyze and evaluate the available information as businesses are facing a huge amount of unstructured data.

The results show that decision-makers might resort to their intuition based on either their cognitive style or lack of understanding the data. Participants have identified other reasons as well such as the abundance of data, time pressure and lack of trusting insights that emerged from data.

Participants also highlighted the importance of team processes as decisions made in teams tend to be faster, benefitting from various expertise and skills of others. However, the responsibility for the decisions is usually attributed to the person in charge of the decision unit. Participants have repeatedly emphasized the importance of information flow in decision making as one of the key factors that are usually more effective within team decision units.

4.2 Dynamic Capabilities

Several subcategories associated with dynamic capabilities were identified, including managerial capabilities, knowledge management, and learning.

Participants highlighted the role of management in improving the attitude of their employees toward data-driven decisions as not everyone necessarily trusts data and the insights that are emerged from it. Another critical factor that was repeatedly mentioned by the participants was the awareness of and understanding of both the business and the data. Cognition was also identified as one of the dimensions of managerial capabilities that varies from person to person based on their perceptions, skills, thought processes, experiences and knowledge acquisition techniques. Social capital is also another vital dimension of managerial capabilities identified as it is concerned with social relations, interactions and networking within a team that facilitates the flow of information and influences the decision-making process.

The concepts of Knowledge Articulation and constructive collaboration have been identified as two of the most important dimensions of knowledge management by which collective knowledge can be made available within the organization. Participants believe that by using collaboration, not only can they share the extant knowledge but also learn from others and update the existing knowledge. Those concepts refer to the articulation of the knowledge which is supposed to be coming from a source. In this sense, knowledge exploration is also mentioned referring to actively seeking new knowledge. Once the knowledge is available and articulated, it might be codified or remain as tacit knowledge in the minds of organizational members.

4.3 Organizational Design

In terms of organizational design, human resource management, organizational structure, and organizational culture were among the important dimensions of the organizational design identified by the participants.

Regular collection of a massive amount of structured and unstructured data has influenced organizational structure in terms of IT infrastructures, collecting, storing, cleansing, analyzing and disseminating data. For example, some participants argued that it is not always possible to collect the data automatically by leveraging AI, so they might need a team of experts to do this, meaning creating new roles and tasks. Similarly, the participants have mentioned the important role of HR in recruiting skilled data analysts and scientists as well. However, as there is a shortage of those skilled experts, therefore this comes with a high price which would influence the resource allocation as well. In this sense, some of the businesses rely heavily on one person's skills in terms of analyzing and getting insight from data.

In terms of organizational structure, the participants have repeatedly mentioned and highlighted formal and informal communication channels. The participants have argued that informal communication channels such as Dialogue could be equally effective as formal communication channels such as formal meetings.

According to the findings, many of the participants identified the importance of culture as one of the key factors that can influence an organization's success in implementing and managing big data analytics. This concept was also mentioned by them several times in their discussions about the subject. For example, some people rely heavily on insights emerging from data in their decisions, although others, regardless of access to the data, are unwilling to change their decision-making styles. This concept has been identified as openness to change.

In the previous sections, the key role of data analysts and scientists was highlighted. Power dynamics is another concept that emerged from data. This concept refers to the impact of big data on organizational power dynamics as data analysts and scientists are gaining more power as the role of data in running businesses is increasing. In other words, those key individuals analyze and get insights from data and provide the necessary support for decision-makers.

5 Discussion and Conclusion

Decision-makers rely on various decision-making styles, such as normative and intuitive styles. The abundance of new data, lack of sufficient data, time pressure and lack of trusting insights emerged from data are among the reasons that cause decision-makers to resort to their instincts in decisions. The results indicate that data-driven decisions are more effective in data-intensive environments. The findings show that wherever directors deal with familiar markets and data, processing the data will be much easier. However, facing the deluge of new data coming from brand new markets, data processing individually might be challenging. So, this would call for collaboration and team processes. Diversity of thoughts, ideas, expertise, and personalities could contribute to better decisions by challenging various ideas and bringing a wider variety of thoughts to the table. Awareness and understanding of the data could also enhance communication, perception, and interpretation of data across the organization.

At the meso level, the findings show that Dialogue has a key role in not only sharing knowledge but also helping individuals with generating new ideas and reflecting on them. A constructive dialogue can be critical in developing new ideas, sharing thoughts, and learning from other members of the organization in a short period of time in comparison with other learning methods. This would contribute to more dynamic decisions and more effective organizational learning.

At a macro level, resource orchestration is also influenced by introducing big data analytics. In other words, organizations should be designed to facilitate the flow of information coming from Big Data. Accordingly, one influential factor in facing new data and information to make decisions is organizational culture (as one of the crucial drivers of evidence-based decision making) because changing decision-making culture and design could influence the power structure, people's roles, and the ways by which people in organizations communicate and share knowledge and information.

The study findings indicate that deriving value from big data is not just a technical issue. It's also a social issue that requires organizations to develop their managerial capabilities and monitor environmental dynamics constantly. The results indicate that managerial capabilities, knowledge management capabilities, and learning abilities, are among the most critical dynamic capabilities. In terms of organizational design, Human resource management (awareness of the needs and talent management), organizational culture (openness to change and analytical culture), and organizational infrastructure are also imperative in improving decision making in data-intensive environments. In a nutshell, this study highlights the importance of all three levels of dynamism in organizations, from the Micro-level (habits and cognitive styles of decision-makers), and the Meso level (referring to interpersonal relations) to the Macro level (concerning resource orchestration) [20]. All of the mentioned levels are interconnected and could contribute to enhancing the information processing and communication within the organization, by tapping the power of big data tools and technologies coupled with natural language processing [21–23] and advance deep learning methods [24–26], resulting in better, faster, and more informed decisions in data-intensive and dynamic environments.

Developing the necessary systems and capabilities to handle big data can help organizations transform their operations and become more agile. This could allow them to respond more effectively to the changes brought about by the new technologies [8]. This has important implications for practitioners as to what factors are key in reaping value from big data. As it is indicated in the results, practitioners can facilitate this process by increasing awareness about the data and its value, enhancing the data-driven culture (mindset), and facilitating formal and informal communication channels. By taking a holistic approach and integrating all the necessary capabilities discussed, organizations can be more resilient in turbulent climates.

References

1. Liu, L., et al.: A framework to evaluate the interoperability of information systems – Measuring the maturity of the business process alignment. Int. J. Inf. Manag. **54**, 102153 (2020)
2. Day, G.S., Schoemaker, P.J.: Adapting to fast-changing markets and technologies. Calif. Manag. Rev. **58**(4), 59–77 (2016)
3. LaValle, S., et al.: Big data, analytics and the path from insights to value. MIT Sloan Manag. Rev. **52**(2), 21–32 (2011)
4. Huber, G.P.: A theory of the effects of advanced information technologies on organizational design, intelligence, and decision making. Acad. Manag. Rev. **15**(1), 47–71 (1990)
5. Waheed, H., et al.: Predicting academic performance of students from VLE big data using deep learning models. Comput. Hum. Behav. **104**, 106189 (2020)
6. Nawaz, R., et al.: Leveraging AI and machine learning for national student survey: actionable insights from textual feedback to enhance quality of teaching and learning in UK's higher education. Appl. Sci. **12**(1), 514 (2022)
7. Hassan, S.-U., et al.: Leveraging deep learning and SNA approaches for smart city policing in the developing world. Int. J. Inf. Manag. **56**, 102045 (2021)
8. Rialti, R., et al.: Big data and dynamic capabilities: a bibliometric analysis and systematic literature review. Manag. Decis. (2019)
9. Felin, T., Powell, T.C.: Designing organizations for dynamic capabilities. Calif. Manag. Rev. **58**(4), 78–96 (2016)
10. Dixon, S., Meyer, K., Day, M.: Building dynamic capabilities of adaptation and innovation: a study of micro-foundations in a transition economy. Long Range Plan. **47**(4), 186–205 (2014)
11. Teece, D.J., Pisano, G., Shuen, A.: Dynamic capabilities and strategic management. Strateg. Manag. J. **18**(7), 509–533 (1997)
12. Shamim, S., et al.: Role of big data management in enhancing big data decision-making capability and quality among Chinese firms: a dynamic capabilities view. Inf. Manag. (2018)
13. Mikalef, P., Krogstie, J.: Big data governance and dynamic capabilities: the moderating effect of environmental uncertainty. In: Twenty-Second Pacific Asia Conference on Information Systems: Japan (2018)
14. Intezari, A., Pauleen, D.J.: Conceptualizing wise management decision-making: a grounded theory approach: conceptualizing wise management decision-making. Decis. Sci. 1–66 (2017)
15. Troisi, O., et al.: Growth hacking: Insights on data-driven decision-making from three firms. Ind. Mark. Manag. **90**, 538–557 (2020)
16. Gupta, M., George, J.F.: Toward the development of a big data analytics capability. Inf. Manag. **53**(8), 1049–1064 (2016)
17. Visvizi, A., et al.: Think human, act digital: activating data-driven orientation in innovative start-ups. Eur. J. Innov. Manag. (2021)
18. Urquhart, C., Lehmann, H., Myers, M.D.: Putting the 'theory' back into grounded theory: guidelines for grounded theory studies in information systems. Inf. Syst. J. **20**(4), 357–381 (2010)
19. Urquhart, C., Fernandez, W.: Using grounded theory method in information systems: the researcher as blank slate and other myths. J. Inf. Technol. **28**(3), 224–236 (2013)
20. Salvato, C., Vassolo, R.: The sources of dynamism in dynamic capabilities. Strateg. Manag. J. **39**(6), 1728–1752 (2018)
21. Rahi, S., et al.: Citation classification using natural language processing and machine learning models. In: Conference 2019, Name. Springer (2019)

22. Iqbal, S., et al.: A decade of in-text citation analysis based on natural language processing and machine learning techniques: an overview of empirical studies. Scientometrics **126**(8), 6551–6599 (2021)
23. Hassan, S.-U., et al.: Deep context of citations using machine-learning models in scholarly full-text articles. Scientometrics **117**(3), 1645–1662 (2018)
24. Hassan, S.-U., et al.: Sentiment analysis of tweets through Altmetrics: a machine learning approach. J. Inf. Sci. **47**(6), 712–726 (2021)
25. Safder, I., et al.: Sentiment analysis for Urdu online reviews using deep learning models. Exp. Syst. e12751 (2021)
26. Mahmood, Z., et al.: Deep sentiments in roman Urdu text using recurrent convolutional neural network model. Inf. Process. Manag. **57**(4), 102233 (2020)

Performance Measurement and Management Systems in Local Government Networks: Stimulating Resilience Through Dynamic Capabilities

Luca Mazzara, Gennaro Maione, and Giulia Leoni

Abstract Local government networks often develop in unpredictable environments and, as a consequence, their abilities and resources have to be prepared for flexible responses, the so-called "dynamic capabilities". One of the most desirable capacities they might reach is resilience, understood as the skill to cope with unpredicted dangers after they become real. This paper reviews literature and conceptual outcomes resulting from the analysis and contextualization of the Dynamic Capabilities (DCs) Theory, providing a contribution to an effective improvement of resilient governance for performance measurement and management systems (PMMS) within local government networks. The fusion of the concepts of resilience, governance, and DCs applied to PMMS offers both theoretical and practical implications. Regarding the theoretical implications, the presence of DCs in resilient inter-municipal governance might help sense, shape and seize opportunities, as well as enhance, combine and reconfigure assets, not only for the single local government but also for the whole community. Concerning the practical implications, the work suggests that DCs applied to resilient governance al-low and facilitate the overcoming of bureaucratic resistances typical of public sector organizations through the networking of local governments that pursue compatible objectives.

Keywords Performance measurement and management system (PMMS) · Local government networks · Resilience · COVID-19 · Dynamic capabilities

---

L. Mazzara · G. Leoni
University of Bologna, 40126 Bologna, Italy
e-mail: luca.mazzara@unibo.it

G. Leoni
e-mail: giulia.leoni10@unibo.it

G. Maione (✉)
University of Salerno, 84084 Fisciano, Italy
e-mail: gmaione@unisa.it

A. Visvizi et al. (eds.), *Research and Innovation Forum 2022*,
Springer Proceedings in Complexity, https://doi.org/10.1007/978-3-031-19560-0_45

1 Introduction

According to the shift from New Public Management to New Public Governance, sub-national government collaboration focused on interdependent horizontal relationships rates increased among several countries, especially in terms of cooperation among Local Governments (LGs) [1]. Such a theoretical mutation has also led to a change in a practical setting since over the last decades there has been a delegation of competencies from the State to LGs according to a model based on the creation of stable cooperative relationships among same-level institutions to increase the capacity, efficacy, and efficiency in delivering local public services. The collaboration among a multiplicity of LGs may increase resilience by fostering a more efficient resource measurement and management. Within the inter-organizational context, Performance Measurement and Management Systems (PMMSs) are considered the most common approach used for network control. Indeed, inter-institutional PMMSs within LG networks are increasingly playing a decisive role in the decision-making process by supporting monitoring, coordination, and trust among a coordinated plurality of actors [2]. However, to make the jump from measurement to management, performance data requires interpretation and usage [3]. This performance management movement intends to generate a "purposeful" employment of information [4], thus directed to improve efficiency and effectiveness and to support decision and action in a multi-dimensional logic. Through the analysis and contextualization of the Dynamic Capabilities (DCs) Theory, this paper provides a contribution for an effective improvement of resilient governance for PMMSs within LG networks. Specifically, the work aims to answer the following research question (RQ): How can dynamic capabilities help to improve resilient governance for PMMSs within LG networks?

2 Dynamic Capabilities as a Conceptual Framework to Study

The phrase "dynamic capabilities" was first developed by Teece et al. DCs reflect the development of new competencies in response to discontinuous and unpredictable changes of internal and external conditions [5], giving the organization a long-term competitive advantage. When the environment evolves rapidly and unpredictably, organizations may ensure a continuous adaptation through a constant development of resources [6] and routines [7]. The concept of DCs summarizes the two key elements for achieving competitive advantage: the term "dynamic" refers to the ability to renew competencies in changing environments; the term "capabilities" emphasizes the ability of strategic management to redefine and integrate resources and competencies with those held by other organizations [8]. The DCs can be qualified as emerging structures defined at a local level [6], rooted in organizational learning, and designed to mitigate the problem of excessive rigidity generated by using ordinary capabilities [9]. However, the concept of DCs has changed over time undergoing numerous interpretations, not always coincident [10] but the prevailing orientation in the literature goes back to the Teece et al.'s view, recognizing the existence of three interdependent stages [11]: resource coordination/integration (a static concept); learning (a dynamic concept); and asset reconfiguration (a transformational concept).

3 Inter-municipal Collaboration and Resilient Governance

Over the last decades, many countries have undergone sub-national governance, such as inter-municipal collaboration (IMC) with the aim to promote territorial resilience. IMC can be defined as a fulfillment of public municipal task/s (volunteering or mandatory), by two or more municipalities jointly or by a third legal entity, serving at least two municipalities [12, 13]. The formation of IMC represents a great opportunity to promote a resilient governance. It is "resilient" due to the leverage of scarce resource and capabilities across different organizations and the flexible and adaptive state-social relationships [14–17]. It is "governance" because these interactions are formulated by the actions of various LGs and social actors in general with diverse and potentially conflicting aims [18]. Thus, this highlights how these relationships cannot be directly controlled by the government. Literature on resilience is far from new and deals with the ability to survive to unpredicted dangers or shocks [19]. Thus, organizational resilience is the capacity to cope with unpredicted dangers after they become real. The management literature questions whether resilience is an outcome or latent capability [20]. However, we adopt the resilience as "latent capability" approach, meant as potentially capable to make an effective improvement for the PMMS. This is in line with reference to network governance studies where "there seems to be some reluctance among many who study networks to discuss formal mechanisms of control" [20]. However, the public management literature mainly focuses on resilience of single organizations, such as single LGs [21], rarely considering the potentiality of a LG network context and its governance. In this vein, collaborative public management requires the sharing of information, offering the possibility to manage uncertainty [22]. The pandemic seems to have given an impetus to the importance of redesigning the relationship among citizens and public administration [23], underlining the benefits deriving from increased public value in the processes of public services delivery and use [24, 25]. This effect is felt not only at the level of single LGs but also within inter-municipal contexts, where the existence of shared policies fosters a vision and an organic mission of governance. Pooling resources and capabilities makes possible to concretely support the networks of LGs–especially the small ones which are less able to manage COVID-19 outbreak [26]–often undersized in terms of personnel, with competitive deficits in terms of intangible infrastructures, organizational resources, and IT skills [27]. Thus, it is claimed how IMC may lead towards several advantages in terms of better shock absorption. In this way it is possible to encourage resilient governance and to promote the well-being of the whole community [28, 29]. Thus, it is important that all institutions, associations, and municipalities dialogue horizontally to adopt effective and timely measures capable of providing valid support to the inter-municipal context [30]. In resilient inter-municipal governance, the main decisions require choices about the policy-making lifecycle and tools to be explored over a wide available range, since promoting engagement with citizens becomes strategically and operationally important.

4 PMMSs Within Inter-municipal Contexts

As managers are alike responsible for the performance of the organization when public services are delivered through the network, great emphasis is placed on the development of shared PMMSs, necessary for a common vision and strategy, plan objectives, targets, measurements, and evaluation processes. According to the PMMSs approach, every public sector organization should measure, evaluate, and report their performance by producing different documents (performance measuring and evaluating system, performance plan, performance report and so forth). Despite the innovative elements and pragmatic approach, the implementation of this reform by LGs, and their form of cooperation, has proved to be a big challenge. Indeed, most LGs still do not understand the value of PMMS, especially for a multi-level performance analysis system. The criticisms are mainly based on the different organizations and cultures of the cooperating partners, which also differentiate the design and use of PMMSs, and which cannot address the wide range of information required by the LG networks. Indeed, the provision of sufficient information and knowledge necessary to allow for a correct decision-making process becomes difficult without a well-articulated definition of information and performance indicators [31] capable of allowing interoperability among the cooperating partners. Furthermore, the negative effects in terms of low importance and consideration of PMMSs might derive from the differences between the partners (for example, in terms of size there could be small municipalities with few human resources), especially whether the activity is politically disinterested [32]. This leads to measurement difficulties and limited availability of performance information to provide feedback for internal management and public accountability purposes. In this regard, Bryson et al. [33], recognizing the effects that PMMS produce on partners' behaviors and on the structure of their relationships, affirm that governance as a set of coordination and monitoring activities must take place for collaborations to survive. In this sense, the use of a PMMS increases the level of communication, trust, commitment, and participation of the actors involved, recognizing that these are fundamental levers for the success of any network [34].

5 Dynamic Resilient Governance to Improve PMMSs in Inter-municipal Contexts

In light of what has been discussed so far, dynamic resilient governance takes shape as a critical success factor for public sector organizations [35, 36], as governments are continually asked to face changing expectations on their responsiveness and the effectiveness of the services provided. In the inter-municipal context, resilient governance assumes an even more decisive role [37] as it allows the dynamic management of multilateral relations among governments at various levels. Therefore, the use of the DCs framework applied to resilient governance can prove useful to help improve the results of inter-municipal networks and facilitate the achievement of the objectives pursued. In this perspective, the application of DCs to resilient governance may favor an improvement of PMMSs in the inter-municipal context. The fusion of the concepts of resilience, governance, and DCs applied to PMMS offers implications under a dual profile, theoretical and practical. Regarding the theoretical implications, DCs applied to resilient inter-municipal governance could be conveniently disaggregated into three organizational levels: *(i)* sensing and shaping opportunities; *(ii)* seizing opportunities; and *(iii)* enhancing, combining, and reconfiguring the assets of both single LGs and the network as a whole. Concerning the practical implications, the work suggests that DCs applied to resilient governance allow and facilitate the overcoming of bureaucratic resistances typical of public sector organizations through the networking of LGs that pursue compatible objectives. Within the inter-municipal context, dynamic resilient governance implies collaborative interactions aimed at triggering new ideas and their implementation through the coordination and integration of the resources [38]. Thus, being dynamic means that resilient multi-actor governance allows for the regeneration of skills and the exploitation of all relevant resources in terms of knowledge, imagination, creativity, courage, transformative capacity, and political authority among the actors involved [39, 40], such as LGs. To perceive and model emerging opportunities both theoretically and practically, inter-municipal networks should constantly explore the possibility of promoting resilience. Once an opportunity has been identified, it is necessary to outline a shared strategy to achieve the compatible objectives pursued. Finally, a reconfiguration of systems, procedures, routines, structures and know-how is necessary to solve complex problems by adapting the inter-municipal network to the unstable environment in which many LGs operate.

6 Concluding Remarks

This paper focuses on the DC theory contextualized in the COVID era as a lens to understand the possible influence on the issues in LG networking regarding PMMSs. In response to the RQ–How can dynamic capabilities help to improve resilient governance for PMMSs within LG networks?–, the findings represent that a LG network would be more resilient to shocks because collaboration helps municipalities leverage of scarce resource and capabilities across different organizations. Moreover, we provide recommendations useful for both scholars and practitioners (managers and policymakers) on how to achieve an effective performance measurement and management improvement through a resilient governance. The propositions rely on relevant literature with the hope that future research will develop empirical evidence of IMC experiences. Empirical works would offer valuable insights, allowing understanding whether DCs provide concrete support to resilient governance in addressing PMMS challenges within IMC contexts. In this regard, it is worth pointing out that some scholars affirm that the use of DCs theory may prove to be dangerously vague and tautological, theorizing only an indirect effect those capabilities on performance. This is also the limit of this paper: whether on the one hand the DCs theory may prove useful for dealing with environmental changes, on the other, it is unclear how this result may be achieved [41, 42].

References

1. Hulst, J., van Rudie, A.J.G.M.: Montfort: Institutional features of inter-municipal cooperation: cooperative arrangements and their national contexts. Public Policy Adm. **27**(2), 121–144 (2012)
2. Moynihan, D.P., Fernandez, S., Kim, S., LeRoux, K.M., Piotrowski, S.J., Wright, B.E., Yang, K.: Performance regimes amidst governance complexity. J. Public Adm. Res. Theory **21**(suppl_1), i141–i155 (2011)
3. Moynihan, D.P.: Through a glass, darkly: Understanding the effects of performance regimes. Public Perform. Manag. Rev. **32**(4), 592–603 (2009)
4. Alojairi, A., Akhtar, N., Ali, H.M., Basiouni, A.F.: Assessing Canadian business IT capabilities for online selling adoption: a net-enabled business innovation cycle (NEBIC) perspective. Sustainability **11**(13), 3662–3678 (2019)
5. Ludwig, G., Pemberton, J.: A managerial perspective of dynamic capabilities in emerging markets: the case of the Russian steel industry. J. East Eur. Manag. Stud. 215–236 (2011)
6. Teece, D.J., Pisano, G., Shuen, A.: Dynamic capabilities and strategic management. Strateg. Manag. J. **18**(7), 509–533 (1997)
7. Eisenhardt, K.M., Martin, J.A.: Dynamic capabilities: what are they? Strateg. Manag. J. **21**(10–11), 1105–1121 (2000)
8. Vicari, S.: Brand Equity. Il potenziale generativo della fiducia. Milan (1995)
9. Easterby-Smith, M., Lyles, M.A., Peteraf, M.A.: Dynamic capabilities: current debates and future directions. Br. J. Manag. **20**, S1–S8 (2009)
10. Ambrosini, V., Bowman, C., Collier, N.: Dynamic capabilities: an exploration of how firms renew their resource base. Br. J. Manag. **20**, S9–S24 (2009)
11. Kalim, K., Arshad, M.A.: Towards high performance strategy: context of Pakistan's public sector organisations. Int. J. Public Sector Perform. Manag. **4**(4), 498–515 (2018)
12. Steiner, R.: The causes, spread and effects of intermunicipal cooperation and municipal mergers in Switzerland. Public Manag. Rev. **5**(4), 551–571 (2003)
13. Fedele, M., Moini, G.: Cooperare conviene? Intercomunalità e politiche pubbliche. Rivista italiana di politiche pubbliche **1**(1), 71–98 (2006)
14. Oliver, C.: Determinants of interorganizational relationships: integration and future directions. Acad. Manag. Rev. **15**(2), 241–265 (1990)
15. Scholten, K., Schilder, S.: The role of collaboration in supply chain resilience. Supply Chain Manag. Int. J. **20**(4), 471–484 (2015)

16. Barasa, E., Mbau, R., Gilson, L.: What is resilience and how can it be nurtured? A systematic review of empirical literature on organizational resilience. Int. J. Health Policy Manag. **7**(6), 491–503 (2018)
17. Kim, K., Andrew, S.A., Jung, K.: Building resilient organizations: organizational resilience as a network outcome. Int. J. Public Adm. **44**(15), 1319–1328 (2021)
18. Yao, Z., Li, B., Li, G., Zeng, C.: Resilient governance under asymmetric power structure: the case of Enning Road Regeneration Project in Guangzhou, China. Cities **111**(1), 1–17 (2021)
19. Wildavsky, A.B.: Searching for Safety. Transaction Publishers, Piscataway (1988)
20. Linnenluecke, M.K.: Resilience in business and management research: a review of influential publications and a research agenda. Int. J. Manag. Rev. **19**(1), 4–30 (2017)
21. Provan, K.G., Kenis, P.: Modes of network governance: Structure, management, and effectiveness. J. Public Adm. Res. Theory **18**(2), 229–252 (2008)
22. Fitzgerald, A., Lupton, R.: The limits to resilience? The impact of local government spending cuts in London. Local Gov. Stud. **41**(4), 582–600 (2015)
23. McGuire, M.: Collaborative public management: assessing what we know and how we know it. Public Adm. Rev. **66**, 33–43 (2006)
24. Christensen, T., Lægreid, P.: Balancing governance capacity and legitimacy: how the Norwegian government handled the COVID-19 crisis as a high performer. Public Adm. Rev. **80**(5), 774–779 (2020)
25. Osborne, S.P., Nasi, G., Powell, M.: Beyond co-production: value creation and public services. Public Adm. **99**(4), 641–657 (2021)
26. Maione, G., Sorrentino, D., Kruja, A.D.: Open data for accountability at times of exception: an exploratory analysis during the COVID-19 pandemic. Transform. Gov. People Process Policy, Ahead-of-print (2021)
27. Troisi, O.: Governance e Co-creazione di valore nella PA: una rilettura in ottica Service-Dominant Logic. Giappichelli Editore, Turin (2016)
28. Polese, F., Troisi, O., Carrubbo, L., Grimaldi, M.: An integrated framework toward public system governance: insights from viable systems approach. In: Cross-Sectoral Relations in the Delivery of Public Services. Emerald Publishing Limited, Bingley (2018)
29. Visvizi, A., Lytras, M.D.: Government at risk: between distributed risks and threats and effective policy-responses. Transform. Gov. People Process Policy **14**(3), 333–336 (2020)
30. Teles, F.: Local governance and intermunicipal cooperation. Springer, Berlin (2016)
31. Bleyen, P., Klimovský, D., Bouckaert, G., Reichard, C.: Linking budgeting to results? Evidence about performance budgets in European municipalities based on a comparative analytical model. Public Manag. Rev. **19**(7), 932–953 (2017)
32. Barretta, A., Busco, C.: Technologies of government in public sector's networks: In search of cooperation through management control innovations. Manag. Account. Res. **22**(4), 211–219 (2011)
33. Bryson, J.M., Crosby, B.C., Stone, M.M.: The design and implementation of Cross-Sector collaborations: Propositions from the literature. Public Adm. Rev. **66**(1), 44–55 (2006)
34. Kaplan, R.S., Norton, D.P., Rugelsjoen, B.: Managing alliances with the balanced scorecard. Harv. Bus. Rev. **88**(1), 114–120 (2010)
35. Troisi, O., Visvizi, A., Grimaldi, M.: The different shades of innovation emergence in smart service systems: the case of Italian cluster for aerospace technology. J. Bus. Ind. Market. Ahead-of-Print (2021)
36. Visvizi, A., Troisi, O., Grimaldi, M., Loia, F.: Think human, act digital: activating data-driven orientation in innovative start-ups. Eur. J. Innov. Manag. Ahead-of-Print (2021)
37. John, P.: The great survivor: the persistence and resilience of English local government. Local Gov. Stud. **40**(5), 687–704 (2014)
38. Bowman, A.O.M., Parsons, B.M.: Vulnerability and resilience in local government: assessing the strength of performance regimes. State Local Gov. Rev. **41**(1), 13–24 (2009)
39. Chandler, D.: Resilience: The Governance of Complexity. Routledge, Abingdon-on-Thames (2014)

40. Lebel, L., Anderies, J.M., Campbell, B., Folke, C., Hatfield-Dodds, S., Hughes, T.P., Wilson, J.: Governance and the capacity to manage resilience in regional social-ecological systems. Ecol. Soc. **11**(1), 19–41 (2006)
41. Platts-Fowler, D., Robinson, D.: Community resilience: a policy tool for local government? Local Gov. Stud. **42**(5), 762–784 (2016)
42. Wheeler, B.C.: NEBIC: A dynamic capabilities theory for assessing net-enablement. Inf. Syst. Res. **13**(2), 125–146 (2002)

Leveraging Artificial Intelligence (AI) to Build SMEs' Resilience Amid the Global Covid-19 Pandemic

Mandy Parkinson, Jackie Carter, and Raheel Nawaz

Abstract As a result of Covid-19 and the ensuing sharp decline in economic activity, business leaders had to quickly adapt to a disruptive marketplace, changing consumer behaviour, and new internal processes. The crisis accelerated adoption of new technologies such as Artificial Intelligence (AI) which could help to improve business processes and evolve new business models, products and services. This exploratory research reports on a preliminary survey with 81 SME owners from various sectors who undertook a seven-week programme to build foundational knowledge on the opportunities of AI for their business and sector. Of those surveyed, 49% turned to AI due to the impact of Covid-19 on their business, sector, and economy. Our results demonstrate the extent of changes small businesses have made as a result of the pandemic, for example, 64% expanded existing services and developed new products and services, with 45% expanding existing product lines. We anticipate the research will directly contribute to existing knowledge by challenging prevailing beliefs about productivity and the role of digital technology adoption. We show that by contributing to the limited literature on micro-enterprise digital technology adoption and demonstrating how a dual approach of implementation of cutting-edge technologies and new management practices can help overcome repercussions of global crises. Our findings suggest that SMEs are currently facing a wide variety of business challenges. However, the integration of a newly developed AI innovation may enable them to overcome these challenges, if adopted with the right level of support.

Keywords Covid-19 · Artificial intelligence · SMEs

M. Parkinson (✉) · J. Carter · R. Nawaz
Manchester Metropolitan University, Manchester M15 6BH, UK
e-mail: M.Parkinson@mmu.ac.uk

J. Carter
e-mail: j.k.carter@mmu.ac.uk

R. Nawaz
e-mail: R.Nawaz@mmu.ac.uk

A. Visvizi et al. (eds.), *Research and Innovation Forum 2022*,
Springer Proceedings in Complexity, https://doi.org/10.1007/978-3-031-19560-0_46

1 Background

Small to medium sized enterprises (SMEs) employ more than 87 million people across the EU and are considered the strength of the EU economy. Furthermore, in 2019 there are 5.6 million micro-businesses in the UK accounting for 96% of all businesses, micro-businesses have 0.9 employees and sit in the lower quadrant of small businesses [1]. However, support targeted at micro-businesses to assist in their increased productivity and subsequent continued contribution to the economy is often fragmented and not always relevant to the challenges they face [2]. As a result of "Covid-19" and the sharp drop-in economic activity which followed, business leaders had to quickly adapt to a disruptive marketplace, changing consumer behaviour, and new internal processes. The crisis accelerated the adoption of new technologies such as Artificial Intelligence (AI) which could help to improve business processes, enabling them to create new business models, products, and services. However, with adoption comes further challenges, for as Beckinsale and Ram [3] highlighted failure to successfully implement and exploit digital technologies are due in part to management limitations.

This limitation can result in the organisation not viewing digital technologies in a strategic manner and thus implementation will be ad hoc and unformulated [3–6]. In addition, Beckinsale and Ram [3] believe that within a micro-enterprise the failure to successfully implement and exploit digital technologies are due in part to management limitations and with new technological adoption comes new challenges especially for resource-constrained businesses who lack the internal technological expertise, finances, and time to leverage and maximise the value of AI.

Dandridge and Levenburg [7] and Fink and Disterer [8] all maintain that research surrounding digital technologies adoption and in particular digital within small businesses is reasonably well established, however, they also dispute that the same can be said for literature in relation to digital technology adoption and the micro-business. Simmons et al. [9, 10] believe that although limited in relation to small and micro-businesses the importance of digital adoption has an increased standing in literature within the arena.

2 Research Gap

This exploratory research aims to investigate small firm owners, and in particular micro-businesses response to the Covid-19 pandemic and technology adoption in terms of where they perceive AI to fit within their digital strategy and potentially contribute toward their recovery strategy. Importantly, we explore the resources required for SMEs to leverage this cutting-edge technology. Building on the work of Troisi and Grimaldi [11] the long-term impact of forced adoption due to the COVID-19 pandemic needs to be considered to ascertain whether the forced adoption of technologies has had a positive long-term effect on the businesses concerned.

The lack of research available in this area has resulted in practitioners, whilst understanding the importance of digital technologies and AI in particular, feeling ill equipped to maximise on the opportunities these platforms have to offer. The lack of guidance within this arena places considerable pressure on micro businesses to adopt the latest technologies even when they are unclear as to how to maximise these Kalakota and Robinson [12] and Jeyaraj et al. [13]. Furthermore, the sectoral differences can have a detrimental impact when imposing a one-size fits all digital strategy [14].

Existing literature highlights that micro-business have not exploited digital technologies to its maximum potential because they were either incapable, unable or unwilling. Moreover, there were very few examples of effective adoption and application [15]. Had the adoption been the result of a planned strategy of adoption rather than in response to competitive pressures and 'media hype' then the results may have been different [16]. Fitzgerald et al. [17] believe that digital technology adoption occurs in isolated incidents within an organisation, and not usually seen holistically. Moreover, Simmons et al. [9] argue that by adopting digital solutions the micro business can leverage its strengths and thus create competitive advantage. However, as the attitudes of the owner manager have been cited as having a direct impact on adoption and productivity, both positive and negative does adoption of technology automatically ensure competitive advantage? Gray [18] would argue that an understanding of the micro-business owners work motivations is essential in order to ensure this. This was further supported by the work of Baccarani and Golinelli [19] and Troisi et al. [20] which highlighted the importance of business owners adopting an innovation mind-set when pursuing alternative solutions.

It could be argued that the skills-shortages regarding digital solutions is a critical factor in successfully adopting such technologies [12, 21, 22]. This is especially the case within micro-businesses, as they frequently have minimal or no training in utilising digital technologies nor are they fully briefed in the capabilities of such systems and thus do not utilise them fully [23, 24].

Indeed, capacity and capability to embrace digital technologies such as AI, has a direct correlation with those organisations that have demonstrated a stimulus for innovation, which in turn encourages aspiration for growth and potentially increases commercial activity [1]. However, it is important to remember that there are obstacles to growth for the micro-businesses, which can include the lack of business skills and the lack of access to adequate information [25].

It has long been argued that the lack of managerial capabilities, especially in relation to specialist functions such as AI, can hamper successful adoption and positive impact on growth [26]. This lack of managerial capabilities in relation to technology adoption, coupled with the forced nature of technology adoption that was enforced due to Covid-19, had the potential to both positively or negatively impact on the micro-businesses performance during this period.

It can be argued that there is a distinct difference between business and technical knowledge, for the owner-manager there subject knowledge is often the reason the business was created but in order to expand they will need to either develop or secure business knowledge and skills. Often the lack of business skills can result in the owner-manager being risk averse to adoption of new technologies. Furthermore, Williams et al. [27] argued that the fear of failure and negative growth expectations, added a further constraint for micro-business owners. These two components were particularly prevalent during Covid-19 for many businesses.

Williams et al. [27] state that "lack of growth ambition, fuelled by fear of failure, risk aversion, negative growth expectations, and a focus on non-economic objectives, represents an important constraint for the owner-managers centric MBs but also a challenge for policy makers"

The lack of digital technology skills within micro businesses is the biggest barrier to exploiting the technologies, it is often out of the reach of the micro businesses to employ an AI specialist and therefore they are at a disadvantage to their larger competitors. Inaddition, micro businesses may invest in a technology and when this does not automatically return results, assume the system is at fault, when in fact they are not utilising or putting the time and training into it.

Throughout the literature review, it became apparent that literature on micro-business and digital adoption is highly fragmented and little is known about the intersection of increased productivity and digital technology. Also, it has been argued that technical knowledge alone is not sufficient to increase productivity and thereby growth within a business; knowing how to relate this to business capabilities is what truly propels productivity and thus financial growth. This clearly raises the issue of the need to explore the impact of new technology adoption on productivity within micro businesses, in order to truly close the gap between research and practice in this area and thus provide the support that micro business practitioners' require.

3 AI-Related Readiness, Knowledge, and Skills

We conducted a preliminary, exploratory survey with 81 SME businesses from a range of sectors, these businesses were predominantly micro businesses. This sample was chosen because the businesses recruited had expressed an interest in participation in the Greater Manchester AI Foundry project, during the period October 2020 to December 2021 which was at the height of the Covid_19 global pandemic. GM AI Foundry is an ERDF funded R&I programme for small businesses in the Greater Manchester Local Enterprise Partnership (LEP) area. This collaborative project between 4 universities in the North West of England provides SME business with a seven-week programme to build foundational knowledge on the opportunities of AI for their business and sector.

The survey was conducted online and at the time of the survey the businesses had not yet received any training or support from the AI Foundry programme. The questionnaire included a combination of Likert scale questions, open ended text questions and multiple-choice selection questions. The questions shown in Table 1 were used to assess delegates AI-readiness with regard to having the appropriate infrastructure to support AI integration, the suitable management and governance mechanisms to sustain AI solutions, the ability to maximise the value AI could provide, and whether their team of employees are ready to implement and work with AI.

The results in Fig. 1 indicate that, despite having limited resources and mechanisms in place to support AI technology at present, the majority of respondents have the ability to maximise the value that AI brings.

Table 1 Questionnaire design

Question	Question response options
Q10.1 The impact of Covid-19 on small firms motivated me to explore the opportunities of AI for my business	5-point Likert scale (Strongly Agree, Agree, Neither Agree nor Disagree, Disagree, Strongly Disagree)
Q10.2 The impact of Covid-19 on small firms motivated me to seek support on the opportunities of AI for my business	5-point Likert scale (Strongly Agree, Agree, Neither Agree nor Disagree, Disagree, Strongly Disagree)
Q11 How has the Covid-19 pandemic positively impacted your business?	Open text
Q12 How has the Covid-19 pandemic negatively impacted your business?	Open text
Q13 As a result of the Covid-19 pandemic, my company has: (select all that apply)	Multi-choice selection with options: • Taken on external finance • Expanded existing product lines • Expanded existing services • Developed entirely new products and/or services • Adopted new technology • Identified new skills gaps within my business that need to be filled. Please state which skills gaps below: • Other, please specify below
19_1My Business has the appropriate infrastructure and interfaces to support AI technology	5-point Likert scale (Strongly Agree, Agree, Neither Agree nor Disagree, Disagree, Strongly Disagree)
19_2 My business has suitable management and governance mechanisms to sustain AI solutions	5-point Likert scale (Strongly Agree, Agree, Neither Agree nor Disagree, Disagree, Strongly Disagree)
19_3 My business has the ability to maximise the value it gets from AI	5-point Likert scale (Strongly Agree, Agree, Neither Agree nor Disagree, Disagree, Strongly Disagree)
19_4 My team of employees are ready to implement and work with AI technology	5-point Likert scale (Strongly Agree, Agree, Neither Agree nor Disagree, Disagree, Strongly Disagree)

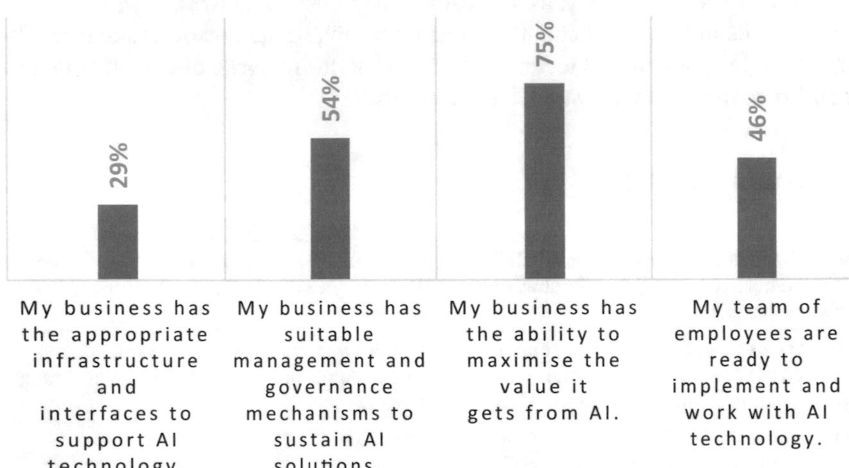

Fig. 1 AI readiness of participants, their knowledge, and skills

4 External Impacts: Covid-19

The external economic impact of Covid-19 on beneficiaries' motivation to adopt AI and seek support on how to implement this technology was investigated. The results highlighted below suggest that almost half of participants turned to AI due to the impact of Covid-19 on their business, sector, and economy, see Table 2.

Of those surveyed, the results highlighted that 49% beneficiaries turned to AI due to the impact of Covid-19 on their business, sector, and economy. The analysis of the findings provides a strong indication of the changes small businesses have made as a result of the pandemic, for example 64% expanded existing services and developed new products and services, with 45% expanding existing product lines, see Fig. 2.

Moreover, we asked beneficiaries to express both the positive and negative impacts the Covid-19 pandemic has had on their business. Responses are highlighted in Table 3. Findings suggest that small firms are currently facing a wide variety of business challenges. However, the integration of a newly developed AI innovation may enable them to overcome these challenges, if adopted with the right level of support. This finding will be further tested by adopting a mixed method case study

Table 2 Summary of results from the survey

Conclusion 1	Conclusion 2
The impact of Covid-19 on small firms motivated **49%** of beneficiaries **to explore the opportunities of AI for their business**	The impact of Covid-19 on small firms motivated **40%** of **beneficiaries to seek support on the opportunities of AI for their business**

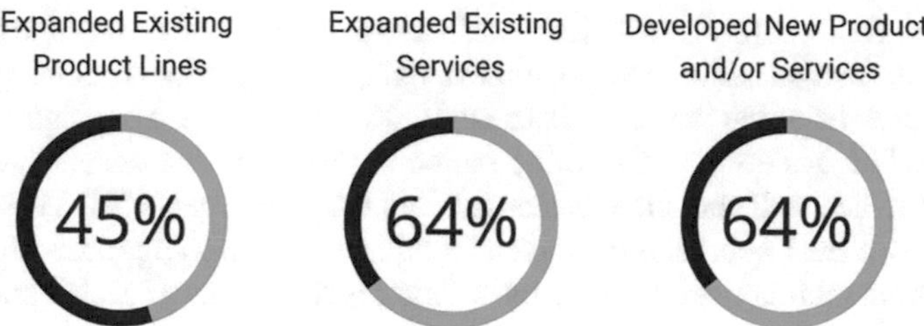

Fig. 2 Results of preliminary, exploratory survey with 81 small firm owners from various sectors

Table 3 Beneficiaries responses to express both the positive and negative impacts the Covid-19 pandemic

Positive	Negative
Increased product/service demand due to changing consumer behaviour	Decreased product/service demand due to changing consumer behaviour
Provided small firm owners the time to start up and launch their company	Disrupted the supply chain
Ability to work more flexibly including in remote locations without a full office in operation This has led to increased technical skills across the workforce due to adoption of remote technology and platforms	Limited the opportunities for networking with others in the industry and demoing
Time saved due to lack of travelling to meetings. More efficient use of time including more time spent on products refinement	Reduced employee productivity and interaction including meeting face-to-face
Technology adoption	Limited investment opportunities
Increased employment number	Decreased turnover
Personal development including increased focus	Decreased advertising spend budgets and spend due to limited return on investment (e.g., in the travel industry)
Improved sustainability	Delayed progress, momentum, and impacted other partners

approach to examine whether AI has effectively supported them to improve their operations and functioning, working toward recovery in the current economic crisis.

5 Contribution to Knowledge

We anticipate the research will directly contribute to existing knowledge by challenging prevailing beliefs about productivity and the role of adoption of digital technologies. Not only will this research build on the work of Jones et al. [21], by contributing to the limited literature on micro-business digital technology adoption by Fink and Disterer [8], it will also contribute to the body of knowledge that exists in this arena. Also, in addition to quantitative data, we plan to incorporate open ended questionnaires from the participants to apply advance natural language processing [28, 29] techniques and deep learning methods [30] to better understand learning behaviour of the small and micro businesses and their sentiments [31, 32] by demonstrating how a dual approach of implementation of cutting-edge technologies and new management practices can help overcome repercussions of global crises.

6 Limitations and Future Work

The survey result does not provide a statistically robust sample size in order to inform any future policy decisions in this arena, in particular on the impact of forced adoption on productivity. This coupled with the fact these particular companies pro-actively expressed an interest in understanding AI provides the potential for survey bias.

For future studies, in order to address this limitation, it would be prudent to expand the survey reach to a wider variety of small businesses and compare the different attitudes to technology and limitations of management capabilities of the owner-managers on the impact of adoption. The expanded survey should incorporate businesses in our region that have not pro-actively expressed interest in such technology.

In addition, a longitudinal study should also be considered in order to review the impact of the forced technologies post COVID-19 pandemic, which could be compared with those of similar studies in other regions of the world.

References

1. Pickernell, D., Jones, P., Packham, G., Thomas, B., White, G., Willis, R.: E-commerce trading activity and the SME sector: an FSB perspective. J. Small Bus. Enterp. Dev. **20**(4), 866–888 (2013)
2. Greenbank, P.: Training micro-business owner-managers: a challenge to current approaches. J. Eur. Ind. Train. (2000)
3. Beckinsale, M., Ram, M.: Delivering ICT to ethnic minority businesses: an action-research approach. Eviron. Plann. C. Gov. Policy **24**(6), 847–867 (2006)
4. Rodgers, J.A., Yen, D.C., Chou, D.C.: Developing e-business; a strategic approach. Inf. Manag. Comput. Secur. (2002)
5. Nath, R., Akmanligil, M., Hjelm, K., Sakaguchi, T., Schultz, M.: Electronic commerce and the Internet: issues, problems, and perspectives. Int. J. Inf. Manag. **18**(2), 91–101 (1998)
6. Darch, H., Lucas, T.: Training as an e-commerce enabler. J. Workplace Learn. (2002)
7. Dandridge, T., Levenburg, N.M.: High-tech potential? An exploratory study of very small firms' usage of the Internet. Int. Small Bus. J. **18**(2), 81–91 (2000)
8. Fink, D., Disterer, G.: International case studies: To what extent is ICT infused into the operations of SMEs? J. Enterp. Inf. Manag. (2006)

9. Simmons, G., Armstrong, G.A., Durkin, M.G.: A conceptualization of the determinants of small business website adoption: setting the research agenda. Int. Small Bus. J. **26**(3), 351–389 (2008)
10. Simmons, G., Armstrong, G.A., Durkin, M.G.: An exploration of small business website optimization: enablers, influencers and an assessment approach. Int. Small Bus. J. **29**(5), 534–561 (2011)
11. Troisi, O., Visvizi, A., Grimaldi, M.: The different shades of innovation emergence in smart service systems: the case of Italian cluster for aerospace technology. J. Bus. Ind. Market. (2021). ahead-of-print, no. ahead-of-print. https://doi.org/10.1108/JBIM-02-2020-0091
12. Kalakota, R., Robinson, M.: "e-Business," Roadmap for Success (2000)
13. Jeyaraj, A., Rottman, J.W., Lacity, M.C.: A review of the predictors, linkages, and biases in IT innovation adoption research. J. Inf. Technol. **21**(1), 1–23 (2006)
14. Drew, S.: Strategic uses of e-commerce by SMEs in the east of England. Eur. Manag. J. **21**(1), 79–88 (2003)
15. Pickernell, D., Jones, P., Packham, G., Thomas, B., White, G., Willis, R.: E-commerce trading activity and the SME sector: an FSB perspective. J. Small Bus. Enterp. Dev. (2013)
16. Taylor, M., Murphy, A.: SMEs and e-business. J. Small Bus. Enterp. Dev. (2004)
17. Fitzgerald, M., Kruschwitz, N., Bonnet, D., Welch, M.: Embracing digital technology: a new strategic imperative. MIT Sloan Manag. Rev. **55**(2), 1 (2014)
18. Gray, C.: Entrepreneurship, resistance to change and growth in small firms. J. Small Bus. Enterp. Dev. (2002)
19. Baccarani, C., Golinelli, G.M.: "Le parole dell'innovazione" Sinergie Italian. J. Manag. **94**, 9–14 (2014)
20. Troisi, O., Grimaldi, M.: Guest editorial: Data-driven orientation and open innovation: the role of resilience in the (co-)development of social changes. Transform. Gov. People Process Policy **16**(2), 165–171 (2022)
21. Jones, P., Simmons, G., Packham, G., Beynon-Davies, P., Pickernell, D.: An exploration of the attitudes and strategic responses of sole-proprietor micro-enterprises in adopting information and communication technology. Int. Small Bus. J. **32**(3), 285–306 (2014)
22. Simpson, M., Docherty, A.: E-commerce adoption support and advice for UK SMEs. J. Small Bus. Enterp. Dev. **11**(3), 315–328 (2004)
23. Wolcott, P., Kamal, M., Qureshi, S.: Meeting the challenges of ICT adoption by micro-enterprises. J. Enterp. Inf. Manag. (2008)
24. Levy, M., Powell, P., Worrall, L.: Strategic intent and e-business in SMEs: enablers and inhibitors. Inf. Resour. Manag. J. (IRMJ) **18**(4), 1–20 (2005)
25. Business Strategy Review: The next entrepreneurial frontier: micro-entrepreneurs (2011). http://faculty.london.edu/rchandy/microentbsr.pdf. Accessed 24 Sept 16
26. MacPherson, A., Holt, R.: Knowledge, learning and small firm growth: a systematic review of the evidence. Res. Policy **36**(2), 172–192 (2007)
27. Williams, N., Gherhes, C.A., Vorley, T., Vasconcelos, A.C.: Distinguishing micro-businesses from SMEs: a systematic review of growth constraints. J. Small Bus. Enterp. Dev. **23**(4), 939–963 (2016)
28. Rahi, S., Safder, I., Iqbal, S., Hassan, S.U., Reid, I., Nawaz, R.: Citation classification using natural language processing and machine learning models. In: International Conference on Smart Information Communication Technologies, pp. 357–365. Springer, Cham (2019)
29. Iqbal, S., Hassan, S.U., Aljohani, N.R., Alelyani, S., Nawaz, R., Bornmann, L.: A decade of in-text citation analysis based on natural language processing and machine learning techniques: an overview of empirical studies. Scientometrics **126**(8), 6551–6599 (2021)

30. Hassan, S.U., Imran, M., Iqbal, S., Aljohani, N.R., Nawaz, R.: Deep context of citations using machine-learning models in scholarly full-text articles. Scientometrics **117**(3), 1645–1662 (2018)
31. Safder, I., Mahmood, Z., Sarwar, R., Hassan, S.U., Zaman, F., Nawab, R.M., Bukhari, F., Abbasi, R.A., Alelyani, S., Aljohani, N.R., Nawaz, R.: Sentiment analysis for Urdu online using deep learning models. Expert. Syst. **28**, e12751 (2021)
32. Mahmood, Z., Safder, I., Nawab, R.M., Bukhari, F., Nawaz, R., Alfakeeh, A.S., Aljohani, N.R., Hassan, S.U.: Deep sentiments in roman Urdu text using recurrent convolutional neural network model. Inf. Process. Manag. **1;57**(4), 102233 (2020)

Circular Economy Engagement in the Agri-Food Industry During the Covid-19: Evidence from the Twitter Debate

Benedetta Esposito, Daniela Sica, Maria Rosaria Sessa, and Ornella Malandrino

Abstract The Covid-19 pandemic has exacerbated the environmental crisis in which our ecosystem is posed. In this context, the call to reorganize the production and consumption models to implement sustainable economic models is emerging. Accordingly, the Circular Economy paradigm, based on the reduction, reuse and recycling practices, has spurred as one of the best ways to manage this emergency state. The scientific literature has highlighted that, to shift from a traditional linear economic model to a circular economic one, the involvement of the whole supply chain is required, especially in the agri-food sector. In this perspective, the stakeholders' engagement plays a pivotal role in reaching the global goal. The present research aims to explore the stakeholders' perception of messages conveyed through social media on circular economy in agri-food, using a coding framework based on the reclassification of the "Glossary of Circular Economy" according to a 4-R paradigm (reduce, reuse, recycle and reduce). In particular, the study analyses the stakeholders' reactions to Twitter posts focused on agri-food and circular economy from the beginning of the pandemic until now.

Keywords Circular economy · Agri-food sector · Stakeholder engagement · Social media

1 Introduction

The recent pandemic generated by the spread of the Covid-19 virus has compounded our ecosystem's social and environmental emergency state. In this scenario, the need to revise production and consumption models by shifting towards a circular economy (CE) paradigm inspired by the reduction, reuse and recycling actions (3R) has gained momentum as the best way to curb the crisis' adverse consequences [1].

Literature contends that the transition towards CE is required at a supply-chain level rather than an individual company level [2]. This makes stakeholder engagement

B. Esposito (✉) · D. Sica · M. R. Sessa · O. Malandrino
Department of Business Studies Management and Innovation Systems, University of Salerno, 84084 Salerno, Italy
e-mail: besposito@unisa.it

A. Visvizi et al. (eds.), *Research and Innovation Forum 2022*,
Springer Proceedings in Complexity, https://doi.org/10.1007/978-3-031-19560-0_47

a pivotal lever to succeeding in this global goal because any weak supply chain link would make an effort to shift towards CE worthless [3]. Therefore, organizations, policymakers and citizens are asked to act synergistically to foster a resilient transition towards sustainable development based on a CE model [4].

Accordingly, companies implementing CE principles in their business activities devote primary attention to disclosing information about their policies and practices CE-related to increase stakeholders' engagement and—in turn—increase sustainable awareness among them [5]. Attuned, the disruptive effects caused by the pandemic have emphasized the urgency to develop a dynamic and interactive approach toward CE disclosure. So, the interactive potential of social media (SM) tools is being more and more exploited by organizations to enhance the engagement of a more comprehensive forum of stakeholders [6, 7].

With this in mind, the present research explores stakeholders' perception of CE disclosure in the agri-food industry through SM, using a coding framework based on the reclassification of the "Glossary of Circular Economy" according to the circular economy paradigm.

Previous studies have explored the social media debate using Twitter to explore the engagement level on several topics (e.g., [8–10]).

To this end, the study analyses the stakeholders' reactions to Twitter posts focused on agri-food and circular economy from the beginning of the pandemic until now. Moreover, the communication direction and balance level have been examined.

The remainder of the paper is structured as follows. After the introduction, the first section presents the literature review and the theoretical background. The second section describes the research methodology adopted. The third section discusses the principal results. Lastly, the main conclusions, implications and future research directions are provided.

2 Literature Review and Theoretical Background

During the last decades, the concept of CE has been widely explored, leading to the development of several definitions according to the multiple currents of study, such as regenerative design [4]; industrial ecology [11]; industrial symbiosis [12]; green economy [13] and cradle to cradle design [14]. For that reason, the CE concept has been defined as an umbrella concept, which inglobes several issues linked to the regeneration and restoration of the ecosystems [15]. In this proliferation of scientific contributions and definitions, the Ellen MacArthur Foundation has provided a widely recognized meaning according to which CE is "a systems solution framework that tackles global challenges like climate change, biodiversity loss, waste, and pollution" [16]. This definition highlights that the implementation of the CE model could be considered the final goal and the mean due to the operational approach provided by its framework.

Taking into account the importance of CE to foster the Ecological Transition, governments– both at the European and International level—have promoted numerous initiatives and guidelines to catalyze the shift from a linear economy to a circular economy, such as the European Circular Economy Action Plan [17–20], the 2030 Agenda for Sustainable Development, the European Green Deal [21–25] and the UN Climate Change Conference 2021 [26].

In order to activate a circular cycle of materials, resources and energies, scholars have demonstrated the need to involve all the actors of the ecosystem to co-create value [27] and activate virtuous processes for the waste, losses and pollution reduction and—at the same time—the process of self-sustainability [28].

In the wake of CE relevance for all the ecosystem actors, stakeholders have called for comprehensive and reliable information from both public and private organizations on CE initiatives, practices and policies [24]. Furthermore, the organizations have to evaluate the results of CE investments in order to implement correction or improvement actions to create circular value. Thus, information related to sustainable and circular practices is essential to developing efficient circularity strategies [29]. Moreover, information on CE is also crucial for investors, policymakers, and other stakeholders due to the growing interest and sensitivity. Sharing information on CE could reduce information asymmetries which are considered one of the main barriers to implementing a CE model [30].

Furthermore, CE information is vital for customers, increasingly engaged in environmental and sustainable needs. As a result, companies committed to sustainable and circular values are oriented to disclose CE information. This communication choice has a double ratio. On the one hand, companies can build their corporate image and—in turn—reach a competitive advantage. On the other hand, they can engage their stakeholders in their best practices [31].

Considering this topic's relevance, scholars have started investigating CE communication strategies and practices (i.e. [32–34]). In particular, Jakhar et al. [32] have explored the impact of stakeholder pressures on a firm's circular economy actions and their consequential disclosure. Instead, Unal et al. [33] have investigated how firms can generate and grasp value from a circular economy business model and communicate this value outside the organization. Lastly, Scarpellini et al. [34] have explored environmental accounting practices implemented in companies that adopt CE models to engage their stakeholders with the sustainable practices and results obtained from their business models. However, no studies explore the CE engagement phenomena to the best of our knowledge. Aiming to fill this gap, the present research is built on a theoretical perspective that combines the stakeholder theory and the legitimacy theory.

The stakeholder theory is based on the assumption that organizations have to create value in the long term through engagement with their stakeholders [35]. Stakeholders are defined by Freeman [36] as a group or individual who can influence or be influenced by the organization's actions. In this pathway toward a circular transition the agri-food sector is called to engage with its stakeholders to create a virtuous cycle and co-create sustainable value for society [37, 38].

Therefore, "the more an organization engages with its stakeholders, the more accountable and responsible that organization is towards these stakeholders" [39, p. 1]. Thus, agri-food industries are called to fulfil the consumers' request for accountability, especially after the pandemic generated by the spread of the Covid-19 pandemic. The use of Social Media (SM) can support the agri-food sector to enhance stakeholder engagement since it can be considered a tool to improve the dialogue between companies and their stakeholders [40]. In this perspective, academics have considered SM as supporting dialogic instruments for sharing information on the stakeholders' expectations [41]. Accordingly, SM are increasingly becoming strategic in the CE field, where it is progressively essential that agri-food companies take stakeholders' engagement to effectively shift towards CE from a supply chain level [42].

Moreover, the use of SM is useful to engage stakeholders, obtain their collaboration and sharing of circular principles and values, obtain legitimacy by their stakeholders, and become more competitive in the market. The legitimacy theory argues that organizations should adhere to the system of values, norms, and expectations of the social context they operate [43]. Therefore, the agri-food sector is called to undertake activities to reach legitimacy from society and survive in the long term [42].

Drawing from this background, the present paper investigates how agri-food companies engage with their stakeholders.

3 Research Methodology

The present study explores the level of CE disclosure and its related degree of stakeholder engagement on tweets focused on the circular economy and the agri-food sector.

Twitter was chosen for several rations. This SM is considered one of the widely used social networks worldwide [44]. Moreover, Twitter supports scholars in collecting and analyzing data since it is a valid open-source platform [45]. Lastly, despite the character limitation of Twitter's posts, the concision of the content can reach a wider audience, and organizations are likely to publish posts more frequently [46]. Accordingly, Twitter has been considered a suitable social network to build engagement [12].

The research methodology implemented is described in Fig. 1. A content analysis of qualitative-quantitative nature has been performed. All tweets with the hashtags "#circulareconomy AND # agri-food" published from the 9th of March 2020 until the extraction data on the 10th of March 2022 were extracted, purified and analyzed using data mining techniques. The period has been chosen since our research objective was to investigate the level of engagement from Twitter's accounts regarding the topics of circular economy in the agri-food sector during the pandemic and post-pandemic period, considering all the Coronavirus waves in order to explore if the pandemic state has raised awareness among stakeholders regarding the need to collaborate in order to reach the ecological transition and overcome the emergency state that involves the environment, society and the worldwide economy. Nevertheless, our data do not consider Tweets published after the extraction, showing only a one-sided view of the disclosure trend for 2022.

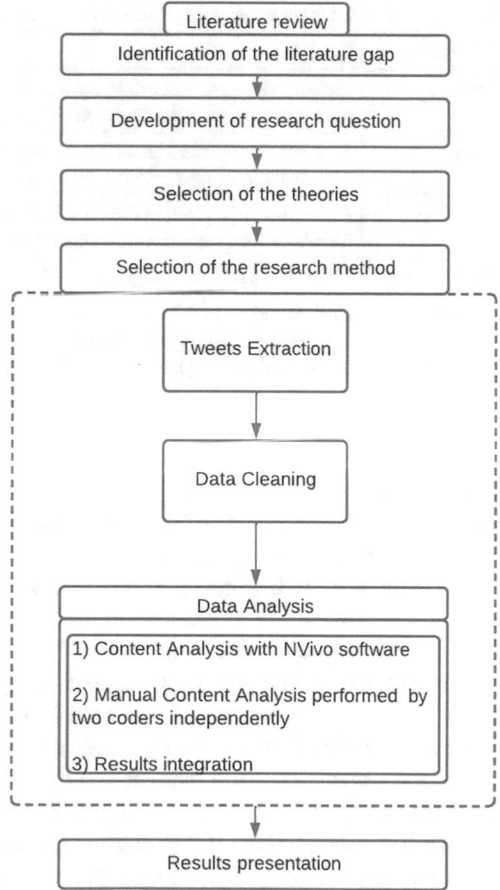

Fig. 1 Research methodology

The data mining procedure was carried out through the NVivo software. In particular, the open-source extension "NCapture" based on Application Programming Interface (API) has been used, which provides access to Twitter's public accounts [47]. The data extracted for each post comprises the Tweet's publication date, the number of "likes", and the number of "retweets". More specifically, the number of likes suggests the users' satisfaction with specific content, while the number of retweets could represent the level of debate among Twitter's users [42].

Following Esposito et al. [42], the data analysis has been developed into three steps (Fig. 1): (1) dictionary-based content analysis through a supervised machine learning technique using the NVivo software, (2) manual content analysis performed by two coders independently, and (3) integration of results.

According to the reduce-reuse-recycle-redesign paradigm, the researchers have developed a dictionary reclassifying the Circular Economic glossary. Moreover, an additional "general" category has been added to englobe all tweets that refer to the circular economy but do not fit the reduce-reuse-recycle-redesign categories. In

Table 1 Circular economy glossary reclassified according to the 4-R paradigm

Categories	Concepts
Reduce	Carbon footprint reduction; environmental impact reduction; raw materials reduction; waste reduction; emissions reduction
Reuse	Alternate materials; disassembly; durability; maintain; redistribute; refurbish; remanufacture; repair; reuse; upcycling; waste diversion
Recycle	Anaerobic digestion; compostable; composting; end-of-life; Radio-Frequency Identification; recyclability; waste conversion; water conservation
Redesign	Dematerialization; design; Raw Material Conversion
General	Circular economy; biodiversity; closed-loop; finite materials; green financing; regenerative production; renewable energy; renewable materials; renewable source; reverse logistics; sharing; virgin materials

Table adapted from Barnabè and Nazir [5] and "Glossary of Circular Economy"

particular, five items have been identified (i.e. reduce items, reuse items, recycle items, redesign items and general CE items). A group of words referring to the CE for each item were selected (Table 1).

The Supervised Machine Learning technique implemented permits the automatic measurement of the occurrences through the use of the NVivo software. Furthermore, with the purpose to see if the content of the Tweets has an informing or interacting character [48], two independent coders have performed a manual content analysis using an empirically grounded approach [49]. The software performs an enhanced qualitative analysis through this process, avoiding the data precoding. More specifically, the researchers have categorized the posts as "informing" if they displayed an activity, a performance or an initiative. On the contrary, the tweets have been classified as "interacting" if they outlined an engagement with stakeholders [42].

Moreover, this study has explored the level of stakeholder engagement by examining the communication direction of each tweet. In particular, the messages that activated a comment by a user were categorized as "two-way communication"; otherwise were classified as "one-way communication" [48]. Furthermore, in order to avoid subjective interpretation and assess the inter-coder reliability, Krippendorff's alpha index (α) was computed. The corresponding coefficient—calculated on the 10th of March 2022 on the first 20% of posts—equals 0.85 is acceptable because it belongs to the range between 1.00 (equivalent) and 0.00 (completely different) [50]. Lastly, the results of both analyses have been integrated and holistically presented in the following section.

4 Results and Discussion

Table 2 summarises the descriptive statistics of "CE" tweets published from 2020 to March 2022. Our findings show that 71.61% of the extracted tweets could be classified as a CE message compared with the total extracted tweets. However, in line with Esposito et al. (2020), our results show that the CE tweets have a higher level of engagement than the non-CE messages (i.e. likes for CE tweets: mean = 72.8; st. dev. = 73.6; retweets for CE tweets: mean = 8.7; st. dev = 9.1). The low value of retweets could be interpreted in the light of the fact that most Twitter users interact with all contents, mainly providing feedback (likes) rather than sharing posts (retweets).

Figure 2 represents the communication trend over time from the beginning of the Covid-19 pandemic until the analysis was carried out. From 2020 to 2022, gradual development of CE communication is shown, with a peak in 2021. These findings are in line with Barnabè and Nazir [5]. In fact, with the rise of the Covid-19 pandemic, there has been a proliferation of tweets and interest in CE issues since the pandemic, which has imposed the need to converse with the agri-food stakeholders. Moreover, these findings suggest that, after the spread of the pandemic, the growing awareness of agri-food companies, policymakers and associations of the need to be proactive in overcoming the crisis are called to disclose CE practices, on the one hand, to obtain legitimacy from its primary stakeholder, on the other hand, to engage with them and involving them for the restart of the whole economic system. The results provided for 2022 are partly considered because the tweets extracted are circumscribed to the date the analysis was carried out (i.e. 10th of March, 2022).

The tweets classified as CE content have been codified following the analytical framework developed according to the 4 R paradigm. The coding process has produced the following results. Table 3 provides the descriptive statistics for each CE dimension (i.e. reduce, reuse, recycle, redesign, general). Our findings suggest that the disclosure of CE in the agri-food sector has been mainly focused on the recycling dimension (i.e. 52.8%). In comparison, only 12.36% is based on reduced issues. These results are in line with previous studies on CE disclosure (e.g., [5, 51]), according to which the attention towards recycling has started before the spread of

Table 2 Classification of Tweets

	CE Tweet extracted		Total Tweet	
	Mean	St. Dev	Mean	St. Dev
Like	72.8	73.6	52.3	53.6
Retweet	8.7	9.1	5.7	6.4
Obs	2750		3840	
Obs%	71.61%		100%	

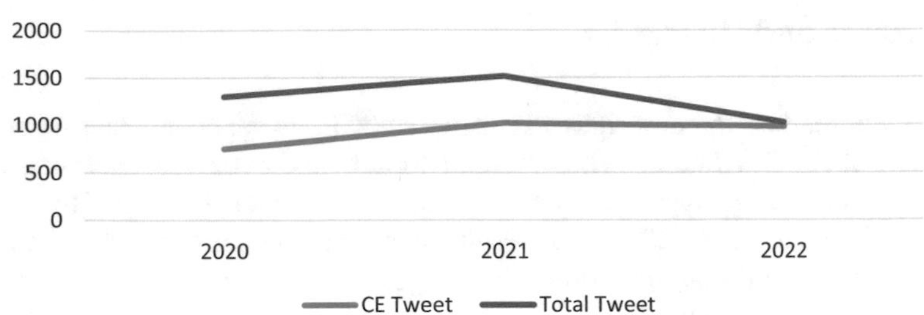

Fig. 2 Evolution of Tweets from 2020 to 2022 (As of the 10th of March 2022)

the CE models, and so there has been a higher awareness on this issues. Considering that recycling practices are more straightforward than reducing, reusing and redesigning, disclosing these issues is easier for companies, no-profit organizations and policymakers.

Furthermore, the average rate of likes shows that recycling, reusing issues and posts that could be considered more generalized posts on CE generate a higher level of engagement (i.e. reuse posts = 32.6; recycling posts = 52.8; general posts = 36.8). The lowest engagement level is detected in the redesign dimension (i.e. 19.6). Nevertheless, posts on general CE issues provide more retweets (i.e. 20.6).

Considering the direction and balance of the communication of the posts, Table 4 provides the total distribution from 2020 to 2021. In general, the highest percentage of CE posts is categorized as two-way communication (i.e. 67.34%). As described in Fig. 3, a higher level of engagement is reached in 2021, confirming our results.

Table 3 Classification of Tweets according to the CE framework

	Reduce		Reuse		Recycle		Redesign		General CE	
	Mean	SD	Mean	SD	Mean	SD	Mean	SD	Mean	SD
Like	26.8	27.2	32.6	33.4	52.8	55.3	19.6	21.2	36.8	38.1
Retweet	15.2	16.5	10.3	11.9	19.3	21.2	9.8	10.2	20.6	21.4
Obs	340		435		890		425		660	
Obs%	12.36%		15.82%		32.36%		15.45%		24.01%	

Table 4 Direction and balance of communication of the CE tweets from 2020 to 2022*

Direction type	Total	
	n	%
One-way communication	898	32.66
Two-way communication	1852	67.34
Informative communication	1253	45.56
Interacting communication	1497	54.44
Total CE tweets		

* As of the 10th of March 2022

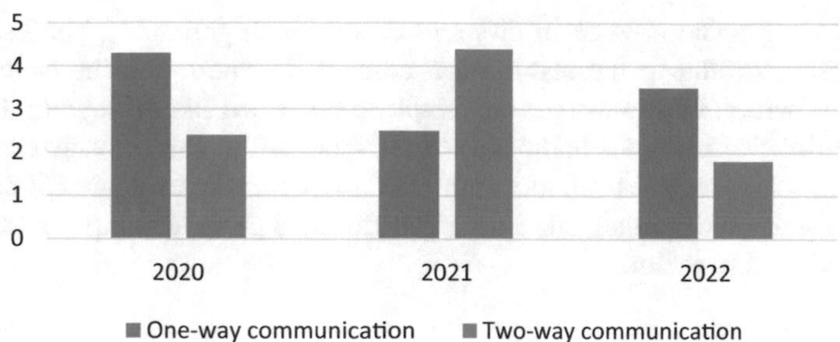

Fig. 3 Evolution of stakeholder engagement on CE from 2020 to 2022

Concerning the balance of communication, Table 4 shows that 54.44% of the CE tweets published have interacting content, while the 45.56% have the purpose of communicating CE practices.

5 Conclusions

In conclusion, our preliminary results have highlighted that social media could be used as communication and disclosure instruments to reveal information about CE practices and policies in the agri-food sector and also to create a dialogue with stakeholders and society at large. This involvement could be useful to raise awareness of the need for the community to be a part of this transition. Indeed, the stakeholder dialogue emerges as an essential issue that allows managers of the agri-food supply chain to include stakeholders' perspectives in their strategies.

This research could be helpful both for academics and managers who can rely on our results to investigate and implement SM disclosure strategies in the CE field to empower stakeholder engagement. Furthermore, social media managers of the agri-food sector could publish interacting content that encourages the digital debate to enhance the engagement level with stakeholders. Lastly, policymakers can develop guidelines and frameworks for CE reporting through SM.

Some theoretical implications could be taken into account in the light of this study. Although the scientific literature on CE engagement and disclosure is still scant, SM could be reputed as the most useful tools to build a dialogue with stakeholders. Accordingly, scholars could deeper investigate this topic in order to provide useful suggestions to the agri-food managers to develop engaging communication strategies.

Furthermore, academics can investigate the CE disclosure through SM in the light of different theoretical approaches and theories.

However, this study is not without limitations. Our study is limited to a specific period, and the Tweets' extraction has been performed using software with some extraction limitations. Accordingly, future studies could use different tools to extract and analyze the data, also developing other analytical frameworks. Moreover, scholars can explore the Tweets from a sample of agri-food companies settled in a specific geographic area or in different countries, highlighting similarities and differences according to the institutional context. Furthermore, the analysis only focuses on Twitter. Future works could explore other SM like Instagram, Facebook and LinkedIn, evaluating the findings obtained from different disclosure channels. In conclusion, scholars could conduct multiple case studies to evaluate CE communication and stakeholder engagement among different agri-food companies according to their commodity sector.

References

1. Merli, R., Preziosi, M., Acampora, A.: How do scholars approach the circular economy? A systematic literature review. J. Clean. Prod. **178**, 703–722 (2018)
2. Esposito, B., Sessa, M.R., Sica, D., Malandrino, O.: Towards circular economy in the Agri-food sector. A systematic literature review. Sustainability **12**(18), 7401 (2020)
3. Gupta, S., Chen, H., Hazen, B.T., Kaur, S., Gonzalez, E.D.S.: Circular economy and big data analytics: a stakeholder perspective. Technol. Forecast. Soc. Chang. **144**, 466–474 (2019)
4. Stahel, W.R.: The circular economy. Nat. News **531**(7595), 435 (2016)
5. Barnabe, F., Nazir, S.: Investigating the interplays between integrated reporting practices and circular economy disclosure. Int. J. Product. Perform. Manag. **70**(8), 2001–2031 (2020)
6. Lytras, M.D., Visvizi, A.: Big data and their social impact: preliminary study. Sustainability **11**(5067), 1–18 (2019)
7. Troisi, O., Fenza, G., Grimaldi, M., Loia, F.: Covid-19 sentiments in smart cities: the role of technology anxiety before and during the pandemic. Comput. Hum. Behav. **126**(106986), 1–16 (2022)
8. Mora-Cantallops, M., Sánchez-Alonso, S., Visvizi, A.: The influence of external political events on social networks: the case of the Brexit Twitter Network. J. Ambient. Intell. Humaniz. Comput. **12**, 4363–4375 (2021)
9. Visvizi, A., Jussila, J., Lytras, M.D., Ijäs, M.: Tweeting and mining OECD-related microcontent in the post-truth era: a cloudbased app. Comput. Hum. Behav. **107**(105958), 1–7 (2020)
10. Alkmanash, E.H., Jussila, J.J., Lytras, M.D., Visvizi, A.: Annotation of smart cities Twitter microcontents for enhanced citizen's engagement. In: IEEE Access **7**, 116267–116276 (2019). (August 2019)
11. Graedel, T.E.: On the concept of industrial ecology. Annu. Rev. Energy Env. **21**(1), 69–98 (1996)
12. Boons, F., Chertow, M., Park, J., Spekkink, W., Shi, H.: Industrial symbiosis dynamics and the problem of equivalence: Proposal for a comparative framework. J. Ind. Ecol. **21**(4), 938–952 (2017)
13. Mohan, S.V., Nikhil, G.N., Chiranjeevi, P., Reddy, C.N., Rohit, M.V., Kumar, A.N., Sarkar, O.: Waste biorefinery models towards sustainable circular bioeconomy: critical review and future perspectives. Biores. Technol. **215**, 2–12 (2016)
14. Braungart, M., McDonough, W., Kälin, A., Bollinger, A.: Cradle-to-cradle design: creating healthy emissions—A strategy for eco-effective product and system design. J. Clean. Produc. **15**(13), 247–271 (2012)
15. Blomsma, F., Brennan, G.: The emergence of circular economy: a new framing around prolonging resource productivity. J. Ind. Ecol. **21**(3), 603–614 (2017)
16. Ellen MacArthur Foundation. Towards the Circular Economy, vol. 1: An Economic and Business Rationale for an Accelerated Transition; Ellen MacArthur Foundation: Cowes, UK (2013)
17. European Commission. Brussels, Belgium, European Commission. Communication from the commission to the European parliament, the council, the European economic and social

committee and the committee of the regions. In: Closing the Loop—An EU Action Plan for the Circular Economy (2014)

18. European Commission: Brussels, Belgium; Ellen MacArthur Foundation. Towards the Circular Economy: Opportunities for the Consumer Goods Sector, (2015)
19. European Commission: A European Agenda for the Collaborative Economy (No. COM(2016) 356 Final). Brussels (2016)
20. European Commission: Ecodesign Working Plan 2016–2019 (No. COM(2016) 773 Final). Brussels (2016)
21. European Commission. Report on the Implementation of the Circular Economy Action Plan. Brussels (2019)
22. European Commission: Sustainable Products in a Circular Economy-towards an EU Product Policy Framework Contributing to the Circular Economy (2019)
23. European Commission: The European Green Deal, Brussels (2019)
24. European Commission: Circular Economy Action Plan-for a Cleaner and More Competitive Europe. Belgium (2020)
25. European Commission: Statement by President von der Leyen ahead of the G20 Summit and the UN (2021)
26. Climate Change Conference (COP26). https://ec.europa.eu/commission/presscorner/detail/en/STATEMENT_21_5643. Accessed 7 Dec 2021
27. Troisi, O., D'Arco, M., Loia, F., Maione, G.: Big data management: the case of Mulino Bianco's engagement platform for value co-creation. Int. J. Eng. Bus. Manag. **10**, 1–8 (2018)
28. Gusmerotti, N.M., Testa, F., Corsini, F., Pretner, G., Iraldo, F.: Drivers and approaches to the circular economy in manufacturing firms. J. Clean. Prod. **230**, 314–327 (2019)
29. Esposito, B., Sessa, M.R., Sica, D., Malandrino, O.: Exploring corporate social responsibility in the Italian wine sector through websites. TQM J. **33**(7), 222–252 (2021)
30. Testa, F., Gusmerotti, N., Corsini, F., Bartoletti, E.: The role of consumer trade-offs in limiting the transition towards circular economy: The case of brand and plastic concern. Resour. Conserv. Recycl. **181**(106262), 1–11 (2022)
31. Kazancoglu, I., Sagnak, M., Kumar Mangla, S., Kazancoglu, Y.: Circular economy and the policy: a framework for improving the corporate environmental management in supply chains. Bus. Strateg. Environ. **30**(1), 590–608 (2021)
32. Jakhar, S.K., Mangla, S.K., Luthra, S., Kusi-Sarpong, S.: When stakeholder pressure drives the circular economy: measuring the mediating role of innovation capabilities. Manag. Decis. **57**(4), 904–920 (2019)
33. Unal, E., Urbinati, A., Chiaroni, D.: Managerial practices for designing circular economy business models. J. Manuf. Technol. Manag. **30**(3), 561–589 (2019)
34. Scarpellini, S., Marín-Vinuesa, L.M., Aranda-Usón, A., Portillo-Tarragona, P.: Dynamic capabilities and environmental accounting for the circular economy in businesses. Sustain. Account. Manag. Policy J. **11**(7), 1129–1158 (2020)
35. Carroll, A.B.: Corporate social responsibility: perspectives on the CSR construct's development and future. Bus. Soc. **60**, 1258–1278 (2021)
36. Freeman, E.R.: Strategic Management: A Stakeholder Approach. Pitman Publishing Inc., Boston (1984)
37. Visvizi, A., Lytras, M.D., Damiani, E., Mathkour, H.: Policy making for smart cities: innovation and social inclusive economic growth for sustainability. J. Sci. Technol. Policy Manag. **9**(2), 126–133 (2018)
38. Troisi, O., Grimaldi, M., Loia, F.: Redesigning business models for data-driven innovation: a three-layered framework. In: The International Research & Innovation Forum, pp. 421–435. Springer (2020)
39. Greenwood, M.: Stakeholder engagement: beyond the myth of corporate responsibility. J. Bus. Ethics **74**, 315–327 (2007)
40. Bellucci, M., Manetti, G.: Facebook as a tool for supporting dialogic accounting? Evidence from large philanthropic foundations in the United States. Account. Audit. Account. J. **30**(4), 874–905 (2017)

41. Bebbington, J., Brown, J., Frame, B.: Accounting technologies and sustainability assessment models. Ecol. Econ. **61**(2), 224–236 (2007)
42. Esposito, B., Sessa, M. R., Sica, D., Malandrino, O.: Corporate social responsibility engagement through social media. Evidence from the University of Salerno. Adm. Sci. **11**(4), 147 (2021)
43. Suchman, M.C.: Managing legitimacy: strategic and institutional approaches. Acad. Manag. Rev. **20**, 571–610 (1995)
44. Mergel, I.: A framework for interpreting social media interactions in the public sector. Gov. Inf. Q. **30**, 327–334 (2013)
45. Panagiotopoulos, P., Alinaghi, Z.B., Steven S.: Citizen–government collaboration on social media: The case of Twitter in the 2011 riots in England. Gov. Inf. Q. **31**(3), 349–57 (2014)
46. Kim, K.S., Sei-Ching, J., Tien-I, T.: Individual differences in social media use for information seeking. J. Acad. Librariansh. **40**(2), 171–178 (2014)
47. Reyes-Menendez, A.S., Josè, R., Cesar, A.A.: Understanding #WorldEnvironmentDay user opinions in Twitter: a topic-based sentiment analysis approach. Int. J. Environ. Res. Public Health **15**(11), 2537 (2018)
48. Schroder, P.: Corporate social responsibility (CSR) communication via social media sites: Evidence from the German banking industry. Corp. Commun. Int. J. **26**(3), 636–654 (2021)
49. Miles, M., Huberman, A.M.: Qualitative Data Analysis: An Expanded Sourcebook. SAGE, London (1994)
50. Krippendorff, K.: Validity in content analysis, in Mochmann, E. (Ed.), Computerstrategien für die kommunikationsanalyse, pp. 69–112. Campus, Frankfurt, Germany. http://repository.upenn.edu/asc_papers/291
51. Barnabè, F., Nazir, S.: Conceptualizing and enabling circular economy through integrated thinking. Corp. Soc. Responsib. Environ. Manag. **29**, 448–468 (2021)

Data-Driven Management of Material Flows in Circular Economy by Logistics Optimization

Anne-Mari Järvenpää, Jari Jussila, Marianne Honkasaari, Olli Koskela, and Iivari Kunttu

Abstract The aim of the Circular economy (CE) business models is to reuse materials and decrease the need for virgin materials in the value chains. This, in turn, requires close collaboration and information sharing between the value chain stakeholders. For this, digitalization and data play a crucial role. This paper studies how small and medium sized enterprises (SMEs) operating in CE can utilize digitalization and data in managing and optimizing the material flows that are central to their production processes. The paper focuses on four case companies, all operating in CE business in Finland, and analyses how these companies have been able to enhance their material flow management by means of data-driven logistics optimization in a research-based university-industry collaboration. The solutions range from conceptual solutions to mathematical optimization and a tool supported solution concept.

Keywords Data-driven decision-making · Circular economy · Logistics optimization · SMEs

1 Introduction

Circular economy (CE) represents an emerging alternative for traditional linear models, where virgin materials are being used and disposed [1]. The CE incorporates a regenerative system that minimizes the entry and waste of resources, emissions, and expenditure of energy through slowing down, closing, and straightening the energy circuits [2]. As CE aims to reuse materials and decrease the need for virgin materials, there are challenges in the supply chain implementation and coordination [3]. In this manner, the CE business concepts aim at addressing sustainable development needs by minimizing resource input and waste, emissions, and energy leakage without jeopardizing growth and prosperity [3].

A.-M. Järvenpää (✉) · J. Jussila · M. Honkasaari · O. Koskela · I. Kunttu
Häme University of Applied Sciences, 13100 Hämeenlinna, Finland
e-mail: anne-mari.jarvenpaa@hamk.fi

A. Visvizi et al. (eds.), *Research and Innovation Forum 2022*,
Springer Proceedings in Complexity, https://doi.org/10.1007/978-3-031-19560-0_48

Operational requirements related to material flows are significantly higher in CE business than in many other business areas. This is because the challenges related to logistics are emphasized in the management of material flows in CE. Digitalization provides several kinds of opportunities to make the flow of materials more effective, and thus make the processes of CE profitable for the stakeholders. Hence, it is natural that the CE companies show increasing interest in developing and investing in digital solutions that help them to make their material flows more effective [4]. Currently available data on the material supply and need, as well as logistics capabilities provides opportunities for data-driven planning and optimization of the logistics in the value chains of CE.

This paper aims at improving understanding on the logistics optimization in the value chains of small and medium sized enterprises (SMEs) operating on the area of CE. Thus, this paper presents four individual cases and seeks answers to the following research question: *How circular economy SME's can develop their data-driven material flow management by means of optimization and planning?*

We approach this question by means of a case study containing four cases, each representing CE SMEs in Finland. In each case, this paper presents a company specific challenge to which solution has been found by means of a practical development project conducted in collaboration between the company and university. In this manner, each case illustrates a company-specific solution for data-driven optimization of the material flows in the case companies.

2 Optimizing Material Flows in Circular Economy

In the area of CE, SMEs play a key role [5]. They are in the center in the efforts to achieve environmental sustainability and more inclusive growth [6]. However, particularly in the SME domain, new tools and techniques provided by digitalization require education, continuous learning, and innovation. A wide range of technological tools are available to develop digital solutions for circular economy, and data can enable the efficient use of resources and reduce environmental effect. However, there are still challenges in the linear supply chains where data is not adequately shared between the stakeholders [7].

The main challenge in the data-driven management of material flows is the lack of centralized data governance [8]. Data has several owners in the private and public sector with different motives. Currently the platform ecosystem for circular economy in Finland aims to build interoperability between industries and to collect, harmonize and enrich data. The platform ecosystem serves material producers, processors and users, and other partners and platforms [8]. Creating profitable business in circular economy requires data of materials, volumes, qualities, and locations. For example, the target to utilize demolition materials was 70% on 2020, but the actual utilization rate was only 50% [9].

The rapid development of digitalization opens possibilities to data-based approaches to planning and optimization of the management of material flows. This, in turn, often requires open interfaces to share the data related to the material supply and demand, as well as transportation. In this paper, we study how data available from the recycled materials can be utilized in the computational optimization of the material flow management.

3 Methodology

This paper presents a qualitative case study of four CE SMEs, all located in Finland. As illustrated in Table 1, the case companies focus on waste management, recycling services and biogas production. Research data was collected by interviewing company representatives on two interview rounds, conducted in November 2021 and January 2022. The interviews were based on a semi-structured questionnaire, in which the questions were related to logistics challenges and optimization needs.

Information was gathered on logistics challenges and optimization need that are described in the Results section for each case company. In each case, the challenges

Table 1 Case companies

Case	Role of person interviewed	Data collection	Number of employees	Core business area	Solution concept
A	CEO	November 2021 workshop, company interview November 2021 and January 2022	50	Waste management, recycling services and solutions for households and companies	Conceptual solution
B	Service manager	Company interview January 2022	60	Waste management and recycling services for households and companies	Tool supported solution concept
C	CEO	November 2021 workshop, company interview November 2021 and January 2022	30	Glass recycling services and glass products	Conceptual solution
D	CDO	November 2021 workshop, company interview November 2021 and January 2022	90	Waste management and recycling services for households and companies	Mathematical optimization

were analysed in a development project carried out by Häme University of Applied Sciences (HAMK). The solution concepts presented in Results section range from conceptual solutions (case A and case C) to mathematical optimization (case D) and tool supported solution concept (case B).

4 Results

This section summarizes the logistics challenges and needs for optimization in case companies and the conceptual solution for each case. Conceptual solutions developed for the challenges are introduced in the following sub-sections. These challenges relate to the optimization of waste management logistics or the contract areas. The conceptual solutions are based on Internet of Things (IoT), material sharing platform or mathematical modelling.

4.1 *Concept Case A: Optimizing Waste Transportation Routes*

The case company A is interested in how the data alongside the material flow can give value for the company itself and its customers. The question is how to handle raw material acquisition, outgoing products, and deliveries to the end user as efficiently as possible? The specific challenge in this company case is to plan back and forth transportation and try to understand how to utilize data to design sensible routes? The existing data from the vehicles includes spatial information, driving mode and fuel consumption. Sensors installed to the waste containers raise an alarm when the container is getting full. This way, an order is entered into the information system of the company. However, despite the fact that the data is collected, it is not fully utilized. As the objective of the case company is to utilize transportation data to demonstrate and minimize the carbon footprint of individual transport routes, it needs to plan efficient collection route for a certain region. So far, this kind of route planning has been done manually by organizing in the map view either by address or manually drawing a route from one destination to another.

As illustrated in Fig. 1, a solution concept for case A consists of monitoring the filling rate of the different kind of containers with the different sensor technologies. The filling rate data and alerts can be visualised in the map and can be used in the route optimization. The concept for case A consists of four types of scenarios for customer orders: (1) industrial actor contacts case company to get rid of waste, (2) households contacts case company to get rid of waste, (3) IoT based service for households where waste containers are monitored and optimally emptied and (4) IoT based service for industry where waste containers are monitored and optimally emptied.

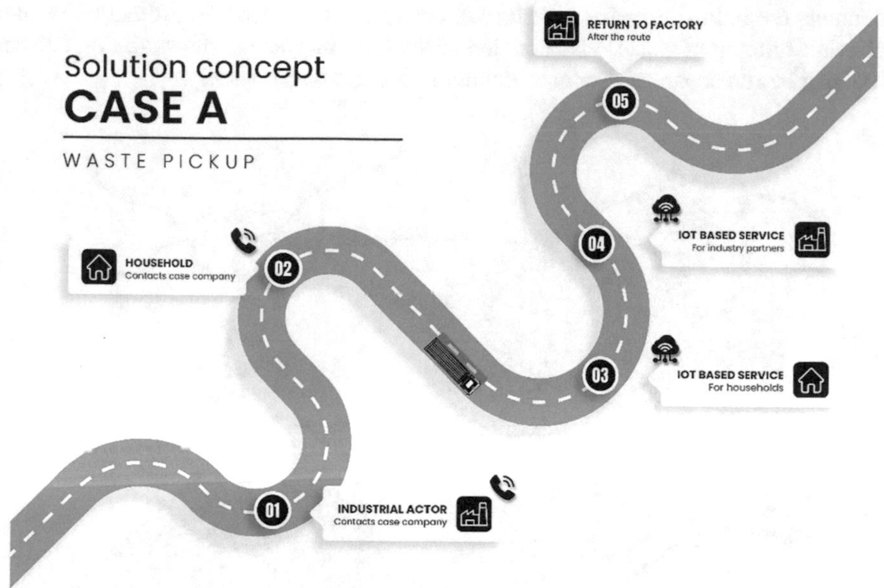

Fig. 1 Case A solution concept

4.2 *Concept Case B: Regional Route Planning*

Case company B needs to be prepared to the changes in law that directs the households waste sorting and their transportation organized by municipalities. It involves publicly procured competition that is to be based on market research and dialogue with prospective service providers. The central questions faced by the company include the following: (1) what kind of collection areas are formed for the contracts, (2) how many vehicles will be needed and (3) how many days the collection takes. There may be multiple contractors by area and by waste, as the law requires the distribution of work. Optimal contract, i.e., collection, areas enable the efficient and flexible collection of waste for the contractor and are accessible for multiple service providers to participate in the tender. Thus, the market mechanism is expected to ensure reasonable pricing for individual households. For the areas of question, there is an existing map-based visualization to support estimations of waste accumulation. Now, company B is interested in quantified indicators of planned routes, waste volume, and capacity in planned contract areas. In addition, company B is interested in route optimization solutions for order-based services without regular routes.

As a solution for case B, a tool was developed that plots customers, i.e., the collection points, on a map and allows the user to draw polygons to specify the desired contract areas. Then, based on these areas, the following indicator metrics were computed: (1) bins within the area, (2) the total number of bins emptying in a year, and (3) the average density of the bins. Using approximations of the bin contents the indicator metrics are further converted to include an estimation of the absolute amount of waste. Visualization of the tool interface is shown in Fig. 2. The tool can be run separately for each contract area and waste variety.

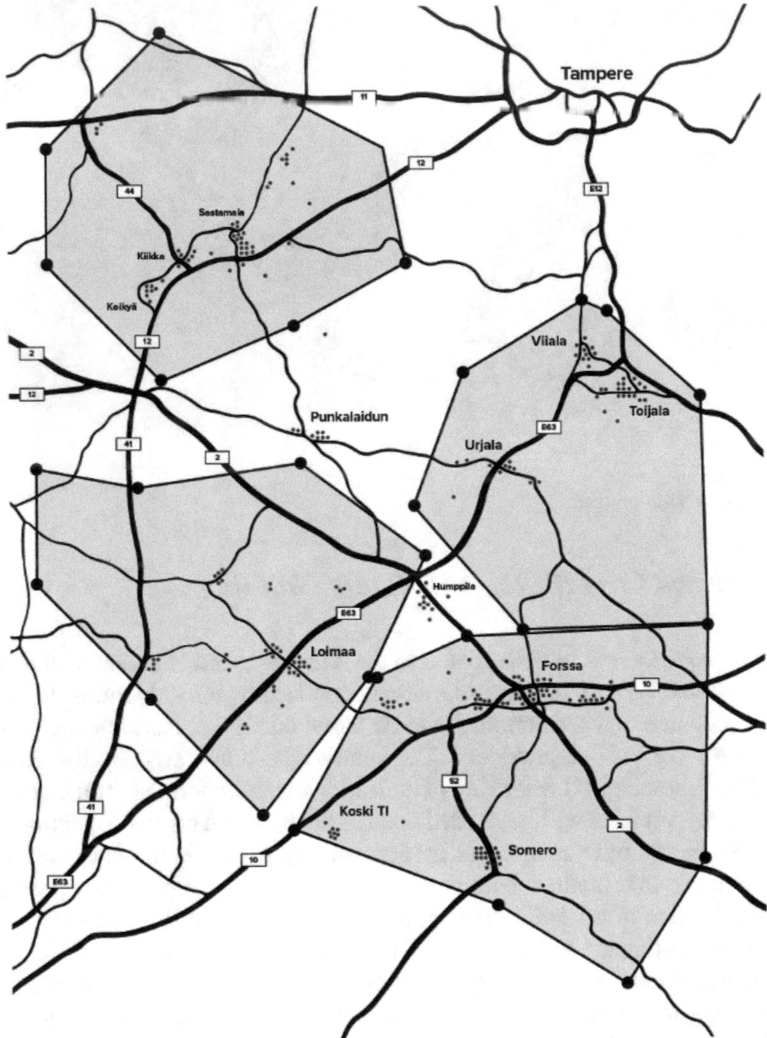

Fig. 2 Illustration of the developed tool in the company case B. Points are shown on the locations where the waste is collected, and polygons are drawn on top of the map to visualize the contract areas

4.3 Concept Case C: Glass Waste Sourcing

Case company C is interested in making a new business concept relying on transportation optimization. The company is making CE business with recycling glass. It transports large quantities and wants to expand operations on the raw material sourcing and export operations. A central question is how to make a long-distance material transport profitable by using sensors, optimization, and smart packaging? In addition, the company is interested in the waste management in construction sites, as there is a lot of glass that is not circulated. Thus, the company is trying to make a solution, where the whole concept competitive in the long run and on the other hand improve their customer service in the raw material procurement.

The solution concept for case C consists of four types of scenarios for glass waste sourcing (Fig. 3): (1) industrial actor generates plate glass as a side stream and contacts case company to get rid of waste, (2) the pre-demolition audit done in construction demolition sites that inform when glass waste is available in demolition sites, (3) waste material availability announcement on the Materiaalitori material sharing platform, (4) IoT based service where waste containers are monitored and optimally emptied (containers that are filled 75% or more).

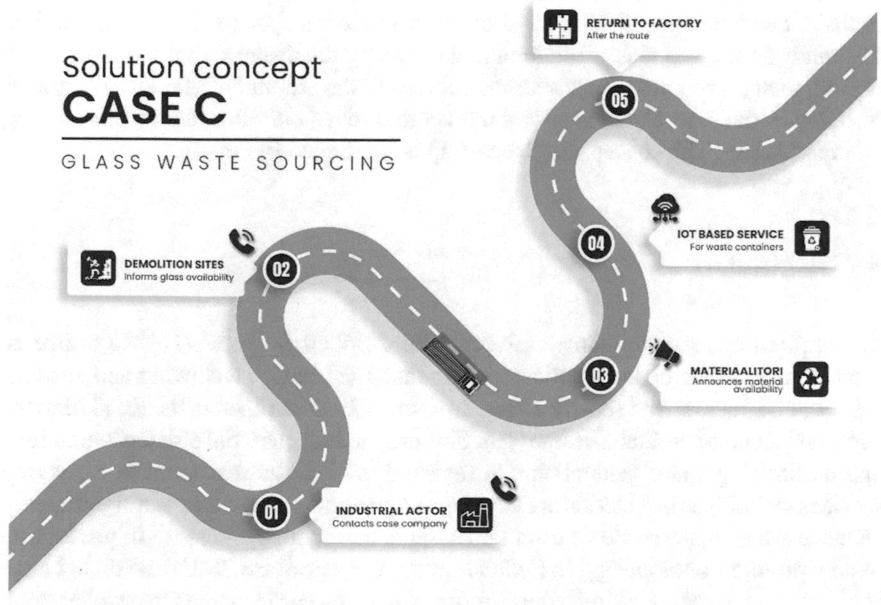

Fig. 3 Case C solution concept

4.4 Concept Case D: Optimizing Locations for Waste Receiving Stations

Company case D explored location options for several new small waste stations in their operating area. Due to a change in the collection of waste management fees from the citizens in a Finnish region, certain waste varieties may now be brought to waste management stations free of charge and their further processing is covered with a mandatory yearly fee. Due to the compulsive nature of waste sorting, the waste receiving stations need to be placed close to the citizens.

In this case, options for possible locations of new waste stations were evaluated based on their average and median distances to the citizens' addresses. In addition, the standard deviation and number of citizens that were considered to be "far away" were used in the estimation. The threshold for "far away" was set to 30 km. The distance was Euclidean, i.e., as the crow flies. Initial situation contained the locations of the current waste stations and the addresses of the citizens in the region. The objective was to create a network of waste stations so that inhabitants have a short distance to the waste stations.

The concept for case D was to explore optimal locations for waste stations by placing potential locations in the operating area and calculating the distances for every resident to the nearest waste station. Locations for new stations were proposed so that all residents would have the shortest possible distance to their nearest station. The analysis showed that, in addition to decreasing the distance to the nearest waste station, well placed new waste stations decreased the amount of "far away" citizens from the initial several thousand to a thousand or even only few hundred depending on the scenario. The concept for the case D is illustrated in Fig. 4.

5 Discussion

The applied research at Häme University of Applied Sciences (HAMK) aims to produce new and creative solutions to increase the vitality, wellbeing and sustainability of businesses and society. Long term goals 2035 outlined in the Road Map for Circular Economy in Kanta-Häme [10], Finland, include material efficient operations and minimizing waste generation. On regional level waste management, recycling services and biogas production are key circular economy actors that can significantly enhance waste material flows from households and industry to reuse in production of new products or as energy. In Finland, many CE actors are SMEs and do not have sophisticated logistics optimization solutions in-house or resources to develop them internally. University-industry collaboration is one means to develop logistics optimization in CE SMEs. The four case studies introduced in the study present solution concepts to regional challenges, however, the way of developing the solutions and the solution concepts themselves can be applied more broadly to CE SMEs in other regions and countries with similar challenges.

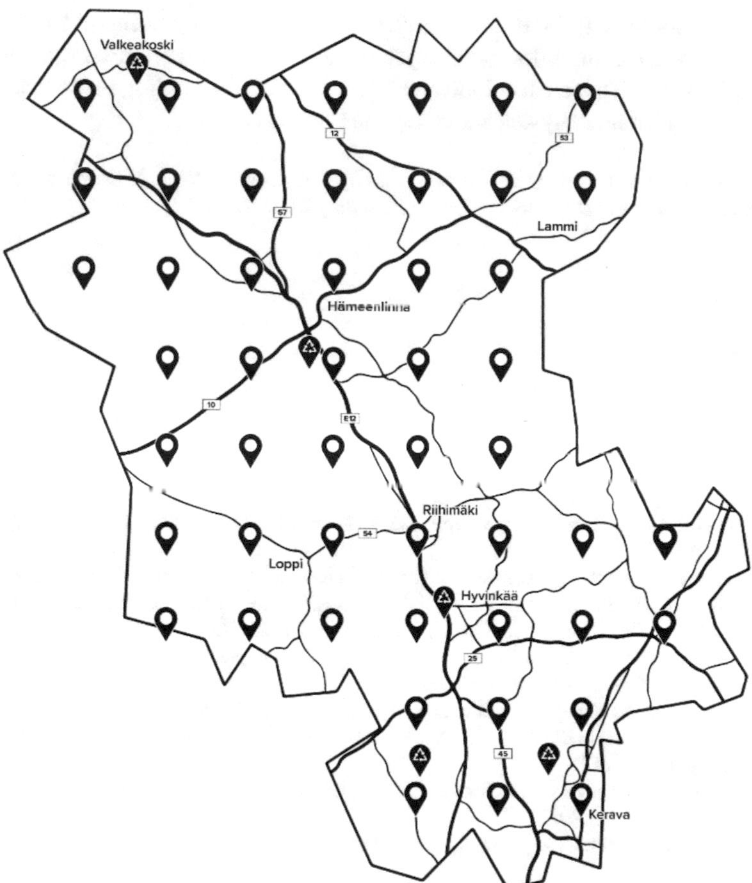

Fig. 4 Case D solution concept

Advanced and continuous logistics optimization requires that data is readily available in a scope that makes optimization possible. For instance, in case B, a tool was developed that plots collection points of material on a map and allows the user to draw polygons to specify desired areas of interest for which all relevant data is computed to support decision-making of company expert. Case D presented a solution concept for determining the optimal placement of new waste stations based on existing data about households. Whereas case A and case C are examples of conceptual solutions that support CE SMEs in sense-making what kind of system can be built to optimize their logistics. As a result of the sense-making process in case A it was decided to launch a student project, where students investigate how historical data of contacts from industrial actors and households could be used to predict new orders for route planning. Case C helped to understand that the case company needs to explore more the available data and opportunities for optimization inside the company, which resulted in recruiting a thesis worker from the university to support the process.

Investigation of CE SMEs problem space also revealed that many CE SMEs in Finland share the same information systems and some of them are based on open source platforms. This clearly indicates opportunities for pooling resources related to logistics information systems and logistics optimization.

Acknowledgements This research was supported by the European Regional Development Project Green Smart Services in Developing Circular Economy SMEs (A77472).

References

1. Suchek, N., Fernandes, C.I., Kraus, S., Filser, M., Sjögrén, H.: Innovation and the circular economy: a systematic literature review. Bus. Strateg. Environ. 1–17 (2021). https://doi.org/10.1002/bse.2834
2. Geissdoerfer, M., Savaget, P., Bocken, N.M.P., Hultink, E.J.: The circular economy–a new sustainability paradigm? J. Clean. Prod. **143**, 757–768 (2017). https://doi.org/10.1016/j.jclepro.2016.12.048
3. Urbinati, A., Chiaroni, D., Chiesa, V.: Towards a new taxonomy of circular economy business models. J. Clean. Prod. **168**, 487–498 (2017). https://doi.org/10.1016/j.jclepro.2017.09.047
4. Järvenpää, A.-M., Kunttu, I., Jussila, J., Mäntyneva, M.: Data-driven decision-making in circular economy SMEs in Finland. Springer Proc. Complex, pp. 371–382 (2021). https://doi.org/10.1007/978-3-030-84311-3_34
5. Järvenpää, A.-M., Kunttu, I., Mäntyneva, M.: Using foresight to shape future expectations in circular economy SMEs. Technol. Innov. Manag. Rev. **10**, 41–50 (2020). https://doi.org/10.22215/timreview/1374
6. OECD: Enhancing the contributions of SMEs in a global and digitalised economy. meet. OECD Counc. Minist. Lev. (2017)
7. Finnish Government: New directions: The strategic programme to promote a circular economy (2021)
8. Kauppila, T., Berg, A., Dahlbo, H., Eilu, P., Heikkilä, P., Hentunen, A., Hilska-Keinänen, K., Horn, S., Ilvesniemi, H., Jenu, S., Karhu, M., Karppinen, T.K.M., Kauppi, S., Kivikytö-Reponen, P., Lavikko, S., Lehtonen, E., Luostarinen, S., Majaniemi, S., Malmi, P., Naumanen, M., Ovaska, J.-P., Pesonen, L., Pesu, J., Pitkänen, H., Pokki, J., Räisänen, M., Salminen, J., Sapon, S., Siivola, E., Sorvari, J., Tanner, H., Tuovila, H., Uusitalo, T., Valtanen, K.: Handbook for a data-driven circular economy in Finland: data sources, tools, and governance for circular design (2022). https://doi.org/10.32040/2242-122X.2022.T401
9. Motiva Oy: Materiaalikiertojen data-alusta ja ekosysteemi [Material cycles data platform and ecosystem]. https://www.motiva.fi/ratkaisut/materiaalitehokkuus/materiaalikiertojen_data-alusta_ja_ekosysteemi. Accessed 12 Mar 2022
10. Regional Council of Häme: Road Map for Circular Economy in Kanta-Häme (2022)

Sustainable Development in the Strategies of Polish Enterprises

Magdalena Tomala

Abstract The purpose of the article is to examine the relationship between corporate revenue and sustainability. The article consists of two parts. The first part analyses the literature on the subject; the second part presents the study of the principles of sustainable development and financial efficiency of enterprises. The survey covered 563 of the biggest industrial and service enterprises operating in Poland (Polska in Największe firmy w Polsce 2021, [1]). This study assesses the contribution that companies have made to the global sustainability concept. It is focused on issues related to the analysis of the dependent variable economic growth and independent variables, i.e.: determinants which described the engagement in sustainable development. In the article, the methods of descriptive statistics have been used. They show the differences in economic growth between different groups of companies engaged in sustainable development. The analysis was completed with selected examples of good practices that companies can use in the future. The article shows, that the implementation of sustainable development at the level of enterprises is not at a satisfactory level. Attention should be paid to the need to introduce state guidelines, which should motivate enterprises to work more intensively in this direction. Therefore, it is necessary to educate and promote the idea of sustainability in order to improve the environmental awareness of consumers. Only their conscious choice of more ecological products can induce companies to work on sustainability.

Keywords Entrepreneurship · Sustainable development · Poland

1 Introduction

In September 2015, world leaders adopted an agenda establishing a set of sustainable development goals to end poverty, protect the planet, ensure the protection of human rights, and guarantee prosperity for all [2]. The adoption of this agenda sets a course of action to reduce economic, social and environmental disparities. It is important to

M. Tomala (✉)
Jan Kochanowski University in Kielce, Kielce, Poland
e-mail: magdalena2828@gmail.com

A. Visvizi et al. (eds.), *Research and Innovation Forum 2022*,
Springer Proceedings in Complexity, https://doi.org/10.1007/978-3-031-19560-0_49

stress that this process fully reflects the European values of social justice, democratic governance and social market economy, as well as environmental protection. These are the goals that should be realized on the regional and state levels, as well as on the level of enterprises.

Following the strategies of companies, it is noticeable that they gradually more often decide to extend the scope of sustainable development goals to other spheres of activity. Corporations more often advertise their activities by emphasizing the advantages of sustainable development on corporate websites and in social media. To what extent does the implementation of the concept of sustainable development in companies contribute to increasing its financial performance? In the literature, researchers note that sustainability is the key to success in business [3–5]. Thus, it can be hypothesized that the implementation of the concept of sustainable development can contribute to increasing corporate profits, establishing the company's position in the market and its development.

The purpose of the article is to examine the relation between corporate revenue and sustainability. The article consists of two parts. The first part analyses the literature on the subject, the second part presents the study of the principles of sustainable development and financial efficiency of enterprises. Polish companies/enterprises refer to entities operating in Poland regardless of ownership.

2 Sustainable Development of Enterprises: Analysis of the Literature on the Subject

The concept of sustainable development is widely discussed in the literature. Researchers pay attention to various aspects of its functioning in enterprises.

The economic aspects of implementing the concept are of primary importance for companies. As Hall points out "entrepreneurs have the potential for creating sustainable economies" [6]. Initially, the idea was considered controversial, but as time goes by, it is increasingly being said that besides profits of tremendous importance are also, for example, the environment and employees. Moreover, the authors point out that a sustainable company brings many benefits. Therefore, Hall mentioned the need to reduce unsustainable business practices as well as to the potential for research development in the indicated area [6].

Initially, the literature on corporate sustainability focused on the fact that environment and economy were conceptualized as separate and competing domains. Researchers mentioned in this context the need to find a compromise between the competing goals of making profits and environmental degradation and resource depletion. McAlister emphasized the need for complex cost–benefit analysis or providing procedural protection [7].

Today, sustainability in business is understood as satisfying all three dimensions: economic, social and environmental. However, the issue of economic growth is still debated because of its contradictions with the goals of capitalism [8]. Nevertheless, the optimal enterprise model should incorporate innovation for sustainable development [9].

Based on the literature analysis, the study of the impact of sustainability on corporate profits has a great research potential [10, 11]. First, there are still doubts about the measurable benefits of implementing sustainability. Second, no analysis of companies in terms of sustainability and possible corporate profits from the implementation of sustainability goals has been found in the literature. In this regard, the present study will make a new contribution to science both in methodological and practical terms.

3 Analysis

The scope of research has been adjusted to the specific nature of the Polish economy and covers the 600 largest companies, classified according to the Money.pl ranking in 2021 as "The biggest." A total of 563 companies were eligible for the survey, with the remainder rejected due to lack of data [12].

The survey covered industrial and service enterprises operating in Poland, while financial institutions (including banks, insurance companies), universities and hospitals were excluded. The survey included companies that represent a variety of industries, such as: manufacturing, fast moving consumer goods, automotive and many others. The largest companies were examined in the analysis, because it is among them that we can expect actions not only focused on obtaining the highest possible profit, but also caring for the environment and employees. It is the largest companies that should be leaders in sustainable development. Among the 563 companies surveyed, we can distinguish five groups of companies (see Table 1).

The companies surveyed can be divided into five groups. The first group includes entities that do not implement or invoke sustainable development. The second group of variables are those that are aware of the importance of sustainable development, which is why they refer to the concept or to environmental and social issues in their

Table 1 Division of enterprises according to the degree of implementation of sustainable development principles

Independent variables	X1: Companies that do not refer to sustainability in their strategy
	X2: Companies which refer in their strategy to sustainable development
	X3: Companies that have a page about sustainable development and corporate social responsibility
	X4: Companies that report on implementation of sustainable development
	X5: Companies that have signed the Global Compact agreement

Source Own Study, based on: [13]

Table 2 Analysis of company profits by sustainability group

Variable	Group	Number	Mean	Sd	Median	Min	Max
X1	0	247	10.13	37.36	4.4	−76.9	303.6
X2	1	51	5.1	17.81	3.9	−33.5	60.2
X3	2	76	5.46	16.5	4.6	−28.1	86.9
X4	3	100	3.74	30.13	2.5	−44.4	236.9
X5	4	89	0.88	18.5	2.8	−50.6	95.7

Source Own study

strategies. The third group are the companies which, in their strategies, refer to the concept of sustainable development and, within their website, devote a special section to informing about the issues of the concept in question. The remaining two groups, i.e. the fourth and fifth, are those which, apart from the theoretical aspect, undertake concrete actions for the implementation of sustainable development in their companies. Both take up the challenge of reporting on sustainability activities, but the fifth group refers specifically to the UN goals and has a signed Global Compact agreement. According to this agreement, it is obliged to produce such reports, referring to the implementation of individual goals (1–17) of sustainable development.

The following table presents basic statistics on corporate profits in relation to each group of variables (Table 2).

The dependent variable is the change in income from total operations, determining the ratio of income in 2021 to 2020. Due to lack of data, some companies were excluded from the study, as they had not indicated in the questionnaire the total revenue for 2020–2021 and, therefore, were not analysed in the further research process. In total, 563 entities were qualified for further analysis. As shown in the table, there is a significant quantitative disproportion between the different groups of variables. The most numerous group consists of enterprises that do not refer in any way to sustainable development. Nearly half of them counterbalance the other groups. The least numerous group invokes only strategy (89 companies). Superficial activities can also be observed. A certain group of companies notes the need to refer to sustainable development, but apart from their rather concise declarations, it is difficult to detect or determine that they are sustainable companies.

It is worth noting that getting to know the direction of structural changes of Polish enterprises in terms of sustainable development is an important task, especially in view of the processes of deepening environmental crisis. Sustainable development in enterprises is one of the links determining the success of the implementation of this concept both at global and national levels. Analysis of the implementation of sustainable development to the company will show the scale of the phenomenon in Poland. I focused on issues related to the analysis of the dependent variable and independent variables using methods of descriptive statistics. This study will assess the contribution that companies have to the global sustainability concept. Furthermore, the analysis was completed with selected examples of good practices that companies can use in the future.

4 Discussion

Analysing the profits of companies operating in Poland, it should be noted that as the involvement in the implementation of sustainable development increases, its profits should increase. The descriptive analysis shows that the opposite trend can be observed in analysed groups. On average, in the years 2020–2021 in the first studied group of enterprises operating in Poland, one can observe an increase of about 10% in total revenues. This is quite a good result, given that many companies in this period were exposed to the crisis related to the Covid-19 pandemic and the economic crisis related to the lock-down introduced at various periods. In all groups, the average is greater than 0, but smallest in the group most committed to sustainable development (less than 1%). Due to the global crisis situation, companies recorded large differences in income. This is evidenced by the minimum and maximum values, which differ significantly from the average for each group. Thus, for example, the largest losses were recorded by Itaka Holdings, representing the travel services industry. Not included in the analysis is Columbus, which experienced unprecedented revenue growth of over 2000%. Columbus [14] operates in the renewable energy market, offering products for households and businesses in areas such as energy production and energy storage. In other words, Columbus Energy S.A. is the leader of micro photovoltaic installations in Poland, which currently consists of 109 companies. Ecological awareness and economics connected with energy production is one of the most important factors of sustainable development. Columbus Energy activities are in line with the goals of this concept.

The unprecedented scale of the company's revenue growth is firstly due to the interest of Polish households in the possibility of producing renewable energy. According to a CBOS (Public Opinion Research Center) report, as many as 22% of single-family house owners in Poland were interested in producing their own energy, which amounts to over 1 million households. RES are rated by far the highest both in terms of safety and prospects of being a prosumer. On the other hand, the lowest ratings, especially in terms of future prospects, are given to oil and coal–only less than two-fifths of respondents see a chance to ensure Poland's energy security in the use of oil, with even fewer people betting on coal in this context [15]. Secondly, the government's policy on changes in the RES market is important. In May 2021, the Polish Ministry of Climate and Environment announced that, with a view to aligning Polish regulations with EU regulations, a change to the prosumer system is planned. Chaos, lack of prudence and lack of public consultation were the main allegations reported by the daily press when analysing the project. In December, the Parliament (Sejm) adopted the amendment to the Renewable Energy Sources Act [16], and the Ministry argues in an advertisement that the proposed changes are very beneficial for the prosumer. They concern the withdrawal of the discount model and the introduction of net-billing. Unfortunately, many experts report negative effects of the changes in the legislation. For example, the PV Industry Association Polska (Stowarzyszenie Branży Fotowoltaicznej Polska) points out that the proposed rules may have negative economic, social, and legal consequences, leading to, e.g., the liquidation of approximately 13.5 thousand companies in the Polish PV market [17]. This is not the only negative effect. Fear and insecurity resulting from the irresponsible behaviour of government representatives will influence citizens' decisions on investing in photovoltaics. It can be assumed that the proposed changes will contribute to changing the trend from increasing to decreasing. This means that fewer and fewer people may decide to invest in renewable energy sources in the coming years. The reason for this will be investment uncertainty, which will hinder the development of this branch of the economy and thus contribute to a decrease in indicators related to sustainable development [18].

5 Conclusions

One of the most important factors for the success of the concept of sustainable development is the trust that society places in individual companies and business as a whole. The higher the level of this trust exists, the greater is the chance for cooperation between stakeholders and business in terms of sustainable development.

Based on the analyses the following conclusions can be drawn. In the process of analysis, the hypothesis on differentiation of profits of Polish enterprises depending on the level of implementation of sustainable development was rejected. Although researchers in the literature pointed to benefits, the analysis showed that the highest increase in profits occurred in those enterprises that addressed sustainable development to the least extent.

Implementing the principles of sustainable development does not depend on the industry. There are examples of companies which comprehensively and professionally implement the principles of sustainable development, and others which do not even refer to this concept (the Bosh Group [19] and LG Electronics [20]). At the same time, there are some companies that did not follow the principles of environmental protection at the beginning. Despite the difficult situation, they managed to reorganise their business in such a way that it can be cited as an example of a good change. Among such companies is Ferrum SA, which until the 1990s used technologies that had to be classified as not energy-efficient and unfriendly to people and the environment. They were therefore placed on the provincial list of environmentally burdensome plants. In 1991, after the steel and foundry divisions were liquidated, they were deleted from the list. The results of these changes are products which meet the world standards. They are manufactured using energy-saving technologies, eliminating the creation of pollution at source. They declare taking into account the principles of the "Global Compact"–an initiative of the UN Secretary General–in their policy and present the effects of their activity in reports since 1990 [21].

In conclusion, we can state that the implementation of sustainable development at the level of enterprises is not at a satisfactory level. Attention should be paid to the need to introduce state guidelines, which should motivate enterprises to work more intensively in this direction. Moreover, it is necessary to educate and promote the idea of sustainability in order to improve the environmental awareness of consumers. Only their conscious choice of more ecological products can induce companies to work on sustainability. That is why sustainability should be a part of universities and schools programmes in Poland. Finally, it should be mentioned that the subject of the analyses could be further investigated with *analysis of variance*, considering the level of implementation of sustainable development and the branch of company.

References

1. Polska, G.W.: Największe firmy w Polsce (2021). https://www.money.pl/ranking-firm/2021/firmy/?page=6. Accessed 24 Nov 2022
2. United Nations: Resolution adopted by the Genaral Assembly on 25 Sept 2015 (2015). https://www.un.org/ga/search/view_doc.asp?symbol=A/RES/70/1&Lang=E. Accessed 23 May 2022
3. Alonso, A.D., Liu, Y.: Old wine region, new concept and sustainable development: winery entrepreneurs' perceived benefits from wine tourism on Spain's Canary Islands. J. Sustain. Tour. **20**(7), 991–1009 (2012)
4. Fotiadis, A.K., Vassiliadis, C.A., Rekleitis, P.D.: Constraints and benefits of sustainable development: a case study based on the perceptions of small-hotel entrepreneurs in Greece. Anatolia **24**(2), 144–161 (2013)
5. Urbaniec, M.: Sustainable entrepreneurship: innovation-related activities in European enterprises. Pol. J. Environ. Stud. **27**(4), 1773–1779 (2018)
6. Hall, J.K., Daneke, G.A., Lenox, M.J.: Sustainable development and entrepreneurship: past contributions and future directions. J. Bus. Ventur. **25**(5), 439–448 (2010)
7. MacAllister, D.M.: Evaluation in environmental planning: assessing environmental, social, economic, and political trade-offs, p. 308, 2. Aufl. The MIT Press, Cambridge, Mass (1982)
8. Balakrishnan, U., Duvall, T., Primeaux, P.: Rewriting the bases of capitalism: reflexive modernity and ecological sustainability as the foundations of a new normative framework. J. Bus. Ethics **47**(4), 299–314 (2003)
9. Wells, P.E.: Business models for sustainability: sustainable business models. Edward Elgar Publishing, Northampton, MA (2013)
10. Visvizi, A.: Artificial intelligence (AI) and sustainable development goals (SDGs): exploring the impact of AI on politics and society. Sustainability **14**, 1730 (2022)
11. Cosimato, S., Troisi, O.: Green supply chain management: practices and tools for logistics competitiveness and sustainability. The DHL case study. TQM J. **27**(2), 256–76 (2015)
12. Polska, G.W.: Największe firmy w Polsce (2021). https://www.money.pl/ranking-firm/2021/firmy/?page=6. Accessed 24 Nov 2021
13. KOS Metodologia rankingu Najwięksi money.pl. https://www.money.pl/gospodarka/metodologia-rankingu-najwieksi-money-pl-6655898851719744a.html. Accessed 24 Feb 2022
14. Columbus Strona główna. https://columbusenergy.pl/. Accessed 22 Feb 2022
15. CEBOS Polacy o źródłach energii, polityce energetycznej i stanie środowiska, https://www.cbos.pl/PL/publikacje/diagnozy/034.pdf+&cd=5&hl=pl&ct=clnk&gl=pl&client=firefox-b-d. Accessed 22 Feb 2022
16. Ministerstwo Klimatu i Środowiska Sejm przyjął nowelizację ustawy o odnawialnych źródłach energii wprowadzającą zmiany do systemu rozliczeń prosumentów. https://www.gov.pl/web/klimat/sejm-przyjal-nowelizacje-ustawy-o-odnawialnych-zrodlach-energii-wprowadzajaca-zmiany-do-systemu-rozliczen-prosumentow. Accessed 10 Jan 2022
17. Teraz-Środowisko Projekt zmiany systemu prosumenckiego. Zagrożone 13, 5 tys. firm i 86 tys. miejsc pracy? https://www.teraz-srodowisko.pl/aktualnosci/system-prosumencki-miejsca-pracy-instalacje-PV-11107.html. Accessed 10 Jan 2022

18. Tomala, M.: Energia odnawialna w Polsce w obliczu nowych wyzwań UE ze szczególnym uwzględnieniem sektora ciepłowniczego. In: Wojcieszak, Ł. (ed.), Bezpieczeństwo energetyczne Polski na początku trzeciej dekady XXI w. Fundacja na rzecz Czystej Energii, Poznań (2022)
19. Bosch w Polsce. https://www.bosch.pl/internet-rzeczy/neutralnosc-klimatyczna.html. Accessed 23 May 2022
20. LG Electronics. https://www.lg.com/pl. Accessed 23 May 2022
21. Ferrum. https://www.ferrum.com.pl/. Accessed 23 May 2022

The Role of Blockchain for Introducing Resilience in Insurance Domain: A Systematic Review

Julio C. Mendoza-Tello, Higinio Mora, and Tatiana Mendoza-Tello

Abstract The insured market requires accuracy and agility to meet the requirements of customers, insurers, and other service providers. Some catastrophic events are impossible to avoid, and their negative effects impede the economic recovery of people and companies. Theft, illnesses, accidents, pandemics, and natural disasters are events that affect the possibility of ruin. In this context, the acquisition of an insurance contract allows the risk to be transferred to an insurance company and reduces uninsured losses. This market needs to maintain an audit trail of transactions, preserving the transparency of the information exchange process and the responsibility for the behavior of its participants. However, this segment faces new limitations related to market policies and risk coverage. Not all people can purchase insurance because they do not have banking services, which impedes the creation of economic resilience in populations that suffer from extreme poverty. Technologies such as blockchain offer a solution to financial exclusion and provide global financial access to anyone, without any discrimination. Blockchain technology provides reliability, transparency, and immutability to transactions using cryptographic techniques and distributed consensus methods, without the need for third parties to testify to the veracity of the events. With this consideration, the paper assesses how technology can address challenges, and identifies the future direction in terms of adoption within the insurance value chain. Finally, conclusions are provided for taking advantage of the versatility of blockchain in this business area.

J. C. Mendoza-Tello
Faculty of Engineering and Applied Sciences, Central University of Ecuador, Quito, Ecuador
e-mail: jcmendoza@uce.edu.ec

H. Mora (✉)
Department of Computer Science Technology and Computation, University of Alicante, Alicante, Spain
e-mail: hmora@ua.es

T. Mendoza-Tello
Insurance Consultant, E3-12, Cristóbal Colón Avenue Quito, Ecuador

A. Visvizi et al. (eds.), *Research and Innovation Forum 2022*,
Springer Proceedings in Complexity, https://doi.org/10.1007/978-3-031-19560-0_50

1 Introduction

The insurance industry faces profound changes related to digitization and new consumer models. The impact of the Covid-19 pandemic has forced insurance companies to reinvent themselves to give continuity to their operations without affecting the service quality offered to customers. During this period, some insurers experienced weak economic growth and increased requests for access to information. The confidentiality of sensitive data implies high management costs, and difficulties with supervision. The insurance business is based on trust between the insurer and the insured. There are still problems related to fraud due to false claims and the unwillingness of insurance companies to compensate claims [1]. In this context, blockchain can improve data exchange, audit the veracity of the facts, reduce transaction processing times, reduce processing time and reliable payment settlement [2]. The integration of a smart contract (as a computer program) incorporates the conditions necessary for the definition and execution of an insurance contract. Thus, the insurance business can expedite the management of identity and processes, which reduces manipulation and forger. In these scenarios, blockchain plays an important role for removing paperwork too, which reduces the risk of Covid-19 infection. This avoids unnecessary transfers of the insured to the insurance company because the parties involved supervise the entire process. With these considerations, the paper provides a scope review that identifies the role and status of blockchain integration for introducing resilience in the insurance domain.

The paper is organized as follows. Section two describes the methodology. Section three provides a brief description of the two research fields. Section four contains a review of insurance-focused blockchain research. In addition, it explains how the versatility of blockchain can influence and introduce resilience to the environment of the insurance market. Finally, the conclusions are given at the end of the paper.

2 Methodology

This research analyzes two study fields, namely: blockchain, and insurance. In this context, an exhaustive search of the blockchain literature focused on insurance was conducted. Then, the review was carried out using two approaches, namely: academic literature, and projects addressed by insurance companies. Based on these considerations, the role of blockchain to introduce resilience is analyzed in each of the insurance processes.

3 Background

3.1 Blockchain and Smart Contracts

Blockchain became relevant because it was initially conceived as the underlying technology for the exchange of cryptocurrency. A relevant consideration in the evolution of blockchain technology is the versatility for registering, confirming, transporting, and transferring any digital token. The adhesion of a blockchain contract allows that (i) the critical aspects of a transaction to be addressed, (ii) legal and operational costs to be reduced, and (iii) transparency and anonymity to be guaranteed. A smart contract is implemented through the publication of a transaction in the blockchain (Fig. 1).

3.2 Insurance

The dynamics of an insurance contract is the payment of a fee (called a premium) from an insured party (which transfers its risk) to an insurer. This contract specifies the benefits, limitations, and exclusions of a set of contingencies or insured events. Next, Table 1 shows a brief description of insurance process.

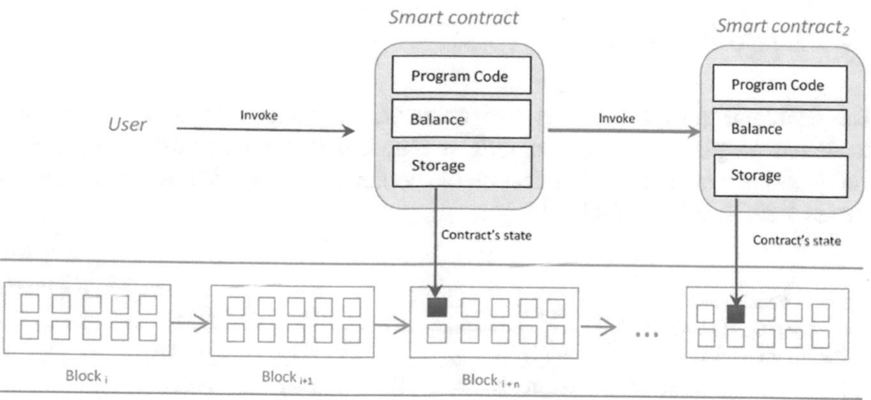

Fig. 1 Invocation of smart contracts

Table 1 Summary of insurance process

Insurance process	Description
Policy sale	Defines the contractual start between the insured party and insurer. Defines possible niche markets
Underwriting	Defines the market price expectation. Sets the subscription cost for the new market. Adjusts the coverage and price of a premium based on the expectation of the market price. Evaluates the risk based on the financial and health status of the insured party. Determines if a person is insurable or not
Reinsurance	Reduces the risk of subscription and transfers it to a reinsurer
Reserve calculation	Predicts the liabilities of future claims based on uncertainty. Evaluates the solvency of the insurer through the actuarial reserve
Fraud detection	Audit, inspect and detect anomalies in claims. Identify dishonest behaviors that cause financial and reputational losses to the insurance company
Incident management	Records the details of the incident and determines responsibility
Claims management	Validates whether a claim is consistent with the policy and terms of a policy in monetary terms. Provides a source of data for fraud management administration

4 The Role of Blockchain for Enabling Resilience in the Insurance Domain

4.1 Literature Review

The literature review involved two approaches, namely: business reports and academic. Regarding business reports, a search was conducted using websites of multinational insurance companies. In this context, 7 projects in test and production phases were identified (Table 2).

Table 2 Blockchain-based insurance projects addressed by the insurance industry

Project	Insurance use cases
B3i [3]	Insurance value chain
Insurer Network [4]	Reinsurance process
Fizzy [5]	Delay flight insurance
Insurwave [6]	Marine hull insurance
Lemonade [7]	Homeowners insurance
Etherisc [8]	Insurance for Crop, flight delay and hurricane protection
Isley [9]	Purchase and sale of insurance contracts in general

Table 3 Prior academic research of blockchain focused on insurance

Research	Findings
Storage and access control scheme [1, 11]	Using IPFS (InterPlanetary File System) and blockchain for security record storage
Usage-based auto insurance platform [12–15]	Insurance premium based on driving time, distance, and area
Incentives offered by auto insurance companies [16, 17]	Algorithm to provide rewards based on driving behavior
Fine-grained transportation insurance [18]	Insurance premium based on vehicle use and driver behavior
Healthcare insurance fraud detection [19]	Framework to support fraud detection
Health insurance claims processing [20]	Smart contracts to manage insurance claims
Billing for insurance companies and healthcare providers [21]	Transfer of billing information from hospitals to insurance companies
Distributed platform for insurance process [22]	Framework to support the execution of the insurance process in hospitals
Fraud detection in health insurance claims [23]	Use of Non-repudiation over Transport Layer Security (TLS-N) in communication between insured, insurer, and medical center
Medical insurance storage system [24]	Scheme for distributed storage of medical insurance

Blockchain is an emerging technology drawing attention in higher education [10]. Regarding the academic review, a search was conducted in prestigious bibliographic databases, such as: IEEE Xplore, Springer, Elsevier, ACM Digital, EBSCO, and Scopus. For this, four sequential tasks were executed. First, search strategy. Keywords were used, namely: blockchain, smart contract and insurance. Search was refined by paper title field. A sample of 75 papers were found; however, most of these studies do not correspond to the central focus of the research. Thus, only 30 papers were extracted in this first stage. Second, identification of exclusion and inclusion criteria. The language of bibliographic references is English. Only book chapters, articles, journals, and proceedings were considered. Third, the selection was studied, and the quality was evaluated. Only full studies were considered; that is, papers that describe the analysis, methodology, and data collection method. In this evaluation, 15 papers were deleted, and 15 papers remained (Table 3). Fourth, data extraction and synthesis. MS Excel and Mendeley were used for this. The review was completed on November 30, 2021.

4.2 *Implications of Blockchain for Insurance*

The characteristics of blockchain technology have the versatility to address business challenges [25] with efficiency and dynamism [26]. In this scenario, the integration

Table 4 The role of blockchain for introducing resilience in the insurance domain

Insurance process	Role of blockchain
Policy sales	– Conditions, benefits, and contractual obligations codified within smart contracts – Decision-making support according to the coverage policy
Underwriting	– Use of shared data eliminates information asymmetry – Identification of patterns and metrics for evaluating the insured risk
Reserve calculation	– Risk estimation for a current insurance contract
Reinsurance	– Support for risk analysis according to an insured value and exclusions
Fraud detection	– Immutable record of events and behavior of participants in the insurance process
Incident management	– Usage of oracles for supervision of external events that automatically activate a smart contract
Claims management	– The establishment of responsibilities is expedited, and the payment is automated through smart contracts

of technology provides opportunities and benefits to insurance activities. Based on researched literature were derived the most important aspects of the use of blockchain in insurance. Table 4 shows the role of blockchain for enabling resilience in the insurance domain.

Policy sales

A contract reflects the sale of a product in the insurance area. In each sale, the insurance companies try to be specific and clear in relation to the conditions, benefits, and contractual obligations of the parties. Policy making could support governance[27], sustainable innovation [28], resilient operations [29], and information management development [30]. That is, the clearer and more detailed a contract is, the less claims and legal disputes it must face. However, the self-validation of the contractual act and the instantaneous reflection of unforeseen errors is a challenge. In this context, blockchain provides opportunities for enabling the sustainable development gear [31] and policy making [32]. A smart contract can reduce inaccuracy and has the versatility to explicitly define the sales policy. These guidelines are stored in a smart contract and include the necessary guidelines to support the decision-making related to the coverage policy and general conditions of the products (or policies) to be sold.

Insurance underwriting

Transparency and collaboration are implicit properties of a blockchain. This makes it possible for independent sources of data to be shared amongst policy underwriters, canceling the asymmetry of the information. That is, they will be able to know market expectations, the offer price, underwriting costs, classification, and risk assessment, before the issuance of the policy. In this context, partnerships between insurers may be required by providing a key utility: shared intelligence management. In this area, blockchain as the underlying technology provides a shared interface, which ensures the validity of the contractual terms and facilitates the automatic verification of the participants.

Reinsurance

Blockchain can strengthen strategic partnerships between insurance companies and data providers (public services, notarial records, amongst others). This collaboration allows valid information from partners and insured parties to be captured. In turn, it improves the accuracy of policy prices through a transparent computation based on cryptographic tasks. Along these lines, this cooperation also allows reinsurance between insurers, retaining the transparency and behavior of their negotiations. In turn, these operations are registered in blockchain, which acts as an audit trail that keeps operations unchanged over time.

Reserve calculation

Initially, insurance premiums are calculated based on the presumption of risks arising from some financial causes (e.g., return on investment, inflation) and demographic aspects (age, type of work, gender, family size, education, income, marital status, religion, birth rate, mortality rate). In this area, actuarial procedures are required. For example, the calculation of a life insurance premium is based on the mortality factor; that is, the probabilities of survival and life of the insured party based on age. As a result, insurers determine the level of risk: high, medium, and low. Over time, these values need valuation, constant updating, and adjustment for a current policy, which is known as reserve determination.

Fraud detection

The incident event involves some participants or agents. Each of them performs billing and settlement activities, which are shared through the blockchain. In this way, each payment receipt constitutes immutable and transparent proof of the settlement of a claim. In turn, the validation of the data and the sending of the payment are automated. In this case, the settlement of claims is automated and the asymmetry of the information in the payment process is eliminated. In addition, each entity has the freedom to collect information and use it for its internal operations. Along these lines, transaction costs and the risk of fraud are reduced because each participant has knowledge of the billing and can validate it [19].

Insurance claims and incident management

Blockchain as a distributed ledger keeps track of the operations and responsibility of each stakeholder within the process. In turn, blockchain provides the basis for the implementation of smart contracts on top of its protocol. This combination allows procedures and codifications to be transported according to a business rule. These instructions need external agents called "oracles", which stimulate their execution. Consequently, the response time by authorized centers is reduced.

5 Conclusions

This research provides some cases of uses that can be exploited for introducing resilience in the insurance business. However, blockchain is in the early stages of adoption in society and therefore, does not yet demonstrate its practical implications within the insurance industry. This raises many concerns regarding its implementation, especially in the management of operational, technological, and legal risks.

As future work, a work agenda between insurance companies and government control organizations is important. Thus, guidelines and standards are obtained that allow the coordination and exchange of information amongst insurers, which has an impact on: decision making, process efficiency and privacy of the insured party's data. In this sense, a legal regulation that defines policies and procedures for the proper use of this technology is necessary.

References

1. Sun, J., Yao, X., Wang, S., Wu, Y.: Non-repudiation storage and access control scheme of insurance data based on blockchain in IPFS. IEEE Access. **8**, 155145–155155 (2020). https://doi.org/10.1109/ACCESS.2020.3018816
2. Mendoza-Tello, J.C., Mora, H., Pujol-López, F.A., Lytras, M.D.: Disruptive innovation of cryptocurrencies in consumer acceptance and trust. Inf. Syst. E-bus. Manag. **17**, 195–222 (2019). https://doi.org/10.1007/s10257-019-00415-w
3. B3i: B3i-The Blockchain insurance industry initiative. https://b3i.tech/
4. IBM: Thai Re launches ASEAN's first reinsurance smart contract platform using IBM's blockchain and hybrid cloud technology. https://newsroom.ibm.com/2021-01-06-Thai-Re-launches-ASEANs-first-reinsurance-smart-contract-platform-using-IBMs-blockchain-and-hybrid-cloud%0A
5. AXA: AXA goes blockchain with fizzy. https://www.axa.com/en/magazine/axa-goes-blockchain-with-fizzy
6. Guardtime: World's First Blockchain Platform for Marine Insurance Now in Commercial Use. https://guardtime.com/blog/world-s-first-blockchain-platform-for-marine-insurance-now-in-commercial-use
7. Lemonade: Forget Everything You Know About Insurance. https://www.lemonade.com/
8. Decentralized Insurance Foundation: Make Insurance Fair and Accessible Insurance. https://etherisc.com/
9. Fidentiax: World's 1st Marketplace for Tradable Insurance Policies. https://www.fidentiax.com/
10. Visvizi, A., Lytras, M.D., Sarirete, A.: Emerging technologies and higher education: management and administration in focus. In: Management and Administration of Higher Education Institutions at Times of Change, pp. 1–11 (2019)
11. Nizamuddin, N., Abugabah, A.: Blockchain for automotive: an insight towards the IPFS blockchain-based auto insurance sector. Int. J. Electr. Comput. Eng. **11**, 2443–2456 (2021). https://doi.org/10.11591/ijece.v11i3.pp2443-2456

12. Lin, W.Y., Lin, F.Y.S., Wu, T.H., Tai. K.Y.: An on-board equipment and blockchain-based automobile insurance and maintenance platform. In: Barolli, L., Takizawa, M., Enokido, T., Chen, H.C.M.K. (ed.) Advances on Broad-Band Wireless Computing, Communication and Applications. BWCCA 2020. Lecture Notes in Networks and Systems, pp. 223–232. Springer, Cham (2021)
13. Wan, Z., Guan, Z., Cheng, X.: PRIDE: A private and decentralized usage-based insurance using blockchain. In: 2018 IEEE International Conference on Internet of Things (iThings) and IEEE Green Computing and Communications (GreenCom) and IEEE Cyber, Physical and Social Computing (CPSCom) and IEEE Smart Data (SmartData), pp. 1349–1354. IEEE (2018)
14. Lamberti, F., Gatteschi, V., Demartini, C., Pelissier, M., Gómez, A., Santamaría, V.: Blockchains can work for car insurance. IEEE Consum. Electron. Mag. **7**, 72–81 (2018). https://doi.org/10.1109/MCE.2018.2816247
15. Demir, M., Turetken, O., Ferworn, A.: Blockchain based transparent vehicle insurance management. In: 2019 6th International Conference on Software Defined Systems, SDS 2019, pp. 213–220. IEEE, Rome, Italy (2019)
16. Singh, P.K., Singh, R., Muchahary, G., Lahon, M., Nandi, S.: A blockchain-based approach for usage based insurance and incentive in ITS. In: TENCON 2019-2019 IEEE Region 10 Conference (TENCON), pp. 1202–1207. IEEE (2019)
17. Palma, L.M., Gomes, F.O., Vigil, M., Martina, J.E.: A transparent and privacy-aware approach using smart contracts for car insurance reward programs. In: Garg, D., Kumar, N., Shyamasundar, R. (ed.) Information Systems Security. ICISS 2019. Lecture Notes in Computer Science, pp. 3–20. Springer, Cham (2011)
18. Li, Z., Xiao, Z., Xu, Q., Sotthiwat, E., Mong Goh, R.S., Liang, X.: Blockchain and IoT data analytics for fine-grained transportation insurance. In: 2018 IEEE 24th International Conference on Parallel and Distributed Systems (ICPADS), pp. 1022–1027. IEEE (2018)
19. Mendoza-Tello, J.C., Mendoza-Tello, T., Mora, H.: Blockchain as a healthcare insurance fraud detection tool. In: Visvizi, A., Lytras, M.D., Aljohani, N.R. (eds.) Research and Innovation Forum 2020: Disruptive Technologies in Times of Change. Springer Proceedings in Complexity, p. 340. Springer, Cham (2020)
20. Thenmozhi, M., Dhanalakshmi, R., Geetha, S., Valli, R.: Implementing blockchain technologies for health insurance claim processing in hospitals. Mater. Today Proc. (2021). https://doi.org/10.1016/j.matpr.2021.02.776
21. Saeedi, K., Wali, A., Alahmadi, D., Babour, A., AlQahtani, F., AlQahtani, R., Raghad, K., Rabah, Z.: Building a blockchain application: a show case for healthcare providers and insurance companies. In: Arai, K., Bhatia, R., Kapoor, S. (eds.) Proceedings of the Future Technologies Conference (FTC) 2019. FTC 2019. Advances in Intelligent Systems and Computing, pp. 785–801. Springer, Cham (2020)
22. Raikwar, M., Mazumdar, S., Ruj, S., Gupta, S. Sen, Anupam Chattopadhyay, K.-Y.L.: A blockchain framework for insurance processes in hospitals. In: 2018 9th IFIP International Conference on New Technologies, Mobility and Security (NTMS), pp. 1–4 (2018)
23. Mohan, T., Praveen, K.: Fraud detection in medical insurance claim with privacy preserving data publishing in TLS-N using blockchain. In: Singh, M., Gupta, P., Tyagi, V., Flusser, J., Ören, T.K.R. (eds.) Advances in Computing and Data Sciences. ICACDS 2019. Communications in Computer and Information Science, pp. 211–220. Springer, Singapore (2019)
24. Zhou, L., Wang, L., Sun, Y.: MIStore: a blockchain-based medical insurance storage system. J. Med. Syst. **42** (2018). https://doi.org/10.1007/s10916-018-0996-4
25. Mora, H., Mendoza-Tello, J.C., Varela-Guzmán, E.G., Szymanski, J.: Blockchain technologies to address smart city and society challenges. Comput. Human Behav. **122** (2021). https://doi.org/10.1016/j.chb.2021.106854
26. Saeedi, K., Almalki, M.D., Aljeaid, D., Visvizi, A., Aslam, M.A.: Design pattern elicitation framework for proof of integrity in blockchain applications. Sustainability **12**, 1–16 (2020). https://doi.org/10.3390/su12208404
27. Polese, F., Troisi, O., Grimaldi, M., Loia, F.: Reinterpreting governance in smart cities: an ecosystem-based view. In: Visvizi, A., Pérez del Hoyo, R. (eds.) Smart Cities and the un SDGs, pp. 71–89. Elsevier (2021)

28. Ciasullo, M.V., Troisi, O., Grimaldi, M., Leone, D.: Multi-level governance for sustainable innovation in smart communities: an ecosystems approach. Int. Entrep. Manag. J. **16**, 1167–1195 (2020). https://doi.org/10.1007/s11365-020-00641-6
29. Visvizi, A., Troisi, O.: Effective management of the smart city: an outline of a conversation. In: Managing Smart Cities, pp. 1–10. Springer, Cham (2022)
30. Lytras, M.D., Visvizi, A., Chopdar, P.K., Sarirete, A., Alhalabi, W.: Information management in smart cities: turning end users' views into multi-item scale development, validation, and policy-making recommendations. Int. J. Inf. Manag. **56**, 102146 (2021). https://doi.org/10.1016/j.ijinfomgt.2020.102146
31. Mora, H., Pujol-López, F.A., Mendoza-Tello, J.C., Morales-Morales, M.R.: An education-based approach for enabling the sustainable development gear. Comput. Human Behav. **107**, 105775 (2020). https://doi.org/10.1016/j.chb.2018.11.004
32. Sicilia, M.A., Visvizi, A.: Blockchain and OECD data repositories: opportunities and policy-making implications. Libr. Hi Tech. **37**, 30–42 (2019). https://doi.org/10.1108/LHT-12-2017-0276

Gig Economy Practices, Ecosystem, and Women's Entrepreneurship: A Theoretical Model

Ali Mohamad Mouazen and Ana Beatriz Hernández-Lara

Abstract The current economic situation in many countries, aggravated by the COVID-19 pandemic, has forced local organizations to downsize the number of employees, reduce working hours or rely on temporary workers to perform the job by means of gig workers. While these conditions could be considered as a threat, some workforce vulnerable groups, like women, grasped this opportunity to develop entrepreneurial behavior and start-up their businesses. This paper aims to investigate the factors that under the explained circumstances encourage women to be entrepreneurs, proposing a theoretical model of relationships between gig economy practices and opportunistic and necessity women's entrepreneurship start-ups.

Keywords Women's entrepreneurship · Gig economy · Gig workers · Crisis · Start-ups

1 Introduction

Contemporary research investigates how entrepreneurial initiatives are articulated, stimulated, and supported by several elements or factors by means of entrepreneurial ecosystem and gig economy [1–4].

Entrepreneurship, understood as an ecosystem in the management field, or a complex collection of actors, elements or pillars that are interconnected, and arranged in a manner to facilitate entrepreneurship and start-ups [5, 6], constitutes a challenge for both, policymakers and scholars [7, 8]. The gig economy takes part in this

A. M. Mouazen
Management and International Management Department, Lebanese International University, Beirut International University, Beirut, Lebanon
e-mail: ali.mouazen@liu.edu.lb

A. B. Hernández-Lara (✉)
Department of Business Management, Universitat Rovira i Virgili, Avinguda Universitat, 43204 Reus, Spain
e-mail: anabeatriz.hernandez@urv.cat

A. Visvizi et al. (eds.), *Research and Innovation Forum 2022*,
Springer Proceedings in Complexity, https://doi.org/10.1007/978-3-031-19560-0_51

ecosystem, and is gradually becoming a relevant setting to understand entrepreneurship, employment, freelancing, digital labor, innovation and technology management [4, 9]. Despite the fact that the literature on management ecosystems, and more specifically on the entrepreneurship ecosystem, has grown in recent years [5], to evolve into a trend [10], there are still significant gaps in the theoretical development of the discipline. These gaps, for instance, are based on the dearth of research on the causal links between the ecosystem factors [3, 5], on the parameters that influence the ecosystem effectiveness [11], and on the influence of culture and context [12]. More recently, Cao and Shi [6] concluded that, researching ecosystems and entrepreneurship in emerging economies would provide additional insights on this topic.

Despite the substantial body of research conducted on entrepreneurship [13–17], less attention has been paid to entrepreneurship decisions adopted by specific vulnerable groups, like women, and specially in developing countries [18]. Under the focus of the gender dimension, the attempting of women to find the right balance between paid work and home responsibilities and family caring, is likely to be particularly influenced by their local country economic conditions. In times of serious economic crisis and recession, the search of new business opportunities and becoming entrepreneurs might be viable alternatives to paid employment [19].

However, despite the relevance of this phenomenon, there is still a dearth of fundamental literature to understand the characteristics of women's entrepreneurship and the factors that stimulate it, especially in developing countries. To our best knowledge, it remains unexplored the female role on the entrepreneurship ecosystem, the practices of gig economy where women take place, and the factors and circumstances that foster and challenge women start-up decisions. Accordingly, this research proposes a theoretical model to better understand the entrepreneurship ecosystem and gig economy practices from the women entrepreneur's perspective.

2 Literature Review and Theoretical Framework

2.1 Entrepreneurship Ecosystem and Women's Entrepreneurship

The concept of ecosystems arose from organic and biological science and popularized in business research and social science since 1980 [20, 21], establishing itself as a key and definitive concept for entrepreneurship start-ups.

Since the early twentieth century, economic research and literature has studied the link between organizations and the areas in which they operate or were established [3]. Initially, businesses would have an advantage by being physically close to other similar organizations, these advantages include but are not limited to, availability of skilled employees, efficiency and decreasing cost, and knowledge and technological spillover [22].

As we relate the environment to entrepreneurial behavior, it is possible to pinpoint to some factors that facilitate the establishment and expansion of start-ups in certain areas or regions. This environment may be defined as a collection of conditions and factors that contribute to the growth of entrepreneurship and include: (I) Financial support; (II) Non-financial support; (III) Government rules and policies; (IV) Socioeconomic conditions; and (V) Business skills and entrepreneurial endeavor [23]. Adding to this, having entrepreneurial and managerial competencies, would foster the establishment of new businesses [23].

Isenberg [3] also acknowledges entrepreneurial ecosystems as a feasible system for promoting economic success via the coordinated action of several actors. Those actors as stated by Isenberg [3] are: (I) Financial capital; (II) Public policies; (III) Markets; (IV) Culture; (V) Human resources and supporting institutions.

Despite the relative spreading of Isenberg's paradigm, it was not subject to empirical investigations that could, in practice, confirm the relationship between the proposed ecosystem actors and their effects on the environment [5, 11]. Furthermore, their approach does not take into consideration evidence gathered directly from current or prospected entrepreneurs [24].

Aiming to fill this gap in the Isenberg [3] model, Foster et al. [24] have created an analytical model that assumes an entrepreneurial ecosystem that is supported by eight main actors, including: (I) Human capital; (II) Financing; (III) Availability and accessibility of markets; (IV) Availability of large universities serving as catalysts; (V) Cultural support; (VI) Training and education; (VII) Regulatory framework and infrastructure availability (VIII) Availability of advisors, mentors and supporting system. availability of financial funding.

Another key aspect of the entrepreneurial ecosystem approach is that it considers entrepreneurship not just as an outcome of the ecosystem, but also as a central contributor (leader) in the system. Based on this view, dislike previous approaches, the role of the government can be classified as an ecosystem "feeder" and not as a "leader".

In addition to the critical role of entrepreneurs in directing the ecosystem development and serving as mentors or advise, Feld [25] emphasized the role of nine actors in the ecosystem, that include: (I) Leadership; (II) Talents; (III) Network density; (IV) Intermediaries; (V) Support services; (VI) Engagement; (VII) Companies; (VIII) Capital; and (IX) Government.

Based on the insights provided in the literature, we developed an entrepreneurial ecosystem value chain model that integrate all actors and elements discussed previously. This model aims to provide more depth on the relationship between ecosystem actors and its output. This model is divided into four layers, in the "forward movement", these layers pinpoints to "entrepreneurial activity leaders", "entrepreneurial activity feeders", that fosters "opportunistic and necessity entrepreneurial activity" leading to "aggregate social and economic value".

In the specific case of women's entrepreneurship, the existence of a proper ecosystem to foster and promote entrepreneurship may act as a catalyst to overcome the difficulties women find to be entrepreneurs and enhance their entrepreneurial initiatives and activities, being our thesis that the positive effect of this ecosystem on women's entrepreneurship would be more evident than in a general context where the gender dimension is not under consideration. Therefore, the following proposition is stated:

Proposition 1 *Entrepreneurship ecosystem value chain model and women's entrepreneurship activity are related in both, necessity and opportunity female entrepreneurship.*

2.2 Gig Economy and Women's Entrepreneurship

The gig economy, understood as a free market system in which temporary, flexible jobs are more common, and companies rely on independent contractors and freelancers (known as gig workers) instead of full-time employees [26]), constitutes a recent and fundamental business model. This economic system relies on smart mobile applications and online platforms to connect prospected clients (organizations) and service suppliers and providers (gig workers), who provide the service that fits their knowledge and qualifications, getting a compensation in return. The contemporary gig platforms provide exceptional work flexibility for gig workers, who may work based on their own day and time schedule. The gig workforce is more prepared to accept non-traditional jobs, improving income and working-hour flexibility [27]. Similarly, the digital economy environmental settings enable the gig economy development and growth; consequently, multi user peer-to-peer design to facilitate businesses, decision making and knowledge sharing [28].

To get a better understanding of the possible implications of gig work on entrepreneurial activities, it is believed that an individual's decision to pursue entrepreneurship vs full-time wage employment is governed by the relative expected returns given by the two options, selecting the agents the choice with the highest projected utility [29–34]. The key assumption underlying decision to enrol in entrepreneurship will be influenced by the potential gains of being able to access gig employment chances in case of a failure. Additionally, the presence of gig opportunities might permit a prospective entrepreneur to establish a firm that would not generate enough revenue without the additional supportive gig revenue. The security provided by the constant availability of gig economy opportunities should be more attractive to a prospective entrepreneur.

However, there is also an opposing view on the effect of gig economy on entrepreneurial activity. The flexibility and low entry cost provided by ad hoc employment may result to an increase in the entrepreneurial activity, since it enables employed or young entrepreneurs to manage their time in order to acquire the necessary resources, to commence a project or run their start-up [35, 36]. However, scholars have also noticed that unemployment and underemployment are key factors influencing entrepreneurial activity. Thus, people may choose to participate in entrepreneurial activities due to their low opportunity costs [37, 38]. In such scenario, gig economy may reduce entrepreneurial activity by presenting other work possibilities for these necessity entrepreneurs [39].

Even if previous research has not analyzed specifically the relationship of women's entrepreneurship and the gig economy, we could expect that the same assumptions could explain the relation between both phenomena, only with some particularities on the intensity of the relationship when overall entrepreneurship is compared to women's entrepreneurship. It means that given the difficulties of women in the labor market and to be entrepreneurs, the existence of gig economy opportunities, may be especially relevant for women, when they try to obtain a better balance of their laboral and personal life. Therefore, we propose the following:

Proposition 2 *Gig economy and women's entrepreneurship activity are related in both, necessity and opportunity female entrepreneurship.*

A summary of the proposed theoretical framework is shown in Fig. 1. This framework aims to relate entrepreneurial ecosystem and its factors, that drive women to pursue entrepreneurial endeavours. The framework also aims to state the relation between gig economy and women's entrepreneurship.

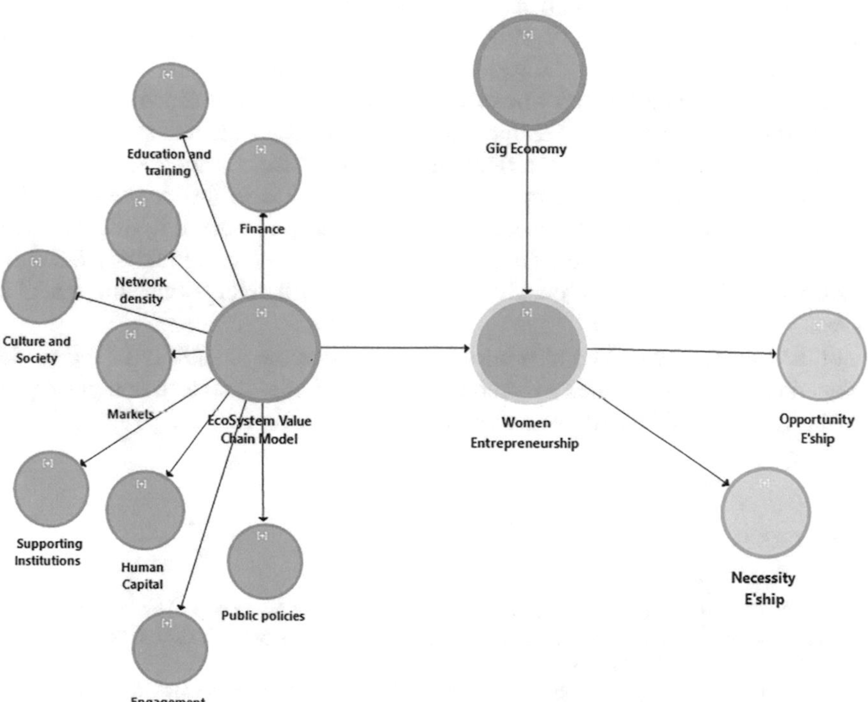

Fig. 1 Theoretical framework: Entrepreneurship ecosystem, gig economy and women's entrepreneurship

3 Conclusions

The main objective of this study was to develop a theoretical framework to understand the female role on the entrepreneurship ecosystem and the influence also exerted by the gig economy.

Despite the huge research body conducted on entrepreneurship [13–17], female entrepreneurship, and more specifically its role in the entrepreneurship ecosystem, as well as its relationship with gig economy, remain unexplored [18]. The proposed framework of relationships aims to cover this gap, contributing theoretically to understand the possible effects and relations of entrepreneurship ecosystem, fostered by its leaders and feeders, on women's entrepreneurship, as well as how the proliferation of the gig economy also impacts on women's entrepreneurship, considering both, the case of necessity and opportunity entrepreneurship. The main thesis proposed in this study is that given the extra difficulties of women to initiate their own businesses and find their success way in the labor market, the existence of a proper entrepreneurial ecosystem and the opportunity of being part of the gig economy can act as a catalyst for female entrepreneurs, more evident than in a general context where gender dimension is not considered, and can act as a means to overcome the difficulties they normally face to gain a proper balance between work and personal life.

In a next step of this investigation, the purpose will be to apply the theoretical framework and demonstrate the validity of its relationship. It will be done in the context of a developing country where, there is still a dearth of evidence to understand the characteristics of women's entrepreneurship and the factors that stimulate it.

This study aims valuable contributions especially from a theoretical point of view. The theoretical model proposed in this study aims to justify the suitability of entrepreneurship ecosystem theory and gig economy, as part of it, to understand women's entrepreneurship, especially in contexts where the difficulties faced by women are outstanding. The practical contributions will emerge once the model is proved.

In the future, it would be also interesting to check the suitability of the model comparing women's and men's entrepreneurship activities, and also comparing the suitability of the model in developed and developing regions.

References

1. Burtch, G., Carnahan, S., Greenwood, B.N.: Can you gig it? An empirical examination of the gig economy and entrepreneurial activity. Manag. Sci. **64**(12), 5497–5520 (2018). https://doi.org/10.1287/mnsc.2017.2916
2. Mason, C., Brown, R.: Entrepreneurial ecosystems and growth oriented entrepreneurship. Final Report to OECD, Paris, vol. 30, no. 1, pp. 77–102 (2014). http://www.oecd.org/cfe/leed/Entrepreneurial-ecosystems.pdf
3. Isenberg, D.: The entrepreneurship ecosystem strategy as a new paradigm for economic policy: Principles for cultivating entrepreneurship. Inst. Int. Eur. Aff. **1**(781), 1–13 (2011)

4. Malik, R., Visvizi, A., Skrzek-Lubasińska, M.: The gig economy: current issues, the debate, and the new avenues of research. Sustainability **13**(9), 5023 (2021). https://doi.org/10.3390/su13095023
5. Alvedalen, J., Boschma, R.: A critical review of entrepreneurial ecosystems research: towards a future research agenda. Eur. Plan. Stud. **25**(6), 887–903 (2017). https://doi.org/10.1080/09654313.2017.1299694
6. Cao, Z., Shi, X.: A systematic literature review of entrepreneurial ecosystems in advanced and emerging economies. Small Bus. Econ. **57**(1), 75–110 (2021). https://doi.org/10.1007/s11187-020-00326-y
7. Parker, S.C.: The Economics of Entrepreneurship. Cambridge University Press (2018)
8. Dale, A.: Self-employment and entrepreneurship: notes on two problematic concepts. In: Deciphering the Enterprise Culture (Routledge Revivals), Routledge, pp. 55–72 (2015)
9. Tsujimoto, M., Kajikawa, Y., Tomita, J., Matsumoto, Y.: A review of the ecosystem concept—towards coherent ecosystem design. Technol. Forecast. Soc. Chang **136**, 49–58 (2018). https://doi.org/10.1016/j.techfore.2017.06.032
10. Brown, R., Mason, C.: Looking inside the spiky bits: a critical review and conceptualisation of entrepreneurial ecosystems. Small Bus. Econ. **49**(1), 11–30 (2017). https://doi.org/10.1007/s11187-017-9865-7
11. Mack, E., Mayer, H.: The evolutionary dynamics of entrepreneurial ecosystems. Urban Stud. **53**(10), 2118–2133 (2016). https://doi.org/10.1177/0042098015586547
12. Audretsch, D.B., Belitski, M.: Entrepreneurial ecosystems in cities: establishing the framework conditions. J. Technol. Transf. **42**(5), 1030–1051 (2017). https://doi.org/10.1007/s10961-016-9473-8
13. Szaban, J., Skrzek-Lubasińska, M.: Self-employment and entrepreneurship: a theoretical approach. J. Manag. Bus. Adm. Cent. Eur. **26**(2), 89–120 (2018). https://doi.org/10.7206/jmba.ce.2450-7814.230
14. Shin, J., Kim, J.: Study on entering self-employment of young workers. Asia-Pacific J. Bus. Ventur. Entrep. **15**(1), 247–257 (2020)
15. Sallah, C.A., Caesar, L.D.: Intangible resources and the growth of women businesses. J. Entrep. Emerg. Econ. **12**(3), 329–355 (2020). https://doi.org/10.1108/JEEE-05-2019-0070
16. Fields, G.S.: Self-employment and poverty in developing countries. IZA World Labor. (2014). https://doi.org/10.15185/izawol.60
17. Block, J.H., Wagner, M.: Necessity and opportunity entrepreneurs in Germany: characteristics and earning s differentials. Schmalenbach Bus. Rev. **62**(2), 154–174 (2010). https://doi.org/10.1007/BF03396803
18. Simoes, N., Crespo, N., Moreira, S.B.: Individual determinants of self-employment entry: What do we really know? J. Econ. Surv. **30**(4), 783–806 (2016). https://doi.org/10.1111/joes.12111
19. Low, S.A., Weiler, S.: Employment risk, returns, and entrepreneurship. Econ. Dev. Q. **26**(3), 238–251 (2012)
20. Porter, M.E., Schwab, K., Sala-i-Martin, X., Lopez-Claros, A.: The global competitiveness report 2007–2008. Citeseer (2007)
21. Kilduff, M., Tsai, W.: Social Networks and Organizations. SAGE Publications Ltd., 6 Bonhill Street, London England EC2A 4PU United Kingdom (2003)
22. Marshall, A.: Princípios de economia: tratado introdutório, vol. 1. Nova Cultural São Paulo (1985)
23. Gnyawali, D.R., Fogel, D.S.: Environments for entrepreneurship development: key dimensions and research implications. Entrep. Theory Pract. **18**(4), 43–62 (1994). https://doi.org/10.1177/104225879401800403
24. Foster, G., et al.: Entrepreneurial ecosystems around the globe and company growth dynamics. World Economic Forum Geneva, Switzerland (2013). https://www3.weforum.org/docs/WEF_EntrepreneurialEcosystems_Report_2013.pdf
25. Feld, B.: Startup Communities: Building an Entrepreneurial Ecosystem in Your City, 2nd ed. Wiley (2020)

26. Donovan, S.A., Bradley, D.H., Shimabukuru, J.O.: What does the gig economy mean for workers? (2016)
27. Gleim, M.R., Johnson, C.M., Lawson, S.J.: Sharers and sellers: a multi-group examination of gig economy workers' perceptions. J. Bus. Res. **98**, 142–152 (2019). https://doi.org/10.1016/j.jbusres.2019.01.041
28. Kaine, S., Josserand, E.: The organisation and experience of work in the gig economy. J. Ind. Relat. **61**(4), 479–501 (2019)
29. Bewley, T.F.: Knightian decision theory and econometric inferences. J. Econ. Theory **146**(3), 1134–1147 (2011). https://doi.org/10.1016/j.jet.2010.12.012
30. Lucas, R.E.: Asset prices in an exchange economy. Econometrica **46**(6), 1429 (1978). https://doi.org/10.2307/1913837
31. Kihlstrom, R.E., Laffont, J.-J.: A general equilibrium entrepreneurial theory of firm formation based on risk aversion. J. Polit. Econ. **87**(4), 719–748 (1979). https://doi.org/10.1086/260790
32. Evans, D.S., Jovanovic, B.: An estimated model of entrepreneurial choice under liquidity constraints. J. Polit. Econ. **97**(4), 808–827 (1989). https://www.jstor.org/stable/1832192
33. Bewley, T.F.: Market innovation and entrepreneurship: a Knightian view. In: Economics Essays, pp. 41–58. Springer (1989)
34. Bewley, T.F.: Knightian decision theory and econometric inference. J Econ Theory **146**(3), 1134–1147 (1988). https://elischolar.library.yale.edu/cowles-discussion-paper-series/1111%0A%0A
35. Douglas, E.J., Shepherd, D.A.: Self-employment as a career choice: attitudes, entrepreneurial intentions, and utility maximization. Entrep. Theory Pract. **26**(3), 81–90 (2002). https://doi.org/10.1177/104225870202600305
36. Agrawal, V.V., Bellos, I.: The potential of servicizing as a green business model. Manag. Sci. **63**(5), 1545–1562 (2017). https://doi.org/10.1287/mnsc.2015.2399
37. Acs, Z.J., Armington, C.: Entrepreneurship, geography, and American economic growth. Cambridge University Press (2006)
38. Fairlie, R.W., Fossen, F.M.: Defining opportunity versus necessity entrepreneurship: two components of business creation. Emerald Publishing Limited (2020)
39. Block, J., Koellinger, P.: I Can't get no satisfaction-necessity entrepreneurship and procedural utility. Kyklos **62**(2), 191–209 (2009). https://doi.org/10.1111/j.1467-6435.2009.00431.x

Challenges of Agricultural Innovation Ecosystems—The Case Study of Central Europe's First R&D Purpose Vertical Farm

Klaudia Gabriella Horváth and Ferenc Pongrácz

Abstract An analysis of the professional experiences of Central Europe's first R&D purpose vertical farm is presented in the paper, which was formed through the Triple Helix innovation model with a collaboration of a wide variety of partners in Hungary. The aim of the article is threefold. Firstly, since sustainable food production is a worldwide problem, the scientific and practical knowledge of potential business models of agricultural innovations are further expanded in the paper using Porter's generic strategies framework. Secondly, whereas agricultural innovations are becoming more valuable in the future, most of these innovations are yet unprofitable. Therefore, the challenges of transforming R&D results to profitable commercial products and services from an ecosystem management point of view is analyzed. Thirdly, the research provides interesting insights on how Covid-19 influenced the diffusion of vertical agriculture technology. The research was conducted using qualitative methods, the findings of the case study is based on twenty-six semi-structured interviews.

Keywords Innovation ecosystem · Agricultural innovation · Vertical farms · Case study

1 Introduction

Today, the agricultural sector is facing increasingly pressing challenges as a result of the Earth's growing population, dwindling resources and the constantly deteriorating natural environment. Solutions to these challenges should be found primarily through more efficient production and, in this context, agricultural innovation [19].

K. G. Horváth
National University of Public Service, Budapest, Hungary
e-mail: horvath.klaudia.gabriella@uni-nke.hu

F. Pongrácz (✉)
Tungsram Operations Kft, Budapest, Hungary
e-mail: Ferenc.Pongracz@tungsram.com

A. Visvizi et al. (eds.), *Research and Innovation Forum 2022*,
Springer Proceedings in Complexity, https://doi.org/10.1007/978-3-031-19560-0_52

Currently, one of the most intriguing innovations in crop production is controlled environment agriculture (CEA), which mainly includes indoor agriculture and vertical farming technology [6]. The aim of CEA technologies are to provide protection from the outdoor elements (e.g. climate, temperature etc.) and maintain optimal growing conditions throughout the cultivating process of crops in a closed system. The technology is currently emerging as a result of the rapidly changing climate conditions and growing interest of alternative food supply options [3, 10].

Due to Covid 19 pandemic and the ongoing war on Ukraine food supply safety and promoting resilience (e.g. energy diversification) have became strategic matters for EU [10, 12]. Hungary is one of the few European countries that have a strong, export oriented primary sector, therefore agricultural innovations—like CEA technologies—are of paramount importance.

The history of Tungsram, a Hungarian multinational company analyzed in the study, goes back more than 120 yrs, starting with the establishment of the United Incandescent Lamp and Electrical Company in 1896. The company was acquired by General Electric (GE) in 1989 as a result of privatisation, and Tungsram returned to the market in April 2018 as a Hungarian technology intensive lighting brand.

In the field of indoor agriculture, the first vertical farm for research and development in Central Europe was inaugurated in May 2021 at Tungsram's headquarters. The farm covers 150 square metres and grows micro vegetables in a hydroponic system with LED lighting developed by Tungsram. As the project was implemented by Tungsram in cooperation with higher education institutions, SME companies and governmental actors, this vertical farm is a good example of an innovation ecosystem based on the so-called Triple Helix model.

In the context of the above, three research objectives were set in this study, each of which was investigated in a case study method using qualitative, semi-structured interviews. (1) First, we sought to answer the question of how the Covid-19 epidemic has affected the expansion of vertical farms in Tungsram's experience. (2) Second, given that vertical farms have very high investment costs and are unprofitable to operate for the time being, we outline business opportunities and models for vertical farming that can be identified in Tungsram's experience, based on Porter's three competitive strategies. (3) Thirdly, we analyse the Triple Helix cooperation with scientific relevance by examining Tungsram's agri-innovation ecosystem.

The article is structured as follows: the Chap. 2 introduces a summary of literature review regarding vertical farming and innovation ecosystems; the Chap. 3 describes the research methodology; the Chap. 4 presents the results of the analysis based on the three research questions above; while the Chap. 5 is a short discussion of the findings and the Chap. 6 is the conclusion of the research where further research directions are outlined.

2 Literature Review and Conceptual Approach

2.1 The Vertical Farm as an Agricultural Innovation

In Despommier's interpretation, vertical agriculture, is essentially the practice of growing plants in vertically stacked layers in an innovative indoor crop production system in an artificially constructed environment [6]. The goal of the vertical farm is to create optimal conditions for high-yielding, resource-efficient crop production while eliminating seasonal variations in yield caused by weather and other external factors.

In terms of geography, 33% of operating vertical farms are located in North America, 30% in Europe and approximately 25% in Asia and the Middle East, according to 2020 data. The market for vertical farms was estimated at $3.24 trillion in 2020, which is projected to exceed $24 trillion by 2030, growing at a rate of around 22% per year [13]. Major European players in the sector are Growx, Seven Steps to Heaven and Future Crops from the Netherlands, Urban Harvest from Brussels and Planet Farms from Italy. In Central Europe, Vertigo Farms from Poland and Tungsram from Hungary are currently the main market players.

On one hand, as the technology of vertical farms is getting more recognition, market players are pressured to choose a possible but at the same time viable business model to reach consumers. The topic has already been analyzed from various aspects for e.g. [2] have constructed a unique Vertical Farm Business Framework; while [16] formulated a typology of vertical farms; and [4, 15] had explored the marketing mix of selling CEA cultivated crops. These researches clearly indicate that there isn't any ultimate business model for vertical farms, because this is still a rather ambigous market area. Therefore a commonly used business strategy framework as Porter's generic strategies may help us to see the current market opportunities "in the big picture".

On the other hand, as vertical farming is a very new but innovation-oriented and therefore costly and complex industry segment, cooperation between related industry players is a prerequisite for market entry.

2.2 The Triple Helix Innovation Ecosystem Model

In the literature, the innovation ecosystem can be seen as a "buzzword" for innovation-related collaborations. According to Granstrand and Holgersson [8], innovation ecosystems are a set of actors, activities, instruments, institutions and the consciously shaped relationships between them that determine the innovative performance and value creation capacity of an economic actor or set of actors. One of the most common models of innovation ecosystems is Triple Helix.

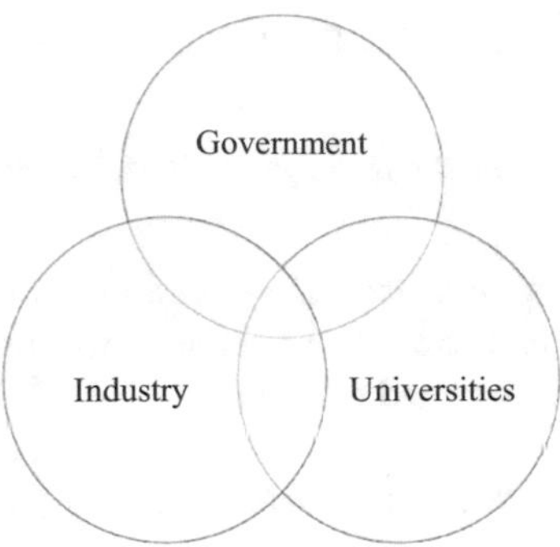

Fig. 1 Triple helix model *Source* Own editing based on [7]

As Fig. 1 indicates, Triple Helix model is based on the premise that strengthening the links between actors can create a dialogue between the academic research community, which holds academic knowledge, the profit-oriented companies that meet market needs, and the public institutions that promote inclusive societies and national economic development [7]

Since the agri-innovation ecosystem of Tungsram, which we analyzed, includes a state partner, two other corporate partners, a foreign and a Hungarian higher education institution, the cooperation can be considered a practical application of the Triple Helix model.

3 Research Methodology

This research can be considered an exploratory and explanatory research. Although the inevitable limitation of case study methodology (especially single case studies) is that no general conclusions can be drawn on the basis of the results obtained, it is still a frequently used methodology in the field of management sciences because it can be used to gain a practical and comprehensive in-depth understanding of scientific problems [20]. Case study approach is also a commonly used method on the field of CEA—see e.g. [4, 5], as practial implications of theories and laboratory research results provide highly valuable feedback for market players and researchers.

The study seeks to answer the following three research questions:

(1) What impact has the Covid-19 epidemic had on the diffusion of vertical farms based on the experience of the Tungsram-led ecosystem?
(2) What business models can be identified based on the experience of the Tungsram-led ecosystem?
(3) What strengths and weaknesses do Tungsram's cooperation partners identify in the agri-innovation ecosystem?

Table 1 Number of interviews by partners

Stakeholders in cooperation	Partner type	Number of interviews conducted
Public institution	Public partner	3
Companies	Leading company	8
	Corporate partners (in Hungary)	7
Higher education institutions	University partners in Hungary	5
	University partner abroad	3

Source Own editing

The questions were examined using a qualitative method, document analysis and 26 semi-structured interviews with experts. The interviews were conducted between October 2021 and February 2022, and the interviews were audio-recorded for later analysis, resulting in 39 h of audio material recorded during the research. Table 1 summarizes the interviews conducted with the participants, while Fig. 2 demonstrates the structure of the ecosystem.

In the following, the results of the research are presented in the context of the research questions above, and in all cases quotations are used with the consent of the respondents.

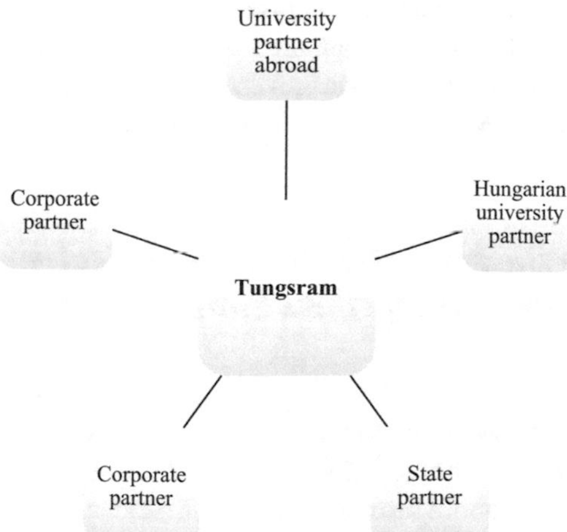

Fig. 2 Tungsram's ecosystem. *Source* Own editing

4 Results

4.1 The Impact of Covid-19 on the Diffusion of Vertical Farms

Although the Covid-19 outbreak highlighted in many ways that even in developed countries, the problem of securing a stable food supply can be problematic, the interviewees' views and Tungsram's market experience suggest that the outbreak situation has not had a significant impact on market demand. The reason for this was seen by interviewees as "*because the economic recovery following the Covid-19 outbreak was accompanied by a large increase in energy prices, even though the outbreak highlighted food security issues and the importance of vertical farms, the energy costs actually made it more expensive to run a vertical farm, which in turn neutralised any potential increase in demand as a result of the outbreak.*" While another respondent said "*Due to the EU's strong agricultural subsidies and regulations (crops grown in the EU without land cannot be certified organic), the growth potential of the European market is artificially limited*". Another point of view is that "*…the ecosystem of vertical farming is definitely incomplete, just like electric cars, tha concept is developed, the exploitation options are quite clear, but neither the regulation and infrastructure, nor the policital mindset enable us to reach the consumers*".

The results of our research therefore show that the Covid-19 epidemic has not had a significant positive impact on the expansion of vertical farms and the European market. However, interviewees has also implicated (specifically corporate partners), that the rising inflation and food shortages, which has a major impact on food production and sales' prices could enhance the attractiveness of vertical farming. Not to mention the moral pressure on governments to move spectatively towards sustainability and domestic food production for self-reliance.

4.2 Identifying Business Models in the Vertical Agriculture Market (Porter's Competitive Strategy Analysis)

Most of the vertical farms currently in operation are loss-making, mainly because of the high investment costs mentioned above. The issue of the market introduction and positioning of vertical farm technology is further complicated by the fact that on the one hand we can talk about farm technology as a technical innovation and on the other hand we have to talk about the issue of the marketing of agricultural products produced on vertical farms. Figure 3 indicates the circular connection with the two above mentioned sides of vertical farm market.

In the context of the above problem, our second research question was to identify market opportunities and business models in the global market for vertical farms

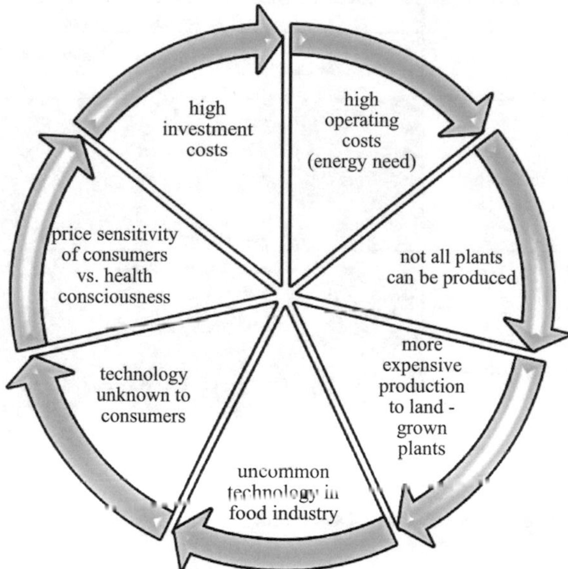

Fig. 3 Linking the technological and food industry challenges of vertical agriculture. *Source* Own editing

based on Tungsram's experience. We chose the well-known Porter's three competitive strategy models as the framework for our analysis [14] The three strategies are very well known, and thus their practical applicability is supported by a large body of empirical research [11], furthermore this analysis has not been applied to the vertical farm market before [2, 19]. Our results are indicated in Fig. 4.

Cost leadership strategy. *Solution-based business model.* The so-called solution-based business model provides for the installation and operation of vertical farms as an integrated service, with the farm being operated by the installation company and its partners.

Technological simplification (type farms). The model consists in replacing vertical farms, which can be used to grow basically any type of crop, with type farms, which can grow only one or two crops with similar needs. The marketability of the business model is supported by the fact that, in any case, economies of scale do not allow for too many crops to be grown on one farm.

Differentation strategy. *Premium health preservation product model.* Since the vitamin and mineral content of farm-grown plants can be maximised due to the artificial growing environment, the health preserving properties of vertical farm-grown greens can be differentiated in the premium segment.

"Skyfarming" model. The phenomenon of skyfarming, which involves small-scale farms on the roofs of skyscrapers or apartment buildings, has long been widespread, particularly in Japan and the USA. These rooftop farms provide a form of recreational,

		Strategic Advantage	
		Low Cost Position	Uniqueness Percieved by the Consumer
Strategic Target	Industrywide	**Cost Leadership** *Solution-based business model* *Technological simplification*	**Differentation** *Premium products* *"Skyfarming"*
	Particular Segment Only	**Focus** *Regional specialisation* *Specialisation in cosmetics and pharmaceuticals* *Grow & Collect business model*	

Fig. 4 Vertical farm business models identified from the expert interviews within the framework of Porter's competitive strategies. *Source* Own editing

green space primarily for residents or workers. This business model is also in close connection to smart cities and villages, because on one hand agricultural innovations enable local communities and stakeholders to reconnect with nature, on the other hand, a more technology intensive primary sector could enhance the attractiveness of rural life [17, 18].

Focus strategy. *Regional specialization.* As the Middle East is highly exposed to climate change and water conservation is a pressing issue in the region, vertical farm sales are expected to boom in this region by 2030.

Cosmetics and pharmaceuticals specialization. Not only can indoor farming technology become increasingly important for beauty and pharmaceutical companies in terms of security of supply, but vertical farms may also offer a solution to the moral problems they face (e.g. animal testing, buying raw materials harvested using child labour, environmental pollution).

The Grow & Collect business model. The Grow and Collect model is all about allowing consumers to buy convenient, easy-to-manage, few-storey farms that can be placed in the home and grow vegetables in a closed, home-based system. Interviews suggest that the B2C strategy may be a particularly good option for companies that are primarily involved in technology development but lack the capital to build commercial vertical farms or consciously do not want to have a food industry tie-in.

From the Porter competition strategy analysis based on the results of the interviews, it can be concluded that in order to exploit the business opportunities offered by vertical farm technology, it is necessary to take into account both the technical and food specificities of vertical farms. The difficulty in the European context is mainly due to the fact that technical technology companies usually lack experience in the food industry (e.g. Tungsram), while the food industry also lacks the necessary know-how to operate farms and the uncertain regulatory environment limits the possibilities for production.

4.3 Practical Experience of the Agri-Innovation Ecosystem Led by Tungsram

As we have mentioned before, innovation cooperations between different market actors are particularly valued in the vertical farm industry segment. Therefore, we will now present the strengths and weaknesses of Tungsram's agri-innovation ecosystem based on the experiences identified during the interviews.

Strengths of ecosystem cooperation. The primary strength of ecosystem cooperation identified from the interviews is the potential to combine complementary resources, as is common to ecosystems in general [1, 5, 8]. As vertical farming requires both technical and food expertise, ecosystem collaboration helps to combine the resources of the different industry actors, enabling each participant to create more value than they could individually.

The trust between participants was also identified by the actors as a strength of the ecosystem and one of its key operational foundations. Based on the interviews conducted, the functioning of Tungsram's ecosystem is largely determined by the fact that the partners know each other, often through previous relationships, and that the person in charge of Tungsram's agri-innovation business is seen by the partners as a credible representative and leader of the ecosystem.

The weaknesses of ecosystem cooperation. Based on the interviews, Tungsram and the partners identified as a fundamental weakness of the ecosystem that the cooperation did not facilitate the acquisition of external funding, mainly venture capital, to take the project forward. The public partner basically provided innovation support for the construction of the farm, which had run out by the time the market entry and production phase was reached. This confirms, at the micro level, the innovation problem that theoretical research suggests is a feature of Europe, namely that EU Member States generally have difficulties in bringing innovations and basic research results to the market, partly due to deficient support structures [9].

A difficulty encountered during the interviews, from the university side, is that the flexibility of the ecosystem is difficult for the university bureaucracy to manage, thus the least direct links in Tiple Helix can be identified in the university-company relationship.

The participants newly perceived the ecosystem's lack of a clear structure as a problem. To address this difficulty, Tungsram released the whole agri-innovation ecosystem as a spin-off company in spring 2022, called Tungsram Food Autnonomy. The experience of this new kind of collaboration will be useful to compare in a future research with the previous experiences of the partners that we have explored.

5 Discussion

This paper analyzed three research questions regarding (1) Impacts of Covid-29 on vertical farm market; (2) business models of vertical farm market; and (3) practical implications of an agri-innovation ecosystem led the Hungarian Tungsram. In the following we outline the most important findings in parallel to the existing literature.

Impacts of Covid-19 on vertical farm market. Based on the interviews conducted, we conclude that the uptake of vertical farms has not been significantly positively affected by the Covid-19 epidemic, mainly because the high investment costs of farms and the high energy demand for their operation make the investment less and less profitable in the long run due to the current energy crisis. The findings are in line with the existing literature stating that the most important hindering component of the diffusion of vertical farming is high investment cost [15]. However, as this aspect of vertical farm market is not widely discussed yet, the rising inflation and food shortages, may enhance the attractiveness of vertical farming in the future, therfore Covid-19 may have positive impacts on the market in the long run.

Business models of vertical farm market. As we have outlined in the literature review, the market and business opportunities of vertical farm technology is relatively discussed in the literature [2, 4, 19], however using Porter's generic strategies was formulated a broader picture about the current market. The results of the case study indicated, that vertical farming is not so different to any other products or services in general, therefore special business frameworks may not needed in the future. However, as vertival farming has a strong technology aspect and an implicit food regulation viewpoint, it is quite essential for market players to choose business models and sales channels visely, or work in close co-operation with each other. Tungsram's experiences suggest that pursuing different business models at the same time is not beneficial due to high cost and narrow market segments.

Practical implications of an agri-innovation ecosystem. The results of the research highlight the potential of innovation ecosystems to exploit the technical and food specificities of vertical farming. On one hand, the strenghts of Tungsram's ecosystem is in line with the current stream of ecosystem literature [1, 5, 8], namely we identified as advantages: combining complementary resources; trust; cooperation without competitors. On the other hand, we concluded some intersting insights on the wekanesses of Tungsram's ecosystem. These are: ecosystem collaboration did not facilitate external capital mobilisation; companies did not trust important tasks for universities due to lack of trust in their capabilities; and last but not least, the lack of clear structure of the ecosystem during product development phase hindered the whole process. Overall, it is an intriguing question for the future, whether these weaknesses are generalizable for other ecosystems as well, or they characterize only agricultural ones.

6 Conclusions and Further Research Implications

Given that the present study analyzed the market for vertical farms in a case study format based on the professional and practical experience of an agri-innovation ecosystem in Hungary, no general conclusions may be drawn from the above results. However, new results are concluded regarding vertical farm market from business point of view.

Firstly, Porter's analysis of competitive strategy can serve as a useful tool for other researchers and industry players to see the market opportunities "in the big picture". Tungsram experiences indicated that no speical business framework is needed for succesful market entry, however it is advisable for companies to choose their business scope before penetrating the market and formulate co-operations with potential suppliers.

Secondly, the functioning of the analyzed ecosystem confirmed the mainstream literature regarding the benefits of ecosystem co-operations, however as for the weaknesses, market players should keep in mind that ecosystem participation may not facilitate external capital mobilisation and a clear structure is preferable in an ecosystem where a lot of different participants are working together.

As ecosystems change and evolve dynamically, we plan to repeat the research in a few years to compare collaborative experiences and explore directions for development. Regarding Covid-19, the research outlined different scenarios regarding vertical farms, we may see which one will be realized in the future.

References

1. Adner, R.: The Wide Lens—A New Strategy for Innovation. Portfolio Penguin, London (2012)
2. Allegasert, D., Wubben, E.F.M., Hegaelaar, G.: Where is the business? a study into prominent items of the vertical farm business framework. Eur. J. Hortic. Sci. **85**(5), 344–353 (2020)
3. Berners-Lee, M., Kennelly, C., Watson, R., Hewitt, C.N.: Current global food production is sufficient to meet human nutritional needs in 2050 provided there is radical societal adaptation. Elem.: Sci. Anthropocene **6**(52), 1–14 (2021)
4. Broad, G.M., Marchall, W., Ezzeddine, M.: Consumer attitudes to vertical farming (indoor plant factory with artificial lighting) in China, Singapore, UK, and USA: A multi-method study. Food Res. Int. **150**(13), 1–17 (2021)
5. De Oliveira, F.B., Forbes, H., Schaefer, D., Syed, J.M.: Lean principles in vertical farming: a case study. Procedia CIRP **93**(9), 712–717 (2020)
6. Despommier, D.: The Vertical Farm: Feeding the World in the 21st Century. Picador, London (2010)
7. Etzkowitz, H., Leydesdorff, L.: Emergence of a triple helix of University-industry-government relations. Sci. Public Policy **23**(5), 279–286 (1996)
8. Granstrand, O., Holgersson, M.: Innovation ecosystems: a conceptual review and a new definition. Technovation **90–91**(2–3), 1–12 (2020)
9. Heder, M.: From NASA to EU: the evolution of the TRL scale in public sector innovation. Innov. J.: Public Sect. Innov. J. **22**(2), 1–24 (2017)
10. Hoobs, J.E.: Food supply chain resilience and the COVID-19 pandemic: what have we learned? Can. J. Agric. Econ. **69**(2), 189–196 (2021)

11. Islami, X., Musata, N., Latkovij, M.T.: Linking Porter's generic strategies to firm performance. Fut. Busin. J **6**(3), 1–15 (2020)
12. Lal, R.: Home gardening and urban agriculture for advancing foos and nutritional security in response to the Covid-19 pandemic. Food Secur. **12**(4), 871–876 (2020)
13. Mordor Intelligence.: Vertical farming market—Growth, trends, COVID-19 impact, and forecasts (2022–2027). https://www.mordorintelligence.com/industry-reports/vertical-farming-market. Last Accessed 02 April 2022
14. Porter, E.M.: Competitive Strategy. The Free Press, New York (1980)
15. Sprecht, K., Weith, T., Swobonda, K., Sieber, R.: Socially acceptable urban agriculture business. Agron. Sustain. Dev. **36**(7), 1–17 (2016)
16. Thomaier, S., Specht, K., Henckel, D., Dierich, A., Siebert, R., Freisinger, U.D., Sawicka, M.: [2014]: Farming in and on urban buildings: present practice and specific novelties of zero-acreage farming (ZFarming). Renewable Agric. Food Syst. **31**(1), 43–54 (2014)
17. Visvizi, A., Lytras, M.D., Mudri, G.: Smart Villages in the EU and Beyond. Emerald Publishing, Bingley, UK (2018)
18. Visvizi, A., Lytras, M.D.: It's not a fad: smart cities and smart villages research in European and global contexts. Sustainability **10**(8), 1–10 (2018)
19. Weidner, T., Yang, A., Hamm, M.W.: Consolidating the current knowledge on urban agriculture in productive urban food systems: learnings, gaps and outlook. J. Clean. Prod. **27**(4), 1637–1655 (2019)
20. Yin, R.K.: Case Study Research and Applications: Design and Methods, 6th edn. Sage, London (2018)

Fresh Food Deliveries to Military Units During the COVID-19 Pandemic

Małgorzata Dymyt and Marta Wincewicz-Bosy

Abstract This paper examines the military supply chains during the Covid-19 pandemic especially as viewed from the perspective of food purchase and delivery processes. Specifically, the objective is to identify the conditions necessary to ascertain the continuity of fresh food supplies and against this backdrop to develop a model of supplying military units with fresh food under conditions of emergency, e.g. a pandemic. The research methods employed included qualitative methods, such as mapping the processes related to the supply of fresh food to military units on a selected case, observation, expert interviews and analysis of literature and documents. As a result of the research, key areas of the supply process exposed to continuity disturbances and their determinants were identified. The conclusions relate to the possibility of improving activities in the field of supplier relationship management, taking into account the specificity of civil-military cooperation. The article concerns civil-military cooperation and disruptions in specific circumstances of the initial phase of a pandemic crisis. It is argued that there is a need for flexible, adaptive shaping of relations in terms of readiness to implement changes in food supplies in crisis situations.

Keywords Supply · Fresh food · Military food supply chain · Collaboration · Covid-19 Pandemic

Article Classification Case study

M. Dymyt · M. Wincewicz-Bosy (✉)
General Tadeusz Kosciuszko Military University of Land Forces, Wroclaw, Poland
e-mail: marta.wincewicz-bosy@awl.edu.pl

M. Dymyt
e-mail: malgorzata.dymyt@awl.edu.pl

A. Visvizi et al. (eds.), *Research and Innovation Forum 2022*,
Springer Proceedings in Complexity, https://doi.org/10.1007/978-3-031-19560-0_53

1 Introduction

The outbreak of the Covid-19 pandemic led to serious disruptions in many areas of social and economic life. Mobility restrictions, orders to limit activities in many spheres disturbed the flow of goods and services. There were significant discrepancies between the changing needs of consumers and the supply and distribution capacity of product suppliers, including food products. Consequently, the coronavirus pandemic has exposed a fundamental lack of resilience in the food supply chain [1]. As a result of the pandemic, food insecurity emerged. The latter is defined as a situation in which people lack secure access to sufficient, safe and nutritious food for normal growth and development and an active and healthy life [2]. It underlines that despite high food production, the pandemic could have a negative impact on the level of food production due to factors such as increased worker morbidity, supply chain disruptions and restrictive measures, as well as individual governments' efforts to limit exports food to meet national needs [3]. In this context, it turned out that the greatest threat to food security is not the availability of food itself, but the problem of delivery-related consumer access to food [4].

In the broader discussion on this issue, four main pillars of food security affected by Covid-19 are identified, such as: availability (adequate food supply), access (ability to meet human food needs), utilisation (possible nutrient intake at a sufficient level) and stability (constancy of access to food) [5]. Indeed, during the Covid-19 pandemic, disruptions in the flow of food were particularly visible in the sector of fresh food. Any delays in deliveries resulted in deterioration of the quality of this type of food, which exacerbated the situation. Against this backdrop, the focus of this paper is directed at the military supply chains of fresh food. In this context the specificity of the military food supply chains under conditions of Covid-driven emergency is examined. Interestingly, while the Covid-19 pandemic disrupted military food supply chains, affecting their continuity and reliability, especially in the area of fresh food, it is also exposed the centrality of civilian suppliers for fresh food purchase and delivery process in the military sector. Hence, the key question that this paper address is what factors determine the continuity of the army's fresh food supply processes carried out in civil-military cooperation, and what actions should be taken to ensure food supplies?

The argument in this paper builds on the following assumptions. First, the efficiency of military supply chains is key to ensuring readiness to act both in times of peace and in times of crisis. Second, due to the nature of the military, unlike civilian systems, the military is less flexible, and so is the ability to respond to crisis situations. Third, to bypass this challenge, it is necessary to look for opportunities to improve civil-military cooperation in the framework of fresh food supplies.

The article is of a conceptual and research nature and aims to analyse the essence of the functioning of the military food supply chains, including fresh food, in relations with civilian suppliers. In the work, the authors ask the following research questions:

The reminder of the discussion is structured as follows. The article consists of five other sections following the introduction, containing a description and analysis of the issue. Section 2 presents the theoretical aspects of managing fresh food supply chains in the "Covid-19" pandemic. The next part of the article describes the research approach and methodology. Section 4 provides a description of the case study, analysis and its results. The last section contains conclusions, main theoretical and management implications, recommendations, limitations and future research perspectives.

2 Theoretical Framework

Food supply chain management (FSCM) is an extremely difficult task due to the complexity of the chain itself, as well as its diversity in terms of factors such as food quality, safety and freshness over a limited period of time [6]. The food supply chain is not just a simple, single sequence of specific entities, but a complex network of interrelated elements that work together to make the products available to consumers [7]. The challenge for food supply chain management is the need to use advanced handling systems to deal with issues such as perishability, unexpected changes in the supply chain and food safety attributes that have a significant impact on environmental sustainability, health and consumer loyalty also supply chain profitability [8].

Relationships in the food supply chain become even more complex when there are specific needs and requirements as well as complicated procedures and principles of cooperation, or conditions of operation. Such specific requirements relate to the supply of fresh food, for which systems and logistical processes have to be designed taking into account time pressures and quality requirements. Fresh products are defined as products without any thermal or other treatment, grown locally without any preservation prior to storage, which, due to the short shelf-life, presents particular challenges for food logistics and need to be part of a cool chain to reduce waste [9]. For perishable and seasonal products, the key issue is the short shelf life which requires considerable effort to keep the product fresh and available [10]. Managing the supply chain of perishable products due to the complexity of processes resulting from the nature of the product, high uncertainty of demand and costs, and increased consumer concern for food and changing environmental conditions (e.g. changes in consumer needs, technological development) requires stakeholder cooperation based on closer relationships within supply chains and networks [11].

In the case of military supply chains, cooperation in the field of food supply requires a relationship between the civil sphere (producers, suppliers) and the military sphere (recipients and consumers). These relationships, due to the specific needs and requirements of the military, are complex, formalised, and are subject to strict procedures of demand planning, supplier selection and implementation of logistics processes. The civil-military cooperation means "*a joint function comprising a set of capabilities integral to supporting the achievement of mission objectives and enabling NATO commands to participate effectively in a broad spectrum of civil-military interaction with diverse non-military actors*" [12] (p. 2.1). This set of capabilities must therefore be adapted to the current needs and nature of the activities carried out.

The purpose of military logistics is planning, preparation and use of means of supply (including the feeding system), as well as the implementation of specialised services in order to maintain adequate combat readiness of the army on the territory of the state and to ensure the functioning of troops performing tasks outside its borders [13]. Safe food and water, including freshly prepared, attractively served, nutritious meals are key to the health, well-being and working capacity of soldiers in peacetime and mission operations [14].

The role and nature of activities carried out by the military, characterised by specific requirements in terms of safety, quality, supplier selection procedure and purchasing organisation, constitute a significant challenge for collaboration within the supply chain. Supply chain collaboration in general is the ability to work across organisational boundaries aimed at building and managing unique, complex and value-added processes to better meet customer needs [15]. Collaboration means the supplier's ability to work closely together and the willingness to share a range of data, from cost structures to planning and logistics, and it should be manifested in attitudes that relate to supplier integrity, reliability, help in cost reduction, synergy and customer service support [16]. The foundation of collaboration is collaborative decision-making and problem-solving, which requires sharing information, risks, rewards and responsibilities [17]. Collaboration means caring for relationships that relate to issues such as mutual understanding, shared vision, shared resources and the achievement of common goals [18].

3 Research Design and Methods

The starting point for the development of the research procedure was the assumption that due to the specificity and importance of the functioning of military food chains, including fresh food, it is necessary to identify factors and actions that are important to ensure continuity of supply in the conditions of disruptions that occurred in connection with the coronavirus pandemic. The problematic situation was caused by the disruptions in the supply of fresh food in connection with the health crisis.

In this context, a research problem was formulated, the essence of which is to search for an answer to the question of what are the conditions and possibilities of improving activities in the field of supplier relationship management, taking into account the specificity of civil-military cooperation, including decision-making processes regarding the selection of suppliers, planning and flexibility of supplies, system dependencies and communication.

In order to solve the research problem and achieve the assumed goal, a multi-stage research procedure was adopted, including various research methods representing the qualitative approach such as: literature study (review and in-depth analysis of scientific literature), document analysis (review and analysis of documents, reports, legal acts), participant observation, individual, partially structured interview with experts, case study and mapping and analysis processes. The initial stage of the research procedure was a review and analysis of scientific literature and articles as well as the analysis of documents enabling the recognition and systematisation of the state of knowledge on the functioning of national food supply chains in the conditions of the COVID-19 pandemic. The main element of the research procedure is the case study method, allowing for in-depth analysis within the selected research facility representing the problem under study. After selecting the case, in-depth, partially structured interviews with specialists were conducted, which allowed to define the structure of military food chains, their participants and their tasks. As a result, a map of the food supply process was developed in the next stage. The process mapping method was helpful for the in-depth analysis, allowing for a detailed understanding of the structure of the system, its goals (results), as well as the identification of entities, their responsibility for the processes and tasks carried out. According to Biazzo, process mapping means building a model that represents the relationships between activities, people, data and objects involved in achieving a specific result and, consequently, enables the improvement and redesign of business processes [19].

The in-depth analysis of the process map has resulted in the identification of the factors of efficient civil-military cooperation in the scope of the functioning of warfare supply chains of fresh food and development of solutions recommended for use in order to ensure continuity of supplies in conditions of disruptions.

4 Case and Findings

The object of research was a military unit which, apart from standard tasks, carries out the process of training and educating military personnel. Due to the nature of the activities, the detailed data has been anonymised. In order to answer the formulated research questions, the identification of critical factors determining the efficiency of military fresh food chains, in particular in the area of relations with civilian suppliers.

The food supply activities carried out in military units require full board for the accommodated soldiers on a 24-h basis. Deliveries of fresh food carried out on the basis of contracts concluded with suppliers after a tender procedure and in accordance with the provisions of the demand plan specifying quantitative and qualitative requirements.

The following elements are present in the overall fresh food supply process:

1. Input data relating to: qualitative and quantitative requirements of the delivered food in accordance with the demand specified in the plan and the principles of conducting the purchasing procedure,
2. Transformation activities including: packaging, quality control, transport, storage, distribution and preparation of meals, and waste management,
3. Output data, i.e. the nutritional needs of soldiers (qualitative and quantitative).

The general model of cooperation includes the following phases of operation: conclusion of the transaction (negotiation, signing of contracts), execution of the transaction as well as control and settlement of the transaction.

In the case under study, there were disruptions in the implementation of food supplies contracted by the military unit, caused by the coronavirus pandemic. Consequently, the necessary modification of standard activities to ensure continuity as well as the security of fresh food supplies to the military unit. It was necessary to introduce a stage including verification activities in the area of processes (transactions), their control and improvement, as well as changes due to external factors (disruptions caused by COVID-19) to the general procedure for supplying a military unit with fresh food.

The identified activities represent process models that are conceptual, descriptive and graphical in the form of process maps. The developed process maps illustrate the individual phases of the fresh food supply procedure, including: conclusion of a transaction (see Fig. 1), execution of a transaction (see Fig. 2), verification of processes and settlement of transactions (see Fig. 3).

The following entities participate in the model process: a military unit, channel partners—wholesalers, food manufacturers and sub-suppliers of packaging.

The food delivery process is carried out on the basis of the regulations on tendering procedures that cover this type of deliveries—in accordance with the statutory requirements of the operation of Military Economic Units and military units. Both food manufacturers and intermediaries can participate in the tender. In the event of any disruptions in the sphere of deliveries, intermediaries constitute a kind of safety buffer, taking over the risk and increasing the reliability of deliveries, ensuring the possibility of substitution of suppliers. At the same time, the participation of intermediaries extends the communication process. Food producers are both direct producers (suppliers of raw materials, e.g. potatoes) and their processors (e.g. dairies, butchers). The contract includes all formal, technical, phyto-sanitary and time conditions related to the delivery.

The concluded supply contract is implemented in parts (it may take the form of cyclical deliveries) according to the agreed delivery schedule, adapted to the nutrition plan and menus. The presented model is a process that is repeated many times within the framework of a specific contract. Food production includes the implementation of phyto-sanitary control processes. Pre-shipment inspection of goods primarily consists in assessing compliance with the quantitative and qualitative (assortment) requirements resulting from the contract. It should be emphasised that fresh food has limited storage options due to its short shelf-life. During the picking process at an intermediary, short-term storage may occur.

Receipt of delivery includes physical food and document inspection processes. Phyto-sanitary control before handing over for use in the process of preparing a meal is carried out in the laboratory of a given military unit. In addition, in accordance with the requirements of the applicable regulations, the collected food samples are transferred to the Sanitary Inspectorate, on the basis of the principles of mass nutrition.

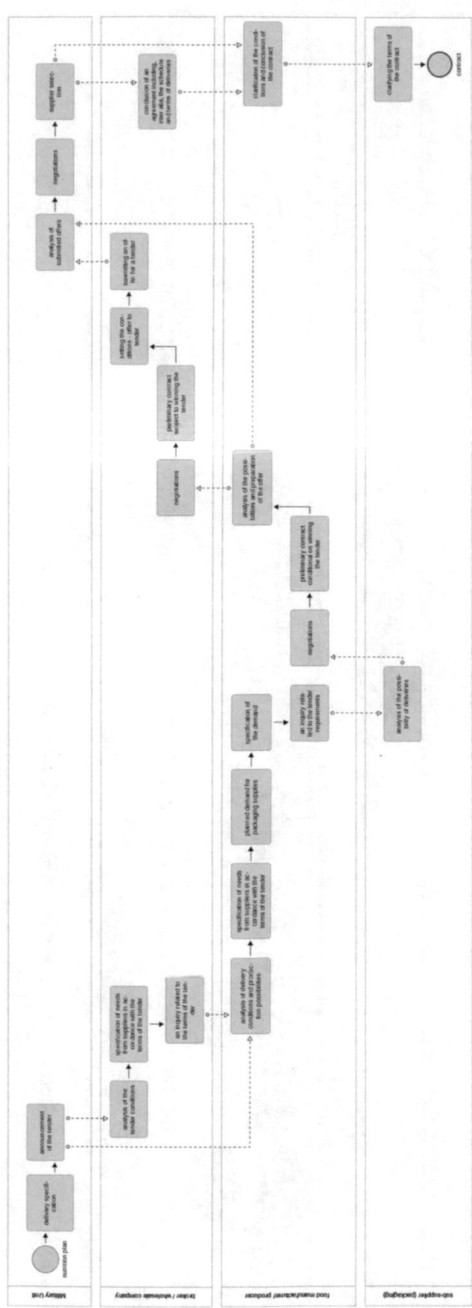

Fig. 1 The map of the fresh food supply process to a military unit—transaction conclusion phase

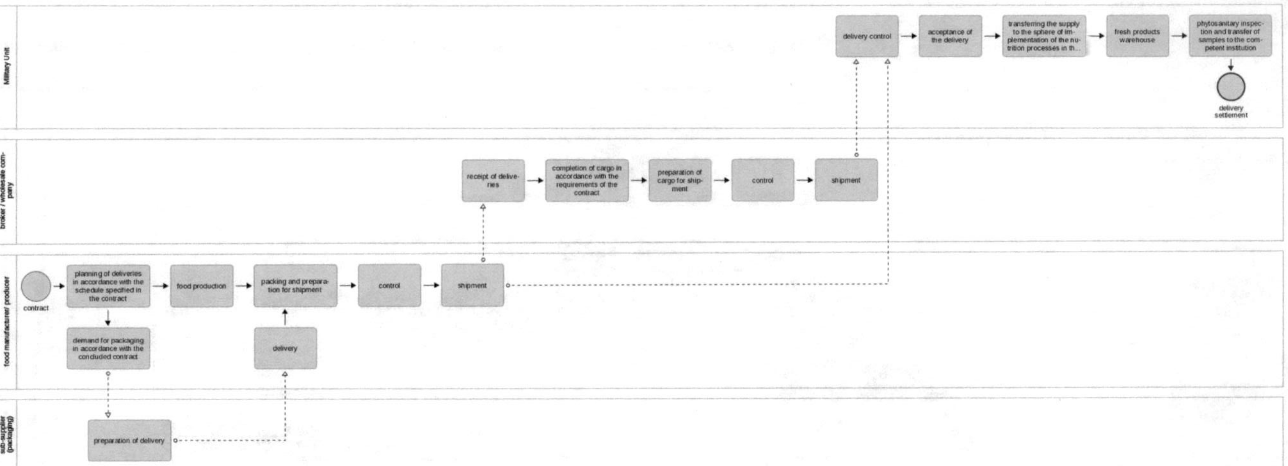

Fig. 2 The map of the fresh food supply process to a military unit—transaction execution phase

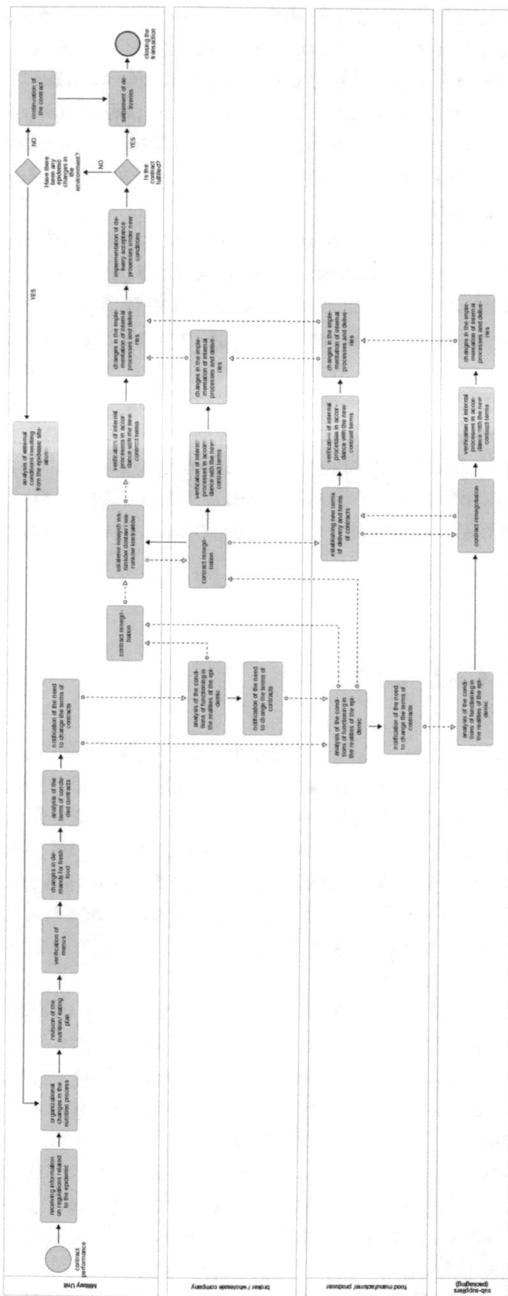

Fig. 3 The map of the fresh food supply process to a military unit—process verification and transaction settlement phase

In the next phase, as a result of various factors, it was necessary to verify the implemented processes, in particular in the field of control and improvement.

The control process consists mainly in determining the compliance of the implemented processes with the provisions adopted in the contract. If any deviations are found, they must be corrected. The adjustments may concern both the implementation of processes and the adopted assumptions. Improving implemented processes refers to changes aimed at ensuring their efficiency and effectiveness. Control processes are carried out both in individual entities and in relation to the entire relationship. The main verifier is a military unit.

The processes are also verified due to the occurrence of external factors (environmental factors), including factors such as the risk of COVID-19 infection.

In the case under study, the processes were verified in two phases. The first phase was related to the emergence of the epidemic during the delivery period, which forced changes in the course of contracts. It was necessary to adapt to formal changes and the health situation on an ongoing basis. The model covers the situation when the contract terms are verified on the initiative of the contracting entity. But due to the epidemic situation, verification may also be forced from the level of suppliers. Many of them limited the scope of their activity, suspended it or even closed it. The second phase of the verification was related to the changes that were included in the terms of the new contracts.

5 Discussion and Conclusions

Logistics efficiency is crucial in managing the food supply chain during a crisis, as any disruptions may have a negative impact on food quality, freshness and safety, and may hinder market access and affordability [20]. Among the main forces influencing the shape of the food chain are: digitisation and the resulting new possibilities for the structure and logistics of the food chain; focus on food waste and productivity; changing consumer demands and interests regarding provenance and traceability [9]. The success of these activities depends on the potential for relationship building and collaboration along the supply chain.

Cao et al. [21] propose a model of collaboration in the supply chain based on seven elements constituting mechanisms of cost and risk reduction, such as: information sharing, goal congruence, decision synchronisation, incentive alignment, resource sharing, collaborative communication and joint knowledge creation. According to Simatupang and Sridharan [22], the collaborative supply chain model includes five features: collaborative performance system; information sharing; decision synchronisation; incentive alignment; and integrated supply chain processes.

These activities are particularly important in the agri-food supply chain, which is characterised by significant barriers that may limit cooperation due to the complex and heterogeneous structure of the industry. The multiplicity of entities in the food supply chain, their diversity in terms of economic size, structure, access to modern technologies, as well as the number of interactions increases the complexity of their relations and may cause problems with the exchange of information, and consequently worsen the intensity of collaboration [23].

In civil-military relations, there are also further barriers that limit the flexibility of cooperation in the field of decision-making, information sharing. The main barriers limiting the management of relations within the framework of civil-military supply chain cooperation include: restrictions on the form and scope of communication, data confidentiality, formalisation of the decision-making process regarding the selection of suppliers and concluding contracts, low level of flexibility in terms of changes in the concluded contract, contract renegotiation, limiting relations, excluding informal relations with suppliers.

Another group of factors determining supplier relationship management results from the specificity of fresh food and the model of its delivery within the concept of short local supply chains. The concept of shortening the supply chain requires a new way of organising and managing the functions performed by entities at different levels of the chain, including: (1) coordination and management function, meaning the ability to design and control relevant aspects of transactions and interactions between different actors; (2) logistic function, expressed in the ability to physically connect producers and consumers in an efficient way reducing distribution costs without increasing agricultural production costs; (3) information function related to the ability to convey complex qualitative attributes relevant to the consumer [24].

Factors determining relationships with suppliers in the context of the health crisis caused by COVID-19 include restrictions in the field of mobility, change of the form of direct contacts to indirect contacts with the use of Internet technologies or traditional communication tools, additional control procedures, the need to implement measures increasing the safety of food products as well as employees.

The applied research methodology based on process mapping can be used to implement decision-making processes at the operational level in modelling relationships with suppliers. The proposed changes in the analysed logistics chain may contribute to the improvement of the fresh food supply chain at the organisational and inter-organisational level during a health crisis, taking into account the specific needs and requirements of military units.

The authors recognize the potential limitations of the proposed research procedure, in particular in terms of the analysis and the failure to identify all the factors conditioning the efficiency of fresh food supplies and the management of civil-military relations within the supply chain. The single case study method used allowed for the development of a universal supply chain model, recognizing that its application is limited by the situational context resulting from national formal and organisational solutions. In the case under study, the focus was on information flows, while material flows were not analysed. However, due to the limitations, the presented analyses, conclusions and proposed recommendations may be useful in the practical, managerial and scientific dimensions, constituting a starting point for broader, in-depth research.

References

1. Chenarides, L., Manfredo, M., Richards, T.J.: COVID-19 and food supply chains. Appl. Econ. Perspect. Policy **00**(00), 1–10 (2020)
2. World Health Organization: Glossary of Health Emergency and Disaster Risk Management Terminology. Geneva (2020)
3. Espitia, A., Rocha, N., Ruta, M.: Covid-19 and food protectionism. In: The Impact of the Pandemic and Export Restrictions on World Food Markets. Policy Research Working Paper 9253 World Bank, Macroeconomics, Trade and Investment Global Practice (2020)
4. The Organization for Economic Co-operation and Development: Food Supply Chains and COVID-19: Impacts and Policy Lessons (2020). http://www.oecd.org/coronavirus/policy-responses/food-supplychains-and-covid-19-impacts-and-policy-lessons-71b57aea/. Last Accessed 30 Dec 2021
5. Laborde, D., Martin, W., Swinnen, J., Vos, R.: COVID-19 risks to global food security. Science **31**, 369 (6503), 500–502 (2000)
6. La Scalia, G., Settanni, L., Micale, R., Enea, M.: Predictive shelf life model based on RF technology for improving the management of food supply chain: a case study. Int. J. RF Technol. **7**(1), 31–42 (2016)
7. Gani, S.: Food Supply Chain Management and Logistics. Wydawnictwo Naukowe PWN SA, Warszawa (2016)
8. Shashi, S.R., Centobelli, P., Cerchione, R.: Evaluating partnerships in sustainability-oriented food supply chain: a five-stage performance measurement model. Energies **11**, 3473 (2018)
9. Collison, M., Collison, T., Myroniuk, I., Boyko, N., Pellegrini, G.: Transformation trends in food logistics for short food supply chains – what is new? Stud. Agricult. Econ. **121**, 102–110 (2019)
10. Eksoz, C., Afshin Mansouri, S., Bourlakis, M.: Collaborative forecasting in the food supply chain: a conceptual framework. Int. J. Prod. Econ. **158**, 120–135 (2014)
11. Pérez-Mesa, J.C., Piedra-Muñoz, L., García-Barranco, M.C., Giagnocavo, C.: Response of fresh food suppliers to sustainable supply chain management of large European retailers. Sustainability **11**, 3885 (2019)
12. North Atlantic Treaty Organization: Nato Standard. AJP-3.19. Allied Joint Doctrine for Civil-Military Cooperation. Edition A Version 1, Allied Joint Publication, November (2018)
13. D-4 (B). Doktryna Logistyczna Sił Zbrojnych Rzeczypospolitej Polskiej. Wersja 2. [Logistics Doctrine of the Armed Forces of the Republic of Poland. Version 2], Ministerstwo Obrony Narodowej, Centrum Doktryn i Szkolenia Sił Zbrojnych, Bydgoszcz (2019)
14. North Atlantic Treaty Organization: NATO Standard AMedP-4.6 Food Safety, Defence, and Production Standards in Deployed Operations, Edition B, Version 1, March (2019)
15. Fawcett, S.E., Magnan, G.M., McCarter, M.W.: Benefits, barriers and bridges to effective supply chain management. Supply Chain Manag. Int. J. **13**(1), 35–48 (2008)
16. Shore, B., Venkatachalam, A.: Evaluating the information sharing capabilities of supply chain partners: a fuzzy logic model. Int. J. Phys. Distrib. Logist. Manag. **33**(9), 804–824 (2003)
17. Min, S., Kim, S., Chen, H.: Developing social identity and social capital for supply chain management. J. Bus. Logist. **29**(1), 283–304 (2008)
18. Skipper, J., Craighead, C., Byrd, T., Rainer, R.: Towards a theoretical foundation of supply network interdependence and technology-enabled coordination strategies. Int. J. Phys. Distrib. Logist. Manag. **38**(1), 39–56 (2008)
19. Biazzo, S.: Process mapping techniques and organisational analysis. Lessons Sociotech. Syst. Theory, Busin. Process Manag. J. **8**(1), 42–52 (2002)
20. Food and Agriculture Organization of the United Nations: Responding to the impact of the COVID-19 outbreak on food value chains through efficient logistics (2020). http://www.fao.org/3/ca8466en/CA8466EN.pdf. Last Accessed 10 Feb 2021
21. Cao, M., Vonderembse, M.A., Zhang, Q., Ragu-Nathan, T.S.: Supply chain collaboration: conceptualisation and instrument development. Int. J. Prod. Res. **48**(22), 6613–6635 (2010)

22. Simatupang, T.M., Sridharan, R.: An integrative framework for supply chain collaboration. Int. J. Logist. Manag. **16**(2), 257–274 (2005)
23. Matopoulos, A., Vlachopoulou, M., Manthou, V., Manos, B.: A conceptual framework for supply chain collaboration: empirical evidence from the agri-food industry. Supply Chain Manag. Int. J. **12**(3), 177–186 (2007)
24. United Nations Industrial Development Organization: Short Food Supply Chains for Promoting Local Food on Local Markets (2020). https://tii.unido.org/node/2879. Last Accessed 10 Feb 2021

Evaluating the Impact of the Covid-19 Pandemic on the Strategic Alignment Competences of Organizations: A Case Study in Logistics

Konstantinos Tsilionis, Yves Wautelet, and Dorien Martinet

Abstract The strategic alignment tenets describe the generic processes/mechanisms for the Information Technology (IT) domain to be impactful to the ascertainment of business strategies within organizations. However, an in-depth reassessment of the level of integration between the organizational IT approach and the business convolutions needs to be performed in times of crisis such as the recent Covid-19 pandemic. For this reason, the present research aims to explore the impact of the pandemic on the state of strategic alignment maturity within the logistics sector. To this end, the Strategic Alignment Maturity Model (SAMM), composed by Luftman, was used to measure the pre-, and post-pandemic strategic alignment maturity levels in four production and distribution facilities in Belgium. This model was chosen due to its capacity to assign strategic alignment maturity scores in a wide spectrum of individual alignment dimensions. For the application of the SAMM, we performed semi-structured interviews to determine an overall strategic alignment maturity score for each company, for the stages before, and after the outset of the pandemic. Despite some fluctuations amongst the companies' alignment maturity trends between the two stages, our results suggest an aggregate increase in the average alignment maturity level for all the surveyed companies after the start of the pandemic.

Keywords Strategic alignment · Business IT alignment · Strategic alignment maturity model · Case study · Logistics · Covid-19

1 Introduction

One of the most recent worldly events, and perhaps the most prominent in terms of its permeated changes in our ways of working and consuming, is the ongoing Covid-19 pandemic. The manifestation of the latter has triggered enterprises and

K. Tsilionis (✉) · Y. Wautelet · D. Martinet
KU Leuven, Leuven, Belgium
e-mail: konstantinos.tsilionis@kuleuven.be

Y. Wautelet
e-mail: yves.wautelet@kuleuven.be

A. Visvizi et al. (eds.), *Research and Innovation Forum 2022*,
Springer Proceedings in Complexity, https://doi.org/10.1007/978-3-031-19560-0_54

organizations to relinquish a series of digital accommodations to their workforce and clientele in the attempt to ensure a minimum level of operational continuity [1, 2]; for example, we can cite (i) the transition to remote-working standards becoming the norm for many businesses since the beginning of the pandemic [3], and (ii) the need for many service-oriented companies to rethink their business models in the attempt to ensure an uninterrupted ideation process for business-related innovations and a steady stream of income during the extended lockdown regimes [4–6]. However, the alteration of working-modes within organizations or the reconceptualization of their business offerings, necessitates a deeper integration and alignment between their major business processes and IT interventions [7, 8].

The notion of Business and IT Alignment[1] (BITA) refers to a complex dynamic process that organizations have to activate in order to enable extensive IT capabilities and achieve their business objectives [9]. This topic has been thoroughly analyzed in the literature (see [10–13]). More recently, Wautelet [14] presented a conceptual modeling-based approach enhancing the alignment evaluation capabilities of business and IT organizational strategies via the use of abstract governance-assisting service entities; Tsilionis and Wautelet [15] and Tsilionis et al. [16] detail a design-driven framework demonstrating the benefits from the in-situ placement of mechanisms/tools that can help organizations evaluate swiftly the alignment of their business and IT strategic convolutions, during the occurrences of business-impacting events (i.e., Covid-19). Pertaining to this topic, Tsilionis and Wautelet [17] place the strategic (and functional) alignment as the impacting factor determining the reconciliation between strategic organizational objectives and (operational) agile-driven software development processes.

The present research is meant to focus on the dynamicity of strategic alignment competences in the logistics sector in the course of the Covid-19 pandemic. There are prior studies that try to condition the impact of the pandemic purely on the operational amplitude of companies activated in the logistics sector (see [7, 8, 18]). Nevertheless, to the best of our knowledge, the universe of discourse has not addressed holistically the evolution of strategic alignment competencies within this sector in the aftermath of the Covid-19 pandemic.

In this setting, we chose to utilize the Strategic Alignment Maturity Model (SAMM) [19, 20] in order to investigate the state of strategic alignment maturity level in four production and distribution facilities in Belgium before, and during the evolution of the pandemic. The SAMM is a renowned evaluation framework designed to assign a strategic alignment maturity score to organizations; indeed, it utilizes six alignment dimensions to determine the phase of strategic alignment progress within organizations [19–21]. These alignment dimensions, along with their incorporated alignment criteria, are visualized in Fig. 1.

There are many studies demonstrating an elevated goodness-of-fit for the SAMM in evaluating the alignment capacities within organizations (see [22–25]). Therefore, a full evaluation of the SAMM is not within the scope of this study; rather, we aim

[1] In this study, BITA is also referred to as 'strategic alignment' or simply as 'alignment'. The terms will be used interchangeably.

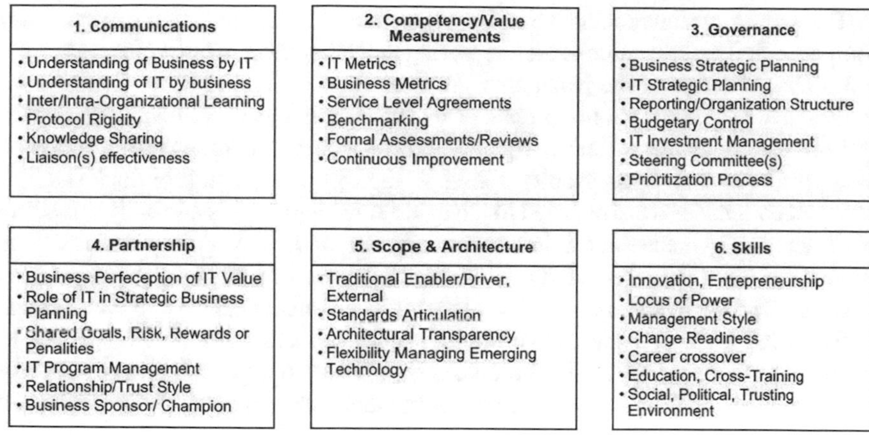

Fig. 1 The six alignment dimensions of the SAMM and their corresponding alignment criteria. *Source* Luftman [19, 20]

to utilize this framework to explore the strategic alignment maturity trends in four production and distribution facilities in Belgium before, and during the evolution of the Covid-19 crisis. Our present exploratory study constitutes the research prelude of a comprehensive descriptive research to be communicated in the future; the latter aims to yield a multilayered analysis, as the one described in the study of Liu et al. [26], to fully trace and analyze some of the pragmatic factors (i.e., behavioral, cultural, etc.,) that influence positive or negative strategic alignment transgressions within these four production and distribution facilities during the pandemic. For the moment, we are merely interested in surveying whether the pandemic has influenced any changes in the strategic alignment maturity trends of these facilities.

The remainder of this paper is the following: Sect. 2 describes the research methodology and the motivation for the choice of the case study which bestows the units of analysis for the exploration of the strategic alignment trends within these four facilities. This section describes also the data collection method whereas some preliminary results are discussed in Sect. 3. Conclusions can be found in Sect. 4.

2 Research Methodology

2.1 Research Approach

Presently, we perform an exploratory case study [27, 28] intended to collect empirical data to help us identify and compare the strategic alignment maturity tendencies within four production and distribution centers in Belgium. We expect the analysis of this data to lead us to the identification of themes/patterns explicating the evolution of these centers' alignment maturities as impacted by the pandemic.

The setting of our research exercise began with the identification of these four companies and the collection of data exploring their alignment maturity scores before, and after the outset of the pandemic. We wanted to capture rich in-content, non-quantitative data in a bottom-up manner by the stakeholders' opinions/experiences within these companies; thereby, a qualitative research methodology [29] is followed with each one of these companies acting as the case study's unit of analysis. The retrieved companies were required to have one subsidiary that is currently operating in Belgium to guarantee a minimum convergence in their working standards. For privacy reasons, the present study will refer these units of analysis as 'Company A, B, C, and D respectively. An overview of their characteristics can be found in Table 1.

Given the space constraints, we do not proceed in a full descriptive analysis to unearth the factors explicating the changes in these strategic alignment maturity trends; this is to be fully instantiated in another study that is meant to be relinquished in a future communication. Presently, the point of emphasis is on the design of a methodology (see Sects. 2.2 and 2.3) to materialize the exploration of any strategic alignment maturity change within the premises of the case study participants pre-, and post-Covid-19.

As it comes to the case study's units of analysis, companies A and B are headquartered in Belgium. Company A is specialized in manufacturing, storing and distributing IT software solutions used in the healthcare industry. Company B owns and manages the operations over three warehouses in Belgium, including two small-scaled distribution centers serving about 60 countries in Europe. Both companies are specialized in road transport, automotive, air and sea freight, logistic services and warehousing. Company C is headquartered in the Netherlands and operates many service hubs in Europe and Asia. Their Belgian offices are specialized in warehousing systems for chemical products such as powders and granulates. The company's main activities are focused on forwarding and shipping, warehousing and distribution, and the operation of breakbulk terminals. Company D is headquartered in Germany and is actively operating all over the world. The company offers transportation and supply chain services for automotive, healthcare, aerospace, and industrial goods. Their supply chain services consist of warehousing, supply chain consulting and order management.

Table 1 Participating companies and their characteristics

Company affiliation	Main business activity	Number of operating warehouses in Belgium
Company A	Manufacturing, warehousing and IT solutions	1
Company B	Warehousing and distribution	3
Company C	Warehousing and distribution	4
Company D	Warehousing and distribution	20

2.2 Data Collection

The nature of our research dictates the use of a purposive sampling technique [30] in order to determine the pool of respondents for each one of the participating companies. We used each company's email directory as our initial sampling frame in order to make our study known to the employees of each company. Our target sample was consisted primarily of executives that would occupy a role in the intersection of business and IT. Our initial email invitation to participate in the survey was accepted by 15 employees (for companies A, B, C and D) occupying varying roles. We arranged a 20-min online prescreening session with these 15 incumbents where we explained the purpose of our study. Out of these 15 employees, 11 were eliminated from the selection process as their roles were entirely focused on commercial and administrative activities (i.e., sales-oriented role, office assistant, etc.,). The remaining 4 candidates (one for each company) were knowledgeable in matters of BITA configurations and they were aware of the SAMM. At a later stage, these 4 executives received a formal email invitation to participate in our survey along with detailed information about the interview process which would take the form of an online individual conversation. Due to privacy reasons, the names of these respondents will not be revealed; Table 2 provides an overview of their background and responsibilities.

For the purposes of our study, we used a semi-structured interview format. Overall, the interview process was split in two phases; the first phase (Phase 0) lasted about 40 min for each respondent and was composed of an in-depth discussion taking place online. Each respondent was initially asked to describe his/her role/responsibilities and provide a short overview of the profile of his/her company and its ongoing strategic alignment processes. Next, the interviewer described the SAMM and its corresponding alignment dimensions to each survey-participant; the alignment criteria and the alignment maturity levels, per alignment dimension, were also explained to each interviewee. The centerpiece of this phase was dedicated to each interviewee utilizing the SAMM format to appraise the state of the alignment maturity level of his/her respective company shortly before the Covid-19 pandemic;

Table 2 Participating respondents and their characteristics

Respondent	Company	Job function	Responsibilities
Respondent 1	A	Vice President and Chief Information Officer	Oversees IT department; participates in the board of directors
Respondent 2	B	ICT Software Development Manager	Oversees the company's software development division
Respondent 3	C	IT Manager	Oversees the software applications division in Belgium
Respondent 4	D	Head of IT Belux Region	Oversees the IT divisions in the Belux region

for this, a mixture of 48 closed and open-ended questions were used. The second phase of the interview (Phase 1) took place right after the end of the first phase and lasted 20 min. For this part, the same set of 48 questions was asked; however, this time the goal was to capture the interviewees' opinions about their companies' state of alignment maturity, during the pandemic. The specifics of the interview protocol (and questionnaire) are given below.

2.3 *Interview Protocol*

To reiterate, for each respondent, the interview process was split in two phases: Phase 0 was meant to focalize the respondents' content on their companies' strategic alignment maturity level before the pandemic. For this phase, the 48 interview questions were divided conceptually into six segments, each one corresponding to one of the six alignment dimensions of the SAMM (i.e., *Communications*, *Competency/Value Measurements*, *Governance*, *Partnership*, *Scope and Architecture*, and *Skills*). Each segment contained a mixture of 8 closed and open-ended questions; the former were meant to allow the respondents to assign a numerical score between '1' and '5' to each of the dimensions' alignment criteria. Each score corresponded to one out of five levels of alignment maturity i.e., *Level 1: Without Process* (no alignment), *Level 2: Beginning Process, Level 3: Establishing Process, Level 4: Improved Process,* and *Level 5: Optimal Process* (complete alignment) [19, 20]. The open questions of each segment were meant to justify these numerical scores and articulate the respondents' perception for their companies' alignment competencies, for that particular dimension. To offer an example, the closed questions addressing the '*Communications*' dimension were meant to allow the respondents rank: (i) the level of understanding between the Business and IT domains within each company (1st criterion), (ii) the level of Inter/Intra-Organizational Learning accommodated between the Business and IT domains (2nd criterion), etc. The reader can retrace the remaining alignment criteria for this dimension in Fig. 1. We have to note that not all alignment criteria (as presented in Fig. 1) were asked to be evaluated by the respondents; some of these criteria were excluded from the interview questionnaire as there were deemed by the research team to be overlapping with others from their respective alignment dimension. The ranking scores and the justification of these scores by the respondents concluded the first phase of the interview.

The next phase of the interview (Phase 1) was meant to focalize the respondents' content on their companies' strategic alignment maturity level after the start of the Covid-19 crisis. The structure of the questionnaire was the same as in Phase 0. Any indicated divergence in the strategic alignment maturity scores between the two phases had to be elaborated. The respondents' scores for the two phases, per company, are presented in Table 3.

3 Preliminary Data Analysis and Results

The discussions taking place during the two phases of the interview session were recorded in audio format after having acquired the consent of each interviewee. These recordings were transcribed into text, analyzed, and codified; this means that parts of the respondents' answers were annotated a code representing a certain opinion/justification/theme etc. Particular attention was given to the justification of the respondents' answers to the open questions; the latter were purposed to capture the respondents' opinions when evaluating the score assigned to a specific alignment criterion for each of the six alignment dimensions of the SAMM. The answers to these open questions were cross-referenced by the research team to determine whether they truly associate to the respondents' given scores to each alignment criterion.

Given the small sample size of our participants in a relatively controlled environment [31], we did not proceed in a statistical (descriptive) realization of the respondent's ranking scores. Contrastingly, we treated the results of the entire exercise under the tenets of analytic generalization [32]. More specifically, we relied on the premises described in Evers and Wu [33] and Yin [34] and considered each one of our sample instances[2] as a transmitter of a cohesive amount of empirical knowledge for the determination of the strategic alignment maturity trend in his/her respective company. The respondents' scores (rankings) for each alignment criterion are summarized in Table 3. The bold-font numbers apposed next to each alignment dimension represent the average of these scores. Finally, the last row presents the average of these alignment dimension scores, per company, for the two phases.

The data, as summarized in Table 3, can yield some primary remarks; first, the interviewees seem to be acknowledging that the partnership maturity, between the business/IT domains of the surveyed companies, has increased after the pandemic. This result can be explicated by the need of the logistics sector's business domain to engage in an advanced state of digitalization awareness due to the transition from conventional ways of working to virtual-working schemes with the aid of cloud-based solutions [35]. Second, the interviewees seem to agree that the workload and range of duties of their IT departments have expanded after the pandemic; this seems to result in an enlarged scope for the logistic sector's IT domain hence the increasing alignment maturity trend amongst all companies in terms of the 'Scope and Architecture' dimension. Third, the sector seems to be embracing a 'pro-change' mentality after the pandemic in order to anticipate (and cope with) change considering the fast-changing safety regulations inflicted by the government (i.e., lockdown imposition, teleworking, etc.,). This constant requirement for flexibility and adaptation forced these production/storage/distribution companies to reprioritize their strategic goals but seems to be the cause for decreased focalization on innovation and encouragement of entrepreneurship within the sector. Fourth, the interviewees are aware of a negative impact on the established communication protocols (between the business and IT domains), resulting from the sudden transition to teleworking norms.

[2] A sample instance refers to each one of the respondents of the survey.

Table 3 Alignment maturity scores for companies A, B, C, D, before, and during Covid-19

	Company A		Company B		Company C		Company D	
	Before Covid-19 (Phase 0)	During Covid-19 (Phase 1)	Before Covid-19 (Phase 0)	During Covid-19 (Phase 1)	Before Covid-19 (Phase 0)	During Covid-19 (Phase 1)	Before Covid-19 (Phase 0)	During Covid-19 (Phase 1)
Communications*	**3,4**	**3,0**	**4,3**	**4,50**	**3,8**	**3,5**	**4.0**	**4,3**
Protocol Rigidity	5	4	5	4	5	4	4	5
Inter-/Intra-Organizational Learning	2,5	2	4	4	4	4	5	4
Understanding of Business by IT	4	4	4	5	3	3	4	4
Understanding of IT by Business	2	2	4	5	3	3	3	4
Governance*	**3,7**	**3,6**	**4,0**	**3,8**	**4,3**	**4,3**	**4,2**	**4,4**
IT Strategic Planning	5	5	4	4	5	5	4	4
Business Strategic Planning	5	5	4	4	5	5	4	4
Reporting Structure	2	2	5	5	n/a	n/a	5	5
IT Investment Management	1,5	1	5	5	2	2	4	4
Steering Committee(s)	5	5	2	1	5	5	4	5
Competency/Value Measurements*	**4,2**	**4,0**	**3,8**	**3,8**	**3,2**	**3,2**	**4,2**	**4,2**
IT metrics	4,5	4	1	1	1	1	2	2
Business Metrics	3,5	3	4	4	4	4	5	5
Service Level Agreements	4	4	4	4	5	5	5	5
Benchmarking	4	5	5	5	2	2	4	5

(continued)

Table 3 (continued)

	Company A		Company B		Company C		Company D	
	Before Covid-19 (Phase 0)	During Covid-19 (Phase 1)	Before Covid-19 (Phase 0)	During Covid-19 (Phase 1)	Before Covid-19 (Phase 0)	During Covid-19 (Phase 1)	Before Covid-19 (Phase 0)	During Covid-19 (Phase 1)
Continuous Improvement	5	4	5	5	4	4	5	4
Partnership*	**3,0**	**3,3**	**4,8**	**5,0**	**4,0**	**4,5**	**4,0**	**4,5**
Business Perception of IT Value	2	2	5	5	3	4	3	4
Relationship between Business and IT	3	4	4	5	3	4	3	4
Business Sponsoring IT Projects	5	5	5	5	5	5	5	5
Role of IT in Business Strategic Planning	2	2	5	5	5	5	5	5
Scope & Architecture*	**3,0**	**3,5**	**3,3**	**4,0**	**3,2**	**3,3**	**3,0**	**3,3**
IT's Role in Organizational Scope	3	3,5	3	4	4,5	4	3	4
Architectural Transparency	2	3	3	3	1	2	2	2
Management of Emerging Technologies	4	4	4	5	4	4	4	4
Skills*	**2,5**	**2,8**	**3,0**	**3,4**	**2,8**	**3,1**	**3,8**	**3,5**
Innovation and Entrepreneurship	3	2	4	4,5	4	3	5	3

(continued)

Table 3 (continued)

	Company A		Company B		Company C		Company D	
	Before Covid-19 (Phase 0)	During Covid-19 (Phase 1)	Before Covid-19 (Phase 0)	During Covid-19 (Phase 1)	Before Covid-19 (Phase 0)	During Covid-19 (Phase 1)	Before Covid-19 (Phase 0)	During Covid-19 (Phase 1)
Locus of Power	2	2	4	4	4	4,5	4	3
Change Readiness	2	4	3	4	2	4	4	5
Career Crossovers	3	3	1	1	1	1	2	3
Average Strategic Alignment Maturity Score*	**3,3**	**3,4**	**3,9**	**4,1**	**3,5**	**3,7**	**3,9**	**4,0**

* Rounded to one decimal place, maximum score: 5

Source Luftman [19, 20]

However, the data suggest that companies that already scored high on their alignment maturity for their communications dimension before Covid-19, the pandemic has impacted their communication protocols only moderately by increasing their formality and by neglecting the distribution of inter-/intra-organizational learning. Finally, the value measurement approach of the logistic sector's IT domain seems not to have been impacted by the pandemic. The interviewees do not recognize a disparity between the utilized pre-, and post-pandemic efficiency-measurements and employee-productivity indicators.

Overall, the last row of Table 3 suggests an increased strategic alignment maturity trend after the pandemic for all the surveyed companies affiliated to the logistics sector. This result should be perceived with some moderation; to reiterate, the present study was meant to explore the existence of any trend-differences in the strategic alignment of these companies before, and after the pandemic. Given that we do notice some alignment changes, the next step of the analysis would be to examine whether these alignment maturity differences, before and during the pandemic, are statistically significant. Therefore, a second stage of interviews with a wider sample base needs to be organized; this stage will also incorporate the analysis of the factors within the logistic sector's business/IT micro-environments that may be the cause of these alignment changes. These analyses are purposed as the future orientation of the present work.

4 Conclusion

The present study was purposed to explore whether an increased state of digital acquiescence in times of the Covid-19 crisis can signify a positive evolution in the strategic alignment maturity of organizations within the logistics sector. In that context, our study has delineated the alignment maturity levels across four production and distribution centers in Belgium before, and during the pandemic. The estimation and appraisal of their alignment maturities were performed with the utilization of the SAMM; the latter, being able to incorporate a variety of well-defined alignment dimensions/criteria, has proven to be a valuable tool. When looking at each alignment dimension individually, we notice an ambivalence amongst these companies in the evolution of their strategic alignment maturities before, and during the pandemic; however, when looking at the results holistically, they seem to be yielding an increasing trend in the average strategic alignment maturity of these companies during the pandemic. The insights gathered from our interviewees are meant to be used as the input for a comprehensive analysis of the factors that instigate a change in these strategic alignment maturities during the evolution of the pandemic.

References

1. Hovestadt, C., Recker, J., Richter, J., Werder, K. (eds.) Digital Responses to Covid-19: Digital Innovation, Transformation, and Entrepreneurship During Pandemic Outbreaks. Springer Nature (2021).
2. Kim, R.Y.: The impact of Covid-19 on consumers: preparing for digital sales. IEEE Eng. Manage. Rev. **48**(3), 212–218 (2020)
3. Seetharaman, P.: Business models shifts: impact of covid-19. Int. J. Inf. Manage. **54**(1), 102173–102173 (2020)
4. Verma, S., Gustafsson, A.: Investigating the emerging Covid-19 research trends in the field of business and management: a bibliometric analysis approach. J. Bus. Res. **118**, 253–261 (2020)
5. Liebowitz, J. (ed.) The Business of Pandemics: The Covid-19 Story. CRC Press (2020)
6. Visvizi, A., Troisi, O., Grimaldi, M., Loia, F.: Think human, act digital: activating data-driven orientation in innovative start-ups. Eur. J. Innov. Manage. (2021)
7. Zito, M., Ingusci, E., Cortese, C.G., Giancaspro, M.L., Manuti, A., Molino, M., .Russo, V.: Does the end justify the means? The role of organizational communication among work-from-home employees during the covid-19 pandemic. Int. J. Environ. Res. Public Health **18**(8), 3933 (2021)
8. He, W., Zhang, Z., Li, W.: Information technology solutions, challenges and suggestions for tackling the covid-19 pandemic. Int. J. Inf. Manage. **57**, 102287 (2021)
9. Dairo, M., Adekola, J., Apostolopoulos, C., Tsaramirsis, G.: Benchmarking strategic alignment of business and IT strategies: opportunities, risks, challenges and solutions. Int. J. Inf. Technol. **13**(6), 2191–2197 (2021)
10. Henderson, J.C., Venkatraman, H.: Strategic alignment: leveraging information technology for transforming organizations. IBM Syst. J. **32**(1), 472–484 (1993)
11. Luftman, J., Papp, R., Brier, T.: enablers and inhibitors of business-IT alignment. Commun. Assoc. Inf. Syst. **1**(1), 11 (1999)
12. Tallon, P.P., Kraemer, K.L.: Investigating the relationship between strategic alignment and information technology business value: The discovery of a paradox. In: Creating Business Value With Information Technology: Challenges and Solutions, pp. 1–22 (2003)
13. Coltman, T., Tallon, P., Sharma, R., Queiroz, M.: Strategic IT alignment: twenty-five years on. J. Inf. Technol. **30**(2), 91–100 (2015)
14. Wautelet, Y.: A model-driven IT governance process based on the strategic impact evaluation of services. J. Syst. Softw. **149**, 462–475 (2019)
15. Tsilionis, K., Wautelet, Y.: A model-driven framework to support strategic agility: value-added perspective. Inf. Softw. Technol. **141**, 106734 (2022)
16. Tsilionis, K., Wautelet, Y., Tupili, D.: Digital transformation and operational agility: love story or conceptual mismatch. In: Proceedings of Workshops Co-Organized with the 14th IFIP WG 8.1 Working Conference on the Practice of Enterprise Modelling, vol. 3031, pp. 1–14. CEUR Workshop Proceedings (2021).
17. Tsilionis, K., Wautelet, Y.: From Service-Orientation to Agile Development by Conceptually Linking Business IT Services and User Stories: A Meta-Model and a Process Fragment. In: 2021 IEEE 23rd Conference on Business Informatics (CBI), vol. 2, pp. 153–162 (2021)
18. Alao, B.B., Lukman, O.G.: Coronavirus pandemic and business disruption: the consideration of accounting roles in business revival. Int. J. Acad. Multidiscip. Res. **4**(5), 108–115 (2020)
19. Luftman, J.: Assessing business-IT alignment maturity. Commun. Assoc. Inf. Syst. **4**(14) (2000)
20. Luftman, J.: Assessing IT/business alignment. Inf. Syst. Manage. **20**(4), 9–15 (2003)
21. De Haes, S., Van Grembergen, W., Joshi, A., Huygh, T.: Enterprise Governance of Information Technology: Achieving Alignment and Value in Digital Organizations, 3rd edn. Cham, Springer Nature Switzerland (2020)
22. Sledgianowski, D., Luftman, J.N., Reilly, R.R.: Development and validation of an instrument to measure maturity of IT business strategic alignment mechanisms. Inf. Resour. Manage. J. (IRMJ) **19**(3), 18–33 (2006)

23. Khaiata, M., Zualkernan, I.A.: A simple instrument to measure IT-business alignment maturity. Inf. Syst. Manage. **26**(2), 138–152 (2009)
24. Ahuja, S.: Strategic alignment maturity model (SAMM) in a cascading balanced scorecard (BSC) environment: utilization and challenges. In: International Conference on Advanced Information Systems Engineering, pp. 567–579. Springer, Berlin (2012)
25. Belfo, F.P., Sousa, R.D.: A critical review of Luftman's instrument for business-IT Alignment (2012)
26. Liu, L., Li, W., Aljohani, N.R., Lytras, M.D., Hassan, S.U., Nawaz, R.: A framework to evaluate the interoperability of information systems - measuring the maturity of the business process alignment. Int. J. Inf. Manage. **54**, 102153 (2020)
27. Brown, T.A.: Confirmatory factor analysis for applied research. New York: Guilford. Organ. Res. Methods **13**(1), 214–217 (2006)
28. Wohlin, C.: Case study research in software engineering - it is a case, and it is a study, but is it a case study? Inf. Softw. Technol. **133**, 106514 (2021)
29. Mack, N.: Qualitative Research Methods: A Data Collector's Field Guide (2005)
30. Saunders, M., Lewis, P., Thornhill, A.: Research Methods for Business Students. 7th edn. Pearson, Harlow (2016)
31. Tsang, E.W.K.: Generalizing from research findings: the merits of case studies. Int. J. Manage. Rev. **16**, 369–383 (2014)
32. Yin, R.: Analytic generalization. In: Mills, A., Durepos, G., Wiebe, E. (eds.) Encyclopedia of Case Study Research, pp. 21–23. SAGE Publications, Inc, Thousand Oaks (2010)
33. Evers, C.W., Wu, E.H.: On generalising from single case studies: epistemological reflections. J. Philos. Educ. **40**(4), 511–526 (2006)
34. Yin, R.K.: Validity and generalization in future case study evaluations. Evaluation **19**(3), 321–332 (2013)
35. Tsilionis, K., Sassenus, S., Wautelet, Y.: Determining the benefits and drawbacks of agile (scrum) and devops in addressing the development challenges of cloud applications. In: The International Research and Innovation Forum, pp. 109–123. Springer, Cham (2021)

Corporate Digital Identity Based on Blockchain

Guillermo Balastegui-García, Elvi Mihai Sabau Sabau, Antonio Soriano Payá, and Higinio Mora

Abstract Digital Identity refers to the set of information that defines an entity, generally a person, which can be presented to other entities of digital world. This information represents different information such as skills, biological, activities and other data that may result useful. There are currently several projects developing digital identity systems for personal use and aiming to standardize its use among the population. Digital Identity finds in blockchain the support technology to enable this decentralized management in a secure and reliable way. Blockchain allows citizens identification to each application, contract or business, by providing only the information requested, and avoiding unnecessarily exposing of other private information about citizens. However, there are no digital identity systems for legal entities. Currently, the digital identity concept for business and corporations is limited to its appearance in the digital environment and their corporate image on internet. The objective of this work is the proposal and development of a digital identity for legal entities based on a blockchain network. The methodology is based on applying the same idea of personal digital identity. In addition, the proposal considers Ethereum blockchain since it is strongly accepted for the deployment of smart contracts. As a result, the proposed corporate digital identity provides companies the control of the information they share with government, people and other companies. In addition, the relationships between entities can be defined, and eventually automate, through smart contracts in the blockchain. The main limitation of this idea is the technology adoption by institutions and public administrations in order to normalizing digital relations between all agent involved: corporates, citizens and administrations.

Keywords Blockchain · Digital identity · Legal entities · e-government

---

G. Balastegui-García · E. M. S. Sabau · A. S. Payá · H. Mora (✉)
Department of Computer Science Technology and Computation, University of Alicante, Alicante, Spain
e-mail: hmora@ua.es

G. Balastegui-García
e-mail: guillermo.balasterri@ua.es

A. S. Payá
e-mail: soriano@ua.es

A. Visvizi et al. (eds.), *Research and Innovation Forum 2022*,
Springer Proceedings in Complexity, https://doi.org/10.1007/978-3-031-19560-0_55

1 Introduction

Currently, the creation and transfer of digital content which occurs in the world are overwhelming. Such a data flow generates opportunities for people who can acquire, process, and understand that information. The Internet has changed the way people interacts in their lives between them and with their environment.

In this context, the need to protect and authenticate information arises in order to provide control over data and trust in information communications between entities. Governments, firms and governments search for comprehensive solutions that enable citizens and identities to identify themselves.

While there is developing a digital identity concept for personal data [1], there is not an equivalent solution for companies yet. Currently, there is not a reliable system that allows a legal entity governs its own data, nor to share and exchange information through a digital environment in a safe, trustworthy, fast, and efficient manner with another entity or regulatory or certifying agency.

This is an important issue for modern societies where digital relations between participants are encouraged in order to increase productivity, reducing costs, and provide a better service to citizens and firms.

A recent example was the pandemic caused by Covid-19. In this situation, personal relations and paper-based document exchange were not recommendable practices [22]. The generalization of an interoperable digital identity among citizens, firms and public administration will allow ubiquitous relations where anyone could be able to make transactions anywhere at any time.

In this work it is proposed the idea about a corporative digital identity system that gives enterprises a tool for sharing information and storing procedures, taking a special focus on integrity and authenticity, making easier communication and operation processes with the agents with it relates (citizens, public administrations, regulatory entities and commissions, and other companies).

The methodology adopted starts from analysing the Digital Identity concept for personal purposes since this tool already has many of the desired features. From this starting point, this work proposes an implementation based on the same technology and includes transaction support for interaction operations.

The structure of this work is as follows. Section 1 gives the introduction and provides the purpose of the work. Section 2 describes the background and related work about digital identities and implementations under blockchain technology. Section 3 introduces the main elements of the proposed idea. Section 4 portrays an example of implementation using Ethereum and smart contracts blockchain and Sect. 5 illustrates the conclusions of the work.

2 Background

Self-sovereign digital identity gives back the user the control of this data, letting him know who and in which terms accesses its information.

2.1 Personal Digital Identity

Digital Identity, also known as Identity 2.0, is all information related to a person which is located in a digital storage, including all the publications in which the person appears, regardless of the publisher, its documentation, and its data in general [2]. On the digital world, this is equivalent to its real-life identity.

Digital identity is increasingly used is many processes related to information sharing in the digital world, such as employee recruitment [3], financial services [4], and also in healthcare [5].

Currently, digital identity implementations based on blockchain provide the desired features of security, data integrity and anonymity. In addition, there is no need of a third-party organization in the middle or in charge of the data communications [6]. Consequently, this technology enables that digital identity becomes the primary source of authentication for digital services and other trust systems in the near future.

Blockchain technology is based on cryptography and distributed ledger systems (DLT), which operate through decentralization and consensus. These pillars provide security to a system that is based on a Blockchain network, making it tamper-proof and fraud-proof. This system offers strong advantages such as privacy, identity self-management, safe communication, security and trust through smart contracts, and immutability of information [7]. Based on these features, blockchain technology provides solutions for data sharing by providing trust to participants without any intermediary organization [19, 20]. These blockchain solutions give people back control over their data, letting them decide who can see or use their information in a precise way [8, 9].

Public and private key pairs must be created and used through "identity wallets" to create a Blockchain-based digital identity system. Those wallets are responsible for calculating the hash of the public key and storing it in a blockchain, which is immutable. The user has control of their identifiers from that moment on, which are decentralized and independent of any higher body.

The identity wallet is also the interface that allows the user to interact with the Blockchain network and the other users, being a simple and usable method. The premise is that as long as you keep your cryptographic keys stored, you remain in control of your information [10]. Meanwhile, user data is stored in the network's cloud, which is distributed and replicated across the various devices in the network.

As shown in Fig. 1, an entity, such as a university, creates a degree that certifies that a person has an education, gives the credential to the owner, and rises this to the

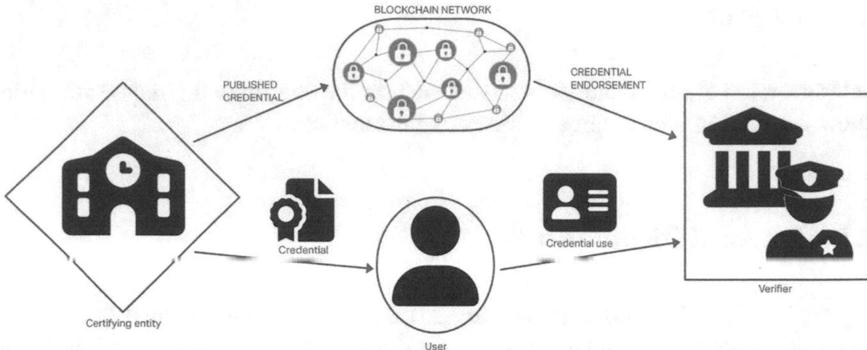

Fig. 1 Blockchain-based digital identity flowchart

Blockchain network, so that when another entity needs to check if the person owns this degree, it will only have to look for the credential, which will be backed by the Blockchain network [11].

Big companies such as Microsoft and various governments including European Union are currently working on diverse projects of decentralized digital identity [12–14].

2.2 *Corporative Digital Identity*

The concept of Corporate Digital Identity (CDI) is currently understood as the image that a company has online through the content that generates, shares and controls on the various platforms on which it is located [15]. This concept of identity shapes a "digital reputation", the opinion that its followers and community have of the company. This identity and this reputation are shaped by the media manager or by marketing decisions [16].

Although those are concepts that go hand in hand, they are not the same, digital identity is more tangible, it can be seen in the image, the colours, the publications, the logo of the brand, and reputation has a more abstract nature, formed by the perception and opinion of people about the brand.

According to Spanish National Cybersecurity Institute (INCIBE), most companies have a digital identity that they actively maintain, which generates a reputation [17].

This reputation can be attacked in different ways, INCIBE itself offers prevention and reaction guidelines for corporations to protect themselves from attacks, although these are based on community relations solutions rather than on technical or technological means.

2.3 *Findings of the Related Work*

SSI—Self-Sovereign Identity and blockchain are currently gaining importance thanks to the advantages they offer to their systems users and various international organizations are already committed to their use.

The digital identity of a company is currently understood as the image the corporation shows to the public through its social networks, its website, its blog, and sometimes through the image offered by its customers when they rate its product and publish their opinions about the company. This concept of corporate digital identity is only related to its image, not as a tool that can be given a utility beyond representing.

The advancement of SSI systems and digital identity should also reach corporate-type profiles enabling companies to take advantage of the SSI benefits supported by blockchain networks.

3 Corporate Digital Identity Model

Companies have a large amount of documentation of different areas keeping all their legal and economic aspects up to date, and they must also be confident that all the information and products they obtain from other organizations are true, complete, and authentic. There is also relevant formal documentation for regulatory entities such as tax entities, employment entities or labour institutions among others, which may vary depending on the region where the company is located, the sector, and the type of services it offers.

Human resources dedicated to keeping the company formalized are costly and usually perform repetitive tasks that consist of receiving, formalizing, and forwarding contracts. All these tasks can be addressed by a blockchain-based Digital Identity system for legal entities that automates the document exchange and verification processes. These legal entities can be firms, societies, institutions, and even governments at different level.

A Blockchain network that interconnects legal entities with each other, allows them to verify the authenticity and originality of their documentations, products and services and their ability to carry out transactions legally, among other operations, improves efficiency in government, industrial, commercial, and service processes.

A priori, digital identity services for organizations can be connected to digital identity services for individuals, thus improving processes in which information provided by a person must be verified or in which information generated by another company about that person must be accessed in a transparent, legal, and secure way. Figure 2 depicts the most overall scheme of this proposal.

This scheme represents the digital relations of corporations with other entities and administrations. In fact, the CDI concept becomes in a tool for enhancing digital relations with any other agent of the society, including citizens, and other firms. The

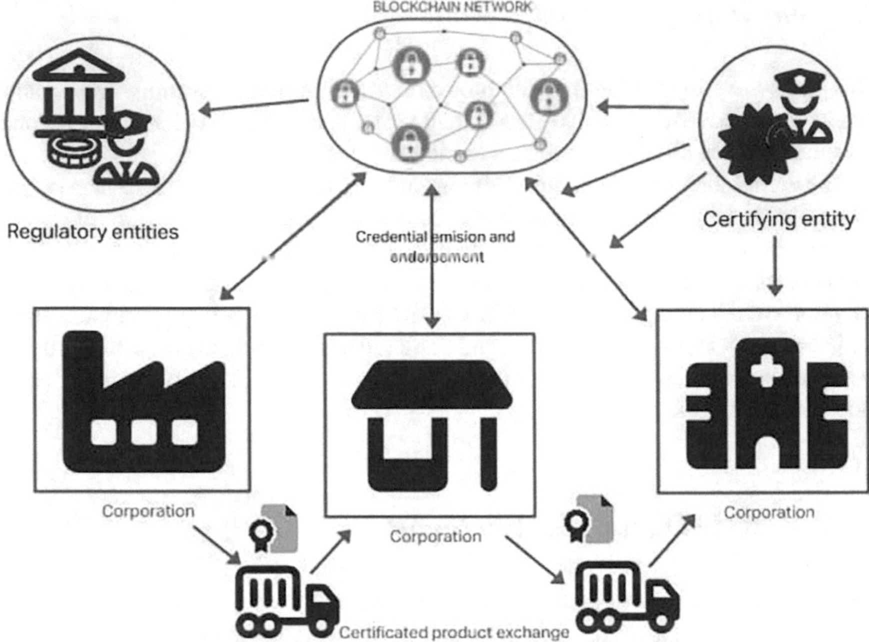

Fig. 2 Corporative digital identity scheme

ability to include in this CDI not only information, but also functions and contracts enhances digital operations to interchange information and make transactions.

The advantages of a corporate digital identity system are of a different nature. Next, the most relevant aspects are listed:

- Agility in the exchange of information between corporations.
- Improved trust during exchanges of goods, services, or information between people.
- Improved security during communication between two or more undertakings.
- Reduction of bureaucratic costs.
- Reduction of regularisation times for companies.
- Transparency in information exchanges, both in the eyes of a company and in the eyes of its customers and partners.
- Traceability of exchanged products.
- Reduction and control of fraudulent sales of a product by persons or organizations external to the product developer.
- Increasing of automation and making digital transactions.

The proposed CDI model provides companies with a tool that allows them to store their information, which may vary depending on the sector and field in which the company is located, but in general terms, it may be composed of:

- *Assigned attributes* (for example): Fiscal and legal data of the company; Identity of the company (name, image, values, mission, etc.).
- *Cumulative attributes* (for example): Patents, products and services; Certificates; Employees; Subsidiaries or branches; Ubication; Current payment certificate.

These are some of the data that the proposed digital identity could contain. In addition, it facilitates an easier relationship with other companies, and not only with companies but also with employees, customers and users using its SSI system, providing confidence to them, as they will be able to see authentic information about the company, thus facilitating recruitment of new human resources.

4 Blockchain-Based Implementation

There are currently many different implementations of blockchain. Not all of these implementations follow the original model of the Bitcoin network as a public, pseudo-anonymous network.

As a proposed implementation, we use the *Ethereum protocol* (https://ethereum.org/en/) and Smart Contracts to develop a system that can manage said attributes for a legal entity based on the blockchain technology.

4.1 About the Blockchain Used

Ethereum is a public blockchain which adds a trust factor over the users who use it, since any transaction, process, or modification of the stored data can be verified by anyone. In addition, Ethereum protocol is open source anyone can develop their own blockchain using this protocol, although it is recommendable to use an already existing and well-populated one, since the more nodes and users a blockchain has, the more solid and secure it becomes.

Smart Contracts (SC) are programs that run over the Ethereum protocol. SC is a software stored on a blockchain that are automatically executed when predetermined terms and conditions are met. This feature guarantees that the contract cannot be modified, and it will always behave in the same way. This immutability in execution allows to program automatic executions avoiding bureaucratic processes that are energy and time-consuming [21].

The programming language used can be an already existing one such as JS, C++ , Rust, Python, or Solidity since there are many libraries for each language that brings the Ethereum ecosystem into the development environment.

Each function calls on a smart contract represents a transaction in the blockchain since the function will execute a logic or change the state of the data stored in the blockchain.

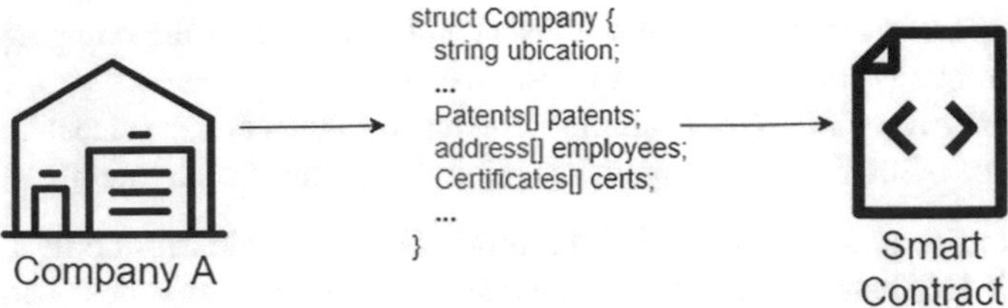

Fig. 3 Example of company data structure

In the Ethereum ecosystem, to prevent infinite loops, speed up transaction time, and to incentivize users to validate the transactions issued by a smart contract, function calls have a fee in "gas". Gas is a unit used to measure the cryptocurrency needs to execute logic on the blockchain, and it varies from time to time depending on a few factors such as the computational load of the blockchain, the storage used, and the complexity of the operations to be computed [18].

4.2 Implementation Details

Smart contracts provide the ability to develop a system that manages the desired attributes of the digital identity for each company, share them specifically with any entity or with anyone, and verify their integrity and validity.

A data model can be defined by using data structures that will define the assigned and cumulative attributes of the company as an entity in the blockchain. Figure 3 shows an example of company data structure.

Certain attributes such as certificates can be hard-bound to accounts, in a way that no one can alter the contents or change the ownership of the attribute. And others such as patents can be traded between companies as assets, for example by turning these attributes into "Non Fungible Tokens" (NFTs). NFTs can also be managed and transferred by approved entities. By this way, the resposability of managing certain assets could be delegated to external companies. Figure 4 illustrates some cases of NFTs management.

And for private data transfers, the only way to share private data without letting other parties read its contents is by encrypting it.

The public and private keys bound to each account can be used to encrypt and decrypt private data shared between entities, by using public-key cryptography and hashing functions for data integrity validation. Figure 5. describes a simple encryption-decryption process. In this way, entities can share private data with other companies by encrypting the data with the receiver's company public key.

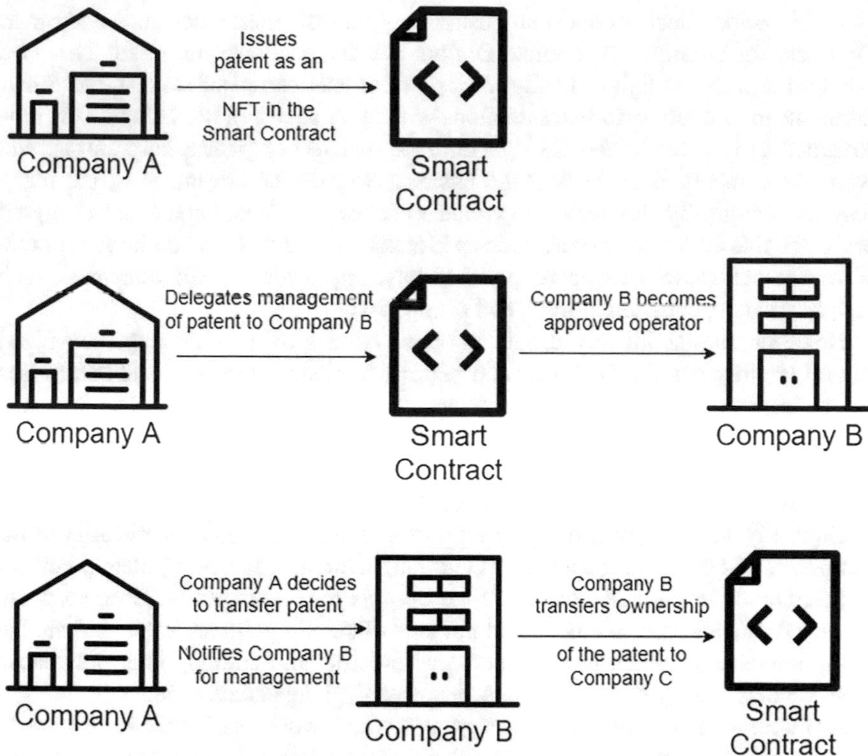

Fig. 4 NFT management examples

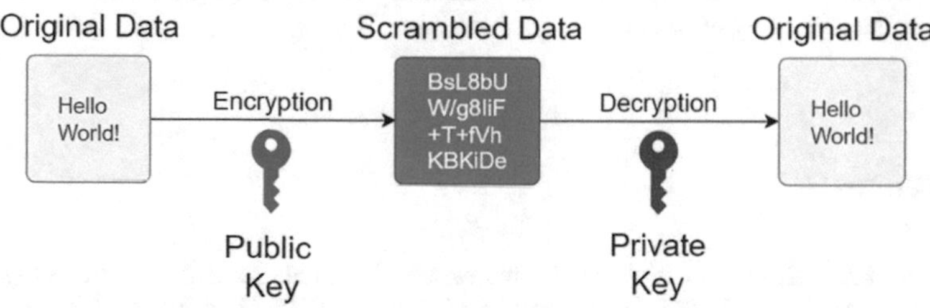

Fig. 5 Encryption and decryption process

5 Conclusions

Self-governed digital identity is emerging as a tool of the future that can change the way we understand information exchange and data management. There is a growing number of public and private projects that aim to implement this technology in people's lives and improve the digitalization of society.

In this work, a technological proposal based on Ethereum blockchain and Smart Contracts for building a Corporate Digital Identity has been proposed. This idea extends the personal digital identity concept to companies and allows integrate Smart Contracts in order to perform transactions with other agents of the society. This self-governed digital identity tool for legal entities, such as companies, corporations and even governments, should allow the business world take advantage of the many benefits brought by this technology, and enhance the development in the digital world. To this end, technologies such as blockchain and SSI play an important role in this transformation. This concept could have application at different areas, such as in healthcare, commerce, sport, and production.

However, digital Identity debate is open. Access to a universally recognized Digital Identity could unlock new and better experiences for users and enterprises as they interact online with businesses, service providers, public administrations and their community. The implications and the expectations are high. The high efficiency of the transactions and the ubiquity of the interactions will boost the development of new kind of social interactions and business.

Currently, technology is ready to support identity value-added services both for companies and citizens, and it is expected that, in the near future, regulatory entities could advance in normalization of SSI proposals for legal entities. Even so, there are also disadvantages. Society could not be sufficiently prepared for assimilate the disruptive changes witch new technologies provide. In addition, other drawbacks exist, such as new cyberattacks forms, technological dependence, etc.

As future work, we aim to explore regulation frameworks at national and European level, in order to propose applications to develop the capabilities and potential of CDI.

Acknowledgements This work was supported by the Conselleria of Participation, Transparency, Coop-eration and Democratic Quality, of the Community of Valencia, Spain, under SDG oriented projects of the University of Alicante.

References

1. Seifert, R.: (2020) Digital identities – self-sovereignty and blockchain are the keys to success. Netw. Secur. **11**, 17–19 (2020). https://doi.org/10.1016/S1353-4858(20)30131-8
2. Feher, K.: Digital identity and the online self: footprint strategies – An exploratory and comparative research study. J. Inf. Sci. (2019). https://doi.org/10.1177/0165551519879702
3. Krishna, S.: Digital identity, datafication and social justice: understanding Aadhaar use among informal workers in south India. Inf. Technol. Develop. **27**(1) (2021). https://doi.org/10.1080/02681102.2020.1818544
4. Jessel, B., Lowmaster, K., Hughes, N.: Digital identity: the foundation for trusted transactions in financial services. J. Financ. Transform., Capco Inst. **47**, 143–150 (2018)
5. Houtan, B., Hafid, A.S., Makrakis, D.: A survey on blockchain-based self-sovereign patient identity in healthcare. IEEE Access **8** (2020). https://doi.org/10.1109/ACCESS.2020.2994090
6. Malyan, R.S., Madan, A.K.: Blockchain technology as a tool to manage digital identity: a conceptual study. In: Singari R.M., Mathiyazhagan K., Kumar H. (eds.), Advances in Manufacturing and Industrial Engineering. Lecture Notes in Mechanical Engineering. Springer, Singapore (2021). https://doi.org/10.1007/978-981-15-8542-5_55

7. Mora, H., Morales-Morales, M.R., Pujol-López, F.A., Mollá-Sirvent, R.: Social cryptocurrencies as model for enhancing sustainable development. Kybernetes **50**(10), 2883–2916 (2021). https://doi.org/10.1108/K-05-2020-0259
8. Mora, H., Mendoza-Tello, J.C., Varela-Guzmán, E.G., Szymanski, J.: Blockchain technologies to address smart city and society challenges. Comput. Hum. Behav. **122**, 106854 (2021). https://doi.org/10.1016/j.chb.2021.106854
9. Mora, H., Pérez-delHoyo, R., Sirvent, R.M., Gilart-Iglesias, V., Management City Model Based on Blockchain and Smart Contracts Technology. In: Visvizi, A., Lytras, M., (eds.), Research & Innovation Forum 2019. RIIFORUM 2019. Springer Proceedings in Complexity. Springer, Cham (2019). https://doi.org/10.1007/978-3-030-30809-4_28
10. Soltani, R., Nguyen, U.T., Aijun, A.: Practical key recovery model for self-sovereign identity based digital wallets. In: IEEE International Conference on Dependable, Autonomic and Secure Computing, https://doi.org/10.1109/DASC/PiCom/CBDCom/CyberSciTech.2019.00066
11. Mora, H., Pujol-López, F.A., Morales, M.R., Mollá-Sirvent, R. (2021). Disruptive Technologies for Enabling Smart Government. In: Visvizi, A., Lytras, M.D., Aljohani, N.R. (eds.), Research and Innovation Forum 2020. RIIFORUM 2020. Springer Proceedings in Complexity. Springer, Cham. https://doi.org/10.1007/978-3-030-62066-0_6
12. Spanish Standard (2020) UNE 71307–1. Digital Enabling Technologies. Decentralised Identity Management Model based on Blockchain and other Distributed Ledgers Technologies. Part 1: Reference Framework.
13. European Parliament, (2021) EPRS, Updating the European digital identity framework, legislative briefing, October 2021
14. DIF - Decentralized Identity Foundation. https://identity.foundation/. Accessed 05 Jan 2022
15. Saran, S.M., Shokouhyar, S.: Crossing the chasm between green corporate image and green corporate identity: a text mining, social media-based case study on automakers. J. Strateg. Mark. (2021). https://doi.org/10.1080/0965254X.2021.1874490
16. Mora, H., Sirvent, R.M.: Analysis of virtual currencies as driver of business marketing. In: Visvizi, A., Lytras, M. (eds.), Research and Innovation Forum 2019. RIIFORUM 2019. Springer Proceedings in Complexity. Springer, Cham (2019). https://doi.org/10.1007/978-3-030-30809-4_48
17. Digital identity: the online reputation of companies (in Spanish), Spanish National Institute of Security (INCIBE). https://www.incibe.es/protege-tu-empresa/blog/infografia-id-empreas. Accessed 05 Jan 2022
18. Werner, S.M., Pritz, P.J., Perez, D.: Step on the gas? A better approach for recommending the Ethereum gas price. In: Pardalos, P., Kotsireas, I., Guo, Y., Knottenbelt, W., (eds.), Mathematical Research for Blockchain Economy. Springer Proceedings in Business and Economics. Springer, Cham (2020). https://doi.org/10.1007/978-3-030-53356-4_10
19. Sicilia, M., Visvizi, A.: Blockchain and OECD data repositories: opportunities and policymaking implications. Library Hi Tech **37**(1), 30–42 (2019). https://doi.org/10.1108/LHT-12-2017-0276
20. Çekani, V.: The role of blockchain technology during Covid-19 in the healthcare sector. In: Visvizi, A., Troisi, O., Saeedi, K. (eds.), Research and Innovation Forum 2021. RIIFORUM 2021. Springer Proceedings in Complexity. Springer, Cham (2021). https://doi.org/10.1007/978-3-030-84311-3_4
21. Sirianni, C.A., Sabbagh, P., Marra, F.: An ethereum-based chain for diagnosis and control of Covid-19. In: Visvizi, A., Troisi, O., Saeedi, K. (eds.), Research and Innovation Forum 2021. RIIFORUM 2021. Springer Proceedings in Complexity. Springer, Cham (2021). https://doi.org/10.1007/978-3-030-84311-3_6
22. De Falco, C.C., Romeo, E.: ICT and social value co-creation during the Covid-19 emergency: the case of Mascherin Amica. In: Visvizi, A., Troisi, O., Saeedi, K. (eds.), Research and Innovation Forum 2021. RIIFORUM 2021. Springer Proceedings in Complexity. Springer, Cham (2021). https://doi.org/10.1007/978-3-030-84311-3_1

Reverse Knowledge Transfer in Service Industry, Towards a New Taxonomy of Service Centers

Francesco Polese and Radosław Malik

Abstract This research bridges two streams of research that have not been linked before: reverse knowledge transfer from subsidiaries to parent companies in service industry and service offshoring evolution from simple tasks to value-creating activities. The review of existing classifications of offshoring service centers shows that research into evolutionary path of their development has not covered intensity and characteristics of reverse knowledge transfer (RKT). This creates a rationale for development of the new taxonomy offshoring service centers. Moreover, the results indicate under-researched area of knowledge transfers, in particular reverse knowledge transfers, in service theory and business practice. The research has important implications for management practices by providing managers with new insight into knowledge creation and diffusion in international corporate networks.

Keywords Reverse knowledge transfer · Service center · Offshoring · Service · Taxonomy · Knowledge transfer · Outsourcing

1 Introduction

The primary purpose of the chapter is to review taxonomies of offshore service centers to verify the extent to which evolution of service offshoring and the intensity of reverse knowledge transfer between offshore service centers and parent companies has been explored in the literature. The research builds on two main lines of scientific investigation which have thus far been advanced separately: reverse knowledge transfer (RKT)—knowledge flows from the subsidiary to the parent company [1]. and the evolution of offshore captive centers—wholly owned near-shore or offshore subsidiaries that provide IT, business process and R&D services to their

F. Polese
Department of Management and Innovation Systems, University of Salerno, Via Giovanni Paolo II 132, 84084 Fisciano, Italy

R. Malik (✉)
SGH Warsaw School of Economics, al. Niepodległości 162, 02-554 Warsaw, Poland
e-mail: rmalik@sgh.waw.pl

A. Visvizi et al. (eds.), *Research and Innovation Forum 2022*,
Springer Proceedings in Complexity, https://doi.org/10.1007/978-3-031-19560-0_56

parent firms [2]. The results will provide new insight into international delivery of service that contributes to evolution of service industry [3], service ecosystems [4] and smart services [5] as important contributor to the modern economies [6] and management [7].

The initial insight from the existing literature reviews and meta-analyses on related subjects of offshoring and shared service centers e.g. [8–12] shows that evolution of service offshoring is a broadly discussed theme. However, offshore service centers have not been analysed in the context of reverse knowledge transfer which constitutes a novelty of the undertaken research. The research uses the integrative literature review, a type of non-systematic literature review, that is utilized to combine various perspectives and can be used to support creation of new conceptual models [13]. The results of this research will be subsequently applied to support the further conceptual investigation and qualitative research linking the intensity of reverse knowledge transfer with evolution of service offshoring.

2 Reverse Knowledge Transfer

Transferring knowledge from headquarters is a major objective of multinational enterprises while setting up or acquiring subsidiaries [14]. Therefore, the majority of studies have been focused on the parent units transferring their knowledge to their subsidiaries [15]. However, it has been noted that subsidiaries could develop considerable strategic independence, strengthen their links to international networks and improve the knowledge intensity and complexity of their work [16]. Consequently, knowledge creation is increasingly viewed as a result of the exploration of multidirectional knowledge connectivity in global networks that connect geographically dispersed knowledge centers [17]. It is, however, yet to be investigated to what extent offshore service centers—as a specific form of subsidiaries, are involved in these processes. Thus, it constitutes an important dimension of the problem to be addressed in the research.

Although, it has been long ago observed that reverse knowledge transfers are probably necessary stepping stones in the evolution of the multinational towards a true distributed innovation network [18], only a few studies have focused on reverse knowledge transfer in an explicit way [19]. Intensity and characteristics of reverse knowledge transfer can be assessed based on four main constructs already used in the previous research on RKT from subsidiaries to parent companies, and these are: the potential to create knowledge, relevance of the knowledge created, the ability to reverse transfer new knowledge and the motivation to reverse transfer new knowledge [20].

Studies on RKT highlight the pivotal role of transforming the location-bound assets in the host countries to galvanize a competitive advantage for the entire multinational enterprise network [21]. The results demonstrated positive effects between the extent and benefits of RKT [22]. Collaboration was found to have a positive influence on both the extent and benefits of RKT, and knowledge tacitness has proved to have a positive impact on the benefits from RKT. Moreover, it has been shown that reverse knowledge transfer is positively influenced by subsidiary capability, knowledge relevance, and the absorptive capacity of the parent company [23]. Thus, recently scholars have called for more research on reverse knowledge transfer [24]. As research on RKT between subsidiaries and parent companies have progressed, it is yet to be verified if the results of the that research apply to knowledge transfer between offshore service centers and their parent companies.

3 Offshoring Evolution

Simultaneously, it has been recognized that offshoring is increasingly a knowledge-seeking phenomena [25] which evolves towards an amplified transfer of high-value, knowledge-intensive activities and can be utilized to relocate advanced tasks, R&D activities and certain segments of the innovation process [26]. As managers become more open to sourcing knowledge-intensive parts of the value chain from abroad, offshoring is increasingly viewed as a promising and legitimate management practice which enables to extract benefits from various knowledge sources [27]. Enhanced legitimacy facilitates allocation resources to projects and reduces the cognitive barriers to utilize knowledge intensive input, such as the "not-invented-here" syndrome [28]. As a result, knowledge-intensive and innovation inputs sourced abroad can be more effectively integrated across borders, resulting in new combinations of ideas and technologies. As new combinations of diverse resources can generate increased innovation output shrinking barriers to knowledge sharing, adoption, and integration that should increase the offshoring effectiveness even in innovative activities [29]. However, it requires further studies to verify if the increased complicity and knowledge intensity of offshoring translates into knowledge development in offshore service centers and their ability to share it in reverse knowledge transfer to parent companies.

More than 75% of Fortune 500 companies have established various models of offshore captive centers with the aim of achieving superior performance, primarily by cost savings and service enhancements [12]. The offshoring evolution towards more complex activities is increasingly motivated by efficiency and knowledge augmentation motives which reinforces the shift in the strategic role that offshore service centers play in supporting the parent's global sourcing strategy as they follow the evolutionary path. Up to the present it has been assumed that a typical evolutionary path for offshore captive centers follows a four-stage pattern: (i) establishing a basic captive center; (ii) evolving to the hybrid model; (iii) evolving to the shared model; (iv) divesting a captive center [30].

4 Reverse Knowledge Transfer in Offshore Service Centers

With the evolution of offshoring and increased importance of reverse knowledge transfers in globally dispersed corporate networks, the exiting consensus on offshore service centers evolution may require an update, especially in the assessment of the final stage of the process. For example, it can be speculated that some offshore service centers may evolve into some form of centers of excellence in corporate networks [31]. This could be reinforced by the proliferation of new technologies such as robotics and process automation which can be utilized and improved in offshore captive centers and transfer to parent companies in a form of reverse knowledge transfer [32].

The results of the recent research seem to show that the extent of knowledge sharing is indeed stronger in a captive mode than in an external sourcing mode, and that structural (tie strength), cognitive (shared understanding), and relational (trust) aspects of social capital mediate the effect of the sourcing mode on the extent of the knowledge sharing [33]. This could potentially reinforce the ability to reverse knowledge transfer from offshore service centers to parent companies, but the subject requires further investigation.

5 Taxonomy of Offshore Service Centers

The studies on offshore service center typology based on governance modes, role in MNCs value networks, knowledge creation capacities and other criteria remain scarce. The distinction between captive offshoring and offshore outsourcing generated by far more attention of the scholars [34] but the insight into typology of offshore service centers—type of captive offshoring remains scant. Based on the literature review undertaken in this study it seems that there is an absence of conceptual research that attempts to define typology of offshore service centers based on their ability to contribute to the reverse knowledge transfer in international production networks of MNCs.

The results of the literate review indicate that scholars have frequently discussed the factors identified as crucial for shared service center (SSC) configurations but have not suggested specific types of SSC structure [12]. Some scholars have compared private and public sector SSCs, highlighting differences, such as the transfer of best practices in implementation and operation of both types [35]. The review of the research on shared service centers shows that the important discrimination area between categories of offshore service centers are types of services that are delivered in these organizations [36]. Recently, novel research direction has emerged exploring the concept of global service centers, thus indicating the geographical scope of service delivery as an important differentiator of the service center type [25].

Shared service model of service delivery is also utilized in public administration and based on the ownership of the service delivery capacities, three types of shared service centers have been researched: horizontal shared service models, vertical shared service models, intergovernmental contracting shared service models [37]. In a rare example of the research that resulted in a development of a conceptual framework, a shared service center business models were discussed based on four dimensions: (i) the governance structure of the SSC, (ii) the strategic rationale behind the SSC, (iii) the nature of the SSC services and (iv) the customer orientation of the SSC [38].

The evolution of workplace institutions, increased role of technology in transformations of work and Covid-19 pandemic have promoted the adoption of remote work modes that have prompted the research into novel types of work arrangement such as globally distributed platform work and virtual companies [27] that may also affect offshore service centers. Moreover, it has been highlighted that the increased robotic process automation in shared service centers will largely affect their operations and their governance modes in the future [39].

6 Conclusions—Towards the New Taxonomy of Offshore Service Centers

With the use of integrative literature review this study attempts to bridge two broad and well-developed streams of research that, up to the present, have progressed in isolation which constitutes the important theoretical contribution of the research. The undertaken research attempts to reconcile the insights from literature on reverse knowledge transfer (RKT), which highlights its increased importance in understanding the performance of contemporary multinationals, with the substantial body of research on the evolving characteristics of offshoring into more knowledge intensive and complex phenomena.

Moreover, the insight into existing taxonomies for offshore service centers, developed as a part of the this research, indicated that the topic of reverse knowledge transfers from offshore service centers to the parent companies has been largely absent from the scientific inquiry. Thus, this under-researched area creates significant research gap that could be addressed in further studies. The outcome of such future research will significantly contribute to an improved understanding of offshore service center performance and its role in multinationals' networks. Moreover, it may also challenge the prevailing perception of offshore service centers as units of relatively standardized and low value-added activities.

Furthermore, the prospective research linking the reverse knowledge transfer with offshore service center typology is likely to have important implications for management practices by providing managers with new insights into knowledge creation and diffusion in the international corporate networks. The taxonomy of offshore service centers, created based on the intensity of reverse knowledge transfer to the parent companies, will allow managers to map offshore service centers and more precisely appraise their strategic importance for knowledge creation in corporate networks.

Moreover, the insight from the updated evolutionary path of offshore service center may support decision makers in estimating the maturity of their offshore service centers and investigate the potential for their further development. Consequently, insight from such research will provide the opportunity to explicitly and methodically review reverse knowledge transfer in global multinational enterprises in order to optimize the knowledge flow and ensues consistency with the strategic imperatives.

Funding Research presented in this paper constitutes a part of the implementation of the following grant: "From invoice factories to knowledge hubs? Investigating intensity of reverse knowledge transfers from offshored captive centers to parent companies." The Polish National Agency for Academic Exchange (NAWA), The Bekker NAWA Programme, BPN/BEK/2021/2/00024/U/00001.

References

1. Yang, Q., Mudambi, R., Meyer, K.E.: Conventional and reverse knowledge flows in multinational corporations. J. Manag. **34**(5), 882–902 (2008). https://doi.org/10.1177/0149206308321546
2. Malik, R.: Key location factors and the evolution of motives for business service offshoring to Poland. J. Econ. Manag. **31**(1), 119–132 (2018). https://doi.org/10.22367/jem.2018.31.06
3. Polese, F., Payne, A., Frow, P., Sarno, D., Nenonen, S.: Emergence and phase transitions in service ecosystems. J. Bus. Res. **127**, 25–34 (2021). https://doi.org/10.1016/j.jbusres.2020.11.067
4. Gummesson, E., Mele, C., Polese, F.: Complexity and viability in service ecosystems. Mark. Theory **19**(1), 3–7 (2019). https://doi.org/10.1177/1470593118774201
5. Malik, R., Visvizi, A., Troisi, O., Grimaldi, M.: Smart services in smart cities: insights from science mapping analysis. Sustainability **14**(11), 6506 (2022). https://doi.org/10.3390/su14116506
6. Lytras, M.D., Visvizi, A., Sarirete, A.: Clustering smart city services: Perceptions, expectations, responses. Sustainability **11**(6), 1669 (2019). https://doi.org/10.3390/su11061669
7. Visvizi, A., Troisi, O., Grimaldi, M., Loia, F.: Think human, act digital: activating data-driven orientation in innovative start-ups. Eur. J. Innov. Manag. **25**(6), 452–478 (2022). https://doi.org/10.1108/EJIM-04-2021-0206
8. Gonzalez, R., Llopis, J., Gasco, J.: Information systems offshore outsourcing: managerial conclusions from academic research. Int. Entrepreneurship Manag. J. **9**(2), 229–259 (2013). https://doi.org/10.1007/S11365-013-0250-Y
9. Lacity, M.C., Khan, S., Yan, A., Willcocks, L.P.: A review of the IT outsourcing empirical literature and future research directions. J. Inf. Technol. **25**(4), 395–433 (2010). https://doi.org/10.1057/JIT.2010.21
10. Schmeisser, B.: A systematic review of literature on offshoring of value chain activities. J. Int. Manag. **19**(4), 390–406 (2013). https://doi.org/10.1016/J.INTMAN.2013.03.011
11. Pisani, N., Ricart, J.E.: Offshoring of services: a review of the literature and organizing framework. Manag. Int. Rev. **56**(3), 385–424 (2016). https://doi.org/10.1007/S11575-015-0270-7
12. Richter, P.C., Brühl, R.: Shared service center research: a review of the past, present, and future. Eur. Manag. J. **35**(1), 26–38 (2017). https://doi.org/10.1016/J.EMJ.2016.08.004
13. Snyder, H.: Literature review as a research methodology: an overview and guidelines. J. Bus. Res. **104**, 333–339 (2019). https://doi.org/10.1016/J.JBUSRES.2019.07.039

14. Luo, Y., Tung, R.L.: International expansion of emerging market enterprises: a springboard perspective. J. Int. Bus. Stud. **38**(4), 481–498 (2007). https://doi.org/10.1057/PALGRAVE.JIBS.8400275
15. Jiménez-Jiménez, D., Martínez-Costa, M., Sanz-Valle, R.: Reverse knowledge transfer and innovation in MNCs. Eur. J. Innov. Manag. **23**(4), 629–648 (2020). https://doi.org/10.1108/EJIM-10-2018-0226
16. Mudambi, R., Navarra, P.: Is knowledge power? Knowledge flows, subsidiary power and rent-seeking within MNCs. In: The Eclectic Paradigm. Palgrave Macmillan, London, pp. 157–191 (2015). doi:https://doi.org/10.1007/978-1-137-54471-1_7
17. Cano-Kollmann, M., Cantwell, J., Hannigan, T.J., Mudambi, R., Song, J.: Knowledge connectivity: an agenda for innovation research in international business. J. Int. Bus. Stud. **47**(3), 255–262 (2016). https://doi.org/10.1057/JIBS.2016.8/FIGURES/1
18. Frost, T.S., Zhou, C.: R&D co-practice and 'reverse' knowledge integration in multinational firms. J. Int. Bus. Stud. **36**(6), 676–687 (2005). https://doi.org/10.1057/PALGRAVE.JIBS.8400168
19. Sinai, A.R., Heo, D.: Determinants of reverse knowledge transfer: a systematic literature review. Int. J. Knowl. Manag. Stud. **13**(2), 213–229 (2022). https://doi.org/10.1504/IJKMS.2022.121876
20. McGuinness, M., Demirbag, M., Bandara, S.: Towards a multi-perspective model of reverse knowledge transfer in multinational enterprises: a case study of coats plc. Eur. Manag. J. **31**(2), 179–195 (2013). https://doi.org/10.1016/J.EMJ.2012.03.013
21. D'Agostino, L.M., Santangelo, G.D.: Do overseas R&D laboratories in emerging markets contribute to home knowledge creation? Manag. Int. Rev. **52**(2), 251–273 (2012). https://doi.org/10.1007/S11575-012-0135-2/TABLES/2
22. Pereira, V., Bamel, U., Temouri, Y., Budhwar, P., Del Giudice, M.: Mapping the evolution, current state of affairs and future research direction of managing cross-border knowledge for innovation. In: International Business Review (In Press), p. 101834 (2021). https://doi.org/10.1016/J.IBUSREV.2021.101834
23. Nair, S.R., Demirbag, M., Mellahi, K., Pillai, K.G.: Do parent units benefit from reverse knowledge transfer? Br. J. Manag. **29**(3), 428–444 (2018). https://doi.org/10.1111/1467-8551.12234
24. Peltokorpi, V., Froese, F.J., Reiche, B.S., Klar, S.: Reverse knowledge flows: how and when do preparation and reintegration facilitate repatriate knowledge transfer? J. Manag. Stud. (Online First) (2022). https://doi.org/10.1111/JOMS.12802
25. Klimek, A.: Offshoring of white-collar jobs: theory and evidence. Int. J. Manag. Econ. **57**(1), 69–84 (2021). https://doi.org/10.2478/IJME-2021-0003
26. Tojeiro-Rivero, D.: What effect does the aggregate industrial R&D offshoring have on you? A multilevel study. J. Int. Manag. **28**(2), 100881 (2022). https://doi.org/10.1016/J.INTMAN.2021.100881
27. Erickson, C.L., Norlander, P.: How the past of outsourcing and offshoring is the future of post-pandemic remote work: a typology, a model and a review. Ind. Relat. J. **53**(1), 71–89 (2022). https://doi.org/10.1111/IRJ.12355
28. Antons, D., Piller, F.T.: Opening the black box of "Not Invented Here": Attitudes, decision biases, and behavioral consequences. Acad. Manag. Perspect. **29**(2), 193–217 (2015). https://doi.org/10.5465/AMP.2013.0091
29. Dzikowska, M., Malik, R.: Determinants of relocation mode choice: effect of resource endowment, competitive intensity and activity character. Europ. J. Int. Manag. (Online First) (2020). https://doi.org/10.1504/EJIM.2020.10022072
30. Oshri, I.: Choosing an Evolutionary Path for Offshore Captive Centers. MIS Quarterly Executive 12(3), 151–165 (2013), Access: https://aisel.aisnet.org/misqe/vol12/iss3/5.
31. De Silva, L.M.H.: Review of the current status of the offshoring industry: Insights for practice. Int. J. Business Manag. **14**(1), 76–93 (2019). https://doi.org/10.5539/ijbm.v14n1p76
32. Anagnoste, S.: Setting up a robotic process automation center of excellence. Manag. Dyn. Knowl. Econ. **6**(2), 307–332 (2018). https://doi.org/10.25019/MDKE/6.2.07

33. Zimmermann, A., Oshri, I., Lioliou, E., Gerbasi, A.: Sourcing in or out: Implications for social capital and knowledge sharing. J. Strateg. Inf. Syst. **27**(1), 82–100 (2018). https://doi.org/10.1016/J.JSIS.2017.05.001
34. Rodríguez, A., Nieto, M.J.: Does R&D offshoring lead to SME growth? D ifferent governance modes and the mediating role of innovation. Strateg. Manag. J. **37**(8), 1734–1753 (2016). https://doi.org/10.1002/SMJ.2413
35. Kamal, M.M.: Shared services: lessons from private sector for public sector domain. J. Enterp. Inf. Manag. **25**(5), 431–440 (2012). https://doi.org/10.1108/17410391211265124
36. Fielt, E., Bandara, W., Miskon, S., Gable, G.: Exploring shared services from an IS perspective: a literature review and research agenda. Commun. Assoc. Inf. Syst. **34**(1), 1001–1040 (2014). https://doi.org/10.17705/1cais.03454
37. Dollery, B., Grant, B., Akimov, A.: A typology of shared service provision in Australian local government. Aust. Geogr. **41**(2), 217–231 (2010). https://doi.org/10.1080/00049181003742310
38. Joha, A., Janssen, M.: Types of shared services business models in public administration. In: Proceedings of the 12th Annual International Digital Government Research Conference: Digital Government Innovation in Challenging Times, pp. 26–35 (2011). https://doi.org/10.1145/2037556.2037562
39. Figueiredo, A.S., Pinto, L.H.: Robotizing shared service centres: key challenges and outcomes. J. Serv. Theory Pract. **31**(1), 157–178 (2021). https://doi.org/10.1108/JSTP-06-2020-0126

Integrating a Digital Platform Within Museum Ecosystem: A New 'Phygital' Experience Driving Sustainable Recovery

Giovanni Baldi

Abstract Managing disruption, resilience and recovery means investing in culture, ICT and networking. Public administration and cultural institutions must also develop efficient networks and innovation ecosystems with an open innovation approach to overcome the crisis. In the cultural field, mainly in museums one, it could bring to light an integrated ecosystem based on a digital platform which creates a new sustainable business model for museum shops, fitting to a new experience for system actors. An experience reaching outside the traditional museum boundaries, that crosses the logic of the eco-museum and through which the user has the feeling of never leaving that place—even when he or she returns home, combining physical and digital resources: *Phygital*. With a conceptual approach, by reviewing the literature on systems, digital eco-systems and retail digitization, we try to suggest an original integration of theoretical patterns that could lead to a sustainable reconfiguration of museum shops to improve visitor engagement.

Keywords Museum ecosystem · Phygital · Sustainable development visitor engagement

1 Introduction

The climatic crisis and the digital revolution are two disruptive and global changes that need to be approached from a unique perspective: digitization is the dominant enabler for sustainability, and no sustainability is possible unless innovation. All investments in technology are viewed through a sustainability lens, and enterprises have a responsibility to understand how technology can deliver new sustainable consumption, business and sales models, as well as products (including services) that positively impact on the triple bottom line of sustainability: people, planet and profit. Thereafter, Covid-19 also provided a substantial challenge to companies, most of which had to adjust as a result of the pandemic's changing conditions in a short time

G. Baldi (✉)
University of Salerno, Via Giovanni Paolo II, 132, 84084 Fisciano, SA, Italy
e-mail: gbaldi@unisa.it

A. Visvizi et al. (eds.), *Research and Innovation Forum 2022*,
Springer Proceedings in Complexity, https://doi.org/10.1007/978-3-031-19560-0_57

[1]. It is crucial that organizations develop solutions which allow them to both respond to changing scenarios and also provide sustainable competitive differentiation to try to *recover* with proven organizational *resilience* [2]. With an exploratory approach, in the following section we will review the literature on Systems, the ecosystem of digital platforms and the digitisation of physical retail.

After that, we have only theoretically combined the two ecosystems, trying to give a new and original application not found in the literature on this topic.

2 Theoretical Framework

2.1 A Systemic Perspective

The role of systems studies is to comprehend and adjust any sort of complexity. In accordance with major systemic theories (such as the VSA, Vital Systemic Approach), a system is a structured layout of self-contained interrelated and interacting entities that is equitable, well-balanced and organised. Taking up the main definitions on the subject of the most well-known authors, we can briefly observe that a system is "a complex of rationally and inescapably connected interacting elements" [3, 4]. A service system is related to provider/client relationships and is consequently seen as an open system able to enhance its equilibrium state by acquiring, sharing and using resources [5]. Service systems are chunks of people, technologies, value propositions and knowledge that allows them to codesign value [6].

The rise of a Science of Service and the ongoing cultural and denominological evolution represent a shift in approach to the understanding of rules, strategies among the actors of the Exchange, relations between the actors of the system and systems. And it lays the solid basis for competitive advantage, thereby contributing to a significant change of perspective with regards to conventional paradigms. The same Actor in the system can therefore simultaneously be a supplier, a user or an intermediary, according to the type of transaction that involves him or her and that is being investigated, as well as the time and context in which it is observed, resulting in Actor-For-Actor (A4A) type relationships where the mutual interest of the actors leads to long-lasting interactions which do not allow for opportunistic and/or speculative behavior [7].

2.2 The Digital Platform Eco-System

Nowadays and in the years to come the entire world will be more and more interconnected, therefore there is a huge attention on learning and innovative processes, on advancement across all fields, most notably in art, education, research and overall culture. In this extremely disciplined field, where heritage management is often in the hands of politics and the government of cities and states, the regulations are always changing. The innovative processes also change the technologies, the reporting protocols/standards, the number of actors, the different levels of interaction. Thus, in this system, the contribution of the end-users increases, and social and ethical implications are more relevant, where culture and art turn into a collective matter. Organizations must therefore develop a range of innovative solutions to maximize the use of available resources in the service delivery process, enhancing their offering. Each Actor (of a system) needs to seek to provide its own personal contribution to the quality enhancement of the national cultural and artistic heritage system, and thus of the whole museum eco-system. The distribution chain is reconceived as a service network and therefore it has a non-prior adjustable layout, but able to be mutable, to fit and develop in relation to the ever-changing contextual conditions [8]. The knowledge input, expertise appliance, continuous remodeling, the willingness to forge long-term relationships with strategic key actors are all part of a systemic mode of being responsive. Companies especially will have to be incrementally more able to reshape and reorganize their entire processes to maintain a constant and sustainable level over time. Everything will be even more interrelated and interconnected in the future, but this is something that also exists in the Now. In the museum sector, the smart logic frequently involves the implementation of digital platforms that provide all the services: from fully virtual tours to AR tours, which are integrated with physical ones.

Reflecting this pattern, Digital Platforms are a new paradigm integrating multifarious issues, technology, actors, concerns and goals, enlarging pre-existing markets and creating new ones [9, 10]. The key project aspect of a digital platform eco-system consists of cohesive and flanking elements, that are both modular and interrelated, and are kept in conjunction by agreed regulations and a holistic value proposition. Hence, in order to be effective, these digital solutions imply coordination of actors with non-aligned interests. E-platforms were identified as packages consisting of digital tools, services or solutions designed and in-house developed by one or more enterprises, constituting a backbone from which other companies are able to build complementary and related products, services and solutions, thereby also generating significant network benefits [11].

Business theory relates Digital Platform to Ecosystem, with regard to the community of Actors who jointly (co)create value with a platform leader [12]. From the systemic standpoint, we can identify four types of actors who shape the e-system, as shown in Fig. 1: the platform sponsor, the platform supplier, the users on the offering side and the users on the demand side [13].

The platform sponsor is the overarching developer and the owner of the intellectual assets in the ecosystem fixing the policies and infras of its virtual environment. Demand-side users are the final recipients. Supply-side users are content and software producers. Supplying and delivering distinctive attributes involving actors in the platform. And, last but not least, the supplier is the liaison of the whole system. Most commonly, the function of platform supplier is covered by platform owner [14]. E-platform ecosystems may generate added value by means of accompanying innovation and/or through their power to quickly bridge the gap between the two sides

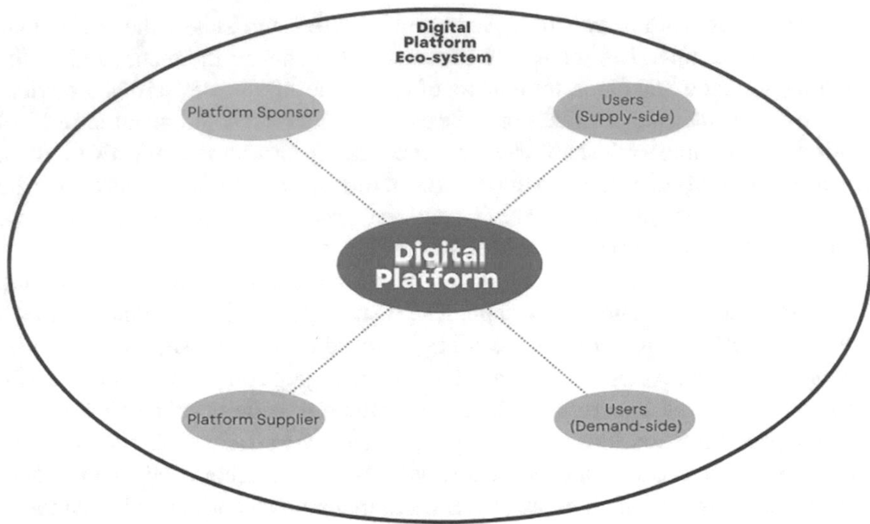

Fig. 1 Digital Platform Eco-system. *Source* Author's elaboration from [10]

of the market and the actors involved in the ecosystem of digital platforms interact and co-create value using various digital algorithms [15, 16]. The concept of eco-system is linked to Sustainable Innovation: Eco-Innovation. Hence, sustainable innovation is more than eco-innovation since it embeds society's interests as well as ties closer to sustainable development. Sustainable innovation is part of a process where sustainability is ingrained in social and entrepreneurial schemes, starting from idea conceiving to implementation [17]. Therefore, it is not a mechanical or straightforward flow, but an eco-system made up of ongoing interplay across the economy, society and the environment [18], an eco-slex of actors and interlinkages whereby co-operation holds an impressive leading role.

2.3 The 'Phygitalization' of Retailing

The wide range of in-store technologies to which customers may have access, enables them to effortlessly search, match, track and purchase products. Alongside all these activities, shoppers are empowered to discover extra information regarding products and services, develop and share their wish lists and shop using easy-to-use platforms. Looking at consumer behavior research, the Technology Acceptance Model (TAM) [19] has been augmented by other drivers such as risk avoidance, trust, hedonic and utility value [20–22]. Indeed, among these, Trust has a critical and significant function under uncertainty and risk situations. In ordinary retailers, the greatest wellspring of trust is the vendor, pooling his or her professional expertise, fairness, goodwill and reliability to sustain the consumer's purchasing behaviour and customer engagement [23, 24] and a number of papers described how hard it is to establish solid connections with the user by using self-service technologies due to the absence of human interaction [25]. In addition, evidence suggests customers are constantly looking for recommendations and feedback in order to make the right decision [26]. In high-end retail, a differentiation can be made between in-store technologies that are fully owned by the shop (touch screens, VR fitting rooms, self-checkouts, tablet-supported personnel, RFID technology, etc.) and hybrid in-store technologies since they are deployed around the retail space using the customer's smartphone (QR scanner, NFC, Bluetooth Beacon, Digital Shop Platforms, etc.). Evidence demonstrates how the majority of customer-focused experiences in retail are hybrid, forming the seamless *omni-channel* continuum [27]. Having both solutions and integrating the shop's own technologies with those of the consumer can facilitate and stimulate the buying process. Not least because smartphones are in any case used for most of the time spent inside the shop, where the phenomenon of 'showrooming' is very common. This consists of looking for or trying something in physical retail, then comparing competitors' prices and buying online—with the risk of buying products from other shops [28]. The in-store experience must be hybrid, i.e., with retailer-owned digital technologies that are seamlessly integrated with consumer technologies such as smartphones, tablets, digital watches, etc. and with a constant human interaction of the staff. It is an Omnichannel In-Store experience to improve the customer engagement that combines the physical and digital worlds in the same place and space: *Phygital* [29].

3 A New Shopping Experience in the Museum Ecosystem

Thinking about combining the museum eco-system with the eco-system of a digital platform implementing Phygital technology within the museum shop may create a fascinating concept towards sustainable development and recovery. The museum ecosystem as shown in Fig. 2 is characterised by the participation of many Actors, both public and private, and these may include: the international arena in which the International Council of Museums (ICOM) stands out, as global and voluntary organization supporting museums and cultural operators; the Public with its Ministries (for example, MiC in Italy) organises and drives the system; associations that foster the stewardship of cultural heritage; the artists; nature and the local territory; the research institutions such as universities and professionals; the educational system; the local community; and visitors [30]. The visitors, as well as enthusiasts, professionals and students, may also be tourists,[1] as the museum complex's part of the tourism environment [32].

[1] UNWTO defines the tourist "as any person who travels to countries other than that in which he or she has his or her habitual residence, outside his or her everyday environment, for a period not exceeding one year and whose main purpose of the visit is other than the pursuit of any kind of remunerated activity within the country visited" [31].

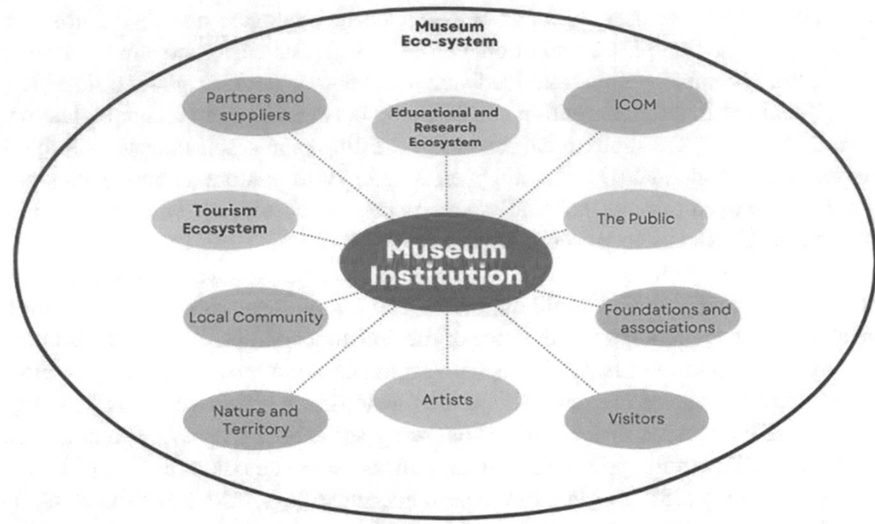

Fig. 2 The museum ecosystem. *Source* Author's elaboration from [30]

Also in the museum system, which is going through a slow crisis, people are starting to think about sustainable innovation, understood not only as respect for the environment and ecology, but also in a broader sense as co-creation and sharing of value in the ecosystem, to facilitate disruptive innovation, whether technological or cultural, and to generate a positive edge. This means pursuing models of business whereby in the process of co-creating value, priority is given to the relationship with the territory and the community to which it belongs, as well as with a digital community, by implementing concrete measures that can improve its social and environmental impact as also found in other studies in this field [33]. In this model, relationships with customers and other stakeholders are strengthened by being educational and actively supporting a social or environmental cause. The challenge is to create a business model for the tourism and culture sector with the lowest transaction costs [34] and the highest value creation for customers [35] while enabling responsible production and consumption according to sustainable development.

The conceptual innovation proposed here is indeed a practical implementation and integration of a digital platform (as museum partner\supplier) [36] for retail 'Phygitalization' within the museum ecosystem. To be more accurate as shown in Fig. 3, it would mean upgrading the retailing system within the museum shop (no longer a mere book-shop) and not just in terms of technology, as seen as a *hybrid*, but also as concerns the business model and the nature of the products offered.

It is rather intuitive to say that the bookshop is the place where the museum public, at the end of a visit through the works or architecture of a museum, gallery, art gallery or archaeological site, usually looks for the prolongation of their experience in the form of an object. The bookshop and museum gadgets are fully included in the category of services that a cultural institution offers, together with ticketing, reception,

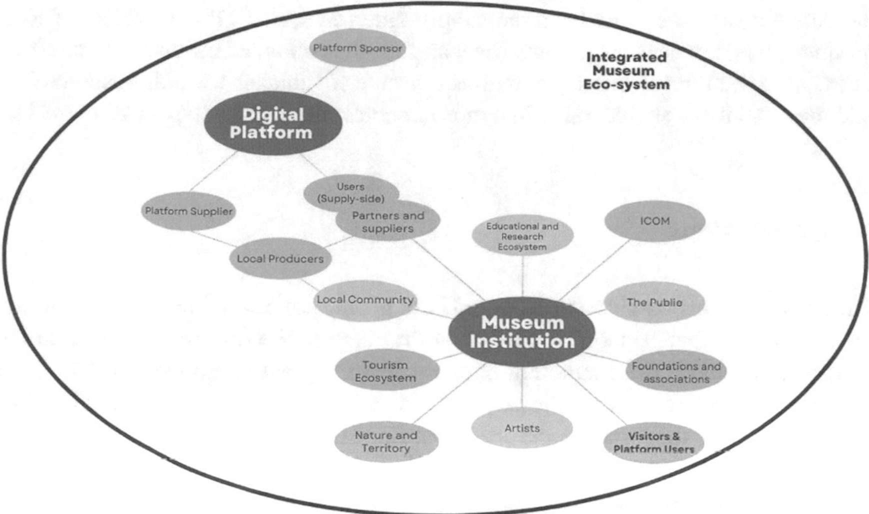

Fig. 3 Integrating an eco-platform into museum ecosystems. *Source* Author's elaboration from [10, 30]

cafeteria or restaurant, guided tours, and surveillance. The bookshop sector, as well as museum merchandising, is an ever-growing sector indispensable for cultural enterprises looking for additional services and products able to generate new financial resources for sustenance [37]. In fact, it is necessary to understand that the public seeks in museum objects a tangible memory of its visit and a cultural institution must be ready to interpret this desire, in order to be more attractive and satisfy not only the cultural need for knowledge and information but also the experiential one, continuing to feel part of that community. The consumption of museum branded objects is also generated by the desire for novelty and change in everyday life; materializing the memory of a previous experience; the instantaneous emergence of a need for a product [37].

The bookshop should not only house the ticketing services or simple gadgets but should be an important touch point between the visitor, the cultural institution, and the territory–which operate, as we have already seen, in the same network. Therefore, the catalogue of products will be linked both to the originality and uniqueness of the visiting experience and to a need to represent the values of the territory. The products have to be exclusive and numbered, coming directly from local producers (Fig. 3). These suppliers have to provide a sample for the *showcase* set-up as first physical touch point with the customers who will be able to acquire the items through a Phygital experience served by a digital platform (eco-system) and have it sent directly home, in a drop-shipping[2] system if they are unable to take it with them. In this way there is a continuous co-operation and co-creation between the users on

[2] Drop-shipping is a business model in which a vendor sells a product to the end user without having it in stock, by having it shipped directly from the manufacturer [38].

the demand side, the users from the supply side (museums), the providers of local products, whether artisans or industries, and all the Actors of the ecosystem, which as we have seen in the literature implies massive advantages towards sustainability intended mainly as shared value in a more general smart city perspective [39–41].

4 Conclusions

The integration of eco-systems proposed here is only conceptual in order to embrace the principles of sustainability. Co-operation between the different actors also has an impact on territorial marketing because it offers visitors an overview of local excellence, understood as traditional products, enhancing them with global visibility, in combination with the nature and artistic-cultural heritage of the territory. The result is a *win–win* for all actors [7]: the museum has enormous benefits and can increase visits, as well as visitors, by pursuing the concept of an extended museum integrated in the territory and its traditions, which goes beyond its borders. As if you had never left it.

Since this work is a conceptual proceeding, it does not present an empirical research, so it has limitations. In the future, the study can be continued by researching museums that already partially or totally adopt this business model, analyzing the benefits related to the actors of the system. Indeed, the implications are manifold both on the research side, lacking on this topic, and on the application side. This may represent a starting point for digital platform developers to match their solutions to these targeted needs, but also in other fields.

In conclusion, this could represent a dynamic way in which communities can conserve, interpret and manage their artistic, cultural, economic and social heritage, and through which museums can improve visitor engagement for a sustainable and *Resilient* development. To ride the wave of the digital transition, *Recovering* from the *Rupture* and getting to an even better level than the pre-pandemic one.

References

1. Kraus, S., Clauss, T., Breier, M., Gast, J., Zardini, A., Tiberius, V.: The economics of COVID-19: initial empirical evidence on how family firms in five European countries cope with the corona crisis. Int. J. Entrep. Behav. Res. **26**(5), 1067–1092 (2020). https://doi.org/10.1108/IJEBR-04-2020-0214
2. Ullah, F., Qayyum, S., Thaheem, M.J., Al-Turjman, F., Sepasgozar, S.M.E.: Risk management in sustainable smart cities governance: a TOE framework. Technol. Forecast. Soc. Chang. **167**, 120743 (2021). https://doi.org/10.1016/j.techfore.2021.120743
3. Von Bertalanffy, L.: General System Theory: Foundations, Development, Applications. George Braziller. Inc., New York (1968)
4. Luhmann, N.: Soziale systeme: Grundriss einer allgemeinen theorie. Suhrkamp (1984)
5. Bassano, C., Barile, S.: The evolving dynamics of service co-creation in a viable systems perspective. In: 13th Toulon-Verona Conference (2010)

6. Spohrer, J., Maglio, P.P., Bailey, J., Gruhl, D.: Steps toward a science of service systems. Computer **40**(1), 71–77 (2007). https://doi.org/10.1109/MC.2007.33
7. Polese, F., Carrubbo, L., Bruni, R., Maione, G.: The viable system perspective of actors in eco-systems. TQM J. **29**(6), 783–799 (2017). https://doi.org/10.1108/TQM-05-2017-0055
8. Barile, S., Polese, F.: Smart service systems and viable service systems: applying systems theory to service science. Serv. Sci. **2** (2012). https://doi.org/10.1287/serv.2.1_2.21
9. Busch, T., Bruce-Clark, P., Derwall, J., Eccles, R., Hebb, T., Hoepner, A., Klein, C., Krueger, P., Paetzold, F., Scholtens, B., Weber, O.: Impact investments: A call for (re)orientation. SN Busin. Econ. 1 (2021). https://doi.org/10.1007/s43546-020-00033-6
10. Calabrese, M., Sala, A.L., Fuller, R.P., Laudando, A.: Digital platform ecosystems for sustainable innovation: toward a new meta-organizational model? Administ. Sci. **11**(4) (2021). https://doi.org/10.3390/admsci11040119
11. Evans, D.: Some empirical aspects of multi-sided platform industries. Rev. Netw. Econ. **2**, 191–209 (2003). https://doi.org/10.2139/ssrn.447981
12. Jacobides, M., Cennamo, C., Gawer, A.: Towards a theory of ecosystems. Strat. Manag. J. **39** (2018). https://doi.org/10.1002/smj.2904
13. Baek, S., Kim, K., Altmann, J.: Role of platform providers in service networks: the case of Salesforce.com app exchange **1**, 39–45. https://doi.org/10.1109/CBI.2014.58
14. Gawer, A., Cusumano, M.A.: Industry platforms and ecosystem innovation. J. Prod. Innov. Manag. **31**(3), 417–433 (2014). https://doi.org/10.1111/jpim.12105
15. Jovanovic, M., Sjödin, D., Parida, V.:. Co-evolution of platform architecture, platform services, and platform governance: expanding the platform value of industrial digital platforms. Technovation 102218 (2021). https://doi.org/10.1016/j.technovation.2020.102218
16. Polese, F., Pels, J., Tronvoll, B., Bruni, R., Carrubbo, L.: A4A relationships. J. Serv. Theor. Pract. **27**(5), 1040–1056. (2017). https://doi.org/10.1108/JSTP-05-2017-0085
17. Charter, M., Gray, C., Clark, T., & Woolman, T.: Review: The role of business in realising sustainable consumption and production. In: System Innovation for Sustainability 1: Perspectives on Radical Changes to Sustainable Consumption and Production, pp. 46–69. https://doi.org/10.4324/9781351280204-10
18. Malerba, F.: Sectoral systems of innovation and production. Innovat. Syst. **31**(2), 247–264 (2002). https://doi.org/10.1016/S0048-7333(01)00139-1
19. Davis, F.D.: Perceived usefulness, perceived ease of use, and user acceptance of information technology. MIS Q. 319–340. (1989)
20. Groß, M.: Mobile shopping: A classification framework and literature review. Int. J. Retail Distrib. Manag. **43**(3), 221–241 (2015). https://doi.org/10.1108/IJRDM-06-2013-0119
21. Pantano, E.: Innovation drivers in retail industry. Int. J. Inf. Manage. **34**(3), 344–350 (2014). https://doi.org/10.1016/j.ijinfomgt.2014.03.002
22. Pantano, E., Priporas, C.-V.: The effect of mobile retailing on consumers' purchasing experiences: a dynamic perspective. Comput. Hum. Behav. **61**, 548–555 (2016). https://doi.org/10.1016/j.chb.2016.03.071
23. Pantano, E., Migliarese, P.: Exploiting consumer–employee–retailer interactions in technology-enriched retail environments through a relational lens. J. Retail. Consum. Serv. **21**, 958–965 (2014). https://doi.org/10.1016/j.jretconser.2014.08.015
24. Gentile, C., Spiller, N., Noci, G.: How to sustain the customer experience: an overview of experience components that co-create value with the customer. Eur. Manag. J. **25**(5), 395–410. (2007). S0263237307000886. https://doi.org/10.1016/j.emj.2007.08.005
25. Kim, H., Suh, K.-S., Lee, U.-K.: Effects of collaborative online shopping on shopping experience through social and relational perspectives. Inf. Manag. **50**(4), 169–180 (2013). https://doi.org/10.1016/j.im.2013.02.003
26. Li, C.-Y.: Persuasive messages on information system acceptance: a theoretical extension of elaboration likelihood model and social influence theory. Comput. Hum. Behav. **29**(1), 264–275 (2013). https://doi.org/10.1016/j.chb.2012.09.003
27. Willems, K., Smolders, A., Brengman, M., Luyten, K., Schöning, J.: The path-to-purchase is paved with digital opportunities: an inventory of shopper-oriented retail technologies. Technol. Forecast. Soc. Chang. **124**, 228–242 (2017). https://doi.org/10.1016/j.techfore.2016.10.066

28. Rapp, A., Baker, T.L., Bachrach, D.G., Ogilvie, J., Beitelspacher, L.S.: Perceived customer showrooming behavior and the effect on retail salesperson self-efficacy and performance. Multi-Channel Retail. **91**(2), 358–369 (2015). https://doi.org/10.1016/j.jretai.2014.12.007
29. Belghiti, S., Ochs, A., Lemoine, J.-F., Badot, O.: The phygital shopping experience: an attempt at conceptualization and empirical investigation 74 (2018). https://doi.org/10.1007/978-3-319-68750-6_18
30. Jung, Y., Rowson Love, A.: Systems thinking in museums: theory and practice (2017)
31. United Nations.: International Recommendations for Tourism Statistics IRTS, 2.13 (2008). https://unstats.un.org/unsd/publication/seriesm/seriesm_83rev1e.pdf
32. Waligo, V.M., Clarke, J., Hawkins, R.: Implementing sustainable tourism: a multi-stakeholder involvement management framework. Tour. Manage. **36**, 342–353 (2013). https://doi.org/10.1016/j.tourman.2012.10.008
33. Russo-Spena, T., Tregua, M., D'Auria, A., Bifulco, F.: A digital business model: an illustrated framework from the cultural heritage business. Int. J. Entrepren. Behav. Res. Ahead-of-print No. ahead-of-print. https://doi.org/10.1108/IJEBR-01-2021-0088
34. Williamson, O.E.: Markets and Hierarchies: Analysis and Antitrust Implications: A Study in the Economics of Internal Organization (1975). University of Illinois at Urbana-Champaign's Academy for Entrepreneurial Leadership Historical Research Reference in Entrepreneurship, Available at SSRN: https://ssrn.com/abstract=1496220
35. Andreassen, T., Lervik-Olsen, L., Snyder, H., van Riel, A., Sweeney, J., Vaerenbergh, Y.V.: Business model innovation and value-creation: the triadic way. J. Serv. Manag. 29. https://doi.org/10.1108/JOSM-05-2018-0125
36. Visvizi, A., Troisi, O., Grimaldi, M., Loia, F.: Think human, act digital: activating data-driven orientation in innovative start-ups. Eur. J. Innov. Manag. **25**(6), 452–478 (2022). https://doi.org/10.1108/EJIM-04-2021-0206
37. Severino, F., (ed.), Comunicare la cultura, vol. 9. FrancoAngeli (2007)
38. Khouja, M.: The evaluation of drop shipping option for e-commerce retailers. Comput. Ind. Eng. **41**(2), 109–126 (2001). https://doi.org/10.1016/S0360-8352(01)00046-8
39. Kashef, M., Visvizi, A., Troisi, O.: Smart city as a smart service system: human-computer interaction and smart city surveillance systems. Comput. Hum. Behav. **124**, 106923 (2021). https://doi.org/10.1016/j.chb.2021.106923
40. Buhalis, D., Taheri, B., Rahimi, R. (Eds.). Smart cities and tourism: co-creating experiences, challenges and opportunities. (2022)
41. Polese, F., Barile, S., Caputo, F., Carrubbo, L., Waletzky, L.: Determinants for value cocreation and collaborative paths in complex service systems: a focus on (smart) cities. Serv. Sci. **10**(4), 397–407. (2018)
42. Taheri, B., Jafari, A., O'Gorman, K.: Keeping your audience: presenting a visitor engagement scale. Tourism Manag. **42**, 321–329 (2014) S0261517713002239. https://doi.org/10.1016/j.tourman.2013.12.011
43. Visvizi, A., Lytras, M. (Eds.).: Smart cities: issues and challenges: mapping political, social and economic risks and threats. (2019)
44. Hiller Connell, K.Y., Kozar, J.M.: Introduction to special issue on sustainability and the triple bottom line within the global clothing and textiles industry. Fashion Text. **4**(1), 1–3. (2017)

Reifying Kintsugi Art in Post-covid Era: A Remote Smart Working Model, Augmented Intelligence-Based, for Antifragile Companies

Andrea Moretta Tartaglione, Ylenia Cavacece, Luca Carrubbo, and Antonietta Megaro

Abstract The aim of this conceptual paper is to understand if augmented intelligence may be considered a driver of antifragility that can be allegorically represented by the Japanese art of Kintsugi, which consists of the use of gold or silver to repair broken objects in ceramic to get a better aesthetic form. Covid-19, like a black swan, represented, for many companies, understood as systems, a complex situation capable of upsetting their equilibrium. It had thus forced them to accelerate the digitization process. Digitalization, based on artificial intelligence (AI) tools, brings in many fields new perspectives, such as new business scenarios and models. By using the Viable System Approach (vSa) lens, we investigated the impact of smart working, widely spread to manage a complex situation (Covid-19), in allowing companies to cope with changes and to be antifragile. A remote smart working model is proposed, as an evolution of smart working, based on a new culture of "doing business" to search for new viable conditions. It can allow companies a more efficient resources management, an endless orientation towards results, but also new synergies in new contexts thanks to new and increased networks, for new collaborations and new forms of interactions, as well as more profitable relationships with employees, based on a strong relationship of trust and on better opportunities for work-life balance.

Keywords Antifragility · Remote smart working · Augmented intelligence · Viable system approach

---

A. M. Tartaglione · Y. Cavacece
University of Cassino and Southern Lazio, Viale Dell'Università, 03043 Cassino, FR, Italy
e-mail: a.moretta@unicas.it

Y. Cavacece
e-mail: ylenia.cavacece@unicas.it

L. Carrubbo · A. Megaro (✉)
University of Salerno, Via Giovanni Paolo II, 132, 84084 Fisciano, SA, Italy
e-mail: amegaro@unisa.it

L. Carrubbo
e-mail: lcarrubbo@unisa.it

A. Visvizi et al. (eds.), *Research and Innovation Forum 2022*,
Springer Proceedings in Complexity, https://doi.org/10.1007/978-3-031-19560-0_58

1 Introduction

Decision-makers often operate in dynamic contexts and their standardized solutions may be inadequate [1] so they are forced to face them with a holistic vision [2].

Covid-19, like a black swan, represented, for many companies understood as systems, a rupture, a complex situation capable of upsetting their viability. Decision-makers always have to look for solutions, also thanks to new technologies, to acquire information and recommendations to improve the outcome of their decision and be antifragile.

Antifragility [3] can well be allegorically represented by the Japanese art of Kintsugi, which consists of the use of gold or silver to repair objects in ceramic because from imperfection, from a wound, from a break, an even greater form of aesthetic perfection can be born.

Covid-19 had thus forced companies to accelerate the digitalization process [4] that has opened up new scenarios in terms of new potential markets to reach, new purchasing experiences for consumers [5], new relationships with the context [6], new service ecosystems [7] and more viable [8], and new organizational patterns, increasingly lean, for increasingly adaptive business models [9]. The digitalization path can improve and augment the decision-making process in a complex situation (such as Covid-19), by stimulating a change in their information variety [10] According to the vSa, complexity can be scaled through knowledge and there is in fact a close connection between decision and knowledge and the knowledge, in a given instant, of a given decision maker or viable system, is characterized by the variety of information composed of information units, interpretative schemes and value categories of the decision maker or viable system [2].

In the literature, the role of technologies in allowing companies to be antifragile has not been sufficiently studied; for this reason, this conceptual paper will try to answer the following research question:

R.Q.: is it possible to consider the augmented intelligence as a driver to make companies antifragile?

After the description of the theoretical background (Sect. 2), we will investigate the role of digitization, based on artificial intelligence (AI) solutions, in making companies antifragile (Sect. 3) by investigating its impact in supporting companies to survive despite Covid-19 and in promoting new business models.

The remote smart working model (AI-based) will be proposed, as an evolution of the smart working practice (Sect. 3.1).

The work ends with non-conclusive considerations (Sect. 4).

2 Theoretical Background

2.1 *Antifragility*

In literature, a distinction between antifragility and robustness has been made [11] and between them and resilience [12]. Robust systems are unable to react to abnormal events and complex situations tend to undermine their equilibrium [11]. Resilience implies the ability of the system to recover equilibrium conditions following the problematic event (which do not necessarily have to be identical to the initial ones) [12]. For antifragile systems, anomalous events, such as "black swans", do not create interruptions but make systems stronger and more creative, more capable of accepting new challenges [3].

Not only that, antifragile systems can take advantage of them by producing better results than the reference level of performance over time, thanks to a series of changes made following a certain level of variation, uncertainty and risk; in fact, antifragile systems grow and thrive if exposed to volatility, randomness, disorder and uncertainty [3].

Antifragility is an attribute of the system that determines the ability of the system to transform limits into opportunities, facing the uncertain, facing it as a possibility of rebirth and improvement [13] rather than as an imposition of resistance.

The system regenerates itself following the anomalous event thanks to its ability to activate learning processes [14] and to change [15].

2.2 *Managing Ruptures by Searching for Viable Conditions: The Viable System Approach Perspective*

The Viable System Approach (vSa) [16] investigates the conditions that allow a system to pursue the ultimate goal of survival, focusing attention on the identification and qualification of relevant actors, who influence decisions in complex contexts [17]. It suggests that the adoption of a systemic vision of phenomenal reality helps to highlight some relevant aspects of complexity, which allow organizations to be governed and managed with greater awareness [2].

The complexity depends on the perception of each observer and is based on his/her interpretative schemes and the need to solve or reduce complexity pushes organizations to develop ever-new interpretative models [18].

A key role is played by the knowledge that implies specific interpretative models, which adapt to changing context conditions and may be useful to manage difficult situations [19].

According to vSa, the evolution of knowledge takes place through the resolution of four problematic situations: chaos, complexity, complication and certainty [20].

The resolution of problem areas depends on the ability of the decision-maker to formulate and formalize new interpretative schemes [2].

Augmented decision-making is a decision-making process in which algorithmic insights are used in an accurate and discriminatory way thanks to the decision-maker's ability to accurately discern both when and when not to integrate the algorithm's judgment into their decision-making process and to supervise the algorithm. to intervene and provide input where necessary [21].

The quality of decisions affects the overall performance of the company [22] and is affected by the quality and quantity of information used. For this reason, it is interesting to evaluate how the decision-maker acquires information and recommendations to improve the outcome of the decision [23].

The vSa explores the role played by AI in the enhancement of decision-making processes. Augmented intelligence supports people in deciding by changing their information variety [24] and can influence the decision-maker potential to make a wise decision in conditions of complexity [25].

Augmented intelligence is a different concept compared to artificial intelligence: the latter is a tool, while the former is the result of an interaction between humans and tools/machines, given by the change in information variety that occurred in humans by using machines or algorithms (based on AI systems) [24].

According to the vSa, augmented intelligence occurs as a result of a variation in the variety of information, in terms of an increase in information units capable of increasing the potential of interpretative schemes for an enhancement of humans' intelligence and their interpretative capacity [10].

3 Digital Technologies as Kintsugi Art Gold in the Post Covid-19 Era

Covid-19 can be considered a digital transition accelerator [26].

Digital technologies have supported companies operating in different sectors that do not stop their activities [27, 28].

Although Covid-19 has dramatically tested the capabilities of health systems globally, forcing them to cope with an unprecedented shortage of resources, other sectors have also been strongly impacted to the point of having to think about a re-design of business models.

Digital technologies must be understood as drivers of a long-term sustainable survival capacity [29]. Bai et al. [30], for example, believe that managers of micro and small businesses need to review their business strategies, incorporating crisis scenarios and business continuity plans based on digital transformation paths.

Digital technologies can enable the augmented intelligence that derives from a collaborative and effective human–machine interaction and integration that enable systems to evolve into wiser configurations, based on rational and emotional components [31], useful in managing complex situations.

3.1 AI to Foster a Remote Smart Working Model

Smart Working is a form of working characterized by the absence of time or spatial constraints and the only burden is ensuring the agreed results [32].

Remote Working allows the worker to independently choose the place from which to provide work services, in compliance with the regulations in force in the office [33].

AI brings in many fields new perspectives [34, 35] and requires the revision of governance methods applied up to now [36], with the potential to improve people's well-being, opening new business scenarios for companies [37].

A business model based on a remote smart working model could be possible thanks to AI tools and would generate a series of benefits for both companies and employees. It could allow companies not only more efficient resource management, and an endless orientation towards results, but also new synergies in new contexts thanks to new and increased networks, for new collaborations and new forms of interactions. As well as, more profitable relationships with employees, based on a strong relationship of trust and on better opportunities for work-life balance, are possible [38].

Illustration case: South working. South working is a social promotion project that stimulates agile work from Southern Italy and marginalized areas. At the base of South working there is the idea of 'giving back' something: the return of skills to the territories where one has chosen to live and work. Remote work is a useful tool to reduce the economic, social and territorial gap in the country, and is able to improve the quality of life of workers, companies and territories. This project is based on Confidence Systems platform (AI-based system). This latter, thanks to AI, allows companies to monitor and arrange their production processes, making sure that they are carried out correctly. The startup Edgemony deals with the training of Italian and international companies in the field of digital innovation by promoting the south working model providing them with all the necessary tools to work remotely. Confidence Systems has allowed the return of people thanks to remote, smart, working paths.

4 Discussion and Conclusion

This example shows how new technologies, based on AI systems, have allowed company decision-makers and others, such as territorial and public actors [39], to face the criticalities and challenges posed by Covid-19, enabling wiser decision-making processes thanks to augmented intelligence [10] and an antifragile behaviour.

Thanks to the review of the business model, now based on a remote smart working model, the company has the opportunity to review not only its physical structure, but also its extended structure, and its defined organizational scheme, from which a new viable system [17] may emerge, also thanks to the identification of new potential supra-systems, potential resources providers, in an A4A perspective [40].

Thanks to this study, we can answer affirmatively to the **R.Q.**: by using this perspective, is possible to consider augmented intelligence (spread by AI use) as an antifragility driver.

The main implication of this work is in the link between vSa and antifragility literature. It fosters companies to adopt a new culture of "doing business" that goes beyond traditional structures, with physical and national boundaries, and superstructures, and to always seek new viability conditions.

In further research, it may be useful to highlight how viability can be affected by antifragility.

References

1. Gummesson, E., Mele, C., Polese, F.: Complexity and viability in service ecosystems. Mark. Theory **19**(1), 3–7 (2019)
2. Barile, S.: Management Sistemico Vitale, vol. 1. Giappichelli (2009)
3. Taleb, N.N.: Antifragile: How to Live in a World We Don't Understand, vol. 3. Allen Lane, London (2012)
4. Troisi, O., Fenza, G., Grimaldi, M., Loia, F.: Covid-19 sentiments in smart cities: the role of technology anxiety before and during the pandemic. Comput. Hum. Behav. **126**, 106986 (2022)
5. Formisano, V., Cavacece, Y., Fedele, M., Tartaglione, A. M., Douglas, A.: Service innovation for customer engagement in the Italian banking sector: a case study. In: Predicting Trends and Building Strategies for Consumer Engagement in Retail Environments, pp. 62–87. IGI Global (2019)
6. Barile, S., Ciasullo, M.V., Troisi, O., Sarno, D.: The role of technology and institutions in tourism service ecosystems: findings from a case study. TQM J. **29**(6), 811–833 (2017)
7. Ciasullo, M.V., Troisi, O., Grimaldi, M., Leone, D.: Multi-level governance for sustainable innovation in smart communities: an ecosystems approach. Int. Entrepren. Manag. J. **16**(4), 1167–1195 (2020)
8. Ciasullo, M. V., Polese, F., Montera, R., Carrubbo, L.: A digital servitization framework for viable manufacturing companies. J. Busin. Indust. Market. (2021)
9. Gigauri, I.: Effects of Covid-19 on human resource management from the perspective of digitalization and work-life-balance. Int. J. Innov. Technol. Econ. **4**(31) (2020)
10. Barile, S., Bassano, C., Piciocchi, P., Saviano, M., Spohrer, J.C.: Empowering value co-creation in the digital age. J. Busin. Indust. Market. (2021)
11. de Bruijn, H., Größler, A., Videira, N.: Antifragility as a design criterion for modelling dynamic systems. Syst. Res. Behav. Sci. **37**(1), 23–37 (2020)
12. Munoz, A., Billsberry, J., Ambrosini, V.: Resilience, robustness, and antifragility: towards an appreciation of distinct organizational responses to adversity. Int. J. Manag. Rev. (2022)
13. Ramezani, J., Camarinha-Matos, L.M.: Approaches for resilience and antifragility in collaborative business ecosystems. Technol. Forecast. Soc. Chang. **151**, 119846 (2020)
14. Größler, A.: A managerial operationalization of antifragility and its consequences in supply chains. Syst. Res. Behav. Sci. **37**(6), 896–905 (2020)
15. Derbyshire, J., Wright, G.: Preparing for the future: development of an 'antifragile' methodology that complements scenario planning by omitting causation. Technol. Forecast. Soc. Chang. **82**, 215–225 (2014)
16. Golinelli, G.M.: L'approccio Sistemico Vitale (ASV) al governo dell'impresa. Cedam (2011)
17. Golinelli, G.M.: L'approccio sistemico (ASV) al governo dell'impresa. Wolters Kluwer (2017)

18. Barile, S., Saviano, M., Polese, F., Di Nauta, P.: Reflections on service systems boundaries: a viable systems perspective: the case of the London Borough of Sutton. Eur. Manag. J. **30**(5), 451–465 (2012)
19. Saviano, M., Caputo, F.: Managerial choices between systems, knowledge and viability. In: Contributions to Theoretical and Practical Advances in Management. A Viable Systems Approach (VSA), vol. 2, pp. 219–242 (2013)
20. Barile, S., Saviano, M.: A new perspective of systems complexity in service science. Impresa, Ambiente, Manag. **4**(3), 375–414 (2010)
21. Burton, J.W., Stein, M.K., Jensen, T.B.: A systematic review of algorithm aversion in augmented decision making. J. Behav. Decis. Mak. **33**(2), 220–239 (2020)
22. Abubakar, A.M., Elrehail, H., Alatailat, M.A., Elçi, A.: Knowledge management, decision-making style and organizational performance. J. Innov. Knowl. **4**(2), 104–114 (2019)
23. Keding, C., Meissner, P.: Managerial overreliance on AI-augmented decision-making processes: how the use of AI-based advisory systems shapes choice behavior in R&D investment decisions. Technol. Forecast. Soc. Chang. **171**, 120970 (2021)
24. Barile, S., Bassano, C., Lettieri, M., Piciocchi, P., Saviano, M.: Intelligence augmentation (IA) in complex decision making: a new view of the VSA concept of relevance. In: International Conference on Applied Human Factors and Ergonomics, pp. 251–258. Springer, Cham (2020)
25. Barile, S., Piciocchi, P., Bassano, C., Spohrer, J., Pietronudo, M.C.: Re-defining the role of artificial intelligence (AI) in wiser service systems. In: International Conference on Applied Human Factors and Ergonomics, pp. 159–170. Springer, Cham (2018)
26. Agostino, D., Arnaboldi, M., Lema, M.D.: New development: COVID-19 as an accelerator of digital transformation in public service delivery. Public Money Manag. **41**(1), 69–72 (2021)
27. Chandra, M., Kumar, K., Thakur, P., Chattopadhyaya, S., Alam, F., Kumar, S.: Digital technologies, healthcare and Covid-19: insights from developing and emerging nations. Health Technol. 1–22 (2022)
28. Quayson, M., Bai, C., Osei, V.: Digital inclusion for resilient post- COVID-19 supply chains: smallholder farmer perspectives. IEEE Eng. Manage. Rev. **48**(3), 104–110 (2020)
29. Kashef, M., Visvizi, A., Troisi, O.: Smart city as a smart service system: Human-computer interaction and smart city surveillance systems. Comput. Human Behav. 106923 (2021)
30. Bai, C., Quayson, M., Sarkis, J.: COVID-19 pandemic digitization lessons for sustainable development of micro-and small-enterprises. Sustain. Product. Consum. **27**, 1989–2001 (2021)
31. Badr, N.G., Carrubbo, L., Ruberto, M.: Responding to COVID-19: potential hospital-at-home solutions to re-configure the healthcare service ecosystem. EALTHINF 344–351 (2021)
32. Ales, E., Curzi, Y., Fabbri, T., Rymkevich, O., Senatori, I., Solinas, G.: Working in digital and smart organizations. In: Legal, Economic and Organizational Perspectives on the Digitalization of Labour Relations (2018)
33. Miller, T.: AI and Remote Working: A Paradigm Shift in Employment. Business Expert Press (2021)
34. Visvizi, A., Bodziany, M.: Artificial intelligence and its context: an introduction. In: Artificial Intelligence and Its Contexts, pp. 1–9. Springer, Cham (2021)
35. Carrubbo, L., Polese, F., Drăgoicea, M., Walletzký, L., Megaro, A.: Value co-creation 'gradients': enabling human-machine interactions through AI-based DSS. In: ITM Web of Conferences, p. 01002. EDP Sciences (2022)
36. Lytras, M.D., Visvizi, A.: Artificial intelligence and cognitive computing: methods, technologies, systems, applications and policy making. Sustainability **13**(7), 3598 (2021)
37. Nickson, D., Siddons, S.: Remote Working. Routledge (2012)
38. Sullivan, C.: Remote working and work-life balance. In: Work and Quality of Life, pp. 275–290. Springer, Dordrecht (2012)
39. Polese, F., Troisi, O., Carrubbo, L., Grimaldi, M.: An integrated framework toward public system governance: insights from viable systems approach. In: Cross-Sectoral Relations in the Delivery of Public Services. Emerald Publishing Limited (2018)
40. Polese, F., Grimaldi, M., Sarno, D., Troisi, O. From B2B to A4A: an integrated framework for viable value co-creation. In: From B2B to A4A: an Integrated Framework for Viable Value Co-Creation, pp. 135–161 (2018)

Managing Risks and Risk Assessment in Ergonomics—A Case Study

Miroslav Bednář, Michal Šimon, Filip Rybnikár, Ilona Kačerová, Jana Kleinová, and Pavel Vránek

Abstract The paper deals with the evaluation of the ergonomic risks of an injection moulding machine which is used to foam components for car interiors. RULA and NIOSH analyses and methods for assessing working positions according to current legislation were performed as part of the risk assessment. The muscle load on the forearm was also measured while working with the upper limbs in compliance with legislative regulations. Integrated electromyography was used to measure local muscle loading. The Tecnomatix Jack software program was used for the analyses. After identifying the problem areas, corrective measures were then proposed and visualised. After the rationalisation, the workplace was re-measured and evaluated. It was found that the proposed corrective measures would enable the company to improve working conditions for their employees, which would, among other things, enable the company to streamline production, eliminate worker sickness and reduce the potential for occupational illness and injury. The case studies were carried out in the Czech Republic.

Keywords Ergonomics · RULA · Musculoskeletal disease

M. Bednář (✉) · M. Šimon · F. Rybnikár · I. Kačerová · J. Kleinová · P. Vránek
Faculty of Engineering, University of West Bohemia, Pilsen, Czech Republic
e-mail: bednarm@kpv.zcu.cz

M. Šimon
e-mail: simon@kpv.zcu.cz

F. Rybnikár
e-mail: rybnikar@kpv.zcu.cz

I. Kačerová
e-mail: ikacerov@kpv.zcu.cz

J. Kleinová
e-mail: kleinova@kpv.zcu.cz

P. Vránek
e-mail: vranek@kpv.zcu.cz

A. Visvizi et al. (eds.), *Research and Innovation Forum 2022*,
Springer Proceedings in Complexity, https://doi.org/10.1007/978-3-031-19560-0_59

1 Introduction

In recent years, a great deal of emphasis has been placed on optimising the working environment. Developments in science and technology have opened up new methods and possibilities to facilitate human work, which is still an integral part of production despite advances in robotics. In this respect, industrial companies are focusing primarily on the organisation of workplaces and the elimination of potential risks that subsequently lead to accidents or occupational diseases. Unsuitable working conditions and the associated inefficient production levels can put many industrial enterprises at a competitive disadvantage. One way of optimising the working environment is to use the principles of ergonomics [1]. The main aim of this work is to analyse workplaces using ergonomic methods to identify problematic areas and subsequently to rationalise workplaces in an industrial enterprise. In addressing this analysis, several questions arose that provide further avenues for research. For example, it is a question of determining the appropriate methodology to determine the correct workplace ergonomics analysis based on the work process. Modern ergonomic risk assessment methods are a tool for postural analysis and allow early identification and comprehensive assessment of the risk of musculoskeletal damage. These modern methods include RULA (Rapid Upper Limb Assessment) [2] and REBA (Rapid Entire Body Assessment) [3]. These methods may become significant components in the prevention of musculoskeletal disorders; they are quick, simple and inexpensive, so it is beneficial and desirable that they are used as standard in ergonomic risk assessment at work. The main reason why companies should be concerned about the health of their employees in the first place is the development of musculoskeletal disorders and the potential for occupational disease [4]. Musculoskeletal disorders (MSDs) are the most common work-related health problems in Europe and accounted for more than 50% of all reported work-related health problems in 15 European Union (EU) countries in 2020. MSDs are among the most common problems related to work activities. They affect millions of workers across Europe and cost employers billions of euros [5]. MSDs are a group of diseases affecting muscles, joints, tendons, ligaments, nerves, bones and the circulatory system. They are a serious problem with widespread societal, health, social and economic consequences [6]. Their incidence is steadily increasing. MSDs are the most common cause of occupational diseases in EU countries [7]. Workers may have health problems that range from minor aches and pains to more serious health problems that cause absence from work and the worker to seek professional treatment [8]. There are also chronic cases that can lead to disability and the need to leave employment [9]. A significant problem in industry is the repetitive nature of movements. Research shows that almost two-thirds of European Union workers say they perform work that requires repetitive upper limb movements. In addition, a quarter of workers say they are exposed to vibrations from work tools [5]. These are the two most important risk factors for work-related neck and upper limb disorders (WRULDs). These disorders affect many workers in a variety of occupations [10]. At the same time, there is emerging evidence that these disorders are also related to psychosocial risk factors, such as high job demands, low levels of autonomy, and low job satisfaction. Musculoskeletal disorders [1, 11]. The prevalence of occupational diseases, which includes the risk of occupational disease, is an important indicator of the health status of a population and working conditions in a country. In order to address the issue of MSDs, according to the latest research, an integrated, multidisciplinary management system is needed, including a comprehensive risk prevention system, the application of appropriate surveillance-based methods and the introduction of good occupational health and safety practices. Implementation of ergonomic principles, i.e. identification of ergonomic risks, their analysis, design and implementation of solutions and subsequent evaluation of the effectiveness of measures, can significantly reduce the number of MSDs [9]. The standard methods used, e.g. integrated electromyography, strain gauging, biomechanical assessment models, questionnaires or observational methods, cannot comprehensively assess all the ergonomic risks and some of them are based on subjective load assessment. Therefore, the introduction of modern methods of ergonomic risk assessment that simplify and facilitate the risk assessment of MSDs and allow for a comprehensive assessment of the different risk factors at work is essential [12]. In RULA and REBA, the positions of individual body parts (arms, forearms, wrists, neck, trunk and lower limbs) are scored with respect to deviation from a neutral position. The so-called 'basic positions' are described for each body part to obtain a basic score. These are different ranges of flexions and extensions that are scored in ascending order as the deviation from neutral increases. There are also descriptions of positions to obtain additional points for the so-called 'variable score' (e.g. rotations and lunges). The weight of the manipulated load (load-force score) and the effect of the static position at work (scores used for muscles, activity score) are also included in the final score [13]. The risk assessment methods are categorical, semi-quantitative and they tell us whether ergonomic risk is present at work and whether preventive action is needed. These methods for the assessment of ergonomic risks at work are not yet sufficiently used in industrial enterprises or by hygiene stations. However, it would be beneficial and desirable to put them into practice, as they are an important component in the prevention of MSDs, and they are quick, simple and inexpensive (only pencil and paper are needed) [14].

2 Methodology

In order to choose an appropriate method of ergonomic workplace measurement, we need to determine what we want to measure. There are several methods that can be used to measure workplace ergonomics. These methods can measure the working position of the whole body or parts of the body. Furthermore, these methods can be used to measure load lifting or material handling. As mentioned earlier, the RULA method was chosen for ergonomic measurement. This method was chosen based on expert estimation, thanks to several previous ergonomic measurements that have been carried out. These include studies such as: Influence of the Upper Limb Position on the Forearm EMG Activity—Preliminary Results, or Ergonomic Design of a Workplace Using Virtual Reality and a Motion Capture Suit [15, 16]. The biggest problem in the workplace was the working position of the upper body. Due to this fact, it was determined that the RULA method would be the most appropriate method for analyzing workplace ergonomics. To evaluate the analysis, it is most appropriate to use one of the software that calculates the total load score of certain parts based on the set positions. Because of previous experience, Tecnomatix Jack software was chosen, which is used by ergonomics experts.

The RULA (Rapid Upper Limb Assessment) method was designed in 1993 in the UK as an ergonomic tool for rapid assessment of workplace risks, the main purpose of which is to determine the risk of loading on the upper limbs and neck. This method was designed as a rapid screening-based tool. In this method, the biomechanical and postural loading of the upper body is assessed [17].

The RULA analysis was developed with the following objectives:

- to assess the muscular strain occurring in jobs characterised by repetitive movements that can lead to excessive muscle strain,
- using a simple scoring method to identify the urgency of changes,
- provide an assessment tool that is user-friendly and requires minimal time and effort.

Benefits of RULA:

- measuring musculoskeletal risk,
- quick and easy to use,
- increasing work efficiency,
- a good information tool for management and workers in high-risk positions.

Disadvantages of RULA:

- does not consider organizational and psychological factors,
- challenging to capture movements using photographs,
- limited to access at one moment. Evaluator must choose the most extreme posture [2].

The basis of this method is observation of the worker in the workplace performing normal work tasks over several work cycles. Next, the assessor selects the postures that are risky, the most frequently performed and the postures that exert the highest force loads. Prior to the subsequent scoring, the evaluator shall determine whether the right or left side will be evaluated. These positions are then appropriately scored.

The positions of the arms, forearms, wrists, neck, trunk and lower limbs are included in the overall assessment. Each of these body parts is given a basic position, which is scored with a basic score. In addition, additional points can be earned which are included in the so-called variable score. The scores are also entered into individual tables. Table A presents the scores for the upper limb positions—wrist, arm and forearm. Table B presents the scores for the neck, trunk and leg positions. The resulting Table C contains the C and D scores [13].

Table 1 RULA score

Category	RULA score	Risk level	Action
1	1–2	None	Not necessary
2	3–4	Low	More investigation needed
3	5–6	Medium	More investigation needed and some changes needed
4	6 +	High	Changes needed immediately

Score C = Table A score + Muscle score + Strength score—Stress score

Score D = Table B score + Muscle score + Strength score—Weight score.

The final output for the evaluation of this method is an overall rating divided into four categories ranging from 1–7. The categories are further specified below (Table 1).

The Tecnomatix Jack program from Siemens serves as the main tool for the practical part of the work. It is a software program invented in the 80-90 s of the twentieth century by NASA. This software serves as a tool for placing a human model in a virtual environment. Data from a population survey database is used to create an accurate biomechanical model of a human. It is also possible to assign precise tasks to the model and subsequently monitor its performance.

A variety of graphics from Computer Aided Design (CAD) software can be inserted into the software to create a copy of the real environment of the workplace, as well as the entire production system. The Tecnomatix Jack software serves, among other things, as a simulation tool in which it is possible to work in real time. The human biomechanical model in the software consists of 69 segments and 68 joints.

Most of the segments are located in the upper body, specifically in the spine (17 segments) and in the arms (16 segments). For this reason, this program is intended to be a suitable software support for the RULA analysis, which, as mentioned in the RULA chapter, deals with the analysis of the load on the upper limbs. The Tecnomatix Jack program allows the evaluation of virtual person performance in many ways, for which additional modules must be used, such as the Occupant Packaging Toolkit or the Task Analysis Toolkit. These add-on modules allow the evaluator to work with a wide variety of analyses, including the NIOSH analysis, REBA, etc. [18].

In this study, the workplace and the worker's work activities needed to be analyzed first. Next, it was necessary to determine by expert estimation the tasks to be analysed and to select an appropriate method of ergonomic measurement. After the tasks to be analysed were selected and the appropriate method was chosen, the entire workplace had to be modelled and imported into a suitable analysis application, in our case Tecnomatix Jack. The next step was the actual analysis using the selected program. The penultimate step after the actual analysis was to create corrective measures that would improve the ergonomics of the workplace. Finally, these corrective actions were implemented into Tecnomatix Jack and an analysis was performed with the corrective actions (Fig. 1).

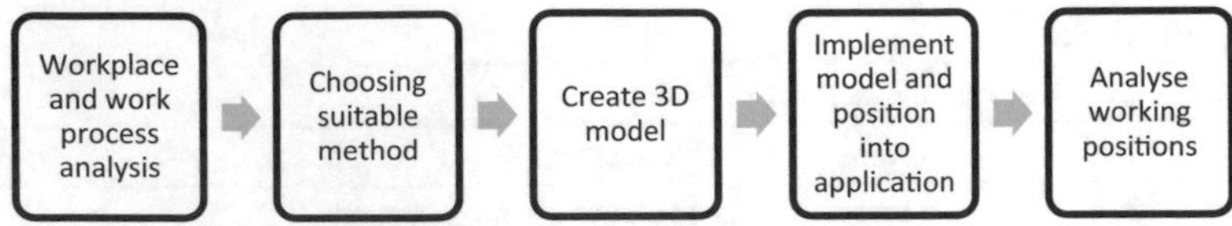

Fig. 1 Process of analyzing

3 Workplace Analysis

The observation and data collection were carried out in an automotive company in Czech Republic, whose main focus is the production of interior components for cars. This company was chosen because of huge cooperation with the University of West Bohemia. The worker operates two workstations. They move freely between the workstations and all the work is performed standing up. The standard production per shift is a total of 200 pieces per worker. The workplace consists of a complex pressing machine, which is equipped with yellow safety cages with barriers. Among other things, there are two metal racks with pre-pressed products which are further processed on these presses. There is also an additional rack for the finished foamed mouldings. A spray gun is used to coat the mould with non-stick lubricant, the blank is inserted, the machine is started and then the blank is removed and placed on a rack (Figs. 2 and 3).

To allow subsequent working in Tecnomatix Jack, it was first necessary to model the workstation with the individual objects so that the 3D model was as realistic as possible. Ergonomic risk assessments were performed for men at the 5th and 95th percentile and in 4 basic work positions.

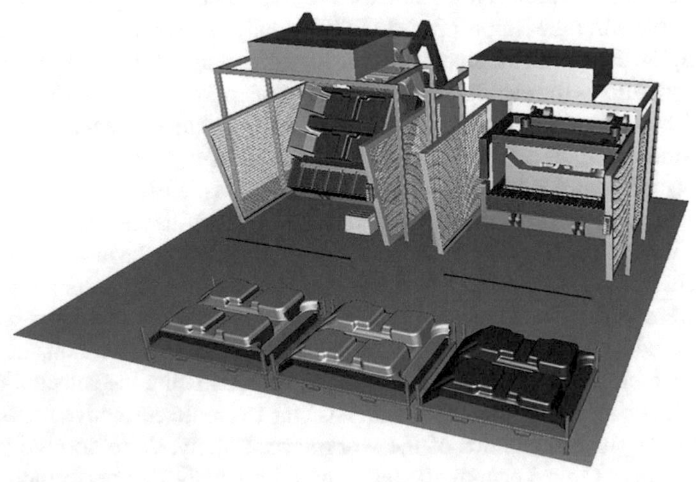

Fig. 2 Workplace

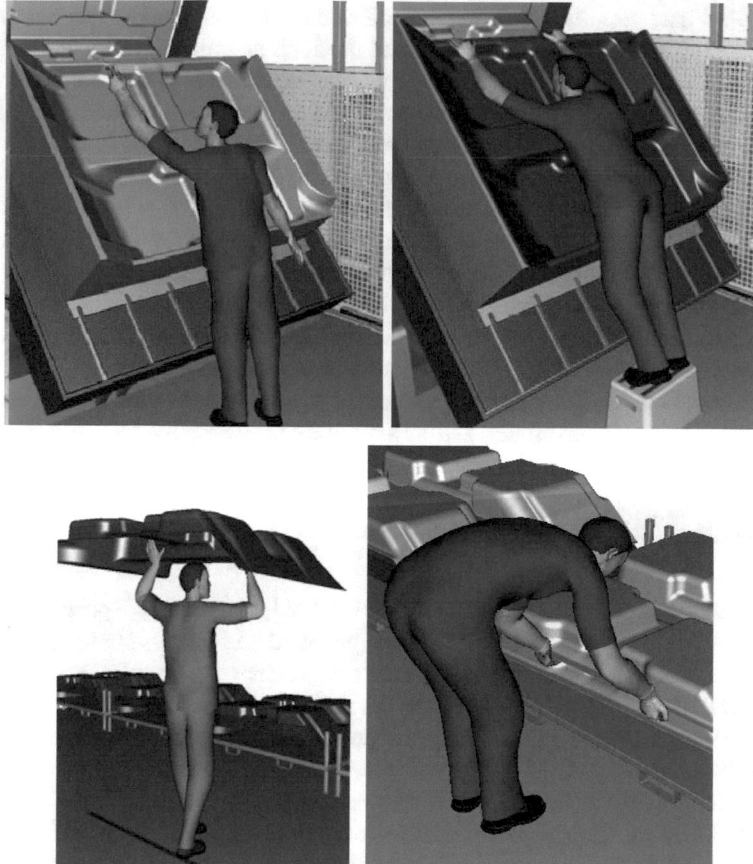

Fig. 3 Work positions

- **Application of degreasing compound**—a position was selected from the process of applying the degreasing compound to the steel mould using a gun, in which the worker holds the application gun in one hand. The other hand is held loosely against the body throughout the process. The worker has the choice of either standing on the ground or on a box. The whole process of applying the mixture takes between 35–40 s and is repeated a total of 200 times per shift. In total, the worker spends approximately 116–133 min per shift on this movement.
- **Removal of the filled part**—the worker removes a foamed black carpet from a steel mould. During observation, it was seen that this operation is always done standing on the box. The total time of the removal process was calculated to be in the range of 7–10 s. The worker was standing on the box and leaning over the mould during the whole process. The worker first removes the upper part of the carpet and after removing the lower part, the whole product is removed. Standing on the unsecured box was found to be a safety risk.

- **Product transfer**—this position was chosen for RULA from the whole process of gripping and transferring the finished product. At this point, the worker holds the product in an overhead grip. The worker repeats this process a total of 200 times per shift, at regular intervals of approximately 2.25 min. The total time from gripping to placing the product on the pallet is between 6–10 s. The worker therefore spends approximately 33 min in this position, out of a total time of 450 min. The total weight of the carpet after foaming is 4–6 kg, depending on the type of carpet. In this position, both shoulders and elbows are at an angle of approximately 45°. This position was found to be ergonomically challenging due to the regular and repetitive strain on the upper limbs and back.
- **Handling the preformed part**—the last position on which RULA was used is the movement in which the worker bends for the penultimate part that is placed on the frame. This position was chosen because of the considerable forward bending, as the frame rail is positioned at approximately knee height.

4 Results

The table below presents an evaluation of RULA for the four selected positions, which in all cases were performed for both selected percentiles of the population. According to the overall RULA score, the worst result was achieved for position 2, which is the removal of the finished product from the steel mould. According to observations and interviews in the plant, it can be confidently confirmed and concluded that this is the most physically demanding position and also very risky from a safety point of view. The results for the 5th percentile were worse for positions 1 and 3 than for the 95th percentile. The most acceptable results of the analysis were obtained for Position 1, where the 95th percentile worker achieved a RULA score of 3, which does not represent a high level of risk. However, there is potential for improvement and therefore actions will be proposed for this position. The results for position 4 are concerning as the overall RULA score for both percentiles was 7 (Table 2).

5 Corrective Actions

Based on the results of the analyses, it is evident that the current state of the workplace is not satisfactory in terms of ergonomics and there are deficiencies. Workers are physically overworked in some positions. In addition, the use of the box to stand on, which is unstable and poses a risk of work-related injury, appears to be a significant safety risk. The aim of this section is therefore to propose appropriate measures to eliminate the identified deficiencies or at least to mitigate the negative effects on the worker. The first remedy is a pedestal that would be placed under the frame. This measure was designed to alleviate the burden on the worker who bends a total of 200 times during a shift for a part that is destined for further processing. The pedestal would be a hydraulic jack on which the frame with the material would be placed. The hydraulic jack would react to the change in weight and raise/lower as the number of pieces increase/decrease. This would mean that the position for the worker does not change when either removing the part or, conversely, putting down the finished product. The hydraulic jack was designed to be 150 cm long and 90 cm high, which

Table 2 RULA results

Position number	Object	5th percentile	95th percentile
1	Upper arm	4	4
	Lower arm	2	3
	Wrist	2	1
	Twisted wrist	1	1
	Neck	5	2
	Trunk	3	1
	Overall RULA	6	3
2	Upper arm	5	5
	Lower arm	2	3
	Wrist	3	2
	Twisted wrist	1	2
	Neck	5	5
	Trunk	3	4
	Overall RULA	7	7
Position number	Object	5th percentile	95th percentile
3	Upper arm	5	4
	Lower arm	3	3
	Wrist	2	3
	Twisted wrist	2	1
	Neck	5	2
	Trunk	4	1
	Overall RULA	7	6

(continued)

Table 2 (continued)

Position number	Object	5th percentile	95th percentile
4	Upper arm	4	5
	Lower arm	3	3
	Wrist	2	2
	Twisted wrist	2	2
	Neck	4	4
	Trunk	4	4
	Overall RULA	7	7

allows the frame to be handled with a pallet truck. The overall height of the hydraulic jack, including the frame, is designed so that the top is always at least 105 cm high (Fig. 4).

The second proposed measure is an ergonomic anti-slip mat. This would be placed in front of the pressing machine. This mat would prevent unwanted movement of the box. The dimensions of the mat were designed for the size of the workstation, specifically 2 m long and 1 m wide. The main advantage of this mat is its ergonomic design which enhances the health, safety at work and productivity of workers, as well as reducing fatigue and pressure on the lower limbs, including the hips and spine (Fig. 5).

Its anti-slip property would reduce the risks of workplace accidents. Another rationalization measure was the extension of the pistol. This activity occupies a significant part of the working time, as the spraying before inserting each part takes between 35–40 s. Overall, this activity accounts for approximately one quarter of the working shift. Therefore, an extension was designed and modelled to extend

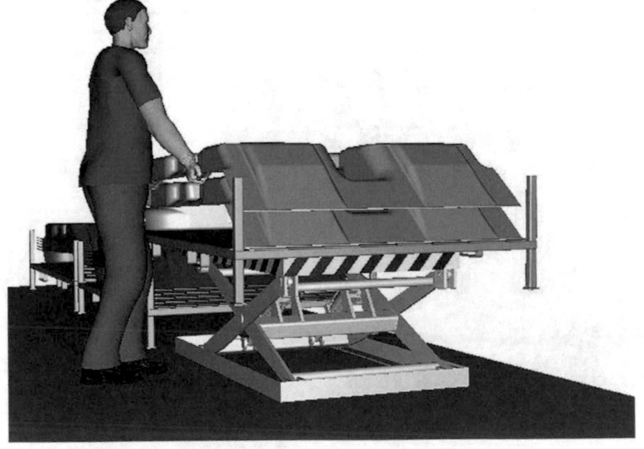

Fig. 4 Corrective action—hydraulic jack

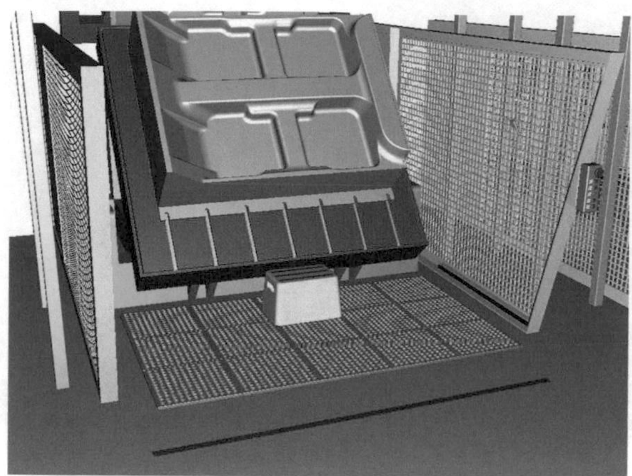

Fig. 5 Corrective action—anti-slip mat

the spray gun handle by 25 cm. The extension would allow the worker to reach all the necessary places without subjecting the upper limb to unacceptable angles. The figure shows a simulation of worker movement for the 95th percentile of the population when spraying with the gun before the measure (left) and with the gun after implementation of the measure (right) (Fig. 6).

One additional measure that seems appropriate is to rearrange the placement of the pallets. The current workstation is arranged so that there are three pallets. Two for storing material ready for foaming and one for storing finished parts. However, for the second machine (on the left) the distance to the frame is approximately double. The proposal is therefore to arrange the frames in such a way that the distances from both machines to the pallets for the finished parts are identical. In the design, the storage rack for the finished parts is in the middle and the other two pallets opposite each

Fig. 6 Corrective action—spray gun extension

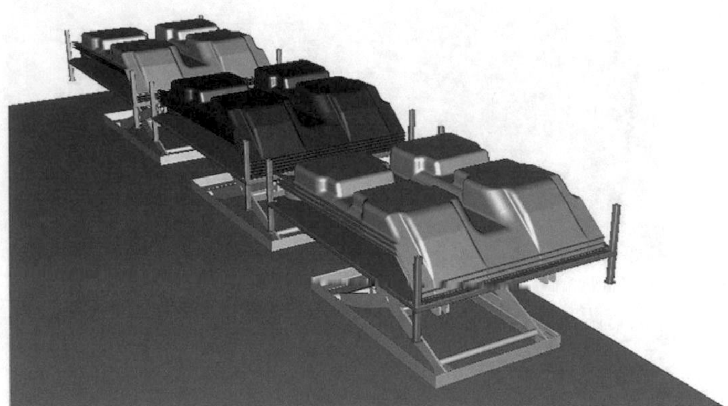

Fig. 7 Corrective action—rearranged storage

machine contain the remaining two frames of pre-prepared material. The advantages of this measure include a reduction in the time it takes a worker to carry a finished part, which is approximately 2 kg heavier after foaming, and the associated reduction in physical strain.

Corrective measures reduce the potential for workplace injury, worker absenteeism or occupational illness, while improving workplace safety and efficiency. These facts increase the turnover rate of production thereby increasing the prosperity of the company (Fig. 7).

6 Discussion

The RULA method is a suitable analysis for investigating workplace ergonomics when focusing on the upper body. With this method, it was possible to detect that the worker spends a large part of the working time in an unnatural position. Due to such poorly set up workplaces, it is possible that workers may develop some of the musculoskeletal disorders. Thanks to previous studies that focused on similar types of tasks but in different workplaces, it was possible to immediately identify which movements would be problematic during the observation. However, it is still up to the observer and their expert judgement to determine what positions and what actions will be investigated. After modifying the workplace and implementing corrective measures, the results of the RULA method were reduced by at least one grade, thus improving the ergonomics of the workplace and increasing the comfort of the worker, which will increase their efficiency and there is no increased likelihood of occupational disease. For future research, it would be useful to develop a methodology for identifying unnatural movements so that even an unknowledgeable consultant can determine what needs to be targeted. This research could focus on possible software support for determining the correct ergonomic analysis for a given work activity, as well as for determining the movement that needs to be analysed and, if necessary, corrective action developed. The biggest problem in developing this methodology will still be the height differences of the population. These differences cannot be easily implemented into various ergonomic methods, as a workplace set up for an average height person may not suit workers occurring in the 5th or 95th percentile.

7 Conclusion

Manufacturing businesses must pay certain fees if workspaces don't follow ergonomic guidelines and employees are overworked. Regular medical check-ups, breaks, employee rotation, and workplace rationalization are some of the measures that can be taken to make up for the money lost. Occupational illnesses already occurring in ergonomically unsuitable workplaces mean significantly higher costs in terms of various forms of compensation for workers' health problems. In order to eliminate these risks, it is essential that workplace conditions are regularly monitored and improved. Improvements lead to increasing comfort of workers in the workplace and, above all, to the elimination of the above-mentioned cost risks.

In the current era of technology development, companies are placing emphasis on improving working conditions and eliminating safety and ergonomic risks in the workplace by using ergonomic rationalisation. This can be used to increase worker safety in the workplace, provide suitable working conditions for the worker and reduce worker fatigue. These measures aim not only to improve ergonomic conditions in the workplace, but above all to reduce the risk of occupational diseases. A secondary aspect of ergonomic rationalisation is often to increase worker performance. To eliminate occupational diseases, it is advisable to focus on ergonomics in the design of the workplace itself. The creation of a 3D model of the workplace and the subsequent transfer of this model into virtual reality can serve this purpose. The ergonomic analysis is then carried out in the virtual world, where the coordinates of the measured body parts are transferred directly to the application using a Motion Capture suit worn by the worker, evaluating the non-compliant positions with respect to the chosen methodology [19].

The principles of ergonomic rationalisation of the workplace can be achieved by using ergonomic analysis and assessment methodologies such as RULA or REBA. These approaches consist of modelling the real world in digital form, identifying critical ergonomic sections and safety hazards and then proposing adjustments leading to their elimination or reduction. If these risks are not addressed, occupational injuries or illnesses can occur.

Funding This work was supported by the Internal Science Foundation of the University of West Bohemia under Grant SGS-2021–028 'Developmental and Training Tools for the Interaction of Man and the Cyber–Physical Production System'.

References

1. Nunes, I.: Ergonomunsic risk assessment methodologies for work-related musculoskeletal disorders: a patent overview. Recent Patents Biomed. Eng. **2**, 121–132 (2010). https://doi.org/10.2174/1874764710902020121
2. Gómez-Galán, M., Callejón-Ferre, Á.-J., Pérez-Alonso, J., Díaz-Pérez, M., Carrillo-Castrillo, J.-A.: Musculoskeletal risks: RULA bibliometric review. Int. J. Environ. Res. Public Health **17** (2020). https://doi.org/10.3390/ijerph17124354
3. Hita Gutiérrez, M., Gómez-Galán, M., Díaz-Pérez, M., Callejón-Ferre, Á.-J.: An overview of REBA method applications in the world. Int. J. Environ. Res. Public Health **17** (2020). https://doi.org/10.3390/ijerph17082635
4. Dul, J., Patrick Neumann, W.: Ergonomics contributions to company strategies. Appl. Ergon. **40**, 745–752 (2009). https://doi.org/10.1016/j.apergo.2008.07.001
5. Bevan, S.: Economic impact of musculoskeletal disorders (MSDs) on work in Europe. Best Pract. Res. Clin. Rheumatol. **29**, 356–373 (2015). https://doi.org/10.1016/j.berh.2015.08.002
6. Hales, T.R., Bernard, B.P.: Epidemiology of work-related musculoskeletal disorders. Orthoped. Clinics North Am. **27**, 679–709 (1996). https://doi.org/10.1016/S0030-5898(20)32117-9
7. Govaerts, R., Tassignon, B., Ghillebert, J., Serrien, B., De Bock, S., Ampe, T., El Makrini, I., Vanderborght, B., Meeusen, R., De Pauw, K.: Prevalence and incidence of work-related musculoskeletal disorders in secondary industries of 21st century Europe: a systematic review and meta-analysis. BMC Musculosk. Disord. **22**, 751 (2021). https://doi.org/10.1186/s12891-021-04615-9
8. Buckle, P.W., Jason Devereux, J.: The nature of work-related neck and upper limb musculoskeletal disorders. Appl. Ergon. **33**, 207–217. https://doi.org/10.1016/S0003-6870(02)00014-5
9. Crawford, J.O., Berkovic, D., Erwin, J., Copsey, S.M., Davis, A., Giagloglou, E., Yazdani, A., Hartvigsen, J., Graveling, R., Woolf, A.: Musculoskeletal health in the workplace. Best Pract. Res. Clin. Rheumatol. **34**, 101558 (2020). https://doi.org/10.1016/j.berh.2020.101558
10. Balogh, I., Arvidsson, I., Björk, J., Hansson, G.-Å., Ohlsson, K., Skerfving, S., Nordander, C.: Work-related neck and upper limb disorders - Quantitative exposure-response relationships adjusted for personal characteristics and psychosocial conditions. BMC Musculosk. Disord. **20** (2019). https://doi.org/10.1186/s12891-019-2491-6
11. Cho, C.-Y., Hwang, I.-S., Chen, C-C.: The association between psychological distress and musculoskeletal symptoms experienced by chinese high school students. J. Orthopaedic Sports Phys. Therapy **33**, 344–353 (2003). https://doi.org/10.2519/jospt.2003.33.6.344
12. Rothmore, P., Aylward, P., Karnon, J.: The implementation of ergonomics advice and the stage of change approach. Appl. Ergon. **51**, 370–376 (2015). https://doi.org/10.1016/j.apergo.2015.06.013
13. Kee, D.: Systematic comparison of OWAS, RULA, and REBA based on a literature review. Int. J. Environ. Res. Public Health **19** (2022). https://doi.org/10.3390/ijerph19010595
14. Hulshof, C.T.J., Colosio, C., Daams, J.G., Ivanov, I.D., Prakash, K.C., Kuijer, P.P.F.M., Leppink, N., et al.: WHO/ILO work-related burden of disease and injury: protocol for systematic reviews of exposure to occupational ergonomic risk factors and of the effect of exposure to occupational ergonomic risk factors on osteoarthritis of hip or knee and selected other. Environ. Int. **125**, 554–566 (2019). https://doi.org/10.1016/j.envint.2018.09.053
15. Kačerová, I., Bures, M., Kaba, M., Görner, T.: Influence of the upper limb position on the forearm EMG activity – preliminary results, pp. 34–43 (2020). https://doi.org/10.1007/978-3-030-20142-5_4
16. Kacerova, I., Kubr, J., Horejsi, P., Kleinova. J.: Ergonomic Design of a Workplace Using Virtual Reality and a Motion Capture Suit. Applied Ssciences-Basel 12. St Alban-Anlage 66, Ch-4052 Basel, Switzerland: MDPI (2022). https://doi.org/10.3390/app12042150
17. Kee, D., Comparison of OWAS, RULA and REBA for assessing potential work-related musculoskeletal disorders. Int. J. Indust. Ergon. **83**, 103140 (2021). https://doi.org/10.1016/j.ergon.2021.103140

18. Pekarcikova, M., Trebuna, P., Kronova, J., Izarikova, G.: The application of software tecnomatix Jack for design the ergonomics solutions. In: . Burduk, A., Chlebus, E., Nowakowski, T., Tubis, A. (eds.), Intelligent Systems in Production Engineering and Maintenance, vol. 835, pp. 325–336 (2019). Advances in Intelligent Systems and Computing. Gewerbestrasse 11, Cham, Ch-6330, Switzerland: Springer International Publishing AG. https://doi.org/10.1007/978-3-319-97490-3_32
19. Polcar, J., Gregor, M., Horejsi, P., Kopeček, P.: Methodology for designing virtual reality applications, pp. 768–774 (2016). https://doi.org/10.2507/26th.daaam.proceedings.107.

Senegalese Fashion Apparels Classification System Using Deep Learning

Adja Codou Seck, Kaleemunnisa, Krishna M. Bathula, and Christelle Scharff

Abstract There is a vested interest in applying Artificial Intelligence (AI) to fashion. However, studies have mainly focused on Western fashion. More inclusive automated fashion classification processes are required to categorize, organize, identify, advertise, recommend, and detect counterfeits for the global fashion industry. African fashion is a $31 billion industry, the second-largest economic sector of the continent after agriculture. This paper focused on a case study that relates to African culture, authenticity, living, and heritage. It presents a small Senegalese fashion dataset and a model capable of classifying various Senegalese apparels, called Boubous and Taille Mames by using transfer learning for image classification with MobileNetV2 as the base model. This paper raises the issue of the need of less western-centric datasets and proposes a preliminary reflection on addressing global AI-related fashion issues.

Keywords AI · CNN · Fashion · Image classification · Senegal · Transfer learning

1 Introduction

The use of AI in fashion is gaining interest in both research and industry, transforming e-commerce and in-store experiences, collection design processes, and fashion shows. Applications go from AI-generated design, identifying fashion trends and virtual try-ons, to recommendation systems, counterfeit detection, and apparel recognition and classification [1, 10, 11, 25, 28, 29]. They have mainly focused on

A. C. Seck
Department of Computer Science, Université Alioune Diop de Bambey, Bambey, Senegal
e-mail: adjacodou.seck@uadb.edu.sn

Kaleemunnisa · K. M. Bathula · C. Scharff (✉)
School of Computer Science and Information Systems, Pace University, New York, NY 10038, USA
e-mail: cscharff@pace.edu

K. M. Bathula
e-mail: kbathula@pace.edu

A. Visvizi et al. (eds.), *Research and Innovation Forum 2022*,
Springer Proceedings in Complexity, https://doi.org/10.1007/978-3-031-19560-0_60

western-centric fashion. This study takes a novel approach by focusing on African fashion, Senegalese fashion in particular.

African fashion is a growing industry, already estimated at $31 billion, the second-largest economic sector of the continent after agriculture, according to Euromonitor International. Senegal is a hotspot of African fashion with influencers reaching the global fashion scene, such as Adama Paris, creator of Dakar Fashion Week, Diarablu, part of Stitch Fix US, Lahad Gueye, champion of the Made In Senegal label, and Sophie Zinga, founder of Dakar Fashion Design Hub. In Senegal, fashion is crucial for prestige and status, and highly influenced by social identity, culture, and heritage [13]. Boubou is the classic Senegalese robe, worn by both men and women all over West Africa, Senegal in particular. It is composed of one single piece of fabric. Boubou (or Kaftan) has declinations with different fabrics (e.g., basin, damask, or wax) and with different sizes. Taille Mame is another traditional dress worn by women that highlights the waistline. There are many other types of apparels including Taille Basse, a fitted two-piece composed of a top and an ankle-length skirt. Recently, Senegalese traditional garments are undergoing transformations with adoption of diverse international inspirations and styles to produce relatively new and unique versions. Social media popularity and e-commerce growth in Africa amplify the need of AI solutions for African fashion and global fashion.

Existing fashion datasets [30] need to be extended to account for non-western fashion. To address this lack of data, a first attempt to create a Senegalese dataset was carried out by focusing on the two popular apparels—Boubou and Taille Mame. The (small) labeled dataset contains 256 images collected from Senegalese designers. The images were preprocessed to contain the garment only, on a white background. Data augmentation [19, 26] was applied to the dataset to generate more data and obtain a more robust dataset.

This study used Transfer Learning to classify these two different apparels. Transfer learning is the process of re-using a pre-trained model as the starting point for solving computer vision and natural language processing (NLP) [15, 21]. This approach is often used when the dataset is small and to speed up the development of the models. The state-of-the-art CNN-based models, VGG19 [27], Xception [2] and MobileNetV2 [24], which are performing well for image classification, were selected as base models to train the Senegalese Fashion dataset. The experimental results show that MobileNetV2, augmented with a Global Average Pooling layer and a Dese layer with SoftMax, achieved the highest score with Accuracy of 96.43, Precision of 93.33, Recall of 1 and F1-score of 96.55.

The paper is organized as follow. Section 2 describes current advances on the use of AI in fashion and state-of-the-art techniques used in image classification. Section 3 describes the experimentation that led to the development of the CNN model to classify Senegalese apparels. Section 4 evaluates the model and, finally, Sect. 5 discuss our contribution and its potential for global fashion.

2 Background

This section provides the theoretical background necessary to understand the methodology used in the following sections. It presents existing fashion datasets, followed by transfer learning as an approach to build models based on pre-trained ones. It also provides the most relevant details of three CNN-architectures that were considered to build a model for the Senegalese Fashion dataset.

2.1 Fashion Datasets

There exists a number of image datasets built to address and benchmark problems related to fashion. They are inspired from ImageNet [3], a dataset of 14 million high resolution quality-controlled, human-annotated, and object-boxed images that was instrumental in advancing deep learning and computer vision research. Fashion-MNIST [30] was built to benchmark Machine Learning (ML) algorithms. With Fashion-MNIST, Zalando, the German fashion and lifestyle e-commerce company, sought to replace the original MNIST [18] dataset containing handwritten digits, considered too simple for benchmarking, and establishing itself as a leader in AI solutions related to fashion. Fashion-MNIST contains 70,000 images of fashion products from 10 categories (pants, sneakers etc.). The image format is itself inherited from the MNIST dataset so that all the ML algorithms are compatible with the fashion images. This dataset comes preloaded in popular packages such as Keras and is used in many fashion research works. Amazon Catalog images (for fashion) [8] and DeepFashion [17] are datasets that achieve state-of-the-art performance on fashion items recognition. They contain 1 million and 800,000 images respectively organized in 43 classes. The Amazon dataset was also used to recognize clothing in movies.

2.2 Transfer Learning

Many studies and research on fashion detection and clothing recognition are effectively conducted using Convolutional Neural Networks (CNN), by cascading multiple CNNs [16], combining Region-CNN (R-CNN) and Support Vector Machine (SVM) [7], and using bi-directional CNNs [32].

Transfer Learning is the process of re-using a pre-trained model as the starting point of building a model for specific computer vision and natural language processing tasks [14, 15, 21, 22]. Methods of transfer learning are designed at the feature representation level or at the classifier level for knowledge transfer from the source domain to the target domain. It permits to speed up model development and, also, to improve the generality of the target model in special cases where the labelled dataset is too small or insufficient to train an original model. CNN models may take days or weeks to train on large datasets. The question of accuracy transfer arises for this type of approach. In the case of image datasets, studies showed that models generalize well across datasets [4, 14, 21, 22]. The generalization of features is, however, not highly effective when data is limited. In this case, Data Augmentation techniques can be used to increase the size of the dataset [19, 26]. They generate additional images based on the available data by applying orientation changes, adding pixel information and feature, making geometric transformations, and adding noise.

This study focuses on the pre-trained models VGG19, Xception and MobileNetV2.

VGG19 [27] is a very deep CNN for large scale image classification (maximum 224×224). It attains state-of-the-art performance on the ImageNet challenge dataset and generalizes well to wide range of datasets and tasks. It consists of 19 weight layers (16 convolution ones and 3 dense ones) using 3×3 convolutional filters. This convolution is followed by 5 MaxPool layers and ends with a SoftMax layer.

Xception (for Extreme Inception) CNN architecture [2] consists of depth wise separable convolutions. It consists of 36 convolutional layers with linear residual connections in-between and is structured into 14 modules. These layers are followed by a logistic regression layer for exclusive investigation of image classification. Xception proved to have best performance for image classification on ImageNet dataset.

MobileNetV2 [24] is a CNN architecture build on a bottleneck depth-separable convolution with residuals. It is composed of an initial layer with 32 filters followed by 19 residual bottleneck layers. The network strategically uses the efficiencies gained by making use of convolution operators with highly optimized matrix multiplications within the layers. MobileNetV2 is a highly efficient mobile model [5]. MobileNetV2 improves performance on image classification and object detection.

3 Methodology

This study permitted to create the Senegalese Fashion dataset and select, experimentally, a CNN-architecture that performs well on this small dataset.

3.1 Senegalese Fashion Dataset

The Senegalese fashion dataset consists of images of Senegalese fashion comprising two categories of apparels named ‘boubou’, and ‘mame’ (women apparels only). Images were gathered from Instagram with the permission of several designers from Senegal. The dataset was prepared in such way that it has an equal number of images of each class. It is composed of 286 images, 143 images of each category. The images were labelled according to their class and validated by domain experts (female students from Senegal). Figure 1 presents a snapshot of the images of the dataset.

Fig. 1 Senegalese fashion dataset snapshot

3.2 Data Preprocessing and Augmentation

The images were preprocessed using Adobe Photoshop to focus on the main region of attention and the important characteristics of the image, including the pattern, design, style, and contour, that actively determine the class label. Any background was removed and replaced by plain white background to only highlight the apparel. Overlapping elements and texts were removed to preserve the uniqueness of targeted classes. These techniques were adopted for best feature extraction of the different apparels.

Each image was resized to 160×160 pixels. To achieve dimensioning at same scale, the images were normalized with the pixel values ranging from −1 to 1. Additionally, morphological operations such as contrast enhancement, sharpening and brightness were performed.

The Senegalese Fashion dataset is rather insufficient for building a generalized model for fashion classification. In particular, we do not have enough images at this time. We adopted data augmentation techniques to enhance the dataset. These techniques included applying zoom and shear on images by a random percentage

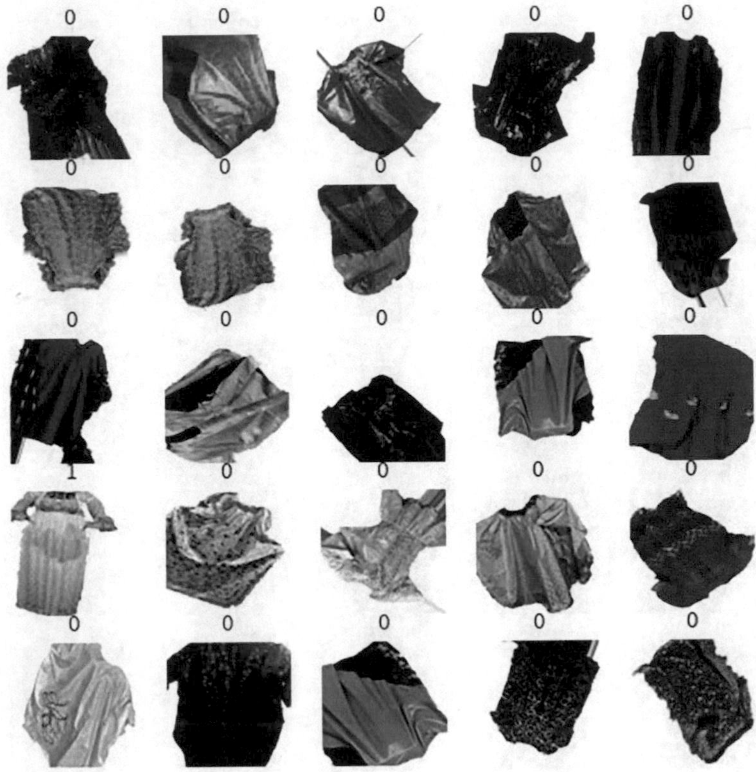

Fig. 2 Data augmentation on senegalese fashion dataset

of up to 20%, shifting images horizontally and vertically by a fraction of 10%, and mirroring and flipping the image on the horizontal and vertical axis. Figure 2 depicts the augmented data.

Once augmented, the dataset was split into training data (70%) to build the model, validation data (10%) to tune the model, and test data (20%) to evaluate the model.

3.3 Model Architecture

We used transfer learning to build the model architecture starting with and comparing performance of state-of-the-art models, MobileNetV2, VGG19 and Xception CNNs. These models were imported along their ImageNet weights from Keras, freezing their base layers and excluding the last classification layer. The base layers of the models were augmented with a Global Average Pooling layer and a Dense layer with Softmax as the classifier function. The models had identical input images, dense layer, classifier and hyperparameters. The model architecture that performed best

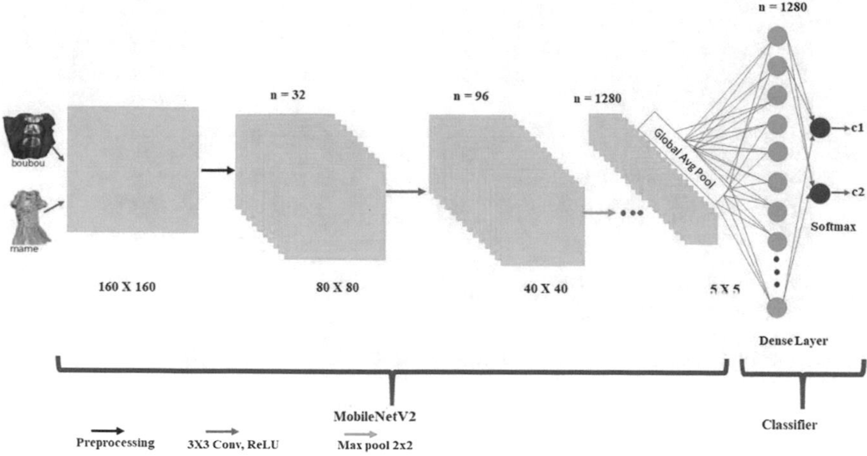

Fig. 3 Final model architecture based on MobileNetV2

on the Senegalese Fashion dataset is based on MobileNetV2 (see Sect. 4 for further explanations). Figure 3 describes the model architecture in detail.

While training the CNN model, images are treated with convolutional operations, in the convolutional layers with a 3 × 3 size kernel. The first convolutional block has 32 filters that produces 80 × 80 feature maps. The subsequent block has 96 filters generating 40 × 40 feature vectors, and, so on until the final block receives 5 × 5 feature maps with 1280 filters. This final convolutional layer is connected to the Global Average Pooling layer followed by the prediction layer, thereby implementing correspondence between feature maps and classes.

In the Global Average Pooling Layer, one feature map is generated for the matching class labels by aggregating the spatial information and finalizing the image classification task. This offers robustness and permits to avoid overfitting of the model at this layer level. A combination of these features defines the region of interest for both the 'boubou' and 'mame' images. The characteristics of the apparels such as texture, contour, grain, designs etc. are determined by these features above pixel information.

Figure 4 provides a summary of the model directly from the Google Colab notebook.

3.4 Experimentation

We experimented with various hyperparameters to evaluate their impact on the performance of the three models. These parameters included Batch Size, Class Mode, Activation Function, Learning Rate and Epochs. The most stable settings of these hyperparameters resulted in the best feature detection and image classification results.

```
# Checking MobileNetV2 model for trainable parameters
MobileNetV2.summary()
```

```
Model: "sequential"
_________________________________________________________________
 Layer (type)                Output Shape              Param #
=================================================================
 mobilenetv2_1.00_160 (Funct  (None, 5, 5, 1280)       2257984
 ional)

 global_average_pooling2d (G  (None, 1280)             0
 lobalAveragePooling2D)

 dense (Dense)               (None, 2)                 2562

=================================================================
Total params: 2,260,546
Trainable params: 2,562
Non-trainable params: 2,257,984
_________________________________________________________________
```

Fig. 4 Description of the model based on MobileNetV2

We settled with a Batch Size of 32, Class Mode set as sparse categorical, Adam as the Optimizer, a Learning Rate of 0.001 and 20 epochs to obtain the best results.

4 Results

The results from the experimentation on the three models show that choosing the right CNN architecture is a complex task to handle. Any minor architectural changes will have high impact on the model's performance. Even a slight change in the number of neurons or layers can result in decreasing accuracy.

Our experimentation and a study of accuracy and loss converged toward MobileNetV2.

With the above-described stable hyperparameters, the training accuracies achieved are 96.74% for MobileNetV2, 85.33% for VGG19 and 95.65% for Xception models respectively. Training accuracy and loss are provided on Fig. 5.

The models are evaluated for their performance and generality on the test dataset where the test accuracies achieved are 96.43% for MobileNetV2, 83.93% for VGG19 and 94.64% for Xception models. Test accuracy and loss are provided on Fig. 6.

Figure 7 presents the confusion matrix plotted for the MobileNetV2 model's prediction summary.

The classification report of Table 1 is generated to evaluate the quality of predictions by the model whether they are true or false predictions. The metrics we used are Accuracy, Precision, Recall and F1-Score. The MobileNetV2 achieved the highest score with Accuracy of 96.43, Precision of 93.33, Recall of 1 and F1-score of 96.55.

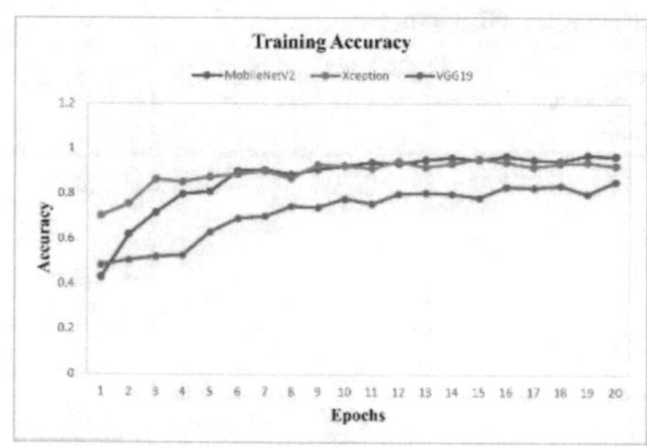

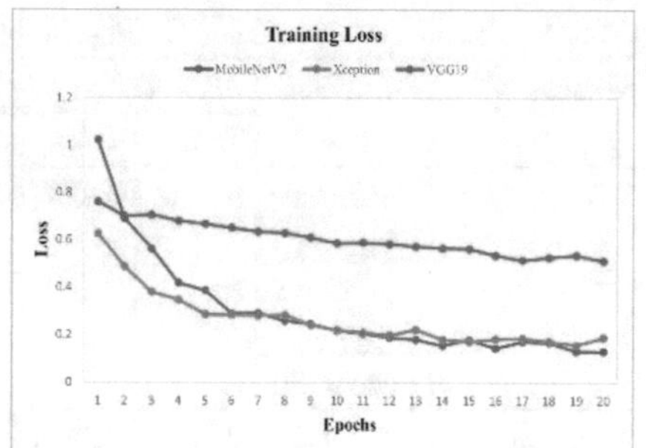

Fig. 5 Training accuracy and loss

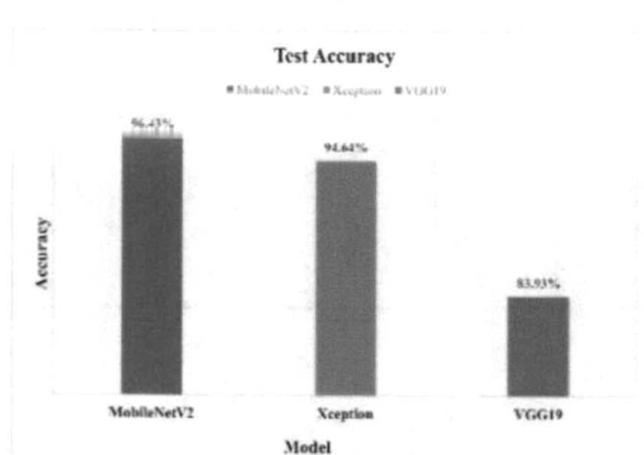

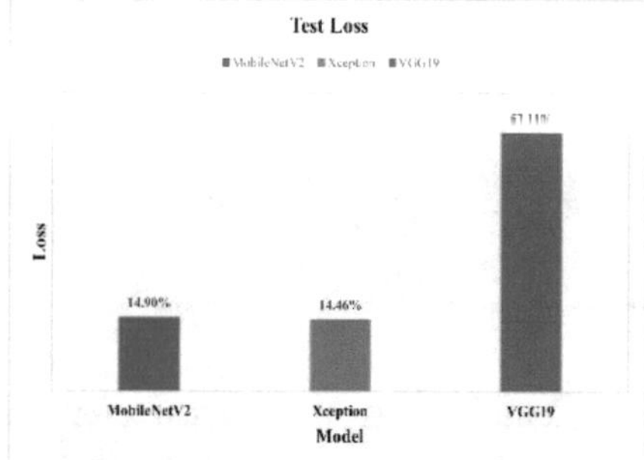

Fig. 6 Test accuracy and loss

Fig. 7 Confusion matrix for MobileNetV2

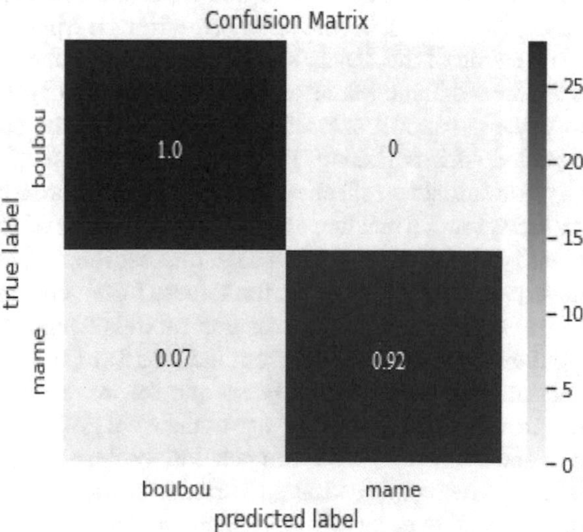

Table 1 Classification report highlighting MobileNetV2 performance

Model	Parameters	Accuracy		Loss		Precision	Recall	F1-Score
		Train	Test	Train	Test			
VGG19	Batch Size: 32	0.8533	0.8393	0.5114	0.5211	0.6279	0.9643	0.7606
	Optimizer: Adam							
	Epochs: 20							
Xception	Batch Size: 32	0.9239	0.9464	0.1884	0.1490	0.9310	0.9643	0.9474
	Optimizer: Adam							
	Epochs: 20							
MobileNetV2	Batch Size: 32	**0.9674**	**0.9643**	**0.1312**	**0.1446**	**0.9333**	**1.0000**	**0.9655**
	Optimizer: Adam							
	Epochs: 20							

5 Conclusion and Future Work

The lack of data is a problem for the development of inclusive AI solutions. Accessing data is difficult in the Global South because of inexistent open data policies and privately-owned data with strong copyright infringements [9]. This paper describes the creation of the Senegalese Fashion dataset that contains 246 images describing Boubou and Taille Mame apparels. We plan to develop a web and mobile interface to make our model accessible to the community, but also to collect more images depicted African fashion. The idea behind the creation of this dataset was to diversify and initiate a reflection on the western-centric aspects of the existing fashion datasets. Dataset creation comes with high levels of responsibility. When a dataset is biased or imbalanced or miss data, the resulting AI models are not reliable and will have detrimental impact on the targeted domain. International fashion is missing from datasets, and existing fashion models cannot be tested on garments such as Boubou from Senegal, Sari from India or Sampot from Cambodia. In the process of dataset creation, the World Economic forum recommends all entities to document the provenance, creation and use of datasets [12]. In [6], datasheets require information about the motivation, composition, collection process (preprocessing, cleaning, labeling), recommended uses, distribution, and maintenance. Such requirements have the potential to ensure ML-dataset integrity and increase transparency and accountability within the ML community. The recommendations should also insist on the generality of the dataset.

Transfer Learning was used to determine the best performing pre-trained model on the Senegalese dataset after data augmentation. The best performing model,

MobileNetV2, is capable of efficiently detecting the features in the images and classify them. Future work will consist of building an incremental model by adding men garments and adding more types of Senegalese apparels such as Taille Basse and Moussor (the Senegalese headscarf).

Acknowledgements This work was partially funded by a Google TensorFlow grant.

References

1. Bossard, L., et al.: Apparel classification with style. In: Computer Vision – ACCV 2012, 2013, pp. 321–335
2. Chollet, F.: Xception: deep learning with depthwise separable convolutions. In: 2017 IEEE Conference on Computer Vision and Pattern Recognition (CVPR) (2017)
3. Deng, J., et al.: ImageNet: a large-scale hierarchical image database. In: 2009 IEEE Conference on Computer Vision and Pattern Recognition (2009)
4. Donati, L., et al.: Fashion product classification through deep learning and computer vision. Appl. Sci. **9**(7), 1385 (2019)
5. Dong, K., et al.: MOBILENETV2 model for image classification. In: 2020 2nd International Conference on Information Technology and Computer Application (ITCA) (2020)
6. Gebru, T., et al.: Datasheets for datasets. Commun. ACM **64**(12), 86–92 (2021)
7. Hara, K., et al.: Fashion apparel detection: the role of deep convolutional neural network and pose-dependent priors. In: 2016 IEEE Winter Conference on Applications of Computer Vision (WACV) (2016)
8. Heilbron, F.C. et al.: Clothing recognition in the wild using the amazon catalog. In: 2019 IEEE/CVF International Conference on Computer Vision Workshop (ICCVW) (2019), pp. 3145–3148 (2019)
9. Heng, S., et al.: Understanding AI ecosystems in the global south: the cases of senegal and Cambodia. Int. J. Inf. Manage. **64**, 102454 (2022)
10. Henrique, A.S., et al.: Classifying garments from fashion-mnist dataset through Cnns. Adv. Sci., Technol. Eng. Syst. J. **6**(1), 989–994 (2021)
11. Hong, C., et al.: Composite templates for cloth modeling and sketching. In: 2006 IEEE Computer Society Conference on Computer Vision and Pattern Recognition - Volume 1 (CVPR'06) (2006)
12. How to Prevent Discriminatory Outcomes in Machine Learning: World Economic Forum (2018). https://www3.weforum.org/docs/WEF_40065_White_Paper_How_to_Prevent_Discriminatory_Outcomes_in_Machine_Learning.pdf
13. Kastner, K.: Making fashion, forming bodies and persons in Urban Senegal. African J. Online, Africa Develop. (2019)
14. Kornblith, S., et al.: Do better imagenet models transfer better? In: 2019 IEEE/CVF Conference on Computer Vision and Pattern Recognition (CVPR) (2019)
15. Shao, L., et al.: Transfer learning for visual categorization: a survey. IEEE Trans. Neural Netw. Learn. Syst. **26**(5), 1019–1034 (2015)
16. Liu, Z., et al.: Fashion landmark detection in the wild. ArXiv.org, 10 Aug (2016)
17. Liu, Z., et al.: Deepfashion: powering robust clothes recognition and retrieval with rich annotations. In: 2016 IEEE Conference on Computer Vision and Pattern Recognition (CVPR), N.p, pp. 1096–1104 (2016)
18. Maaten, L. V. D.: A new benchmark dataset for handwritten character recognition. Semant. Schol. (2009)

19. Mikolajczyk, À., Grochowski, M.: Data augmentation for improving deep learning in image classification problem. In: 2018 International Interdisciplinary PhD Workshop (IIPhDW) (2018)
20. Muhammed, M.A.E., et al.: Benchmark analysis of popular imagenet classification deep CNN architectures. In: 2017 International Conference On Smart Technologies For Smart Nation (SmartTechCon) (2017)
21. Raina, R., et al.: Self-taught learning. In: Proceedings of the 24th International Conference on Machine Learning - ICML '07 (2007)
22. Rohrmanstorfer, S., et al.: Image classification for the automatic feature extraction in human worn fashion data. Mathematics **9**(6), 624 (2021)
23. Russakovsky, O., et al.: Imagenet large scale visual recognition challenge. CoRR abs/1409.0575 (2015)
24. Sandler, M., et al.: MobileNetV2: inverted residuals and linear bottlenecks. In: 2018 IEEE/CVF Conference on Computer Vision and Pattern Recognition (2018)
25. Seo, Y., Shin, K.-S.: Image classification of fine-grained fashion image based on style using pre-trained convolutional neural network. In: 2018 IEEE 3rd International Conference on Big Data Analysis (ICBDA) (2018)
26. Shorten, C., Khoshgoftaar, T.M.: A survey on image data augmentation for deep learning. J. Big Data **6**(1) (2019)
27. Simonyan, K., Zisserman, A.: Very deep convolutional networks for large-scale image recognition. ArXiv.org (2015)
28. Tuinhof, H., et al.: Image-based fashion product recommendation with deep learning. In: International Conference on Machine Learning, Optimization, and Data Science, pp. 472–481 (2019)
29. Wang, W., et al.: Attentive fashion grammar network for fashion landmark detection and clothing category classification. In: 2018 IEEE/CVF Conference on Computer Vision and Pattern Recognition (2018)
30. Xiao, H., et al.: Fashion-MNIST: a novel image dataset for benchmarking machine learning algorithms. ArXiv.org (2017)

Covid-19 and the Society: Resilience and Recovery

Digital Networks and Leadership Collaboration During Times of Uncertainty: An Exploration of Humanistic Leadership and Virtual Communities of Practice

Angela Lehr and **Susie Vaughan**

Abstract Over the past two years, ripple effects from the Covid-19 pandemic have amplified disconnection and loss while also fueling greater human creativity and connection through technological mediums. How can leaders continue maximizing technology and collaboration to best support and foster interpersonal connection and innovation while also navigating the human conditions and emotions present during times of volatility, uncertainty, complexity, and ambiguity (VUCA)? The unpredictable nature of these times requires leaders and change agents to continue learning and refining ways to help their teams, organizations, and communities to flourish. Within the current context of Covid-19 rebuilding and repair, this paper explores how participation in digital or virtual communities of practice (VCoP) and leadership centered on psychological safety and humanistic principles have been experienced by leaders and coaches. An informal discussion was facilitated, and a qualitative survey was conducted to begin exploring questions pertinent to VCoP and humanistic leadership (HL). Practitioner testimonials and key insights are shared. The findings of this exploratory study suggest that the main benefits of VCoP have been leader exposure to diverse industry practices and the bridging of physical and professional disconnection caused by Covid-19 social distancing and remote working requirements. Findings related to HL indicate that the leaders surveyed believe HL is essential in the work they do and that HL and fostering resilience entail both practicing and promoting holistic self-care and relational connection. Meaningful participation in VCoP and HL practices are framed as vital tools for leading and evolving in a post-Covid-19 world.

Keywords Virtual communities of practice · VUCA paradox · Humanistic leadership

A. Lehr (✉)
Creighton University, Omaha, NE 68178, USA
e-mail: angelalehr1@gmail.com

S. Vaughan
Susie Vaughan & Company, McKinney, TX 75070, USA
e-mail: susie@susievaughanandcompany.com

A. Visvizi et al. (eds.), *Research and Innovation Forum 2022*,
Springer Proceedings in Complexity, https://doi.org/10.1007/978-3-031-19560-0_61

1 Introduction

Globally, interpersonal boundaries and professional needs have changed significantly since January 2020. The impacts of the Covid-19 pandemic as well as worldwide sociopolitical unrest have been far-reaching, changing the way people in organizations, governments, and communities communicate and interact [9]. Shifting personal and collective priorities and constant adaptation are critical factors shaping current individual and professional functioning [10]. Concurrently, technological platforms have allowed collaboration and knowledge-sharing to expand beyond geographical constraints allowing for more diverse partnerships [13]. More must be studied in relation to how these fluctuations and adaptations are affecting leaders across sectors.

The exploratory study presented in this paper focused on digital networks and humanistic collaboration related to participating in virtual communities of practice (VCoP) and to humanistic leadership (HL). The lived experiences of leaders were captured through a qualitative survey and offer a springboard for further study. The results and findings of this study on VCoP and HL are presented along with key insights and recommendations. The following sections discuss post-Covid-19 leadership and collaboration considerations related to VCoP and HL principles. The literature reviewed and this study's initial findings support the assertion that engaging and inclusive VCoP provide professional and personal enrichments for leaders, educators, and coaches benefitting progress across sectors [4, 9, 11].

2 Post-covid-19 Leadership and Collaboration

To discuss leadership and collaboration within the post-Covid-19 context necessitates acknowledging that the pandemic is not over and that, along with the pandemic, the current context is also a period of racial, societal, and political reckoning around the world [5]. VUCA, a term coined by the U.S. military to denote contexts or conditions that present volatility, uncertainty, complexity, and ambiguity has been integrated into government and corporate vernacular in recent decades [6]. In light of the events of the past two years, VUCA is now a relevant concept for the education, non-profit, and small business sectors.

The term *vuka* in Swahili and Zulu translates to wake-up and be aware [17, 18]. The meaning of vuka offers poignant food for thought for today's leaders, educators, and coaches. The intensity of working and living in times of VUCA can result in fatigue, reactivity, and inaction [6]. Leaders also have the opportunity to respond to uncertain conditions by deepening their understanding of how VUCA arises in their spheres of influence, and by engaging intentionally and proactively in on-going planning and pivot strategies [2]. The current VUCA context calls on leaders and change agents to remain awake and attentive to what is unfolding at any given moment. The VUCA paradox, as described by General George Casey, calls on leaders to accept that the environment is volatile, uncertain, complex, and ambiguous without allowing external factors to determine their results [2]. To further expand on this, Bob Johansen frames proactive leadership strategies by repurposing the VUCA acronym to represent vision, understanding, clarity, and adaptability [6].

2.1 *The Intersection of Humanistic Leadership and the VUCA Paradox*

The VUCA paradox serves as a reminder that people within collective groups have agency and influence in determining favorable outcomes. HL is a combination of leadership mindsets, behaviors, and attributes [9, 11]. Humanistic principles in leadership emphasize that the holistic well-functioning of individual members directly impacts the well-functioning of organizations, communities, and societies [7]. Leading, teaching, and coaching groups and individuals through mutable and complex circumstances requires leaders to have empathy and understanding for the grief, trauma, and overwhelm people are experiencing [12, 16]. Practices consistent with HL focus on prioritizing people over profit and bureaucracy. Moreover, humanistically navigating losses or difficulties requires actively engaging with and naming challenges rather than giving up or working to the point of exhaustion [11, 15].

The losses and developments since the onset of the pandemic have resulted in both collective growth and collective trauma. The current phase of recovery requires sector leaders and coaches to simultaneously foster hope while holding space for despair and frustration [16]. Practicing self-compassion, patience, and a long-game mentality are valuable humanistic leadership perspectives. Additionally, sharing power within organizations, investing in collaborative learning, and focusing on strong working relationships are HL approaches that promote engagement and mitigate burnout and fatigue [11].

2.2 *Virtual Communities of Practice*

Wenger's social theory of learning and the establishment of communities of practice (CoP) are not new concepts in education [19]. The benefits of learning in groups have been studied and documented for thirty years and highlight enhanced knowledge acquisition, meaning-making, and application [20]. CoP are organized learning groups that share a domain or competency area, pursue common interests together, and develop practitioner skills and resources. CoP occur in person, online, or in a hybrid format [19, 20]. Following the onset of the Covid-19 pandemic, VCoP became a lifeline for many leadership practitioners navigating social distance mandates and socioeconomic ripple effects. Through the connection and learning provided by

VCoP, leaders assisted one another in their evolving roles as change agents and models for leading through crisis and uncertainty [1, 20].

VCoP continue to provide vital online human connection and collaboration points. VCoP offer the most value in promoting open innovation by fostering psychological safety and facilitating meaningful conversations about topics that matter to the collective of individuals [4, 19]. Successful VCoP foster mutually beneficial relationships among cognitively diverse individuals and treat the time spent together as precious [13, 19]. Lastly, VCoP normalize for members of the learning ecosystem the liminal space of not knowing all the answers [19]. In this post-Covid-19 context, VCoP have become more critical to connecting leaders globally for the purpose of sharing knowledge, building competencies, and optimizing open and collaborative innovation.

3 Method

As leadership practitioners, the authors, co-facilitated an informal discussion about HL within their own VCoP, the *June Cohort of Certified Dare to Lead Facilitators*. The semi-structured conversation was held on Zoom and recorded with permission. The conversation revealed that the VCoP members participating in the conversation actively identified as humanistic leaders; some had been practicing HL for most of their careers. Several participants articulated feeling a sense of relief and camaraderie in knowing that there were fellow humanistic practitioners within the greater community. The self-identified newer HL practitioners said they felt inspired to continue leading humanistically as a result of the informal dialogue. The revelations of this conversation served as a springboard for further inquiry. An exploratory study was conducted based on two guiding research questions:

RQ1—What are the experiences of leadership practitioners as they relate to participating in virtual communities of practice (VCoP) in the current post-Covid-19 context?

RQ2—How do leadership practitioners view and practice humanistic leadership (HL)?

The researchers/authors co-created a qualitative survey consisting of ten open-ended, long answer questions. The first half of the survey focused on exploring respondent experiences related to their respective participation in VCoP. The second half focused on exploring HL practices and lived experiences related to leading humanistically as well as examples of non-humanistic leadership impacts. The survey also included demographic questions and one scaled question regarding the importance of humanistic leadership.

Each section included operational definitions of VCoP and of HL to establish a common understanding and language for exploring the topics of this study. A recruitment message explaining the purpose of the study and emphasizing the voluntary and anonymous nature of the survey was developed and shared with potential participants. The recruitment message and survey link was sent to the professional newsletter email list maintained by one of the authors consisting of 125 leaders and executive coaches. The study message and survey link were also posted on the social media accounts of both authors. During the one-week period the survey was open, 11 responses were received. The responses were reviewed and interpreted with converging themes and diverging answers in mind.

Of the 11 participants, there were seven Pre K-12 school administrators as well as leaders in the areas of executive coaching, finance, and green industry/forestry services. Eight respondents identified as female, two as male, nine as white, and one participant identified as Black or African American. Lastly, eight respondents stated they were between the ages of 45–54, one was between 35–44, and one was age 65 or older.

4 Results and Findings

The survey responses highlighted the top benefits of VCoP, ways virtual meetings and engagements can feel disconnecting, specific humanistic and resilience-building practices, and negative experiences related to non-humanistic approaches (see Table 1). The VCoP provided benefits centered on connection and collaboration. Specifically, respondents believed reassurance from peers and exposure to other leaders and practices provided much needed connection. Having online mediums for collaborating and learning with others were also of value. The advantages expressed by study respondents were congruent with the cognitive diversity and knowledge-sharing benefits of social learning cited in the literature [13, 19, 20]. Moreover, the virtual nature of the respondents' professional groups were essential to supporting their work during the onset of the pandemic [1, 9].

Five respondents also discussed elements of VCoP that were not beneficial. Namely, large virtual meetings were described as impersonal, unilateral, and boring. One respondent stated that it is more difficult to decipher non-verbal cues that aid human connection, and another said they disliked that sometimes virtual meetings took the place of face-to-face contact. The expressed barriers to connection also aligned with the literature emphasizing intentional connection through the CoP best practices of engaging in small group discussions, meaningful dialogue, and encouraging varying levels of participation [1, 19].

Regarding HL, every respondent identified that leading humanistically is essential in this post-Covid-19 context. Respondents shared about ways they have practiced HL during the pandemic. Each of the respondents described cultivating individual relationships and being present for the people they work with as fundamental to their HL practices. Additional examples of HL included regular gestures of appreciation, open communication, role modeling kindness and acceptance, and promoting

Table 1 Participant quotes supporting the findings

Related themes	Participant quotes
Post-covid-19 VCoP as collaborative and connecting	"These platforms have helped me stay connected to others both locally, throughout the state, and nationally. These connections and the similar struggles have helped create a community of leaders and learners who have genuine empathy. These same connections have led to deep discussions and practical solutions for the situations that public educators are facing" "While my participation has been limited it has been enlightening to know that I am not alone in this struggle. I have been able to learn and get ideas from others and just have open truthful rich conversations!" "[VCoP] Provided deep connection necessary to combat loneliness"
Drawbacks of participating in VCoP	"When the meeting is large, and you are not an active participant, it is hard to stay attentive" "Virtual meetings in my district need an overhaul if we continue them - they are boring and unengaging"
HL as essential in the post-covid-19 context	"I live and breathe humanistic philosophies. I believe in order to work with others, to be in community with others, we have to look at the whole person -mind, body, spirit. I embed this philosophy daily in my coaching and leadership training" "I am passionate about my role and education, yet I lead with empathy and the mindset that others have the best intentions for our school community. I support and encourage yet, I am honest, kind, and caring in my approach to feedback whether it is positive or somewhat negative"
Examples of non-humanistic leadership	"Organization focused on $ only. Metrics that were about financial gain" "I had a supervisor who was very controlling and emotionally abusive…I was young and thought it was my fault. [It] left me feeling less than and I dreaded work. To this day, I still dream of that job and how unhealthy it was for me"
Promoting resilience through HL	"I continuously encourage personal time, offer additional support through extra resources or removing unnecessary work, celebrating failures that associates can learn from, and encouraging self-reflection" "Being professionally connected to others and promoting both self-care and mutual care"

social emotional learning opportunities. In terms of resilience-building, respondents articulated engaging in and promoting holistic self-care and renewal practices for well-being. Conversely, examples of non-humanistic behaviors experienced by respondents included controlling behavior, limited autonomy, demeaning language, reactive decision-making, and poor communication or transparency. The examples above supported assertions in the literature related to the power of prioritizing mutually beneficial relationships, nurturing the whole person, and enabling autonomy [4, 8, 11, 19].

5 Discussion and Recommendations

The qualitative survey data collected and presented offer insights into the lived experiences of leadership practitioners participating in VCoP. The responses also raise awareness related to HL and the practices leaders deem as valuable and detrimental [7, 8, 12, 19]. The experiential accounts generated from this survey give depth and personalization to the subjects of VCoP and HL as a strength of this study. This study also contributes to the literature by highlighting possible areas of study related to VCoP and HL. The main limitations of this study relate to the low survey response rate resulting in a small sample size. Another limitation is the lack of respondent gender and racial diversity. While the data collected is limited, it offers a rich view into the realties faced by leaders following Covid-19.

During times of uncertainty, individual and organizational performance is impacted more negatively when leaders or cultures are not taking care of their people holistically [7, 15]. Thought leader and researcher, Tim Clark, purports, "Innovation is the process of connected people connecting things" [3]. Employing HL to proactively create psychologically safe and collaborative VCoP ecosystems promotes innovation and cultivates an environment that rewards experimentation and growth. Principles of HL also assist leaders in skillfully acknowledging experiences of grief, loss, inertia, and languishing in the post-Covid-19 workplace [8, 11, 15].

As the findings of this study suggest, VCoP and cultures guided by HL can serve as fertile ground for fostering creativity, sustainability, and transformative growth benefitting leaders personally and within their respective systems. The digital and virtual frontiers must be humanistic for organization members to build trust, embody compassion and empathy, and process grief and resistance in the midst of ongoing change [7, 11, 14]. Investing in human capital and fostering meaningful human connections virtually are Post-Covid-19 necessities [16].

To augment collaboration, professional development, and open innovation, organizations and VCoP can incorporate HL approaches and concrete practices. Activities and considerations that build psychological safety and bolster HL include the following:

- Establish inclusive and thoughtful rituals or exchanges to mark the beginning, middle, and end of meetings or learning session [14].

- Practice active listening and invite equity in participation by offering a variety of engagement options including small group breakout opportunities [1, 20].
- Allow for and embrace all emotions, experiences, and learning paces as valid, and view regrets or setback as catalysts for learning and growth [15, 16].
- Remember that all people are navigating unique and collective challenges. Treat people with dignity, compassion, and patience while maintaining clear and healthy boundaries [8, 14].
- Celebrate progress while also caring for yourself and others intentionally [16].
- Leading humanistically and fostering virtual environments that are responsive and honor the holistic needs of all organization members are critical to rebuilding and evolving in this post-Covid-19 context of VUCA [6, 11, 12, 14].

6 Conclusion

The ripple effects of the Covid-19 pandemic and the sociopolitical strife across the globe have changed the way people work and lead. Volatility, uncertainty, complexity, and ambiguity–VUCA–are conditions now faced by nearly every organizational sector. These conditions also invite greater engagement, awareness, clarity, and compassion [6]. Building resilience through understanding the VUCA paradox, promoting VCoP that optimize thought diversity, and leading humanistically offer pathways to sustainability and open innovation [11, 13].

The exploratory study presented in this paper provides valuable insights into the lived experiences of leaders regarding VCoP and HL. The study also presents possibilities for future research in the areas of VCoP, HL, resilience, humanistic technology, and the relationship between psychological safety and innovation. Promoting and elevating virtual communities of learning and humanistic practices based on mutual respect, human dignity, and thoughtful regard for the challenges organizations face is critical to individual and collective adaptability.

References

1. Bolisani, E., Fedeli, M., DeMarchi, V., Bierema, L.: Together we win: communities of practice to face the Covid crisis in higher education. In: Proceedings of the 17th International Conference on Intellectual Capital, Knowledge Management and Organizational Learning ICICKM (2020)
2. Casey, G.W.: "Living in a VUCA world. https://www.youtube.com/watch?v=yHpHN49__G4. Last Accessed 04 April 2022
3. Clark, T.R.: The 4 Stages of Psychological Safety: Defining the Path to Inclusion and Innovation. Berrett-Kohler Publishing, Oakland, CA, USA (2020)
4. Clark, T.R.: Agile doesn't work without psychological safety. Harvard Busin. Rev. https://hbr.org/2022/02/agile-doesnt-work-without-psychological-safety, last accessed 2022/04/04.
5. Council on Foreign Relations. Global Conflict Tracker. https://www.cfr.org/global-conflict-tracker/?category=us&vm=list. Last Accessed 04 April 2022
6. Glaeser, W.: VUCA world: leadership skills and strategies. VUCA World. https://www.vuca-world.org/ 2021. Last Accessed 04 April 2022

7. Hamel, G., Zanini, M.: Humanocracy: Creating Organizations as Amazing as the People Inside Them. Harvard Business Review Press, Boston, MA, USA (2020)
8. Hicks, D.: Dignity: Its Essential Role in Resolving Conflict. Yale University Press, New Haven, CT, USA and London (2011)
9. Lehr, A., Vaughan, S.: Grief, growth, and silver linings: humanistic leadership during the Covid-19 pandemic. In: Visvizi, A., Troisi, O., Saeedi, K. (eds.) Research and Innovation Forum 2021. RIIFORUM 2021. Springer Proceedings in Complexity. Springer Champ (2021). https://rdcu.be/cInrj
10. Moss, J.: Beyond burned out. Harvard Business Review, http://www.hbr.org/2021/02/beyond-burned-out. Last Accessed 27 Mar 2021
11. Nathanson, C.: Humanistic leadership during Covid. https://drcraignathanson.com/humanistic-leadership-during covid-19/. Last Accessed 04 April 2022
12. Nathanson, C.: Ten ideas to lead, manage, and coach remote teams as a humanistic leader. https://drcraignathanson.com/ten-ideas-to-lead-manage-and-coach-remote-teams-as-a-humanistic-leader/. Last Accessed 04 April 2022
13. Page, S.: The Diversity Bonus: How Great Teams Pay Off in the Knowledge Economy. Princeton University Press, Princeton (2017)
14. Parker, P.: The Art of Gathering: How We Meet and Why it Matters. Riverhead Books, New York (2018)
15. Pink, D.: The Power of Regret: How Looking Backward Moves Us Forward. Riverhead Books, New York (2022)
16. Rock, D.: We need time to rehabilitate from the trauma of the pandemic. Harvard Business Review. https://hbr.org/2022/02/we-need-time-to-rehabilitate-from-the-trauma-of-the-pandemic. 2022. Last Accessed 04 April 2022
17. Vuka. Wiktionary (2022). https://en.wiktionary.org/wiki/vuka. Last Accessed 04 April 2022
18. "Vuka." Xpand.eu., https://xpand.eu/sa/vuca-understanding-volatility/. Last Accessed 04 April 2022
19. Wenger-Trayner, Bev: In-person, online, hybrid: The future. https://wenger-trayner.com/reflections/in-person-online-hybrid-the-future/. Last Accessed 04 April 2022
20. Yarris, L.M., Chan, T. M., Gottlieb, M., Juve, A.M.: Finding your people in the digital age: virtual communities of practice to promote education scholarship. J. Grad. Med. Educ. https://www.ncbi.nlm.nih.gov/pmc/articles/PMC6375332/. Last Accessed 04 April 2022

Digital Divide and Entrepreneurial Orientation in the Global South: Quantifying and Explaining the Nexus

Anna Visvizi, Orlando Troisi, and Mara Grimaldi

Abstract Motivated by the imperatives enshrined in the Sustainable Development Goals (SDG 5, SDG 9, SDG 10) the study applies the concept of digital divide to the willingness of individuals to undertake entrepreneurial activities. The aim is to explore the relationship between digital skills, propensity to adopt technology, and entrepreneurial orientation of individuals in developing countries. By adopting a quantitative approach, this study seeks to identify the weight (and so the impact) of digital skills (therefore, the possible digital gap) in the dual context of entrepreneurship-related technology adoption in entrepreneurial activities. This paper, including its conceptual approach and findings (including the managerial implications), adds to the broader debate on digital divide and ways of addressing it, all in context of the SDG debate.

Keywords Digital divide · Global South · Digital skills · Technology adoption · Entrepreneurial orientation · SDG5 · SDG9 · SDG10

1 Introduction

Advances in information and communication technology (ICT) and the resultant digitalization of all aspects of social interaction, including business and economy, imply that to maintain their competitive edge, the business sector needs to keep pace with digitalization. Nevertheless, research suggests that several barriers hamper the process of successfully exploiting the advantages resulting from digitalization [1, 2]. The phenomenon of uneven spread and adoption of digital technologies, and the

A. Visvizi
SGH Warsaw School of Economics, Al. Niepodległości 162, 02-554 Warszawa, Poland

Effat University, 8482 Qasr Khouzam, Al-Nazlah Al-Yamaniyah, Jeddah 22332, Saudi Arabia

O. Troisi · M. Grimaldi (✉)
University of Salerno, Via Giovanni Paolo II, 132, 84014 Fisciano, Italy
e-mail: margrimaldi@unisa.it

O. Troisi
e-mail: otroisi@unisa.it

A. Visvizi et al. (eds.), *Research and Innovation Forum 2022*,
Springer Proceedings in Complexity, https://doi.org/10.1007/978-3-031-19560-0_62

resultant socio-economic implications thereof, has been dubbed in the literature as the "digital divide" [3, 4]. While typically attributed to developing countries, or in broader terms, to so called Global South [5, 6], digital divide has many facets and is by no means reserved to the developing world. Indeed, the digital divide exists among diverse societal groups, i.e., gender-, age-, income-based [7, 8], across countries and continents, thus including also the Global North. While the literature suggests that digital divide it is a derivative of technological progress [9], it ought to be stressed that it is also conditioned by other variables such as inefficient and ill-targeted economic policies. Moreover, the implications of center-periphery relationship, and – albeit at a different level –cultural factors, education patterns, and managerial practices can also have an impact on digital divide.

The well-being and prosperity-related implications of digital divide on individuals [10, 11] and societies [12, 13] have been well documented in the literature. By means of taking the analysis a step further, it is imperative to ask the question of the relationship between the digital divide and an individual's propensity to engage in entrepreneurial activities. It is argued that a positive correlation exists between an individual's ability to use digital technology and his/her propensity to engage in entrepreneurial activity. Accordingly, the prospect and ability to acquire certain basic threshold of digital skills may increase an individual's attitude toward considering engaging with entrepreneurial activities [14].

Despite the need to explore the different dimensions that can affect digital divide and to clarify the impact of the possession of digital skills on entrepreneurship, future research does not analyze these issues adequately. Hence, this paper seeks to bridge the gap in literature related to the absence of studies that explore digital skills and entrepreneurship in the Global South.

To explore and test the validity of these theses, against the backdrop of an international survey (N = 200), this paper sets on (i) to identify the relevance of the acquisition of the necessary threshold of digital skills on the willingness to adopt technology; (ii) to analyze the impact of technology adoption on entrepreneurial orientation. The argument is structured as follows. In the next section, the research model, methodology and the materials are elaborated on. A careful analysis of the survey results ensues. Discussion and conclusions follow.

2 The Research Model and Key Concepts Employed

2.1 Digital Divide and Attitude Towards Technology

The digital divide is defined as a gap in two strictly interrelated factors: skills and the physical access technology. The implicit assumption underpinning the concept of digital divide is that that without access to technology, it is difficult to develop technical skills [8]. Accordingly, in a world undergoing digital transformation, the lack of access to technology, and the related impossibility to acquire skills to use it becomes an additional source of deprivation and exclusion [15, 16]. In an aggregate view, i.e., at the level of a company or even an entire economy, the right skills, including especially digital skills [17] can drive the competitiveness, innovation and resilience capacity [18].

The enhancement of digital skills can have social, cultural and economic implications. In particular, it can enable the development of business start-ups and entrepreneurship [19, 20]. Not only the acquisition of the skills to use technology efficiently is a necessity for contemporary entrepreneurs but owning digital skills can enhance the proper integration of technology in business strategies and processes.

As global digital transformation unfolds, the issue of digital divide is strictly related to digital inclusion. For this reason, the lack of digital competence can be emphasized especially in the developing countries that are already plagued with social inequalities [21].

Hence, there is the need to investigate if the possession of digital skills can improve technology adoption and, in turn, can stimulate the entrepreneurial orientation in disadvantaged countries. For this reason, the following hypothesis can be formulated:

H1. The possession of digital skills can influence the willingness to adopt technologies in entrepreneurial activities.

2.2 Technology Adoption and Entrepreneurial Orientation

The ability to use technology and to integrate it effectively in business processes can foster entrepreneurial orientation today [22, 9]. Digital technologies can impact on the creation of new business ventures [23], can lower the cost and increase profits [24]. ICT and innovation are the key elements of entrepreneurship for the creation of new opportunities for business [25]. Adopting ICTs wisely in business management can reduce the barriers to the entrepreneurship [26].

In extant research, very few studies explore the impact of ICT on entrepreneurship in rural areas. Venkatesh, who introduced TAM (the technology acceptance model) and UTAUT (the unified theory of acceptance and use of technology), provides an extension of the TAM framework to examine the influence of IT adoption on entrepreneurship [27]. Moreover, in extant research is acknowledged that the intention to use new technologies can be influenced by the possession of the right skills to use technology [28] and of a strategic approach to the use of technology and of data [29].

Thus, the second research hypothesis can be as follows:

H2. Technology adoption can have an impact on entrepreneurial orientation in developing countries.

3 Methodology

3.1 Instruments and Measures

The survey has been developed by adopting validated items for each construct borrowed from previous literature on digital divide, entrepreneurship and management. The measurement items for the construct of digital skills have been borrowed and re-adapted from Area and Guarro [30], Rodríguez-de-Dios et al. [31] and Manco-Chavez et al. [32]. The items for the construct of entrepreneurial orientation have been borrowed from Lumpkin e Dess [33], Bolton and Lane [34] and Taatila and Down [35].

The intention of the entrepreneurs in the sample to accept and use technology has been measured through the classical sub-dimensions of UTAUT (performance expectancy, effort expectancy, social influence, behavioural intention). The items for the UTAUT construct are borrowed and adapted from Venkatesh et al. [36]. The items of all the constructs were measured by using a self-designating 7-point scale ranging between "strongly disagree" (1) and "strongly agree" (7).

Lastly, the survey included questions on basic demographics (gender, age, faculty). The self-administered questionnaire was designed and shared through a web-based survey service and administered to young students aspiring entrepreneurs from Pakistan. To validate the research hypotheses, the technique of multiple linear regression was employed by using SPSS.

3.2 Data Collection

In line with the research objectives, the population in the sample is composed of international businessmen that usually employ technology in their businesses. A convenience sampling approach was adopted, i.e., the researchers randomly contacted some organizational members through online communities (on LinkedIn, Facebook) and through international associations.

Data were collected between February and May 2022. A total of 200 responses was obtained. As Table 1 shows, the gender ratio of the respondents was 27% male and 73% female. 40% of the respondents are managers (19.7% general manager, 7.2% Human resources manager, 6.1% office manager). The majority of individuals in the sample is in the age range from 18–20 and the average age in the sample is 21.

35% of respondents attend a course in social and communication science, whereas the 22% attend a course in science, technology and math.

Table 1 Demographics and description of the sample

Variables	Categories	%
Gender	Female	73
	Male	27
Age	18–20	52
	21–25	31
	26–30	17
Faculty	Social and communication sciences	35
	Science, technology and math	22
	Humanities and education science	18.5
	Engineering	14.5
	Political science	5.5
	Medicine and pharmacy	5
	Arts, literature, philosophy	4.5
	Economics	1
	Law	1

4 Findings

To examine the survey results, factor analysis was employed. The latter is a technique that allows to highlight the existence of a structure of latent traits, the so-called factors or dimensions. These cannot be measured directly, within a set of variables directly observable (sometimes also referred to as indicator variables or instrumental variables) that relate to the latent traits [37]. The factor loadings describe the saturation, that is the strength of the relationship between the factor and the measured variable. Low saturations (values lower than 0.30 or 0.40) are generally used to exclude the relationship between a variable and a factor, thus simplifying the structure. High saturations (from 0.50 onwards) indicate a high degree of concordance between variable and factor.

The factor analysis is employed as a preliminary analysis aimed at validating the goodness of belonging of a set of items to a multi-level construct before carrying out regression analysis between the aforementioned construct and other constructs.

The factor loadings for each item comply with the threshold of 0.7, by showing a high degree of internal coherence in each construct, as shown also by the value of Cronbach Alpha, which is higher than 0.9. Cronbach alpha estimates the reliability of the multi-item instrument of measures: digital skills with 0.902, UTAUT with 0.963, Entrepreneurial orientation with 0.954. All the alpha coefficients are above the cut-off point of 0.7, indicating an acceptable level of reliability for each construct [38] (Table 2).

After the purification of the measurement items employed for each construct, with the subsequent removal of the items with a factor loading lower than 0.7, the regression analysis has been performed. Figure 1 reports the regression coefficients that

Table 2 Factor analysis

Construct	Items	Factor loading	Cronbach α
Digital skills	I know how to bookmark a website I like so I can view it later	0.678	0.902
	I know how to download information I found online	0.835	
	I do not like downloading apps for smartphones as I find difficult to learn how to use them	0.869	
	I know how to deactivate the function showing my geographical position (e.g., Facebook, apps)	0.789	
	I know when I can post pictures and videos of other people online	0.567	
	I know how to change the sharing settings of social media to choose what others can see about me	0.672	
	I know how to compare different sources to decide if information is true	0.863	
	I know how to identify the author of the information to evaluate its reliability	0.798	
	I know how to compare different apps in order to choose which one is most reliable and secure	0.654	
	I know how to detect a virus in my digital device	0.778	
	I know how to block unwanted or junk mail/spam	0.698	
	If something doesn't work occurs while I am using a device (computer, smartphone, etc.), I usually know what it is and how to fix the problem	0.877	
	I find hard to decide what the best keywords are for online searching	0.734	

(continued)

Table 2 (continued)

Construct	Items	Factor loading	Cronbach α
	I find confusing the way in which many websites are designed	0.767	
	Sometimes I find difficult to determine how useful the information is for my purpose	0.843	
	Depending on who I want to communicate with, it is better to use one method over the other (make a call, send a WhatsApp message, an email)	0.912	
	I know how to send any file to a contact using a smartphone	0.765	
	No matter with who I communicate: emojis are always useful	0.834	
Unified theory of acceptance and use of technology (UTAUT)	Using technology would improve my work performance	0.925	0.963
	Using technology would be helpful in my work	0.768	
	Using technology would enhance the effectiveness of business activities in my company	0.923	
	Using technology would increase my productivity at work	0.922	
	Interaction with a new technology would be clear and understandable	0.93	
	It would be easy for me to become skilful at using a new technology	0.851	
	I would find a new technology not easy to use	0.768	
	Learning to operate a new technology would be easy for me	0.789	
	People who influence my behavior think that I should use new technologies in my work	0.754	
	People who are important to me would think that I should use new technologies in my work	0.734	

(continued)

Table 2 (continued)

Construct	Items	Factor loading	Cronbach α
	People around me will take a positive view of me using new technologies in my work	0.723	
	People around me would think that I should not use new technologies	0.589	
	I have the resources necessary to use a new technology	0.622	
	I have the knowledge necessary to use a new technology	0.878	
	I feel comfortable using new technologies in my daily life	0.891	
	I have not problems to use new technologies the next time I study/work/communicate with my friends	0.834	
	I Intend to use a new technology in my work	0.795	
	I will probably use a new technology in my work	0.853	
	I am decided to use a new technology in my work	0.785	
Entrepreneurial Orientation (EO)	I like to take bold action by venturing into the unknown	0.835	0.954
	I am willing to invest a lot of time and/or money on something that might yield a high return	0.869	
	I tend to act "boldly" in situations where risk is involved	0.773	
	I often like to try new and unusual activities that are not typical but not necessarily risky	0.9	
	In general, I prefer a strong emphasis in projects on unique, one-of-a-kind approaches rather than revisiting tried and true approaches used before	0.865	
	I prefer to try my own unique way when learning new things rather than doing it like everyone else does	0.695	

(continued)

Table 2 (continued)

Construct	Items	Factor loading	Cronbach α
	I favour experimentation and original approaches to problem solving rather than using methods others generally use for solving their problems	0.774	
	I usually act in anticipation of future problems, needs or changes		
	I tend to plan ahead on projects	0.823	
	I prefer to "step-up" and get things going on projects rather than sit and wait for someone else to do it	0.567	

Fig. 1 The findings of regression analysis

estimate the relationships between the three constructs inserted in the research model. Results show that there is a positive linear relationship between digital skills and technology adoption (R^2 0.657) and between technology adoption and entrepreneurial orientation (R^2 0.732). Both the coefficients exceed the threshold value of 0.3 which makes a regression coefficient acceptable [39].

5 Discussion

The findings of the study reveal: (i) the high impact of digital skills on the willingness to employ technology in entrepreneurial activities (H1 confirmed); (ii) the positive relationship between the intention to adopt technology and entrepreneurial orientation. Hence, the empirical research demonstrates that a positive correlation exists between the level of digital skills, propensity to use technology and engage in entrepreneurial activity. The results are in line with extant research that shows that the criticalities related to technological evolution, such as the digital divide, can affect entrepreneurship [40, 41]. The assessment of the influence of digital skills in the development of entrepreneurial orientation reveals that that entrepreneurship is a socially embedded process. Moreover, the findings confirm the impact of technology on entrepreneurship and the influence that digital competencies can have on the success of entrepreneurial activities. The research also highlights the dynamic interrelation between social, cultural and technological dimension. social networks and ICT use.

6 Conclusion

The research seeks to bridge the gap in literature related to the absence of studies that explore digital skills and entrepreneurship in the Global South. However, even if it is acknowledged that technology can bring different advantages for entrepreneurship, the downside of technological progress and the digital divide can increase exclusion and poverty not only in the Global South, but also in the peripheries of Global North. Since extant studies tend to focus on the analysis of digital divide in Global South, future research can try to compare the weight of the impact of digital divide on entrepreneurship in Global North and Global South.

However, the research model adopted in this study considers only the technological dimension (the digital skills) as a predictor of technology adoption intention. Obviously, other dimensions, such as culture, education, managerial practices should be taken into account [42, 43]. Thus, to address the limitations of the study, further research can address the problem by assessing the impact of cultural and social variables, gender gap, education on the attitude to employ technology.

References

1. Reddy, R.C., Bhattacharjee, B., Mishra, D., Mandal, A.: A systematic literature review towards a conceptual framework for enablers and barriers of an enterprise data science strategy. Inf. Syst. E-Bus. Manag. 1–33 (2022)
2. Matt, D.T., Pedrini, G., Bonfanti, A., Orzes, G.: Industrial digitalization. A systematic literature review and research agenda. Eur. Manag. J. (2022)
3. Fang, M.L., Sarah, L., Canham,S.L., Battersby, L., Sixsmith, J., Wada, M., Sixsmith A.: Exploring privilege in the digital divide: implications for theory, policy, and practice. Gerontologist. **59** (1), e1–e15 (2019).https://doi.org/10.1093/geront/gny037
4. Lai, J., Widmar, N.O.: Revisiting the digital divide in the covid-19 era. Appl Econ Perspect Policy **43**, 458–464 (2021). https://doi.org/10.1002/aepp.13104
5. Hill, C., Lawton, W.: Universities, the digital divide and global inequality. J. High. Educ. Policy Manag. **40**(6), 598–610 (2018). https://doi.org/10.1080/1360080X.2018.1531211
6. Ragnedda, M., Gladkova, A. (eds.): Digital Inequalities in the Global South. Palgrave Macmillan, Cham (2020) https://doi.org/10.1007/978-3-030-32706-4
7. Van Dijk, J.: The Digital Divide. Wiley (2020)
8. Kularski, C., Moller, S.: The digital divide as a continuation of traditional systems of inequality. Sociology **5151**(15), 1–2 (2012)
9. Millán, J.M., Lyalkov, S., Burke, A., Millán, A., van Stel, A.: Digital divide' among European entrepreneurs: Which types benefit most from ICT implementation? J. Bus. Res. **125**, 533–547 (2021)
10. Jamil, S.: A widening digital divide and its impacts on existing social inequalities and democracy in Pakistan. In: Ragnedda, M., Gladkova, A. (eds.) Digital Inequalities in the Global South.

Global Transformations in Media and Communication Research—A Palgrave and IAMCR Series. Palgrave Macmillan, Cham (2020). https://doi.org/10.1007/978-3-030-32706-4_4

11. Yoon, H., Jang, Y., Vaughan, P.W., Garcia, M.: Older adults' internet use for health information: digital divide by race/ethnicity and socioeconomic status. J. Appl. Gerontol. **39**(1), 105–110 (2020). https://doi.org/10.1177/0733464818770772
12. Solomon, M.E., van Klyton, A.: The impact of digital technology usage on economic growth in Africa. Util. Policy **67** (2020). https://doi.org/10.1016/j.jup.2020.101104
13. Răileanu Szeles, M.: New insights from a multilevel approach to the regional digital divide in the European Union. Telecommun. Policy **42**(6) (2018). https://doi.org/10.1016/j.telpol.2018.03.007
14. Ibrahim, N., Mas'ud, A.: Moderating role of entrepreneurial orientation on the relationship between entrepreneurial skills, environmental factors and entrepreneurial intention: a PLS approach. Manag. Sci. Lett. **6**(3), 225–236 (2016)
15. Visvizi, A., Lytras, M.D., Damiani, E., Mathkour, H.: Policy making for smart cities: innovation and social inclusive economic growth for sustainability. J. Sci. Technol. Policy Manag. **9**(2), 126–133 (2018)
16. Adam, I.O., Alhassan, M.D.: Bridging the global digital divide through digital inclusion: the role of ICT access and ICT use. Transform. Gov. Ment: People, Process. Pol. **15**(4), 580–596 (2021)
17. Gašová, K., Mišík, T., Štofková, Z.: Employers demands on e-skills of university students in conditions of digital economy. In: CBU International Conference Proceedings, vol. 6, pp. 146–151 (2018)
18. Levano-Francia, L., Sanchez Diaz, S., Guillén-Aparicio, P., Tello-Cabello, S., Herrera-Paico, N., Collantes-Inga, Z.: Digital competences and education. J. Educ. Psychol.-Propos. Y Rep-Resentaciones **7**(2), 579–588 (2019)
19. Alderete, M.V.: Mobile broadband: A key enabling technology for entrepreneurship? J. Small Bus. Manag. **55**(2), 254–269 (2017)
20. Visvizi, A., Troisi, O., Grimaldi, M., Loia, F.: Think human, act digital: activating data-driven orientation in innovative start-ups. Eur. J. Innov. Manag. **25**(6), 452–478 (2022)
21. Unesco.: A Global Framework of Reference on Digital Literacy Skills for Indicator 4.4.2. UNESCO Institute for Statistic, Montreal (2018)
22. Chatterjee, S., Gupta, S.D., Upadhyay, P.: Technology adoption and entrepreneurial orientation for rural women: evidence from India. Technol. Forecast. Soc. Chang. **160**, 120236 (2020)
23. Elia, G., Margherita, A., Passiante, G.: Digital entrepreneurship ecosystem: how digital technologies and collective intelligence are reshaping the entrepreneurial process. Technol. Forecast. Soc. Chang. **150**, 119791 (2020)
24. Bharadwaj, A.S.: A resource-based perspective on information technology capability and firm performance: an empirical investigation. MIS Quarterly 169–196 (2000)
25. Yunis, M., Tarhini, A., Kassar, A.: The role of ICT and innovation in enhancing organizational performance: the catalysing effect of corporate entrepreneurship. J. Bus. Res. **88**, 344–356 (2018)
26. Sharma, M., Chaudhary, V., Bala, R., Chauhan, R.: Rural entrepreneurship in developing countries: challenges, problems and performance appraisal. Glob. J. Manag. Bus. Stud. **3**(9), 1035–1040 (2013)
27. Venkatesh, V., Shaw, J.D., Sykes, T.A., Wamba, S.F., Macharia, M.: Networks, technology, and entrepreneurship: a field quasi-experiment among women in rural India. Acad. Manag. J. **60**(5), 1709–1740 (2017)
28. Zaheer, H., Breyer, Y., Dumay, J.: Digital entrepreneurship: An interdisciplinary structured literature review and research agenda. Technol. Forecast. Soc. Chang. **148** (2019)
29. Gupta, M., George, J.F.: Toward the development of a big data analytics capability. Inf. Manag. **53**(8), 1049–1064 (2016)
30. Area, M., Guarro, A.: La alfabetización informacional y digital: fundamentos pedagógicos para la enseñanza y el aprendizaje competente. Rev. Esp. Doc. Científica., (Monográfico), 46–7 (2012)

31. Rodríguez-de-Dios, I., van Oosten, J.M., Igartua, J.J.: A study of the relationship between parental mediation and adolescents' digital skills, online risks and online opportunities. Comput. Hum. Behav. **82**, 186–198 (2018)
32. Manco-Chavez, J.A., Uribe-Hernandez, Y.C., Buendia-Aparcana, R., Vertiz-Osores, J.J., Isla Alcoser, S.D., Rengifo-Lozano, R.A.: Integration of ICTS and digital skills in times of the pandemic covid-19. Int. J. High. Educ. **9**(9), 11–20 (2020)
33. Lumpkin, G.T., Dess, G.G.: Clarifying the entrepreneurial orientation construct and linking it to performance. Acad. Manag. Rev. **21**(1), 135–172 (1996)
34. Bolton, D.L., Lane, M.D.: Individual entrepreneurial orientation: development of a measurement instrument. Educ. + Train. **54**(2/3), 219–233 (2012)
35. Taatila, V., Down, S.: Measuring entrepreneurial orientation of university students. Educ. Train **54**(8), 244–760 (2012)
36. Venkatesh, V., Morris, M.G., Davis, G.B., Davis, F.D.: User acceptance of information technology: Toward a unified view. MIS Quarterly 425–478 (2003)
37. Barbaranelli, C., Natali, E.: I test psicologici. Teorie e modelli psicometrici. Carocci, Roma (2005)
38. Nunnally, B., Bernstein, I.R.: Psychometric Theory. Oxford University, New York (1994)
39. Di Franco, G.: Models and Techniques for Multivariate Analysis. FrancoAngeli, Milano (2017)
40. Ojediran, F., Anderson, A.: Women's entrepreneurship in the global south: empowering and emancipating? Adm. Sci. **10**(4), 87 (2020)
41. Wood, B.P., Ng, P.Y., Bastian, B.L.: Hegemonic conceptualizations of empowerment in entrepreneurship and their suitability for collective contexts. Adm. Sciences. **11**(1), 28 (2021)
42. Troisi, O., Fenza, G., Grimaldi, M., Loia, F.: Covid-19 sentiments in smart cities: the role of technology anxiety before and during the pandemic. Comput. Hum. Behav. **126**, 106986 (2022)
43. Hujran, O., Abu-Shanab, E., Aljaafreh, A.: Predictors for the adoption of e-democracy: an empirical evaluation based on a citizen-centric approach. Transform. Gov. Ment: People, Process. Pol. **14**(3), 523–544 (2020)

Resilience Assessment in Times of Covid-19 in Ecuador

Daisy Valdivieso Salazar, María Luisa Pertegal-Felices, Aldrin Espín-León, and Antonio Jimeno-Morenilla

Abstract Since the 2019 outbreak of the Covid-19 pandemic, Ecuador was challenged by a combination of high health risk and serious economic and social impact. A need emerged to evaluate the resilience of the population in a social situation at the level of health and public safety unprecedented since the term resilience appeared in the scientific literature. The aim of this study is to create an instrument to assess resilience during the period when traumatic events are occurring due to the pandemic. Furthermore, it will consider the fact that the researchers conducting this study are themselves immersed in the same traumatic events provoked by Covid-19. For the development of the instrument, Ungar's Ecological Model has been taken as a reference, which encompasses several theories from the behavioural, cognitive, systemic, and functionalist currents. The instrument consists of 29 items that provide information on 4 dimensions associated with resilience: (1). interpersonal resources, (2). formal support networks, (3). informal support networks and 4. facing the risk situation. The instrument was validated by the "Judges Method" using the Aiken coefficient, in which 10 professionals from various universities in Ecuador participated. As for the reliability of the instrument, internal consistency was assessed using Cronbach's alpha.

Keywords Cultural identity measurement · Preservation of ethnic identity · Amazonia · Waorani community

D. V. Salazar · A. Espín-León
Faculty of Sociology and Social Work, Central University of Ecuador, Quito, Ecuador

A. Jimeno-Morenilla
Department of Computer Technology, University of Alicante, Alicante, Spain

M. L. Pertegal-Felices (✉)
Developmental and Educational Psychology Department, University of Alicante, Alicante, Spain
e-mail: ml.pertegal@ua.es

A. Visvizi et al. (eds.), *Research and Innovation Forum 2022*,
Springer Proceedings in Complexity, https://doi.org/10.1007/978-3-031-19560-0_63

1 Introduction

The Covid-19 pandemic drastically altered people's lives, as well as economic, political, social aspects worldwide whose repercussions affected the living conditions of citizens, generating uncertainty, fear, massive blockades, economic recessions that rearranged supply and demand in the economy [1].

Education was not left untouched by the impact of the crisis; this sector was one of the hardest hit by the pandemic worldwide [2]. Authors such as Hanushek and Woessman [3] have related the impact of school loss to the long-term economic impact, where the loss of one third of an academic year, induced by the Coronavirus, will affect their income by approximately 2–4% during their working life. The magnitude of these social and economic consequences have been compared to economic disasters, revolutions and terrorist attacks, the difference being that they carry a considerable public health risk demanding direct attention to the family [4].

In Ecuador, the impact of Covid-19 has hit families, mainly the most vulnerable. At the beginning of the pandemic, in the first months of 2020, 26,336 cases of people with Covid-19 and 1,063 deaths were identified, a figure that in a relatively small country generated national alarm [5]. Bonilla and Guachamín [6] points out that only 37.23% of households have a computer at the national level and that, in rural areas and slums, the percentage is less than 23.27%, while employment in the first year of primary education is lower than in the second year of secondary education [7]. The employment rate in the first semester of 2020 fell by 11.1–52.8%, which meant an additional 643,420 unemployed, with young people between 15 and 24 years of age being the most affected (25.8% reduction), many of whom are students [8].

Resilience is certainly one of the factors that have become most evident in this pandemic context, in fact, for Intriago et al. [9] resilience is characterised by the human being's condition to push for a goal as opposed to defeat. It is also the engine that drives human beings to achieve the goals they pursue based on the application of their skills and abilities, which exerts the motivation required for the momentum of the achievements and their attainment [10].

For Ungar [11] resilience should not have a dominant view of resilience as something that individuals have but should be seen as a process facilitated by families, schools, governments and communities. In this sense, given that resilience is related to social factors, a social ecological interpretation that recognises people and their relationship with their environment and territory is necessary, as protective factors depend on the social construction of each factor and the meaning attributed to it [12, 13].

Cultural and contextual differences expressed in the resilience of individuals, families and communities reflect in a construct, a postmodern understanding and demonstrate between risk and protective factors, a non-hierarchical and non-systemic relationship [14].

The objective of this research is to create a Social Resilience evaluation instrument to assess the social problems and coping strategies associated with the Covid-19 health emergency in Ecuadorian students, parents and professionals. This instrument will ultimately allow for the construction of a resilient social intervention model based on life experiences and the recognition of coping strategies for the Covid-19 pandemic.

2 Literature Review

There are methods and instruments to measure resilience. Jew et al. attempt to measure levels of individual resilience in three factors: optimism, skill acquisition and risk taking, while Doll et al. [15] measure interpersonal relationships through four factors: frequency of social interaction, ability to resolve minor disagreements, frequency of prosocial behaviours and ability to resolve conflicts. On the other hand, based on two factors: personal competence and self-acceptance, Wagnild and Young [16] elaborated a scale of resilience in five aspects: perseverance, self-confidence, meaning in life, philosophy of life and equanimity. Other authors such as Hjemdal et al. [17] in a 14-item instrument, define the "resilient self" as an individual's abilities to self-manage, so that an individual adapts when faced with new situations.

In Latin America, Salgado [18], based on Peru's own reality, developed an instrument to assess five factors of resilience: Self-esteem, Empathy, Self-nomy, Humour and Creativity, while in Mexico, Valadez [19] evaluated the so-called "Escala Mexicana", which consists of a 43-item questionnaire arranged in four factors: Social Competences, Family Support, Social Support and Structure, likewise Palomar and Gómez [20] developed a measurement scale of 25 items that are grouped into five dimensions: personal competence, high standards and tenacity, self-confidence, tolerance of negative situations and coping with the effects of stress, secure relationships and acceptance of change, control, and spiritual influence.

In the literature reviewed, no studies were found that addressed this new situation and its coping through resilience. In particular, these methods and instruments were not developed under this new nor an equivalent reality of a global pandemic and do not included the particular realities of Ecuadorian peoples and indigenous peoples. For this reason, the need to analyze this "new" social problem from the perspective of resilience became evident in Ecuadorian universities. The study of the living conditions of the Ecuadorian population became a priority for the Social Work degrees of 11 universities in the country. These conditions are generating traumatic situations in the population that are strictly related to the threat to life, collective and individual health and the type of coping of those individuals who are impacted by this crisis [21, 22]. In this scenario of social-educational transformation it was proposed to create telematic support networks. For Aranda and Pando [23] these transformations take place within the framework of social interrelationships, and are identified with care and protection "for the other" using support networks, while for Hernández et al. [24] it is the informal support networks where the strengthening of their identity is also constructed.

3 Methodology

In the absence of methods and instruments that measure resilience under the reality of a global pandemic, many resilience measures from various disciplines are transferable to address the pandemic, however, to include realities from the cosmovision of the Ecuadorian peoples [25], experts from 11 Ecuadorian universities (social workers, sociologists, psychologists and anthropologists), using the Delphi method and the existing theoretical review, developed a measurement instrument with resilience factors (dimensions) and items during the pandemic caused by Covid-19.

3.1 *Participants*

The different stages and phases of this research were carried out exclusively with the participation of students, authorities, administrative staff and members of the academic community of 11 Ecuadorian universities. Thus, the survey was applied to 7441 people over 18 yrs of age throughout the national territory (except for the inhabitants of the Galapagos Islands), including students, parents and working and non-working professionals, of whom 2947 were men and 4494 women.

3.2 *Procedure*

The surveys were conducted using a computer application in which informed consent was given (in accordance with ethical standards for the application of non-experimental human subjects' surveys) in which the individual agreed or did not agree to participate in the study and the social-resilience questionnaire detailed below.

3.3 *Instrument*

A multi-theoretical social resilience questionnaire of 29 items grouped into 4 dimensions was developed (see Table 1). Each of the 29 items [26] is assessed on a Likert-type scale 1–5. The scores of the 4 dimensions were standardised so that each value was between 0 and 100.

Table 1 Multi-theoretical resilience survey

Variable	Dimensions	Indicators	Items	Data type
Resilience social	Interpersonal resources	– Maintaining social and friendly networks – Restoring self-esteem and self-image in difficult times – Being assertive – Empathy with others	1–10	Survey Resilience Social Qualitative Nominal Likert scale 5 response options
	Family resources	– Adaptation family – Participation family – Resources personal – Family effect – Resources relatives	11–15	
	Sociocultural resources	– Community resources and social support – Credibility in the institutions – Confidence in state institutions – Construction of national identity and patriotic values	16–25	
	Situation of risk	– Perception of risk (contagion) – Personal configuration of the situation	26–29	

3.4 Reliability and Content Validity

To determine whether the items are reliable, we first applied Cronbach's Alpha reliability coefficient for the 29 items of the 7441 surveys, which obtained an acceptable value of 0.791.

Content Validity refers to the degree to which an instrument reflects a specific content domain of what it measures [27]. Whereas Content-related evidence determines the extent to which the items of an instrument are representative of the variables it is intended to measure (degree of representativeness) [28].

Validity was determined by expert judgement. The instrument was sent for analysis to 10 professionals from the fields of Social Work, Sociology, Anthropology and Psychology with master's degrees and PhDs. Once the validation documents were received from the judges, the content validity method was used, by means of the expert judgement technique, using Aiken's V coefficient [29]. The results of the V coefficient of 0.96 were obtained, categorising each of the items as valid in terms of relevance, pertinence, and clarity.

Aiken V-coefficient

$$V = \frac{S}{(n(c-1))}$$

where

S: sum of whether
n: number of judges
c: number of values in the rating scale

4 Results

The results obtained (see Table 2) show that Interpersonal resources and Family resources are more resilient, while Coping with the risk situation and Socio-cultural resources are less resilient.

To determine whether there are differences between the male and female population groups, a t-test for independent samples was performed. The results show that there are significant differences in 3 of the 4 dimensions. There is no significant difference for the dimension of Resilience socio-cultural resources. (Fig. 1 and Table 3).

Table 2 Descriptive statistics

	Gender	Mean	Std deviation	N
1. Interpersonal resources	Man	73.4	11.84	2947
	Woman	75.9	10.71	4494
	Total	74.9	11.24	7441
2. Family resources	Man	73.5	21.28	2947
	Woman	75.6	20.48	4494
	Total	74.8	20.83	7441
3. Socio-cultural resources	Man	50.8	18.90	2947
	Woman	51.6	17.87	4494
	Total	51.3	18.28	7441
4. Coping with the risk situation	Man	52.1	16.95	2947
	Woman	50.8	16.03	4494
	Total	51.3	16.41	7441

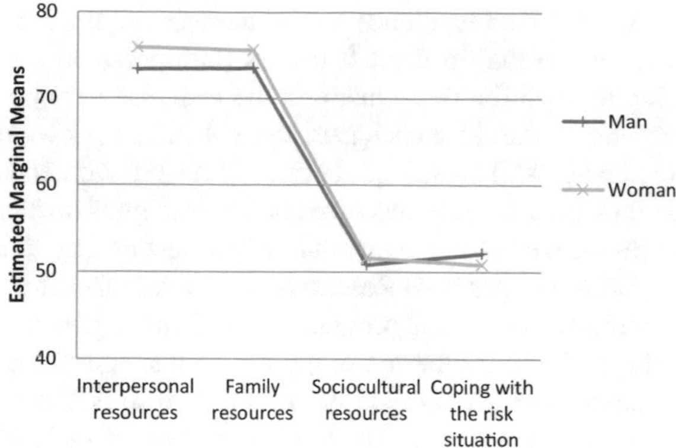

Fig. 1 Means of resilience factors between men and women

Table 3 Independent samples test (4 dimensions of Resilience)

	T-test for equality of meanst		
	t	Sig	Difference of means
1. Interpersonal resources	9.423	0.000*	2.49
2. Family resources	4.305	0.000*	2.12
3. Socio-cultural resources	1.819	0.069	0.79
4. Coping with the risk situation	−3.31	0.001*	−1.29

*Significant at 0.05 level

5 Discussion and Conclusions

Covid-19, being a new social phenomenon, generated dramatic consequences at global, regional, and local levels; however, in the case of Ecuador, under this new scenario, no updated resilience studies were found, nor the social, economic, and developmental impact on students and their families.

Of all the resilience factors analysed, the highest values fall on the items of the Interpersonal Resources factor (average 74.91) and the Family Resources factors (average 74.81). This result could be due to the characteristics of worldview [30] (ancestral vision, the relationship between the self and its natural environment), social values and cultural values of Latin American families, which together make up a protective factor for its members.

The dimensions with the least resilience are Coping with the risk situation (average 51.33) and Sociocultural values (average 51.34). The first factor would be related to the ability to access effective protective factors for academic continuity, subject to the lack or limitation of effective protection mechanisms that expose them to academic dropout. In the context of the pandemic, socio-cultural resources are less different from other factors, probably because families and the community developed coping strategies based on cultural values in response to the need for health services, food, housing and working conditions among their members.

In 3 of the 4 resilience factors there are significant differences between men and women, so that in these 3 factors (Interpersonal resources, Family resources and Coping with the risk situation), the response to cope with problems or the level of accepting or not the change to new risk situations would have a difference according to gender, while resilience between men and women would be considered as similar in the Socio-cultural resources factor as there is no significant difference. This could be associated with the fact that individuals belong to the same society and culture.

Both Interpersonal Resources and Family Resources have high resilience values. For this reason, social programs should not neglect attention to family ties and social relationships. Furthermore, the analysis suggests that resilience should be further strengthened and worked on with regard to Socio-cultural Resources and Coping with the risk situation. The implementation of a tele-care system by the universities' student welfare departments would be the first action.

Although the Ecological Model in the resilience of an individual proposes incorporating their families, educational centres, communities and their environment, the social factors of influence and protection, in the case of Ecuador, which has a diversity of ethnicities and cultures, each construct may contain particular items that respond to the realities of each people.

This study shows important shortcomings in relation to "Socio-cultural resources" and "Coping with the risk situation". These results suggest generating, from a political and social perspective, mechanisms to minimise the impact of pandemics or serious crises, to strengthen the role of the family in the students, to generate spaces for discussion and social participation from the academy, to promote return strategies, to improve the rate of permanence in the university or to make recommendations to the authorities to increase the level of resilience.

References

1. Urzúa, A., Vera-Villarroel, P., Caqueo-Urízar, A., Polanco-Carrasco, R.: La Psicología en la prevención y manejo del COVID-19. Aportes desde la evidencia inicial. Ter. Psicológica. **38**, 103–118 (2020)
2. Nicola, M., Alsafi, Z., Sohrabi, C., Kerwan, A., Al-Jabir, A., Iosifidis, C., Agha, M., Agha, R.: The socio-economic implications of the coronavirus pandemic (COVID-19): a review. Int. J. Surg. **78**, 185–193 (2020)
3. Hanushek, E.A., Woessmann, L.: The economic impacts of learning losses. OECD Education Working Papers, Paris (2020). https://doi.org/10.1787/21908d74-en
4. Prime, H., Wade, M., Browne, D.T.: Risk and resilience in family well-being during the COVID-19 pandemic. Am. Psychol. **75**, 631 (2020)
5. Chen, X., Zhang, S.X., Jahanshahi, A.A., Alvarez-Risco, A., Dai, H., Li, J., Ibarra, V.G.: Belief in a COVID-19 conspiracy theory as a predictor of mental health and well-being of health care workers in Ecuador: cross-sectional survey study. JMIR Public Health Surveill. **6**, e20737 (2020)
6. Bonilla-Guachamín, J.A.: Las dos caras de la educación en el COVID-19. CienciAmérica. **9**, 89–98 (2020)
7. INEC: Instituto Nacional de Estadística y Censo.: https://www.ecuadorencifras.gob.ec/institucional/home/. Last Accessed 12 Jan 2022

8. Esteves, A.: El impacto del COVID-19 en el mercado de trabajo de Ecuador. Mundos Plur—Rev. Latinoam. Políticas Acción Pública. **7**, 35–41 (2020)
9. Intriago, J.O.V., Pinargote, C.L.B., Mendoza, E.M.C.: Niveles de resiliencia y la presencia de síntomas depresivos en los estudiantes universitarios. Rev. Ecuat. Psicol. **4**, 134–142 (2021)
10. Lacal, P.L.P.: Teorías de Bandura aplicadas al aprendizaje. Málaga (2009)
11. Ungar, M.: The Social Ecology of Resilience: A Handbook of Theory and Practice. Springer Science & Business Media (2011)
12. Ungar, M.: The social ecology of resilience: addressing contextual and cultural ambiguity of a nascent construct. Am. J. Orthopsychiatry. **81**, 1 (2011)
13. Libório, R.M.C., Ungar, M.: Resiliência oculta: a construção social do conceito e suas implicações para práticas profissionais junto a adolescentes em situação de risco. Psicol. Reflex. E Crítica. **23**, 476–484 (2010)
14. Ungar, M.: A constructionist discourse on resilience: multiple contexts, multiple realities among at-risk children and youth. Youth Soc. **35**, 341–365 (2004). https://doi.org/10.1177/0044118X03257030
15. Doll, B., Murphy, P., Song, S.Y.: The relationship between children's self-reported recess problems, and peer acceptance and friendships. J. Sch. Psychol. **41**, 113–130 (2003). https://doi.org/10.1016/S0022-4405(03)00029-3
16. Wagnild, G.M., Young, H.M.: Development and psychometric. J. Nurs. Meas. **1**, 165–17847 (1993)
17. Hjemdal, O., Friborg, O., Stiles, T.C., Martinussen, M., Rosenvinge, J.H.: A new scale for adolescent resilience: grasping the central protective resources behind healthy development. Meas. Eval. Couns. Dev. **39**, 84–96 (2006). https://doi.org/10.1080/07481756.2006.11909791
18. Salgado Lévano, A.C.: Métodos e instrumentos para medir la resiliencia: una alternativa peruana. Liberabit. **11**, 41–48 (2005)
19. Valadez, D.C.: Propiedades psicométricas de la escala de resiliencia mexicana en población del norte de México. Enseñ. E Investig. En Psicol. **21**, 78–83 (2016)
20. Palomar Lever, J., Gómez Valdez, N.E.: Desarrollo de una escala de medición de la resiliencia con mexicanos (RESI-M). Interdisciplinaria. **27**, 7–22 (2010)
21. Arciniega, J. de D.U.: La resiliencia comunitaria en situaciones catastróficas y de emergencia. Int. J. Dev. Educ. Psychol. **1**, 687–693 (2010)
22. Páez, D., Fernández, I., Beristain, C.M.: Catástrofes, traumas y conductas colectivas: procesos y efectos culturales. Catástrofes Ayuda En Emerg. Estrateg. Eval. Prev. Trat. 85–148 (2001)
23. Aranda, C., Pando, M.: Conceptualización del apoyo social y las redes de apoyo social. Rev. Investig. En Psicol. **16**, 233–245 (2013)
24. Hernández, M.G., Carrasco, G.M.R., Rosell, C.F.: Evaluación de las principales redes de apoyo informal en adultos mayores del Municipio Cerro. GeroInfo. **5**, 1–11 (2010)
25. López, M.E.D.: Sumak Kawsay o Buen Vivir, desde la cosmovisión andina hacia la ética de la sustentabilidad. Pensam. Actual. **10**, 51–61 (2010)
26. Pertegal-Felices, M.L., Valdivieso-Salazar, D.A., Espín-León, A., Jimeno-Morenilla, A.: Resilience and academic dropout in ecuadorian university students during Covid-19. Sustainability. **14**, 8066 (2022)
27. Esther, C.: La Validez. Univ. Alicante. 6–13 (2011)
28. Martínez, M., March, T.: Caracterización de la validez y confiabilidad en el constructo metodológico de la investigación social. REDHECS. **20**, 107–127 (2015)
29. Mayaute, L.M.E.: Cuantificación de la validez de contenido por criterio de jueces. Rev. Psicol. **6**, 103–111 (1988)
30. Montero, G.: La cosmovisión de los pueblos indígenas. https://www.uv.mx/blogs/iihs/investigadores/199-2. pp. 105–126 (2016)

Resources, the Environment and Energy Security: Unpacking the Challenge of Blackouts

Wojciech Horyn

Abstract One of the areas of care for environmental protection is the reduction of greenhouse gases. The production of electricity primarily based on conventional energy sources, which are high emitters, significantly contributes to their formation. The usc of renewable energy meets these needs. Along with environmental security, the security of individuals, society, and the state is paramount, and we are talking about energy security. In the article, the literature analysis and case study method were used. Due to the threats that appear there, selected cases of the blackout were presented, and it was shown that a complete transition to renewable energy sources, despite the introduction of modern technological solutions, may still be the cause of the blackout. Therefore, technological solutions should be sought, energy storage should be built, and local security systems should be established to ensure energy security.

Keywords Energy security · Blackout · Renewable energy sources

1 Introduction

Environmental protection has always played a significant role in the functioning of society. However, in recent decades, humankind has realised that the lack of proper care for the environment in the long term may lead to extinction. In the twenty-first century, we are witnessing the so-called greenhouse effect. The temperature is rising, glaciers are melting, and the temperature of the oceans and seas is also growing. The high emission of carbon dioxide, creating the ozone hole, is considered the primary cause of the greenhouse effect. International actions are being taken to limit carbon dioxide emissions to prevent that. These include introducing modern technologies in the industry, moving away from cars powered by traditional fuels to electric cells, or moving away from coal power to alternative energy sources. There is a change in the philosophy underpinning the approach to energy sources in that

W. Horyn (✉)
The General Kosciuszko Military University of Land Forces, 51-147 Wroclaw, Poland
e-mail: wojciech.horyn@awl.edu.pl

A. Visvizi et al. (eds.), *Research and Innovation Forum 2022*,
Springer Proceedings in Complexity, https://doi.org/10.1007/978-3-031-19560-0_64

renewable energy sources are emphasized today [1]. These include: solar radiation (solar energy), wind energy (wind power), water fall energy (hydropower), biomass (energy from the combustion of plants), geothermal energy (energy of hot deep water), energy of sea tides, and temperature differences between surface and deep water. The article presents the development of renewable energy sources and their importance for the energy industry internationally and in Poland. It was indicated that blackout phenomena are often caused by disruptions in the proper functioning of the transmission of energy produced from renewable energy sources.

Most often in the literature, blackout phenomena are described as a technological problem with transmission lines and distribution systems. On the other hand, less attention is paid to the interrelationships between individual energy sources, i.e. the causes resulting in a blackout. The purpose of the article is to point out the complexity of the process of using energy only from renewable sources. Given the experience of using renewable energy sources, blackout has occurred several times. In order to prevent this, it is necessary to conduct research on improving the use of "green energy" and seek new technological solutions. Since there is no return to fossil energy sources for environmental reasons, the issues presented in the article should still be pursued in research.

1.1 Environmental Protection and the Transition to Renewable Energy Sources

The development of the use of renewable energy sources is carried out in three areas [2], including electricity from renewable energy sources, heat and cold from renewable energy sources, biocomponents used in liquid fuels and liquid biofuels. In this paper the first area, namely, electricity and renewable energy sources, is addressed. Environmental protection issues influence the policy pursued on the international arena, especially the European one. Representatives of the European Union (EU) Member States have decided to reduce greenhouse gases, which involves moving away from coal-based energy towards renewable energy sources [3–11].

By 2050, greenhouse gases are to be reduced by using renewable energy as one of the solutions. The Paris Agreement aims to limit global warming by 2 °C and further reduce it by 1.5 °C [12]. Under the Agreement, countries will strive to achieve carbon neutrality by 2050 through low-carbon solutions [13]. China has committed to carbon neutrality [14]. The President of the United States of America also declared his acceptance of the Paris Agreement in January 2021 [15]. The EU has adopted an ambitious target of climate neutrality by 2050 and a 32% share of renewable energy sources by 2030 [16].

Renewable energy sources are an alternative to power generated from conventional fossil fuels, which [2] still form the basis of the global energy system. Currently, about 81% of the world's total primary energy production comes from fossil fuels, with oil accounting for 31.5%, coal 26.9%, natural gas 22.8%, biofuels 9.3%, nuclear power

4.9%, and hydropower 2.5%. Only 2.1% of the world's primary energy comes from solar, wind, geothermal, biomass, or other alternative energy sources [17, 18].

Globally, there is a shift away from conventional energy towards renewable energy, exemplified by the annual increase in the share of renewable energy in global energy production. Renewable energy generation capacity in 2020 was 2799 GW, of which hydropower was 1211 GW, wind energy 733 GW, solar energy 714 GW, bioenergy 127 GW, geothermal energy 14 GW, and marine energy 500 MW [19].

Renewable energy capacity increased by 261 GW (+10.3%), of which solar energy by 127 GW (+22%), wind energy by 111 GW (+18%), hydropower by 20 GW (+2%), bioenergy by 2 GW (+2%), and geothermal energy by 164 MW. Solar and wind energy accounted for 91% of all renewable energy sources. Investment in renewable energy is diverse. The most considerable investments are in North America and Europe, as shown in Table 1.

1.2 *Investing in Renewable Energy Sources in Poland*

A similar situation is also taking place in Poland, where the government implements the European Union's recommendations. Through programs involving subsidies for installations based on renewable energy sources, Poland has seen a dynamic increase in the capacity and number of installed devices.

Generators in small RES installations (765 entities) produced over 340 GWh of energy in 2020. Compared to 2019, the most significant number of facilities using solar power increased by 33%, including a 41% increase in capacity. Table 2 shows the capacity of installed installations with a breakdown by source type [20].

In addition to small installations, 459 168 micro-installations were installed in Poland (as of the end of 2020) [21]. Their total installed capacity was over 3 GW. The micro-installations using solar radiation energy (PV) were the greatest in terms of the number (458 675) and installed capacity (3 015.4 MW). It is displayed below in Table 3.

Even though the above summary of developments covers only one European country, it shows the direction of the dynamic changes taking place in energy policy approaches.

2 Methodology

When considering the issue of energy security in the context of a blackout, the following research question was posed: to what extent can renewable sources be the cause of an energy emergency and lead to a blackout?

A literature review and qualitative analysis were carried out to answer the question. The literature was selected based on time and subject censorship, which included an analysis of publications, reports, and articles covering the use of renewable energy

Table 1 Renewable generation capacity by region

	North America	Central America and the Caribbean	South America	Europe	Middle East	Africa	Euroasia	Asia	Oceania
Capacity	422 GW	16 GW	233 GW	609 GW	24 GW	54 GW	116 GW	1286 GW	44 GW
Global share	15%	1%	8%	22%	1%	2%	4%	46%	2%
Change	+32.1 GW	+0.3 GW	+9.2 GW	+34.3 GW	+1.2 GW	+2.6 GW	+6.2 GW	+167.6 GW	+6.9 GW
Growth	+8.2%	+2.1%	+4.1%	+6.0%	+5.2%	+5.0%	+6.0%	+15%	+18.5%

IRENA [19]

Table 2 RES installations entered in the register of small-scale energy producers by source type (as of the end of 2020)

Type of RES installation	Number of installations	Installed capacity [MW]
Using hydropower (WO)	343	51,96
Using solar thermal (PV)	328	66,86
Using biogas (BG)	117	32,10
Using wind energy (WI)	108	31,71
Using biomass (BM)	2	0,47
Total	**898**	**183,10**

Electricity generation in Poland in small RES installations [20]

Table 3 RES micro-installations by type of renewable energy source (as of the end of 2020)

RES micro installation type	Number of micro-installations [pcs.]	Total installed capacity [MW]
Using biogas other than agricultural biogas	20	0,1
Using agricultural biogas	30	0,8
Using biomass	18	0,3
Using solar radiation	458 675	3 015,4
Using solar radiation/biogas other than agricultural biogas	1	0,0
Using solar radiation/wind	44	0,5
Using solar radiation	4	0,1
Wind-based	83	0,5
Hydro-based	293	8,0
Total	**459 168**	**3 025,8**

Energy Regulatory Office [21]

sources and describing blackout phenomena. Four events covering the year 2003—USA/Canada, Italy, Scandinavia; 2006—Europe; 2016—Australia; 2018—China were selected for analysis. The study used is based on a qualitative, case study approach, which is commonly used in security sciences. Due to the framework of the article and the lack of need to replicate the findings described in the articles, the study details the main issues related to the research question and identifies the sources to familiarise the reader with the details of the events described.

The study is based on the presentation of the importance of renewable energy sources in the energy economy and environmental protection, the problems that may arise in the production and transmission of energy from renewable sources and extreme situations—blackouts, the cause of which, often indirectly, are renewable energy sources. The methodology applied in this way made it possible to find answers to the research question and present them in the form of conclusions.

3 Technical Problems in the Power Production and Distribution and the Blackout Phenomenon

In today's world, energy issues must be viewed from a broad, global perspective. Although electricity is produced in individual countries, a more comprehensive range of consumers takes advantage of it. Individual power transmission lines are interconnected to form an energy supply system. The European grid covering the area from Lisbon to Istanbul is divided into two regions: the northwest and southeast. Moreover, it appears imperative that transmission networks have a frequency of 50 Hz to operate smoothly since any deviation could damage the connected equipment. If frequency fluctuations were not reduced within a few minutes, it could cause damage to the entire European high-voltage grid, potentially causing power cuts for millions of customers. Any disruption to the grid can cause a cascading phenomenon, which translates into the occurrence of a blackout.

The causes of a blackout include:

A. failures,
B. network overloads caused by the excess current in the network or bad distribution (scheduling of power demand due to weather conditions, morning start-up, etc.).

Failures

Failures can occur from many factors, including unpredictable ones, such as damage to the hardware or software responsible for producing and distributing electricity. Other causes may include fires, earthquakes, and tsunamis. Failures can also occur in the power plant itself, where a team of specialists makes decisions to shut down individual power units on an ongoing basis, and the resulting shortages are made up from external suppliers. Since the power system is a network of connections between electricity producers, there is a real possibility of meeting energy shortages. It requires sufficient flexibility in the power system operation.

Weather anomalies can be another reason for failure, but these can be anticipated, and measures can be taken to protect the power system. Cyclones have a significant impact on the operation of wind farms, hydroelectric power plants or photovoltaics and power lines. For example, a storm or gale caused by a typhoon can lead to a temporary increase in wind power, a wind turbine stop, or even damage to the wind turbine. If the electricity system cannot consume the sudden growth in wind power over time, wind abandonment will occur, further decreasing the energy efficiency of power systems. On the other hand, if the reserve generation capacity or transmission capacity is insufficient, wind power will suddenly decrease and load loss will occur. In the meantime, a large amount of rainfall occurs, which will affect the performance of hydropower plants. Thus, wind strength and speed, and heavy rainfall under the influence of a hurricane, can affect the output relationship of each power source and load the system at a highly unfavorable level for safe, stable, and economical electricity generation and distribution [22]. China alone experiences an average of 7.2 typhoons in the southeastern coastal areas each year, resulting in a direct loss of approximately 0.4% of GDP [23, 24].

The difference between unpredictable and weather-related failures is that in the case of weather anomalies, smart grids can be built and the parameters and functionality of the power generation equipment can be selected to optimally utilize the phenomena occurring (e.g. wind intensity in a time interval, amount of precipitation, etc.).

Network overloads, cascading, caused by overcurrent in the network or lousy distribution (scheduling of power demand due to weather conditions, morning start-up, etc.)

As mentioned earlier, the current international trend is directed towards increasing the amount of energy produced from renewable sources and away from conventional fossil fuels. It makes the renewable energy generated dependent on variable and intermittent factors such as wind and sun. Thereby, it creates the need to develop the energy system flexibility. In the literature, flexibility is distinguished into three main categories: planning, operational, and exported. Planning flexibility is mainly concerned with long-term planning related to the transmission of the system design [25]. Operational flexibility is related to the hardware of the generation system and its real-time response to power changes through optimized controllability [26]. Exported flexibility means the operational flexibility that can be offered to a neighboring network through tie points [27].

Renewable energy generation (solar, wind) is characterized by variability and uncertainty. The frequent and natural fluctuations of wind and photovoltaic power pose a challenge to conventional domestic generators. It is dictated by the need for fast, sudden and significant ramp-ups and frequent start-ups. Reserves appear necessary due to the inevitable errors between forecast and actual wind and PV output. Hence, to adequately accommodate large amounts of wind and photovoltaic energy, the system must be flexible enough to keep up with the variability and uncertainty [28]. Too much variation in energy production can cause disturbances in the grid's transmission frequency stability, thus resulting in a blackout [29].

The load on the energy system results from economic and technological factors and is determined by supply and demand. The energy produced from renewable sources often meets the needs of energy producers (prosumers), and the surplus is fed into the grid. Nonetheless, when there is a lack of energy from renewable sources (no wind, night-time), the energy must be produced by other energy sources or used from energy storage. Therefore, for power production and distribution, scheduling of energy production is an essential element. For example, during the morning hours, the industry and other places commence work, and the demand for electricity increases. In many electricity systems, such rapid changes occur in the morning when demand can change by about 30% within a few hours [30, 31].

Conventional power plants, mainly coal-fired, are used to generate energy to compensate for shortfalls during the peak period.

4 Examples of Blackouts—A Case Study

Blackouts have already occurred many times in the twenty-first century. The year 2003 was a special one, where within two months, power systems failed in the USA/Canada on 14 August 2003, in Italy on 28 September 2003, and in Scandinavia on 23 September 2003. These cases have been widely described in the literature, among others, the articles "General Blackout in Italy", "Blackouts in North America…", and "Blackout in the US/Canada…" [32–34] present a detailed analysis of the blackout cases. The analysis shows that the main reasons for power system failures were overvoltage, overload, and voltage problems in the network. They were most often caused by situations that did not foretell such events. These included repairs of power units, maintenance of power equipment, and the economic factor—the decision to purchase (especially in the peak period) a certain amount of energy from a neighboring country.

The above factors brought about disturbances in the operation of power grids and primarily in the transmission of energy with adequate capacity. Each difference and direction of energy flow caused jumps in voltage and transmission capacities. For technological reasons, with too large or uncontrolled power fluctuations, protection systems were activated, which separated the system due to voltage instability and collapse. The occurrence of such an event in one power area may affect the next one, thus, causing a cascade of events.

Subsequent blackouts indicate existing risks directly or indirectly related to renewable energy sources. The blackout phenomenon occurred in Europe on 4 November 2006. As a result, 15 million European citizens in Austria, Belgium, the Netherlands, France, Germany, Italy, Portugal, and Spain were left without electricity. The outage meant that about 15 GW of electricity was not supplied to consumers for almost 1.5 h [35] (Table 4).

The event, which took place in the evening on 4 November 2006, indicates technological and communication problems between network operators and the speed of the occurrence. The planned disconnection of a high-voltage line on the Ems River in the Netz network circuit forced a 13,700 MW energy shortfall to be made up. Part of the new energy came from renewable sources—3200 MW from wind power stations.

Table 4 Load shedding of some TSO

Country/TSO	Load shed (GW)	Country/TSO	Load shed (GW)
Austria/APG	1.5	Italy/TERNA	1.5
Austria/Tiwag	0.04	Netherlands/TenneT	0.4
Belgium/Elia	0.8	Portugal/REN	0.5
France/RTE	5.2	Apanish/REE	2.1
Germany/E.ON	0.4	Slovenia/ELES	0.1
Germany/RWE	2		

Li et al. [35]

Due to the weather conditions, there was a decline in the wind power produced. The E.OW flow of the Netz system was combined with the neighboring RWE system to compensate for the shortfall in transmitted energy. The resulting excessive grid instability, underloading, and then overloading caused electricity to seek alternative routes within an astonishing 14 s, and a cascade of power line outages spread across Germany. In another 5 s, the outage cascaded as far east as Romania and Croatia in the southwest.

The above example shows that renewables had a part to play in making the wrong decisions, resulting in the blackout.

Another case is the occurrence of typhoon-induced power disruptions. Typhoon Rammasun destroyed 220 k transmission lines in Guangdong province in China in July 2014, and super typhoon Mangkhut, which occurred on the coast in Guangdong province on 18 September 2018, led to the shutdown of more than 200 lines and deprived nearly 6 million users of electricity [36, 37]. According to researchers, such a phenomenon will appear cyclically, and counteracting it necessitates developing new ways of modelling the functioning of renewable energy in extreme situations.

The blackout in Australia on 28 September 2016 indicates a close link between the event that occurred and the production of energy from renewable sources [38]. On that day, a severe storm crashed the South Australia (SA) grid and damaged several remote transmission towers resulting in the SA grid losing about 52% of its wind generation within minutes. That deficit had to be compensated for by importing power from neighboring Victoria (VIC) via the Heywood AC interconnection. About 1.7 million people were left without power [7, 39]. At the time of the blackout, of the approximately 2900 MW of load, wind power accounted for about 1600 MW and photovoltaic (PV) generation 730 MW. Before the blackout, almost 50% of the total was provided by wind and PV as demand in SA, while conventional synchronous generation accounted for only 17.6%, with the rest coming from interconnection. The energy production is shown in Table 5. The resulting overvoltage redistribution, the drop and subsequent increase in frequency caused the separation and cascading of the event [40] Despite modern protections, too rapid changes in current flow caused the equipment to be unable to react in time, and the resulting delay in equipment operation led to separation and the blackout.

Table 5 Pre-event power generation profile in South Australia

Source	Generation (MW)
Synchronous generators	330
Wind	883
PV (behind-meter)	50
Import from Victoria via Heywood	500
Import from Victoria via Murraylink	113
Total load (including losses)	1826

Yan et al. [40]

5 Conclusions

The above analysis has shown that renewable energy sources are gaining ground in energy production, which is only the beginning of the transition away from fossil energy sources. Also, as the recent case in Australia shows, renewable energy sources will continue to contribute to the blackout phenomenon. Conventional energy sources most often compensate for energy shortfalls. Nonetheless, the basis for the efficiency of the system is the production of modern equipment to secure the energy transmission and skillful management of the power grid so that decisions taken in situations of disturbance in the production and distribution of energy are quick and the algorithms developed meet actual needs. By switching to renewable energy sources, there is a need to change the philosophy of using such energy. At present, the shortcomings in energy compensation that result from its intermittency (wind, sun) are compensated for by supplementing energy from other suppliers, sometimes thousands of kilometers away from the user. However, it raises technological and logistical problems. The best solution is self-sufficiency, i.e., planning the production of renewable energy so that it is used for current and local needs. Surplus energy from energy production should be stored at the production stage and used by the producer. However, the analysis made by Narayanan et al. [41] indicates that green energy is not able to satisfy 100% of energy demand. The study results showed that renewable energy sources could produce 63% of the energy needed on a global basis with no breakdown by day or week to secure the 75,000 inhabitants city of Kortrijk in Belgium. It seems that renewables cannot be the only solution for the future. Regardless of the environmental costs, there is still room for traditional energy sources at the current technological stage, all the more so if we want to prevent blackouts.

One of the solutions is the use of artificial intelligence for the skillful management of energy sources, e.g., in cities, but this is another area of possible research [42].

Furthermore, for blackouts to occur less frequently, it is necessary to:

- upgrade protection devices to make them more sensitive to power and frequency fluctuations,
- build modern transmission networks,
- introduce software into energy systems that will make optimal decisions in real-time,
- develop low-cost energy storage technology so that the economic factor does not interfere with the decision to invest in energy storage,
- create local security systems based on the generators using the maximum amount of energy,
- Introducing the above proposals will increase the security of using renewable energy and reduce the likelihood of blackouts.

There is a need for continued scientific research to improve the management of transmission lines and the distribution system to prevent further blackouts. Artificial intelligence can be used in this regard. Another area of research should be the search for alternative energy sources (e.g., hydrogen) to fill the gap in an unstable energy system.

References

1. Lewandowski, W.M.: Proekologiczne odnawialne źródła energii [Pro-environmental renewable energy sources], Wydawnictwa Naukowo-Techniczne, Warszawa (2006)
2. Kopczewski, M., Narloch, J.: Bezpieczeństwo energetyczne. Odnawialne źródła energii [Energy Security. Renewable energy sources], Wydawnictwo Akademii Wojsk Lądowych, Wrocław (2019)
3. Report from the Commission to the European Parliament, The Council, The European Economic and Social Committee and the Committee of the Regions. Renewable Energy Progress Report, COM 952 final (2020)
4. European Commission, European Climate Pact, ML-03–20–815-EN-N. https://doi.org/10.2834/43590
5. European Commission, Feasibility and scoping study for the commission to become climate neutral by 2030. Final report, ML-02–20–693-EN-N. https://doi.org/10.2834/020283
6. European Commission, State of the Union 2020. National energy and climate plans: Member State contributions to the EU's 2030 climate ambitions, ML-04–20–490-EN-N. https://doi.org/10.2834/046416
7. European Commission, In-depth report on the results of the European Climate Pact open public consultation. Final report, ML-06–20–097-EN-N. https://doi.org/10.2834/008230
8. European Commission, Going climate-neutral by 2050. A strategic long-term vision for a prosperous, modern, competitive and climate-neutral EU economy, ML-04–19–339-EN-N. https://doi.org/10.2834/02074
9. European Commission, Climate change, Report. Special Eurobarometer 459, March 2017
10. European Commission, Climate change, November 2015. https://doi.org/10.2834/90050
11. Renewables 2021. Global Status Report, Paris 2021
12. United Nations. Paris Agreement. 2015. Available online: https://unfccc.int/process-and-meetings/the-paris-agreement/the-umowaparyska. Accessed 20 May 2021
13. Connolly, D., Lund, H., Mathiesen, B.V.: Smart Energy Europe: The technical and economic impact of one potential 100% renewable energy scenario for the European Union. Renew. Sustain. Energy Rev. **60**, 1634–1653 (2016)
14. The Guardian. China Pledges to Become Carbon Neutral Before 2060. Available online: https://www.theguardian.com/environment/2020/sep/22/china-pledges-to-reach-carbon-neutrality-before-2060 (2020). Accessed 2 May 2021
15. Joseph, R., Biden, J.R.: Acceptance on Behalf of the United States of America. Available online: https://www.whitehouse.gov/briefing-room/statements-releases/2021/01/20/paris-climate-agreement/ (2021). Accessed 10 March 2021
16. Makešová, M., Valentová, M.: The concept of multiple impacts of renewable energy sources: a critical review. Energies **14**, 3183 (2021). https://doi.org/10.3390/en14113183
17. International Energy Agency (IEA). Statistics Report. IEA World Energy Balances 2020: Overview. Available online: https://www.iea.org/reports/world-energy-balances-overview#world (2020). Accessed 12 May 2021
18. Komarnicka, A., Murawska, A.: Comparison of consumption and renewable sources of energy in European Union Countries—sectoral indicators, economic conditions and environmental impacts. Energies **14**, 3714 (2021). https://doi.org/10.3390/en14123714
19. IRENA, International Renewable Energy Agency, Renewable capacity highlights, 31 March 2021. https://www.irena.org/publications/2021/Aug/Renewable-energy-statistics-2021. Accessed 12 June 2021
20. Electricity generation in Poland in small RES installations. Report of the President of ERO for 2020. (Legal basis: Article 17 of the Renewable Energy Sources Act), Warsaw, April 2021
21. Energy Regulatory Office, Report containing summary information concerning electricity generated from a renewable energy source in a micro-installation (including by prosumers) and introduced into the distribution grid in 2020 (Article 6a of the RES Act), Warsaw, March 2021

22. Qian, M., Chen, N., Chen, Y., Chen, C., Qiu, W., Zhao, D., Lin, Z.: Optimal coordinated dispatching strategy of multi-sources power system with wind, hydro and thermal power based on CVaR in Typhoon environment. Energies **14**, 3735 (2021). https://doi.org/10.3390/en14133735
23. Luo, Y.L., Sun, J.S., Li, Y., Xia, R.D., Du, Y., Yang, S., Zhang, Y.C., Chen, J., Dai, K., Shen, X.S., et al.: Science and prediction of heavy rainfall over China: research progress since the reform and opening-up of new China. J. Meteorol. Res. **34**, 427–459 (2020)
24. An, Z., Shen, C., Zheng, Z.T., Liu, F., Chang, X.Q., Wei, W.: Scenario-based analysis and probability assessment of sub-synchronous oscillation caused by wind farms with direct-driven wind generators. J. Mod. Power Syst. Clean Energy 7, 243–253 (2019)
25. Lannoye, E., Flynn, D., O'Malley, M.: Transmission, variable generation, and power system flexibility. IEEE Trans. Power Syst. **30**, 57–66 (2015)
26. Yang, J., Zhang, L., Han, X., Wang, M.: Evaluation of operational flexibility for power system with energy storage. In: Proceedings of the 2016 International Conference on Smart Grid and Clean Energy Technologies (ICSGCE), Chengdu, China, 19–22 October 2016, pp. 187–191 (2016)
27. Bucher, M.A., Chatzivasileiadis, S., Andersson, G.: Managing flexibility in multi-area power systems. IEEE Trans. Power Syst. **31**, 1218–1226 (2016)
28. Mladenov, V., Chobanov, V., Georgiev, A.: Impact of renewable energy sources on power system flexibility requirements. Energies **14**, 2813 (2021). https://doi.org/10.3390/en14102813
29. Dreidy, M., Mokhlis, H., Mekhilef, S.: Application of meta-heuristic techniques for optimal load shedding in islanded distribution network with high penetration of solar PV generation. Energies **10**, 150 (2017). https://doi.org/10.3390/en10020150
30. Mielczarski, W.: Impact of Energy Storage on Load Balancing. In: Proceedings of the 2018 15th International Conference on the European Energy Market (EEM), Lodz, Poland, 27–29 June 2018, pp. 1–5
31. Chudy, D., Leśniak, A.: Advantages of applying large-scale energy storage for load-generation balancing. Energies **14**, 3093 (2021). https://doi.org/10.3390/en14113093
32. Corsi S., Sabelli, C.: General blackout in Italy Sunday September 28, 2003, h. 03:28:00, IEEE Power Engineering Society General Meeting, 1691–1702 (2004). https://doi.org/10.1109/PES.2004.1373162
33. Makarov, Y.V., Reshetov, V.I., Stroev, V.A., Voropai, N.I.: Blackouts in North America and Europe: analysis and generalization. IEEE Russia Power Tech. 1–7 (2005). https://doi.org/10.1109/PTC.2005.4524782
34. Bialek, J.W.: Blackouts in the US/Canada and continental Europe in 2003: Is liberalisation to blame? IEEE Russia Power Tech. 1–7 (2005). https://doi.org/10.1109/PTC.2005.4524781
35. Li, C.L.,et al.: Analysis of the blackout in Europe on November 4, 2006. In: 2007 International Power Engineering Conference (IPEC 2007), pp.939–944 (2007)
36. Feng, L., Hu, S., Liu, X.T., Xiao, H., Pan, X., Xia, F., Ou, G.H., Zhang, C.: Precipitation microphysical characteristics of typhoon Mangkhut in southern China using 2D video dendrometers. Atmosphere **11**, 975 (2020)
37. Ma, X.C.: Impact of winter meteorological disasters on wind farms. In: Proceedings of the 2018 5th International Conference on Key Engineering Materials and Computer Science, Vancouver, BC, Canada, 10–12 January 2018
38. Todd, M.J., Yıldırım, E.A.: On Khachiyan's algorithm for the computation of minimum-volume enclosing ellipsoids. Discret. Appl. Math. **155**(13), 1731–1744 (2007)
39. Kumar, P., Yildirim, E.A.: Minimum volume enclosing ellipsoids and core sets. J. Optim. Theory Appl. **126**(1), 1–21 (2005)
40. Yan, R., Masood, M., Saha, T.K., Bai, F., Gu, H.: The anatomy of the 2016 South Australia blackout: a catastrophic event in a high renewable network. IEEE Trans. Power Syst. **33**(5), 5374–5388 (2018). https://doi.org/10.1109/TPWRS.2018.2820150

41. Narayanan, A., Mets, K., Strobbe, M., Develder, C.H.: Feasibility of 100% renewable energy-based electricity production for cities with storage and flexibility. Renew. Energy **134**, 698–709 (2019)
42. Chui, K.T., Lytras, M.D., Visvizi, A.: Energy sustainability in smart cietes: artificial intelligence, smart monitoring, and optimization of energy consuption. Energies **11**, 2869 (2018). https://doi.org/10.3390/en11112869

Design Thinking, Soft Skills Development, and the Digital Society

Marina-Paola Ojan and Pablo Lara-Navarra

Abstract The Covid-19 pandemic reinforced changes that had been taking place over the past decade. These changes were triggered by the Fourth Industrial Revolution (4IR). The 4IR does affect the industry and the human being as we know it. In this way, it is not hard to envision a society where technology is so developed and integrated into our lives that it increases humanity's capacity and threshold for human intelligence, cognition, and physical abilities. This research proposes a teaching–learning framework for the techno-humanist ecosystem by developing the demanded soft skills with the design process methodology. The first step of this research is to understand the essential elements to conceptualise a framework. To reach this objective, we started with a systematic review of the literature to analyse, and the second item is understanding the evolution of job offer asking for soft skills. We are investigating soft skills, design thinking and digital humanism, and the job market evolution. Then, considering the state of the art, this research will study how those ideas converge and affect each other. Finally, the primary outcome of this research will be a visual matrix representing the point of junction of these concepts. It will help to build the initial elements of a framework of a teaching–learning model that can help adapt to and face the challenges of a human-digital society.

Keywords Soft skills · Digital humanism · Design thinking

1 Introduction

The Fourth Industrial Revolution (4IR) [1] is a term used to study the impact of emerging technologies on human development at the beginning of the twenty-first century. It is transforming the social, economic and technological rules that we have

M.-P. Ojan (✉)
Istituto Europeo di Design. Biada, 11, 08012 Barcelona, Spain
e-mail: m.ojan@ied.es

P. Lara-Navarra
Universitat Oberta de Catalunya. Tibidabo, 39, 08035 Barcelona, Spain
e-mail: plara@uoc.edu

A. Visvizi et al. (eds.), *Research and Innovation Forum 2022*,
Springer Proceedings in Complexity, https://doi.org/10.1007/978-3-031-19560-0_65

been endowing since the nineteenth century. Therefore, it is due to the hybridisation of physical, digital, and biological elements such as quantum computing and nanotechnology, providing new approaches to the complex interaction between humans and technology, provoking new ways of perceiving, acting and being. This concept analyses the impact of emerging factors on the economy, politics, law, technology and the environment. In this sense, we can affirm that these changes impact the way of working, consumption habits, communicating and informing ourselves, and our way of life.

The technological change of the 4IR is a relevant engine of transformation in all industries and society. It demands jobs requiring higher education and technical studies, leaving aside jobs involving physical or routine tasks. The WEF's Future of Jobs 2016 Report surveys top HR executives and predicts that future jobs will increasingly require complex problem-solving, social and systems skills.

Considering the challenges posed by the 4IR about education and employability, this work presents a new educational research framework based on design thinking and soft skills and the study of the use of soft skills in the labour market. Next, the methodology developed for analysing sources of information in bibliographic databases is presented with the study of the profiles of job offers in the Spanish portal Infojobs. Finally, the principal results and conclusions of the proposed research are established.

2 Methodology

This research is an exploratory work based on two analyses. First, a search of studies that support the theoretical framework of the use of design thinking for the teaching–learning of soft skills in a digital humanistic context. The databases selected to carry out the searches were Scopus, ProQuest Central and Factiva-Reuters. Specific queries were established mixing techno-humanism, digital-humanism, design, design thinking and soft skills.

On the other hand, a qualitative analysis of the Spanish job offers on a selected job search platforms (Infojobs) was made. This platform allows searches filtered by keywords, geographical area and date of publication of the offer.

Different search equations were established once the information source was selected to perform the data extraction for its subsequent analysis. For the research, we choose to stay within the framework of the Spanish territory, including offers throughout the country from local and foreign companies. We focus on the most recent proposals at the time frame level, published in 15 days.

3 Evolution of Design Thinking

Studying the evolution of design thinking, we detected two theoretical branches. Both are human-centred and oriented to collaborative innovation management, but the first focus on a problem-solving process, and the second is oriented toward a project management methodology.

The best promoter of this second approach is the global design and innovation company IDEO, through David Kelly, Tim Brown and Jane Fulton. Fulton [2] considers Human-Centred Design and Design Thinking a set of "corporate ethnography" methods. Brown [3] defines Design Thinking as a methodology that permeates the entire spectrum of innovation activities with a spirit of Human-Centred Design and a creative management process for companies and social projects.

Brown's approach to Design Thinking focuses on presenting the creative exercise as a design activity centred on people, which includes versatile tools that stimulate collaborative innovation; and working with indeterminate or "wicked" problems [3, 4]. Thus, as of 2008, Design Thinking began to sound like business management and design service management strategy [5].

From this approach, Design Thinking is usually assumed as a set of linear and iterative creative-collaborative methodologies to respond to a design challenge through different phases of work.

From project management [6], use the following four questions to identify four critical stages of a project: "What is it?" refers to the current scenario; "What if?" refers to the vision of the future; "What enthusiasm?" refers to user empathy; "What works?" refers to market implementation.

On the other hand, in 2005, Jane Fulton published a book on Design Thinking as an ethnographic process to develop creative intuition.

The main characteristics of his approach are the following: Human-Centred Design should be assumed as an ethnographic research process called "corporate ethnography"; Radical innovation is achieved through the development of intuition in the process of design and the observation of the thoughtless acts of people; Design is a social situation.

The human-centred perspective of Design Thinking is crucial for this research to connect this methodology with the skills required for the job market in the 4IR society.

4 Basics of Soft Skills

Until the twentieth century, most manufacturing companies based their production on technical aspects linked to methodologies, systems, and facilities. Therefore, they needed to select technically prepared workers capable of performing specific tasks. In recent decades, however, the rapid progress and rise of digital technologies and artificial intelligence have created a world that is more interconnected and complex in many ways [7–9]. Before this paradigm shift, workers were asked for certain technical skills obtained through specific formal (or non-formal) studies that allowed them to perform in a job: the so-called hard skills. Today, these skills have been joined by others that are not linked to a specific task but that companies, organisations and institutions consider strategic elements.

To cite one of the large consultancies researching the future of professional figures, the McKinsey Global Institute has analysed the type of jobs lost and created with the consolidation of automation, AI, and robotics. In addition, this analysis has identified the kind of high-level skills that will become increasingly important: the demand for basic manual, physical and cognitive skills will decrease, but the need for more advanced technological, social, emotional and cognitive skills will increase.

Going back to the literature, since the '90 s of the last century [10], attempts have been made to delimit, define, relate and standardise names and descriptors of the skills or competencies that fall outside the framework of the hard skills. Even international organizations and research teams [7–13] have begun to link the new economic and industrial context with the demand for skilled workers who also have non-technical skills, such as attitudes and additional skills. Many studies show the difficulty of differentiating a psychological attitude towards a task from a competence-based on abilities and skills.

Research on "Soft Skills in Higher Education" [14] highlights how the record of soft skills includes a wide range of attributes that changes depending on the study or intervention. It is a central problem with skills today: their relevance depends on the historical moment and the context, with variables such as the occupation, lifestyle, and society of the person who evaluates the skills or who is in demand coming into play. Moreover, it implies that it is difficult to define it in a standard way: the names, descriptors and even indicators are completely different over time and the organization that states them.

Our research shows that most of the literature on the concept of 'skill' emphasizes that all skills can be learned and developed through appropriate (and observable) training of particular types of activities and tasks combined to achieve complex results.

According to some writers, skills can be thought of as behaviours that apply knowledge, skills, and personality traits. Others say they constitute the corpus of knowledge, procedures, skills, abilities, and attitudes needed to carry out various activities with a certain degree of quality, efficiency, independence and flexibility.

Kingsley [15] distinguishes between Self-Oriented/Intrapsychic (what the person must understand and develop on their own) and Other-Oriented/Interpersonal Skills (what the person can develop by interacting with other people). This distinction can also be made in terms of Personal Skills (mainly corresponding to Cognitive Skills, such as knowledge and thinking skills) and Social Skills (referring to relationships with other people) [16].

Some authors define soft skills as self-oriented or intrapersonal and other-oriented or interpersonal skills, considered socio-emotional skills essential for personal development, social participation and job success [15, 17]. They include skills such as communication, the ability to work in multidisciplinary teams, adaptability, resilience among others.

The interpersonal can be divided into two groups: individual and social. In addition, systemic competencies are considered organisational, business and leadership skills [18]. Delving into the development of the concept of soft skills is crucial in this research to identify those most related to the human-centred perspective of design thinking. For example, suppose that design thinking is a lever for strategic thinking and the engine of creative change necessary for a techno-humanistic society. In that case, soft skills are the vehicle for people to manage these changes.

Another concern that we find when analysing soft skills is their evaluation. There is indeed talk of the need to enhance these competencies so that future workers possess them; however, their evaluation is highly complicated: "many of these attributes can only be assessed subjectively, that is, there are no objective tests for, say, skills interpersonal and managerial. Furthermore, several attributes that make up the soft skills taxonomies refer to dispositional traits that may change very little over the years of higher education and are known to affect academic grades" [14]. Perhaps this is one of the reasons why formal and non-formal education finds it more challenging to include soft skills training programs in their curricular designs. We took into consideration that: there is no catalogue of specific competencies with explicit descriptors (rubrics) to train them; skills change depending on the need for each training/employment cycle; the ways of measuring them are subjective or outdated, and many of these more "recent" tests are from the end of the twentieth century.

5 Results

From the queries done in the database and the platform, we did not find results crossing soft skills and digital humanism, while we had a better result with design and digital humanism. In any case, we can affirm that there is a lack of study connecting digital humanism, design and soft skills.

Then, we focused on Infojob using more general keywords such as "competencies" and "skills". Again, it gave us a high result of 1603 and 5069 offers.

Entering the offers, we saw how the number of requests that meet this criterion is higher due to the degree of generalisation of the terms. It was associated with specific technical skills of the position and other more generic ones that respond to our profiling. Refining the search on the concept of ability and focusing on interpersonal skills, the result passed from 5069 to 195 results.

Finally, we tried to apply the concept of 'soft skills and 'soft skills and their incorporation into the criteria of the offers. We found eight answers using the "soft skills" versus 181 results with "soft skills". We divided, in turn, the job offers that

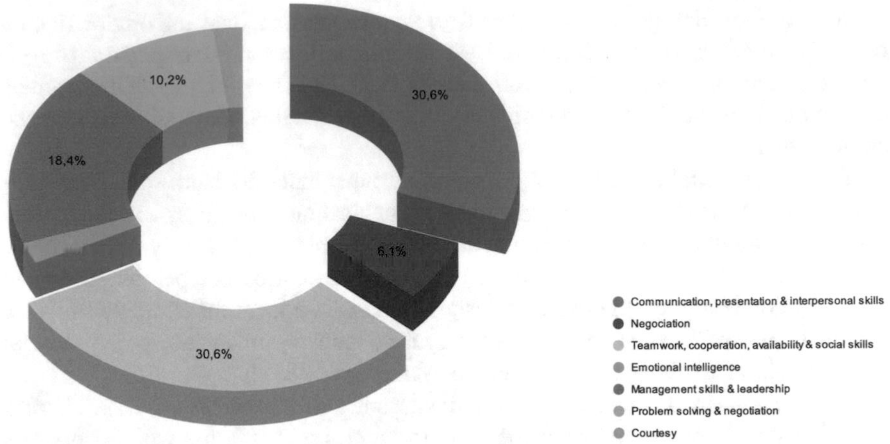

Fig. 1 Other-oriented or inter-personal skills demanded in the selected job offers

included soft skills either among the selection requirements or as a value proposition within the training opportunities offered by the job applicant company.

We focused on the offers that valued soft skills as a requirement or an element to be appreciated. Of the 181 offers, 16.6% (a total of 30 offers) specify the soft skills required for the position, and we focus on which ones were included among these skills. At the same time, we look at which ones focus on 'other-oriented' or inter-personal skills and 'self-oriented' or intra-personal skills.

Among the skills that stood out, we associate '*other-oriented*' *or inter-personal skills* (see Fig. 1).

We associate the '*self-oriented*' *or intra-personal* (see Fig. 2).

6 Conclusions

The proposed research opens a critical path to establishing soft skills as a lever for change in the 4IR. In this sense, the role of design thinking and digital humanism seems to be the most plausible scenario to produce a paradigm shift in teaching for the new labour demands. However, there are many uncertainties to be resolved.

The results from the exploratory research yield a series of conclusions classified into three groups: open theme, soft skills market and classification of soft skills.

First, we can affirm that there is a lack of study connecting digital humanism, design and soft skills. Currently, the field of digital humanism is in the analysis phase, the existing works related to the subject are still open, and academic literature is lacking. However, on the other hand, the design seems to be a methodology that can rely on digital humanism to produce changes. Both concepts have the person at the centre of the model, and, lastly, soft skills is a concept that can help reorient the role of education more specifically in the postulates of Life Long Learning.

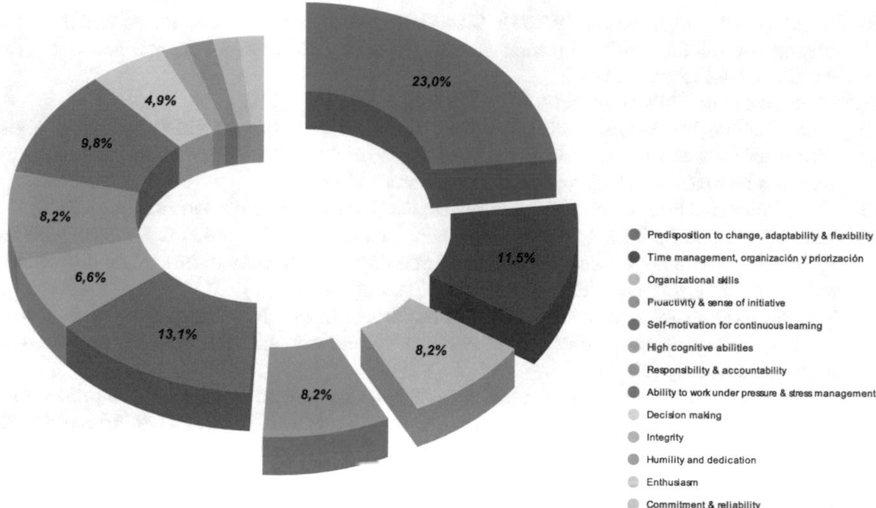

Fig. 2 Self-oriented or intra-personal skills demanded in the selected job offers

We have verified that there is still a lack of maturation in the soft skills market. Although they are necessary for change in this 4IR, very few companies still request them in job offers and, in addition, the meaning of requesting them in specific jobs or why having soft skills can help the worker in their job growth is not defined. In this sense, the projection of the soft skills in a professional career is not seen in the offers studied. Lastly, the lack of consensus in classifying soft skills does not help a correct use. Sometimes they are confused, repeated, or found in sections unrelated to soft skills, which shows the need for studies on their applicability.

References

1. Oxman, R.: Thinking difference: theories and models of parametric design thinking. Des. Stud. **52**, 4–39 (2017)
2. Fulton, J.: ¿Thoughtless acts? Observations on Intuitive Design. Chronicle, Books LLC, San Francisco (2005)
3. Brown T.: Change by Design: How Design Thinking Transforms Organization and Inspires Innovation. Harper Collins/NY (2009)
4. Kimbell, L.: Rethinking Design Thinking (part. 1). Design and Culture (2011a)
5. Kimbell, L.: Designing for service as one way of designing services. Int. J. Design (2011b)
6. Liedtka, J., Ogilvie, T.: Designing for growth: a design thinking tool kit for managers. Columbia University Press (2011)
7. Accenture: An inclusive future of work: A call to action (2018)
8. World Economic Forum (WEF): The future of jobs report: 2020. World Economic Forum (2020)
9. World Economic Forum (WEF): Towards a Reskilling Revolution: Industry-Led Action for the Future of Work (2019)

10. World Health Organization (WHO): Life skills education: planning for research as an integral part of life skills education development, implementation and maintenance (No. MNH/PSF/96.2. Rev. 1) (1996)
11. Accenture: New skills now: inclusion in the digital economy (2017)
12. United Nations Educational, Scientific, and Cultural Organization (UNESCO): Digital skills critical for jobs and social inclusion. UNESCO. https://en.unesco.org/news/digital-skills-critical-jobs-and-social-inclusion. Accessed 15 March (2018b).
13. World Economic Forum (WEF): Operating Models for the Future of Consumption (2018)
14. Chamorro-Premuzic, T., Arteche, A., Bremner, A.J., Greven, C., Furnham, A.: Soft skills in higher education: importance and improvement ratings as a function of individual differences and academic performance. Educ. Psychol. **30**(2), 221–241 (2010)
15. Kingsley, B.: Self Awareness and Emotional Intelligence. Speech at "Soft Skills and their role in employability—New perspectives in teaching, assessment and certification", workshop in Bertinoro, FC, Italy (2015)
16. Engelberg, E., Limbach-Reich, A.: The role of empathy in case management: a pilot study. Soc. Work. Educ. **34**(8), 1021–1033 (2015). https://doi.org/10.1080/02615479.2015.1087996
17. Kechagias, K.: Teaching and assessing soft skills. MASS Project (2011)
18. Poblete Ruiz, M.: How to teach and develop Soft Skills. Evolution of generic skills (Soft Skills) following the declaration of Bologna 1999. Speech at "Soft Skills and their role in employability—New perspectives in teaching, assessment and certification", workshop in Bertinoro, FC, Italy (2015)

Mobility Intentions of Latvian High-School Graduates Amid Covid-19 Pandemic and Beyond

Zane Varpina and **Kata Fredheim**

Abstract During the Covid-19 pandemic, time had become legato, if not stationary for many. This included secondary school students who were about to finish high school and transition to a new phase in their life, be that work, higher education or other activities. Many feared they are missing out or lost opportunities. In this paper, we explore how Latvian secondary school graduates perceive their mobility opportunities and intentions using survey data gathered during years 2019, 2020 and 2021, i.e., the year before Covid-19 and during two years of pandemic. This will provide insight into Generation Z students' plans for the future as well as how they adopt to a world that is freer of restrictions but not what it used to be.

Keywords School graduates · Youth mobility · Generation Z · Migration intentions · Covid-19 · Latvia

1 Introduction

Latvia, a European country with a population of 2 million people in the north-east of Europe, has seen one of the greatest rates of population loss in recent history, with a population of 19% smaller in 2020 than in 2000 [1]. This is due in part to the population's ageing and in part to emigration. Young people leave Latvia for a variety of reasons, but most importantly for studying and working overseas. Two and a half thousand people aged 15–24 leave Latvia every year [2], more than 1% of the total population in this age group, and close to 20% of all emigrants. Hence, this constitutes an important moment of brain drain with potential long-term

This paper aims to shed light on trajectories of graduates after they exit the high-school door. We present evidence on mobility plans of youngsters and allow quantitative estimation of loss of human capital towards Western European countries.

Z. Varpina (✉) · K. Fredheim
Stockholm School of Economics in Riga, Strelnieku Iela 4a, Riga 1010, Latvia
e-mail: Zane.Varpina@sseriga.edu

Baltic International Centre for Economic Policy Studies, Strelnieku Iela 4a, Riga 1010, Latvia

A. Visvizi et al. (eds.), *Research and Innovation Forum 2022*,
Springer Proceedings in Complexity, https://doi.org/10.1007/978-3-031-19560-0_66

consequences for productivity, country's human capital, innovation and demographic development in general.

Migration is a selective phenomenon characteristic specifically to young ages. The youth group of 15–29 years is especially mobile—in Europe it was the most mobile part of population group of all people in years 2013–2017, with peak between 25–29 years. Although young generation, now Generation Z, people constitute majority of all migration flows, relatively little is known about their mobility intentions and aspirations, especially the gap is evident in literature on Central and Eastern Europe (CEE). Further, the Covid-19 pandemic that hit globally in early 2020 brought about new circumstances with regards to international mobility and studies abroad.

Migration intentions develop throughout adolescence [3]. The graduation from secondary school is a focal moment in one's life when the young person is to decide what to study and where to live. In this research we focus on mobility intentions of young people at the moment when they exit the school door and set off to independent life. We assess their plans and aspirations to study abroad and live abroad. We also explore the role the pandemic has potentially played in the decisions. There are indeed two separate questions that we address: What are the study mobility intentions of the Latvian adolescents and the reasoning behind? And what are the lifetime mobility aspirations?

The focus of this study is on the moment of transition from school to life: secondary education to higher education; never worked to employment, and, more generally, 'childhood' to 'adulthood', which all happens for this cohort, Generation Z, during the Covid-19 global pandemic. We explore how the young people's international mobility intentions are formed at this complicated moment.

2 Background

2.1 *Definitions, Life Cycle and Youth Mobility*

The first distinction is to be made between migration or mobility. Migration is moving from one place to another followed by staying in the destination for a certain period. Mobility however is understood as a more dynamic movement without the 'set' time and can involve moving back and forth between places (countries) or transitioning further to new locations. Concerning youth, at the age of 19, it is arguable how well an adolescent can project the moves beyond one year, if even that far ahead, so indeed we are speaking about both terms here without being able to distinguish between them. Also more generally in Europe, researchers notice a convergence of the two terms, linked with transnationalism and the fluid nature of migration.

King et al. [4] note a shift in terminology by EU, IOM and UNDP. They report that the institutions tend to replace the term 'migration' with the more elastic 'mobility'. 'Migration' implies that migrants will stay for some time, perhaps for good, while 'mobility' signals that people will not stay but move on, either back to their home country or onwards to another one.

More and more commonly, migratory movements are not permanent, hence gaining the name "liquid migration". Such things as multiple identities, regular commuting across borders, living on regular basis or shifting between different societies have become more widespread [4–6]. This system has become known as transnationalism, also known as circular migration which is particularly true in younger ages.

To be precise, we should also define the concept of "youth" since it can differ from culture to culture. UN and EC define youth as population aged 15–24. OECD uses the same age group, while occasionally expanding the age limit up to 34. In other studies, for example, King et al. [4] it is the age 18–34. In this study we concentrate the moment around high-school graduation which in Latvia takes place the year when individual achieves the age of 19, sometimes 18, which defines our age brackets for youth/adolescents.

Migration is a selective process as not all population subgroups have equal mobility prevalence. Age is the prime determinant of likelihood of mobility. From a human life cycle perspective, migration is closely linked to important events in the life of every individual. In the early stages of life the moves are tied with parents or caregivers and individuals have little if any choice in making mobility decisions, however reaching age of adolescence and maturity (that can be at different ages in different countries), the individuals start to make moves that are associated with changes in own lifecycle. Most pivotal life course events triggering migration typically happen in young ages (18–34) hence majority of migrants are young. This period typically also covers the transition from school to higher education institution, from studying to employment and from living with caregivers to living independently. Also according to Fassman et al. [6], the age-specific distribution of migrants rises sharply after age 18 when secondary education is completed which in most of the European countries is between 16 and 20 years of age, and peaks between 23 and 25 years when tertiary education is started or when entering the labour market. In this study we are concerned with the very early stage of the independent migration, and looking to assess migration intentions as the best predictor for the actual behaviour. Even though youth is the most mobile population group there are still gaps regarding the factors that drive actual and intended generation Z migration.

2.2 *Theoretical Perspectives Relevant to Youth Migration*

The usual framework in migration research is the 'push and pull' model by Lee [7] that establishes two sets of factors driving the individual migration at micro level: the push factors—all circumstances that the individual dislikes or is not satisfied in the country/place of origin, and pull factors—the more attractive circumstances in the other locations where the individual is attracted to. This theoretical basis remains valid nowadays and for youth migration likewise. It is a simple and still very accurate theoretical basis to build upon. The model has been revisited and updated, and one of the revisions in the 'push–pull plus' model by Van Hear et al. [8]—stating the relationship between determinants of migration that are embedded in "economic, social, political, cultural, and environmental context, and more immediate factors". The author suggests that it is helpful to differentiate between four different sets of factors: predisposing, proximate, precipitating, and mediating.

Fluid and fragmented migration trajectory is a characteristic of youth migration across countries. Instead of moving strictly from one place to another and staying there, young people flow and move depending on circumstances, which means that mobility alters with 'set' periods. As Cairns [5] put it: "Young people's migration is characterized by temporality, flexibility, fluidity and open-endedness, and perhaps a sense of placelessness and social disembeddedness". Further, for youngsters the cost of migration is also generally low, meaning that returns to investment in education or a career abroad is potentially high [9].

Youth choices can also be grounded in an extra-economic logic [10, 11]. Instead, the driver is the wish to become part of global world and happening, desire to acquire cosmopolitan dispositions, or to follow lifestyles that revolve around international life and fun. "The increasing mobility of young people is not only due to life course transitions but is also linked to the fact that young people are growing up in an increasingly mobile world", Veale and Dona [12] write. For globally oriented youngsters the very experience of spending time abroad is an important step in transition to adulthood [13]. This increasingly mobile world is not encountered explicitly in push, pull or individual factors, however it does not contradict the economic sources of migration either.

Youth mobility acts as an 'eye-opener' [5] for "new experiences, broadening awareness of a wide range of future possibilities, with the value of the experience emerging not immediately but rather many years later". The migration of young people also comes along with the process of individualization, that manifests as self-realisation, life-style choices and a "thirst for adventure" [9]. And finally, mobility and foreign experience if sometimes perceived as asset that one can capitalise on later in the labour market.

In recent years an increasing number of studies has addressed migration intentions and their determinants rather than actual migration (for example, see, [14–17]). Intentions may not result in a real move, but they are both related to the same determinants. Because of that, migration intentions can be a good predictor for future moves. Especially regarding youth, it is beneficial to study mobility aspirations while the cohort is still reachable in the country as post-migration it is more complicated to reach the emigrants and virtually impossible to approach them as a cohort. As a result, the subjective reasoning and incentives are not known. As noted by Varpina et al. [16], if the sending country is willing to retain the population, as is the case for Latvia, it is usually too late to act when the population has left the country. "From a policy perspective, exploring the migration incentives of those still residing in country of origin is just as important as investigating migration motives of people who already moved," the authors note.

Intentions help to predict future behaviour. Adolescent plans may well be unstable and change quickly, but even then, the sentiment with respect to living abroad is a useful indicator of what might be expected. Further, investigating the population before mobility helps to avoid self-selection bias that is known to be present in all migrant data, and as a result quantitative estimates are more reliable and valid.

Migration intentions in Latvia have been studied before. As part of cross country and comparative research we can mention [17] who analysed Eurobarometer data with respect to migration plans in selected European countries, incl. Latvia. Authors report that the main factors that positively contribute to willingness to live abroad are prior foreign experience, stagnating economy and employment in simple manual work. Another study based on Eurobarometer data [18] showed that for Latvia and other CEE countries perceived alienation and other subjective factors are key determinants of high intentions for migration in the future. Low life satisfaction was found proved to be a determinant of emigration plans by Otrachshenko and Popova [19]. Also social networks and more generally migration culture are positively linked to migration intentions as showed by Docquier et al. [20] in a cross-sectional study in 138 countries based on Gallup survey data.

Migration research in Latvia is well developed, but mostly covers actual migration. For example, see a recent collection of articles in the book by Kasa and Mierina [21]. However, less is known about migration intentions and specifically with respect to young Generation Z.

The research on Covid-19 as a factor in mobility or migration decisions has not yet been fully established. The first immediate effects such as halting travel and other mobility were there as the first response to pandemic, that have been lifted and reintroduced in response to virus infection rates. The more far-reaching implications are yet to be seen. There are suggestions Covid-19 has acted as a catalysator for mobility decisions, at the same time many flows have stopped for good. How the situation plays out on young generation mobilities is yet to be seen.

3 Methodology

This research uses survey approach to answer the research question and explore migration intentions of Latvian adolescents amid Covid-19 pandemic. Three rounds of secondary school graduates survey were run online during spring periods 2019–2021.

In line with Varpina et al. [16] analysis that originates from the same research project, we use two different indicators for intended migration: first, we record the intended country of further studies for the adolescents that plan to start higher education in the year of observation. If the respondent plans to go abroad, we also record the country of destination. Further, we inquire if graduates would like to live abroad for "some period of their life" with answer options 'Definitely yes', 'Probably yes', 'Might or might not', 'Probably not' and 'Definitely not'. The question does not specify a timeframe, because we believe that 18–19 year old persons are unlikely to be able to answer in a precise time window. For us it was more important to understand the feelings and thoughts with respect to life abroad. The indistinct time frame approach has been utilised in literature elsewhere (see, for example, [22, 23]).

The survey was run online in Latvian secondary schools, targeting final year students, soon to be graduates in 2019, 2020 and 2021, one or two months before graduation. We picked the last moment when all the cohort is still in school and not busy with final exams. This approach also made it possible to use the digital platform 'e klase' used by every.

We consider the achieved sample to be representative of the country in general, as well as of all regions, urban and rural, variety of school sizes, language of instruction (Latvian or Russian), as well as quality of schools (gymnasiums and regular secondary schools). The total sample for analysis was 1972 responses with the following distribution among years: 2019—406 respondents, 2020—1074 respondents, 2021—492 respondents. We apply descriptive statistics analysis methods. For illustrations we also use Sankey diagrams.

4 Results

4.1 Intentions to Study Abroad

A large majority of school leavers in Latvia intend to carry on with education in the year following graduation from secondary education—84% of individuals across the three years of observation said they will study. Further, a large share—44%—intended to become employed. In Latvia work in parallel to studies in higher education is a widespread habit, which is seen also in our sample that 47% of the graduates contemplate both working and studying. Apart from the undecided 13%, there is a small proportion of graduates that would choose alternative paths—a gap year, use the year to travel and see the world, try out different things, but also establish or develop own business or do professional sports. The after-school careers seem appear rigid and see little variation. However, as expressed by youth mobility experts (Latvijas [24]), it is becoming more common than in the previous decade to take a break after secondary school and do voluntary work abroad instead of rushing into higher education.

Based on the work by Varpina et al. [25] that uses the same database as the current work, it is known that Covid-19 has influenced the future plans of adolescents. The authors report that graduates, due to uncertainty, had somewhat different effects on their intended plans for the year after graduation. While 44% say they are less likely to live abroad, 11% express the contrary—that they are more likely to leave. Same is true for alternative plans—studying is less affected for the majority however there are similar shares of adolescents (14%–16%) for whom the plans have been positively or negatively affected. Also, for business and entering labour market—effects diverge.

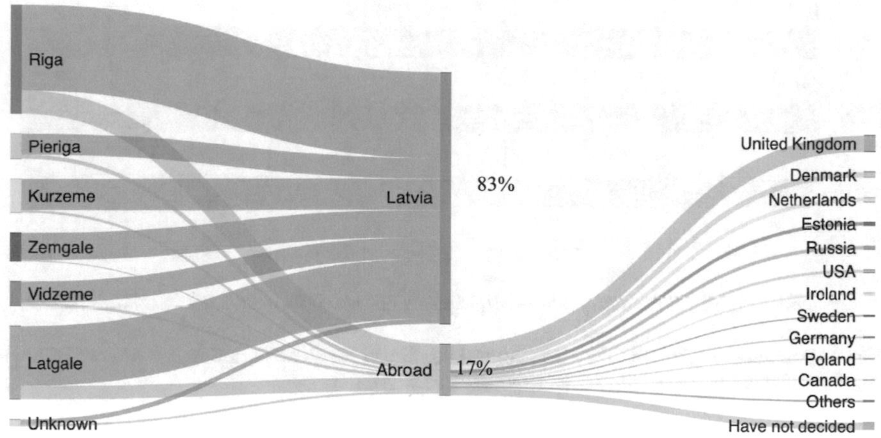

Fig. 1 Study plans, % of individuals planning to study *Source* Authors calculations based on Graduate survey 2019–2021

Concerning one of the primarily interests of this paper—graduate intentions to acquire higher education abroad, we notice that for the pooled sample as much as 83% of secondary school leavers who intend to continue studies in the year of observation, would remain in Latvia, but the other 17% consider studying abroad (Fig. 1). Relative to other European countries the proportion of abroad studies is low, however in Latvian context where wider access to studies abroad only opened after approximately 2010, the one fifth of internationally driven students is notable. For students from Latvia, United Kingdom (even after Brexit) is the main target country, and it is followed by Denmark and the Netherlands, as well as Estonia, Russia, USA, Ireland and Sweden. We do not attribute precise shares to the countries as the number of observations in each cell is too small to be able to draw certain conclusions, however the top destinations are the ones mentioned in media and available from qualitative studies.

Further, there are regional differences with respect to international studies intentions. The students from metropolitan Riga are more abroad oriented as indicated by the share of graduates with foreign studies ambitions.

4.2 Migration Intentions

Regarding overall migration aspirations, the data show that almost two thirds (59% across three years) of school leavers think positive regarding living in another country some time in the future. (see Fig. 2). Half of these positively minded adolescents (29%) would definitely go and live for some time period, and the other half (30% of all population) were less sure, but still positive. There are just 15% of adolescents

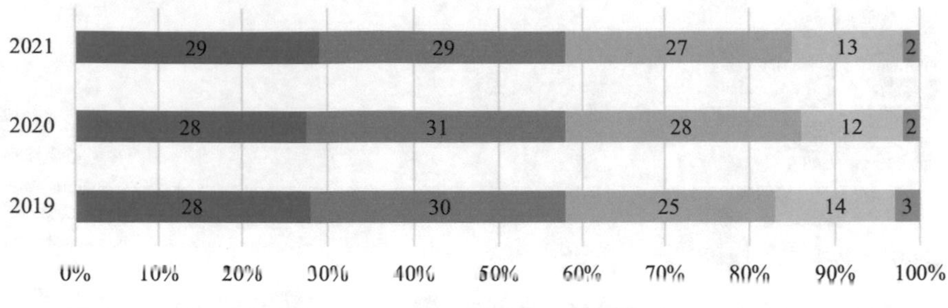

Fig. 2 Answers to question "Do you plan to live abroad for some period in the future?" *Source* Authors calculations based on Graduate survey 2019–2021

in Latvian population of 19 year-olds that are definitely (2%) or probably (13%) negative. Thirty percent of adolescents had not decided when asked.

Perhaps the most surprising finding is the robustness of answers over the three years. To remind, in spring of 2020 the first wave of Covid-19 started, the uncertainty of which could have affected the responses. Schools were closed and moved to remote mode along with most of economy that was under tough restrictions. The spring of 2021 in Latvia also featured strict limitations, while people had got used to them. Still, we see that Covid-19 did not seem to affect youngsters' desires and plans to live abroad in the future.

Graduates' stated reasons behind willingness to live abroad reveal what we have seen developing in the literature regarding youth migration elsewhere—the adolescents in Latvia are looking for the global experience (Fig. 3). Among their motivations dominate the non-economic reasoning—such as new experiences and different cultures. The graduates realise that the globalised world might give them more opportunities as compared to the small home country, too small for one's ambitions and lacking dynamics that is intrinsic to large scale economies and global metropolises.

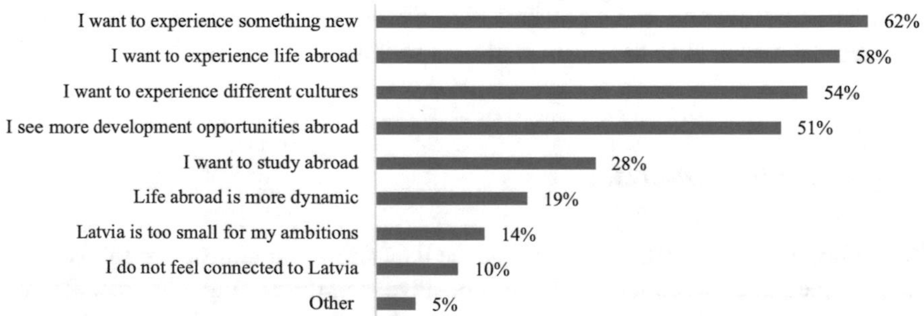

Fig. 3 Answers to question "Why do you want to live abroad?" (reasons) (%, multiple responses possible) *Source* Authors calculations based on Graduate survey 2019–2021

Quite alarmingly a tenth of youngsters do not feel connected to Latvijas and Varpina et al. [16, 24] explain that majority of these individuals are for non-Latvian ethnicity. Only when we turn to 'other' reasons, the factor of better pay comes up. Still other reasons mentioned by several individual graduates include better climate in other countries, Latvian depressive society, LGBT+ rights and safety.

5 Conclusions

We intended look into secondary school graduates' mobility aspirations during the time of global pandemic. More specifically, we asked what the study mobility intentions are and what are the life mobility aspirations of adolescents in Latvia.

Regarding the studies, we learn that nine in ten graduates plan to continue studies in higher education in the year following secondary school, and as high as 17% would apply for studies abroad. The 83% (expected) transition rate to higher education is seen as very high in comparison to other countries and is attributable to three factors. First, higher education in Latvia is often compatible with part time or even full-time studies, and this is what many graduates suggest they will do. Second, the legacy of soviet and post-soviet education career where the so-called gap years were not common. The principle would be either continue into higher levels of education without a pause or start work career. Third, Covid-19 had broken some of the plans; and during higher uncertainty about future education appears to be a value to hold on to. For those graduates who aspired for studies abroad education in European countries is most attractive, especially UK—for the well-established reputation of schools, and Netherlands and Denmark that offer high quality tuition free education.

Concerning life mobility aspirations, two thirds of adolescents, across years, express their wish to live abroad in the future, from the other side, only one in six adolescents had no interest or strictly rejected possibility to live abroad in the future. The reasons for living aboard are grounded almost exclusively in non-economic factors such as lifestyle, globalisation, wish to learn new cultures and experiences, gain independence and 'find himself'. Yet, we also observe the reasoning that Latvia is too small to achieve the life objectives and having better opportunities abroad.

The drive to experience life abroad in the form of studies or unconnected to education for Latvian adolescents is very high. Covid-19 has not affected the general perceptions and overall plans, if only shifted priorities in short term. This phenomenon has important implications for the national higher education system. Local universities struggle to compete for enrolment with foreign ones even if the quality of education and offered opportunities are on par or better. It is impossible to limit youth emigration if such was the aim to prevent depopulation and brain-drain. Instead, the decision makers should shift attention to attracting the people after they have gained their share of global world. For that more attention should be paid to what a returnee would find back home—how open and flexible are the labour markets, how is the state of security, corruption, civil society, how much of social services are available—just to mention a few. Higher education institutions in

Latvia would appear more attractive to the incoming Generation Z if they offered more and emphasized the international cooperation opportunities, such as exchange semesters, and welcomed more international students and faculty.

Acknowledgements This research was supported by Latvian Council of Science, Funding number: lzp-2018/1-0486, acronym: FLPP-2018-1

References

1. UN Population Division database, ND. https://esa.un.org/unpd/wpp/Download/Standard/Population/. Accessed 12 August 2022
2. Oficiālais statistikas portāls, ND. https://data.stat.gov.lv/pxweb/lv/OSP_PUB/START__POP__IB__IBE/IBE030/table/tableViewLayout1/. Accessed 12 Aug 2022
3. Lulle, A., Jurkane-Hobein, I.: Strangers within? Russian-speakers' migration from Latvia to London: a study in power geometry and intersectionality. J. Ethn. Migr. Stud. **43**(4), 596–612 (2017)
4. King, R., Lulle, A., Morosanu, L., Williams, A.: International Youth Mobility and Life Transitions in Europe: Questions, Definitions, Typologies and Theoretical Approaches'. Working Paper No. 86. University of Sussex. Sussex Centre for Migration Research (2016)
5. Cairns, D.: The Palgrave Handbook of Youth Mobility and Educational Migration. In: The Palgrave Handbook of Youth Mobility and Educational Migration (2021). https://doi.org/10.1007/978-3-030-64235-8
6. Fassmann, H., Gruber, E., Nemeth, A.: Conceptual framework for the study of youth migration in the Danube region (2018)
7. Lee, E.S.: A theory of migration. Demography **3**(1), 47–57 (1966)
8. Van Hear, N., Bakewell, O., Long, K.: Push-pull plus: reconsidering the drivers of migration. J. Ethn. Migr. Stud. **44**(6), 927–944 (2018)
9. Gruber, E., Schorn, M.: Österreichische Gesellschaft für Europapolitik (ÖGfE) | Rotenhausgasse 6/8–9 | A-1090 Wien | europa@oegfe.at | oegfe.at | +43. 1, 4999. Retrieved from: http://atlas.espon.eu/ (2019)
10. Krzaklewska, E.: Youth, mobility and generations—the meanings and impact of migration and mobility experiences on transitions to adulthood. Studia Migracyjne - Przegląd Polonijny **171**(1), 41–59 (2019). https://doi.org/10.4467/25444972smpp.19.002.10252
11. Skrbis, Z., Woodward, I., Bean, C.: Seeds of cosmopolitan future? Young people and their aspirations for future mobility. J. Youth Stud. **17**(5), 614–625 (2014). https://doi.org/10.1080/13676261.2013.834314
12. Veale, A., Dona, G.: Child and Youth Migration. Mobility-in-migration in an era of Globalization. Palgrave Macmillan, London (2014)
13. Beck, U., Beck-Gernsheim, E.: Individualization. Institutionalized Individualism and its Social and Political Consequences. Sage, London (2002)
14. Esipova, N., Srinivasan, R., Ray, J.: Global desires to migrate. Adjusting to a World in Motion: Trends in Global Migration and Migration Policy, 21–57 (2016)
15. Migali, S., Scipioni, M.: Who's about to leave? a global survey of aspirations and intentions to migrate. Int. Migr. **57**(5), 181–200 (2019)
16. Varpina, Z., Fredheim, K., Krumina, M.: Who is more eager to leave? Differences in emigration intentions among Latvian and Russian speaking school graduates in Latvia. In: society. Integration. Education. Proceedings of the International Scientific Conference (vol. 6, pp. 207–218) (2021)
17. Williams, A.M., Jephcote, C., Janta, H., Li, G.: The migration intentions of young adults in Europe: a comparative, multilevel analysis. Popul. Space Place **24**(1), e2123 (2018)

18. Van Mol, C.: Migration aspirations of European youth in times of crisis. J. Youth Stud. **19**(10), 1303–1320 (2016)
19. Otrachshenko, V., Popova, O.: Life (dis) satisfaction and the intention to migrate: evidence from Central and Eastern Europe. J. Socio Econ. **48**, 40–49 (2014)
20. Docquier, F., Peri, G., Ruyssen, I.: The cross-country determinants of potential and actual migration. Int. Migr. Rev. **48**(1 suppl), 37–99 (2014)
21. Kaša, R., Mieriņa, I.: The Emigrant Communities of Latvia: National Identity, Transnational Belonging, and Diaspora Politics (p. 298). Springer Nature (2019)
22. Cairns, D.: "I wouldn't stay here": economic crisis and youth mobility in Ireland. Int. Migr. **52**(3), 236–249 (2014)
23. Van Dalen, H.P., Henkens, K.: Explaining low international labour mobility: the role of networks, personality, and perceived labour market opportunities. Popul. Space Place **18**(1), 31–44 (2012)
24. Latvijas, R.: Aptauja: Lielākā daļa vidusskolas absolventu izvēlas mācības turpināt Latvijā. https://lr1.lsm.lv/lv/raksts/gimenes-studija/studet-arzemes-uz-kurieni-ved-latvijas-jauniesu-celi.a149597/. Accessed 12 August 2022 (2021)
25. Varpina, Z., Fredheim, K., Krumina, M.: Implications of the Covid-19 pandemic on high school graduates plans and education paths. In: Visvizi, A. (ed.) "Moving Higher Education Beyond Covid-19: Innovative and technology-enhanced approaches to teaching and learning". Emerald Publishing (2022)

The Covid-19 Pandemic's Impact on Migrants' Decision to Return Home to Latvia

Kata Fredheim and Zane Varpina

Abstract The Covid-19 pandemic restricted people's movement but also changed their course of life. For some migrants, this meant re-evaluating opportunities abroad and back home. This paper uses findings from interviews with those who returned to Latvia during the pandemic to gain insight into the ways the pandemic influenced their decision to return. We find that the pandemic impacted how people think of return. It was both a reason and a catalyst, accelerating life events and leading to decisions to return. For some who contemplated return the pandemic accelerated decision, motivated by missing people, loneliness, and missing community. The pandemic and its immediate consequences also directly affected migrants; livelihood and work; some returned quickly. For some of these migrants, the pandemic also acted as a barrier to leaving again soon after a return. Circular migration journeys of coming back and leaving again feed into the narrative that for many migrants returning is more a stop in their journey than the destination itself. The much anticipated great wave of return, it seems was more like a tide. People moved back and forth between borders, seeking safety and community in times of uncertainty while trying to maintain their work and studies.

Keywords Latvia · Return migration · Covid-19 · Reasons for return · Decision to return

1 Introduction

Latvia experienced the high rate of depopulation in the first decade of the 2000s [1]. Migration was a key contributor; after the accession the European Union, Latvians increasingly opted for opportunities abroad [2]. Attracting return migrants has been

K. Fredheim (✉) · Z. Varpina
Stockholm School of Economics in Riga, 4a Strelnieku Iela, Riga 1010, Latvia
e-mail: kata.fredheim@sseriga.edu

Z. Varpina
e-mail: Zane.Varpina@sseriga.edu

Baltic International Centre for Economic Policy, 4a Strelnieku Iela, Riga 1010, Latvia

A. Visvizi et al. (eds.), *Research and Innovation Forum 2022*,
Springer Proceedings in Complexity, https://doi.org/10.1007/978-3-031-19560-0_67

a priority for policymakers and businesses in Latvia, as in other countries that joined the European Union after 2004 [3]. Return migration has been linked to increase in entrepreneurship [4] and brain gain, investment, and skill transfer by return migrants [5]. In recent year, only a third of emigration is return migration [6] and the government launched messaging campaigns and support initiatives including grants and dedicated coordinators to assist with return and resettlement [7, 8].

What makes people return is a key question to design effective policies, especially returnees' reasons to migrate back differs from that of other voluntary migrants. Family reasons and seeking growth weigh more than economic reasons for returnees [9, 10]. This is the case for Latvia also, homesickness, patriotism and family were found to be the most important reasons for returning to the homeland [11–13]. At the same time, seeking opportunities for a better lifestyle [14] or entrepreneurship [4] also influenced returnees' decision-making.

At the start of the COVID-19 pandemic, many popular theories circulated about how the pandemic could impact migration and return migration. Some speculated that return migration will increase since Latvians will want to be home during these difficult times, others feared that all forms of migration will come to a halt due to borders closing. Migration flows decreased due to travel restrictions and border closures [15]. Efforts to attract and support returns intensified [16].

At least temporarily the world has changed: there were more restrictions, life seemed to have slowed for many, and unemployment across Europe soared. At the same time, barriers to mobility increased. As countries closed their borders, many feared staying abroad or could not leave home again. Migration to Latvia dropped, and almost half of the migrants to Latvia were Latvian nationals [15].

Because it is a recent phenomenon, understanding of migrants' narratives of return is still lacking. In this paper we explore how did the Covid-19 pandemic influence Latvian return migrants' decision to return. Rather than generalising, we seek to understand the influence of the pandemic on return, considering existing knowledge on reasons to return. Based on the interviews we argue that for some the pandemic acted as catalyst, speeding up the time between contemplating return to decision; while for others it triggered an immediate decision.

2 Background

2.1 Why Do Migrants Return?

Studies conducted in recent decades show that decisions of return migration vary. Push and pull factors to a large degree explain why people migrate and how they choose where to Lee [17]. However, return migration differs from other forms of migration not only in that people return to a place known to them. Thus, the reasons for return differ also.

Attachment to host and home countries matters [18], yet they are not mutually exclusive [19, 20]. Strong social ties in the home country, opportunities, family, and motivations related to one's identity such as longing for home are determinants of return intentions [21] but so is failure in the host country [22]. What increases the intention not to return is work, education, and job satisfaction [21–25]. For return migrants, social ties and families play an even greater role than in other forms of migration, while economic factors matter less [9, 10, 26, 27]. Stark's mapping of reasons for return migration provides one of the most comprehensive lists of reasons for return in recent literature and includes economic reasons, growth and one's place in society, and family reasons [28]. These motivations and reasons interact and result in complex decisions and analyses [18].

Latvian returnees' reasons for return align with reasons for return in other studies. The main factors that influence return migration to Latvia are homesickness, patriotism, and family ties [11, 29]. Latvia pro-actively sought to help its citizens return and integrate into society, with new policies, laws, and messaging as well as with support from remigration coordinators in all regions (e.g.: Migration Policy Conception, Diaspora law). In a recent study, coordinators discussed the complexity of reasons for return but agreed that patriotism and family were the two most important factors for Latvians when deciding to return [11]. For young Latvians the main reasons for return were homesickness, taking care of family, and wanting their children to grow up in the home country [28]. The reason for return is rarely economic [12, 13]. Failure is a theme, and unsuccessful integration in the host country also contributes to return [12]. Professional factors count; these include applying skills learned abroad [11], and starting a business [4]. Skills learned abroad however are not always acknowledged by employers [30] and even if they are considered it is mainly so in the capital, Riga [31]. Life events such as starting a family, and professional opportunities expedited such return intentions and turned them into a decision to return [14]. Younger diaspora return migrants seek a better life in Latvia [32]. For them, the return may not be the end of their migration journey but part of the seeking new opportunities and experiences [26]. But what happens during a crisis like the pandemic when life as we know it pauses in many ways. What are the reasons for return then? In this paper, we ask not only how the pandemic influenced decisions to return but also what the reasons for return were.

These reasons do not come at once; in fact, migrants' often think about belonging, ideas of the self and homeland, often juxtaposed with life abroad. Voluntary migration is not an immediate process for the individual, and decisions do not take place on a uniform schedule due to the complexity of feelings, thoughts, and factors. Instead, migration definitions can be described in three phases. First, they are is considered, then planned, and finally executed [33–36]. While many migrants go through the first two stages, it is in the last one that arrangements facilitating return are made. Often, the gap between considering mobility and actual mobility is large, some may consider returning for a long time but never return [37–39]. Preparedness to return refers both to their willingness to return and their readiness to return [22]. How the Covid-19 pandemic impacted these phases for Latvian returnees has not been explored in detail yet.

2.2 *Pandemic's Impact on Migration*

The Covid-19 pandemic is a relatively recent crisis As a result, published studies remain limited. It impacted so many aspects of people's social, economic, and personal lives. Migration too was an impacted area. The pandemic, with lockdowns, restrictions, and border closures increased the immobility of populations from one moment to another [40]. This was recognized by policymakers; the Latvian governments' efforts to reach and attract back Latvians intensified during the pandemic [16]. The efforts acknowledged the diversity of the diaspora [7, 8].

At the beginning of the pandemic, there were assisted returns when the government helped those abroad to come home. As economies shrunk, unemployment grew. This was especially relevant for the service sector and temporary employment, areas where many migrants work in. Migrants globally faced economic problems, not only losing employment but sometimes not being paid for the work they completed [41–43].

Understanding changes in one's lives may be a way to understand migration more during the pandemic. During the pandemic, health, employment and social life changed [44, 45]. Changes in one part of one's life often lead to changes in other parts of life, and trajectories also [46–48]. Exploring the role of live events and changes has been used to study mobility and return migrants' journeys [49–53]. Examples of such life stages include caring duties such as having school-aged children or elderly parents [52] or family planning [54, 55]. Yet in migration decisions, identity formation, access to the labour market, or feelings for the 'homeland' still play a role [52]. Thus, family, professional and historical reasons are layered upon each other [46, 56]. The decision is then made prompted by a change, or rather transition in one's life [57]. The pandemic brought many changes in people's lives, including new caring duties, or professional crossroads.

While life changed during the pandemic, demand in certain sectors did not. A recent article by Paul highlights that for many EU13 migrants' opportunities remained as they were in high demand for agricultural and service sector jobs abroad [58]. In Europe, when restrictions were the strictest, government negotiations led to exceptions that allowed CEE migrants to fill essential positions abroad to avoid a food crisis. Migrants also proactively returned when they could to their jobs abroad [58]. Thus, return migration was often short-lived for CEE migrants, many returning abroad to work. Return migration was closely linked to other areas of life seemingly fast-tracked by the pandemic, such as hybrid and remote work. For Slovenian migrants considering return combining working online and living at home, with a higher quality of life was a significant factor [41].

Hybrid or remote work increasingly made lifestyle migration, for instance earning a Western European salary but settling in CEE, possible. This was already a consideration for Estonian migrants before the pandemic [53]. Moving abroad, or returning are also not singular events but for part of their migration journey. In this sense, transitional migrants, migrants continually negotiate their identities between home and host countries, maintaining strong links to their home communities [22] Thus, brain circulation, repeat, and circular migration better describe migration flows than the limited dual point A to B dynamic [26].

However, the ways reasons for the return and people's decision-making to migrate back to Latvia may have changed during the pandemic have not been explored, there are no current studies on return migration on Covid-19 in Latvia.

3 Methods

We conducted 74 in-depth interviews with return migrants in the three Baltic countries between 2018–2022 to find out more about their thoughts on migration, reasons for return, and life upon return. This article is based on part of this dataset, 12 interviews conducted with Latvian return migrants who returned during the Covid-19 pandemic.

For this study we use the definition of a return migrant as someone who spent at least three months abroad and then returned to their home country, in line with similar research in the Baltics [59].

The interviews were a mix of face-to-face, phone, or video conference. Most interviews were conducted in Latvian. The interviews were semi-structured. In interviews, all returnees were asked about their migration background, return journey, resettlement, and plans for the future. Interviewees come from different professions, regions, age groups, ethnicity, and gender. Since the population and most companies in Latvia are centred in Riga and major regional centres, these make up the majority of locations.

Potential interviewees were contacted after searching on registers and social networks (LinkedIn), personal networks of the research team, and employment networks. There was also an opportunity to register on the research projects' website if someone was interested in being interviewed.

All potential interviewees received information on the research and consented to the interviews. Interviews were recorded, and anonymous and all interviewees reccived pseudonyms. The analytical approach was thematic, looking for patterns and relationships between them. The sample is not representative, but it captures perceptions of returnees in these two years about how they see the reasons for returning to their home country, Latvia.

4 Results

This section presents two different ways the pandemic was weaved into returnees' narratives of return. For each pattern, we not only discuss the way the pandemic influenced returns but also how this influence was connected to the main reasons to return in participants' stories.

4.1 Pandemic as a Catalyst

This group of returnees in the sample have been thinking about return, but they have not thought of the specifics of when. This staged approach aligns with different stages of migration decision, first contemplated and only finally realized [14].

As borders, public spaces, workplaces shut down some returnees still living abroad, and living alone described feelings of loneliness and missing people. The pandemic and the restrictions had an even bigger impact on those living alone or being isolated from society [60–62]. Migrants, who did not have family abroad or who did not (yet) integrate into the host society relatively fully faced challenges of loneliness [63, 64]. Contrasting being able to interact with people daily with isolation, some migrants chose to return to be with family. Rita described how the pandemic influenced her decision to return: "*I spent several months in Geneva home alone. In the beginning, I enjoyed it, but the more it lasted, the less I liked it. I understood that actually I would like to be in Latvia. I had the feeling that the pandemic won't disappear from a to another and that it will begin again in autumn. And I thought that I would not like to be again home alone. I preferred to be in Latvia*" (Rita). Rita only recently arrived in Geneva and felt increasingly isolated. She contrasted her isolation with the opportunities she would have for socialising in Latvia. Edgars felt the same way, he did not want to 'hermit abroad' especially since he never thought of not returning. "*Well because one of the reasons is Latvia is the place I want to stay and that was just a source of higher income. That was in no way a long-term situation. So that's why I really didn't feel like a returning migrant. I just returned home*" (Edgars).

Other returnees reflected on whether they would rather stay abroad during the pandemic. Latvia did comparatively well especially at the beginning of the pandemic when much of Western Europe was in a strict lockdown. Thus, migrants saw opportunities to live without isolation, work and even find new opportunities. Sandis retrospectively argued "*I am happy that in our country it is more or less well… If I would be abroad now, I would definitely would like to come home.*" (Sandis). As the pandemic continued, these differences evaporated somewhat. By 2020 autumn, Latvia too had lockdowns. Yet, migrants' feelings of loneliness did not disappear though became more of the status quo.

The decision to return was not always with permanence in mind. Edgars too came home but was able to continue his work remotely. His initial plan of coming back for a short time only was extended repeatedly and he now thinks he returned for good: "*Well, there was a chance I could work online and move back to Latvia. The main reason was that I didn't know how long the border closings would last. In the beginning, I was thinking of staying there during Covid-19. But as rumors of border closings were shuffling and the duration would be unknown, I didn't want to hermit there for a few years… The positive was also that I got back to Latvia and met my family and friends and don't regret the decision.*" (Edgars). Toms, who moved abroad to start an enterprise had to return quickly.

These feelings, and the opportunity to return, played out against the backdrop of having thought of return before. For these migrants, the pandemic facilitated or rather sped up the decision to return. At the core of the reasons, such missing people and the home country align with the reasons in literature before the pandemic. However, loneliness and longing intensified and catalised the decision-making process from contemplation to action.

4.2 Pandemic as a Reason to Return

For the second group of returnees, the pandemic had such a direct impact on their livelihood, work, or studies that they returned without contemplating returning significantly before that. This was primarily true at the beginning of the pandemic when the economy halted.

Unemployment across Europe soared at the beginning of the crisis with the most impacted sector being the service sector. As the world shot down, people lost their roles. Since CEE migrants filled many service sector jobs in Western Europe, they were disproportionally impacted. For migrants not having a job resulted not only in stress but also in economic difficulties. For Una and her family, the pandemic also meant losing their jobs, which prompted them to return quickly, they moved from consideration to execution within days: "*Things are closing there and so on, like one of the bookshops that I worked in. And my husband's job kind of closed. We had a flight on March 26th*" (Una). Una felt they had no choice but to return as life abroad without income was more expensive. Her family's decision was made quickly.

Students studying abroad faced a unique dilemma. As their studies shifted online, social life came to an abrupt halt. Anna, who studied in the Netherlands quickly decided to return to Latvia, on the second to last repatriation flight when the pandemic hit. "*It was the 25th of March, just when all that situation started. There were a lot of people flying to Latvia. Honestly, the majority were students who were studying abroad. From the ones I know, rarely anyone stayed. I know that in the beginning, I thought I would stay in the Netherlands myself but then I realised that they are closing borders and everything, then I got the feeling "Oh Shit, it's getting real*" (Anna). Anna then spent the next two years flying back for exams or when studies were on-site, only to be in quarantine or self-isolation for weeks on end. Often, these trips would be decided from one day to the other since changing restrictions left little time for planning. In the end, she did not return to the Netherlands but stayed in Latvia despite previous plans.

For some who left quickly, and without planning it before. the pandemic also acted as a barrier. Returnees contemplated leaving again, but they couldn't leave. Una, who had to return because she and her husband's job ended during the pandemic was waiting for the earliest opportunity for the borders open to leave "*All the borders are closed until the end of July... I hope that the situation will change until January.*"

(Una). For students, this often became costly, especially as modes of studies, regulations, and flights changed from one day to the other. "*Then I returned to the Netherlands for an exam...the next day we received an email that said: "If you live abroad, you can stay in your home countries and not go to the Netherlands. Don't worry, the exam will be held online.*" *And then I thought "damn, wtf". The thing is I couldn't immediately fly back (to Latvia) because I would have to do another Covid test. It was expensive and tickets were expensive for the last moment flights. Then I was stuck in the Netherlands writing a test and after a month they stated that all studies are going online again. All are bad. Go back (to home country)*" (Anna). Toms, who moved to Spain three months before the pandemic hit Europe to start a business drove back 120 km per hour across Europe once he realized borders are closing. He then left again only to realise the second wave is there "*And then the second wave hit. And then we realised we couldn't predict how long this one will last. Either we head back to Latvia now or we stay and just run out of money. So, we decided to head back before zeros and not with debt*" (Toms). He returned with a positive balance sheet and stayed in Latvia. The feeling of being stuck therefore can be described not only as a one-off experience but a reoccurring frustration for migrants moving back and forth between home and abroad.

For this group of migrants, who returned early in the pandemic, what they do every day was not an option. Thus, they decided to move back to Latvia quickly. They also started to contemplate leaving Latvia quickly. However, the environment remained uncertain as lockdowns, quarantines and regulations changed. Leaving was not easy and costly in terms of time and funds also.

5 Conclusions and Discussion

A few recent events had a larger impact on people's lifestyles, relationships, or even jobs than the Covid-19 pandemic. This paper aimed to answer the question of how the Covid-19 pandemic influenced Latvian migrants' decision to return to Latvia. While searching for the reason for return, we found that returnees' decision-making process, consideration, planning, and execution were impacted. Interviews reveal that the pandemic's influence may be described in two ways.

Firstly, for a group of participants, the pandemic was a catalyst allowing migrants to move from contemplating a return to returning quickly. The pandemic impacted how people thought of return. For them, missing people, loneliness, and missing community were some of the main reasons for return. These reasons are about social ties and home, very much aligning to reasons Latvians migrated home during the pandemic. The return journey of this group of migrants highlights not only the vulnerability of migrants in times of crisis but also re-emphasises the importance of integration for successful migration journeys. Integration and strong social ties provides resilience in times of crisis. At the same time, it is an example how the gap between considering mobility and mobility shrunk when faced with a crisis, uncertainty, and isolation abroad.

Secondly, the pandemic and its immediate consequences may be seen as reasons for return. For this group of returnees, loss of employment and online studies were the most significant reason. Yet, there is no sense that these experiences are linked to failure, often sighted as a reason to return. Instead, they are rooted in uncertainty and a need for safety. These returnees moved quickly. For some of these migrants, the pandemic also acted as a barrier to leaving again soon after a return. Circular migration journeys of coming back and leaving again, feed into the narrative that for many migrants returning is more a stop in their journey than the destination itself.

The much-anticipated great wave of return, it seems was more like a tide. People moved back and forth between borders, seeking safety and community in times of uncertainty while trying to maintain their work and studies. Yet, lessons can be learned from this experience.

It was migrants living alone, who contemplated return prior to the pandemic who decided to return and stayed. This highlights not only the vulnerability of migrants abroad but also the importance of the consideration phase. In this consideration phase policymakers wishing to attract returnees have the opportunity to remind and emphasise migrants about the opportunities and values of the home country which then can play a large role in the decision to return.

For these transnational migrants returning and leaving again, coming back is not an end-point. They successfully maintained links to both their home and host country, and what attracted them to leave remains valid even in times of crisis.

Acknowledgements This work was supported by the National Research Program Project grant number VPP-IZM-2018/1-0015 and by the Latvian Council of Science, project No. lzp-2018/1-0486.

References

1. Martin, R., Radu, D.: Return migration: the experience of Eastern Europe 1: return migration: the experience of Eastern Europe. Int. Migr. **50**(6), 109–128 (2012)
2. Hazans, M.: 'Migration experience of the baltic countries in the context of economic crisis'. In: Kahanec, M., Zimmermann, K.F. (eds.) Labor Migration, EU Enlargement, and the Great Recession, pp. 297–344. Springer, Berlin Heidelberg (2016). https://doi.org/10.1007/978-3-662-45320-9_13. Accessed 16 Feb 2022
3. Boros, L., Gábor, H.: 'European National Policies Aimed at Stimulating Return Migration'. In: Nadler, R., Kovács, Z., Glorius, B., Lang, L. (eds.) Return Migration and Regional Development in Europe, pp. 333–357. Palgrave Macmillan, London, UK (2016). https://doi.org/10.1057/978-1-137-57509-8_15. Accessed 19 Feb 2022
4. Varpina, Z., Kata, F., Anders, P., Marija, K.: 'Back for Business: The Link between Foreign Experience and Entrepreneurial Activity in Latvia'. Forthcoming (2022)
5. Wahba, J.: 'Who Benefits from Return Migration to Developing Countries?' IZA World of Labor. https://wol.iza.org/articles/who-benefits-from-return-migration-to-developing-countries (2021). Accessed 19 Feb 2022
6. Latvian Central Statistical Office database, code IBE060; IBR040
7. Birka, L., Kļaviņš, D.: Diaspora diplomacy: nordic and baltic perspective. Diaspora Stud. **13**(2), 115–132 (2020)

8. Birka, I.: Engaging the diaspora for economic gain: what can latvia expect? J. Baltic Stud. **51**(4), 497–511 (2020)
9. Markowitz, F., Stefansson, A.H. (eds.): Homecomings: Unsettling Paths of Return. Lexington Books, Lanham, Md (2004)
10. Sussman, N.M.: Return Migration and Identity: A Global Phenomenon, a Hong Kong Case. Hong Kong University Press, Hong Kong (2011)
11. Prusakova, L., Bērziņš, M., Apsīte-Beriņa, E.: 'Institutionalisation of return migration in latvia: the case of regional coordinators.' Soc. Integr. Educ. **6**, 185–195 (2021)
12. Kļave, E., Šūpule, I.: Return migration process in policy and practice. In: The emigrant communities of Latvia, pp. 261–282. Springer, Cham (2019)
13. Šūpule, I.: Intentions to stay or to return among highly skilled Latvians in the EU: who is more likely to return? J. Baltic Stud. **52**(4), 547–563 (2021)
14. Kata, F., Varpina, Z.: There is a reason why: Baltic return migrants' reasons for return. Forthcoming (2022)
15. Official Statistics of Latvia.: More Rapid Decline in Population. https://stat.gov.lv/en/statistics-themes/population/population-number/press-releases/6935-number-population-latvia-2020 (2021). Accessed 24 March 2022
16. Lāce, A.: "Working from Home": Government Initiatives to Promote Returning to Latvia Amidst the Covid-19 Pandemic. In: Visvizi, A., Troisi, O., Saeedi, K. (eds.) Research and Innovation Forum 2021. RIIFORUM 2021. Springer Proceedings in Complexity. Springer, Cham (2021). https://doi.org/10.1007/978-3-030-84311-3_49
17. Lee, E.S.: A theory of migration. Demography **3**(1), 47–57 (1966)
18. Carling, J., Pettersen, S.V.: Return migration intentions in the integration–transnationalism matrix. Int. Migr. **52**(6), 13–30 (2014)
19. Snel, E., Engbersen, G., Leerkes, A.: Transnational involvement and social integration. Global Netw. **6**(3), 285–308 (2006)
20. Oeppen, C.: A stranger at 'home': interactions between transnational return visits and integration for Afghan-American professionals. Global Netw. **13**(2), 261–278 (2013)
21. De Haas, H., Fokkema, T., Fihri, M.F.: Return migration as failure or success? J. Int. Migr. Integr. **16**(2), 415–429 (2015)
22. Cassarino, J.P.: Theorising return migration: The conceptual approach to return migrants revisited. Int. J. Multicul. Soc. (IJMS) **6**(2), 253–279 (2004)
23. Waldorf, B.: Determinants of international return migration intentions. Prof. Geogr. **47**(2), 125–136 (1995)
24. Wang, W.W., Fan, C.C.: Success or failure: Selectivity and reasons of return migration in Sichuan and Anhui. China. Environ. Planning A **38**(5), 939–958 (2006)
25. Skrentny, J.D., Chan, S., Fox, J., Kim, D.: Defining nations in Asia and Europe: a comparative analysis of ethnic return migration policy 1. Int. Migr. Rev. **41**(4), 793–825 (2007)
26. Zaiceva-Razzolini, A., Zimmermann, K.F.: Returning Home at Times of Trouble? Return Migration of EU Enlargement Migrants during the Crisis (IZA Discussion Paper No. 7111) (2012)
27. Stark, O.: The Migration of Labor. Cambridge, Mass., USA; Oxford, UK: B. Blackwell (1991)
28. Stark, O.: Behavior in reverse: reasons for return migration. Behav. Public Policy **3**(2), 104–126 (2019)
29. Krisjane, Z., Apsite-Berina, E., Sechi, G., Bērziņš, M.: 'Juxtaposed Intra-EU Youth Mobility: Motivations among Returnees to Latvia'. Belgeo (3) (2018). http://journals.openedition.org/belgeo/21167. Accessed 15 Feb 2022
30. Fredheim, K., Varpina, Z.: What a Manager Wants? How Return Migrants' Experiences Are Valued by Managers in the Baltics. In: Visvizi, A., Troisi, O., Saeedi, K. (eds.) Research and Innovation Forum 2021, pp. 577–587. Springer International Publishing, Cham (2021)
31. Mieriņa, I., Bela, B.: 'Can return migration revitalise latvia's regions? facilitators and barriers to human capital gains.' Soc. Integr. Educ. Proc. Int. Scient. Conf. **6**, 142–159 (2021)
32. Lulle, A., Krisjane, Z., Bauls, A.: 'Diverse Return Mobilities and Evolving Identities among Returnees in Latvia'. Loughborough University. https://hdl.handle.net/2134/36068 (2019)

33. De Jong, G.F., et al.: Migration intentions and behavior: decision making in a rural Philippine province. Popul. Environ. **8**(1–2), 41–62 (1985)
34. Fawcett, J.T.: Migration psychology: new behavioral models. Popul. Environ. **8**(1–2), 5–14 (1985)
35. Kley, S.A., Mulder, C.H.: Considering, planning, and realizing migration in early adulthood. the influence of life-course events and perceived opportunities on leaving the City in Germany. J. Housing Built Environ. **25**(1), 73–94 (2010)
36. Kley, S.: Explaining the stages of migration within a life-course framework. Eur. Sociol. Rev. **27**(4), 469–486 (2011)
37. Lu, M.: Analyzing migration decisionmaking: relationships between residential satisfaction, mobility intentions, and moving behavior. Environ. Planning A Econ. Space **30**(8), 1473 1495 (1998)
38. Fang, Y.: Residential satisfaction, moving intention and moving behaviours: a study of redeveloped neighbourhoods in inner-city Beijing. Hous. Stud. **21**(5), 671–694 (2006)
39. de Groot, C., Mulder, C.H., Manting, D.: Intentions to move and actual moving behaviour in the Netherlands. Hous. Stud. **26**(3), 307–328 (2011)
40. Martin, S., Bergmann, J.: (Im) mobility in the age of COVID-19. Int. Migr. Rev. **55**(3), 660–687 (2021)
41. Valentinčič, D.: Brain circulation and return migration in slovenia before and during the Covid-19 pandemic. In: The International Research & Innovation Forum, pp. 549–566. Springer, Cham (2021)
42. Foley, L., Piper, N.: Returning home empty handed: examining how COVID-19 exacerbates the non-payment of temporary migrant workers' wages. Global Social Policy **21**(3), 468–489 (2021)
43. Zeeshan, M., Sultana, A.: 'Return Migration to Pakistan during COVID-19 Pandemic: Unmaking the Challenges'. Pakistan Perspect. **25**(1) (2020)
44. Gangopadhyaya, A., Garrett, A.B.: 'Unemployment, Health Insurance, and the COVID-19 Recession'. SSRN Electr. J. (2020). https://www.ssrn.com/abstract=3568489. Accessed 27 May 2022
45. Xiao, S., Luo, D., Xiao, Y.: Survivors of COVID-19 are at high risk of posttraumatic stress disorder. Global Health Res. Policy **5**(1), 29 (2020)
46. Elder, G.H.: Time, human agency, and social change: perspectives on the life course. Soc. Psychol. Quart. **57**(1), 4 (1994)
47. Heinz, W.R., Marshall, V.W. (eds.): Social Dynamics of the Life Course: Transitions, Institutions, and Interrelations. Aldine de Gruyter, New York (2003)
48. Kulu, H., Milewski, N.: Family change and migration in the life course: an introduction. Demogr. Res. **17**, 567–590 (2007)
49. Bettin, G., Cela, E., Fokkema, T.: Return intentions over the life course: evidence on the effects of life events from a longitudinal sample of first-and second-generation Turkish migrants in Germany. Demogr. Res. **39**, 1009–1038 (2018)
50. Kley, S.A., Mulder, C.H.: Considering, planning, and realizing migration in early adulthood. The influence of life-course events and perceived opportunities on leaving the city in Germany. J. Housing Built Environ. **25**(1), 73–94
51. Bloem, B., Van Tilburg, T., Thomése, F.: Residential mobility in older Dutch adults: influence of later life events. Int. J. Ageing Later Life **3**(1), 21–44 (2008)
52. Erdal, M.B., Ezzati, R.: 'Where are you from' or 'when did you come'? Temporal dimensions in migrants' reflections about settlement and return. Ethn. Racial Stud. **38**(7), 1202–1217 (2015)
53. Saar, M., Saar, E.: Can the concept of lifestyle migration be applied to return migration? The Case of Estonians in the UK. Int. Migr. **58**(2), 52–66 (2020)
54. Corcoran, M.: Global cosmopolites: issues of self-identity and collective identity among the transnational Irish elite. Etudes Irlandaises **28**(2), 135–150 (2003)
55. Duda-Mikulin, E.A.: Should i stay or should i go now? exploring polish women's returns "home." Int. Migr. **56**(4), 140–153 (2018)
56. Elder, G.H., Jr.: Family history and the life course. J. Fam. Hist. **2**(4), 279–304 (1977)

57. Brettell, C.: Gendered lives: transitions and turning points in personal, family, and historical time. Curr. Anthropol. **43**(S4), S45–S61 (2002)
58. Paul, R.: Europe's essential workers: migration and pandemic politics in central and Eastern Europe during COVID-19. Eur. Policy Anal. **6**(2), 238–263 (2020)
59. Hazans, M.: *Return to Latvia: Remigrants' Survey Results [Atgriešanās Latvijā: Remigrantu Aptaujas Rezultāti]*. Rīga: LU Diasporas un migrācijas pētījumu centrs (2016)
60. Salman, D., et al.: Impact of social restrictions during the COVID-19 pandemic on the physical activity levels of adults aged 50–92 years: a baseline survey of the CHARIOT COVID-19 rapid response prospective cohort study. BMJ Open **11**(8), e050680 (2021)
61. Langenkamp, A., Cano, T., Czymara, C.S.: My home is my castle? the role of living arrangements on experiencing the COVID-19 pandemic: evidence from Germany. Front. Sociol. **6**, 785201 (2022)
62. Xu, Z., et al.: Does it matter who you live with during COVID-19 lockdown? association of living arrangements with psychosocial health, life satisfaction, and quality of life: a pilot study. Int. J. Environ. Res. Public Health **19**(3), 1827 (2022)
63. Spiritus-Beerden, E., et al.: Mental health of refugees and migrants during the COVID-19 pandemic: the role of experienced discrimination and daily stressors. Int. J. Environ. Res. Public Health **18**(12), 6354 (2021)
64. COVID-19 Crisis Puts Migration and Progress on Integration at Risk.: OECD. https://www.oecd.org/migration/covid-19-crisis-puts-migration-and-progress-on-integration-at-risk.htm (2020). Accessed 31 March 2022

How the Covid-19 Pandemic Affects Housing Design to Adapt With Households' New Needs in Egypt?

Rania Nasreldin and Asmaa Ibrahim

Abstract As of late, "STAY AT HOME" is the main slogan; household needs are constantly changing for many reasons, such as the change in the human life cycle, the shift to smart cities, and adopting new modern technologies to reduce the risks. However, while moving to a smart solution, many forgotten social dimensions are being interpreted into the design of many services, including housing. Accordingly, this study aims to explore housing flexibility through a review of relevant literature and how housing design will change to accommodate new needs through quarantine and spread of Corona virus Disease 2019 (Covid-19) to formulate new design codes for stakeholders and real-estate developers to consider in the future. It examines the impact of quarantine on personal household priorities, house design, and how they innovate in their interior design to suit their new needs by conducting a wide online social survey. The research uses an online survey to evaluate the importance of the new arrangement of household requirements using quantitative analysis tools and techniques. The findings display housing guidelines to apply housing design flexibility to cope with any external or internal changes that may happen in the next period and affect household needs in smart cities and others.

Keywords Housing flexibility · Basic needs · COVID-19 · Quarantine · Hoffice · Online survey

R. Nasreldin (✉)
Department of Architecture, Faculty of Engineering, Cairo University, Giza, Egypt
e-mail: raniaibrahim@eng.cu.edu.eg

A. Ibrahim
Department of Architecture, College of Architecture and Design, Effat University, Jeddah, Saudi Arabia
e-mail: asibrahim@effatuniversity.edu.sa

A. Visvizi et al. (eds.), *Research and Innovation Forum 2022*,
Springer Proceedings in Complexity, https://doi.org/10.1007/978-3-031-19560-0_68

1 Introduction

A house is a place for human activities; it cannot be viewed independently from its context and users and the external crisis that may happen at any time. The "Covid-19" crisis may essentially change our relationship with public space also. In the current period, it will be a must to study, highlight and measure the impact of changes on architecture, housing, and open spaces in order to inform real-estate developers, urban planners and designers in a post-Covid world. Several research studies talk about "Covid-19" lockdown and its impact on mental health [1] and on health overall; other researches have addressed housing attributes as concrete determinants of various macro-level and micro-level factors, Covid-19 also has an impact on housing prices and housing market performance all over the world [2]. This paper focuses on studying the impact of Covid-19 on housing design during the quarantine periods and the emergence of new needs, such as staying at home, working from home, and many other needs. As a result of the experience with Covid-19, the lines between life, work, learning, and play will gradually blur.

2 Household Needs Life Cycle

Housing and spatial needs differ by the change in family life cycle, as a family passes through four different stages, as shown in Fig. 1. In this cycle, studio is more appropriate at the beginning of 20–30 years old. After marriage, at age of 30–40 years, there is a need for a separate bedroom, living room, and services. Stage three is from age 40–60 years old and is considered the peak in family size and requirements, where the need to provide separate bedrooms for girls and boys becomes evident (in most cultures). After this cycle, family size begins to decline, and household needs also shrink in the last cycle before death [3].

3 Maslow's Human Needs Hierarchy Versus Housing Basic Requirements

The house is the main settlement in helping to fulfill basic physiological needs and helping to keep up with life requirements. Maslow presented his pyramid of needs and classified the basic human needs for life [4].

According to many studies, one of the primary roles of houses is to fulfill various levels of wishes, expectations, and changing needs. Banham [5] and Oliver [6] suggested that residential spaces can have three different notions; shelter, house, and home. Unlike a house, a home is more than the meaning of a building; Shelters are provided to protect people, while the definition of a home represents deep socio-cultural relationships and structures. It is obvious that the variations aren't

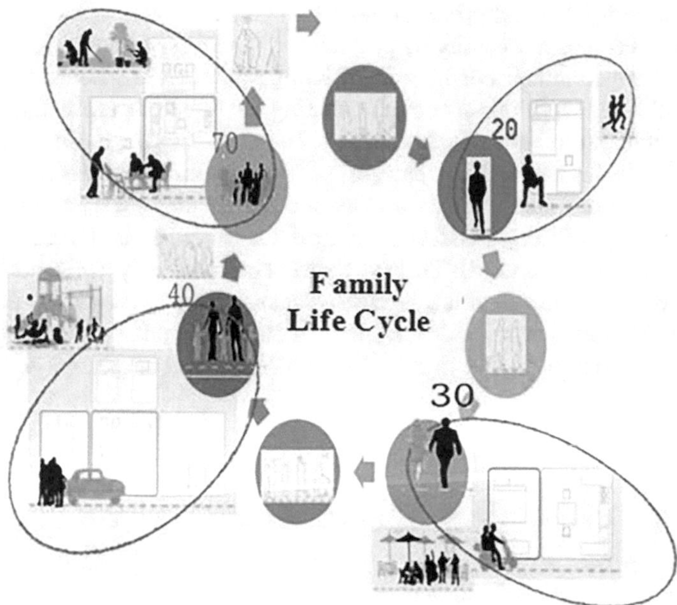

Fig. 1 Family life cycle and housing needs. *Source* Authors

inconse-quential or romantic but essential. According to Olive [6] 'house' is the concept of denotative, which is a 'small dwelling' describing a building or physical construction, while 'home' represents the concept of connotative.

Maslow's hierarchy of needs was developed by Israel [7]. This section tries to categorize basic household needs and add the new requirements during the spread of "Covid-19" with reference to the further use of the house as quarantine.

By applying these needs to housing and how to fulfill them, we find that these basic needs are subject to be modified due to many reasons, such as personal priorities and the impact of external factors imposed by the built and natural environment, such as housing affordability, household with low income can only accommodate basic needs (physiological needs).

There is a direct correlation between affordability and the fulfillment of all human needs; as the affordability increase, the more housing and human needs are fulfilled. When the gap between housing demand and supply decreases, households can choose the best housing design that responds to all their needs (indirect relation). For physiological needs, the first definition of a House is "shelter", as it protects people from external circumstances. It also provides the required space to do all necessary human activities properly, such as sleeping, eating, clothing, etc. Such activities should be conducted in a house that provides suitable ventilation, thermal comfort, and sufficient lighting.

Housing could offer a sense of safety from the external world and circumstances. Structural safety, visual and sound privacy are all factors that give the household a sense of security. Now, after the Covid-19 pandemic, health security and healthy housing have become a necessity by providing a flexible housing design that would suffice during quarantine. For love and belonging needs, the house is the living environment in which people spend time sharing the same space with their families, which aids in the formation of relationships. Esteem needs are basic human needs; Maslow classified esteem into two categories: first, esteem for oneself as self-respect, success, mastery, and independence and second, the desire for respect from others as status and prestige. Finally, satisfaction of aesthetic needs. An individual's home can tell others a lot about them [8]. There is a direct correlation between the housing's concern with the fulfillment of the psychological needs of the owner and the owner's self-esteem. Households can reflect a person's character and needs by the ability to change and redesign the house to achieve the highest level of creativity.

4 Needs/Priorities

4.1 Quarantine Protocol

The World Health Organization (WHO) published guidelines for home quarantine [9]. The most important points that affect the housing requirements and standards for home isolation are:

- Allocate a separate room with adequate ventilation at home.
- The patient should stay in a separate room or be parted from the other family members.
- Other family members should use a separate bathroom.

4.2 Working from Home (WFH)

Due to quarantine, several people are now working from home. That is why a home office might become a vital space in some houses. Several spaces can serve as an office, such as a living room that can accommodate work at the home area with at least a desk, chair, and a filing cabinet, dining table, or bar in the kitchen is also an alternative for a laptop. For more efficiency, and if the housing unit area is enough, preferably find a space that can be permanently used as a home office, as it will be able to work more professionally [10].

Due to the recent residential trend to reduce unit sizes in the name of efficiency, just enough space for a living room furniture design and arrangement, perhaps enough room for a dining table, a small kitchen, and a sleeping zone. Though this concept will continue, designers must consider how a home office might be part of the mix. For working from home, the household needs a separate private space that includes at least a desk, chair, suitable lighting, and built-in storage, or a two-seat dining table might help as alternative [11] , also enrich digital culture by removing in this way the barriers to the use of technology and study the psychological dimensions of working from home [12].

4.3 Interact with Nature

In the past, balconies were considered unused space. In contrast, now and after spending much time at home during the pandemic, balconies become significant spaces of community, present a semi-public place to look out, communicate with neighbors, and interact with nature [13].

5 Flexibility and Adaptable Housing Design

Many studies use the word "flexible" for physical changes, but the word "adaptable" is used for non-physical changes [14]; flexibility can be defined as the 1st level of adaptability that the user makes to fulfill his own renewable needs, and this means the housing elements cannot modify themselves [15].

While adaptable housing can be defined as accessible housing that doesn't look different from other housing and has a design and features that can be adjusted, added, or removed in a concise time as required to suit the householders' needs [16].

6 Indicators of Flexibility and Adaptability

Table 1 shows different indicators of flexible/adaptable housing, including a clear indicator definition, unique requirements for application, and a sketch as an example for each indicator. Most of these indicators depend on the flexibility of the structural system and building legislation that allow households to renovate their house, and adapt the building envelop.

7 Empirical Part

In order to trace and highlight the changes happening in household needs, the following section highlights how householders improve their home design to accommodate new requirements because of quarantine or because of staying at home for long periods.

Table 1 Flexibility Indicators

Type		Indicator	Definition	Requirements	Source
Fixed surface area	V1	Convertibility	Allowing for changes in use within the building  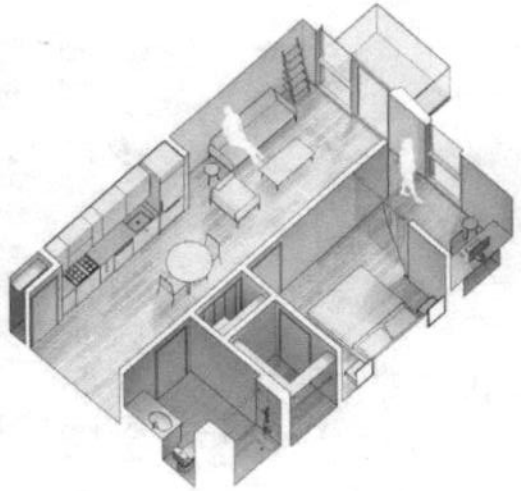 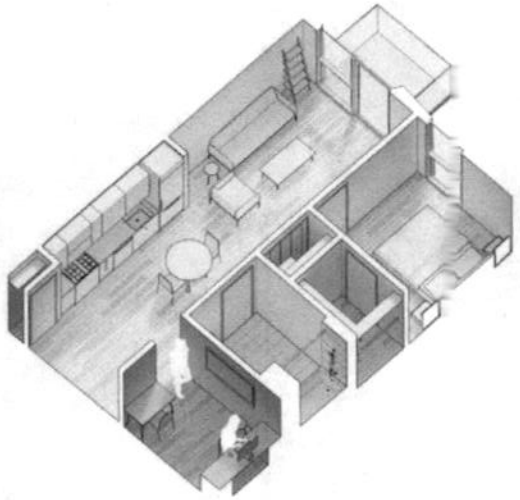	– Flexible structure system – Adjustment and adaptability of the building envelop	[6,1]
	V2	Expanda bility	Facilitating additions to the quantity of space in a building 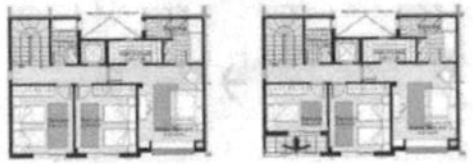	– Flexible structure system – Regulations	[6]

(continued)

Table 1 (continued)

Type		Indicator	Definition	Requirements	Source
	V4	Divisibility	(Dividing up) The potential to divide a larger unit 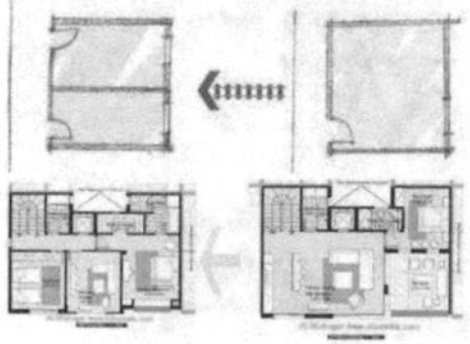	– Flexible structure system – Space area and proportions – Opening places (doors and windows) – Use mobile equipment (equip walls, cabinets, or pre-fabricated modular interior partitions)	[1] [7,8, 1]

(continued)

Table 1 (continued)

Type		Indicator	Definition	Requirements	Source
	V5	Multi-functionality	…is the possibility of using or deploying space, construction or installation components for several functions 	– Space area – Furniture size	[9,1]
	V6	Rearrangeability	Change the layout of spaces 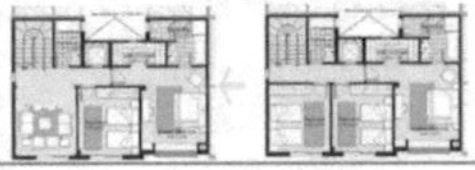	– Furniture flexibility and size	[7]
	V7	Versatility	Capability of being turned 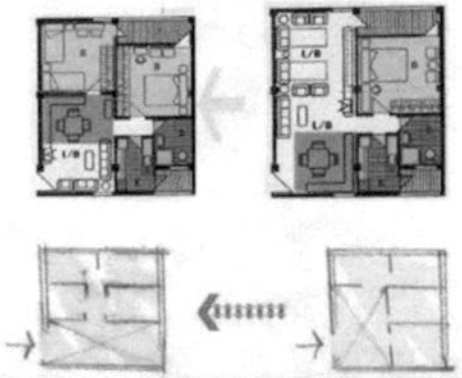	– Flexible structural system	[10,1]
Non fixed surface area	V3	Expansion/extendibility	This factor concerns to which extent the use surface of a building can be increased in the future (horizontal and/or vertical) 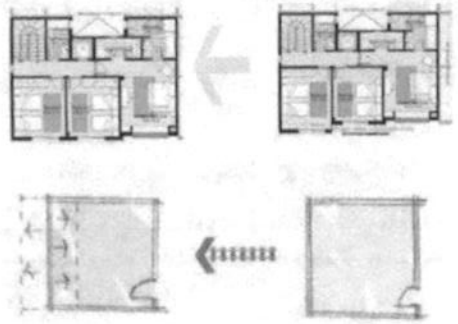	Building technology	[1]

7.1 Methodology

According to Covid-19 circumstances, an online survey is the safest method for collecting data. Both qualitative and quantitative methods are used for data collection.

7.2 Population and Sampling Technique

A cross-sectional web-based online survey was carried out between September 3 and October 11 2020, in Egypt, which included 102 participants. A multistage sampling method was adopted in which several rounds of cluster sampling were carried out prior to establishing the accessible population. The whole population was subdivided into clusters according to their housing prototype, and random samples were then collected from each group.

7.3 Questionnaire

The link of an online survey is promoted through various techniques, such as social media or online discussion platforms, emails, and potential survey participants are invited to fill the survey. The questionnaire's format was simple for the participants to browse, and they should only need a minimum of computer skills to participate as it could be completed by using smartphones or computers. To facilitate for participants, the questions were designed to be short, clear and easy to read.

The online survey is designed to avoid multiple responses. It is designed to include a feature that enables registering interested participants (through their email or IP address). The questionnaire consists of 5 different sections; the first section assesses the socio-demographic details of the participants as age and family size.

The second section concentrates on the house description as housing prototype, type of tenure, housing level, number of rooms, area, number of bathrooms, and balconies. Section three examines the concept of quarantine and whether the house is used as a quarantine or not, and if the participant or one of the family members got 'Covid-19' or not, to know if the house was capable of fitting with quarantine requirements or not.

The fourth section covers the concept of "staying at home" and how staying at home for an extended period affects household needs such as working from home, the need to connect safely with nature, storage area, gym, or any other requirement. The last section examines the definition of flexibility in housing renovation by asking about seven different indicators and if the household uses one of these indicators or not to fit with the new needs through the duration of quarantine and 'Covid-19'.

7.4 Data Collection

Collecting data with an online survey has the potential to collect large amounts of data in a timely and efficient way with fewer errors due to the lack of transferring written data to a computer and lower cost as it requires little human resource efforts while collecting or managing data [17, 18]. Google forms were formed with closed-ended questions in Arabic for the online survey.

A pilot test was conducted before the questionnaire was shared with the target group. The workability and communicability of the questions were tested in a pilot study with 10 participants. Following the completion of the forms, data was imported into a Microsoft Excel spreadsheet and analyzed with IBM SPSS software to generate a correlation matrix between different variables to investigate the relationship between variables such as unit area and needs, human needs' priorities, and their arrangements. The sample included 102 inputs. The study used maximal variation and the snowball sampling technique to include householders with different backgrounds and who live in different house prototypes and may have different opinions and evaluations.

8 Results

8.1 Sample Description

One hundred two persons filled the online survey. The results showed that the mean age of the participants was 20–40 years old, and the average family size is five persons.

8.2 Housing Unit's Description

Apartment area varies between less than 60 m^2 , which presents only 1% of the sample, and the most popular are the units with an area more than 120 m^2 and less than 200 m^2 , which present 42% of the sample with average of three bedrooms (49% of the sample) one bathroom (45%) and one balcony (36% of the sample), (49%) live in new cities, and more than half of the sample live in multifamily housing (64%), middle income housing (65%), 81% of the sample are owners, while 13% are in new rental law housing and 6% are in old rental law housing.

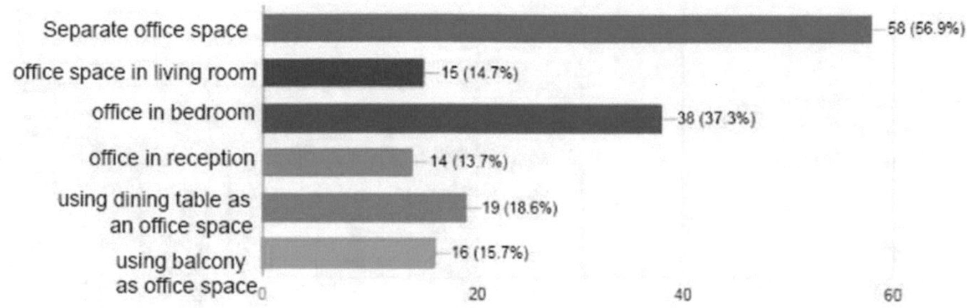

Fig. 2. Hoffice and the best design alternatives

8.3 "Stay at Home" and Required Needs

Most of the sample (87.3%) think that the apartment area can be enough to accommodate new needs during the period of staying at home to interact with nature, while 63.7% use a balcony to interact with nature, while 21.6% used it for studying and working as outdoor space. At the same time, 11% of the sample did not use the balcony. Despite this, 72.5% of the sample thinks that the area of the balcony is enough for activities. But 65.7% of the sample thinks that a balcony is not enough to interact with nature. Only 17% of the sample used the roof for activities because of regulations and safety that may not allow working/studying from home: 73.5% reported that working from home has a significant effect on housing design. 57% said the best zone for working or studying is the private office room, while 13.7% said reception is the best zone (see Fig. 2).

8.4 Housing Renovation

By asking about the arrangement of housing renovation priorities, adding a separate office room became the first priority. The second priority was increasing the number of bathrooms (see Fig. 3). Housing tenure has a significant effect on housing renovation as shown in Fig. 4.

8.5 Housing and Quarantine

17% of the samples have been infected with Coronavirus, while 22% have changed their house to quarantine or confinement during the Coronavirus disease 2020 outbreak. When asked about problems in-home quarantine and how suitable your house design is, answers vary according to the unit area; housing with area 120 to above 200 m^2 have no problems, units with areas from 90 to 120 m^2 mentioned that

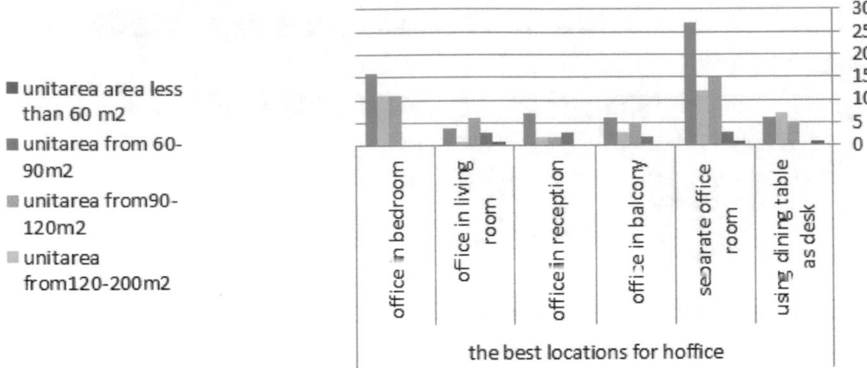

Fig. 3 Unit area and Hoffice priorities alternatives

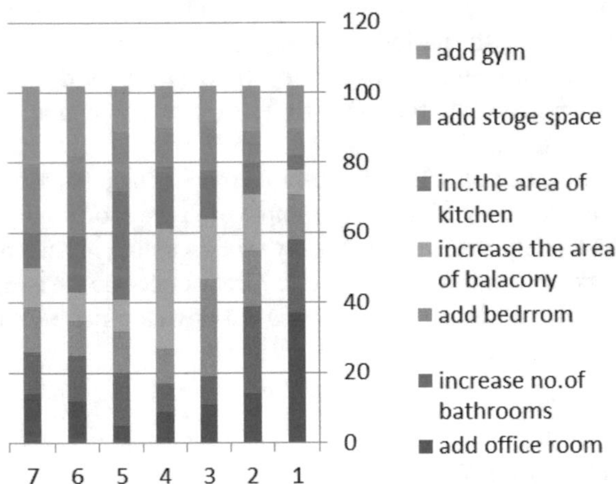

Fig. 4 Housing renovation priorities

the most severe problem is the number of bathrooms, while units from 60 to 90 m^2 mentioned many problems such as the number of rooms and bathrooms which are not enough for quarantine (see Fig. 5).

8.6 *Flexibility*

Type of tenure has a significant impact on housing renovation level, new or old rent law household has a limited level of the renovation as they can rearrange furniture or change finishing colors, multi-functionality or versatility. Still, they cannot change

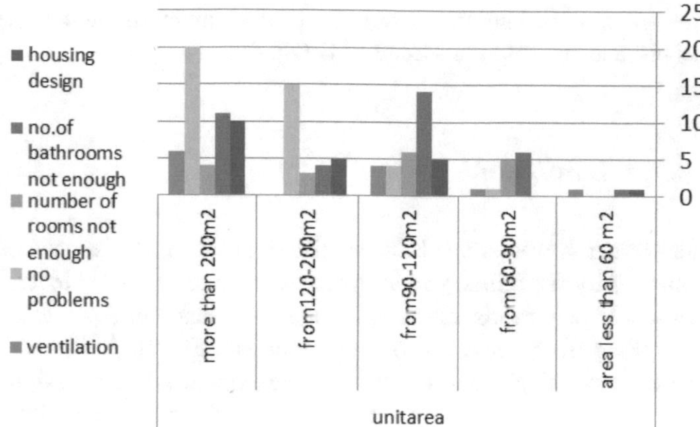

Fig. 5 The relation between housing area and the problems through quarantine

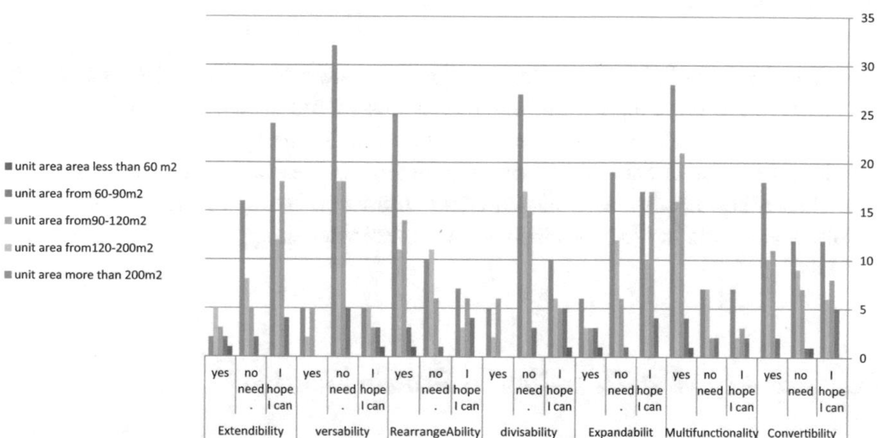

Fig. 6 The relation between housing unit area and flexibility level

the house design by division, or extendibility, or expandability. In addition, house area affects flexibility level (see Fig. 6).

8.7 Discussion

The research aims to examine the impact of quarantine on personal household priorities, house design, and how they innovate in their interior design to suit their new needs by using a broad online social survey in Egypt. The results highlight some housing guidelines that can be applied to housing design standards and flexibility to cope with any external or internal changes that may happen in the next period and affect household needs due to the spread of Covid-19.

8.8 Research Limitation

Although this research was created during the duration of the spread of 'Covid-19', data collected by the online survey from September 2020 to November 2020 highlight household new needs after and through the quarantine and after the long duration of working from home and online education. The study has several limitations. For instance, using a cross-sectional questionnaire-based design does not allow confirming the needs in rural areas which has limited use of the internet.

9 Conclusion

Based on the early indications stated in the literature review and online survey, the paper tried to explore significant impacts on household needs, identify key factors that should be considered for better housing design. Most importantly, designers and developers of residential resilient projects in smart cities will need to consider how households' new needs can be integrated into these new settings during and following a pandemic. Guidelines can be classified at different levels applied on housing design scale as;

9.1 For Applying Flexibility in Housing Design

- Considering the horizontal and vertical expansion of housing areas based on users' changeable needs without negatively impacting the elevation design.
- The spatial distribution of rooms and the proportions of each room and its entrances allow the division or merging of spaces to satisfy the varying needs of residents.
- Providing a suitable structural system (as a flat slab) to allow users to rearrange spaces.
- Design openings in suitable ways that can allow splitting up space division, rearranging, or combining different spatial units without the change in elevations.
- Using multi-use/portable furniture.

9.2 Working from Home

Through and after the COVID-19 pandemic, our homes are anticipated to remain venues for both learning and working; therefore, designers and housing developers may start admitting that these spaces will need to enhance health and productivity; housing design should be adapted and be used in a diversity of ways to meet the changing role of the home. Architects may also design new treatments to overcome acoustic issues by using technical solutions such as sound insulators, soft and porous materials, or separation between rooms, especially for home-office spaces.

- Increase the number of closed rooms to be used as a separate office to provide privacy be more suitable for working or learning.
- In units with an area of 120–200 or more than 200 m^2, separate office space is the best option, the units with area <120 m^2, office space can be merged into a bedroom to be more quiet and private, rather than the living room
- Use the dining table as a temporary desk in small units.
- Divide the living room and reception area into separate spaces; One of them can be locked to suit the requirements of converting it to a workplace.
- Enhance design requirements such as good lighting and ventilation, movable furniture, internet connection, and access to nature to decrease stress and improve users' mental health during working or learning.

9.3 Living Space

- Ensure the efficient design of distribution elements (such as corridors and lobby) and service elements (toilets and kitchen).
- Use flexible, portable furniture that can suit temporal activities such as living space, playing, working area and sometimes sleeping sofa at night.
- Increase the area for living room from 10 to15 m^2 at least to adapt new activities.

9.4 Connection to Nature

- Findings indicated that residents prefer small private open spaces than balconies, but in multifamily housing, the best preference is balconies rather than the roof.
- Balconies provide users with a private open space that can be used for socializing, working, and relaxing, but it is necessary to consider the implications of direction and climatic data when determining the ideal balcony design and use.
- Roofs are less safe and less private when it comes to fulfilling users' needs and connection to nature, it can ideally be used in the penthouse or single-family housing prototypes.

– Users prefer a small garden area to ensure privacy, while outdoor semi-public spaces between residential buildings have been less popular among users than balconies or roofs due to the shared usage between residents and lack of privacy.
– Housing tenure type dramatically affects the ability for adjustments as it is challenging to change wall placements or alter open spaces. However, combining activities or replacing furniture is more convenient to obey their needs.

9.5 Converting House to Quarantine

Minimum housing requirements that should be added to housing design codes:

– At least one master bedroom with a private bathroom, plus one to two bathrooms for other family members.
– At least 2 balconies to provide the patient with a private open space where he/she can connect with nature safely or a personal open space on the roof or garden.
– Enhance natural ventilation and sunlight access, especially in the bedrooms.
– Raise awareness about the concept of healthy buildings is a central part of the international WELL building institute (IWBI)'s mission by engaging individuals in the housings project and highlighting the WELL features pursued.
– Provide a storage space for food and other housing products, and gym space in a separate room or part of the living room or balcony.

References

1. Amerio, A., Brambilla, A., Morganti, A., Aguglia, A., Bianchi, D., Santi, F., Costantini, L., Odone, A., Costanza, A., Signorelli, C., Serafini, G., Amore, M., Capolongo, S.: COVID-19 lockdown: housing built environment's effects on mental health. Int. J. Environ. Res. Pub. Health (2020). https://doi.org/10.3390/ijerph17165973
2. Apergis, N.: The role of housing market in the effectiveness of monetary policy over the Covid-19 era. Econ. Lett. 109749 (2021). ISSN 0165-1765, https://doi.org/10.1016/j.econlet.2021.109749
3. Nasreldin, R.I.: Residential mobility in Egypt, as a solution for middle income housing. Doctoral dissertation, Cairo University, Faculty of engineering, Architecture department, Egypt (2015)
4. Maslow, A.H.: A theory of human motivation. Origin. Publ. Psychol. Rev. **50**, 370–396 (1943)
5. Banham, R.: A home is not a house. In: Miller, B. (ed.) Housing and Dwelling: Perspective on Modern Domestic Architecture, pp. 54–61. Routledge. New York (2007)
6. Oliver, P.: The cultural context of shelter provision. In: Oliver, P. (ed.) Built to Meet Needs: Cultural Issues in Vernacular Architecture, pp. 185–197. Architectural Press, Italy (2006)
7. Israel, T.: Some Place Like Home: Using Design Psychology to Create Ideal Places. Academy Editions/Wiley, Chichester (2003)
8. Venngage Homepage. https://venngage.net/p/74670/maslows-hierarchy-of-needs-housing. Accessed 24 Sept 2020

9. WHO Homepage. https://www.who.int/docs/default-source/searo/whe/coronavirus19/the-guideline-for-home-quarantine---quarantine-in-non-health-care-settings-is-intended-for-any one-who-believes-they-have-been-exposed-to-covid-19-and-are-required-to-be-home-quaran tined-to-prevent-community-trans.pdf?sfvrsn=1bc12565_4. Accessed 24 Sept 2020
10. Bahadursingh, N.: 8 ways covid-19 will change architecture. Architizer (2020). https://archit izer.com/blog/inspiration/industry/covid19-city-design/
11. Peters, T., Halleran, A.: How our homes impact our health: using a COVID-19 informed approach to examine urban apartment housing. Archnet-IJAR: Int. J. Archit. Res. **13**(1)–**14**(3) (2020). https://doi.org/10.1108/ARCH-08-2020-0159
12. Troisi, O., Fenza, G., Grimaldi, M., Loia, F.: Covid-19 sentiments in smart cities: the role of technology anxiety before and during the pandemic. Comput. Hum. Behav. **126**, 106986 (2022). ISSN 0747-5632, https://doi.org/10.1016/j.chb.2021.106986
13. Maiztegui, B.: Green balconies: gardens with altitude. Arch Daily (2020). https://www.archda ily.com/937886/green-balconies-gardens-with-altitude?ad_source5sea
14. Estaji, H.: A review of flexibility and adaptability in housing design. New Arch Int. J. Contemp. Archit. **4**, 37–49 (2007)
15. Lelieveld, C.M.J.L., Voorbij, A.I.M., Poelman, W.A.: Adaptable Architecture. Building Stock Activation, pp. 245–252. TAIHEI Printing Co., Tokyo (2007)
16. Bostrom, J.A., Mace, R., Long, M.: Adaptable Housing: A Technical Manual for Implementing Adaptable Dwelling Unit Specifications. Diane Publishing Company, USA (1987)
17. Kevin, B.W.: Researching Internet-based populations: advantages and dis-advantages of online survey research, online questionnaire authoring software packages, and web survey services. J. Comput.-Mediat. Commun. **10**(3), 1 April 2005. https://doi.org/10.1111/j.1083-6101.2005.tb00259
18. Lytras, M.D., Visvizi, A.: Information management as a dual-purpose process in the smart city: collecting, managing and utilizing information. Int. J. Inf. Manage. (2021). https://doi.org/10.1016/j.ijinfomgt.2020.102224

Exploring Reliability of TEA CAPTIV Motion Capture Suit for a Virtual Reality Workplace Ergonomic Design

David Krákora, Jan Kubr, Petr Hořejší, Ilona Kačerová, and Marek Bureš

Abstract Occupational diseases and specifically musculoskeletal disorders are great issues even in modern factories. Consequences (i.e. occupational diseases or generally lower worker performance) cost a lot of financial and mental funds of employees and employers. The correct way is the prevention using ergonomic analyses of workplaces. We have developed a unique methodological base using modern technologies for an effective and efficient way for quick and precise validation of a particular workplace. The procedure can also be used for employee training. We involve virtual reality (next VR) headset, motion capture, and self-developed VR software (next SW). The validation can be also performed for a virtual (artificial) workplace, thus the proposed technological framework can be used also for learning purposed. We have also proposed a method of validation. We have limited the original research on a representative assembly station and the use of the Perception Neuron suit. This paper extends the original paper. In this research, we are using a different motion capture suit TEA CAPTIV in the scope of the same experimental methods, conditions, and limitations as in the original study. The result is extrapolated to the scope of the full working shift and then compared with the original results.

Using of a different Motion Capture suit is reliable, but there are aspects that still needs deeper research. The process of measuring and software processing differs. Those issues are discussed in the paper.

Keywords Ergonomics · Motion capture · Virtual reality · Workplace · Learning

1 Introduction

Nowadays, the elements and methods of the 4th industrial revolution are already very common in industrial enterprises. Automation, digitalization, robotization and virtualization have been implemented in several processes across the entire industrial sector [1]. Manufacturing industries are one of the most demanding industries as they

D. Krákora (✉) · J. Kubr · P. Hořejší · I. Kačerová · M. Bureš
Faculty of Engineering, University of West Bohemia, Pilsen, Czech Republic
e-mail: krakorad@kpv.zcu.cz

A. Visvizi et al. (eds.), *Research and Innovation Forum 2022*,
Springer Proceedings in Complexity, https://doi.org/10.1007/978-3-031-19560-0_69

require systematic productivity improvements. Proper efficiency and optimization of individual tasks can take a company to a new level, increase its capacity, reduce costs and enable better competitiveness [1–3]. Due to current situations, such as Covid-19 and the various legislative constraints associated with it, it has resulted in the situation that we find some enterprises hastening their transition to digitalisation. This does not mean that enterprises have switched to complete robotization and tried to create an AI-only production line. Elements of human labour and skills will always be needed, it's just that different professions and their responsibilities may change [4].

The same approach (work this means the use of new technologies and measurement innovations) can be understood from an ergonomic perspective. Generally speaking, ergonomics is concerned with the technical solution of requirements in terms of human needs. Its aim is to find a balance solution that best suits the requirements of the individual and the capabilities of society. This is no different in terms of the workload of employees in industries. In every country, there are many standards, regulations, laws or even recommendations that can protect workers, make their work more pleasant and prevent occupational diseases [5, 6]. Nowadays, it is very common for workplaces to have a rotation of workers with different physical proportions. This creates anthropometric parameters that should be met by individual workplaces to facilitate the work of new and existing employees. It is these elements that improve the quality of individual activities, have the right psychological impact on people and eliminate the occurrence of work-related accidents [7].

Therefore, in addition to digitalisation, today more and more emphasis is being placed on elements of workplace rationalisation, ergonomic analysis of individual activities and elements of occupational health and safety. For these studies and measurements, it is again possible to use elements of modern technology to make individual measurements and thus improve the workplace [7, 8]. One possible tool is virtual reality. This type of virtualization can be encountered in workplaces when training individual activities and tasks. So-called virtual training courses are nowadays very often used for staff training. This type of training can effectively transfer the necessary skills to the employee [9]. From an ergonomics and rationalisation perspective, virtual reality can be used to test and measure existing workplaces and design new workplaces to meet all standards. In workplace inspection/design, the use of virtual reality offers several advantages. Some of the basic ones include reducing costs, not stopping the workplace, not disfiguring parts, and making changes in real time. It is therefore possible to fine-tune and test the workplace several times without having to stop operations or reduce capacity. [8, 10, 11] Motion capture technology can be used to speed up ergonomic measurements. This technology is widely used in the film industry or in animation production because it can capture motion in real time. Some of these suits are already focused on stand-alone ergonomics, where in addition to sensors, there is also a program to evaluate and measure the different positions of the limbs. Thus [8, 12], in combination with virtual reality, one can test the workstation and usability of each suit for these measurements and compare their success rate in accuracy with others.

This paper is a pilot study inspired by the article "A Method for a Workplace Validation Using Virtual Reality and a Motion Capture Suit" [13]. The main focus of the article is an innovative methodological basis that explores the use of Motion Capture technology suits and so-called HMD (head mounted display) glasses that project virtual reality from a selected process. The procedure of the experiment was such that first the anthropometric data of the proband was measured, then training on the virtual reality was done to learn the controls and to repeat the whole work process. It is a multi-phase method that involves 2 measurements using VR and MoCap—before (VR_B) after (VR_A) ergonomics workplace optimization, followed by a verification measurement using MoCap or conventional means on a real workplace.

Testing can be repeated using a different type of interaction in VR (using hand-tracking). The same procedure was proposed in the conclusion of [13]. Other possibilities for future research are to link to local muscle loading of the upper limbs, which can be calculated thanks to a mathematical model, or using Motion Capture suits that support this technology, such as TEA CAPTIV or Xsens. This paper follows the use of the TEA CAPTIV suit for ergonomic workplace measurements using virtual reality. This is an original study that will test the capabilities of using the MoCap suit TEA CAPTIV and compare the results with the Noitom suit. The rationale for using this suit was to have custom software for motion evaluation and more efficient sensor manipulation (faster deployment and calibration) [13]. It is possible to follow up this pilot with further studies (e.g. comparing it with other motion capture suits Xsens, Noitom).

2 MoCap Suits Comparison

CAPTIV is a load and motion measurement device from the French company TEA Ergo. This device uses different types of wireless sensors (the currently available ones will be presented below). These sensors work with computer software that records data in the form of a virtual model of the person. From this model, the data can then be analysed and evaluated. The virtual model can be linked to video footage, leading to greater clarity and a faster analysis process. For the evaluation itself, all limits corresponding to the legislation in force must be set.

CAPTIV uses various types of wireless sensors that measure, for example, muscle activity, heart rate, skin conductivity due to sweating or the movements of the person being monitored. The currently available kit includes 6 sensors that measure movement and 4 EMG sensors that measure muscle activity [13] (Fig. 1).

T-Sens Motion sensors record the movements of the person being monitored by measuring changes in angles relative to themselves. For example, if we place the sensors on the forearm and biceps, it is possible to track movements in the elbow.

Before the actual measurement, the suit needs to be calibrated with the software. Compared to the hardware device used in the original article [13], only a simple I pose is required. Subsequently, a visual inspection of the movement and synchronization with the computer SW is performed. During the actual measurement, it is important to switch the computational software on/off during the different cycles of the square (Fig. 2).

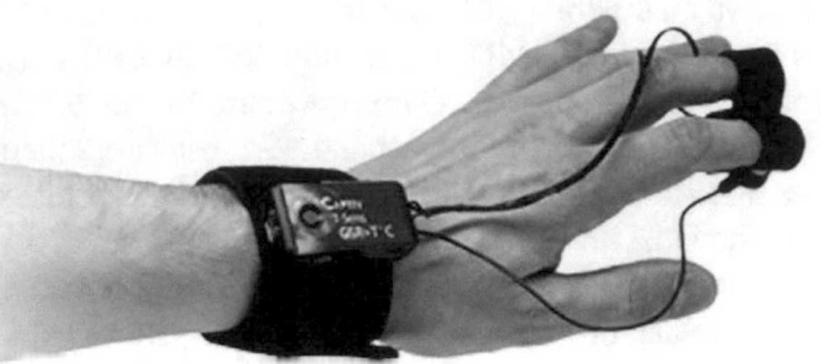

Fig. 1 TEA CAPTIV sensor

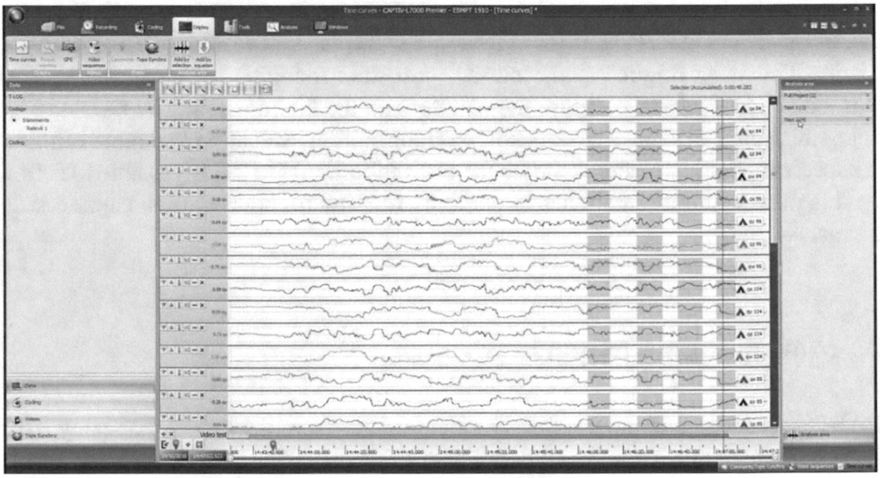

Fig. 2 SW TEA ergo excerpt

3 Reference workplace

For our study, we chose a standardized workplace of an assembly line for the production of door parts in the automotive industry, this is the workplace that was used in the previous study. The VR application used is functionally identical to the application used in the previous study as well as the workstation mentioned [14]. The work was measured on a virtual workstation of an automotive door production line. The worker performs the activity standing at the workbench, parts are taken from the tray or KLT box located above the workbench. The part is then taken to the welding line where the automatic welding process takes place. Both men and women work on the work line. The work plane is static. The workplace cannot be individualized.

Table 1 Rated working positions

	Accaptable	conditionally acceptable	Non-acceptable
Upper limb	0°–40°	40°–60°	Over 60°
Torso flexion	0°–40°	40°–60°	Over 60°
Head rotation	0–25°	Over 25°[a]	Over 25°[b]
Head flexion	0–25°	Over 25°[a]	Over 25°[b]

[a] at a movement frequency of less than 2/min
[b] at a frequency of movements greater than 2/min

4 Method of Ergonomics Evaluation

The ergonomic evaluation was carried out on the basis of the current Czech legislation (Government Regulation No. 361/2007 Coll.), which evaluates the positions according to acceptability into acceptable, conditionally acceptable and unacceptable (This legislation is stricter than other European legislation. Any interpretations can therefore be used within the scope of other legislations.). The legislation determines the maximum time a worker can spend in a given position, acceptability is determined by the angles of the part. In the case of a conditionally acceptable position, the maximum possible time is 160 min per average 8-h shift, in the case of an unacceptable position, the maximum possible time is 30 min out of an average 8-h shift. In the study, the analysis focused on the most problematic parts of the human body—upper limb (shoulder) position, trunk position and head position. The angles determining whether a position is conditionally acceptable or unacceptable are very similar to the RULA (Rapid Upper Limb Assessment) analysis [15, 16] (Table 1).

In the ergonomic assessment, only the unacceptable position was assessed in the case of the head.

The actual evaluation of the acceptability of the positions was carried out in the special TEA Ergo software available with the TEA CAPTIV suit. The outputs can be seen in the figure below. Each of the probands was filmed during the measurements, the videos with the curves were synchronized and the positions were evaluated based on these curves (Fig. 3).

5 Results

TEA Ergo evaluates the working positions either in the form of a percentage distribution (green areas = acceptable positions, orange areas = conditionally acceptable positions, red areas = unacceptable positions) or lists the times in each position for the given measurements (Figs. 4 and 5).

This is an initial pilot study that was conducted on a group of 5 people who made pilot measurements. The group was in the age range of 25–40 years. They

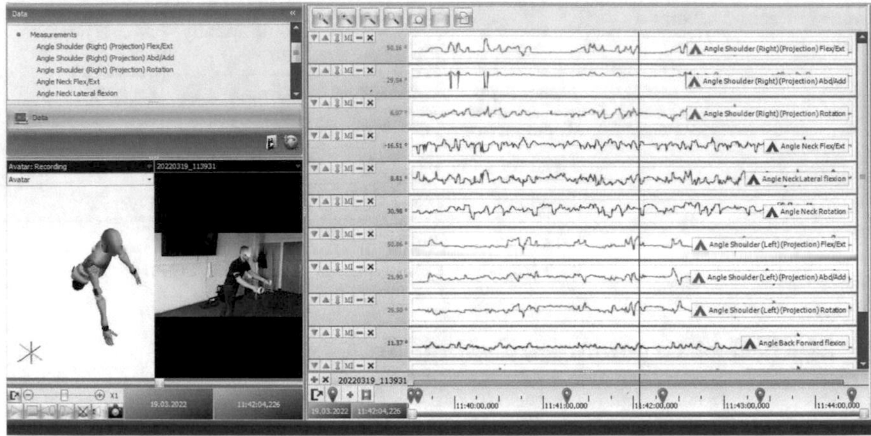

Fig. 3 Part of evaluation

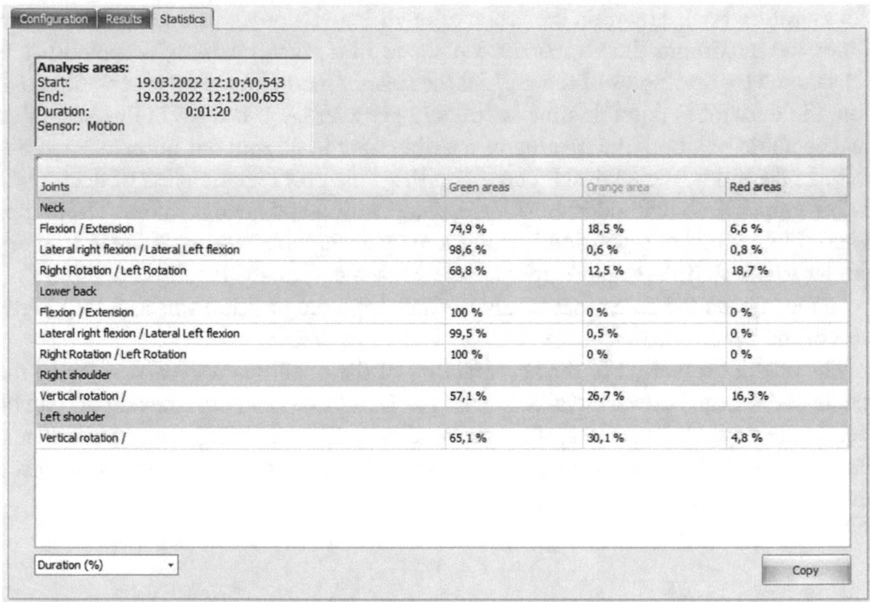

Configuration | Results | Statistics

Analysis areas:
Start: 19.03.2022 12:10:40,543
End: 19.03.2022 12:12:00,655
Duration: 0:01:20
Sensor: Motion

Joints	Green areas	Orange area	Red areas
Neck			
Flexion / Extension	74,9 %	18,5 %	6,6 %
Lateral right flexion / Lateral Left flexion	98,6 %	0,6 %	0,8 %
Right Rotation / Left Rotation	68,8 %	12,5 %	18,7 %
Lower back			
Flexion / Extension	100 %	0 %	0 %
Lateral right flexion / Lateral Left flexion	99,5 %	0,5 %	0 %
Right Rotation / Left Rotation	100 %	0 %	0 %
Right shoulder			
Vertical rotation /	57,1 %	26,7 %	16,3 %
Left shoulder			
Vertical rotation /	65,1 %	30,1 %	4,8 %

Duration (%) | Copy

Fig. 4 Measured values percentage

were 4 men and one woman who were almost all completely healthy, with no movement disorders, health problems or hand surgery. The vast majority were university students or administrative workers who have sedentary jobs and spend most of their time working with computers. One proband had a history of lumbar spine surgery. If the health problem was more severe and obvious during the measurement, the result would not be included in the statistical evaluation. The probands had different

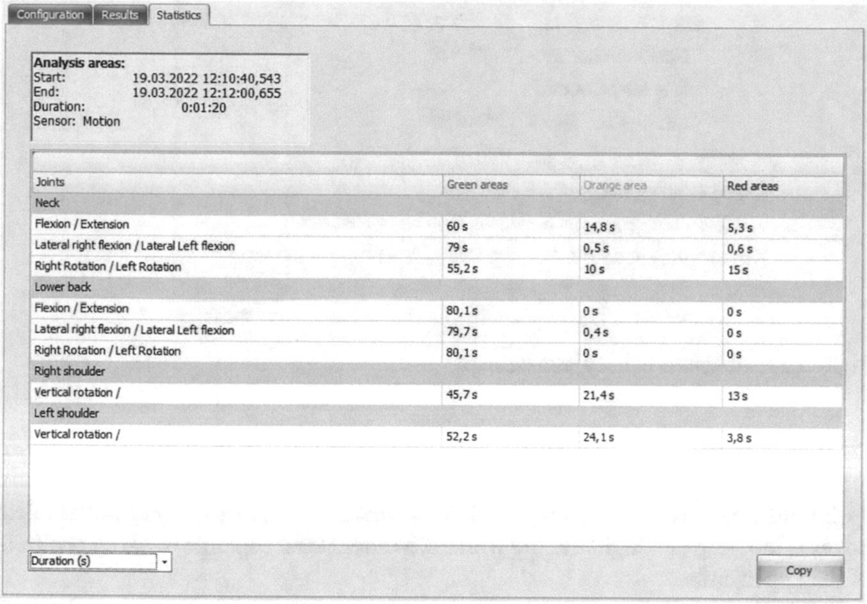

Configuration | Results | Statistics

Analysis areas:
Start: 19.03.2022 12:10:40,543
End: 19.03.2022 12:12:00,655
Duration: 0:01:20
Sensor: Motion

Joints	Green areas	Orange area	Red areas
Neck			
Flexion / Extension	60 s	14,8 s	5,3 s
Lateral right flexion / Lateral Left flexion	79 s	0,5 s	0,6 s
Right Rotation / Left Rotation	55,2 s	10 s	15 s
Lower back			
Flexion / Extension	80,1 s	0 s	0 s
Lateral right flexion / Lateral Left flexion	79,7 s	0,4 s	0 s
Right Rotation / Left Rotation	80,1 s	0 s	0 s
Right shoulder			
Vertical rotation /	45,7 s	21,4 s	13 s
Left shoulder			
Vertical rotation /	52,2 s	24,1 s	3,8 s

Duration (s) Copy

Fig. 5 Measured values in time distribution

physiological body builds and different physiques. All probands were between 168 and 178 cm tall, falling within the average European population. All probands were of right laterality.

The actual measurement took place with each participant in turn. The participants were explained how to work in a virtual reality environment, were introduced to the workplace and the process itself. They were then dressed in a Motion Capture suit. Subsequently, the calibration in I pose was performed and the actual measurement began. First, the participant performed a work cycle taking products from the top floor of the rack, followed by the middle floor and then the lowest floor. The measurement process was similar for the modified workstation. Then the virtual reality goggles were removed and the suit was taken off and transferred to the next research participant. The training itself took about 5 min, and the measurement then took about 10 min. As part of the evaluation, the workplace was first measured in virtual reality before rationalisation (VRb), then the workplace was adjusted and re-measured based on the findings (VRa) (Fig. 6).

First, ergonomics in virtual reality was evaluated in a virtual workplace. After the analysis, it was found that the workers have an overloaded right upper limb (shoulder joint), the position of the upper limb exceeds the recommended limits by more than 10 min. This result indicates a violation of the ergonomics of the workplace and indicates a necessary modification of the workplace. Another problematic area was the head position - especially the rotation above 15°, which was calculated at 245 min for an average 8-h shift. However, this phenomenon was observed probably due to

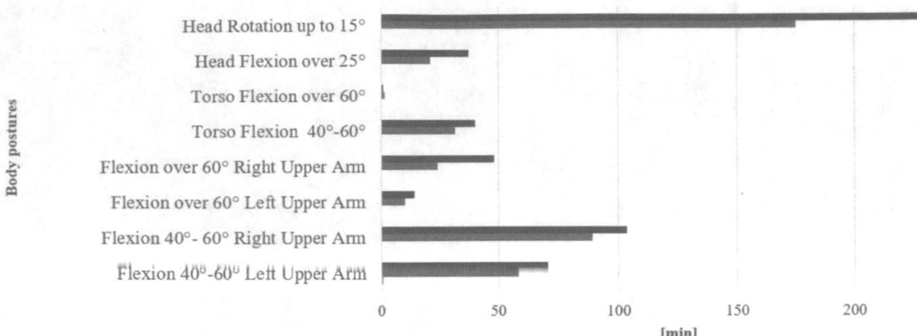

Fig. 6 Final evaluation of body postures

the lack of orientation in the virtual environment. After the analyses, a workstation modification was recommended. The placement of the KLT boxes with material within the comfort zone was adjusted. The current design of the placement of the boxes is too deep and high, forcing workers to adopt unacceptable working positions of the upper limbs.

The rationalisation has saved time in unacceptable working positions by half. After rationalisation, the position of both the right and left upper limbs is within the limit and there is no overloading. Even after rationalisation, the head position (rotation) is still over the limit, but this overrun is not due to the workload itself, but to orientation in the environment.

6 Conclusion and Discussion

In the Table 2 compares the outputs from the previous study with the measured values from the current study. Compared to the previous study [13], the results are different, especially in the area of head and neck position head position was recorded for an average of around 25 min in the previous study, and in the new study it was recorded for 245 min, a percentage increase of 980%. Compared to the previous study, the results may differ mainly because the test group of this research. Probands who were not intimately familiar with the production process and the working environment that realistically corresponds to a manufacturing company were measured. As for other parts of the body - for example, the unacceptable working position of the right upper limb was measured in the previous study for 32 min in an average 8-h shift (VRb), in the new study, the results were again higher at 49 min, an increase of 53%. To increase the correlation of the whole research, the number of probands tested would need to be increased.

Table 2 Results comparison

	VRb–Before rationalization TEA Captive [min]	VRb–Before rationalization perception neuron [min]	VRa–After rationalization TEA Captive [min]	VRa–After rationalization perception neuron [min]
Flexion 40°–60° Left Upper Arm	71	10	58	5
Flexion 40°–60° Right Upper Arm	103	12	89	8
Flexion over 60° Left Upper Arm	14	7	10	4
Flexion over 60° Right Upper Arm	48	32	24	15
Torso Flexion 40°–60°	40	13	32	11
Torso Flexion over 60°	1	16	2	12
Head Flexion over 25°	38	20	21	14
Head Rotation up to 15°	237	26	175	21

The TEA CAPTIV suit measurement method is more suitable for laboratory conditions due to the high susceptibility to ambient magnetic field and the high calibration requirements of the whole set. Another indisputable disadvantage of this solution is the necessity to purchase individual sensors to measure multiple body parts. The main advantage of this solution is the automatic evaluation of the whole ergonomic process directly in the native environment supplied with the suit. From a practical point of view, the great advantage is the speed and versatility of wearing individual suit segments for sensor clamping. This is a small pilot study and the results are not statistically significant, based on the results, it's suspicious the suit TEA CAPTIV could be reliable for performing measurements according to the methodology proposed in [13]. An idea for further research is to expand the sample of probands to a larger group, especially to include workers with different anthropometric characteristics. Validation of the results using conventional methods could follow as the next part, as was done in the aforementioned paper.

Funding This work was supported by the Internal Science Foundation of the University of West Bohemia under Grant SGS-2021–028 'Developmental and Training Tools for the Interaction of Man and the Cyber–Physical Production System'.

References

1. Benanav, A.: Automation and the future of work. (2020)
2. Rosa, C., Silva, F. J.G., Ferreira, L. P.: Improving the quality and productivity of steel wire-rope assembly lines for the automotive industry. Procedia Manuf. **11**, 1035–1042 (2017). Elsevier B.V. https://doi.org/10.1016/J.PROMFG.2017.07.214
3. Abolhassani, A., Harner, E. J., Jaridi, M.: Empirical analysis of productivity enhancement strategies in the North American automotive industry. Int. J. Prod. Econ. **208**, 140–159 (2019). Elsevier. https://doi.org/10.1016/J.IJPE.2018.11.014
4. Sorgner, A.: The automation of jobs: A threat for employment or a source of new entrepreneurial opportunities? Foresight and STI Governance. **11**, 37–48 (2017). National research university, Higher school of econoimics. https://doi.org/10.17323/2500-2597.2017.3.37.48
5. Baraldi, E. C., Kaminski Paulo C. 2011. Ergonomic planned supply in an automotive assembly line. Hum. Factors Ergon. Manuf. **21**, 104–119 (2011). John Wiley & Sons, Ltd. https://doi.org/10.1002/hfm.20228
6. Malý, S., Král, M., Hanáková, E.: ABC ergonomie, vol. 386. Professional Publishing (2010)
7. Rodrigues, J., Probst, P., Cepeda, C, Guede-Fernandez, F., Silva, S., Gamboa, P., Fujao, C., Quaresma, C. R., Hugo , G.: MicroErgo: a concept for an ergonomic self-assessment tool. In: Proceedings of 2021 IEEE 7th international conference on bio signals, images and instrumentation, ICBSII 2021. Institute of electrical and electronics engineers Inc, (2021). https://doi.org/10.1109/ICBSII51839.2021.9445156
8. Vermeulen, B., Psenner, E.: Exploiting the technology-driven structural shift to creative work in regional catching-up: toward an institutional framework. Routledge (2022). https://doi.org/10.1080/09654313.2022.2028737
9. Broekens, J., Harbers, M., Brinkman, W. P., Jonker, C. M., Bosch, K. V. D, Meyer, J. J.: Virtual reality negotiation training increases negotiation knowledge and skill. Lecture notes in computer science (including subseries lecture notes in artificial intelligence and lecture notes in bioinformatics) 7502 LNAI, pp. 218–230. Springer Verlag (2012). https://doi.org/10.1007/978-3-642-33197-8_23/COVER/
10. Hugget, C. 2018. *Virtual Training Basics*. American Society for Training & Development
11. Ma, D., Fan, X., Gausemeier, J., Grafe, m. ed.: 2011. Virtual reality & amp; augmented reality in industry. Springer, Berlin Heidelberg (2011). https://doi.org/10.1007/978-3-642-17376-9
12. Failes, Ian.: What Mocap suit suits you?—VFX voice magazine VFX voice magazine. VFC Voice, (2019)
13. Fingertracking—EST, Engineering systems technologies GmbH & Co. KG
14. Kačerová, I., Kubr, J., Hořejší, P., Kleinová, J.: Ergonomic design of a workplace using virtual reality and a motion capture suit. Appl. Sci. **12**, 2150 (2022). MDPI AG. https://doi.org/10.3390/APP12042150
15. Gómez-Galán, M., Callejón-Ferre, A. J., Pérez-Alonso, J., Díaz-Pérez, M., Carrillo-Castrillo, J.A.: Musculoskeletal risks: RULA bibliometric review. Int. J. Environ. Res. Pub Health. **17**, 1–52 (2020). Multidisciplinary Digital Publishing Institute (MDPI). https://doi.org/10.3390/IJERPH17124354
16. 361/2007 Sb. Nařízení vlády, kterým se stanoví podmínky ochrany zdraví při práci

Using DES to Improve the Efficiency of a Covid-19 Vaccination Centre

Saikat Kundu, Muhammad Latif, and Petr Hořejší

Abstract Many research and development teams around the world have developed and continue to improve Covid-19 vaccines. As vaccines are produced, preparedness and planning for mass vaccination and immunization has become an important aspect of the pandemic management. Mass vaccination has been used by public health agencies in the past and is a viable option for Covid-19 immunization. To be able to rapidly and safely immunize a large number of people against Covid-19, mass vaccination centres are accessible in the UK. Careful planning of these centres is a difficult and important job. Two key considerations are the capacity of each centre (measured as the number of patients served per hour) and the time (in minutes) spent by patients in the centre. This paper discusses a simulation study done to support this planning effort. In this paper, we explore the operations of a vaccination centre and use a simulation tool to enhance patient flow. The discrete event simulation (DES) tool outputs visually and numerically show the average and maximum patient flow times and the number of people that can be served (throughput values) under different number of patient arrivals (hourly). With some experimentation, the results show that marginally reducing the hourly arrival rate, patient congestion reduces enabling good patient service levels to be achieved.

Keywords Mass vaccination centre · Covid-19 vaccination · Discrete event simulation · Capacity planning

S. Kundu · M. Latif
Department of Engineering, Manchester Metropolitan University, Manchester, UK

P. Hořejší (✉)
Faculty of Mechanical Engineering, University of West Bohemia, Pilsen, Czech Republic
e-mail: tucnak@kpv.zcu.cz

A. Visvizi et al. (eds.), *Research and Innovation Forum 2022*,
Springer Proceedings in Complexity, https://doi.org/10.1007/978-3-031-19560-0_70

1 Introduction

In the early phase of the year 2020, a novel virus outbreak led to a worldwide pandemic with millions of confirmed cases [1] that caused large proportions of the world population to be in temporary lockdown. With non-essential travel discouraged and everyone but key workers staying at home the world economy came to a sudden pause [2]. The containment of the virus, a novel coronavirus named Covid-19, required quick resource re-allocation on a large scale and was prioritised on every level of healthcare delivery, first identified in East Asia. As the outbreak continued the epicentre shifted to Europe and the Middle East, and eventually affected the Americas [3]. It led to restrictions on public life previously unimaginable during times of peace [4]. Schools were closed, work from home was strongly encouraged, and non-essential travel was forbidden; some regions, and even countries, were entirely locked down for weeks or months [5].

The UK Covid-19 vaccines delivery plan [6] published on Monday 11 January 2021 to coincide with the opening of seven new regional vaccination centres, said England would have capacity to vaccinate at least two million people per week by the end of January. This will be delivered across 206 hospital sites, 50 vaccination centres, and 1200 local vaccination sites run by primary and community care teams, it says.

The expansion of capacity means that everyone will live within 10 miles of a vaccination centre, or, in the case of a small number of highly rural areas, have access to a mobile unit delivering vaccinations. In this paper, we utilise DES to improve the operational efficiency of a typical vaccination centre. The vaccination centre was located in a sports/community facility that provides spacious accommodation to support high patient flow for advance-booked individuals. In the UK, the population is vaccinated in order of priority group, based mainly on age but also accounting for underlying health conditions and employment as a health or care worker [7].

2 Methodology

Witness Horizon software was used to demonstrate the value of DES computer modelling in supporting operational planning of Covid-19 vaccination centre. This study followed a standard simulation study methodology, consisting of the following steps: vaccination centre operations (scope of study), data collection, analyse data, model building, model testing, results/experimentation, discussion, and conclusion.

The scope of the simulation study was limited to the vaccination centre operations and the key performance measures of capacity and time-in-system. Arrival of patients to the centre is assumed to be by car or walk-ins on the basis of 60:40. Data collection relied upon observational data of a vaccination centre and secondary data from related literature. At the time of the study, the Pfizer-BioNtech(PZ) vaccine was being administered.

3 Vaccination Centre Operations

Vaccination centres require careful planning and implementation and are governed by National Health Service (NHS) England guidelines [8]. The correct number of staff must be assigned to roles when the centre operates. Two key considerations are the capacity of the centre (measured as the number of patients served per hour) and the time (in minutes) spent by patients in the centre (this is known as the flow time or throughput time). Centre capacity affects the number of centres that must be opened, and the total time needed to vaccinate the population. The flow time affects the number of patients who are inside the centre. More patients require more space as they wait to receive treatment. If throughput is too high then unsustainable queues will form, compromising social distancing and impairing patient experience (important for ensuring a repeat visit for any additional dose). On the other hand, if throughput were too low, then this would lead to an uneconomic use of available resources. The balance between centre capacity and flow time is very subjective in mass vaccination and regularly tweaked to meet operational targets. Additional considerations relate to the optimal allocation of activity-level resources to ensure balanced server utilisation and the incorporation of sufficient 'slack' in pathway capacity to ensure any 'shocks' can be readily absorbed (such as staff sickness or a number of patients arriving all at once).

There is, however, very little information and learned experience to guide managers through these considerations. Events of such magnitude have simply not occurred in recent times and so, beyond the limited number of national and regional level emergency preparedness roles, there is little existing knowledge and expertise within the local frontline entities tasked with setting up the vaccination centres. Most previous studies on pandemics and vaccination centres use the discrete-event approach to stochastic simulation, given its capacity of capturing the modelling requirements of service systems with individual entities (such as patients) which flow through a care pathway, competing for resources such as appointment slots [9].

Vaccination centre operates on a pre-booked appointment basis. This means that slots are available on-line for patients to book. The centre receives patients as walk-ins or by car. Either case the patients enters the site via a car park. Cars are directed onto the car park in a controlled manner by marshals (volunteers) who limit car arrivals on to site. Once parked the patient walks to the building and usually join a patients queue that forms at the entrance. Walk-in patients also join the same entrance queue. The entrance queue moves slowly enabling patients to enter the building containing the centre. Upon building entry patients have a temperature check done whilst in a moving queue. Patients follow an orderly queue which meanders along the entrance corridor to enter the main hall. Within the main hall, the patients first stop is Registration that involves confirming basic personal details and collecting a personal data sheet. After Registration, the patient's second stop is Clinical Assessment where the patients' medical condition is evaluated. The patient then follows the snake like queue and is directed to the next available Vaccination cubicle. Within the Vaccination cubicle, the personal data sheet is collected, information is given, and the vaccine administered.

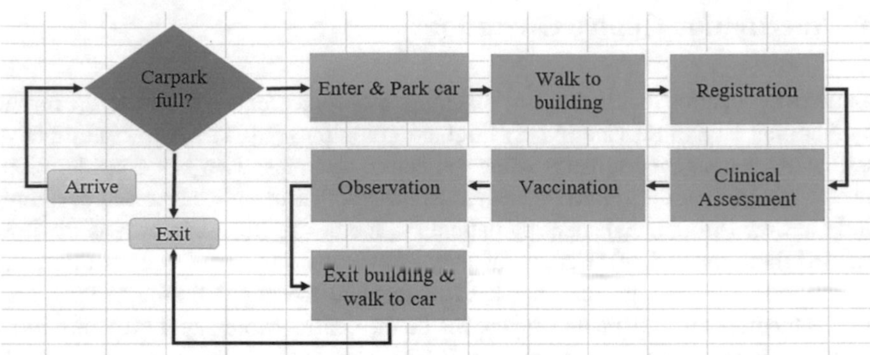

Fig. 1 Patient flow for a drive-in patient

The patient is then directed to take a seat in the Observation area. Volunteers manage the Observation area and allow patients to exit after a 15-min stay. Patients leave the building through a separate exit and walk through the car park and either drive through or walk through the site gates. A simple flowchart representing the drive-in patient is shown in Fig. 1.

For the vast majority of patients, the procedure described earlier reflects their experience. However, some patients are likely to leave the centre with or without vaccination because they have been un-successful at any of the stations. These patients have not been considered in this study because observational evidence suggests the failures are negligible.

4 Data Collection

Operations of the vaccination centre were observed by following the patient flow through the various stations in addition to secondary data [10]. The observations included patient arrivals and mode, queue lengths, walking speeds, distance and capacity of stations, staffing levels, and service times. The various stages involved:

- Arrival either by car or walk-in through the car park
- Walking to the building entrance, essentially joining a patient queue to enter the centre
- Registration
- Clinical Assessment
- Vaccination cubicle
- Observation area
- Exit via the car park.

Although the data collection was carefully planned, the data collected was not complete and may have included some inaccuracies due, in part, to the limited number of people, time, and equipment available to conduct the time study. Missing data was estimated from secondary data [11, 12]. Computer simulation was used to model vaccination centre operations. An analytical solution was not attempted due to limitations on variability. The data was sufficient for constructing a valid simulation model.

5 Data Analysis

Raw data was collected from the centre, mostly by observations and some by estimates. Ultimately the data was collated and manipulated to determine how long a patient spent at each station and ultimately the total time in the centre.

It was determined to separate the conveyance timings from activity timings. This enabled value-added data to be independent of non-value-added data. It was deemed appropriate to use a triangular distribution for conveyance and service times due to the limited data and wide variability. Tables 1, 2 and 3 depicts key parameters and operational timings of the centre.

Table 1 Operational parameters

Patient arrivals	
Number of arrivals expected per hour	100
Operational hours per day	12
New cars allowed on car park at any one time	5
Ratio of car arrivals to walk-ins	60%
Resources	
Registration staff	3
Clinical assessment staff	8
Vaccinators	6
Observation spaces	25
Queue capacity	
From entrance to registration	100
From registration to clinical assessment	10
From clinical assessment to vaccination	10
From vaccination to observation	2
Car park capacity	100

Table 2 Activity timings (in minutes)

1. Registration	
Minimum	0.5
Mode	1
Maximum	1.5
2. Clinical Assessment	
Minimum	4
Mode	5
Maximum	6
3. Administer Vaccine	
Minimum	2
Mode	3
Maximum	4
4. Observation	
Minimum	15
Mode	15
Maximum	15

Table 3 Conveyance timings (in minutes)

1. To Park a Car		2. Walk: Entrance to Registration		3. Walk: Registration to Clinical		4. Walk: Clinical to Vaccination	
Minimum	1	Minimum	0.5	Minimum	0.5	Minimum	0.5
Mode	2	Mode	1	Mode	0.5	Mode	0.5
Maximum	4	Maximum	2	Maximum	0.5	Maximum	0.5
5. Walk: Vaccination to Observation		**6. Walk: Observation to exit building**		**7. Walk: to Car/Site exit**		**8. Car to Site exit**	
Minimum	0.5	Minimum	1	Minimum	1	Minimum	1
Mode	0.5	Mode	2	Mode	1.5	Mode	1.5
Maximum	0.5	Maximum	3	Maximum	3	Maximum	3

6 Simulation Model

Discrete event simulation (DES) is a method of simulating the behaviour and performance of a real-life process, facility or system. DES is being used increasingly in health-care services [12] and the increasing speed and memory of computers has allowed the technique to be applied to problems of increasing size and complexity.

DES models the operation of a system as a (discrete) sequence of events in time. Each event occurs at a particular instant in time and marks a change of state in the system. DES assumes no change in the system between events. DES is used to characterize and analyse queuing processes and networks of queues where there is an emphasis on use of resources. The core elements are:

- Entities: objects that flow through the processes and have work done on them e.g. patients
- Resources: objects that are used in the workflow to process entities e.g. health care services
- Events: important and specific moments in the system's lifetime e.g. vaccination
- Queues: waiting lines.

DES is particularly suitable for models of systems of patient care where the constraints on resource availability are important. This type of study allow patients to have individual attributes and to interact with resource provision. Due to the superior balance of functionality and ease of use, Witness Horizon software was used to develop a model of the vaccination centre. A table of operational parameters were developed, based on a combination of observation data and discussions with management of a vaccination centre. The operational parameters and their values are depicted in Table 1.

A mapping activity produced Table 4, enabling the real world elements to be mapped to Witness Horizon elements.

A DES model was developed and iteratively refined to credibly represent the operations at the target vaccination centre. Figure 3 displays the vaccination centre after one day (12 h) of simulation. Some of the key drivers was to establish performance and patient service levels. National guidance in the UK is that a single vaccination station should deliver 260 vaccinations per 12-h operating period. Secondly, locally agreed patient service level was defined by two criteria:

- Avg patient flow time around 40 min
- Max patient flow time not to exceed 50 min.

To ensure variability and realism, the patient arrival rate (hourly) was implemented using an exponential inter-arrival time.

Table 4 Element mapping

Mapping to witness elements	
Description	Witness element
Patient	Entity
Park car	Activity
Walk to: registration/clinical assessment/vaccination	Queue
Registration/clinical assessment/walk to observation	Activity
Observation area	Queue
Drive/walk off site	Activity
Patient ID/mode of arrival	Attribute
KPI display	Variable array

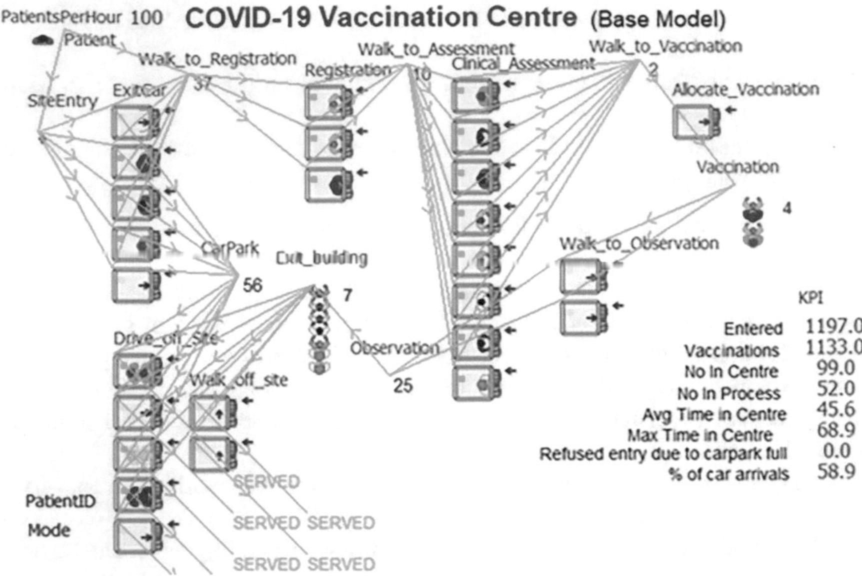

Fig. 3 Base simulation model @12 h

7 Model Testing

To be useful, the model must logically represent the process flow observed in the vaccination centre, known as verification. This was accomplished by techniques such as a structured walkthrough of the model code, test runs and checking of the animation display.

Before the model results were recorded, model behaviour was checked to ensure the model is providing valid results. Validation is about ensuring that model behaviour is close enough to the real-world system for the purposes of the simulation study. The model validation process consisted of comparing simulation results with actual statistics to determine the correctness was a critical step before performing what-if analysis. To our initial surprise, in many cases, our simulated results did not closely replicate the performance of the vaccination centre. This was largely due to the absence of some data to model accurately the operations. The model was then fine-tuned to adjust some of the model parameters to match closer with the performance of the vaccination centre [10].

Finally, demonstration of the model between interested parties provided a forum for communication of model behaviour and helped identify any anomalies. The credibility of any model is dependent on reliable data, which are not always readily available in the British Health Service.

Once the model had been validated, it was run over a set time period and results collected. At this stage, the model is simply reproducing the behaviour of the current process. This "as-is" model provided a visual representation of the whole process, which was important to provide a consensus that the model provides a convincing representation of the process.

8 Results and Experimentation

An arbitrary single simulation run of 12 h of operations is displayed in Fig. 3. However due to the use of random sampling of statistical distributions, a single replication is not representative and some experimentation was necessary.

Full results were obtained by performing multiple replications (500) of the simulation, each with a different random number seed used to generate the timing of patient arrivals, service times, and conveyancing time. In accounting for realistic conditions at the beginning of the operating period, each simulation starts empty and with no warm-up period. Table 5, illustrates the simulated results for the base model.

The key performance indicators (KPIs) in Table 5 show that desired throughput and service levels were not achieved by the configuration of the base model. This finding can actually be derived without modelling—1560 daily vaccinations are not achievable with a patient arrival rate of 100 per hour.

Recognising that the current configuration of the vaccination centre is not achieving the desirable outputs then changes need to be considered. However, the present configuration is operational and works safely. Ultimately, the aim of the project is to maximise throughput, but this must be done under Covid-19 safety rules. If throughput is too high then unsustainable queues will form, compromising social distancing and affecting patient experience. The base model configuration is showing a very stable behaviour across the operating period.

To explore alternative configurations, permissible operational changes were reviewed with management. The outcome was to increase the patient arrival rate (hourly) and resources. However, the service times and conveyancing times could not be changed and hence maintained. The changes transpired as: (1) increase the patient arrival to 130 patient/hr; (2) The staffing levels for the activity levels were Registration (3–5), Clinical Assessment (8–12), Vaccination (6–10); Observation seating (25–40).

An improved simulation model was explored using Witness Horizon's Experimenter and appropriate permissible changes. Initially as a result of the changes, 1200 scenarios were generated. This was subsequently revised. With careful planning of the step size we reduced the optimisation problem to 180 scenarios. Each scenario was run for 50 iterations with a focus on meeting the desired KPIs. An

Table 5 Base model results (time in minutes)

Scenario name	No. of vaccination	Avg. flow time	Max. flow time
Base model	1128.074	48.129	70.682

extract of the results obtained are shown in Fig. 4. Analysis of the results enabled an optimum configuration to be selected. The selection criteria was to maximise throughput, reduce the patient flow time and minimise staffing levels. An adaptive simulated annealing algorithm was used for the optimisation.

From the results shown in Fig. 4, scenario 43 is seen as the best configuration as it meets the desired patient service levels and maximises the patient output. The pinch points of the configuration were the Clinical Assessment capacity and the seating capacity of the Observation area. The latter change is seen an easy fix with little financial implications as the spacious venue could easily accommodate the change. The selected configuration was developed as the improved model as shown in Fig. 5.

Scenario	No of Vaccinations	Registration .Quantity	Clinical_Assessment .Quantity	Vaccination .Capacity	Observation .Capacity	Avg Flow Time (mins)	Max Flow Time (mins)
40	1398.000	3	11	6	40	61.907	92.156
41	1180.750	3	11	8	25	84.699	117.554
42	1414.000	3	11	8	30	58.794	88.862
43	1528.100	3	11	8	35	37.605	47.910
44	1528.100	3	11	8	40	37.603	47.949
45	1180.750	3	11	10	25	85.141	118.020
46	1414.000	3	11	10	30	58.800	89.116
47	1528.250	3	11	10	35	37.558	47.822
48	1528.250	3	11	10	40	37.558	47.915

Fig. 4 Experimentation results

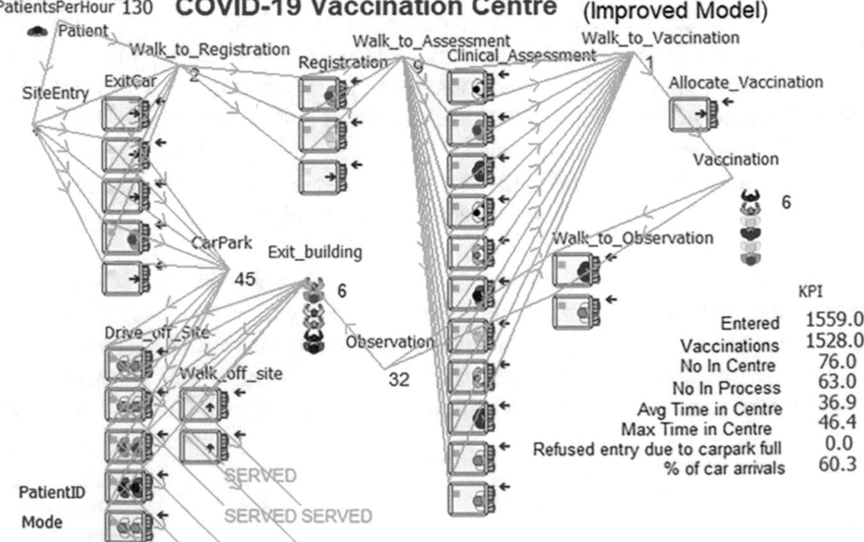

Fig. 5 Improved simulation model @12 h

The improved model was further experimented upon to understand its behaviour. A comparison of the two models for the number of patients in service and in queue was undertaken. The simulated results are shown in Table 6. A noticeable marked decrease in queue size at Registration and Clinical Assessment are predicted using 95% confidence interval. The impressive vaccination output of an additional 400 patients has serious performance impact on the mass vaccination programme for that community.

9 Discussion

We have introduced a simulation tool for evaluating a Covid-19 vaccination centre. Such tools can help with enhancing the service level and operational performance of such facilities. We used Witness Horizon simulation software to develop the model as it provides opportunities for more effective functionality and visualization (2D and 3D) capability over other available tools.

The results presented in Fig. 3 were generated by one realization of the simulation for demonstration purposes. The parameters here have been set using observation data from the vaccination centre in the as-is state. The results of our single simulation were rather different to those observed due to stochastic elements. This was rectified through experimentation. Our goal was to explore operational changes that could improve patient throughput to 1560 daily vaccinations and also meet patient service levels. Improvements were conducted using Witness Horizon Experimenter enabling good confidence levels to be achieved through hundreds of replications. A 30% increase in patient arrivals (hourly) was instigated coupled with minimal increase in resources. The best configuration was determined using adaptive simulated annealing algorithm. The results shown in Fig. 4 indicate that capacity at Clinical Assessment and that of the Observation area are very critical. Utilising this configuration enabled the improved model to be developed. Experimentation with the improved model demonstrated a daily vaccination output of 1528 patients. Whilst the target of 1560 vaccination is not achieved, the simulated output of 1528 is certainly a substantial improvement to the current operations.

Modelling has influenced the decision to increase staffing levels by +3 at Clinical Assessment and +2 at Vaccination as the productivity gains far outweigh the financial implications. The required increase of Observational capacity by +10 has no financial implications but interestingly very critical in the patient flow characteristics. Resource management is like a dynamic commodity and very challenging, whilst additional resources could have been utilised the optimisation algorithm has not shown to have an adverse performance impact (Fig. 4).

Simulation results suggest that the improved configuration will noticeably reduce the patient queues at Registration and Clinical Assessment stations (Table 6). Management are certainly very supportive of trialling these changes and achieving a good balance of high throughput, queue reduction whilst maintaining site safety.

Table 6 Simulation results of improved model after 12 h

	Vaccination output (per day)	Mean number (95% CI) of patients in queue			
		Registration	Clinical assessment	Vaccination	Assessment
Baseline model	1128	20.516 (19.661–21.371)	9.225 (9.161–9.289)	0.805 (0.803–0.808)	0.782 (0.782–0.783)
Improved model	1528	4.149 (3.977–4.321)	6.268 (6.092–6.445)	1.091 (1.089–1.094)	1.060 (1.059–1.061)

10 Conclusion

This study and the simulation models can provide some insights regarding different vaccination centre parameters in terms of patient arrival rate (hourly), staff levels, queue capacities, vaccination cubicles, car park capacity. It is not possible to sensitively validate site performance since real-life operations involve intermittent shutting down of and re-opening of various service channels. This variance in many ways is not fully appreciated by the model.

An important point is that the patient arrival rates impacts the results significantly [13]. In the improved model, we used a fixed arrival rate of 130 patients per hour. However, the arrival rate does not have to be fixed and can vary during the day depending on demographic and environmental factors. The arrival rate has significant impacts on the number of people being vaccinated. It has been assumed that pre-registration has been done and all patients have pre booked appoint slots. According to previous studies, the registration stage contributes most to the formation of bottleneck in mass vaccination systems. The project has provided some insight in just some of the ways in which modelling and simulation can improve vaccination centre operations.

Public health agencies can use our simulation model to examine how many people can be vaccinated for a given number of days, shifts, and working hours per shift. Moreover, the model can help decision makers to have an estimate of how many vaccination centres would be needed to achieve a certain number of immunizations in a specific time period.

As with any type of modelling study, the various assumptions and simplifications can easily contribute to a number of limitations. The service times used in our model have come from observation data and published data [8]. One limitation of this simulation that demands further work is consideration for additional behavioural and user needs, such as the ability of the simulation to allow people who change their mind after they enter the vaccination line to leave, people who need further recovery time and might even need to be taken care of in a caregiving area, patients need for washrooms, etc.

Although this study is based on observational data, measuring the actual impact of proposed interventions on patient arrival rates and patient flow times requires a real-world implementation. Such efforts, however, require financial support. As such, this is another limitation of the current study.

Given that, Covid-19 is unlikely to completely disappear; mass vaccination centres are probably going to become more common, therefore, there are many opportunities for future work in this problem area. In this study we assumed service times were not alterable, however combining Clinical Assessment and Vaccination into a single station is an approach worthy of consideration. We also have not considered the impact of shocks in the system and how readily they can be absorbed such as staff sickness or a number of patients arriving all at once. Seasonal variations have not been considered, extra layer of clothing worn during winter as compared to summer clothes. Magnitude and effects of "no shows" has not been considered.

Funding This work was supported by the Internal Science Foundation of the University of West Bohemia under Grant SGS-2021-028 'Developmental and Training Tools for the Interaction of Man and the Cyber–Physical Production System'.

References

1. Kluge HHP on behalf of the World Health Organization. WHO announces COVID-19 outbreak a pandemic. http://www.euro.who.int/en/health-topics/health-emergencies/coronavirus-covid-19/news/news/2020/3/who-announces-covid-19-outbreak-a-pandemic. Accessed 23 Jun 2021
2. World Economic Forum. It could take three years for the US economy to recover from COVID-19. https://www.weforum.org/agenda/2020/03/economic-impact-covid-19/. Accessed 23 Jun 2021
3. Johns Hopkins Coronavirus Resource Centre. Coronavirus Covid-19 Global Cases by the Center for Systems Science and Engineering. https://coronavirus.jhu.edu/map.html. Accessed 23 Jun 2021
4. UK Government. Coronavirus-19: what you need to do. https://www.gov.uk/coronavirus. Accessed 23 Jun 2021
5. BBC News. Coronavirus: the world in lockdown in maps and charts. https://www.bbc.co.uk/news/world-52103747. Accessed 23 Jun 2021
6. Department of Health and Social Care. UK covid-19 vaccines delivery plan. 11 January 2021. https://assets.publishing.service.gov.uk/government/uploads/system/uploads/attachment_data/file/951284/UK_COVID-19_vaccines_delivery_plan.pdf. Accessed 23 Jun 2021
7. Department of Health and Social Care. Priority groups for coronavirus (COVID-19) vaccination: advice from the JCVI, 30 December 2020. https://www.gov.uk/government/publications/priority-groups-for-coronavirus-covid-19-vaccination-advice-from-the-jcvi-30-december-2020
8. NHS England, UK. COVID-19 Vaccination centres: Operating Framework. Version 1.1, 20 Jan 2021. https://www.england.nhs.uk/coronavirus/wp-content/uploads/sites/52/2021/01/C1034-operating-framework-information-and-guidance-on-operating-vaccination-centres-v1.1-20-january-21.pdf. Accessed 23 Jun 2021
9. Pitt, M., Monks, T., Crowe, S., Vasilakis, C.: Systems modelling and simulation in health service design, delivery and decision making. BMJ Qual. Saf. **25**(1), 38–45 (2016). https://doi.org/10.1136/bmjqs-2015-004430
10. The Strategy Unit, NHS, UK. Strategy Unit releases opensource model for planning vaccine centre capacity. 3rd February 2010. https://www.strategyhunitwm.nhs.uk/news/strategy-unit-releases-opensource-model-planning-vaccine-centre-capacity. Accessed 23 Jun 2021
11. Wood, R.M., Moss, S.J., Murch, B.J., Davies, C., Vasilakis, C.: Improving COVID-19 vaccination centre operation through computer modelling and simulation, Paper in collection COVID-19 SARS-CoV-2 preprints from medRxiv and bioRxiv, March 2021. https://doi.org/10.1101/2021.03.24.21253517. Accessed 23 Jun 2021
12. Zhang, X.: Application of discrete event simulation in health care: a systematic review. BMC Health Serv. Res. **18**, 687 (2018)
13. Hassan, I., Bahalkeh, E., Yih, Y.: Evaluating intensive care unit admission and discharge policies using a discrete event simulation model. Simul. Trans. Soc. Model. Simul. Int. **96**(6), 501–518 (2020)

NATO's Resilience in Response to the Covid-19 Pandemic

Marlena Rybczyńska

Abstract The importance of threats such as terrorist attacks, cyberattacks, hybrid warfare and natural disasters is growing. The North Atlantic Treaty Organization (NATO) must adapt and respond to these different types of challenges, which could severely impact societies and collective defense. Therefore, the subject of research in this article is building the allies' resilience understood as the first line of defense. The purpose of the research was to explore the significance, role and development of resilience particularly in the light of the Covid-19 pandemic. The results of the research show that coordination of tasks and solidarity improved NATO's response to the Covid-19. The conclusion is that NATO should develop its crisis management capabilities. The need of building resilience to non-military threats in the military and civil spheres of allies was indicated. The issues raised in the article may complement the debate on role and tasks of NATO in responding to the emerging challenges, especially in the context of the newly adopted strategic concept.

Keywords Resilience · NATO · Crisis management capability · Covid-19 · Pandemic

1 Introduction

New security challenges, such as cyberattacks, hybrid warfare, as well as natural disasters and biological threats cannot be deterred by the threat of military retaliation. To respond these non-traditional threats NATO is developing new policies, structures and capabilities, including enhancing resilience.

The aim of the article is to identify factors increasing NATO's resilience to new non-military challenges such as the Covid-19 pandemic. The utilized goal is to propose recommendations and conclusions for Alliance in the event of such threats. The aim defined in this way necessitate the formulation of the main research problem, which took the form of the question: How has the pandemic contributed to building

M. Rybczyńska (✉)
The General Tadeusz Kościuszko Military University of Land Forces, Wroclaw, Poland
e-mail: marlena.rybczynska@awl.edu.pl

A. Visvizi et al. (eds.), *Research and Innovation Forum 2022*,
Springer Proceedings in Complexity, https://doi.org/10.1007/978-3-031-19560-0_71

the resilience of Allies and setting new tasks for NATO to reduce the effects of such threats in the future? Due to the complex nature of own research, the main problem has been explicated into specific problems: (1) What conceptual and institutional solutions have been implemented to increase NATO's resilience (2) What undertakings has NATO accomplished to prevent the effects of the Covid-19 pandemic? (3) What solutions adopted by NATO are useful in the event of future threats? Among the research methods, an important role in the research process was played by source analysis, including content analysis and analysis of secondary data contained in research reports and public registers. The primary method, and at the same time inspiring to undertake the research, was passive observation, which gave the basis for specifying the subject of the research and defining the problem situation, which is NATO's resilience to the effects of the Covid-19 pandemic.

In the first part of the study, the definitions of resilience were presented. For this purpose, the literature on the subject were reviewed and analyzed.

In the third section, on the basis of NATO's primary documents as well as media statements, it was explored how the Alliance's resilience concept has developed in response to emerging new security challenges.

In the fourth section, the author analyzed what efforts NATO has made to improve its response to Covid-19 and which of them increased resilience during the second wave of the pandemic. For this purpose, situation reports, press materials and official websites of the NATO institutions and programs were studied.

The last part of the study presents the conclusions that NATO should learn from the pandemic and indicates the possibility of using them to respond to the future crises.

2 Definitions of Resilience

The concept of resilience has been defined in various forms and used in many contexts. The term extended beyond its initial ecological use. It is now applied at different levels (individual, community, state), in various fields such as psychology, economy, organizational management, community studies and security studies. Drawing on terminology from sociology, resilience is defined as "the capacity of a system, enterprise, or person to maintain its core purpose and integrity in the face of dramatically changed circumstances" [1]. The concept can also be presented as process. Walker and Salt characterize resilience as "a capacity of a system to constantly evolve and adapt to disturbance while maintaining its basic function and structure" [2]. Ran et al. make the following conclusion about resilience: "The ability of a system, community or person to prepare for, cope with, recover from and adapt to a hazard or hazardous event" [3].

NATO defines resilience as follows: "Each NATO member country needs to be resilient to resist and recover from a major shock such as a natural disaster, failure of critical infrastructure, or a hybrid or armed attack. Resilience is a society's ability to resist and recover from such shocks and combines both civil preparedness and military capacity (…) NATO supports Allies in assessing and enhancing their civil preparedness" [4]. In line with this resilience is primarily a national responsibility. Each Ally must be strong enough and able to adapt to deal with the full spectrum of crises foreseen by the Alliance [5, 6]. NATO help countries achieve the required resilience and provide benchmarks against which to assess the state of civilian preparedness.

Resilience is not a new task for the Alliance. The principle of the concept is anchored in Article 3 of the Alliance's founding treaty [4]. However, at that time, resilience was associated with deterrence and was intended to reduce the likelihood of an attack against NATO. Currently the term refers to a wider range of threats and challenges—especially non-military ones. It has also become the subject of analyzes and researches by a team of experts for NATO [7, 8].

There are many scientific articles dealing with resilience [9], but a new aspect—enhancing resilience of alliance member states in the face of the Covid-19 pandemic—so far was unexplored and is the subject of this article's research.

3 Enhancing Resilience in NATO Documents and Declarations

Resilience was barely mentioned in the NATO Strategic Concept of 2010. It was only related to ensuring the safety of communication, transport and transit routes by the Allies [10].

The Wales Summit in 2014 was held when Russia "annexed" Crimea and ISIS's activities expanded. Therefore, the intensification of cyberattacks, information warfare and terrorist threats in the Alliance's environment became the catalyst for talks on increasing the resilience of NATO member states. The Summit took steps to counter these security hazards and strengthen the Alliance's defense capabilities as well as enhencing resilience [11].

Due to the growing situation of uncertainty and instability on the periphery of NATO, resilience turned out to be the principal concept of the Warsaw Summit in July 2016. The final Communiqué mentions resilience 16 times and states that: "Civil preparedness is a central pillar of Allies' resilience and a critical enabler for Alliance collective defence. While this remains a national responsibility, NATO can support Allies in assessing and, upon request, enhancing their civil preparedness. We will improve civil preparedness by achieving the NATO Baseline Requirements for National Resilience, which focus on continuity of government, continuity of essential services, security of critical civilian infrastructure, and support to military forces with civilian means" [12]. Accordingly, Allied Leaders decided to strengthen NATO's resilience to the full spectrum of threats and continue developing individual and collective capacity to resist any form of attack. Moreover, heads of state and governments participating in the Summit issued a Commitment to Enhance

Resilience, which states that "resilience is an essential basis for credible deterrence and defence and effective fulfilment of the Alliance's core tasks" [13].

At the NATO summits in Brussels (2017, 2018) and London (2019), the concept of enhancing resilience was upheld but not developed. Then NATO member states had to face new challenges presented by emerging communications technologies, such as 5G, as well as the impact and implications of the Covid-19 pandemic. Therefore, Allied members have begun work on updating their baseline requirements.

Resilience was the subject of statements by key NATO officials. During a virtual press conference in April 2020, NATO secretary general Jens Stoltenberg stated that: "NATO has developed a baseline requirements for civilian resilience, for our societies and we need to look to them to see if there is any need for further developing them in the light of the Covid-19 crises" [14]. During the panel discussion in December 2020, NATO Deputy secretary general Mircea Geoană said that: "Resilient societies are our first line of defense" [15, 16].

Important for making resilience priority was the postulate made by Daniel S. Hamilton to the NATO Parliamentary Assembly in May 2021 [17]. In his Testimony, there was even a proposal to make comprehensive resilience NATO's fourth core task in the next Alliance Strategic Concept.

Another significant event in creating the framework for enhancing resilience was Brussels Summit in June 2021. The final Communiqué mentions resilience 27 times and concerns such issues as: hybrid threats, acts of terrorism, cyber threats, critical infrastructure, climate change, energy security and supply chains. NATO's leaders declared that "Allies will develop a proposal to establish, assess, review and monitor resilience objectives to guide nationally-developed resilience goals and implementation plans" [18]. At the Summit Allied leaders also agreed a Strengthened Resilience Commitment that sets out further steps [19].

Much attention was paid to the resilience, compared to the Lisbon strategic document, in The NATO 2022 Strategic Concept adopted in June at the Madrid Summit. More robust, integrated and coherent approach to building national and Alliance-wide resilience against threats and challenges to security were declared. NATO members committed to building resilience in many areas to which health systems have been added [20].

4 NATO Efforts to Improve Its Response to the Covid-19 Pandemic

Since the beginning of the Covid-19 NATO has taken steps to maintain its capacity to counter traditional threats and to coordinate the efforts of member and partner countries in the fight against the pandemic. Mircea Geoană stated that: "NATO's main task during the pandemic is to make sure the health crisis does not become a security crisis" [21]. To achieve this, NATO has implemented crisis response mechanisms to improve logistics and coordinate the procurement of necessary medical equipment and resources.

The Euro-Atlantic Disaster Response Coordination Centre (EADRCC) has turned out to be a key tool in NATO's response to the pandemic [22, 23]. Since the start of the crisis, EADRCC has coordinated 26 requests for international assistance in response to Covid-19, including 1 from the UN OCHA, 1 from KFOR and 24 from NATO and partner countries garnering dozens of responses in return.[1]

After analyzing the EADRCC reports, it can be concluded that the most common forms of assistance were: the distribution of medical supplies, air transport of patients, construction of field hospitals, creation of quarantine facilities and triage centers, assistance with decontamination. In many cases, bilateral financial assistance was also provided.

Logistics played a huge role in the NATO response to the Covid-19 pandemic [24]. The EADRCC coordinated offers for medical equipment and supplies as well as their delivery to countries in need. To complete these tasks Strategic Airlift International Solution (SALIS) and Strategic Airlift Capability (SAC) were used. These programs have been established by some NATO member and partner countries to rapidly deploy forces and resources to where they are needed. The Alliance also used the Rapid Air Mobility mechanism to accelerate all state approvals for important flights organized as part of NATO capabilities and individual undertakings of the member states.

Moreover, the NATO Support and Procurement Agency (NSPA) has helped Allies and partners achieve economies of scale in purchasing Covid-19 relief material. The NSPA has also provided Allies and partners in transporting urgent relief items to countries in need through mentioned above programs and commercial chartered flights.

The Alliance quickly began working to increase its preparedness and thus its resilience. It took steps to strengthen its response capacity during the second wave of the Covid-19 pandemic. Through the NSPA medical supplies has been stocked to be able to provide immediate relief to Allies or partners in need. In addition, the NATO Pandemic Response Trust Fund has been established to enable rapid procurement of medical materials and services [25].

5 NATO's Lessons Learned from the Pandemic and Conclusions

Learning the lessons of each crisis and looking ahead to imagine the next is an essential part of maintaining the security of NATO. Therefore, some steps have been taken at NATO level to learn lessons from the Covid-19 pandemic. For this purpose, debates, expert groups have been established and guidelines developed [26, 27].

The results of the own research indicate that the pandemic contributed to the intensification of talks about NATO's resilience against non-military threats. It has prompted Allies to make far-reaching commitments to building individual and collective resilience. The scope of resilience has expanded to aspects of supply

[1] Data as of April 9, 2022.

chains and health systems which were included in the new Strategic Concept of 2022. The Euro-Atlantic Centre for Resilience was also established to increase NATO's resilience [28].

Moreover, it can be concluded that information exchange, aid distribution and Allied solidarity have improved NATO's response to Covid-19. EADRCC's successful logistical coordination during the pandemic underscored its unique value. The Centre's activities should be developed as they may increase the resilience in future crises. Strategic airlift capabilities have also proved to be crucial in responding to the pandemic and should be advanced. Pandemic experiences have shown that the Just-In-Time approach to supply chains has failed when the whole world is simultaneously crying out for medical supplies. Therefore, it turned out to be essential to establish their stocks and a trust fund for the purchase of urgently needed items. This could be also a universal solution to future crises.

The effectiveness of overcoming non-military crises increases through join action with various entities. Facing new challenges and threats as well as building resilience requires working with partners and other international organizations. Crisis response is often led by civilians with the support of military resources, so civil-military cooperation is essential. Additionally, regarding environmental security, there is a need for a more systematic networking with the scientific community to identify technical and scientific trends.

The Covid-19 pandemic has forced the Alliance to adapt its operations and instruments to the challenges once again to face global threats. This pandemic highlighted the importance not only of having a strong military but also of creating resilient societies. The Russian invasion of Ukraine in 2022 and the ongoing conflict since the annexation of Crimea, also show that European governments must be prepared not only for military aggression, but also for hybrid warfare as well as they have to build their resilience in various areas. Enhancing resilience and crisis response capabilities should be developed in parallel to effectively counteracting classic military threats. Such approach is intended to prevent future situations in which the omission or inability to respond to various threats will weaken the Alliance's defense potential.

References

1. Langeland, K.S., Manheim, D., McLeod, G., Nacouzi G.: Definitions, characteristics, and assessments of resilience. In: Langeland, K.S., Manheim, D., McLeod, G., Nacouzi G. (eds.) How Civil Institutions Build Resilience, pp. 5–10. RAND Corporation (2016). https://doi.org/10.7249/j.ctt1btc0m7.8. Accessed 10 Apr 2022
2. Walker, B.R., Salt, D.: Sustaining ecosystems and people in a changing world. In: Boon, N.A., Colledge, N.R., Walker, B.R. (eds.) Resilience Thinking. Island Press, London (2006)
3. Ran, J., MacGillivray, B.H., Gong, Y., Hales, T.C.: The Application of Frameworks for Measuring Social Vulnerability and Resilience to Geophysical Hazards within Developing Countries: A Systematic Review and Narrative Synthesis. Science of Total Environment, vol. 711 (2020)
4. NATO. Resilience and Article 3. https://www.nato.int/cps/en/natohq/topics_132722.htm. Accessed 10 Apr 2022

5. Visvizi, A., Lytras, M.D.: Government at risk: between distributed risks and threats and effective policy-responses. Trans. Gov. People Process Policy **14**(3), 333–336 (2020)
6. Visvizi, A., Stępniewski, T. (eds.): Poland, the Czech Republic and NATO in Fragile Security Contexts, IESW Reports, Lublin (2016)
7. Linkov, I., Palma-Oliveira J.M.: An introduction to resilience for critical infrastructures. In: Linkov, I., Palma-Oliveira J.M. (eds.) Resilience and Risk: Methods and Application in Environment, Cyber and Social Domains, 1st ed. Dordrecht: Springer Netherlands: Imprint: Springer (2017)
8. Hodicky, J., Özkan, G., Özdemir, H., Stodola, S., Drozd, J., Buck, W.: Dynamic modeling for resilience measurement: NATO resilience decision support model. Appl. Sci. **10**(8), 1–10 (2020)
9. Dunn Cavelty, M., Kaufmann, M., Søby Kristensen, K.: Resilience and (in)security: Practices, subjects, temporalities. Secur. Dialogue **46**(I), 3–14 (2015)
10. NATO. Strategic Concept for the Defence and Security of the Members of the North Atlantic Treaty Organization. https://www.nato.int/nato_static_fl2014/assets/pdf/pdf_publications/20120214_strategic-concept-2010-eng.pdf. Accessed 10 Apr 2022
11. NATO. Wales Summit Declaration. https://www.nato.int/cps/en/natohq/official_texts_112964.htm. Accessed 10 Apr 2022
12. NATO. Warsaw Summit Communiqué. https://www.nato.int/cps/en/natohq/official_texts_133169.htm. Accessed 10 Apr 2022
13. NATO. Commitment to enhance resilience. https://www.nato.int/cps/en/natohq/official_texts_133180.htm. Accessed 10 Apr 2022
14. NATO. Coronavirus response to top NATO Ministerial agenda. https://www.nato.int/cps/en/natohq/news_174785.htm?selectedLocale=en. Accessed 10 Apr 2022
15. NATO. Building transatlantic resilience: Why critical infrastructure is a matter of national security. https://www.nato.int/cps/en/natohq/opinions_180067.htm. Accessed 10 Apr 2022
16. Roepke, W.D., Thankey, H.: Resilience: the first line of defence. NATO Review. https://www.nato.int/docu/review/articles/2019/02/27/resilience-the-first-line-of-defence/index.html. Accessed 10 Apr 2022
17. Hamilton, D.S.: Democratic Resilience is Foundational to the Alliance. Testimony by before the NATO Parliamentary Assembly May 2021
18. NATO. Brussels Summit Communiqué. https://www.nato.int/cps/en/natohq/news_185000.htm. Accessed 10 Apr 2022
19. NATO. Strengthened Resilience Commitment. https://www.nato.int/cps/en/natohq/official_texts_185340.htm. Accessed 10 Apr 2022
20. NATO. The 2022 NATO Strategic Concept. https://www.nato.int/nato_static_fl2014/assets/pdf/2022/6/pdf/290622-strategic-concept.pdf. Accessed 04 Jul 2022
21. NATO. NATO Deputy Secretary General: NATO is ready for whatever challenges it might face. https://www.nato.int/cps/en/natohq/news_179207.htm?selectedLocale=en. Accessed 10 Apr 2022
22. NATO. Euro-Atlantic Disaster Response Coordination Centre (EADRCC). https://www.nato.int/cps/en/natohq/topics_117757.htm. Accessed 10 Apr 2022
23. Rybczyńska, M.: Euroatlantycki Ośrodek Koordynacji Reagowania w przypadku Katastrof jako mechanizm zarządzania kryzysowego NATO wobec pandemii Covid-19. Rocznik Bezpieczeństwa Międzynarodowego **15**(1), 269–295 (2021), https://doi.org/10.34862/rbm.2021.1.12. Accessed 10 Apr 2022
24. NATO. NATO's Response to the Covid-19 Pandemic. Factsheet. https://www.nato.int/nato_static_fl2014/assets/pdf/2020/4/pdf/200401-factsheet-COVID-19_en.pdf. Accessed 10 Apr 2022
25. NATO. NATO's Response to the Covid-19 Pandemic. Factsheet. https://www.nato.int/nato_static_fl2014/assets/pdf/2020/10/pdf/2010-factsheet-COVID-19_en.pdf. Accessed 10 Apr 2022

26. NATO. Alliance scientists study the military impact of Covid-19. https://www.nato.int/cps/en/natohq/news_182281.htm. Accessed 10 Apr 2022
27. NATO. NATO Deputy Secretary General addresses Covid-19 response at Lessons Learned Conference. https://www.nato.int/cps/en/natohq/news_182227.htm. Accessed 10 Apr 2022
28. Euro-Atlantic Centre for Resilience. https://ue.mae.ro/en/node/1551. Accessed 10 Apr 2022

Customs and Their Role in Cultural Heritage Protection: The Case of Poland

Jacek Dworzecki and Izabela Nowicka

Abstract This paper explores the role of the Polish National Revenue Administration (NRA) in cultural heritage protection in Poland. In this context, the challenges the civil servants and officers employed in NRA face are examined. The key question that this paper addresses is: To what extern and how organizational, legal and technical features of the NRA influence the scope and effectiveness of the NRA in the domain of combating and preventing illegal trade of art works, smuggling of art works, and other illegal activities related to cultural heritage? While the literature of the subject focuses mostly on legal issues in heritage protection, the organizational and technical challenges are only rarely discussed. Against the backdrop of the literature on the subject, legal acts analysis, expert interviews, this paper adds to the discussion.

Keywords Polish National Revenue Administration · Protection of cultural heritage · Poland

1 Introduction

Social and economic changes which have occurred both nationally and internationally raise new phenomena and expectations. In this context, the role of cultural heritage in the development of a country and in the creation of the society of tomorrow also appears as an integrating factor in the cultural diversity of the world [1]. What is today a sign of the present day, tomorrow might be considered as a national heritage, that is, everything will be a national heritage regardless of whether a respective community regards something as a cultural asset or not at a particular time, but also notwithstanding the social and temporal construct [2]. The society have recognized the protection of heritage so vital that requiring a specific strategy of both

J. Dworzecki
AMBIS University, Prague 180 00, Czech Republic

I. Nowicka (✉)
Military University of the Land Forces, 51 147 Wroclaw, Poland
e-mail: izabela.nowicka@awl.edu.pl

A. Visvizi et al. (eds.), *Research and Innovation Forum 2022*,
Springer Proceedings in Complexity, https://doi.org/10.1007/978-3-031-19560-0_72

global and local nature. This protection is shaped by the prism of legal and organizational structures [3]. A tangible and intangible heritage is recognized by the international community as a prodevelopment stimulus for increasing significance in modern societies [4]. All democratic countries with an internationally established position perceive the need to protect their own cultural heritage, especially that within the framework of globalization and free movement of people and goods, many cultural boundaries become blurred. On the example of the European Union (EU) it can be pointed out that in virtually any member state of the community there are ministries responsible for the protection of national heritage and cultural goods.

2 Methodology

The research results presented in this publication were based on the examination of the available subject literature, legal acts, media reports and several expert interviews, but, in particular, on statistical and descriptive data referred to illegal trade in art works, smuggling of works of art or other illegal activities related to cultural heritage and the role of officials of the tax administration at Tax Chamber and Customs and Tax Offices for the years 2017–2021. An in-depth process concerning the effects, i.e. the analysis of the effectiveness of the actions of the NRA in the field of cultural heritage protection, covered: legal provisions regulating the tasks of the NRA indications of practice in this area and postulates regarding changes to the existing legal status. Difficulties in the interpretation of legal constructions have been shown. An expression of this was the need to properly define the areas, directions and methods of carrying out tasks in the field of the protection of cultural goods.

3 Organizational, Legal and Technical Condition of Polish NRA

3.1 Organizational and Legal Condition

The performance of Polish NRA is predominantly based on the Act of 16 November 2016 on the Polish NRA (Journal of Laws of 2016, item 1947), i.e. of 3 March 2020 (Journal of Laws of 2020, item 505) (Law on the Polish NR A). As a single entity, Polish NRA subordinate to the minister competent for public finances, it was established on 1.3.2017 and replaced three independent administrative branches dealing with the collection of budget revenues and the protection of State Treasury's property interests, namely the tax administration (chambers and tax offices), creating a joint tax and customs authority. NRA performs many functions in the Polish tax administration. It is primarily a specialized government administration entrusted with responsibilities in the area of implementation of income from taxes, customs duty rates, fees and non-tax budget receivables, protection of the interests of the State Treasury, as well as protection of the customs territory of the European Union, additionally ensuring the service and support of a taxpayer and a payer in a proper performance of tax obligations and the service in addition to support of an entrepreneur in an appropriate realization of customs obligations.

NRA is headed by the Minister of Finance and the body supervising its activities is the Head of the NRA. A very significant change is the establishment of a new controlling body which is represented by the head of the customs and tax office. It is in charge of a new type of audit, i.e. customs and tax control in entities selected after conducting a risk analysis. The provisions of the Act set out the details of the tasks of The NRA (art. 2 of the Law on NRA) such as: the implementation of customs revenue and other charges related to the importation and exportation of goods; the performance of customs policy resulting from membership of the EU; including customs procedures and regulating the situation of goods referred to the importation and exportation of goods; conducting information and educational activities in the field of tax and customs law; identifying, detecting and combating crimes and offences concerning violations of the provisions on goods whose trade is subject to prohibitions or restrictions under Polish law, European Union law or international agreements, preventing these crimes and offences as well as prosecuting their perpetrators, if they have been disclosed by the Customs and Tax Service; disclosure and recovery of assets in connection with the offences referred to in points 13–16 or Articles 33 § 2 of the Act of 10 September 1999—Penal Fiscal Code (Journal of Laws of 2020, item 19); carrying out tasks resulting from the provisions of EU law regulating statistics on trade in goods between the Member States of the EU (INTRASTAT) and the trade in goods between the Member States of the EU with other countries (EXTRASTAT) and conducting investigations in the field of INTRASTAT performing tasks resulting from prohibitions and restrictions on trade in goods with foreign countries established, in particular, for the protection of human and animal life, health, plants, environment and public safety, consumer protection, international security, national heritage, intellectual property rights and trade policy measures.

An example of activities that should be emphasized in the context of the protection of cultural heritage is the so-called customs and tax control [5]. Tax Ordinance (i.e. Journal of Laws of 2020, item 1325, as amended), which is a procedure of the so-called hard control nature. Its main task is to detect and release irregularities on a wide scale—fraud, extortion and other crimes causing significant damage to public finances, and thus significantly affecting the financial security system of the state. Customs and tax control will include, among others, the activities of organized criminal groups, tax carousels resulting in VAT fraud, as well as attempts to import exports, trade in works of art, etc. The activities of NRA are a fundamental instruction allowing to obtain information on the elements of facts whether the audited entity has lawfully fulfilled its tax obligations which as a consequence will continue in the form of making the assessment of receivables to be paid or reimbursed in the framework of the subsequent audit proceedings.

The focus of NRA activities in the field of combating and preventing tax offenses in relation to the protection of cultural heritage is, among others, on prohibited acts that have been penalized in the Criminal Fiscal Code (most often committed: art. 86 of the c.f.c—customs clearance; Art. 87 of the c.f.c—customs fraud) or in the Act on the Protection and Care of Monuments of 23 July 2003 (most often committed: art. Art. 109—export of antiques without permission).

They may be procedural in nature, but the importance of operational and exploratory activities should be emphasized. Irrespective of those powers, the officers of the NRA are provided with instruments to ensure the safety of persons performing audit activities as well as to detain and bring persons who commit acts detrimental to the financial interests of the public authorities. The powers of the powers of the NRA presented above, in a nutshell, clearly indicate that it is not an entity of which a predominant duty is the protection of cultural heritage. However, this single institution realizes significant tasks for the protection of cultural heritage. The NRA cooperates, among others, with the Ministry of Culture and National Heritage and its subordinates: the National Institute of Museology and Collections Protection, the National Heritage Institute, Museum Institutes, Provincial Historical Monuments Protection Offices, but also the Police and Border Guard.

3.2 *Technical Condition*

The implementation of responsibilities of the NRA is supported with modern technology. Since 17 March 2020 the NR A has an innovative IT system. It was prepared with the use of a computer vision based on deep neural networks and provides the opportunity for further learning.

Contemporary IT tools based on artificial intelligence increasingly take on an offensive character. The operational capabilities of such instruments are inevitably heightened, which on the one hand allows to improve the effectiveness of police detection activities which relate to, for example, the recognition and illegal transactions of works of art etc. [6].

In their day-to-day work, the NRA officers use a number of national and international databases and IT systems for two essential purposes: checking the legality of exports and imports of works of art and analyzing emerging trends related to crime against historic monuments in order to be able to prevent them effectively. The following shall remain at their disposal: the national list of monuments stolen or exported illegally, maintained by the National Institute of Museology and Collections Protection and the catalogue of war losses maintained by the Ministry of Culture and National Heritage, listing the objects lost during World War II, the database of stolen works of art made available by the Interpol Secretariat, the Customs Enforcement Network database and other [7].

The officers of the National Tax Administration carry out risk analysis on the basis of indicated databases, which makes it possible to eliminate fraud by properly targeting customs controls. Controls are carried out in proportion to the level of risk, which means the concentration of forces and measures of the National Tax Administration on the actual threats and supporting legal entities [8].

4 Ways of Enhancing NRA's Effectiveness

The practical expression of cooperation NRA and other is signing an agreement on cooperation within combating illegal exports abroad or imports of monuments from abroad on 7 February, 2020 by the Head of the NRA, the Minister of Culture and National Heritage, the Commander-in-Chief of the Police and the Commander-in-Chief of the Border Guard. The framework of the agreement shapes the form and manner of cooperation in the field of mutual assistance in audit, exchange of information, training and experience. An additional, important part of this document presents the guidelines for acting jointly in international operations against theft and illicit trafficking of cultural goods, such as Pandora and Athena, coordinated by Interpol and the World Customs Organization (WCO). Cooperation with other entities is not solely based on national entities. In the area of regional activities, collaboration within the Police and Customs Cooperation Centers on the border with the Czech Republic, Slovakia and Germany is of great importance. Specialized Units of the National Tax Administration also maintain contacts and conduct training cooperation with groups from the customs service of the Czech Republic—the Operational Intervention Group SON (Skupina Operativniho Nasazeni Celni Správy ČR) and the Criminal Office of the Financial Administration of the Slovak Republic—JSZ KUFS (Jednotka služobných zákrokov Kriminálneho úradu finančnej spravy). This results in effective actions connected with various categories of crime and in the specificity of organized crime. Mutual exchange of information between these international entities which brings effects in the rapid search for wanted persons and stolen goods and vehicles also takes place at the level of border posts, such as the Joint Centre in Chotebuz. It should be remembered that cultural heritage is global and is not provided only for one nation. This has also consequences in a global responsibility to protect it against any possible threat presented above, in a nutshell, clearly indicate that it is not an entity of which a predominant duty is the protection of cultural heritage. However, this single institution realizes significant tasks for the protection of cultural heritage. The cooperates, among others, with the Ministry of Culture and National Heritage and its subordinates: the National Institute of Museology and Collections Protection, the National Heritage Institute, Museum Institutes, Provincial Historical Monuments Protection Offices, but also the Police and Border Guard.

The control activities accomplished by the are limited due to the abolition of customs posts at the borders with the countries belonging to the EU. Therefore, on the territory of Poland, checks are carried out by mobile groups by controlling vehicles at random. If a vehicle is stopped, unless there is a specific indication, it is likely that mainly large items such as furniture, statues, architectural fragments or paintings will be found. However, this does not mean success as they must be assured, based on evidence, that the found monument was to be exported abroad.

In this respect, the intention to export is more likely to be demonstrated by random checks in intra-EU trade carried out at airports, seaports or postal traffic. Practice shows that the risk analysis related to trade outside the customs territory of the EU allows for typical shipments depending on the country to which the goods are shipped; notification of the shipment by an entity that has previously attempted to illegally export cultural goods; the size and shape of the shipment indicating that it may contain cultural goods (e.g. in the case of paintings, large flat parcels or tubes).

Another issue that undoubtedly affects the effectiveness of the activities of the Polish N R A concerning the protection of cultural heritage is knowledge and experience. The customs officer must assess whether the object meets the age and value criterion in order to be able to leave without a permit. Therefore, it depends on the knowledge whether the item will be exported abroad. The conducted analysis of regulations, statistical data, websites shows a certain "training" image of the officers of the Polish.

NRA with reference to the protection of cultural heritage. In accordance with Article 14 Sect. 1 point 2 of the Law on the Polish NRA, the Head of the institution is responsible for shaping the Policy of training in organizational units of NRA. The cooperation of the NRA with other entities (Police, Border Guard, etc.) in the area of the protection of cultural heritage is indicated by the following statistical and descriptive data on the illegal trade in works of art, their smuggling or other illegal activities related to cultural heritage, in which the officers of the tax administration of this organizational unit participated. The data were made available by all organizational units of the NRA. The Head implements this on the basis of the directions of activities and development of the National Tax Administration (Ordinance of the Minister of Development and Finance of 25 July 2017 on directions of actions and development of the national tax administration for the years 2017–2020, Ordinance of the Ministry of Finance, regional policy funds of 23 December 2020 on the determination of directions of action and development of the NRA for the years 2021–2024—pursuant to art. 7 par. 1 of the Act of 16 November, 2016 on the NRA (Journal of Laws, item 1947, as amended). There is no doubt that cooperation with other institutions competent in the field of the protection of cultural heritage can be beneficial for each of the parties. Data for 2017 were analyzed, i.e. since the year of the creation of the National Tax Administration to the present day (to the 2021).

5 Discussion and Conclusions

There is also no doubt that legal basis not only of the acts of statutory rank or implementing acts, but also of agreements between the institutions, allows to carry out tasks effectively by the officers of the NRA to cooperate with other entities. The literature on the subject shows the basic activities of the NRA. Only new issues reveal the problems faced by this cooperation. Nevertheless, the officers emphasize the need for training, especially in view of the limitations and difficulties of the modern world and the ingenuity of the perpetrators supported by modern means of technology. They are not isolated in this, the border guards and police officers cooperating with them in the subject matter have a similar opinion. This is all the more difficult as it covers different categories of cultural heritage. It is necessary to undertake discussions with regard to the effectiveness of cooperation between the described entities. The high quality of activities translates into better protection of cultural goods. The conducted research should outline a strategy for the practical termination of programs, rapid response to current or real situations, while understanding and recognizing the far-reaching effects of these responses. A correctly formulated strategy should help to maximize the opportunities and strengths of the NRA The doctrine indicates that changes in the NRA will be effective when they take into account, apart from completely new solutions (innovations), the improvement of processes already taking place in it, with simultaneously changing external factors. NRA is one of the entities that constitute one element of the system for the protection of cultural heritage. Although the stated protection is not the primary task of this formation, it has forces and means which allow to prevent and combat violations against the cultural heritage. Technology has become a huge support in all activities of entities responsible for the protection of cultural heritage. There is also no doubt that legal basis not only of the acts of statutory rank or implementing acts, but also of agreements between the institutions, allows to carry out tasks effectively by the officers of the NRA and to cooperate with other entities. The conducted research clearly highlighted that activities are under- taken jointly with other entities, but unfortunately this does not apply to all units of the IAS and UCS ranking NRA hese studies, based on data from all organizational units of NRA, and the analysis of selected events, are an added value in the discussion on the effectiveness of such agendas, in particular in cooperation with other entities.

References

1. BItušíková, A.: Kultúrne dedičstvo a globalizácia: Príbeh jednej locality. Museol. Cult. Herit. **2**(1), 9–17 (2014)
2. Jensen, U.J.: Cultural heritage, liberal education, and human flourishing. In: Avrami, E.C., Randali, M., De La Torre, M. (eds.) Values and heritage conservation: research report, p. 38. Getty Conservation Institute, Los Angeles (2000)
3. Dworzecki, J., Nowicka, I., Urbanek, A., Kwiatkowski, A.: Protection of national heritage in the light of the applicable law and the actions provided in this area by police in Poland. Museol. Cult. Herit. **8**(4), 177–198 (2020)
4. Sapanhzna, O.: Museums in the new model of culture: concerning the issue of training professionals in museum education. Museol. Cult. Herit. **6**(1), 7–11 (2018)
5. Act on the National Revenue Administration and the Act of 29 August 1997
6. Dworzecki, J.: The practical use of police databases of stolen works of art in the protection of national heritage in selected European Union countries. Museol. Cult. Herit. **9**(2), 91–101 (2021)

7. Díte, T.: Limity digitalizácie v ochrane kultúrneho dedičstva spravovaného múzeami. Museol. Cult. Herit. **1**(2), 87–100 (2013)
8. Ogrodzki, P.: Krajowy wykaz skradzionych lub wywiezionych nielegalnie z prawem zabytków. In: Karpowicz M., Ogrodzki, P. (eds.) Międzynarodowa współpraca służb policyjnych, granicznych i celnych w zwalczaniu przestępczości przeciwko zabytkom, WSPOL, Szczytno (2005)

Military Involvement in the Evacuation of Nursing Home Patients in the COVID-19 Pandemic

Małgorzata Dymyt, Marta Wincewicz-Bosy, and Robert Kocur

Abstract The subject of the article concerns the involvement of the armed forces in the processes of evacuating patients from nursing homes in the event of contracting coronavirus. The analysis concerned the evacuation of patients from social welfare homes, a Covid-19 carried out in the first stage of the study as part of the "Resistant Spring" action coordinated by the Polish Army. These activities required external resource support. The aim of the article is to develop a model procedure for evacuating patients from care centres in a pandemic situation involving many entities, including military units. The research methods employed included qualitative methods, such as analysis of literature and documents, expert interviews a case study, observation and process mapping. It was found that it is necessary to develop a model evacuation procedure, its implementation among all participants, as well as regular practical exercises increasing the efficiency of the process and communication. In addition, planning and organisational conditions for the efficiency of processes under civil-military cooperation for the purposes of preventing the spread of the Covid-19 were identified. The article concerns civil-military cooperation in special circumstances by considering Conid-19 conditions. The results suggest that it is necessary to consciously shape civil-military relations in terms of readiness to cooperate in crisis situations. The article presents the socio-organisational changes caused by Covid-19, in particular in terms of practical implications at the operational level, as well as the directions of further theoretical considerations regarding decision-making and cooperation between the civil and military spheres.

Article classification Case study.

Keywords Patient evacuation · Transport process · Civil-military cooperation · Covid-19 pandemic

---

M. Dymyt (✉) · M. Wincewicz-Bosy · R. Kocur
General Tadeusz Kosciuszko Military University of Land Forces, Wroclaw, Poland
e-mail: malgorzata.dymyt@awl.edu.pl

M. Wincewicz-Bosy
e-mail: marta.wincewicz-bosy@awl.edu.pl

R. Kocur
e-mail: robert.kocur@awl.edu.pl

A. Visvizi et al. (eds.), *Research and Innovation Forum 2022*,
Springer Proceedings in Complexity, https://doi.org/10.1007/978-3-031-19560-0_73

1 Introduction

The Covid-19 pandemic caused a health crisis was particularly severe for the health care system, both for employees and, above all, for patients. The significant consequences of the crisis caused by the coronavirus pandemic in the sphere of health concern: challenges in diagnostics, quarantine and treatment suspected or confirmed cases, a significant burden on the functioning of the health care system, neglect in treating patients with other diseases and health problems, overloading doctors and other healthcare professionals who are at a very high level of risk, high protection requirements, overloading medical stores and disrupting the medical supply chain [1].

One of the groups of patients, particularly affected in the first phase of the Covid-19 development, were the elderly, residents of social care homes. As a result of mass illnesses of medical and care personnel, it was necessary to relocate nursing home patients to other centres, including hospitals, in order to ensure continuity of medical care. The disaster response has become necessary. This concept is by World Health Organisation (WHO) as actions taken directly before, during or immediately after a disaster in order to save lives, reduce health impacts, ensure public safety and meet the basic subsistence needs of the people affected [2]. It took outside resources to evacuate patients. In Poland, the public authorities have entrusted such tasks to the armed forces, including the Territorial Defence Forces, as well as students of military universities.

The analysis concerns the evacuation of nursing home patients carried out in April 2020 in Poland during the first phase of the Covd-19 as part of the "Resistant Spring" action coordinated by the Polish Army with the participation of soldiers and cadets.

The aim of this article is to develop a universal model of the patient's evacuation process, ensuring their safety in the conditions of a pandemic threat.

In accordance with applicable law, public utility institutions are required to develop safety procedures, including evacuation plans. These plans were based on the implementation of the process with the institution's own resources. However, the Covid-19 has shown that this assumption is unrealistic.

In the case of the health crisis, the cooperation of many organisations and institutions turned out to be necessary. It is particularly important for entities related to the implementation of tasks in the field of care for sick people and people with limited physical and mental condition. One of such entities are nursing homes. The patients staying there are elderly people with reduced mobility, often with multiple diseases.

As a result of the infection, medical and care personnel were subjected to isolation, which made it impossible to evacuate in accordance with the earlier assumptions. State entities reporting to various ministries were asked for support. Therefore, it has become necessary to develop a common procedure based on cooperation and coordination.

The article is of a conceptual and research nature and aims to analyse the essence of effective civil-military cooperation in the area of evacuating patients in situations of a health crisis caused by the coronavirus pandemic.

Our research filled the research gap in the area of conditions and principles of interinstitutional cooperation in civil-military relations in health crisis situations in terms of the implementation of patient evacuation processes.

The research question of this study are: (1) what actions should be taken within the framework of civil-military cooperation to ensure the efficiency of the evacuation processes of patients in a health crisis situation and (2) what factors determine the effectiveness of patient evacuation processes carried out in civil-military cooperation?

The article is organised as follows. First, the theoretical foundations related to the evacuation of patients in the situation of the Covid-19 were explained. Then the principles of civil-military cooperation were presented. The next part of the article describes the design of the study and the methodology used for the analysis, including a case study.

The results section includes a description and analysis of the patient evacuation process carried out as part of the "Resistant Spring" operation and the identification of key actions determining the efficiency of patient transfer. The last part of the article contains an overview, the main theoretical and managerial implications and recommendations, as well as conclusions, limitations and future research perspectives.

2 Conceptual Framework

Patient transfer is a logistical challenge that requires the implementation of many processes involving various entities responsible for the sphere of managing the flow of transported people.

The process of inter-hospital patient transport is complex and burdened with a number of risk factors, including: (a) technical factors related to the equipment and its use, (b) human factors related to the transport team (e.g. lack of training, lack of supervision), (c) collective factors related to the organisation of transport, including: insufficient communication and coordination between the driving and receiving teams, and (d) patient-related factors (including clinical instability) [3].

Therefore, it is necessary to ensure patient safety and minimize the risk of adverse events through proper planning of the logistic aspects of patient flow.

In the logistic process of patient transport, it is necessary to implement the following activities: deciding on the transfer and communication, stabilising and preparing the patient before the transfer, selecting the appropriate means of transport, involving the personnel accompanying the patient, protecting equipment monitoring the patient's condition required during the transfer, preparation of documentation and transfer of the patient to the receiving facility [4]. However, in the event of a threat to health or even life due to coronavirus infection, immediate action, consisting in the rapid evacuation of patients, has become crucial.

Due to the risk of infection, the transport of patients with Covid-19 (including patients in a critical condition) is a unique challenge, especially during an emergency, when it is necessary to evacuate patients immediately. According WHO evacuation means moving people and assets temporarily to safer places before, during or after the occurrence of a hazardous event in order to protect them [2]. The evacuation of a facility is a complex socio-physical process carried out under extreme conditions, during which individuals interact with each other and with the built environment, with the different mobility characteristics and needs of evacuees, especially people with disabilities or limitations (in particular with mobility disabilities) make this process difficult [5].

In order to ensure patient safety and minimize the risk of disease transmission to transport personnel, communication and advance planning are needed for issues such as: ideal destination and type of unit as well as transport time or out-of-hospital time, appropriate use and type of protective equipment, patient care procedures during transport and decontamination after transport, as well as maintaining the appropriate types and amounts of personal protective equipment and barriers for the patient [6].

During the evacuation of a person requiring medical or additional assistance, implementing protective measures in hospitals and residential care homes can be difficult due to the greater potential for Covid-19 impact [7]. With regard to nursing homes, three main difficulties have been identified, which are related to: the specific nature of nursing homes (insufficient resources, competences, organisational and logistical capacity); challenges related to the nature of SARS-CoV-2 (problems with identifying the infection in the elderly); and challenges related to the nature of service recipients (presence of other diseases) [8].

From the first wave of the Covid-19, military personnel were involved in the command and control systems of civil emergency planning, both as a liaison and to increase analytical and planning capabilities (including medical intelligence) [9]. The military response to Covid-19 has been varied and has shown that the military is capable of complementing civilian efforts in health and logistics, as well as welfare and humanitarian aid [10].

The armed forces and military health systems have played an important role in supporting national health systems, in particular in activities such as: recognition of health security threat from Covid-19 spread, detection and announcement of first military cases, invocation of national crisis plans (including announcing of military involvement), information on typologies of military support (how support was provided to specific interventions), dealing with rumours and modifying internal and external routine military activities to accommodate changes posed by the Covid-19 [11].

In Poland, the potential of the Polish Armed Forces (PAF) was involved in counteracting the Covid-19 and supporting the non-military system, in particular in the scope of such activities as: providing assistance to various civilian entities, including local authorities, hospitals, nursing homes and non-governmental organisations with their own specialised abilities [12]. The Polish Ministry of National Defence, as part of the operation "Resistant Spring", indicated common assumptions for cooperation of both the local government, soldiers and sanitary services. The purpose of the operation was to alleviate the effects of the crisis and to strengthen the resilience of local communities to the crisis. Polish cadets and university personnel supported medical staff, people subject to quarantine, veterans and veterans of operations abroad, the elderly, the disabled and single parents.

3 Research Design and Methods

The premise for the development of the research procedure was the assumption that due to the specificity of evacuating nursing home patients in the conditions of the coronavirus pandemic, it is necessary to look for solutions supporting decision making and organisational processes. In this context, it is necessary to identify the key actions and factors for the efficiency of civil-military cooperation in the field of patient evacuation.

The multi-stage research procedure, based on various research methods, made it possible to achieve all the objectives adopted in the article. The following research methods were used to achieve the adopted goals: literature study (a review and in-depth analysis of scientific literature), document analysis (review and analysis of documents, reports, legal acts), participant observation, individual semi-structured interview with experts, case study and process mapping and analysis.

The methodology of the undertaken research focuses on a qualitative approach, starting with assumptions based on the content of articles in the area of patient transfer and evacuation, as well as the role of the armed forces in the implementation of social and health tasks. This methodological approach has been planned in several phases. In the first stage, an analysis of literature and documents was carried out in order to identify and systematise the state of knowledge about the evacuation processes of patients, including the processes of elderly people in the conditions of the coronavirus pandemic.

The core of the research procedure is a single case study method that provides in-depth study and rich description [13]. A single case study allows for a deeper understanding of the explored topic, not only due to the possibility of a richer description of the phenomenon, but also the possibility of questioning the existing theoretical relations and exploring new ones as a consequence of conducting more detailed research [14].

The starting point for conducting in-depth analyses using the method of a single case study was the selection of a research object—a system (case) representing the issue under study. At this stage, the focus was on the evacuation of nursing home patients as part of the "Resistant Spring" operation carried out in April 2020 in Poland with the participation of cadets—students of military universities.

The analysis of documents relating to evacuation procedures in individual entities participating in the evacuation allowed for the identification of differences that were critical points in the actually implemented operation. To confirm the conclusions of the analysis, in-depth, semi-structured interviews were conducted with the participants, who were observers of the activities. Based on the interviews, a map of the evacuation process was created.

As a result of the analysis of the evacuation process, it was possible to develop a map of the complete evacuation process of patients in a pandemic, taking into account the participation of various institutions.

4 Case and Findings

Due to the dynamic spread of the SARS-CoV-2 virus, it was necessary to take measures to support local authorities and sanitary services in the fight against the pandemic. Especially the elderly, lonely people required extra help. The Armed Forces of the Republic of Poland actively participated in the fight against the virus, using their forces and means, supporting the society, providing the necessary formal and organisational (structural) support [15]. The university participated in the implementation of tasks in the field of preventing and combating the coronavirus, including the operation "Resistant Spring", involving a total of 1205 people (1,085 cadets and 120 professional soldiers). One of the undertakings in which the cadets took part was the evacuation of nursing homes.

In order to support the evacuation and restore the original capacity of social welfare homes, from April 20, 2020, Crisis Intervention Teams operated in each territorial defence brigade. These groups are supported by specialist units of the General Command of the Armed Forces. The teams were directed to act by the Command of the Territorial Defence Forces at the request of the Chief Sanitary Inspector or the appropriate voivode, and were involved in the event of evacuation of residents and staff of social care homes.

The analysis concerns the case of the evacuation of patients from the Nursing Home in Kalisz, which was attended by cadets—students of a military university with soldiers of the Territorial Defence Forces. The evacuation concerned 26 people who were transported by military ambulances to the hospital in Kalisz. Most of the patients had to be transported to ambulances on a stretcher. In this case, a particular difficulty was the shortage of staff and the unavailability of the management of the facility due to the coronavirus infection.

Based on the observation of activities carried out during the evacuation as part of the case studied, key stages of the process and entities responsible for the implementation of individual tasks and activities were identified. The identified activities are illustrated in the process model. The model is conceptual, descriptive, and graphically represented in the form of process maps (see Figs. 1, 2 and 3).

An important input assumption of the model are legal limitations and restrictions that define the scope of responsibility of institutions participating in the evacuation processes in a crisis situation. The efficient course of the evacuation of patients in a pandemic emergency requires the implementation of a number of activities that can be classified into the following stages:

1. preparatory phase: decision-administrative, including procedural and subjective arrangements;

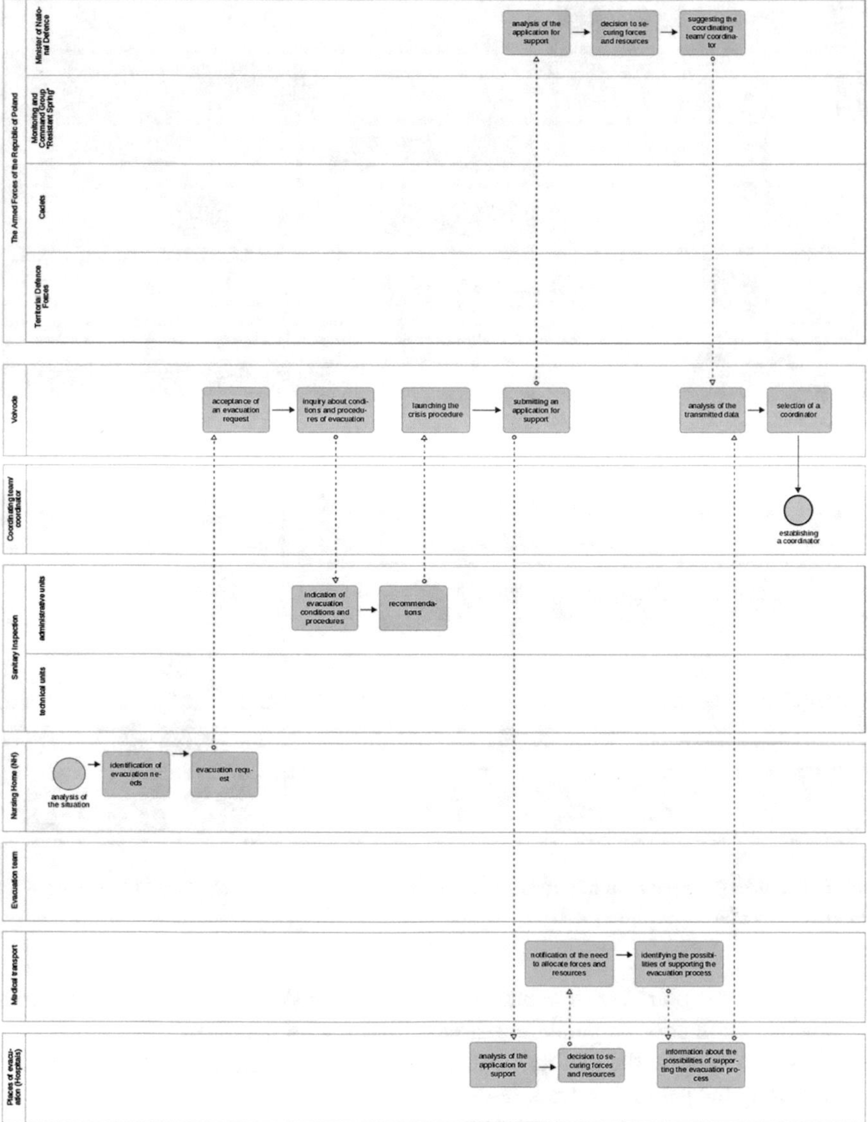

Fig. 1 A model of evacuation of nursing home patients as part of civil-military cooperation—the preparatory phase

2. organisational phase: securing forces and resources, organising the command centre, establishing executive procedures;
3. implementation phase: carrying out the evacuation and handing over the evacuated patients to designated entities;

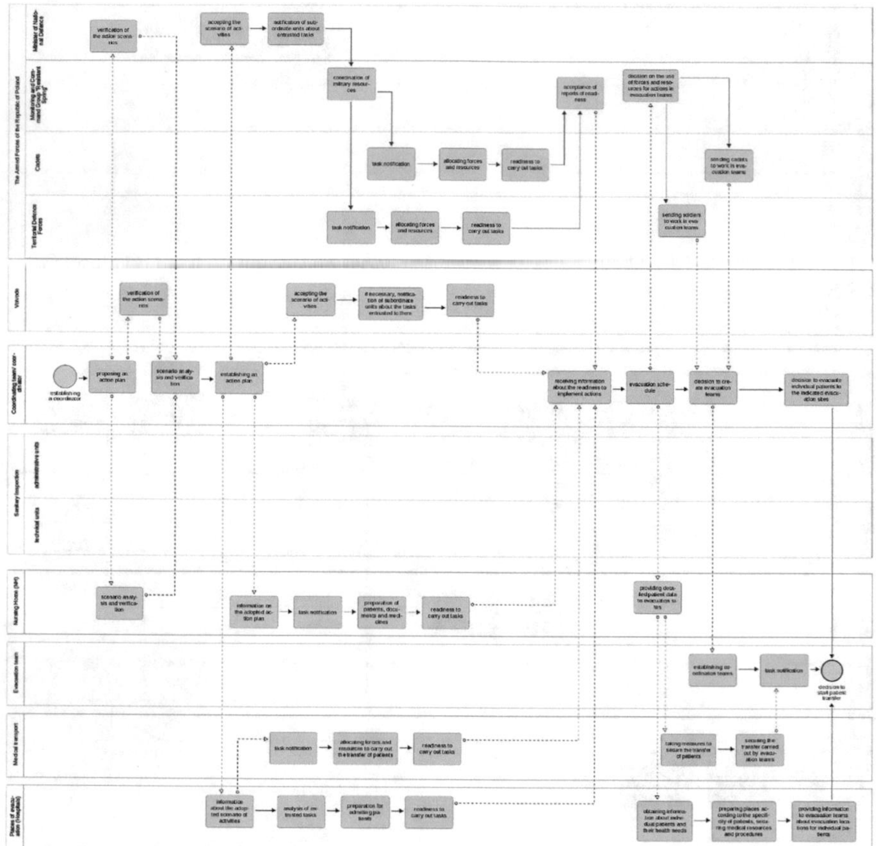

Fig. 2 A model of evacuation of nursing home patients as part of civil-military cooperation—the organisational phase

4. neutralisation phase: with regard to the evacuation site—procedures for neutralisation, elimination of pandemic threats, isolation of personnel;
5. phase of restoring the facility's operational readiness: adjusting the facility and preparing the patient's return, re-evacuation.

The implementation of these tasks requires the involvement of many entities. In the analysed process, two basic groups of entities were identified: the civil and the military sector.

As part of the civilian sector, care homes can be distinguished—initiating the need for support, their founding bodies, local authorities, state authorities (government, voivode), sanitary institutions, and health care system institutions: hospitals, ambulance service, medical transport, as well as other organisations, including non-profit social organisations that can support specific activities. The military sector is

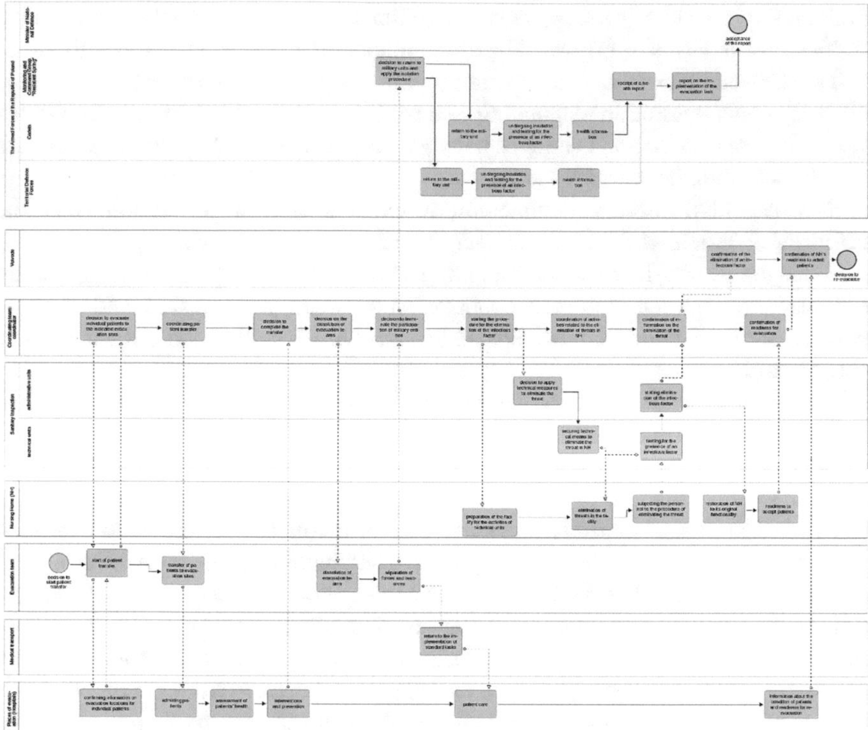

Fig. 3 A model of evacuation of nursing home patients as part of civil-military cooperation—the implementation phase, the neutralisation phase, the phase of restoring the facility's operational readiness

the Ministry of National Defence and its subordinate units, including the Territorial Defence Forces, military universities and crisis response teams.

The interaction of these two sectors requires coordination, and therefore it is necessary to establish a coordinating team composed of representatives of the civilian and military sectors, responsible for substantive and organisational arrangements. In addition, it is necessary to appoint evacuation teams responsible for the implementation of activities during the evacuation of patients.

5 Discussion and Conclusions

The developed model of the evacuation process of nursing home patients is a proposal for comprehensive multi-entity cooperation in connection with the occurrence of a threat caused by a health crisis such as the coronavirus pandemic. In the face of a wave of numerous infections, nursing homes need the support of external entities that ensure professional and immediate action, taking into account the needs and limitations of the elderly and the sick. The military is the professional support for the health sector in terms of organisation and implementation of evacuation activities.

It is stressed that one of the most important lessons that can be learned from the difficulties encountered in responding to a pandemic is the need for a society-wide approach to resilience and synergies between civilian and military actors to address current and future health crises [16].

However, civil-military cooperation requires a number of actions and the fulfilment of certain conditions necessary to achieve success (Table 1).

Managing Covid-19 patients, including patient transport, applies to the entire healthcare system and requires cooperation between pre-hospital and hospital services, some of which are usually unfamiliar with logistics and patient transfer mechanics [17].

Therefore, preparatory activities are of key importance, as are the development of standard scenarios of activities, taking into account various conditions and restrictions.

Emergency preparedness refers to "the long-term planning of activities to strengthen the overall capacity and capability of a country" to efficiently respond to all types of hazards in civil society [18]. The challenges for state services that arose during the Covid-19 indicate the need to develop a strategy for civil-military cooperation aimed at maintaining constant readiness to respond to crises, including health crises. In this context, strategic resilience becomes extremely important. The coronavirus pandemic has highlighted the role of resilience to strategic shocks (such as pandemics) as an element of national and regional security, particularly in terms of the capacity of national health systems, supply chains of critical materials (e.g. personal protective equipment), strategic communication and disinformation, and cybersecurity [19].

The WHO identifies the following key elements for an effective civil-military cooperation in health to develop essential national capabilities for the prevention,

Table 1 A model of evacuation of nursing home patients—actions and factors

Actions necessary during the evacuation process	Factors of the effectiveness of evacuation processes
Development of essential national capabilities for the prevention, detection, response and recovery of health emergencies cooperation between pre-hospital and hospital services Development of standard scenarios of activities Establishment of decision-making structures, an appropriate communication system and conduct training, including practical training, in nursing homes	A society-wide approach to resilience synergies between civilian and military actors Clearer roles before a crisis occurs reduces the amount of coordination needed in the response The readiness and ability of security institutions to fulfil any or all of the roles assigned to them by the civil management Maintaining constant readiness to respond to crises, including health crises Ensuring legal conditions, professional support and adequate resources (material and information)

Source own analysis

detection, response and recovery of health emergencies: (a) establishing a strategic cooperation plan for health emergency preparedness; (b) recognising the differences between the public health sector and the military health service; (c) the definition of technical areas for cooperation based on the national core capacity for health preparedness; (d) institutionalisation of civil-military cooperation in the field of health; and (e) jointly building and training civil-military capabilities for health crisis preparedness [20].

The developed strategies, operational action plans and scenarios must be disseminated among entities involved in evacuation processes. It is therefore important to establish decision-making structures, an appropriate communication system and conduct training, including practical training, in nursing homes.

According to Jensen and Hertz [21], developing clearer roles before a crisis occurs reduces the amount of coordination needed in the response, leading to better outcomes. Efficiency in fulfilling roles and missions includes the readiness and ability of security institutions to fulfil any or all of the roles assigned to them by the civil management, which means meeting three basic requirements, such as: planning with a strategic vision or roadmap for the role of security forces in society, structures and processes for formulating plans and reviewing the implementation of these plans, commitment of resources in the form of political, financial and personnel capital, and any other resources necessary for the assigned roles and missions [22].

In conclusion, it should be stated that environmental threats and health crises revealed the need to strengthen the role of civil-military cooperation at various levels, systemic (state decisions), inter-sectoral (medical and military sectors, logistics) and, above all, local and inter-organisational (implementation of activities). Effectiveness requires professional support and ensuring adequate resources in a specific threat situation, which means that it must be coordinated in a way that allows the flow of information and resources in the inter-entities dimension.

The considerations undertaken in the article led to the formulation of an answer to the accepted research questions and the achievement of research goals. Regardless of the limitations, the presented analyses, conclusions and proposed recommendations may be useful in the practical, managerial and scientific dimensions, constituting a starting point for broader, in-depth research.

References

1. Haleem, A., Javaid, M., Vaishya, R.: Effects of COVID-19 pandemic in daily life, Letter to the Editor. Curr. Med. Res. Pract. **10**, 78–79 (2020)
2. World Health Organization: WHO Glossary of Health Emergency and Disaster Risk Management Terminology, Geneva (2020).
3. Sethi, D., Subramanian, S.: When place and time matter: how to conduct safe inter-hospital transfer of patients. Saudi J. Anesth. **8**, 104–113 (2014)
4. Kulshrestha, A., Singh, J.: Inter-hospital and intra-hospital patient transfer: recent concepts. Indian J. Anaesth. **60**, 451–457 (2016)
5. Haghpanah, F., Ghobadi, K., Schafer, B.W.: Multi-hazard hospital evacuation planning during disease outbreaks using agent-based modelling. Int. J. Disaster Risk Reduct. **66**, 102632 (2021)

6. Brown, A.S., Hustey, F.M., Reddy, A.J.: Inter hospital transport of patients with COVID-19: Cleveland Clinic approach. Clevel. Clin. J. Med. (Jun 2020)
7. FEMA: COVID-19 Supplement for Planning Considerations: Evacuation and Shelter-In-Place (Sep 2020)
8. Ochi, S., Murakami, M., Hasegawa, T., Komagata, Y.: Prevention and control of COVID-19 in imperfect condition: practical guidelines for nursing homes by Japan environment and health safety organization (JEHSO). Int. J. Environ. Res. Public Health **18**, 10188 (2021)
9. Meyer, C.O., Besch, S., Bricknell, M.: Briefing. How the COVID-19 crisis has affected security and defence-related aspects for the EU. Directorate-General for External Policies. Policy Department. EP/EXPO/SEDE/FWC/2019--01/Lot4/1/C/06 EN July 2020-PE 603.510, European Union (2020)
10. Lațici, T.: Members' Research Service PE 649.401–April 2020. European Parliamentary Research Service. European Union (2020)
11. Gad, M., Kazibwe, J., Quirk, E., Gheorghe, A., Homan, Z., Bricknell, M.: Civil–military cooperation in the early response to the COVID-19 pandemic in six European countries. BMJ Mil Health **167**, 234–243 (2021). https://doi.org/10.1136/bmjmilitary-2020-001721
12. Ćwik, Z., Nykiel, D., Pietrzak, R., Hajost, E., Knasiak, G.: Polish Armed Forces Response to COVID-19. Observations, Insights, Lessons. Pandemic outbreak and lockdown phase. Centrum Doktryn i Szkolenia Sił Zbrojnych. Bydgoszcz (2020)
13. Darke, P., Shanks, G., Broadbent, M.: Successfully completing case study research: combining rigour. Relevance Pragmat. Inf. Syst. J. **8**, 273–289 (1998)
14. Gustafsson, J.: Single case studies vs. multiple case studies: A comparative study. Sociology (2017)
15. Decision No. 84 DWOT of the Minister of National Defence of March 16, 2020 on the support by the Polish Armed Forces of activities related to the prevention of the spread of SARS-CoV-2 virus (2020)
16. Garriaud-Maylam, J.: Committee on the Civil Dimension of Security (Cds) enhancing the resilience of allied societies through civil preparedness. Preliminary Draft General Report, 011 CDS 21 E, 3 March (2021)
17. Bredmose, P.P., Diczbalis, M., Butterfield, E., Habig, K., Pearce, A., Osbakk, S.A., Voipio, V., Rudolph, M., Maddock, A., O'Neill, J.: Decision support tool and suggestions for the development of guidelines for the helicopter transport of patients with COVID-19. Scand. J. Trauma Resusc. Emerg. Med. **28**, 43 (2020)
18. World Health Organization: Risk Reduction and Emergency Preparedness: WHO Six-Year Strategy for the Health Sector and Community Capacity Development. Geneva (2007)
19. European Union: How the COVID-19 crisis has affected security and defence-related aspects of the EU In-Depth Analysis. Policy Department, Directorate-General for External Policies, EP/EXPO/SEDE/FWC/2019-01/Lot4/1/C/07 EN. January 2021-PE 653.623 (2021)
20. World Health Organization: National Civil–Military Health Collaboration Framework for Strengthening Health Emergency Preparedness: WHO Guidance Document (2021)
21. Jensen, L.-M., Hertz, S.: The coordination roles of relief organizations in humanitarian logistics. Int. J. Log. Res. Appl. **19**(5), 465–485 (2016)
22. Matei, F.C., Halladay, C., Bruneau, T.C. (eds.): The Routledge Handbook of Civil- Military Relations, 2nd edn. Routledge (2022)

Author Index

A. Visvizi et al. (eds.), *Research and Innovation Forum 2022*, Springer Proceedings in Complexity, https://doi.org/10.1007/978-3-031-19560-0

Printed in the United States
by Baker & Taylor Publisher Services